THE GEOTECTONIC DEVELOPMENT OF CALIFORNIA

W. W. Rubey (1898–1974)

W. G. ERNST
Editor

THE GEOTECTONIC DEVELOPMENT OF CALIFORNIA

Rubey Volume I

Prentice-Hall, Inc., Englewood Cliffs, New Jersey 07632

Library of Congress Cataloging in Publication Data
Main entry under title:

The Geotectonic development of California.

(Rubey; vol. 1)
"Dedicated to the late W. W. Rubey."
Based on a series of lectures delivered in 1978 at UCLA in a course entitled Rubey Colloquium.
Bibliography: p.
Includes index.
1. Geology—California—Congresses. 2. Rubey, William Walden, 1898-1974. I. Ernst, Wallace Gary, 1931- II. Rubey, William Walden, 1898-1974. III. Series.
QE89.G46 557.94 80-20658
ISBN 0-13-353938-5

Editorial/production supervision
and interior design by Karen Skrable
Cover design by Edsal Enterprise
Manufacturing buyer: Anthony Caruso

Printed in the United States of America

10 9 8 7 6 5 4 3 2 1

Prentice-Hall International, Inc., *London*
Prentice-Hall of Australia Pty. Limited, *Sydney*
Prentice-Hall of Canada, Ltd., *Toronto*
Prentice-Hall of India Private Limited, *New Delhi*
Prentice-Hall of Japan, Inc., *Tokyo*
Prentice-Hall of Southeast Asia Pte. Ltd., *Singapore*
Whitehall Books Limited, *Wellington, New Zealand*

CONTENTS

PREFACE

The Geotectonic Development of California attempts to synthesize the geology of California in terms of plate tectonic evolution. It is based on a series of lectures delivered by the authors of the various chapters during winter quarter 1978 at UCLA in a course entitled the Rubey Colloquium. That symposium, and this volume (the first of a series) are dedicated to the late W. W. Rubey, Career Scientist with the U.S. Geological Survey since 1920, and Professor of Geology and Geophysics at UCLA from 1960 until his death in 1974.

Bill Rubey's scientific contributions received international recognition. They include: the systematics of stream hydrology; sedimentation, stratigraphy, compaction, and origin of sedimentary rocks; areal geology of parts of the midcontinent, the Great Plains, and the northern Rocky Mountains (especially the Black Hills and western Wyoming); the origin of the atmosphere, seawater, and chemical differentiation of the earth; mechanisms of overthrust faulting and mountain building; and factors influencing the release of seismic energy. He was a brilliant generalist, studying the interconnections of geologic phenomena in virtually all the problems he addressed; he brought rigor and quantification to many subjects that hitherto had seemed virtually intractable.

Bill's 1951 presidential address to the Geological Society of America dealt with the origin of seawater, and was subsequently amplified in Geological Society of America Special Paper 62, which appeared in 1955. In this article he demonstrated that the atmosphere and hydrosphere accumulated gradually near the earth's surface over the course of geologic time as a consequence of the outgassing of the deep interior. This is perhaps Bill's best-known and most frequently cited work, although papers co-authored with M. King Hubbert on the importance of aqueous fluid pressures in reducing frictional resistance to allow slip along major, low-angle, thrust faults may be equally famous. Clarification of the previously enigmatic mechanism of overthrust faulting represents a major advance in structural geology. The solution to this perplexing problem grew out of Bill's curiosity concerning the origin of imbricate thrust structures like those he was mapping in western Wyoming; it shows how he applied his considerable understanding of physics to a general, first-order problem of tectonics. A related study involved the correlation of microseismic activity in the Denver, Colorado area with fluid injection at the Rocky Mountain Arsenal, which in turn introduced a possible method of controlling release of earthquake energy in tectonically active areas.

Bill Rubey received many honors. He was elected to membership in the National Academy of Sciences in 1945 and the American Philosophical Society in 1952. He was awarded the U.S. Department of the Interior Award of Excellence in 1943, the Distinguished Service Award in 1958, and received the National Medal of Science from President Lyndon B. Johnson in 1965. Bill Rubey was president of the Geological Society of America in 1949–1950 and received that society's highest honor, the Penrose Medal, in 1963. He was also president of the Geological Society of Washington in 1948 and of the Washington Academy of Sciences in 1957. Bill was a member, fellow, or councillor of more than twenty learned societies. He received honorary degrees from Yale University, the University of Missouri, Villanova University, and UCLA.

Bill Rubey's counsel was held in exceptionally high esteem as reflected by his appointment to many official advisory committees and panels—both governmental and academic—and as testified to by the innumerable requests from scientific colleagues for his thoughts on broadly ranging subjects. Bill served as trustee of the Science Service Corporation (1956–1964), of the Carnegie Institution of Washington (1962–1974), and of the Woods Hole Oceanographic Institution (1966–1974). He was a visiting professor at UCLA in 1954, at Cal Tech in 1955, at Johns Hopkins in 1956, and was Silliman lecturer at Yale in 1960. In addition, he participated on numerous university visiting committees. His services in these many advisory capacities were sought out not only because of his well-recognized breadth of experience, knowledge, and insight, but also because it was well known that he "did his homework" and came to meetings prepared for in-depth analysis of the problems at hand.

Bill Rubey influenced the course of earth science in yet another fashion. During his many years of productive associations in Washington, D.C., with the U.S. Geological Survey, and later at UCLA, he hosted informal monthly seminars at his home that brought together earth scientists interested in a variety of first-order problems. A keynote speaker would introduce a subject of mutual interest—such as the chemical evolution of the earth's crust or the origin of life—and protracted but spirited discussion would follow. Judging from the comments of those who were privileged to attend, these "think-tank" sessions were both provocative and stimulating and in no small measure provided impetus for the development of contributions to various aspects of geology.

It was during such an informal "brain-storming" session that the idea for Project Mohole was hatched. Although this rather audacious, but visionary scheme failed scien-

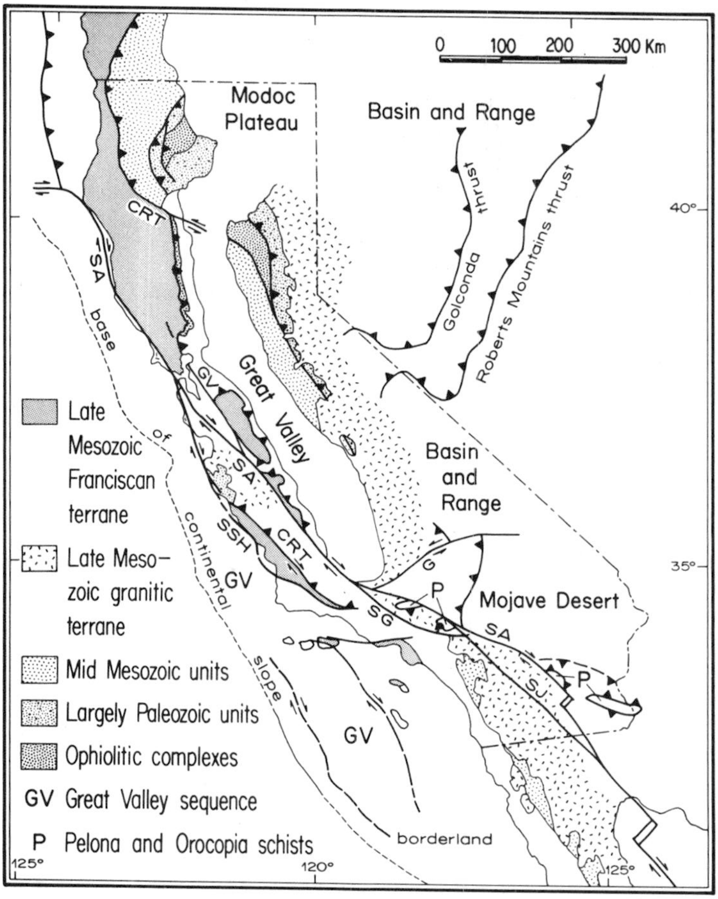

Fig. 1. Aspects of the generalized regional geology of California, in part after Hamilton (1978). Labeled major faults are: CRT = Coast Range thrust; G = Garlock; SA = San Andreas; SG = San Gabriel; SJ = San Jacinto, SSH = San Simeon-Hosgri. Other abbreviations include: GV = Great Valley Sequence; P = Pelona + Orocopia schists.

tifically, largely due to technical, political, and contracting difficulties, the idea was sound. The subsequent phenomenal success of the Joint Oceanographic Institutions for Deep Earth Sampling deep-sea drilling program has amply vindicated the concept. The JOIDES evidence demonstrating the validity of plate tectonic processes is well known today. Thanks to such advances in knowledge which Bill had a hand in, we have been able to set out the plate tectonic history of California in the following chapters.

The Geotectonic Development of California is aimed at the experienced college senior as well as first and second year graduate students. In addition, it is hoped that this book will provide a useful synthesis and regional review for Californian professional geologists. Moreover, it documents a well-studied case history of an evolving Phanerozoic continental margin, and as such should be of interest to all students of orogenic belts and plate tectonic evolution.

Individual chapters are written by authorities in the particular regions. To provide a comprehensive synthesis, an attempt has been made to select authors representing as broadly ranging interest and interpretations as possible, but time and volume constraints provided some inevitable arbitrary limitations.

The regional geology of California is best referred to employing the 1:750,000 State Geologic Map (Jennings, 1977). However, for the reader's convenience, a small-scale interpretive synthesis showing the present-day disposition of lithotectonic belts is presented as Fig. 1, after Hamilton (1978).

Financial support both for this book and for the Rubey Colloquium have been provided by UCLA. I wish to thank all the authors for their enthusiastic cooperation in producing this first Rubey Volume, and the UCLA Earth and Space Sciences staff for their help in preparing the manuscript for publication.

W. G. Ernst

THE GEOTECTONIC DEVELOPMENT OF CALIFORNIA

William R. Dickinson
Department of Geosciences
University of Arizona
Tucson, Arizona 85721

1

PLATE TECTONICS AND THE CONTINENTAL MARGIN OF CALIFORNIA

ABSTRACT

The California continental margin has undergone four main stages of tectonic evolution: (1) a rifted Atlantic-type margin evolved through the late Precambrian and early Paleozoic, (2) a complex Japanese-type margin with offshore island arcs developed in the late Paleozoic and early Mesozoic, (3) an active Andean-type margin, with a trench along the edge of the continent, existed in the late Mesozoic and early Cenozoic, and (4) the present Californian-type margin, dominated by strike slip on the San Andreas transform fault system. The miogeoclinal Precambrian to Devonian succession of the eastern Cordillera was deposited along a passive continental margin following late Precambrian rifting that pursued a jagged course. Prominent marginal offsets in central Idaho and southern California delimited the Nevada segment of the rifted margin where Paleozoic sedimentary facies were most fully developed. Interactions of offshore island arcs with the continental margin led to the thrust emplacement of allochthonous oceanic facies by partial subduction and crustal collision during the Devono-Carboniferous Antler Orogeny and the Permo-Triassic Sonoma Orogeny. Extensive island arc terranes were thus accreted to the continental margin by mid-Triassic time. By the beginning of Jurassic time, subduction had begun beneath the expanded continental margin along the Sierra Nevada Foothills. Further accretion of intraoceanic Jurassic island arc terranes along the Foothills suture belt induced subduction to shift into the Coast Ranges by the end of Jurassic time. The Cretaceous arc-trench system included the Franciscan subduction zone in the Coast Ranges, the Great Valley forearc basin, the Sierra Nevada batholith representing the roots of the magmatic arc, and the backarc Sevier fold-thrust belt. An episode of plate descent at a shallow angle below the Cordillera during the Paleogene led to the Laramide Orogeny inland while subduction continued in the Franciscan terrane. Neogene evolution of the San Andreas transform system offset coastal slices of the continental block in complex patterns, and led to termination of arc volcanism inland where slab descent did not occur adjacent to the transform. Pervasive effects of plate tectonics on the geologic history of the continental margin include changes in sedimentation owing to shifts in paleolatitude and changes in base level owing to eustatic fluctuations in the volume of ocean water or ocean basins.

INTRODUCTION

The architecture of the earth's surface harbors only two main components, continental blocks and oceanic basins, although gradational features intermediate between the two exist. Consequently, the topic of continental margins deals with the interfaces between oceanic and continental regions. Continental margins can have only four basic tectonic configurations (Fig. 1-1):

1. Atlantic-type margins, the passive margins, are formed by rifting along a divergent plate juncture whose later evolution develops a midoceanic rise or spreading ridge

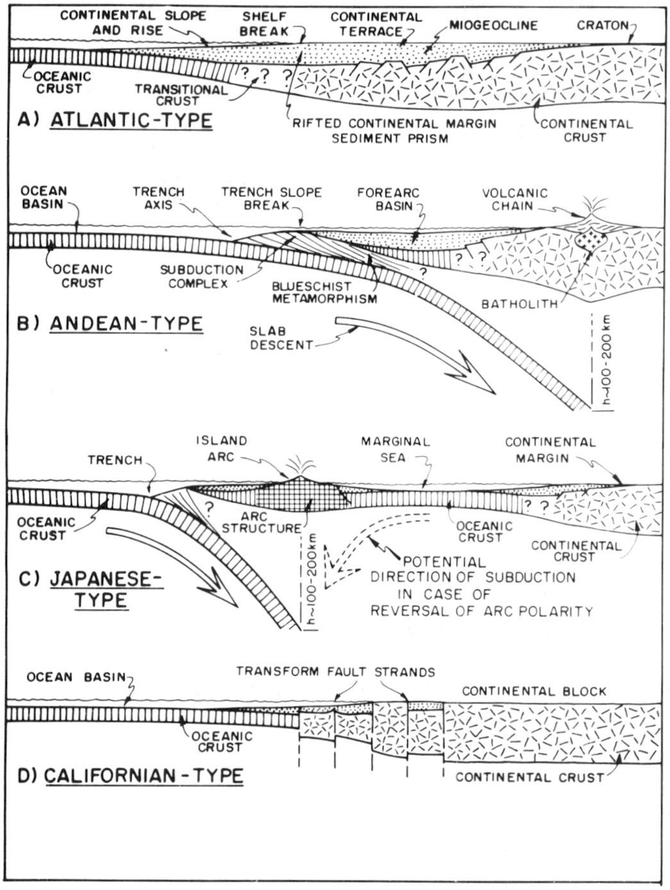

Fig. 1-1. Tectonic configurations of continental margins (after Dickinson, 1976). See text for discussion.

within the adjacent ocean basin; the ocean typically widens with time, although the passive margin may persist until the ocean begins to narrow by subduction elsewhere.

2. Andean-type margins, the classic active margins, form where subduction of oceanic lithosphere occurs at a trench along or near the edge of a continental block; the resulting arc volcanoes stand on the edge of the intact continental block, which extends unbroken across the backarc region.

3. Japanese-type margins are the most complex morphologically, because subduction occurs beneath offshore island arcs that are separated from the main continental block by small ocean basins of the marginal seas; in the classic case, marginal seas are formed by backarc spreading, but may also be older ocean floor merely trapped between arc and continent by seaward stepping of the subduction zone.

4. Californian-type margins are marked by relative lateral displacement of ocean

basin and continental block along a transform fault zone whose main fault s located within or near the continent-ocean interface (a belt of finite width).

The California continental margin has experienced all four of these kinds of t regime during the course of recorded geologic history (Fig. 1–2):

1. Following rifting in late Precambrian time, an Atlantic-type margin pers. through the early Paleozoic until near the Devonian-Carboniferous time boundary; du this time span, the classic Cordilleran miogeocline evolved in an eastern belt of t Cordillera extending into the Mojave block of southeastern California.

2. From Late Devonian to Late Triassic times, a Japanese-type margin with offshore island arcs was either persistent or recurrent; during this time span, much of the eugeosynclinal terrane in the western Cordillera, including most of the Sierra Nevada block, was accreted to the continent during the Antler and Sonoma orogenies.

3. From near the Triassic-Jurassic time boundary until the onset of the Neogene at the Oligocene-Miocene time boundary, an Andean-type margin was continuous as part of the circum-Pacific subduction system; effects included the Nevadan, Sevier, and Laramide orogenies, and accreted terranes include those of the Sierra Nevada Foothills, much of the Klamath block, and most of the Coast Ranges.

4. During the Neogene, plate interactions along the coast gave rise to the San Andreas transform system that now dominates the present Californian-type margin; Neogene intraplate deformation also formed the Basin-and-Range province of the intermountain region, including easternmost California and parts of adjacent states.

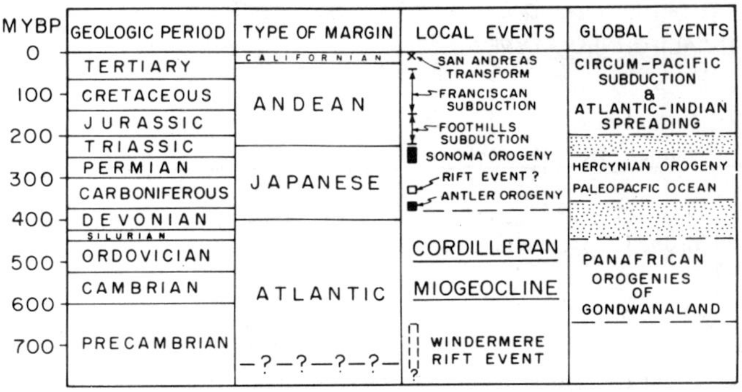

MYBP
GEOLOGIC PERIOD
TYPE OF MARGIN
LOCAL EVENTS
GLOBAL EVENTS
0
100
200
300
400
500
600
700
TERTIARY
CRETACEOUS
JURASSIC
TRIASSIC
PERMIAN
CARBONIFEROUS
DEVONIAN
SILURIAN
ORDOVICIAN
CAMBRIAN
PRECAMBRIAN
CALIFORNIAN
ANDEAN
JAPANESE
ATLANTIC
–?–?–?–?–
SAN ANDREAS TRANSFORM
FRANCISCAN SUBDUCTION
FOOTHILLS SUBDUCTION
SONOMA OROGENY
RIFT EVENT ?
ANTLER OROGENY
CORDILLERAN
MIOGEOCLINE
WINDERMERE RIFT EVENT
CIRCUM-PACIFIC SUBDUCTION
&
ATLANTIC-INDIAN SPREADING
HERCYNIAN OROGENY
PALEOPACFIC OCEAN
PANAFRICAN OROGENIES OF GONDWANALAND

Fig. 1-2. Major phases and events of Cordilleran history shown in relation to key global trends. Time scale generalized after Lambert (1971).

These evolutionary tectonic stages in the history of the California margin can be discussed either in the order in which they happened, from ancient to modern, or in the order in which they must be reconstructed, from modern to ancient. Certain technical advantages lie in both directions. For clarity, I will follow the flow of time from ancient to modern. Bear in mind, however, that my homework was completed in the opposite fashion, from modern to ancient, before I could begin my exposition.

The reconstruction of past plate interactions depends upon the ability to interpret the significance of key petrotectonic assemblages, which are associations of rock masses diagnostic of specific plate tectonic settings (Dickinson, 1972). For the four main types of continental margins, the following petrotectonic assemblages are the critical indicators or signals in the rock record:

1. For Atlantic-type margins, an elongated wedge of shelf sediments, which rests unconformably on older basement rocks, thickens laterally toward the old ocean basin; where the adjacent ocean basin has been closed subsequently by subduction, the sediments thicken instead toward a tectonic join against a parallel belt of deformed but coeval oceanic strata.

2. For Andean-type margins, an intricately deformed belt representing a subduction complex, which formed by detachment of surficial layers from a descending plate at a trench and is composed of oceanic sediments, ocean-floor lavas, and ophiolitic scraps of oceanic crust, lies outboard and parallel to a coeval belt of metavolcanics and batholiths representing the igneous suite of a magmatic arc. Commonly, an intervening belt of relatively undeformed and unmetamorphosed sedimentary rocks was deposited within fore-arc basins of the arc-trench gap, while deformation and metamorphism were underway in the flanking subduction zone and magmatic arc.

3. For Japanese-type margins, metavolcanic belts representing island arcs are separated from rock masses formed along the edge of the continental block by coeval oceanic assemblages that formed within the intervening marginal sea. Once the marginal sea is closed by subduction, the oceanic assemblages are crumpled by severe deformation along the suture belt that formed by later crustal collision (disscussed later) between the exotic rock masses of the offshore island arc and the indigenous ones of the continental margin.

4. For Californian-type margins, disparate sedimentary facies and basement terranes are juxtaposed in elongate belts that have been shuffled complexly by strike slip parallel to the continental margin. Lateral displacement of sediment piles with respect to sediment sources is contemporaneous with deposition and is cumulative with time.

LATE PRECAMBRIAN AND EARLY PALEOZOIC

The initial stage in the evolution of the California continental margin was the deposition of the miogeoclinal sediment wedge along a passive continental margin of Atlantic type (Stewart, 1972). The rifting that initiated deposition was accompanied or preceded by basaltic volcanism at about 850 m.y.b.p. (Dickinson, 1977). Continental separation was probably not completed, however, until about 650 m.y.b.p. (Stewart and Suczek, 1977), or not long before the start of the Cambrian. Sedimentation was thereafter essentially continuous along the passive continental margin until the Late Devonian, when the onset of the Antler Orogeny marked the advent of a wholly different tectonic regime (Dickinson,

1977). At that time, regional thrusting carried eugeoclinal oceanic facies eastward as a deformed allochthon across the miogeoclinal strata (Stewart and Poole, 1974).

Any extrapolation of plate-tectonic concepts into the Precambrian involves some degree of inference. In the present instance, the degree is minimal and of the same order as that required for the application of plate tectonics to ideas about the Paleozoic. Moreover, there is a logic that forces the extrapolation. Facies relationships of lower Paleozoic strata in Nevada document the existence of a passive continental margin of Atlantic type (Lowell, 1960). Those strata pass conformably downward in the Death Valley region into the latest Precambrian strata (Stewart, 1970). Thus, the Cordilleran continental margin either was created by rifting of some kind during the late Precambrian or had existed since the time that crustal blocks first formed early in the Precambrian.

If the Cordilleran margin had long existed, then some igneous and metamorphic record of an active continental margin, or some long sedimentary record of a passive continental margin, should be present within the region. However, neither is the case. There is certainly no hint that any coherent age belt of Precambrian basement rocks lies parallel to the Cordilleran trend. Quite to the contrary, the transcontinental trends of the Precambrian age belts that sweep across the continent are sharply truncated by the trend of the Cordilleran miogeocline (see Fig. 1-3). Nor does the sedimentary record of the miogeocline extend far back into the Precambrian. Although the basal strata do lie below the lowest known Cambrian fossils, the inland flank of the sediment wedge rests with marked angular unconformity across the whole array of Precambrian age belts.

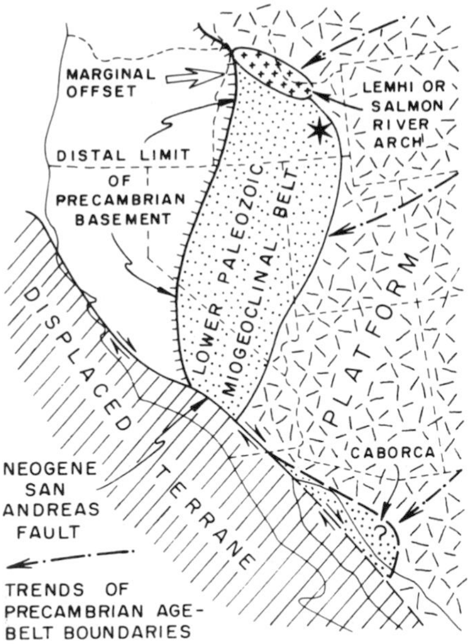

Fig. 1-3. Principal tectonic elements of Cordilleran rifted continental margin of the early Paleozoic (post-Windermere, pre-Antler). Modified after Stewart (1972), Stewart and Poole (1974, 1975), and Suczek (1977a, b). Asterisk denotes Precambrian Bannock Volcanics near Pocatello.

Precambrian Rifting

The early history of the continental margin was somewhat different in the northern and southern parts of the Cordillera. In Canada, initial rifting probably preceded Belt-Purcell deposition that began about 1450 m.y.b.p. (Gabrielse, 1972). This sedimentation apparently followed pre-Grenville continental separation of a crustal block that did not extend south of the type Belt-Purcell basin lying athwart the international boundary (Dickinson, 1977). Farther south, the earliest continental separation leading to the inception of miogeoclinal deposition was evidently a post-Grenville event (Stewart, 1972). In the north, the old Beltian margin was doubtless modified considerably by the later event, which triggered Windermere deposition, for Belt and Windermere strata are separated by a regional unconformity. The Belt and Windermere sequences are both thick wedges of clastic sediments that thicken westward.

The ultimate test of these or other ideas about continental rifting along the Cordillera will be the ability to detect the existence of the detached continental fragments elsewhere in the world. The various crustal blocks separated by suture belts within the composite modern continent of Eurasia (including Alaska) are prime prospects for correlation. Secondary prospects are those crustal blocks that were present within the supercontinent of Gondwana following the Panafrican orogenic events near the Precambrian-Paleozoic time boundary. Because of the Precambrian ages of the postulated rifts, correlation must consist of matching templates defined not by fossils but by radiometric age belts and paleomagnetic polar paths. Unfortunately, the key targets for study lie within interior Eurasia in places that are largely inaccessible politically to field geologists from elsewhere.

Dating of the rifting event associated with Windermere deposition is not yet satisfactory. Interbedded basaltic volcanics near the international boundary were erupted roughly 850 m.y.b.p. However, continued terrestrial deposition above the lavas locally suggests that final continental separation may have been much later. A key component of the Windermere succession along the whole length of the Cordillera is a unit of tillites and associated glaciomarine turbidites. These strata may ultimately provide unique constraints on the timing and paleolatitude of the rifting, although better data are needed to proceed further.

The tillite sequence probably reflects deposition either at high paleolatitudes or within and near highlands presumably formed by thermotectonic uplift along the rift belt prior to and during continental separation. Thus, if the tillites are truly correlative in a time-stratigraphic sense, either a uniformly polar position for the rift belt or simultaneous rifting may be implied. Independent paleomagnetic data potentially can resolve such a question. If the tillites are not coeval, but occur in stratigraphic association with comparable depositional phases of the Windermere rift sequence, then diachronous rifting would be suggested. Diachronous rifting would also be indicated if the tillites are coeval but associated with disparate depositional phases of the Windermere rift sequence. In the general case, of course, plate movements would be expected to generate some diachroneity of both glaciation and rifting across any region as large as the whole Cordillera.

Paleozoic Miogeocline

Between southeastern California and southeastern Idaho, the strike of the Cordilleran miogeocline is generally continuous across Nevada and Utah in a belt now 250 to 500 km wide (Stewart and Poole, 1974). The aggregate thickness of pre-Antler strata increases regularly from perhaps 1000 m at a hinge line on the southeast to about 10,000 m near the Paleozoic orogenic front on the northwest. The shape and dimensions of the sediment wedge are similar to those present beneath the continental terraces or shelves that border the Atlantic Ocean today. Moreover, the rate of sedimentation of shelf deposits within the Cordilleran miogeocline declined logarithmically through time in a manner compatible with the inferred thermotectonic subsidence of a rifted continental margin (Stewart and Suczek, 1977). A basal clastic phase of latest Precambrian to mid-Cambrian sandstones with associated shales and conglomerates attains local thicknesses of as much as perhaps 7500 m, whereas overlying mid-Cambrian to mid-Devonian carbonates and shales nowhere exceed 5000 m in thickness. Some of the initial clastics were probably shed from relict highlands associated with the rift belt that delineated the Cordilleran edge of the continent, but most were dispersed westward from the stable craton of the continental interior. Carbonate sediment was generated along the miogeoclinal trend itself, within a belt of organogenic productivity in the carbonate factory at shelf depth in marine waters (Matti and McKee, 1977).

The overthrust Roberts Mountains allochthon of the Antler orogen contains complexly deformed lower Paleozoic turbidites, argillites, cherts, and greenstones. These strata are interpreted as the deposits of the continental rise and ocean basin that lay adjacent to the rifted Cordilleran margin during early Paleozoic time; transitional facies probably represent continental slope deposits (Stewart and Poole, 1974).

Marginal Offsets

The failure of the Cordilleran miogeoclinal assemblage to continue along strike into central Idaho or into the modern Pacific Ocean off California requires explanation. Several authors have suggested that the miogeoclinal trend was truncated by a rifting event that carried away its southwestern extension during late Paleozoic or early Mesozoic time (e.g., Hamilton, 1969; Burchfiel and Davis, 1972). However, this type of explanation cannot apply to its truncation on the northeast against a Precambrian projection of the continental interior. Perhaps neither termination of the Nevada segment of the miogeocline reflects a postdepositional rifting event.

In plan view, most modern Atlantic-type continental margins display a jagged shape that reflects a compound origin as short rift segments linked by marginal offsets. The latter are oriented parallel rather than perpendicular to the spreading direction, and hence undergo transform faulting during continental separation and hinge faulting during later thermotectonic subsidence. Severe gashing or even rupture of the continental block may occur at the marginal reentrants where paired rift segments and marginal offsets meet. Moreover, the sedimentary history of rifted-margin segments and marginal-offset segments

of the same continental margin may be distinctly different. Consequently, the regional distribution of miogeoclinal sedimentary units can be strongly affected by the locations of marginal offsets inherited from earlier rifting events.
affected by the locations of marginal offsets inherited from earlier rifting events.

Thus, the tectonic discontinuities that terminate the Nevada segment of the miogeocline in central Idaho and southern California are interpreted here as the record of marginal offsets of Precambrian age (see Fig. 1-3). Their trends together define the direction of rift movements during the continental separation that initiated the miogeocline. On the northeast, the trend of the offset is given generally by the elongation of the Lemhi or Salmon River Arch. The same orientation is suggested also by a line connecting major centers of Windermere volcanism in southeastern Idaho near Pocatello and along the international boundary between northeastern Washington and southeastern British Columbia. These volcanics may mark the position of eruptions akin to those that occur along marginal fracture ridges or within marginal reentrants. On the southwest, the trend of the offset is given generally by a line joining the western side of the Mojave block, where the miogeoclinal belt is last seen in California, with the area where thick Paleozoic strata reappear near Caborca in Sonora. Offsets within the much younger San Andreas transform system of similar trend have clearly spoiled any chance to view straightforward relationships along the older structure. For example, the western fringe of the Mojave block has now been faulted far to the northwest as part of the Salinian block in the Coast Ranges.

LATE PALEOZOIC AND EARLY MESOZOIC

The following stage in the evolution of the California continental margin saw the development of a complex Japanese-type morphology with offshore island arcs and marginal seas that influenced the evolution of the continental margin itself. In detail, the system was probably more complex than the Japanese system of the northwest Pacific and may have more closely resembled the Melanesian system of the southwest Pacific (cf. Churkin, 1974b). In particular, the polarity of the offshore island arcs was in part or at times reversed, with the trench on the continental flank of the island arc. This configuration allowed crustal collisions (Fig. 1-4) to occur between active island arcs and the passive continental margin inherited from the previous Atlantic-type stage of evolution. The Devono-Carboniferous Antler Orogeny and the Permo-Triassic Sonoma Orogeny both probably reflect collisions of this type along the continental margin where the Cordilleran miogeocline had been deposited (Dickinson, 1977). Thus, the Japanese-type configuration of the continental margin apparently prevailed from roughly mid-Devonian to mid-Triassic times.

During both the Antler and Sonoma events, oceanic assemblages of turbidites, cherts, argillites, and greenstones were thrust as allochthons across the western flank of the Cordilleran miogeoclinal belt. The Roberts Mountains allochthon of the Antler orogen contains only lower Paleozoic strata of Cambrian through Devonian age. The Golconda allochthon of the Sonoma orogen contains only upper Paleozoic strata of Car-

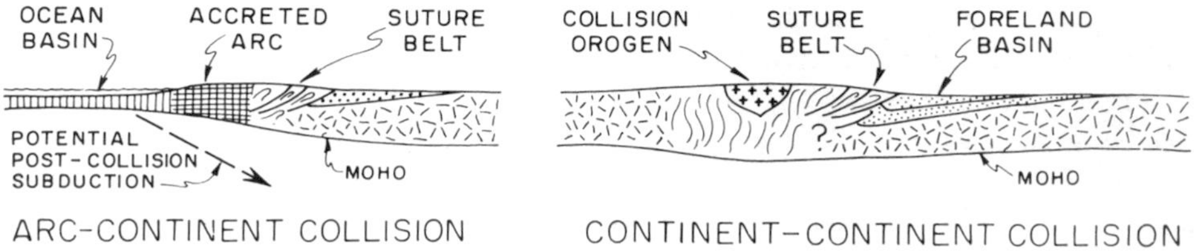

Fig. 1-4. Scenarios for crustal collision. Suture belts mark locations of former ocean basins closed by subduction. See text for discussion.

boniferous and Permian age. The internal structure of both allochthons ranges from complicated to chaotic. Recumbent isoclines, imbricated thrust sheets, schuppen zones of lenticular fault slices, sheared melange bands, and massive olistostromal layers are all present locally. Such tectonic complexity involving oceanic rocks is the mark of a subduction complex formed by the structural telescoping and tectonic stacking of successive increments of oceanic strata peeled by detachment off the top of an oceanic slab descending into the mantle at a subduction zone.

To emplace such a subduction complex upon a previously passive continental margin seemingly requires that the continental block be drawn partly beneath the subduction complex during crustal collision between the continental margin and an arc-trench system, which might be an intraoceanic island arc or might lie along an active continental margin (Fig. 1-4). Without the kind of tectonic contraction and plate descent implied by crustal collision, it is not clear how the stratified fill of the adjoining ocean basin can be either deformed or uplifted enough to cover the edge of the adjacent continental block with a complex allochthon.

The chief alternative for the emplacement of the two allochthons is the concept of backarc thrusting behind an offshore island arc. Although backarc thrusting is well known in association with continental-margin arcs, its occurrence has not been reported in connection with intraoceanic arcs. For an intraoceanic arc, behavior that included backarc thrusting would presumably be transitional to polarity reversal, whereby the main subduction zone shifts from one flank of the island arc to the other. The degree of tectonic dislocation and stratal disruption encountered in the two allochthons is clearly greater than that observed within typical backarc fold-thrust belts, such as the younger Cordilleran belt discussed later. Consequently, the crustal collisions responsible for the Antler and Sonoma events may well have stemmed from polarity reversal of offshore island arcs (cf. Moores, 1970; Speed, 1977), but not from minor backarc thrusting without polarity reversal.

Marginal Seas

With available data, the origins of the late Paleozoic marginal seas remain uncertain (Burchfiel and Davis, 1972; Silberling, 1973; Churkin, 1974a). They apparently were not simply interarc basins formed by backarc spreading, because there are no calc-alkalic igneous assemblages that could represent remnant arc terranes associated anywhere with the strata of the miogeoclinal belt that flanked the marginal seas (Dickinson, 1977). They were perhaps most likely to have been trapped marginal seas formed when initiation of subduction within a preexisting ocean basin formed intraoceanic island arcs not far from the Cordilleran continental margin. The possibility cannot yet be excluded, however, that they were residual marginal seas with only transient existence as remnants of large ocean basins. This idea holds that each wide ocean shrank in response to continued subduction beneath the flank of an exotic island arc that finally collided with the continental margin after the intervening ocean had closed completely (cf. Moores, 1970). In any event, the initial widths of the marginal seas are unknown at present.

Interpretations hinge critically upon whether the late Paleozoic island arc terranes of the present Klamath Mountains and Sierra Nevada stood close to the Cordilleran margin throughout their evolution. If so, prominent pulses of Devono-Carboniferous and Permo-Triassic volcanism along the Sierran-Klamath trend invite correlation with the Antler and Sonoma events of similar timing farther east, and make some brand of backarc thrusting both an attractive and plausible suggestion (e.g., Burchfiel and Davis, 1972). To date, however, there is no paleomagnetic confirmation of the late Paleozoic position of these terranes with respect to the Cordilleran margin. In the overthrust allochthons farther east, there is a surprising paucity of volcanic debris except in direct association with pillow lavas and pillow breccias regarded as fragments of Paleozoic ocean floor. Unless some record of volcaniclastic detritus can be discerned in turbidites of the overthrust oceanic assemblages, it seems unlikely that the strata could represent the fill of marginal seas bounded on one side by fringing island arcs of normal polarity, with subduction downward toward the continent as for modern Japan. Even the presence of island arcs of reversed polarity should be reflected by the occurrence of ash layers within the sediment sequence for all times during which the island arcs stood close to the Cordilleran margin, unless prevailing winds blew consistently or intermittently offshore. A careful census of the volcaniclastic detritus within the Roberts Mountains and Golconda allochthons thus deserves high priority. It is my impression that little if any calc-alkalic debris is present in either assemblage.

Antler Orogeny

The Antler Orogeny is the more mysterious of the two Paleozoic orogenies, for we really cannot be sure what entity collided with the Cordilleran margin then. For all we actually know, it could have been a continental-margin arc on the edge of another continent, although an intraoceanic arc is ordinarily postulated. The reason for the uncertainty is that only the overthrust subduction complex of the Roberts Mountains allochthon subsequently remained permanently attached to the Cordillera. Other elements of the Antler Orogen were evidently rifted away during Carboniferous time, for late Paleozoic ocean floor formed adjacent to the Cordilleran margin between Antler and Sonoma time, just as early Paleozoic ocean floor had formed offshore between Windermere and Antler time.

As a result of the Antler Orogeny (Fig. 1-5), an asymmetric foreland basin formed

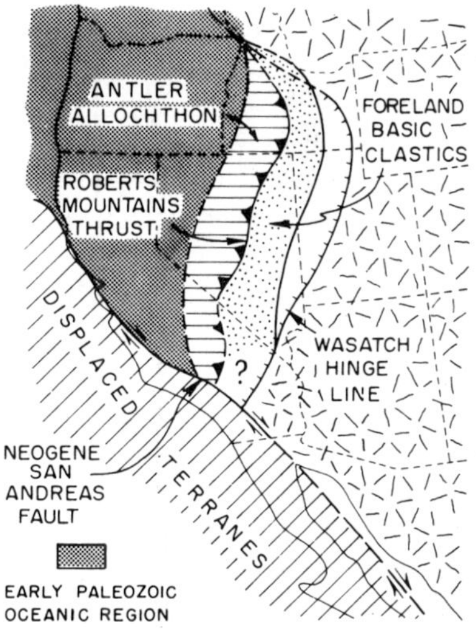

Fig. 1-5. Principal tectonic elements of Devono-Mississippian Antler Orogeny along mid-Paleozoic Cordilleran margin. Modified after Burchfiel and Davis (1972, 1975), Poole (1974), Poole and Sandburg (1977). Wasatch hinge line is the cratonal margin of the miogeoclinal belt (see Fig. 1-3).

east of the Roberts Mountains allochthon where the weight of the allochthon induced a sort of isostatic moat, which subsided by flexure of the continental lithosphere. The keel of the foreland basin lay in the central part of the formerly miogeoclinal belt. Clastic wedges dumped into the foreland basin from the Antler Orogen included both subsea fans and deltaic complexes (Poole, 1974; Nilsen, 1977). In mid-Carboniferous time, when erosion had reduced the Antler Orogen to a level subject again to marine transgression, deep basins began to form in several places along the eastern flank of the Antler foreland where only fine clastics had been deposited previously. These basins included the huge Permo-Carboniferous Oquirrh-Sublette basin of Utah and Idaho, as well as poorly understood troughs in the Death Valley area where varied carbonate turbidites and associated strata of similar age are quite thick locally (Dickinson, 1977). Perhaps these post-Antler, pre-Sonoma sedimentary basins reflect subsidence following extensional deformation of the Cordilleran margin at the time that some form of rifting broke up the Antler Orogen.

Sonoma Orogeny

As a result of the Sonoma Orogeny (Fig. 1-6), extensive island arc terranes exposed now in the Klamath Mountains, northern Sierra Nevada, and northwestern Nevada were accreted to the continental margin (Speed, 1977). The time of suturing was roughly mid-Triassic (Dickinson, 1977), although the final incorporation of accreted terranes may have been slightly earlier or later from place to place along the irregular continental margin. At the same time, the Golconda allochthon was thrust partway across the eroded remnants of the Antler Orogen, but no prominent foreland basin was formed farther east. Instead, local but rapidly subsiding basins along the trend of the suture belt received thick clastic accumulations such as those in the Auld Lang Syne Group of Nevada.

The Roberts Mountains and Golconda allochthons of the Antler and Sonoma oro-

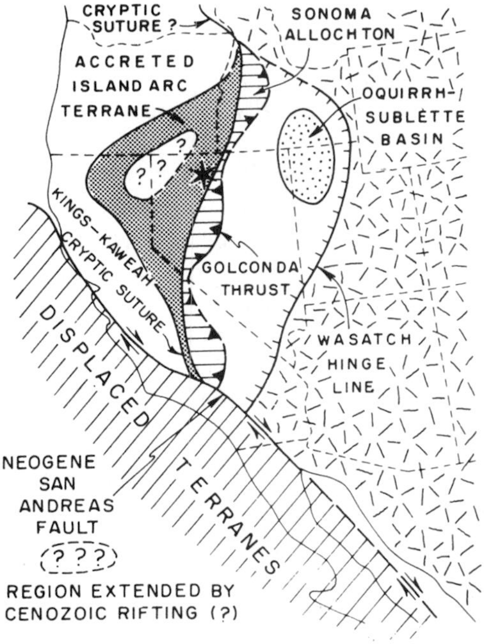

Fig. 1-6. Principal tectonic elements of Permo-Triassic Sonoma Orogeny along accretionary mid-Triassic Cordilleran margin. Modified after Silberling (1973), Burchfiel and Davis (1972, 1975), and Speed (1977). Asterisk denotes location of Auld Lang Syne sutural basin. Blank area between thrust belt and hinge line was largely submerged shelf.

gens are best preserved along the Nevada segment of the Cordilleran miogeocline within a marginal reentrant of the early Paleozoic continental margin (see Figs. 1-3, 1-5, 1-6). To the north, the allochthonous overthrust terranes narrow and eventually pinch out between accreted island arc terranes exposed along Hells Canyon of the Snake River and the projection of Precambrian basement cut by the younger Idaho batholith in central Idaho.

To the south, the overthrust belts strike into the central Sierra Nevada, and are among the tectonic elements that may have been truncated by rifting, together with the parallel miogeoclinal belt. However, the same irregular continental margin and marginal offsets inferred previously to explain the distribution of Paleozoic miogeoclinal facies can also be used to explain the distribution of allochthonous Paleozoic oceanic facies.

The key feature is an isolated remnant of the allochthons located near Garlock Station beside the Garlock fault in the Mojave desert. This Garlock or Mojave remnant of oceanic facies lies well inland from the western limit of Precambrian basement rocks in southern California. It is also much closer to the Wasatch hinge line (see Figs. 1-5 and 1-6), which marks the eastern limit of the miogeoclinal belt, than are comparable klippen farther north. The relations of the Garlock or Mojave allochthon remnant thus suggest important contrasts in the emplacement of thrust sheets along the continental margin during the Antler and Sonoma orogenies. The contrasts can be ascribed as follows to differences in the behavior of rifted-margin and marginal-offset segments of the Paleozoic continental margin during the thrusting associated with later collision events:

1. Simple extensional rifting during inception of the Nevada segment of the rifted margin caused attenuation of continental crust across a wide belt, which subsided to ac-

cumulate the well-developed band of miogeoclinal strata (see Fig. 1-3); during the later emplacement of allochthonous oceanic facies by partial subduction of the continental margin, this broadly attentuated segment was well adapted to extensive preservation of the allochthons, especially along the oceanic flank of the miogeoclinal belt far from the hinge line marking the continental limit of major subsidence.

2. In central Idaho and southern California, however, projections of the continental platform extended anomalously far to the west adjacent to marginal offsets in the initial continental margin; during later thrust emplacement of the allochthons, these relatively intact projections of the continental block could not be subducted as readily as could the attenuated Nevada segment. Consequently, the allochthons either were not emplaced as widely or were eroded more generally in central Idaho and central California than in the intervening ground; where locally present, however, as in the Garlock or Mojave locality, the allochthons extend closer to the stable craton of the continental interior.

In brief, then, marginal reentrants like the Nevada segment underwent relatively mild crustal collision owing to relative ease of partial subduction, and display widespread preservation of oceanic allochthons; by contrast, continental projections induced much stronger deformation during crustal collision, which involved more pronounced telescoping of crustal blocks and more marked uplift leading to general erosion of allochthons (e.g., Dewey and Burke, 1974).

Crucial for a correct interpretation of Sonoma orogenic events are the geologic relations of the complex fault zone termed the Kings-Kaweah suture in the southern Sierra Nevada (Schweickert and others, 1977; Saleeby, Chapter 6, this volume). Commonly viewed as a post-Sonoma transform associated with rift truncation of Paleozoic orogenic trends, the feature may instead be part of the arc-continent suture created by crustal collision during the Sonoma event.

LATE MESOZOIC AND EARLY CENOZOIC

The next stage in the tectonic evolution of California was the inception and maturation of an arc-trench system with a terrestrial volcanic chain standing along an active continental margin of Andean type (Hamilton, 1969). The earliest magmatism clearly related to this tectonic regime was latest Triassic or earliest Jurassic in age (Schweickert, 1976b, in press). The Cordilleran arc-trench system of later Mesozoic and Cenozoic age (Dickinson, 1976) was the local expression of widespread circum-Pacific subduction that began at a similar time along most parts of the Pacific rim. It is surely no accident that circum-Pacific subduction has thus been coeval with expansion of the Atlantic and Indian oceans (Dickinson, 1977). Prior to Atlantic and Indian opening, sea-floor spreading within an expanding paleo-Pacific realm was probably paired in time with the late Paleozoic closure of oceans along suture belts within and marginal to Laurasia during the final assembly of Pangaea. The pre-Jurassic floor of the paleo-Pacific has been wholly subducted beneath the circum-Pacific orogenic system, and the floor of the modern Pacific is entirely Jurassic or younger.

The Cordilleran arc-trench system was probably initiated by flip or reversal of arc polarity following the culmination of arc-continent collision during the Sonoma Orogeny (see Fig. 1-4). Accretion of the eugeosynclinal arc terrane in the Sierra Nevada and adjacent areas terminated westward subduction beneath the island arc and induced eastward subduction beneath the accretionary edge of the expanded continent. Associated volcanics and plutonics that date near the Triassic-Jurassic time boundary by both paleontologic and radiometric methods form the oldest linear belt of fully indigenous igneous rocks along the Cordilleran margin (see Fig. 1-7). This magmatic suite was erupted and emplaced partly within the recently accreted eugeosynclinal terrane of upper Paleozoic island-arc and ocean-floor assemblages, but also crosses the older miogeoclinal belt to punch up through continental basement rocks of Precambrian age in the Death Valley and Mojave regions. Volcanic detritus was first shed into the Colorado Plateau area of the intermountain region within the Upper Triassic Chinle Formation, a fluvial unit that tapped volcanic sources in highlands near the modern Mogollon Rim in Arizona.

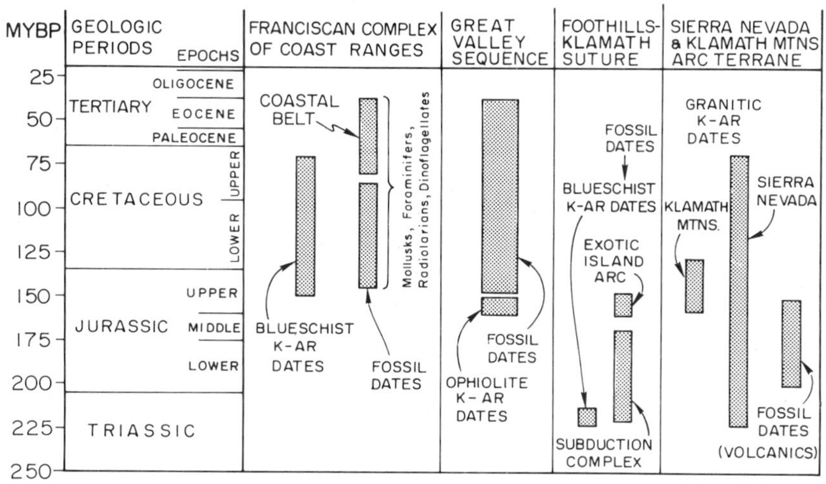

Fig. 1-7. Summary of pertinent age relationships among tectonic elements of Cordilleran arc-trench system. Key data from Armstrong and Suppe (1973), Blake and Jones (1974), Evitt and Pierce (1975), Hotz and others (1977), Irwin and others (1977), Jones and others (1976), Lanphere (1971), Lanphere and others (1978), and Suppe and Armstrong (1972).

Foothills Subduction

Prior to the Late Jurassic (Fig. 1-8), the subduction zone lay along the Sierra Nevada Foothills and within the Klamath Mountains (Schweickert and Cowan, 1975). The subduction complex is a terrane of intricately folded and faulted argillite, ribbon chert, and greenstone within which melangelike and olistostromal units are common, and pods of

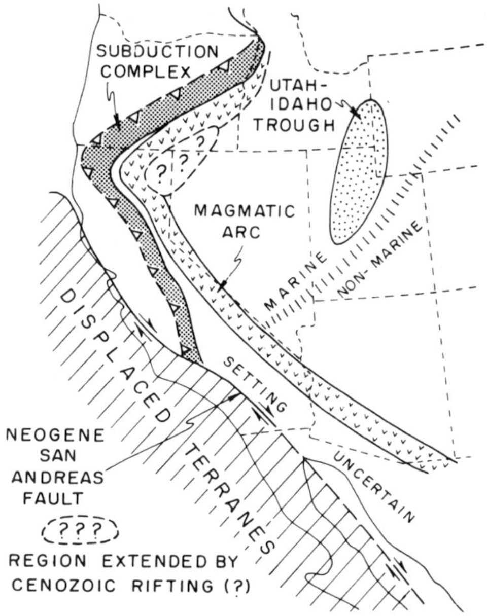

Fig. 1-8. Principal tectonic elements related to Late Triassic to Late Jurassic Foothills subduction zone and suture belt. Modified after Dickinson (1976).

serpentinite and other ophiolitic rocks are widespread. A metamorphic overprint has converted large segments of the belt to slate or phyllite, metachert, and greenschist or amphibolite, as in the Calaveras of the Sierra Nevada. Until recently, all the diagnostic fossils from these oceanic rocks were Paleozoic forms from minor limestone bodies. Dating of radiolarian cherts from the Klamaths has now shown, however, that both Triassic and Jurassic strata are also abundant (Irwin and others, 1977). Probably, the Paleozoic limestones were deposited on seamounts and other sea-floor prominences when the ocean floor was young and still at relatively shallow depth near midoceanic spreading ridges of the paleo-Pacific realm. The younger cherts and argillites were then deposited later after increments of the paleo-Pacific sea floor had subsided prior to insertion into the circum-Pacific subduction system during the Mesozoic. In general, young ocean floor stands relatively shallow and can receive calcareous sediment in favorable latitudes, but old ocean floor cannot receive calcareous sediment once it has subsided below the carbonate compensation depth.

The subduction complexes of combined Paleozoic and Mesozoic oceanic strata are caught now within the Foothills suture belt of the Sierra Nevada and along an analogous suture belt traversing the heart of the Klamath Mountains. East of the suture are the previously accreted eugeosynclinal terranes or segments of the preexisting miogeoclinal belt, and west of the suture are arc volcanics of largely Middle Jurassic age. Structural and stratigraphic relations across the suture belt are everywhere complex and not yet fully resolved. In general, however, the mid-Jurassic rocks west of the suture are less deformed and less disrupted than those of the subduction complex along the suture belt. Apparently, a Mesozoic island arc of intraoceanic origin was drawn into the Cordilleran subduction zone and accreted to the continental margin as a unit block or sliver; the crustal collisions involved apparently occurred within the Late Jurassic in California (Schweickert and Cowan, 1975). Farther north along the Cordillera, similar but even larger island-arc terranes of intraoceanic origin were added to the continental margin by analogous crustal collisions between mid-Jurassic and mid-Cretaceous time (Dickinson, 1976).

Franciscan Subduction

When the Foothills subduction zone was choked with the buoyant and exotic island arc and converted to a crustal suture belt, subduction along the expanding continental margin stepped oceanward into the Coast Ranges. Thus was generated the late Mesozoic arc-trench system (Fig. 1-9). The Franciscan assemblage was the subduction complex, the Great Valley Sequence accumulated in an elongate forearc basin within the arc-trench gap, and the Sierra Nevada batholith represents the intrusive roots of the magmatic arc (Dickinson, 1970). From Late Jurassic to Late Cretaceous, radiometric and paleontologic dates indicate that both deposition and metamorphism of the Franciscan Complex, deposition of the Great Valley Sequence, and emplacement of the Sierra Nevada plutons were coeval (Fig. 1-7). The blueschist metamorphic rocks of the Franciscan and the metamorphic wallrocks of the Sierra Nevada together form paired metamorphic belts of the classic kind linked elsewhere to processes within arc-trench systems (Miyashiro, 1967). No other geologic machine known is capable of generating simultaneously the high pressures and high temperatures required in such closely spaced subparallel belts.

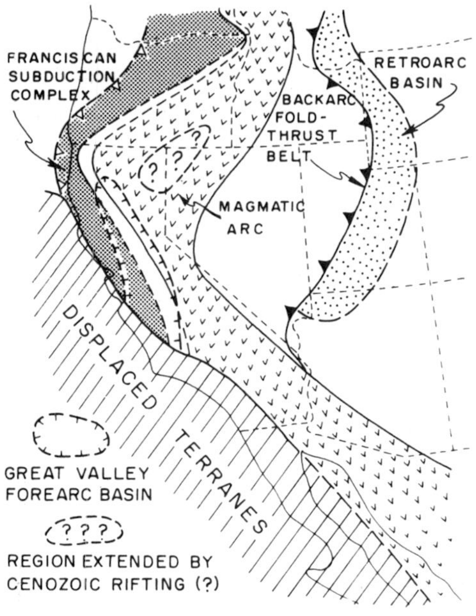

Fig. 1-9. Principal tectonic elements related to Late Jurassic to mid-Tertiary Franciscan subduction in California Coast Ranges. Modified after Dickinson (1976).

Lithologically, the Franciscan Complex is composed principally of lithic-rich and quartzofeldspathic turbidites, but also includes minor greenstone, chert, and limestone (Bailey and others, 1964). The varied facies present suggest that the Franciscan terrane is a vast collage of sea-floor pillow lavas, oceanic pelagites and hemipelagites, seamount volcanics, insular carbonate platforms, archipelagic sediment aprons, abyssal plain turbidites, trench fill turbidites, slope basin deposits, and shaly slope facies. All were presumably shuffled together somehow in the tectonic setting of the trench inner slope. The grade of blueschist metamorphism increases eastward toward the arc terrane (Ernst, 1975), reflecting successively greater depths of subduction.

In detail, several aspects of the Franciscan Complex remain enigmatic (e.g., Blake and Jones, 1974). The question of whether the characteristic lentiform structure of shear surfaces in the disrupted melanges stems from strictly tectonic slicing or from deformation of already isolated blocks floating in olistostromal matrix is still under debate. The juxtaposition of thrust slices of contrasting metamorphic grade is still difficult to reconcile with any plausible geometric scheme for internal deformation within the complex (Suppe, 1972). Perhaps difficulty in understanding stems ultimately from geologic unfamiliarity with the physical conditions that likely exist within a subduction zone. Cool rocks are carried quickly to great depths and deformed there under high strain rates. Unusual conditions of subterranean fluid pressure may develop through deferred dehydration of clays and delayed dewatering of compacting sediment. Pore pressures may thereby greatly exceed hydrostatic and approach lithostatic. Abnormally brittle strain in response to tectonic stress may then be characteristic under such conditions of low temperatures and great depths.

The sedimentary evolution of the Great Valley Sequence conforms well to expected trends (Dickinson and Seely, 1979). Where thickest, it rests depositionally on an ophiolite sequence representing oceanic crust of Late Jurassic age. The oceanic substratum evidently represents a slice of ocean floor that was caught between the subduction zone and the arc massif. As this sliver of ocean floor lay immediately west of the Jurassic island arc that lodged in the Foothills subduction zone (see preceding discussion), it likely was generated by backarc spreading just prior to Late Jurassic crustal collision along the Foothills suture belt (Schweickert and Cowan, 1975). The steep western flank of the regional synclinorium within which the Great Valley Sequence is preserved is underthrust by the eastern part of the Franciscan subduction complex (Ernst, 1970), and imbricated thrust sheets of Great Valley strata are locally present farther west. Components of the accretionary Franciscan assemblage generally young westward on a broad scale (Berkland and others, 1972). An eastern Franciscan belt contains Upper Jurassic and earliest Cretaceous strata, a central Franciscan belt contains mostly Cretaceous strata, and a coastal Franciscan belt contains latest Cretaceous and Paleogene strata. The gentle eastern flank of the Great Valley synclinorium rests depositionally on eroded metamorphic and plutonic rocks of the Sierra Nevada arc massif, across whose eroded roots the basin margin transgressed through Cretaceous time. During the same time span, the axis of plutonism marking the magmatic front retreated eastward from the Foothills belt to the region of the present range crest (Evernden and Kistler, 1970). Thus, the arc-trench gap and the Great Valley forearc basin both widened with time in California.

Within the Great Valley proper (Fig. 1-10), Upper Jurassic strata are largely slope facies that initially prograded into an essentially starved forearc basin ponded behind an incipient subduction complex offshore. Lower Cretaceous strata are mainly turbidites deposited on a deep trough floor by longitudinal flow parallel to the structural grain defined by the basin margins, which were located at the flank of the arc massif on the east and at the position of the accretionary subduction complex on the west. Upper Cretaceous strata are principally subsea-fan turbidites that prograded into the forearc trough from sources in the arc massif. Slope and shelf facies associated with deltaic and shoreline complexes had begun to prograde into the forearc basin by latest Cretaceous time and were dominant within the Great Valley during the Paleogene.

Petrofacies within the Great Valley Sequence consistently reflect sediment derivation from igneous and subordinate metamorphic rocks of the arc massif (Dickinson and Rich, 1972). Not only are plutonic contributions present, but also volcanic debris from cover rocks that have since been almost entirely removed from the source by erosion. Lithic sandstones rich in volcanic rock fragments are dominant near the base of the sediment pile, whereas feldspathic sandstones rich in arkosic detritus derived from granite dominate near the top of the pile. Intermediate horizons contain varied mixtures of volcanic and plutonic materials whose proportions reflect the interplay between magmatism and erosion at different stages in the long evolution of the arc. Key petrologic trends in the igneous suite are faithfully recorded in the sedimentary suite. For example, potassium feldspar first becomes prominent in each at about mid-Cretaceous time. Compositionally, the Great Valley Sequence thus records the changing nature of the adjacent magmatic arc through time. The potassium content of the granitic plutons increases monotonically eastward (Bateman and Dodge, 1970), away from the subduction zone, in harmony with the transverse petrologic trends observed across modern arcs (Dickinson, 1975).

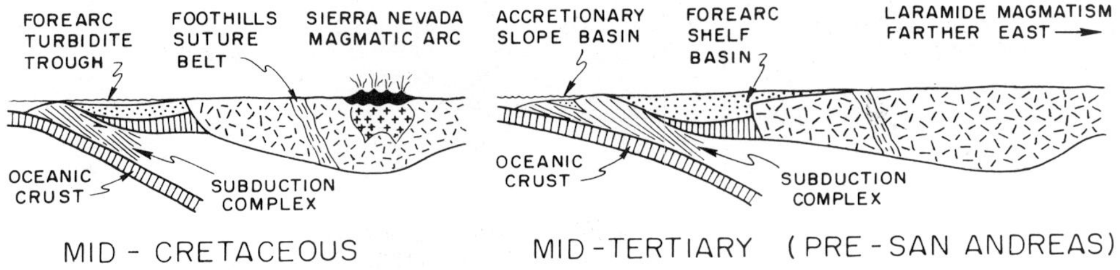

Fig. 1-10. Successive stages (schematic) in the evolution of the Mesozoic-Cenozoic forearc region in northern California. Adapted from Dickinson and Seely (1979).

Backarc Events

Behind the Mesozoic magmatic arc, a backarc fold-thrust belt was prominent along the Sevier belt in the region of the present Great Basin (Burchfiel and Davis, 1975). A prominent retroarc foreland basin occupied the area of the present Colorado Plateau and parts of the Rockies in the late Mesozoic (Fig. 1-9). By the end of Cretaceous time, however, arc magmatism had been extinguished within the Sierra Nevada, and the locus of inland deformation had shifted eastward from the thin-skinned Sevier belt into the classic Laramide region of the central Rocky Mountains (Armstrong, 1974). Laramide deformation involved wholesale crustal buckling of the interior foreland to form fault-bounded and basement-cored uplifts. The Laramide Orogeny in the eastern Cordillera was accompanied by a prominent Paleogene hiatus in arc magmatism covering the whole Great Basin in the western Cordillera. The presence of voluminous Paleogene clastic material in the Coastal Belt Franciscan suggests, however, that subduction continued unabated along the coast. By analogy with reported relations in the modern Andes (Fig. 1-11), the Paleogene magmatic null and the Laramide deformation can be attributed jointly to an episode of shallow subduction beneath the Cordillera (Dickinson and Snyder, 1978). Where plate descent is steep and the descending slab penetrates the asthenosphere, arc magmatism is prominent. Where plate descent is so shallow that the descending plate scrapes along beneath the overriding plate, arc magmatism is suppressed, and internal deformation is widespread across the dormant arc massif.

The Paleogene episode of shallow subduction is perhaps also recorded by the following additional geologic effects: (1) Beneath the granitic Salinian block of the central California coast, geophysical data may imply the presence of underthrust Franciscan rocks in

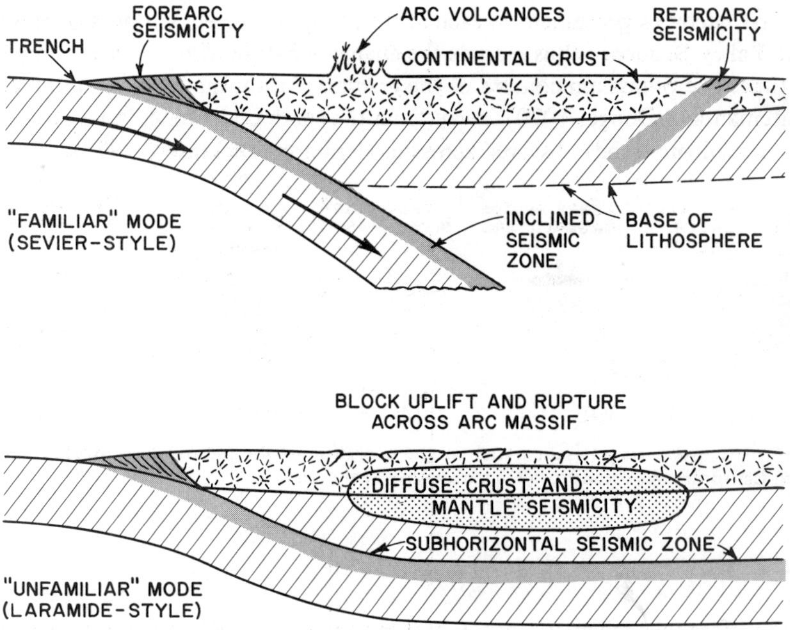

Fig. 1-11. Schematic diagrams to illustrate concept of steep (above) versus shallow (below) subduction to account for contrast between late Mesozoic Nevadan (batholithic)-Sevier (thrust-belt) style of orogeny and Paleogene Laramide Orogeny. After Dickinson and Snyder (1978).

the lower crust below a depth of about 10 km (Stewart, 1968). (2) At various places along and near the San Andreas and Garlock faults where they bound the Mojave block, exposures of Franciscan-like rock occur in the Pelona, Orocopia, and Rand schist terranes. These Franciscan-like terranes may be locally upfaulted through the granitic crust of the Mojave block from an underthrust Franciscan terrane like that present beneath the Salinian block (cf. Burchfiel and Davis, 1972, Fig. 6). The latter tract was presumably once a westward continuation of the Mojave block before offset along the San Andreas fault system. (3) The presence of blocks of Franciscan-type eclogites in Tertiary diatremes of the Colorado plateau (Helmstaedt and Doig, 1975) suggests that some of the underthrust Franciscan Complex was carried well inland without reaching depths much below the base of the overriding lithosphere. These inland occurrences of subducted Franciscan materials, erupted through the continental crust, were probably derived from parts of the subduction complex that had been entrained between the overriding and underthrust plates of the Laramide subduction system (Helmstaedt and Doig, 1975).

LATE CENOZOIC AND PRESENT TIMES

The final stage in the evolution of the California continental margin was the growth of the San Andreas transform system to create a Californian-type margin (Atwater, 1970). The Laramide deformation inland ended by about the end of the Eocene. During the Oligocene, arc magmatism was gradually revived throughout the region of the Paleogene magmatic gap in the Great Basin. By the beginning of the Miocene, a continuous volcanic arc again stood parallel to the whole Cordilleran margin south of Canada. Shortly thereafter, however, coastal subduction began to be terminated by the encounter between the Pacific-Farallon spreading ridge within the ocean and the Farallon-American trench along the Cordilleran continental margin. As increments of the Farallon plate were successively consumed beneath the continental margin, the Pacific and American plates came into contact along the progressively lengthening San Andreas transform. During the Neogene, arc magmatism was gradually interrupted over a larger and larger region where no slab has descended beneath the segment of the Cordillera that is adjacent to the transform. Arc volcanism continues in the Cascades north of the region where the coastal transform system has supplanted the subduction zone.

The San Andreas transform and the San Andreas fault are not strictly equivalent terms. The San Andreas fault is the specific zone along which most transform slip occurs today. In the past, however, other strands of the San Andreas system have served from time to time as the main slip surface of the San Andreas transform. The initial position of the transform was probably offshore close to the continental slope. In general, motion was largely offshore on one or more fault strands during the Miocene, but has been mainly onshore, and chiefly along the present San Andreas fault, during the Pliocene. The peninsula of Baja California began to separate from the mainland near the Miocene-Pliocene time boundary.

Coastal Faults

The correct restoration, for various times in the past (Fig. 1-12), of California coastal slices now lying west of the San Andreas fault can be ambiguous because of fault splays within the transform system. A key example is the Salinian block of granitic and metamorphic rocks along the central California coast. The northern end of this block seemingly has been displaced by about 570 km along the San Andreas fault from counterparts in the Tehachapi Mountains near the Transverse Ranges. However, this inference fails to allow for about 115 km of Neogene offset along the subparallel San Gregorio-Hosgri fault trend (Graham and Dickinson, 1978a, b), which branches off the San Andreas fault near the Golden Gate. Other data constrain the Neogene motion on the San Andreas fault proper in central California to about 305 km. An estimated 50 km of additional net offset can be accommodated by minor displacements along subparallel faults within the elongate Salinian block. About 100 km of the apparent basement offset cannot be explained by

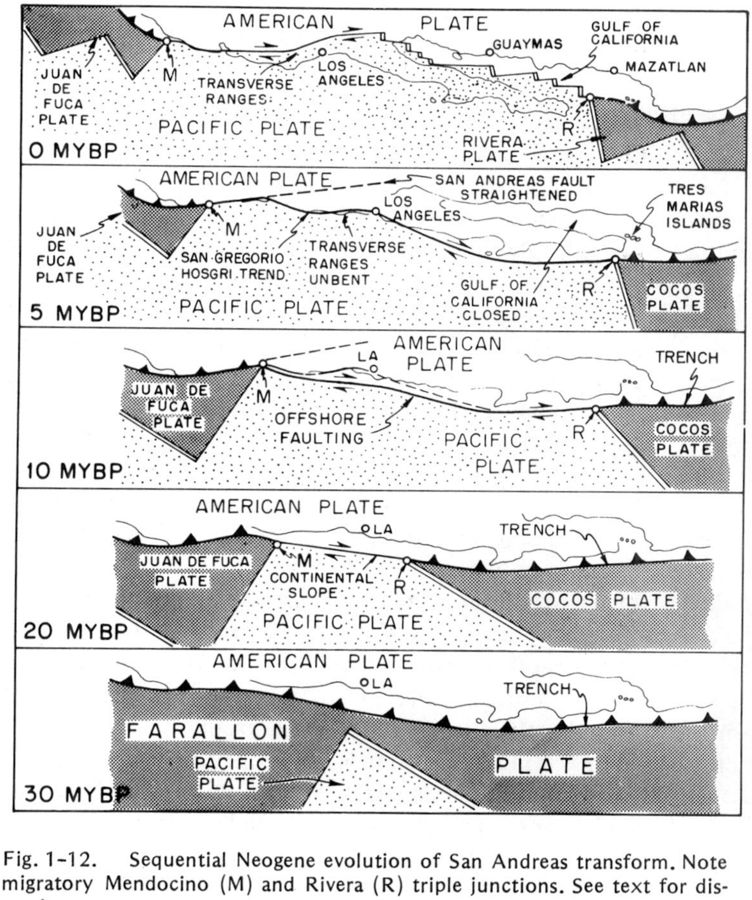

Fig. 1-12. Sequential Neogene evolution of San Andreas transform. Note migratory Mendocino (M) and Rivera (R) triple junctions. See text for discussion.

Neogene slip on any known faults. Presumably, this residual offset reflects subordinate Paleogene strike slip along a proto-San Andreas structure during oblique subduction in the Paleogene. Little is known of this tectonic regime, however, and even less can be said now about the likely magnitude and timing of aggregate lateral motions between the Salinian block and tectonic slices like the Nacimiento block lying still farther to the west. During Paleogene time, however, the Salinian block apparently was part of a fragmented continental borderland including multiple basins and positive blocks (Nilsen and Clarke, 1975).

Coastal Deformation

Neogene tectonics throughout the coastal region of California have been closely related to the San Andreas transform. Widespread structural features of characteristic morphology reflect subordinate contractional or extensional components of motion along the transform (Fig. 1-13). Contractional effects include the development of en echelon wrench folds (Wilcox and others, 1973), especially in the Coast Ranges of central California. Extensional effects include the development of pull-apart basins in southern California (Crowell, 1974a, b), especially within and east of the Peninsular Ranges onshore and within the fragmented continental borderland offshore. Transient and local episodes of basinal subsidence and local volcanism have also occurred in response to the passage of the migratory triple junctions at each end of the growing San Andreas transform (Dickinson and Snyder, 1979).

Most strands of the San Andreas fault system appear to be flexed in the region of the Transverse Ranges. Fault slices that may once have been oriented parallel to the general trend of the dextral San Andreas transform may now be twisted into crude parallelism with the sinistral faults of the Transverse Ranges. Thus, the dextral offset along the San Gregorio-Hosgri fault trend may well be shunted inland north of Point Concepcion to be partly overprinted by sinistral offset along the Santa Ynez fault. If so, the Paleogene turbidite trough of the coastal Santa Ynez Range may now be out of position with respect to the coeval turbidite trough of the inland Sierra Madre Range. Although subparallel now, perhaps the two troughs are offset segments of the same feature. Accordingly, they may once have been aligned as continuous segments of an elongate forearc basin adjacent

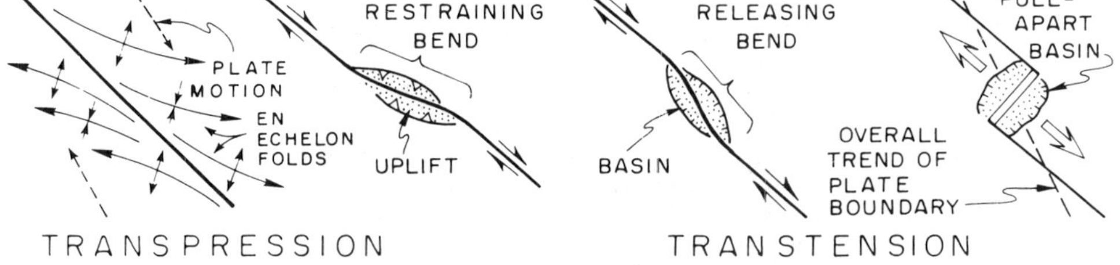

Fig. 1-13. Overall kinematics of contractional deformation from transpression and extensional deformation from transtension along transform fault trends (cf. Crowell, 1974a, b).

to part of the Franciscan subduction complex that forms a structural high in the presently intervening San Rafael Range. Such a postulated dislocation within the Transverse Ranges would have major implications for the geologic history of the Los Angeles and Ventura basins. A proper understanding of structural and stratigraphic relationships within the offshore continental borderland and in the inland Mojave block also hinges critically upon correctly gauging such aspects of the geologic history of the Transverse Ranges.

Arc Switchoff

The evolution of the coastal transform system also had major implications for the course of arc magmatism inland. Arc magmatism depends upon the presence of a subducted slab in the mantle (Lipman and others, 1971). No oceanic slab is subducted along a

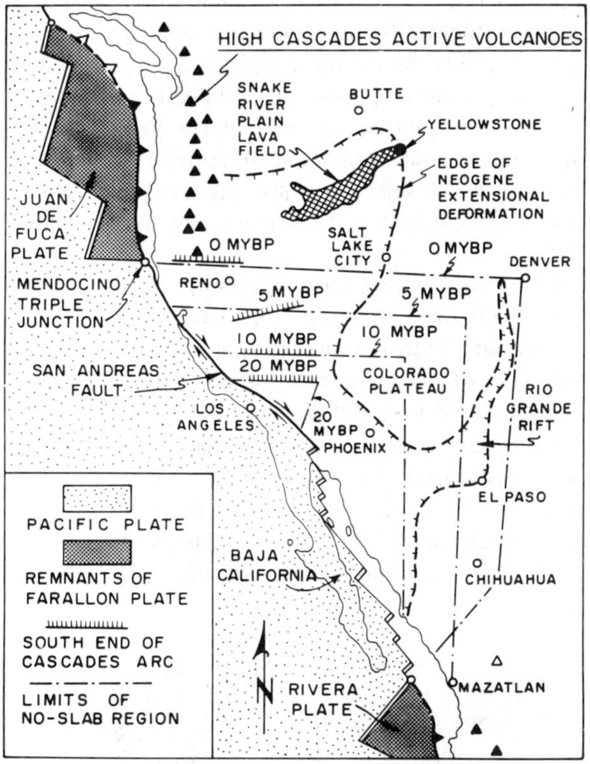

Fig. 1–14. Sketch map illustrating inferred growth of no-slab region above triangular slab-window in subducted slab of lithosphere beneath the Neogene Cordillera. Note apparent relationship to timing of northward retreat of southern termination of Cascades magmatic arc and to areal extent of Basin and Range province of widespread extensional deformation. Adapted from Dickinson and Snyder (1978).

Californian-type transform margin. Consequently, as the transform lengthens, a triangular slab-window develops in the subducted slab, which is inclined beneath the continent; accordingly, a corresponding no-slab region of slightly foreshortened triangular shape develops within the continent (Dickinson and Snyder, 1979). Where the no-slab region exists above the slab-window, arc magmatism is interrupted. The progressive Neogene switchoff of arc magmatism in the southwestern Cordillera was in harmony with the predicted expansion of a slab-window and no-slab region in time and space (Fig. 1-14). Surprisingly, the extent of the no-slab region is also nearly, though not exactly, coincident with the Basin and Range province of extensional tectonics when the latter is taken to include the Rio Grande Rift. Upwelling of hot asthenosphere to fill the expanding slab-window in the descending slab of cool lithosphere is a potential form of mantle diapirism that may have contributed to thermal uplift and stretching of the Cordilleran lithosphere in the Basin and Range province. The late Cenozoic bulk uplift of the Colorado Plateau may stem from the same effect.

PALEOLATITUDE CHANGES AND EUSTATIC FLUCTUATIONS

The role of plate tectonics in the geologic history of the Cordilleran margin has not been restricted to the direct tectonic effects described in preceding sections. In addition, there are two indirect factors that are worldwide in impact. These are (1) changes in paleolatitude as the plates move and (2) fluctuations in mean sea level, relative to the surfaces of the continental blocks, as either the volume of the ocean basins or of the ocean waters changes through time. Both processes exert subtle influences on inland weathering, coastal morphology, and offshore sedimentation that combine to work profound control over the regional nature of the stratigraphic record. The tectonic history of the Cordillera cannot be read properly unless these other, more pervasive signals of shifting paleolatitudes and eustatic changes are filtered correctly out of the geologic data.

Paleolatitude Changes

The general effects of changing paleolatitude are easy to understand. As the plate in which the continental block rides shifted position on the surface of the earth, the Cordilleran margin may have changed both its orientation and its position with respect to paleolatitude. In general, movement across polar, temperate, and tropical zones could change the degree of weathering on land, the direction of the prevailing coastal winds, and various oceanographic parameters of temperature, currents, and tides offshore. These and related factors essentially controlled the nature of shelf sedimentation, which may thus have responded to strong secular trends unrelated to local tectonics along the Cordilleran margin itself.

Available paleomagnetic data allow the broad outlines of such secular trends in sedimentation to be inferred without ambiguity. Paleomagnetic data are ordinarily cast as polar wander paths, which can be plotted on single maps and constitute a continuous inverse record of the movement of continental blocks with respect to the poles through time. An alternate mode of representation that is more ponderous, but more graphic for our purposes here, is a series of maps on which small circles of paleolatitude are drawn separately on each continental block for different intervals or moments of time. From the latter, graphs can be constructed to show the changing paleolatitude and orientation of a given continental margin. This has been done crudely here for the Cordilleran margin near Las Vegas (Fig. 1-15). Generally, tropical paleolatitudes prevailed during much of the Paleozoic while the carbonate-rich sequence of the miogeocline was deposited. A transit of the mid-latitude desert belt of the doldrums occurred during the early Mesozoic when the great dune fields and redbed successions of the Colorado Plateau were formed. Subsequent history has unfolded within the temperate zone.

Glacial Eustatics

The global distribution of continental blocks with respect to paleolatitude can also generate eustatic changes in sea level by promoting or discouraging continental glaciation, which reduces the volume of ocean waters. The transit of the composite Gondwana supercontinent across the South Pole during the Carboniferous and Permian is seen clearly now as the trigger for widespread Permo-Carboniferous glaciation on the southern continents (Crowell and Frakes, 1970). The polar transit, as inferred from the polar wander path for the assembled fragments of Gondwana, was clearly reflected in the gradual shift of glacial maxima from western to eastern Gondwana. Surely the strikingly cyclic coastal sedimen-

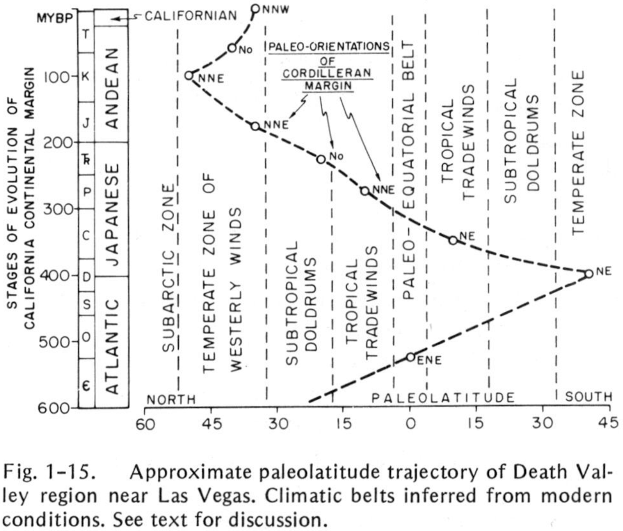

Fig. 1-15. Approximate paleolatitude trajectory of Death Valley region near Las Vegas. Climatic belts inferred from modern conditions. See text for discussion.

tation that prevailed at low latitudes among the Laurasian continental masses during the Permo-Carboniferous reflected the waxing and waning of contemporaneous glaciers in Gondwana (Wanless and Shepard, 1936). Between mid-Mississippian and mid-Permian time, there are about 60 pronounced cyclothems in the mid-continent region adjacent to the Hercynian-Appalachian foreland basin. Cyclic patterns of deposition in Permo-Carboniferous carbonates and shelf clastics of the Cordilleran miogeoclinal belt must harbor similar evidence for glacially controlled global fluctuations in sea level akin to those we know in such detail for the Pleistocene record.

Oceanic Eustatics

Eustatic changes of larger magnitude and longer wave length can be ascribed to changes in the global patterns of sea-floor spreading and subduction. To a first approximation neglecting sediment cover, the depth of the ocean floor is known now to be a simple logarithmic function of its age. The negative elevation below the rise crests, which lie typically at minus 2000 to 3000 m, is proportional to the square root of the age of the igneous substratum below the ocean floor; very roughly, depth equals $(2500 + 350t^{1/2})$ meters, where t is the age of the ocean floor in millions of years (Parsons and Sclater, 1977). Any process that operates to modify, through time, the mean age of the existing ocean floor (in the sense of length of time since its generation) can thus alter the aggregate volume of the ocean basins. As long as water volume remains constant, the freeboard of the surfaces of the continental blocks above mean sea level will also change in harmony, even magnified somewhat by isostatic feedback.

Several workers have suggested that systematic variations in the net global rate of sea-floor spreading would be reflected by changes in the volume of the ocean basins. Doubtless so, except that the implied pulsing of global heat loss would require some sort of thermal hysteresis on an improbably grand scale. Global heat production is itself geared to a rigorously continuous exponential decay governed by the laws of radioactivity. Moreover, the pace of observed eustatic draining and flooding of the continental blocks seems to occur at a tempo that cannot be explained by the thermal decay of static rise systems or the renewal of spreading in a manner much short of explosive (Berger and Winterer, 1974).

Alternately, the mean age of remaining ocean floor can be varied simply by changing the mean age of ocean floor that enters subduction zones. Given the complex global pattern of plate boundaries, this process can promote seemingly random eustatic fluctuations in arbitrarily irregular sequence. If subduction zones in aggregate consume an older and older sample of existing oceanic crust, then the average age of preserved oceanic crust becomes younger, and the volume of the ocean basins decreases. On the other hand, if subduction zones consume ocean floor closer and closer to spreading midoceanic ridges, then the average age of preserved oceanic crust becomes older, and the volume of the ocean basins increases. Continental freeboard increases or decreases accordingly.

The Phanerozoic pattern of major eustatic events is still known only roughly on a worldwide scale. The most pronounced long-term eustatic event during the Cenozoic was clearly the Oligocene drawdown (Ingle and others, 1976). All tectonically stable continental shelves bordering the Atlantic and Indian oceans underwent an episode of erosion or regression that probably was centered on a time early in the Oligocene. Prominent unconformities between Eocene and Miocene beds are widespread beneath many modern shelves, and a prominent hiatus or reduced section is characteristic elsewhere. In California, the Oligocene redbeds of the Sespe and related formations appear within many marine sections without any marked record of local deformation that might have caused uplift.

Perhaps the subduction of the east flank of the ancestral East Pacific Rise beneath the Cordilleran subduction zone allowed sufficiently older oceanic crust to persist elsewhere to cause global draining of the continental shelves. In this fashion, the generation of the San Andreas transform may have left an imprint on worldwide eustatics. This notion should be tested quantitatively, for the eustatic record on the continental blocks is one of the few potential ways in which aspects of the history of pre-Jurassic oceans can be inferred from the stratigraphic record. Unless the correct meaning of the eustatic signal can be discovered for Cenozoic events, the record of older events cannot be interpreted properly. Perhaps California geology, which has played such a key role in the evolution of plate-tectonic theory, can also provide the insight to break the eustatic code!

ACKNOWLEDGMENTS

No one can worry effectively alone about matters like those discussed here. Although none can be held responsible for the thoughts I have presented, each of the following has been generous of especially pertinent ideas in many past discussions with me: R. L. Armstrong, Tanya Atwater, B. C. Burchfiel, P. J. Coney, D. S. Cowan, J. C. Crowell, G. A. Davis, T. W. Dibblee, W. G. Ernst, S. A. Graham, W. B. Hamilton, D. G. Howell, R. V. Ingersoll, D. L. Jones, T. E. Jordan, E. M. Moores, T. H. Nilsen, B. M. Page, E. I. Rich, R. A. Schweickert, N. J. Silberling, W. S. Snyder, J. H. Stewart, and C. A. Suczek. My own research efforts in California have been sustained in recent years by the Earth Sciences Section of the National Science Foundation with NSF grants DES 72-01728 and EAR 76-22636. Thoughtful reviews by W. G. Ernst, D. W. Hyndman, and E. M. Moores significantly improved the manuscript, even though none of the three would subscribe to all the statements and interpretations included.

William P. Irwin
U.S. Geological Survey
Menlo Park, California 94025

2

TECTONIC ACCRETION OF THE KLAMATH MOUNTAINS

ABSTRACT

The Klamath Mountains geologic province is a west-facing arcuate region at the boundary between northwestern California and southwestern Oregon. It consists predominantly of marine arc-related volcanic and sedimentary rocks of Paleozoic and Mesozoic ages. However, ultramafic and other ophiolitic rocks are also important components and are significant to structural and paleotectonic interpretation of the region. Granitic plutons intruded many parts of the province during Jurassic time. Structurally, the

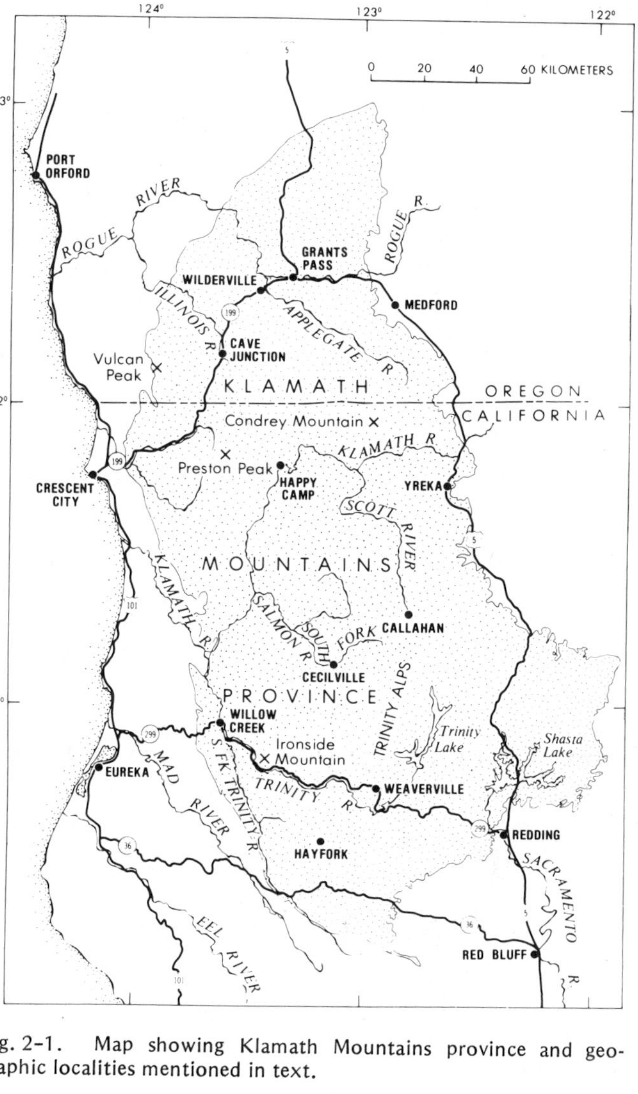

Fig. 2-1. Map showing Klamath Mountains province and geographic localities mentioned in text.

province consists of a series of slices of ancient crust that form an imbricate eastward-dipping sequence.

The province is the product of tectonic accretion of fragments of oceanic crust and island arcs. Paleozoic rocks of the eastern Klamath region formed a nucleus against which other tectonic slices later accreted. The nucleus was a long-standing arc, built on a dominantly ultramafic base, and shows evidence of intermittent volcanism ranging from early Paleozoic into the Jurassic. A layer of amphibolite and mica schist developed beneath the ultramafic substratum of the eastern Klamath region during the Devonian, probably as the result of subduction of the more westward oceanic rocks. Although the record of volcanism in the eastern Klamath region suggests that subduction took place during the late Paleozoic and early Mesozoic, no accretion to the eastern Klamath nucleus seems to have occurred between the Devonian and Jurassic. The various tectonic slices of the western Klamath Mountains were swept against the Paleozoic nucleus only during Jurassic time.

INTRODUCTION

Geologic exploration of the Klamath Mountains province (Fig. 2-1) was begun nearly a century ago by J. S. Diller of the U.S. Geological Survey. His classic field studies particularly in the eastern Klamath region, have provided a foundation for all who followed. Since Diller's work began (ca. 1884), more than 300 geologists have written a total of nearly 1000 abstracts and papers pertaining to the Klamath Mountains. Many of these reports can be attributed to a growing interest in continental borderlands and ophiolitic terranes since the late 1950s, when interest in the geology of the province increased sharply (Fig. 2-2). An early overview of the Klamath Mountains in terms of

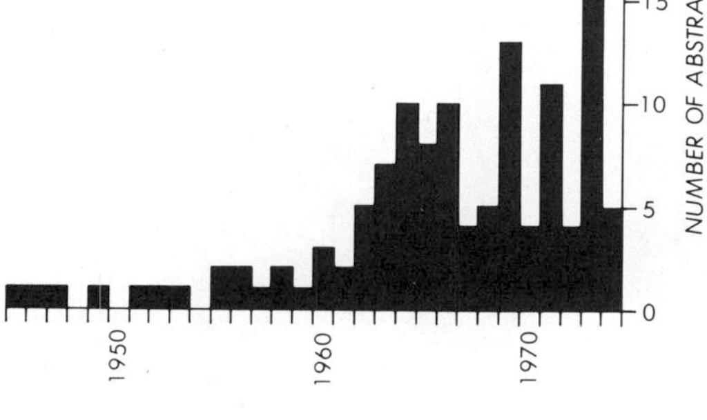

Fig. 2-2. Graph of published abstracts of talks on the geology of the Klamath Mountains for the period 1945 to 1974.

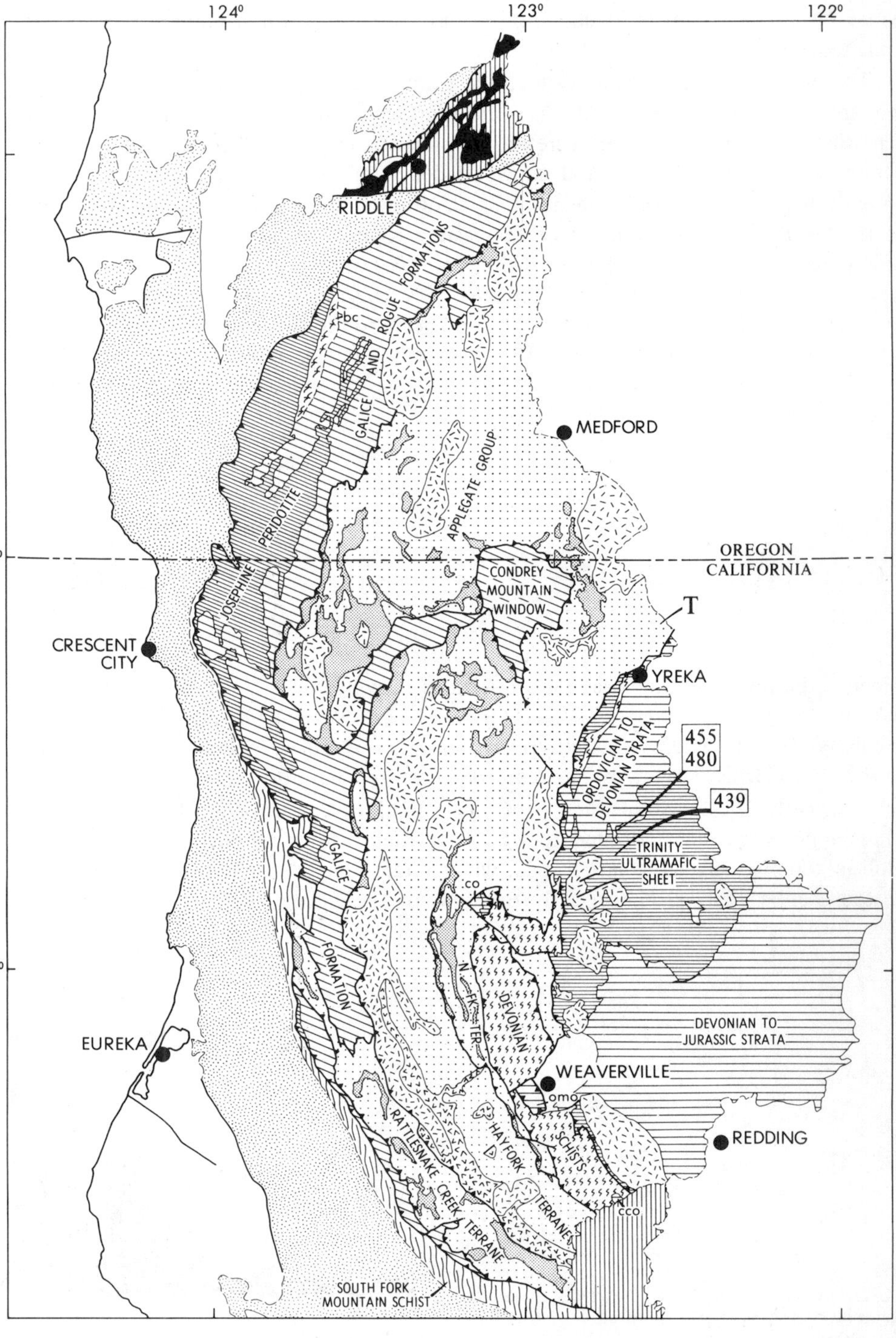

Fig. 2-3. Geologic map of Klamath Mountains province.

Fig. 2-3. (cont.)

EXPLANATION

Tertiary and younger strata; includes minor Upper Cretaceous in California Coast Ranges

Great Valley sequence in California and Myrtle Group in Oregon; Upper Jurassic and Cretaceous; includes Upper Jurassic ophiolitic rocks(oph)

Mesozoic rocks of Coast Range province; mostly Franciscan and Dothan Formations, but includes klippen of Klamath Mountains rocks and Great Valley sequence in Oregon; South Fork Mountain Schist(s) is shown only adjacent to the Klamath Mountains province and includes some semischistose metagraywacke

Western Jurassic plate: mainly Galice and Rogue Formations; includes Josephine Peridotite and related ophiolitic rocks(oph), amphibolite of Briggs Creek(bc), and schist of Condrey Mountain

Western Paleozoic and Triassic plate: mainly rocks of North Fork, Hayfork, and Rattlesnake Creek terranes in southern part of province, undivided rocks in central part, and Applegate Group in Oregon; includes ophiolitic rocks(oph); Hayfork Bally Meta-andesite(ma) shown in Hayfork terrane

Central metamorphic plate: Abrams Mica Schist and Salmon Hornblende Schist of Devonian metamorphic age

Eastern Klamath plate: early Paleozoic strata in the Yreka-Callahan area; Devonian to Jurassic in the Redding area; includes Trinity ultramafic sheet of ophiolitic rocks(oph); klippen of eastern Klamath plate on central metamorphic plate include Cecilville outlier(co), Oregon Mountain outlier(omo), and Cottonwood Creek outlier(cco)

Calc-alkaline plutonic rocks; Jurassic age

455 Radiometrically dated samples of ophiolitic rock

T Permian limestone of Tethyan faunal affinity

a plate tectonic model is in Hamilton's (1969) classic paper on underflow of the Pacific crust and mantle. The present paper will review the lithologies of the various structural slices of the Klamath Mountains, speculate on the ages of associated ophiolitic rocks and melanges, and examine the available data for timing of accretion and suturing.

SUBDIVISION OF THE PROVINCE

In the mid 1950s, when our work in the Klamath Mountains began, most of the western part of the province in California had not yet been mapped. During a project of reconnaissance mapping and compilation of the geology, the province was subdivided into several lithotectonic units that from east to west are called the eastern Klamath belt, the central metamorphic belt, the western Paleozoic and Triassic belt, and the western Jurassic belt (Fig. 2-3; see Irwin, 1960). Some prefer the term "subprovince" for these units instead of "belt" (Davis, 1966). Additional study led to division of the southern part of the western Paleozoic and Triassic belt into three parallel subunits called terranes. From east to west these are the North Fork, Hayfork, and Rattlesnake Creek terranes (Irwin, 1972). The belts and terranes, which are tracts of rock distinguished by a particular association of geologic features, are thought to be thrust plates that form a generally eastward-dipping imbricate sequence. In this report they are called "plates" when referred to as structural units.

Subdivision of the province into lithotectonic units was strongly influenced by the distributional pattern of ultramafic rocks. The reconnaissance during the 1950s revealed that the ultramafic bodies commonly separate rocks of different ages and of different sedimentary and metamorphic facies. These relations and the arcuate linear distribution of the ultramafic rocks are obviously of structural and tectonic significance. The ultramafic rocks were thought to intrude the surrounding rocks, perhaps as cool plastic masses squeezed tectonically along the fault boundaries of plates during regional thrusting (Irwin, 1964). In the past few years, however, our understanding of the evolution of the province has been enhanced by recognition that most of the ultramafic bodies are probably parts of dismembered ophiolite sequences rather than intrusive masses, and that the ophiolites probably represent former oceanic crust and upper mantle. Thus the ultramafic rocks can be viewed as parts of systematic structural-stratigraphic units rather than as randomly intrusive igneous rocks.

EASTERN KLAMATH PLATE

The eastern Klamath plate consists of the Trinity ultramafic sheet at the base, overlain by volcanic and sedimentary strata that range from early Paleozoic to Jurassic in age. Lower Paleozoic (Ordovician to Devonian) strata are exposed in the Yreka-Callahan area at the northern part of the plate. Devonian to Middle Jurassic strata (the Redding section) form a regionally eastward-dipping stratigraphic sequence in the southern and eastern part of the plate. The strata of the Yreka-Callahan area are separated geographically from those of the Redding section by a broad expanse of the Trinity ultramafic sheet. The Paleozoic and Mesozoic strata of the eastern Klamath plate are nearly everywhere separated from the central metamorphic plate to the west by the Trinity ultramafic sheet, except for along the northwest side of the Yreka-Callahan area. Klippen of eastern Klamath plate rest on the central metamorphic plate at three localities.

Yreka-Callahan Area

The lower Paleozoic rocks of the Yreka-Callahan area consist of the Gazelle and Duzel formations. The Gazelle occupies the southeastern part of the Yreka-Callahan area, whereas the Duzel constitutes most of the larger northwestern part. The stratigraphy and structure of these rocks have been much revised during recent studies (Lindsley-Griffin and others, 1974; Hotz, 1977; Potter and others, 1977), and only a brief summary can be given here. The Gazelle is subdivided into several stratigraphic units consisting of shale, volcanic and arkosic wackes, chert, chert-pebble conglomerate, some spilite, and limestone. A conspicuous subunit, the Payton Ranch Limestone Member, is a nodular and thin-bedded limestone overlain by thick cliff-forming limestone.

The Duzel Formation of Wells and others (1959) has been subdivided into six juxtaposed units that may be individual fault slices thrust over the Gazelle (Hotz, 1977). These units are Sissel Gulch Graywacke, Duzel Phyllite, Antelope Mountain Quartzite, Schulmeyer Gulch sequence, Moffett Creek Formation, and limestone of Duzel Rock. They consist of graywacke, phyllite, quartzite, calcareous and sandy siltstone, shale, quartz arenite, chert, and discontinuous bodies of limestone. The prominent landmark, Duzel Rock, is limestone with minor interbedded chert, shale, and basalt. According to Condie and Snansieng (1971), the clastic sedimentary rocks of the Duzel Formation are derived mostly from a plutonic-metamorphic terrane, in contrast to those of the Gazelle, which are from a calc-alkaline (mainly andesitic) volcanic source.

Most of the subdivisions of the Duzel Formation of Wells and others (1959) are not well dated paleontologically, but are thought likely to include both Ordovician and Silurian rocks (Potter and others, 1977). Chert in the Moffett Creek Formation was recently found to contain Ordovician or Silurian radiolarians (Irwin and others, 1978). The age of the Gazelle ranges from Early Silurian to Early Devonian, as indicated by abundant graptolites, corals, brachiopods, and conodonts (Hotz, 1977).

The Duzel Formation along the northwest side of the Yreka-Callahan area is separated by a narrow band of amphibolite from an ultramafic belt that seems to be a northern continuation of the Trinity sheet. The amphibolite yields Devonian isotopic ages (Hotz, 1977) and is considered part of the central metamorphic belt based on age and lithology. However, the amphibolite is not in normal structural sequence with the Duzel and Trinity sheet if it is part of the central metamorphic plate. Its juxtaposition may have resulted from complex thrust faulting (see Fig. 1 in Hotz, 1973). Additional factors relating to the structural complexity of the region are pointed out by Cashman (1977 a,b), who considers the Duzel equivalent in part to the Abrams Mica Schist of the central metamorphic belt, though of somewhat lower metamorphic grade.

Redding Section

Volcanic and sedimentary strata exposed in the Redding section include representatives of all systems of the Paleozoic and Mesozoic from Devonian through Middle Jurassic. Their total exposed stratigraphic thickness is more than 10 km. Lower Paleozoic strata equivalent to those of the Yreka-Callahan area are probably not present below the Devonian of the Redding area and are not shown in the schematic cross section (Fig. 2-4).

The strata of the Redding section are subdivided into sixteen formational units (Fig. 2-5). For a detailed description of these units the reader is referred to the definitive studies of Diller (1905, 1906), Kinkel and others (1956), Albers and Robertson (1961), and Sanborn (1960). The dominantly volcanic units are mainly andesite (Copley Greenstone and Dekkas Andesite) and soda rhyolite (Balaklala Rhyolite and Bully Hill Rhyolite). The sedimentary units are commonly graywacke and shale, and include pyro-

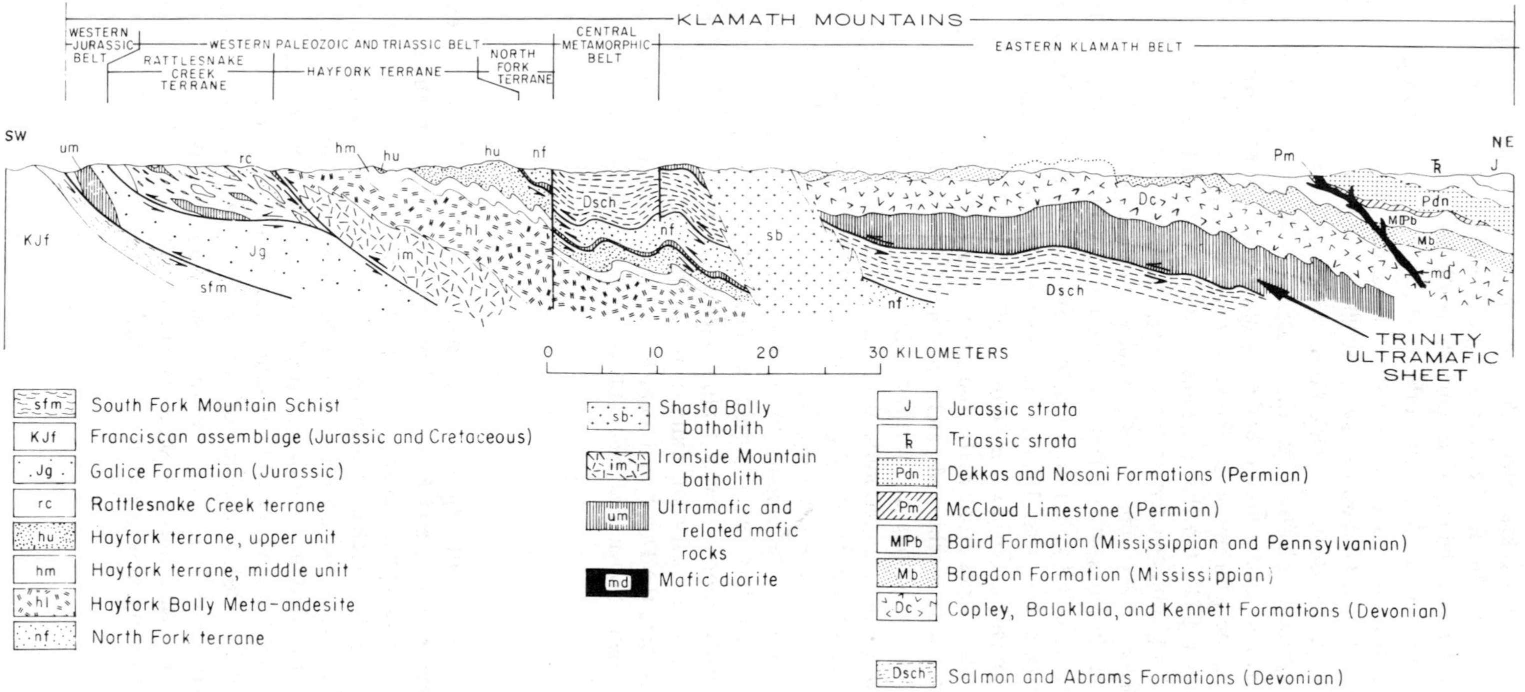

Fig. 2-4. Vertical section trending SW-NE across southern Klamath Mountains, showing major formations and tectonic plates. Modified after Irwin (1977, Fig. 5), with permission of Pacific Section, Society of Economic Paleontologists and Mineralogists.

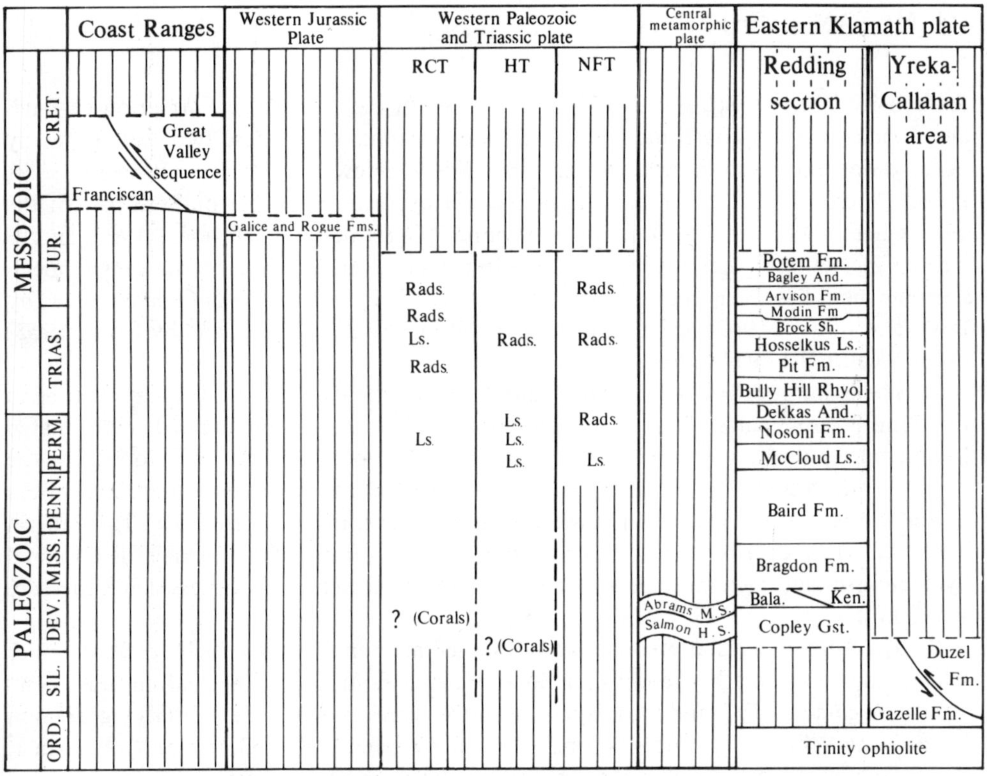

Fig. 2-5. Comparative stratigraphic columns of tectonic plates of the Klamath Mountains and adjacent Coast Ranges.

clastic beds, minor chert, and thin calcareous layers. The conspicuous McCloud Limestone (Lower Permian) locally attains a thickness of a least 800 m. Limestones of the Kennett Formation (Middle Devonian) and the Hosselkus Limestone (Upper Triassic) are thinner and less extensive.

Relations among some of the formations in the Redding section are described as mildly unconformable, but the overall impression is one of general conformability, as shown in Figure 2-4, rather than of discordance. This impression may be misleading, however, for the section seems unlikely to have the extensive continuity of layering shown in the schematic cross section (Fig. 2-4) if the strata represent a volcanic arc that was active during much of the Paleozoic and Mesozoic. Furthermore, an unusual diorite dike (Fig. 2-4, md) intrudes a fault for many kilometers along the trend of the McCloud Limestone and engulfs much of the limestone. The fault may be of greater tectonic importance than commonly believed, because the stratigraphically higher rocks east of the McCloud are generally less deformed than the stratigraphically lower rocks to the west.

Salmon Hornblende Schist and Abrams Mica Schist (Grouse Ridge Formation of Davis and Lipman, 1962) constitute the central metamorphic belt. These two units generally occur along the western side of and structurally below the Trinity ultramafic sheet for an outcrop length of 160 km. Their greatest exposure is in the southern part of the belt, where their width of outcrop commonly ranges between 10 and 20 km. In the Yreka-Callahan area their average width is less than 1 km. Mapping of the central metamorphic belt has been mainly by Hotz (1973, 1977) in the Yreka-Callahan area, by Davis and others (1965) and Cox (1967) in the Trinity Alps region, and by Irwin (1963) in the Weaverville quadrangle.

The Salmon Schist is the structurally lower of the two units and is best exposed in the Trinity Alps region. It is most commonly a fine- to medium-grained, well-foliated hornblende-epidote-albite schist, with a moderately good lineation developed by the parallel orientation of hornblende prisms. Along much of its lower contact with rocks of the western Paleozoic and Triassic belt it is phyllonitic and retrograde. The Salmon Schist was formed from mafic volcanic rocks whose stratigraphic thickness is obscured by tectonism and metamorphism, but whose present structural thickness could measure 1 or 2 km.

The Abrams Schist is predominantly metasedimentary rock and includes quartz-mica schist, micaceous marble, and minor intercalated amphibolite. The marble most commonly occurs as discontinuous layers near the contact with the underlying Salmon Hornblende Schist. The Abrams probably reaches a structural thickness of 2 or 3 km at the trough of a synform in the main area of schist south of Weaverville, but it is tectonically thinned along the eastern side of the belt where it dips eastward beneath the Trinity ultramafic sheet.

The isotopic age of the Salmon and Abrams is Devonian. Rb-Sr ages of approximately 380 m.y. were measured on samples of Abrams from the Weaverville area and from the Coffee Creek area of the Trinity Alps east of Cecilville (Lanphere and others, 1968). K-Ar ages of 390 and 399 m.y. were measured on the amphibolite (Salmon?) along the western side of the Yreka-Callahan area (Hotz, 1977). The Salmon and Abrams are considered to be cometamorphic. There is no evidence for the stratigraphic age of either the Salmon or Abrams, unless one considers the Duzel to be an equivalent of the Abrams, as suggested by Cashman (1977a).

Klippen of Eastern Klamath Plate

Klippen of rocks that correlate with formations of the eastern Klamath plate rest on the central metamorphic plate at three widely spaced localities (Fig. 2-3). One is just north of the South Fork of the Salmon River near Cecilville, a second is the Oregon Mountain outlier west of Weaverville, and the third is the Cottonwood Creek outlier at the southern limit of the province. Ultramafic rocks associated with the klippen at all three localities presumably are detached segments of the Trinity ultramafic sheet. The strata of the Cecilville outlier are chert, phyllite, and sandstone that are correlative with the Duzel Formation of the Yreka-Callahan area (Davis, 1968). Those of the Oregon Mountain outlier are mostly correlative with Bradgon Formation of the Redding section, based on distinctive sandstones and conglomerates, but other formations may be included (Irwin, 1963). The Cottonwood Creek outlier is slate with thin beds of grit and conglomerate and is considered questionably Bragdon (Murphy and others, 1969).

These klippen are of considerable interest for they tend to place limits on the original structural thickness of the underlying central metamorphic plate. It should be noted that the principal broad area of exposure of the central metamorphic plate is a large structural flap (nearly a klippe) resting on rocks of the western Paleozoic and Triassic plate. The flap is detached from its root zone to the east beneath the eastern Klamath plate, except in the region southeast of Weaverville.

WESTERN PALEOZOIC AND TRIASSIC PLATE

Within the structurally complex area west of the central metamorphic plate, a variety of generally undivided rocks is included in the western Paleozoic and Triassic plate. This plate is the most extensive unit in the province, having an exposed length of 300 km and a width that generally ranges between 40 and 80 km. The southern part of the plate is subdivided into the North Fork, Hayfork, and Rattlesnake Creek terranes (Irwin, 1972). In the Cecilville-Trinity Alps region the plate includes mildly metamorphosed strata of the Stuart Fork Formation of Davis and Lipman (1962), and in Oregon the rocks of the plate are called Applegate Group (Wells and others, 1949). The original assignment of these rocks to the Paleozoic and Triassic was based mainly on paleontologic ages of sparsely scattered, fossiliferous limestone bodies (Irwin, 1972; Irwin and Galanis, 1976). However, radiolarians in widespread cherts and tuffs show that the plate also includes strata as young as Jurassic (Irwin and others, 1977 and 1978).

North Fork Terrane

The North Fork terrane occupies a narrow zone along the western side of the central metamorphic plate, ranging generally between 2 and 10 km in width for a length of nearly 100 km. Dismembered ophiolite, including ultramafics, gabbro, diabase, pillow basalt, and red radiolarian chert, forms a selvage along the western side of the zone and is succeeded to the east by siliceous tuff, chert, mafic volcanics, minor lenses or pods of limestone, phyllite, and locally pebble conglomerate. At four localities south of the Trinity River the limestone contains fusulinids and foraminifers, and on the basis of these fossils the North Fork terrane was earlier considered upper Paleozoic. However, recent study of radiolarian-bearing cherts and tuffs has shown that the North Fork terrane at the latitude of Hayfork is mainly Mesozoic rather than Paleozoic. The red chert associated with the ophiolite contains Late Triassic or Early Jurassic radiolarians, and the overlying siliceous tuff and chert contain Early or Middle Jurassic radiolarians

(Irwin and others, 1977). The limestone bodies are now thought to be exotic blocks from an older terrane.

Hayfork Terrane

The Hayfork is the most extensive of the three terranes. It has been traced northward from the Cretaceous overlap at the southern end of the province to beyond the Salmon River, and ranges generally between 15 and 25 km in width. As mapped in the Hayfork quadrangle, the terrane consists mainly of a layered structural sequence of three units and includes the Ironside Mountain batholith. The lowest of the three units is a thick coherent volcanic formation, the Hayfork Bally Meta-andesite (Irwin, 1977), which along much of its known length of more than 80 km is intruded by the remarkably elongate Ironside Mountain batholith and related Wildwood pluton (Fig. 2-3). The meta-andesite grades upward into a middle unit of interlayered chert, argillite, and minor limestone. The upper unit consists of mafic volcanic rocks, siliceous volcanic and sedimentary rocks, chert, phyllite, and minor limestone. Structural disorganization and the presence of seemingly exotic blocks of epidote amphibolite and other metamorphic rocks suggest that the upper unit is a melange. The nature of the boundary between the middle and upper units is not clear, but the presence of thin lenses of serpentinite at some localities suggests that the boundary is a major fault.

Late Paleozoic fossils, mostly foraminifers of Permian age, are found at fourteen localities in limestone of the Hayfork terrane (Irwin and Galanis, 1976). At two of these localities the fossils may have Tethyan faunal affinities. Coarse clasts of limestone in conglomerate at the Mueller mine about 6 km southeast of Hayfork are abundantly fossiliferous, and some of them likely came from the McCloud Limestone of the eastern Klamath plate. In contrast to the Paleozoic limestones, the cherts sampled in the Hayfork terrane are Mesozoic (Irwin and others, 1978). At four localities the chert contains late Triassic or Jurassic radiolarians.

Rattlesnake Creek Terrane

The Rattlesnake Creek terrane crops out for a length of 140 km and a width that ranges commonly between 10 and 15 km. For most of its length it is bordered to the east by the Hayfork terrane and to the west by flysch of the western Jurassic belt. In its northern part, however, it is separated from the Hayfork terrane and is entirely within the flysch of the western Jurassic belt. The terrane consists of a wide variety of rocks, including abundant ultramafic rock, gabbro, diabase, pillow basalt, chert, various mafic volcanic rocks, quartz keratophyre and albite granite, discontinuous lenses and pods of limestone, phyllite, sandstone, and conglomerate. The rocks are greatly disrupted by folding and faulting, and their original relations are even further obscured by widespread slope failure and landsliding. However, the lithologies suggest that much of the Rattlesnake Creek terrane is a dismembered ophiolite. Intermixing of the various rocks, some of which are now known to be of widely differing ages, indicates that the terrane is a melange.

Limestone bodies have yielded paleontologic ages at ten localities in the Rattlesnake Creek terrane. The ages are based on Devonian (?) coral, Carboniferous and Permian foraminifers, and Permian or early Mesozoic foraminifers. Late Triassic ammonites were found at one locality. The presence and extent of Paleozoic rocks other than limestone are not certain. The heterogeneous nature of the terrane, suggested by the various ages of limestone, is further documented by the results of sampling the radiolarian cherts. The cherts were found to contain Mesozoic (Triassic and Jurassic) radiolarians at fifteen localities (Irwin and others, 1978).

Undivided Western Paleozoic and Triassic Plate

The central and northern latitudes of the western Paleozoic and Triassic plate are not subdivided into the three terranes of the southern part. Similar rocks are present, though somewhat more metamorphosed in the central part. In the Cecilville-Trinity Alps region, the eastern part of the belt is mapped as the Stuart Fork Formation, which is also exposed in windows and half-windows in the lip of the central metamorphic plate (Davis and others, 1965; Davis, 1968; Hotz, 1973; Seyfert, 1974). Ando and others (1976) consider the Stuart Fork an equivalent of the North Fork. They also consider the North Fork and Hayfork to be equivalents, based on their mapping along the Salmon-Trinity divide, the names having been applied to the same unit but to opposite limbs of a steeply overturned antiform with an ophiolite core. A northern extension of the Hayfork terrane in the remote Salmon Mountains includes unusual chert-argillite breccia, which is interpreted to be a submarine slide deposit (Cox and Pratt, 1973). A few blocks of limestone are found in the breccia, two of which contain rugose corals of Silurian or Devonian age. To the north, on the Salmon River, rocks correlated with the Hayfork terrane are described in part as tectonic melange (Cashman, 1974).

Only three localities of paleontologically dated limestones are known in the vast area of the western Paleozoic and Triassic plate between Salmon-Trinity Divide and the California-Oregon border. Corals in limestone that is questionably from the main fork of Knownothing Creek, 16 km northwest of Cecilville, are identified as Silurian or Devonian (Merriam, 1961). Foraminifers in thin-bedded limestone on Elk Creek, about 4 km south of Happy Camp, are late Paleozoic (Irwin and Galanis, 1976). Permian fusulinids of Tethyan faunal affinity occur in a small limestone body near the Cretaceous overlap northeast of Yreka (Elliott and Bostwick, 1973).

In the Applegate Group, that part of the western Paleozoic and Triassic belt that extends into Oregon, limestone was early described by Diller and Kay (1909) as occurring in four northeast-trending zones. The two western zones were thought to contain Devonian limestone, and the two eastern zones probably Carboniferous or Triassic. However, these early age designations are now considered to be generally erroneous, and evidence favors a Mesozoic age for most of the fossiliferous limestone. The best locality is a large limestone quarry, about 5 km south of Wilderville, where abundant fossils including Late Triassic corals are found. The only locality of pre-Mesozoic limestone now known in the Applegate is 3 km northeast of Cave Junction, where the limestone contains late Paleozoic foraminifers (Irwin and Galanis, 1976).

Radiolarians were obtained from chert in the undivided western Paleozoic and Triassic plate at several localities along the South Fork of Salmon River, at Quartz Hill near Scott River, and along the Klamath River north of Yreka (Irwin and others, 1978). All are Triassic or Jurassic except for Permian radiolarians and conodonts from red chert near the mouth of Matthews Creek on the South Fork of Salmon River, 8 km northwest of Cecilville. The localities north of Yreka are in the same general area as the Permian limestone of Tethyan faunal affinity (Elliott and Bostwick, 1973). A sample of chert from the Applegate Group in the far western part of the belt contains Jurassic radiolarians.

WESTERN JURASSIC PLATE

The western Jurassic plate is exposed along virtually the entire 350-km length of the western limit of the province. It consists of a thick stratigraphic section of flysch and volcanic-arc deposits of the Galice and Rogue formations, and includes the Josephine Peridotite and related ophiolitic rocks. On the west it is bordered in California mainly by the South Fork Mountain Schist, a narrow belt of regional blueschist of Early Cretaceous metamorphic age (Lanphere and others, 1978), and in Oregon mainly by the Dothan Formation of Late Jurassic and Early Cretaceous (?) age. The Galice consists of slaty shale and graywacke, with some pyroclastic interlayers, and is thousands of meters thick. At some places in the lower part of the section, the volcanic rocks greatly predominate and are known locally as the Rogue Formation (Wells and Walker, 1953). The Rogue is mainly pyroclastic meta-andesite, but includes metabasalt and metarhyolite, and is as much as several kilometers thick where well developed. The presence of the pelecypod *Aucella concentrica* in the Galice indicates a Late Jurassic (Oxfordian to middle Kimmeridgian) age (Imlay, 1959), but datable fossils are not found in the Rogue. Thin-bedded tuff in the Galice, on the Rogue River near its type locality, contains Late Jurassic radiolarians (Irwin and others, 1978).

OPHIOLITIC TERRANES

The largest exposures of ultramafic rock in North America are in the Klamath Mountains province. The general distribution of much of the ultramafic rock was outlined as part of a study of chromite deposits during World War II by F. G. Wells and others of the U.S. Geological Survey. The tabular nature and structural importance of the ultramafic bodies were recognized by the 1960s as an outgrowth of the reconnaissance mapping of the 1950s, but it was not until the early 1970s that the modern ophiolite concept was applied to the province. Most of the ultramafic rocks of the Klamath Mountains are now regarded as parts of allochthonous slabs of ophiolite.

The ophiolitic rocks of the Klamath Mountains are of particular interest because of their unusually wide diversity in age, the geometry of their distribution, and their structural succession. They are important constituents of all the major lithotectonic belts of the province except for the central metamorphic belt. The ophiolitic rocks of the various belts seem to be successively younger slices of oceanic crust and mantle, ranging from early Paleozoic in the eastern part of the province to Early or Middle Jurassic in the western part. The adjacent Coast Range ophiolite is Late Jurassic, and although not in the Klamath Mountains province, it completes the full sequence of ophiolites at this latitude of the Pacific Coast region.

The age of formation of an ophiolite is most directly determined by radiometric dating of its gabbroic or plagiogranitic components, but this does not give the age of the associated ultramafic rocks, nor does it take into consideration the intervals of time that may exist between other lithic elements of the constructional pile. In some instances, the upper part of an ophiolite can be dated paleontologically if the pillow lavas include or are capped by radiolarian chert. An upper limit to the age of some ophiolites is determined by the paleontologic ages of overlying or associated strata, which commonly are flysch or arc-related volcanic rocks; but these ages are interpreted as times of plate convergence at some distance from the spreading center, rather than as times of formation of the ophiolite. Metamorphic rocks such as amphibolites and blueschists also form during plate convergence, and in some instances these also contribute to our interpretation of the age of an associated ophiolite. In addition, regional unconformities and intrusive plutonic rocks are helpful in placing upper limits to the age. Despite inadequate data for precise age assignments, the ophiolite belts described in the following provide a systematic framework for the guidance of further research.

Lower Paleozoic Ophiolite

The Trinity ultramafic sheet is the easternmost ophiolite of the Klamath Mountains province and is the oldest and perhaps largest in the Pacific Coast region. It crops out from beneath the lower Paleozoic and younger strata of the eastern Klamath plate, forming the arcuate western lip of the plate for a length of 160 km, and overlies the central metamorphic plate to the west. The contact with the central metamorphic plate is folded on a large scale, in some places nearly isoclinally, and is offset locally by cross faults (Irwin, 1963; Davis and others, 1965). The fold axes tend to be parallel to the regional trend of the contact, and both compositional layering and foliation in the ultramafic rock are parallel to the contact (Irwin and Lipman, 1962; Lipman, 1964). The exposed lip of the ultramafic sheet ranges in thickness from a few meters to a kilometer or more, and at some places pinches out entirely. Eastward the sheet disappears beneath the Paleozoic strata.

The Trinity sheet consists mostly of tectonitic harzburgite and dunite, with gabbro, pyroxenite, diabase, and plagiogranite. This lithic association led to consideration of the

Trinity sheet as an ophiolite (Mattinson and Hopson, 1972; Irwin, 1973; and Lindsley-Griffin, 1973). Where studied in detail at the south end of the Yreka-Callahan area (Lindsley-Griffin, 1977), the tectonitic peridotite is overlain by a sequence of cumulate gabbro and diorite cut by diabase dikes and sills. The diabase zone seems to grade upward into keratophyric and spilitic volcanic rocks containing secondary calcite and red chert. Overlying the ophiolite is a melange, some clasts of which are peridotite, gabbro, and diorite that may have come from the ophiolite (Lindsley-Griffin, 1977). One of the diorite (trondhjemite) clasts from the melange gave a Pb/U isotopic age of 455 m.y., and a fault slice of gabbro thought to be part of the ophiolite gave an age of 480 m.y. (Mattinson and Hopson, 1972). Gabbroic rocks from farther south in the Trinity sheet gave K-Ar isotopic ages as old as 439 m.y. (Lanphere and others, 1968). Both isotopic ages and structural relations to the Devonian metamorphic rocks are compatible with the idea that the Trinity ophiolite is early Paleozoic (Ordovician?) in age, and that the ophiolite was the substratum on which early Paleozoic and younger strata of the eastern Klamath plate were deposited.

Upper Paleozoic (?) and Triassic Ophiolite

Ophiolitic rocks of probable Triassic age are important constituents of the western Paleozoic and Triassic plate. They occur both as blocks in melange and as large, more coherent slabs. Viewed in the North Fork terrane at the latitude of Hayfork, the ophiolitic rocks are the structurally lowest exposed part and are succeeded upward by mafic volcanic rocks containing interlayered chert, siliceous tuff, and minor limestone. The ophiolite includes serpentinized peridotite, gabbro, diabase, plagiogranite, pillow basalt, and red radiolarian chert. North Fork ophiolite near the latitude of the South Fork of Salmon River is described by Ando (1977) as the core of an antiform that is overturned to the west. A typical ophiolite sequence is not seen there; however, some cumulate gabbro that may be in the upper part of the ophiolite grades upward a few meters into diabase. Tectonitic harzburgite is present between the diabase and overlying pillow basalt. Dikes and sills are absent. Ando (1977) notes the difficulty in comparing the North Fork rocks to "classical" ophiolites.

The Rattlesnake Creek terrane along the western side of the western Paleozoic and Triassic plate is a melange that contains large, chaotically dislocated slabs of peridotite and related ophiolitic rocks, in addition to other volcanic and sedimentary rocks. Near Ironside Mountain, volcaniclastic strata correlative with Hayfork Bally Meta-andesite are reported to rest depositionally on Rattlesnake Creek ophiolite and are interpreted to represent a submarine fan (Charlton, 1978). In the Preston Peak area, farther north along the western edge of the western Paleozoic and Triassic plate, the ophiolite is described by Snoke (1977) as consisting of a basal sheet of serpentinized tectonitic peridotite and minor pyroxenite, overlain and intruded by a diabase complex that is in turn overlain by metabasalt and metavolcanic rocks. The ophiolite here rests in thrust fault contact on rocks of the western Jurassic plate, and is overlain by volcanic and sedimentary strata of the western Paleozoic and Triassic plate.

The Triassic age tentatively assigned to the ophiolites of the western Paleozoic and Triassic plate is based mainly on the presence of Triassic radiolarians in the red chert associated with the ophiolitic rocks at a few places in the North Fork and Rattlesnake Creek terranes (Irwin and others, 1977 and 1978). However, the red chert near the mouth of Matthews Creek contains Permian radiolarians and suggests upper Paleozoic and Triassic plate. Other dated cherts and siliceous tuffs in the plate are Triassic and Early or Middle Jurassic. These scanty radiolarian data seem to be the best available evidence for the minimum age of the ophiolitic rocks. Radiometric ages of the ophiolitic rocks are not available. However, the many examples of late Paleozoic limestone now found in the plate probably formed along island arcs, and it would not be surprising to find in the same plate some tectonic slices of Paleozoic ophiolite on which the arcs were probably built.

Lower or Middle Jurassic Ophiolite

The Josephine Peridotite is exposed for 150 km along the western border of the province, and rivals the Trinity ultramafic sheet in size. Just north of the California-Oregon boundary, where described by Vail and Dasch (1977), the peridotite is associated with cumulate gabbro, diabase, and spilite. There the major element compositions of the rocks are similar to those of ophiolites elsewhere, and the abundance patterns of titanium, zirconium, and rare-earth elements show affinities to oceanic ridge rocks. Slate and graywacke of the Galice Formation depositionally overlie the ophiolite, and although there is no bedded chert at the contact, there are minor amounts of chert between the pillows in the underlying volcanic rocks. Vail and Dasch interpret the ophiolitic rocks as oceanic crust that formed in a marginal basin, near a continent and behind an island arc represented by the Galice and Rogue formations. According to Ramp (1975), the Rogue may represent the sheeted-dike part of an ophiolite at some localities where it is intruded by a multitude of diabase, gabbro, and diorite dikes.

In the Vulcan Peak area farther north, part of the Josephine Peridotite is described by Loney and Himmelberg (1977) as being in fault contact with a terrane of amphibolite and clinopyroxene-bearing ultramafic rock intruded by hornblende gabbro. The ultramafic rock of the terrane overlies the amphibolite, possibly with cumulate magmatic contact. The amphibolite and ultramafic rock were intruded by gabbro during Middle Jurassic time and were then overthrust by the Josephine Peridotite along a south-dipping fault. Both units now structurally overlie the Upper Jurassic or Lower Cretaceous (?) Dothan Formation along the eastward-dipping thrust fault boundary of the Klamath Mountains province. Loney and Himmelberg suggest that rocks of the Vulcan Peak area are parts of an ophiolite, but that relations are somewhat different from an ideal ophiolite if the ultramafic cumulate was deposited on the amphibolite.

Ophiolitic rocks at the northern extent of the Josephine Peridotite near the Rogue River are described by Garcia (1976) as consisting of tectonitic harzburgite and dunite, clinopyroxenite, lherzolite, troctolite, banded anorthosite and gabbro, and quartz diorite. These were folded and metamorphosed prior to the intrusion of 140 m.y. old granodiorite. To the east they are in fault contact with amphibolite of Briggs Creek, which consists of metamorphosed mafic igneous and sedimentary rocks, including bedded chert. The amphibolite is in turn in fault contact with the Galice and Rogue formations. Garcia interprets the ophiolitic rocks and the amphibolite as fragments of oceanic crust on which a volcanic arc was built.

The Josephine Peridotite has not been dated isotopically, but there are isotopic dates on associated rocks that may bear indirectly on its age. Dikes that intrude the peridotite west of Cave Junction give isotopic ages having a strong modal concentration near 150 m.y. (Dick, 1973). The amphibolite of Briggs Creek (Coleman and others, 1976) has an isotopic age of 128 m.y., but this age may have been reset by later intrusion. Metagabbro near Rogue River has an isotopic age of 150 m.y. and is cut by quartz diorite that has an isotopic age of 151 m.y. (Hotz, 1971, and oral communication, 1977).

The significance of the age of the Galice in relation to the Josephine Peridotite is not clear, because the physical relations between the two formations are controversial. Vail and Dasch (1977) state that the Galice overlies the ophiolitic rocks depositionally, a relationship that would place an upper limit of Late Jurassic on the age of the peridotite. Others (e.g., Dick, 1973) consider the Galice and Rogue to be in fault contact with the peridotite, and Harper (1978) interprets the Josephine Peridotite as a regionally overthrust slab lying in fault contact on the Galice. However, considering the available data and the absence of any known older rocks in the plate, it seems permissible to consider the Josephine Peridotite to be a slice of upper mantle or oceanic crust whose tectonic emplacement is closely related in time to the formation of the amphibolite, the igneous intrusions, and to the deposition of the Rogue and Galice.

TIMING OF ACCRETION OF THE KLAMATH PLATES

The various tectonic plates of the Klamath Mountains are remnants of oceanic crust and island arcs that accreted sequentially from east to west. The sutures between the plates are complex eastward-dipping fault zones along which the eastern plates are underthrust successively by the western plates. The timing of underthrusting of the various plates can be ascertained within fairly close limits by considering the ages of rocks that constitute the plates and the ages of metamorphism that occurred in the lower plate as a result of the underthrusting (Fig. 2-6).

The easternmost suture separates the eastern Klamath plate from the central metamorphic plate and is called the Bully Choop thrust fault (Irwin, 1963; same as Trinity thrust zone of Davis, 1968). The thrusting is assigned a Devonian age corresponding to the isotopically dated schists of the central metamorphic plate, because the metamorphism probably occurred when protoliths of the schists were thrust beneath the eastern Klamath plate. The Devonian isotopic dates indicate that the metamorphism occurred early in the development of the island arc that was the nucleus of the Klamath Mountains province. However, there are unresolved complications in this simple relation if part of the early Paleozoic section of the Yreka-Callahan region is indeed a less meta-

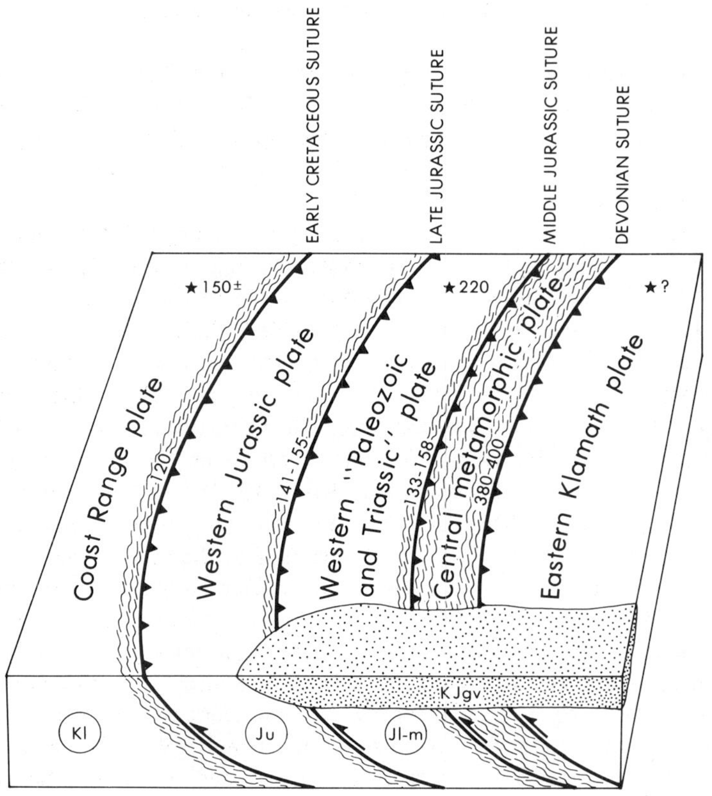

EXPLANATION

155 Isotopic age of rock regionally metamorphosed during underthrusting.

★220 Isotopic age of blueschist knockers.

(Ju) Paleontologic age of youngest strata in underthrust (lower) plate. Jl-m (Lower or Middle Jurassic); Ju (Upper Jurassic); Kl (Lower Cretaceous - Valanginian)

KJgv Great Valley sequence: deposited on accreted plates of Klamath Mountains.

Fig. 2-6. Schematic diagram showing sequential ages of suturing of accretionary plates of the Klamath Mountains and adjacent Coast Ranges.

morphosed equivalent of rocks of the central metamorphic plate, as suggested by Cashman (1977a).

The next major suture to the west is the contact between the Devonian metamorphic rocks and the western Paleozoic and Triassic plate and is called the Siskiyou thrust zone by Davis (1968). West of this suture there is no clear record of subduction occurring between Devonian and Jurassic, although intermittent volcanism in the eastern Klamath plate during the Paleozoic and Mesozoic suggests at least episodic synchronous subduction.

Radiolarian cherts in the western Paleozoic and Triassic plate, involved in collision with the Devonian metamorphic rocks, indicate that the collision culminated during Middle or late Jurassic time. Weakly metamorphosed rocks of the western Paleozoic and Triassic plate near the suture yield K-Ar isotopic ages of 133 and 158 m.y. (Lanphere and others, 1968).

The western Paleozoic and Triassic plate is underthrust from the west by the western Jurassic plate. The time of underthrusting is narrowly constrained by the late Jurassic (Oxfordian to middle Kimmeridgian) age of the Galice and by the Early Cretaceous (Valanginian) age of strata of the Great Valley Sequence that unconformably overlie the Galice. Isotopic ages of about 150 m.y. were measured on semischistose metagraywacke of the Galice (Lanphere and others, 1978), which likely developed its tectonitic fabric with mild recrystallization during thrusting beneath the western Paleozoic and Triassic plate in the vicinity of Condrey Mountain, far to the east of the main area of exposure of the western Jurassic plate (Klein, 1977), and have isotopic ages of 141 m.y. (Lanphere and others, 1968) and 155 m.y. (Suppe and Armstrong, 1972). The isotopic ages of metamorphism of the Galice are similar to the ages of dikes and plutons that intruded the Josephine Peridotite and its associated amphibolite, and perhaps reflect the general contemporaneity of these various aspects of regional underthrusting.

The western boundary of the Klamath Mountains plate is a fault separating the western Jurassic plate from a narrow selvage of South Fork Mountain Schist and related Franciscan rocks to the west. The fault is equivalent to the Coast Range thrust that in the Coast Ranges to the south separates the Great Valley Sequence and its ophiolitic substratum from underlying Franciscan and related rocks. The South Fork Mountain Schist is a regionally developed blueschist-facies rock whose isotopic age of metamorphism is about 120 m.y. (Lanphere and others, 1978). The metamorphism occurred as the Franciscan was thrust eastward under the Klamath Mountains and Great Valley Sequence plates during Early Cretaceous time. The South Fork Mountain Schist is neither intruded by igneous rocks nor depositionally overlain by strata of the Great Valley Sequence.

RELATIONS OF BLUESCHIST KNOCKERS TO SUTURE ZONES

Isolated tectonic blocks of blueschist, now known almost universally as "knockers," occur in three different tectonic settings in northern California: in the Duzel Formation of Wells and others (1959) of the eastern Klamath plate, in the western Paleozoic and

Triassic plate, and in Franciscan melange of the Coast Ranges plate (Fig. 2-6). The metamorphic age of blueschists that form knockers in the Duzel is not known, but the age of those in the western Paleozoic and Triassic plate is about 220 m.y. (Hotz and others, 1977) and in the Franciscan about 150 m.y. (Lanphere and others, 1978). Blueschist knockers are thought by many to indicate the site of a former subduction zone and the age of blueschist metamorphism to date the time of subduction. However, the metamorphic ages of blueschists that occur as knockers in northern California, where known, are significantly older than the indicated times of suturing (subduction) of the tectonic plates containing the knockers. This relation is most evident near the boundary between the Coast Ranges and the Klamath Mountains, where blueschist that occurs as knockers in the Franciscan is Late Jurassic (about 150 m.y.) in age, and where the regional blueschist (South Fork Mountain Schist) formed as a selvage along the suture during the Early Cretaceous (120 m.y.), an age difference of 30 m.y. The knockers in the western Paleozoic and Triassic plate are Triassic (220 m.y.) blueschist, but the time of suturing was Middle or Late Jurassic (about 158 m.y.), a difference of about 60 m.y. In both of these examples the knocker blueschists seem to predate suturing by several tens of millions of years. This difference in age in an actively convergent system suggests that the knocker blueschists formed at sites far removed from the suture zones they now occupy. One might speculate that if the blueschist knockers of the Duzel also follow this pattern, and if the Duzel is a tectonic equivalent of the Abrams Mica Schist, the knockers of the Duzel ultimately should give isotopic ages older than the Devonian age (380 m.y.) obtained on the Abrams. Although blueschist knockers have captured the interest of many geologists, their role in the tectonic development of northern California remains an unsolved geologic puzzle.

B. C. Burchfiel
Department of Earth and Planetary Sciences
Massachusetts Institute of Technology
Cambridge, Massachusetts 02139

Gregory A. Davis
Department of Geological Sciences
University of Southern California
Los Angeles, California 90007

3

TRIASSIC AND JURASSIC TECTONIC EVOLUTION OF THE KLAMATH MOUNTAINS-SIERRA NEVADA GEOLOGIC TERRANE

ABSTRACT

Pre-Mesozoic igneous, sedimentary, and metamorphic rocks in what are now the eastern Klamath Mountains and the northeastern Sierra Nevada constituted an island arc complex that evolved independently of the North American continent. During the Permo-Triassic Sonoma orogeny this complex became accreted to western North America after the oceanic basin that had separated it from the continent narrowed, then closed. No definitive plate models exist for the geometry of closure, but volcanic activity in the eastern Klamath Mountains was not, apparently, interrupted by the collision of Paleozoic arc and continental margin farther east. This relation argues for eastward subduction of oceanic lithosphere beneath the Klamath arc before, during, and after the collisional Sonoma orogeny.

The southwestern United States was affected by a tectonic event in earliest Triassic time that truncated abruptly all northeast-trending Paleozoic tectonic elements and stratigraphic trends (among them the Sierran portion of the Klamath-Sierran arc) and produced a new, northwest-trending continental margin. This event is postulated herein to have been the consequence of left-lateral transform faulting that shifted offset structural and stratigraphic elements to the southeast, presumably into northern Mexico. It is likely that the line of truncation did not extend as far northward as the Klamath Mountains, where the history of latest Permian through Middle Triassic volcanic activity appears to be essentially continuous.

By Late Triassic time the geologically diverse western margin of the continent from the southern Sierra into British Columbia had become the site of oblique (?) plate convergence, eastward subduction of Pacific Ocean lithosphere, and widespread arc magmatism. Oceanic rocks of late Paleozoic-early Mesozoic age and "Calaveras" ("Cache Creek" in Canada) lithology (cherts, argillites, and their ophiolitic basement) were accreted to the continental margin as a consequence of their incomplete subduction along it. A Late Triassic blueschist-grade assemblage of these rocks in the Fort Jones area of the Klamath Mountains probably predates the more extensive accretion of similar but lower-grade rocks of the Rattlesnake Creek and Hayfork-North Fork assemblages farther south in the Klamaths.

Some workers believe that western portions of the Klamath Mountains and Sierra Nevada are comprised of multiple Middle to Late Jurassic island arcs that were "swept" into and accreted against the continental margin during Jurassic plate convergence. Alternatively, we propose that the Middle and Late Jurassic igneous rocks of the two regions might have constituted portions of a *single* evolving arc that was constructed across the sutured plate boundary between the continent and allochthonous rocks of "Calaveras" type. Hence, eastern portions of the Jurassic arc were constructed on a basement of older continental arc rocks, whereas western portions were deposited on previously accreted rocks of oceanic affinity. Internal Middle (?) and Late Jurassic disruption and imbrication of this arc by strike-slip and thrust faulting is believed to have been an intraplate response to continued plate convergence to the west, not a direct expression of the collision and accretion to the continent of multiple arc complexes foreign to North America. The trench and east-dipping subduction zone responsible for Jurassic arc magmatism lay to the west of the present Klamath-Sierran terrane, but it is doubtful that either has been preserved in the geologic record.

INTRODUCTION[1]

The pre-Mesozoic development of the Klamath Mountains-northern Sierra Nevada geologic terrane appears to have been geographically independent of continental North America until the time of the Permo-Triassic Sonoma orogeny. Prior to that time, igneous, metamorphic, and sedimentary rock assemblages in what are now the eastern Klamath Mountains and the northeastern Sierra Nevada comprised a Paleozoic ensimatic island arc that lay an unknown distance from the continental margin in central Nevada (Fig. 3-1). Boucot and Potter (1977) comment that some Ordovician, Devonian, and Permian faunal

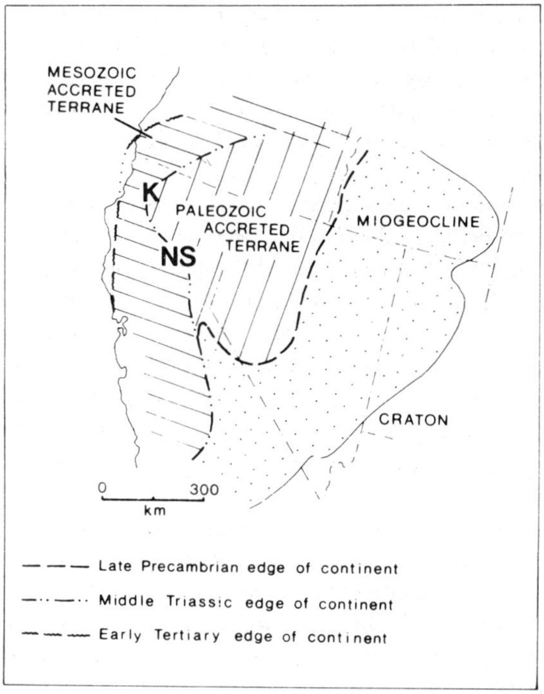

Fig. 3-1. Location of Klamath Mountains (K) and northern Sierra Nevada (NS) with respect to Paleozoic and Mesozoic accreted terranes in the southwestern United States.

[1]Portions of this paper were taken from a geographically more extensive treatment of Mesozoic Cordilleran tectonics by Davis and others (1978).

assemblages of the eastern Klamath Mountains have North American affinities, a relation that suggests relative proximity of the arc to the continent during its development. The Sonoma orogeny represents the collision and accretion of this offshore arc (or arcs) to the continent (Burchfiel and Davis, 1972; Silberling, 1973; Speed, 1974, 1977; see Dickinson, 1977, for a review of possible collisional geometries).

With accretion of the Klamath-Sierran arc by early Early Triassic time, the continental margin shifted abruptly westward from central Nevada to the western or outer side of the former island arc terrane (Fig. 3-1). Metavolcanic and metasedimentary rocks of the Klamath central metamorphic subprovince (Salmon Hornblende Schist and Grouse Ridge Formation) had previously been thrust (subducted) eastward beneath the eastern Klamath arc and its Trinity ophiolitic (?) basement in Devonian time (Burchfiel and Davis, 1972). Thus, following arc-continent collision at the time of the Sonoma orogeny, the outer edge of the arc (the new continental margin) lay somewhere west of presently exposed rocks of the central metamorphic subprovince.

In the eastern Klamath Mountains of northern California, sedimentary and volcanic rocks of Late Triassic (Karnian) and Middle Triassic age are described as lying in a conformable section above Late Permian and Early Triassic volcanic rocks (Dekkas Andesite, Bully Hill Rhyolite); the latter units, in turn, sit with only probable disconformity on sedimentary and volcanic rocks of Middle Permian age (Nosoni Formation). One can, therefore, infer essentially continuous magmatic activity in an eastern Klamath volcanic archipelago before, during, and after the Permo-Triassic orogeny in areas to the east. The tectonic accretion of the Paleozoic arc to the continent in Nevada did not apparently influence the volcanic history of the eastern Klamath subprovince. If the colliding arc and the Klamath-northern Sierran arc were one and the same, the continuity of volcanic activity in the Klamaths argues for *eastward* subduction of oceanic lithosphere beneath the Klamath arc from late Paleozoic into Jurassic time (for contrary interpretations, see Dickinson, Chapter 1, and Schweickert and Snyder, Chapter 7, this volume).

Burchfiel and Davis (1972, 1975) concluded that Permo-Triassic volcanic rocks in the eastern Klamath Mountains (Dekkas Andesite, Bully Hill Rhyolite) were the products of eastward subduction of ocean lithosphere beneath the arc, a geometry of ocean-arc convergence inherited at least from Siluro(?)-Devonian time. They concluded (erroneously) that the probable expression of this convergence is the east-dipping Siskiyou thrust fault, which separates upper plate rocks of the central metamorphic subprovince from lower plate "western Paleozoic and Triassic" units. The Siskiyou thrust fault is now known, however, to be of Jurassic age on the basis of recent field studies (Ando and others, 1977) and the occurrence of Jurassic radiolaria in lower plate rocks of the North Fork terrane (Irwin and others, 1978).

At the present time no candidate structures compatible with east-directed Permo-Triassic subduction beneath the Klamath segment of the arc are known, but the western edge of the arc where such structures might have been present has apparently not been preserved. Late Paleozoic-early Mesozoic (?) sedimentary and volcanic rocks of oceanic affinity that once lay outboard of the Paleozoic arc are probably present in the Fort Jones-Yreka area of the east-central Klamath Mountains, but such rocks do not appear to be widespread in the Klamath region. In the Fort Jones area a presently fault-bounded assemblage of chert, argillite, and mafic volcanic rocks was metamorphosed at considerable depth in Late Triassic time (214 to 222 m.y.a., K-Ar blueschist facies, Hotz and others, 1977). The assemblage may, therefore, be older than lithologically similar rocks exposed to the south in the central Klamath Mountains that Irwin and others describe as having been deposited on an ophiolitic basement of Late Triassic age (North Fork terrane; Irwin and others, 1977). In this latter area, exotic blocks of late Paleozoic carbonates, some containing a Tethyan fauna, are disconcertingly abundant in the Triassic-Jurassic chert-argillite sequence of the North Fork terrane. The Paleozoic source terrane(s) for these blocks is not known. It can be argued that the amount of Jurassic telescoping of Klamath rocks along the overlying Siskiyou thrust fault is extreme, and that the missing late Paleozoic oceanic assemblage (and adjacent trench?) at this latitude lies hidden to the east below the Siskiyou thrust plate. Alternatively, the former outer margin of the Permo-Triassic Klamath arc and rocks of the Pacific ocean basin adjacent to it may have been removed by the rifting or transform faulting that modified the early Mesozoic continental margin to the south (see following discussion).

CONTINENTAL TRUNCATION (POST-SONOMA OROGENY)

The northwest-trending early Mesozoic margin south of the Klamath Mountains (Fig. 3-1) appears not to coincide with the late Paleozoic outer edge of the accreted Klamath-Sierran arc. In central and southern California this margin truncates at a high angle the southwestward projections of all Paleozoic depositional ("eugeosynclinal," miogeoclinal, cratonal) and tectonic (Antler, Sonoma) trends (Figs. 3-2 and 3-3). This geometric relationship argues strongly for Early Mesozoic continental truncation by rifting or transform faulting, as originally proposed by Hamilton and Myers (1966) and schematically represented by Burchfiel and Davis (1972, their Fig. 7).

Schweickert (1976a) proposed that the truncated continental margin is defined by the Melones fault zone of the western Sierra Nevada, although as discussed later we consider the Melones to be a younger fault unrelated to the truncating fault zone. Nevertheless, in the northwestern Sierra, late Paleozoic (Devonian) volcanic rocks of the Klamath-Sierran arc and an older, metasedimentary basement (Shoo Fly Formation) lie directly east of the Melones fault (Fig. 3-3). To the west of the fault and bodies of ultramafic rock that are present along it, lies the "Calaveras Formation," a structurally disarrayed terrane of metacherts, phyllites, metavolcanic rocks, and serpentinized ultramafic rocks that Davis (1969) has correlated with lithologically similar rocks in the Klamath Mountains ("western Paleozoic and Triassic" subprovince). Here, then, the Melones fault does lie between rocks of the Paleozoic Klamath-Sierran arc and a western assemblage of oceanic rocks that is probably at least in part of younger age.

Within the central Sierra Nevada, however, the boundary between pretruncational rocks to the east and accreted rocks to the west is less clear. It certainly lies west of pendants in the eastern Sierran batholith (Log Cabin, Ritter Range, Mount Morrison, and

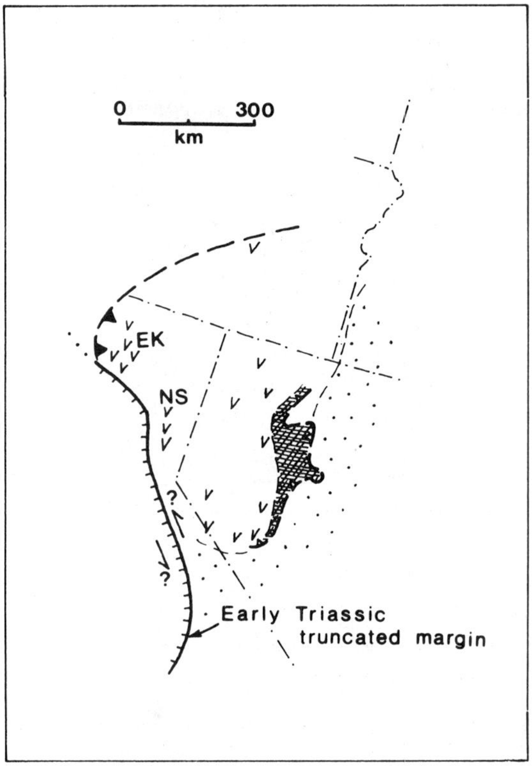

Fig. 3-2. Distribution of Upper Paleozoic rock assemblages with respect to Early Triassic truncated continental margin: clastic deposits related to Devono-Mississippian Antler orogeny (dots); ocean basin or marginal basin assemblage (cross-ruled pattern); volcanic arc rocks (V). EK, eastern Klamath Mountains; NS, northern Sierra Nevada.

Bishop Creek) that can be correlated with Paleozoic terranes of western Nevada (Speed and Kistler, 1977). The margin does not, however, appear to be coincident with the Melones fault, since here "Calaveras" rocks lie east of that fault, not west (Fig. 3-3). Chert-argillite units of the "Calaveras Formation" on both sides of the fault contain limestone lenses that have yielded scarce Permo-Carboniferous fossils, including Permian Tethyan fusulinids (Schweickert and others, 1977). We suggest that *all* "Calaveras" rocks must lie west of the truncated continental margin, and that in the central Sierra Nevada that margin lay east of the Melones fault zone (which must be a younger, unrelated structure) in a region now occupied largely by the Sierran batholith (Fig. 3-3). The "Calaveras Formation" is, in our opinion, a composite assemblage of remnants of one or more oceanic terranes. Older portions of the "formation" were carried into a subduction zone along the truncated margin and accreted to it in Middle or Late Triassic time. This exotic

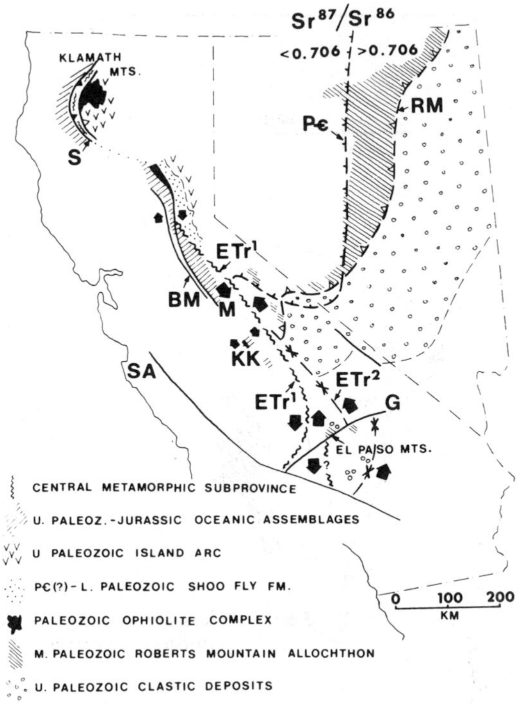

Fig. 3-3. Geologic relations across line of Early Triassic continental truncation. ETr^1, line of continental truncation and hypothesized transform fault with left slip; ETr^2, hypothesized inland fault related to ETr^1; P€, western limit of Precambrian crystalline basement rocks; RM, Roberts Mountains thrust fault of Devono-Mississippian age; S, Siskiyou thrust fault; BM, Bear Mountains fault zone; M, Melones fault; KK, Kings-Kaweah area; SA, San Andreas fault; G, Garlock fault.

oceanic assemblage includes the dismembered Kings-Kaweah ophiolite east of Fresno (Saleeby and others, 1978), which has been dated from plagiogranites as late Paleozoic (250 to 300 m.y., U-Pb, Saleeby, 1979).

The position of the line of truncation farther south is geologically even less definite, again in large part due to extensive Mesozoic plutonism. The plutons, however, themselves provide a clue to its location. Following Kistler and Peterman (1973), the west-to-east change in ^{87}Sr-^{86}Sr initial ratios in granitic rocks of the batholith from $\leqslant 0.704$ to $\geqslant 0.706$ appears to delineate the eastern edge of the accreted "eugeosynclinal" terrane and the westernmost extent of Precambrian crystalline basement. The position of this isotopic boundary, which is coincident in the Southern Sierra with ETr, is shown on Fig. 3-3. Early Paleozoic "eugeosynclinal" rocks that occur in the El Paso Mountains east of the boundary (and north of the Garlock fault) pose a problem to this interpretation. But very similar rocks lie east of the 0.706 line in central Nevada as well—in the Roberts Mountains thrust plate. We postulate that tectonic slicing along a truncated margin of transform type can explain these relations. The allochthonous (?) lower Paleozoic rocks and the upper Paleozoic clastic wedge sedimentary rocks deposited on them (Poole, 1974) may occur within a fault-bounded sliver east of the main transform boundary (Fig. 3-3). Offset of units in the hypothesized sliver from their initial position east of the zone of truncation is left-lateral. Perhaps the late Precambrian-early Paleozoic miogeoclinal section of the Caborca area, Sonora, Mexico, lies within another sliver, for it too is "out of place" in a left-lateral sense with respect to strikingly similar rocks in the Death Valley area. This suggestion is in accord with an earlier hypothesis of Silver and Anderson (1974), although the early to middle Mesozoic left-lateral fault that they proposed crosses the central Mojave Desert and is not coincident with that shown in Fig. 3-3. Van der Voo and others (1977) and Pilger (1978) present geologic arguments pertaining to the Gulf of Mexico region that are supportive of early Mesozoic sinistral fault displacement between the southwestern United States and Mexico. Both papers suggest that the truncated Antler orogenic belt may have been offset from southern California into mainland Mexico.

LATE TRIASSIC AND EARLY JURASSIC HISTORY

Different segments of the earliest Mesozoic continental margin of North America had different evolutionary histories. The margin appears to have been essentially accretional in areas north of the Klamath Mountains, but in southern areas the accreted Paleozoic arc and the continent itself had been subjected to faulting, apparently of transform type, that produced a truncated continental margin. Beginning in Late Triassic time, however, this geologically diverse margin became the site of similar magmatic, sedimentary, and metamorphic phenomena along its entire length, the consequence of oblique (?) convergence and subduction of Pacific Ocean lithosphere beneath the continent (Hamilton, 1969; Monger and others, 1972; Burchfiel and Davis, 1972). Convergence is recorded by widespread arc magmatism with initial ages of circa 220 to 195 m.y. (Late Triassic to Early Jurassic if the Triassic-Jurassic boundary is taken at 212 m.y.a.; R. Armstrong, in press), and by equally widespread and coeval ages for rocks of blueschist facies that lie west of the arc in oceanic rock assemblages (Fig. 3-4). The western margin of the continent was largely submerged, and the volcanic arc appeared as islands that supplied clastic detritus to adjacent basins. Marine basins that lay within or east of the arc were of epicontinental type. To the west of the volcanic archipelago lay a consuming trench, beyond which lay an ocean basin floored by upper Paleozoic and lower Mesozoic rocks of "Calaveras" lithology ("Cache Creek" in Canada).

Klamath Mountains, Oregon and California

Eastern Klamath arc Late Triassic through mid-Jurassic (Bajocian) volcanic arc activity is well preserved in the stratigraphic section of the eastern Klamath Mountains of northwestern California (Figs. 3-4, 6). Broadly correlative volcanic and sedimentary strata

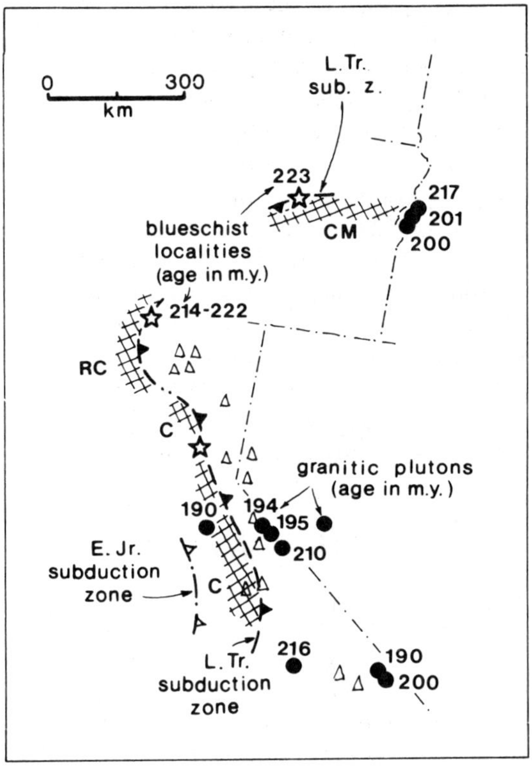

Fig. 3–4. Distribution of late Paleozoic through Early Jurassic oceanic rock assemblages of "Calaveras" lithology (cross-ruled pattern), and Late Triassic-Early Jurassic blueschist localities (stars), volcanic arc rocks (triangles), and granitic plutons (dark circles). C, "Calaveras" assemblages; RC, Rattlesnake Creek, Hayfork-North Fork, and Fort Jones area assemblages; CM, Canyon Mountain Complex, Burnt River Schist, Elkhorn Ridge Argillite.

occur to the south in the "eastern" belt of the Sierra Nevada (Schweickert and Cowan, 1975), where voluminous plutonic intrusion has obscured many initial stratigraphic relations. As mentioned previously, volcanic and sedimentary rocks of Late (Karnian) and Middle Triassic age in the eastern Klamath section sit with apparent conformity on Permo-Triassic rocks (Irwin, 1966). McMath (1966), however, reports that shelf-type carbonates of Late Triassic age (Norian; Hosselkus Limestone, Swearingen Formation) lie with angular unconformity on Permian and probable Permian pyroclastic and volcaniclastic strata in the Taylorsville area of the northern Sierra Nevada.

Rocks in the Klamath-Sierran arc are lithologically diverse and include andesitic lava, tuff, and breccia, siliceous pyroclastic rocks and ignimbrite, argillite, shallow-water limestone, and tuffaceous sandstone. No Late Triassic to mid-Jurassic granitic plutons occur in the Klamath portion of the arc, but a northwest-trending belt of 190 to 210 m.y. old plutons lies along the eastern margin of the Sierran batholith and extends into the eastern Mojave Desert (Fig. 3-4). The discontinuous belt parallels the truncated early Mesozoic margin of the continent and crosscuts older, northeast-trending stratigraphic and structural elements (Burchfiel and Davis, 1972).

Western oceanic assemblages An open ocean basin lay somewhere west of the eastern Klamath arc in Late Triassic and Early Jurassic time. Eastward subduction of oceanic lithosphere beneath the arc is implied by the magmatic activity of the arc and by the occurrence along its present northwestern margin of an allochthonous blueschist terrane of appropriate age (222 to 214 m.y., Hotz and others, 1977). The present structural setting of the blueschists near Fort Jones may or may not, however, date from the time of Late Triassic convergence. Field relations in areas to the south (subsequently treated) suggest that major thrust faulting in the Fort Jones area may be of Jurassic age.

Triassic oceanic crust west of the eastern Klamath and central metamorphic subprovinces may be represented by at least two ophiolitic sequences in the "western Paleozoic and Triassic" subprovince. Identified by Irwin (1972) as the North Fork and Rattlesnake Creek terranes (Fig. 3-5), they lie along the eastern and western margins of the subprovince respectively in areas south of 41°15′N latitude. Irwin has also defined a centrally located lithologic assemblage, the Hayfork terrane, which Ando and others (1977) consider to be largely correlative with sedimentary and volcanic rocks of the North Fork terrane. The Preston Peak ophiolite (Snoke, 1977) lies along the western edge of the subprovince just south of the Oregon border. It may be a northern equivalent of either the ophiolitic Rattlesnake Creek or North Fork terrane, although its age and geologic relations with these terranes is not known.

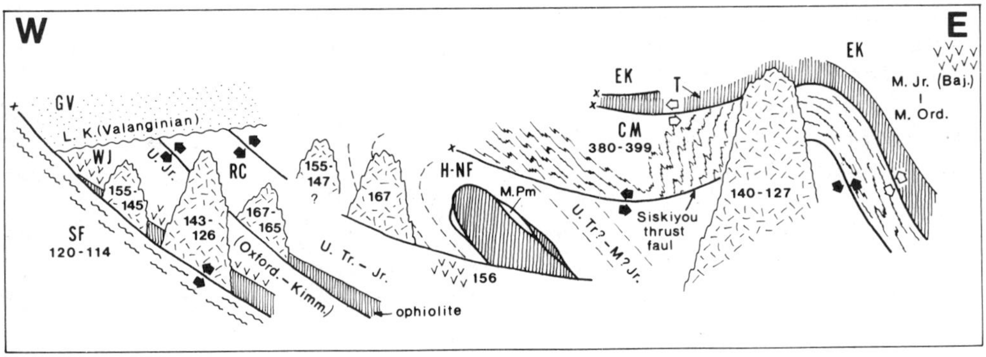

Fig. 3-5. Diagrammatic representation of stratigraphic and radiometric age controls on thrust faulting in the Klamath Mountains, California and Oregon. Length of section is approximately 200 km. SF, South Fork Mountain Schist; GV, Great Valley Sequence; WJ, western Jurassic subprovince; RC, Rattlesnake Creek terrane; H-NF, Hayfork-North Fork terrane; CM, central metamorphic subprovince, T, Trinity ophiolite; EK, eastern Klamath subprovince. Thrust faults with black relative motion arrows are post-Paleozoic in age. Radiometric age data from Hotz (1971), Dick (1973), Young (1974), Lanphere and others (1975), Snoke (1977), and Irwin (1977).

The North Fork ophiolite is tectonically dismembered and lacks the sheeted dike complex characteristic of many other ophiolites. Gabbro exposed in the core of a regional north-plunging antiform (between the Salmon and Trinity Rivers) grades abruptly upward into fine-grained diabase of hypabyssal character (Ando, *in* Ando and others, 1977). Sheetlike units of serpentinized peridotite (North Fork and Twin Sisters bodies) are tectonically interleaved between core gabbro and diabase and structurally higher pillow basalt, chert, and volcaniclastic rocks. Irwin and others (1977) report that radiolarians from red chert "associated with the ultramafic-mafic rocks of the ophiolite" in an area south of the Trinity River are probably Late Triassic. However, southwest of Cecilville, mid-Permian (Wardian) radiolarians and conodonts were found in a block of red chert that lies structurally between the ophiolitic core of the North Fork antiform and the stratigraphic sequence that constitutes its eastern flank (Fig. 3-5; Irwin and others, 1978). The age of the North Fork ophiolite in this area is, therefore, open to question.

North Fork strata in the eastern flank of the antiform include altered mafic volcanic and pyroclastic rocks that appear to lie depositionally above rhythmically bedded cherts and are, in turn, overlain unconformably by shallow-water limestones. Rounded cobbles of metabasalt are present in basal layers of the limestone, which is itself overlain by more rhythmically bedded chert (Davis, 1968). Samples apparently collected from this upper chert unit yield radiolarians of Late Triassic age (Irwin and others, 1978, localities 8, 9, and 10). This stratigraphic sequence is indicative of the growth on the ocean floor of a seamount, its emergence and erosion as an oceanic island, and its subsequent subsidence, all before the deposition of Late Triassic cherts. Exotic blueschist blocks with a lawsonite-glaucophane-(jadeite) mineralogy (E. Ghent, written communication, 1977) occur in a poorly exposed chaotic (olistostromal?) zone (Davis, 1968). The blocks are undated, but some resemble mineralogically the Fort Jones area blueschists, and a 220 m.y. initial age for them seems likely. They are present on both flanks of the North Fork antiform (Fig. 3-5), in the North Fork section to the east and the Hayfork section to the west. This relation supports the equivalency of the "two" terranes (Ando and others, 1977), although a regional east-dipping thrust fault between Hayfork units and the ophiolitic core of the North Fork antiform cannot yet be disproved.

The relation of the North Fork-Hayfork and Rattlesnake Creek ophiolitic terranes to each other is also not clear, although both appear to be essentially coeval (cherts in both have yielded Late Triassic and Jurassic radiolaria). Irwin (1972), however, has previously indicated that the fauna from carbonate pods in the two terranes (now interpreted as exotic blocks) are markedly different and almost mutually exclusive; the Rattlesnake Creek terrane is characterized by a coral-like chaetetid fauna, the North Fork-Hayfork by a fusulinid fauna. Possible paleogeographic relations between the eastern Klamath arc and the North Fork and Rattlesnake Creek oceanic terranes are discussed later, since faults that presently separate these lithologic assemblages are probably all of Late Jurassic age.

Sierra Nevada

The "Calaveras Formation" of the western Sierra Nevada appears to be in part a southern counterpart of the North Fork-Hayfork and Rattlesnake Creek Mesozoic ophiolitic terranes (Davis, 1969). Irwin and others (1978) report the occurrence of Triassic or Jurassic radiolaria in cherts mapped as "Calaveras" in the northern Sierra Nevada. However, the presence of the Late Paleozoic Kings-Kaweah ophiolite in the southwestern Sierra Nevada (Fig. 3-3) upon which Calaveras-type olistostromes were "deposited" (Saleeby and others, 1978) indicates that pre-Mesozoic oceanic rocks are also a component of the accreted terrane. In this southern area a Late Permian Tethyan fusulinid fauna has been collected from the "Calaveras" rocks, specifically from a limestone block in a chert-argillite olistostromal unit (Schweickert and others, 1977). This unit is overlain by continent-derived quartzitic strata of Late Triassic and Early Jurassic age, which are in turn overlain by felsic volcanic rocks. Metamorphic tectonites derived from mafic members of the ophiolite yield K-Ar ages of 190 m.y., an Early Jurassic age considered by Saleeby to be the likely age for emplacement of the ophiolite against the continent. Geologic relations described by Saleeby lead us to the following conclusions: (1) the original truncated margin lies east of all "Calaveras" rocks in the Kings-Kaweah area (Fig. 3-3); (2) late Paleozoic "Calaveras" rocks and their ophiolitic basement were brought against that continental margin following the initiation of Middle to Late Triassic plate convergence along it; (3) subsequent shifting of the zone of convergence to the west of the accreted oceanic terrane was followed, in Early Jurassic time, by its internal disruption along a major strike-slip fault zone (the Kings-Kaweah "suture"); (4) this zone was parallel to an active trench to the west, and its existence implies oblique convergent motion along that plate boundary (Saleeby and others, 1978); and (5) Late Triassic and Early Jurassic sedimentary and volcanic rocks that depositionally overlie western "Calaveras" units (and were also offset by transcurrent faulting) may conceal the older truncational boundary in areas to the east.

Despite convergent accretion of some or all of the "Calaveras" terrane in the Triassic period, Upper Triassic and lowest Jurassic strata in the Sierra Nevada appear to record a time of relative arc inactivity, in contrast to the eastern Klamath Mountains. A number of Sierran stratigraphic sequences of this age are nonvolcanic or, at best, contain only minor contributions from volcanic sources (Schweickert and Cowan, 1975). This relationship is surprising, since scattered eastern Sierran plutons have ages that fall in a Late Triassic-Early Jurassic time period (220 to 190 m.y.) and do attest to at least sporadic magmatic activity within the newborn arc (Fig. 3-4). During later Early Jurassic time, Andean-type magmatism in the Sierran region was evidently more widespread. Thick sequences of volcanic rocks dating from this time are present in the northeastern and east-central Sierra Nevada (Stanley and others, 1971, Fig. 4). Saleeby (1977a) has suggested that in the southern Sierra Nevada, arc volcanism spread far enough westward that Lower Jurassic volcanic rocks and sediments derived from the older "Calaveras" terrane were deposited on top of the ophiolitic complex of the Kings-Kaweah area. Thus, by Early Jurassic time subduction at this latitude had apparently stepped westward from its Middle to Late Triassic position.

The volcanic rocks that lie west of the Melones fault zone in the Sierran foothills have generally been regarded as Middle and Late Jurassic in age (Schweickert and Cowan, 1975), but the studies of Saleeby cited previously and Morgan and Stern (1977) suggest that older volcanic (and plutonic) rocks are also present in the western Sierra Nevada. Morgan and Stern (1977) report that a small pluton near Sonora yields a U-Pb age of

190 m.y. (Early Jurassic). It intrudes alpine-type ultramafic rocks and unconformably overlying Peñon Blanco volcanic rocks (Fig. 3-4) that occur west of the Melones fault zone and east of the Bear Mountains fault zone. Although the mafic Peñon Blanco volcanic rocks (lava, breccia, tuff) may be oceanic in affinity rather than an expression of arc activity, the pluton that intrudes them is indicative of surprisingly early plutonism in this terrane.

The ages of metavolcanic and metatuffaceous rocks (Franklin Canyon and Duffey Dome formations) that occur between the Melones and Bear Mountains fault zones in the northernmost Sierra Nevada and lie unconformably above deformed "Calaveras" cherts, phyllites, and discontinuous limestone bodies are not known (Fig. 3-3). Hietanen (1973) tentatively correlates them with late Paleozoic units of the Klamath-Sierran arc to the east of the Melones fault. We believe that a Mesozoic age assignment is more likely, since these metavolcanic rocks reportedly overlie unconformably an accreted "Calaveras" terrane in a manner comparable to relations described by Saleeby far to the south, and because we consider it unlikely that *any* rocks of the Paleozoic Klamath-Sierran volcanic arc are preserved west of the Melones fault. Hietanen (1977) does report that lower-grade metavolcanic rocks of variable composition (mafic augite basalt, andesite, dacite, soda-rhyolite) and probable Late Jurassic age (Bloomer Hill Formation) overlie phyllites of unknown age north of Lake Oroville. The phyllites are extensions of an undifferentiated "Calaveras" unit mapped by Creely (1965) in the Oroville quadrangle to the south, where a Late Triassic (?) ammonite was collected from overlying strata that includes metavolcanic rocks.

MIDDLE AND LATE JURASSIC HISTORY

Klamath Mountains

The Klamath Mountains of northwestern California and southwestern Oregon are a geologic continuation to the northwest of the Sierra Nevada (Davis, 1969; Hamilton, 1969). This point has been discussed previously in terms of correlative early Mesozoic and late Paleozoic arc and oceanic rock assemblages in the two geographic areas (Figs. 3-2 through 3-4). Most of the Klamath Mountains province west of the Siskiyou thrust fault and much of the Sierra Nevada province west of the Melones fault consist of Jurassic sedimentary and igneous rocks with both arc and oceanic affinities. With the advent of plate tectonics, an increasing number of authors have interpreted the Jurassic rocks of the western Klamath Mountains and Sierra Nevada as exotic elements of the Cordillera, that is, units alien to the North American plate but carried to it atop Pacific Ocean lithosphere that was subducted in Jurassic time along the western edge of the continent.

Hamilton (1978a, p. 46), for example, states that the Klamath Mountains west of the central metamorphic subprovince consist of "variably aggregated island-arc fragments, perhaps analogous to the southern Philippine Islands" (where he interprets a tectonic assemblage of at least six different Cenozoic island-arc systems). The multiple Klamath island arcs are considered by him to have been "individually active during Middle and

Late Jurassic time, and probably earlier" (ibid.). Similarly, Dickinson (1976, p. 1275) concludes that the western Sierra Nevada is a Jurassic subduction complex consisting of "scraps of intraoceanic island arcs of early Late Jurassic and older (?) age with melanges and ophiolitic slices of uncertain but presumably similar age." Both Hamilton (1978a) and Dickinson (1976), among others (e.g., Moores, 1972; Schweickert and Cowan, 1975), regard Middle and Late Jurassic magmatic activity in the western part of the Klamath-Sierra terrane as distinct from and geographically separated from coeval Jurassic magmatic activity in the eastern Klamaths and Sierra Nevada.

It is our opinion that such hypotheses are increasingly difficult to defend as mapping studies in the western Klamath and Sierran regions progress. In the following sections on the two areas we attempt to develop an alternative concept. Specifically, we propose that the Middle and Late Jurassic igneous rocks of the two regions constitute portions of a *single* complex arc that was constructed across a previously sutured [Middle Triassic to Early (?) Jurassic] plate boundary between western oceanic rocks of "Calaveras" type and the continent. We present evidence that major Jurassic faults within this Klamath-Sierra arc are not former sutures or plate boundaries along which multiple exotic arcs collided, but are instead intraplate faults that developed within the evolving arc. The test of our alternative hypothesis, as well as that of contrasting interpretations, will come from additional field, paleontologic, and paleomagnetic studies in these important terranes.

As Figures 3-5 and 3-6 illustrate, the record of Middle and Late Jurassic volcanic and plutonic activity in the Klamath Mountains and northern Sierra Nevada is widespread. In the eastern Klamath Mountains, Mesozoic volcanic activity continued until at least the Bajocian (Middle Jurassic); younger Mesozoic volcanic units, if once present, may lie hidden beneath an eastern cover of Cenozoic Cascade volcanic rocks. Callovian and younger (?) strata in the northeastern Sierran Nevada are present and indicate continuation of Mesozoic volcanic arc activity there into the Late Jurassic. The basement for this eastern, arc-related activity was the ensimatic island arc of Paleozoic age that had been accreted to the continent during the Sonoma orogeny.

Farther west, Charlton (1978; personal communication, 1978) reports that hemipelagic sediments and volcaniclastic rocks correlative with the Hayfork Bally Meta-andesite of Irwin (1977) lie depositionally on pillow basalts that he tentatively assigns to the Rattlesnake Creek ophiolite. Nearby meta-andesites are intruded and deformed by the Ironside Mountain batholith (165 to 167 m.y., Lanphere and others, 1968).

Still farther west, Late Jurassic rocks [Callovian (?), Oxfordian, and Kimmeridgian] in the Klamath Mountains are well represented by the Galice Formation, a unit of phyllite, slate, and semischist (Fig. 3-5) of the western Jurassic subprovince. At most localities the metasedimentary rocks are intercalated with and overlie metavolcanic and metapyroclastic rocks of intermediate to silicic composition (the Rogue Formation in northernmost California and southwestern Oregon). Near O'Brien, Oregon, Vail and Dasch (1977) report that Galice slate and graywacke depositionally overlie pillowed spilite of the Jurassic Josephine ophiolite. Late Jurassic plutons ranging in age from 155 to 140 m.y. intrude the Rogue and Galice formations and mafic and ultramafic rocks assigned to the Josephine ophiolite in both Oregon (Hotz, 1971; Dick, 1973) and California (Young, 1974). At the present time, all these Jurassic units (sedimentary, volcanic, plutonic) lie in the upper

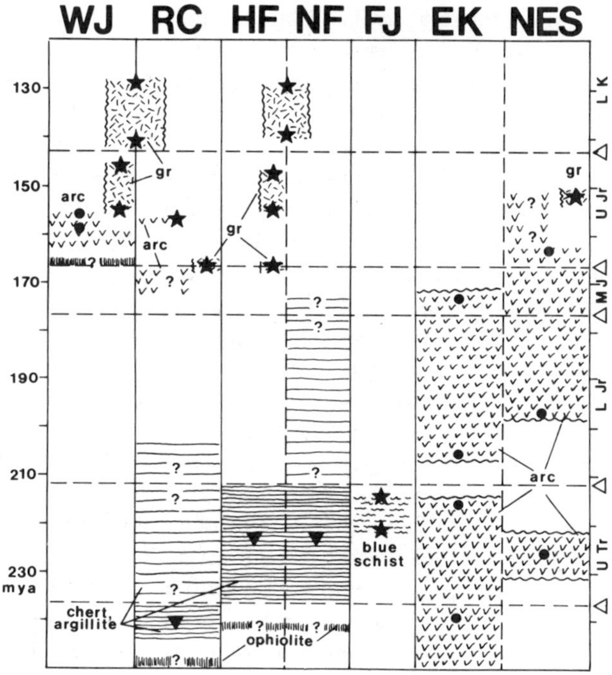

Fig. 3-6. Mesozoic paleogeographic assemblages and intervals of plutonic activity, Klamath Mountains and northeastern Sierra Nevada. Geologic terranes represented by columns, from west to east: WJ, western Jurassic subprovince; RC, Rattlesnake Creek terrane; HF, Hayfork terrane; NF, North Fork terrane; FJ, Fort Jones area blueschist terrane; EK, eastern Klamath subprovince; NES, northeastern Sierra Nevada. Age control: radiometric dates (stars); megafossils (closed circles); radiolaria (closed triangles). References for radiometric dates listed in caption for Fig. 3-5.

plate of a regional thrust fault above Tithonian and younger Franciscan rocks (in Oregon, Dothan) of the Coast Ranges.

The western Jurassic, Rattlesnake Creek, North Fork-Hayfork, and eastern Klamath volcanic and sedimentary assemblages are now all separated from each other by major east-dipping thrust faults (as diagrammatically represented in Fig. 3-5). Geologic and geochronologic data bracket each of the three separating faults rather closely. The lower two (between western Jurassic, Rattlesnake Creek, and North Fork-Hayfork terranes) are almost certainly Late Jurassic in age. The higher Siskiyou thrust fault could conceivably be somewhat older (Middle or Late Jurassic) because of the uncertainties concerning the age of the youngest rocks to be found in the underlying North Fork terrane.

If, as Hamilton (1978a) has suggested, the Middle and Late Jurassic magmatic activity of the western Jurassic, Rattlesnake Creek, and eastern Klamath regions occurred within independent, separately evolving arcs (the first two oceanic), then the fault boundaries that now separate them must represent sutures between major plate elements.

None of these three faults, however, can be identified with certainty as a convergent plate boundary at the time of its formation, and at least one, the Siskiyou thrust, is an intraplate fault that crosscuts older structures in both upper and lower plates (Fig. 3-5; Davis, 1968; Ando and others, 1977).

Geologic relations suggest that the Middle and Late Jurassic volcanic rocks and coeval plutons in the Klamath Mountains could be components of a single, evolving magmatic arc that was constructed across a previously sutured ocean-continent boundary and was related to eastward subduction along a western trench that has not been preserved. Although an east-dipping Late Triassic or Early Jurassic subduction zone probably once separated the North Fork oceanic terrane from the continental plate to the east, the presently intervening Siskiyou thrust fault does not appear to be a direct expression of such an inferred plate boundary. As mentioned previously, the Siskiyou thrust is younger than episodes of low-angle faulting, major antiformal folding, greenschist facies regional metamorphism, and, possibly, Middle Jurassic granitic intrusion into rocks of the North Fork-Hayfork lower plate.

It might be argued that a consuming plate boundary is represented by the thrust fault that separates the North Fork-Hayfork and Rattlesnake Creek terranes, the latter with its Ironside Mountain batholith–Hayfork Bally Meta-andesite arc components (as interpreted by D. Charlton). However, geologic relations favor the interpretation that this fault, too, is the consequence of intraplate shortening. The hanging wall Hayfork-North Fork terrane is intruded by a Middle Jurassic pluton (Forks of Salmon) that is identical in age (167 m.y., K-Ar) and similar in composition (syenodioritic) to the footwall Ironside Mountain pluton (Hotz, 1971). Thus coeval plutonism appears to have affected both of the ophiolitic terranes prior to the period of thrust faulting that now juxtaposes them. The possibility that detritus from the Rattlesnake Creek–Hayfork Bally Meta-andesite "arc" is present in Hayfork strata (D. Charlton, personal communication, 1978) may be supplemental evidence that the two terranes were in proximity during Early to Middle Jurassic time.

The paleogeographic setting of the Rogue-Galice assemblage is admittedly uncertain, but the presence of heavy detrital minerals in Galice graywackes (including glaucophane, hornblende, garnet, tourmaline, epidote, and zircon; Harper, 1978) may indicate that Galice sediments were derived from older Klamath crystalline rocks, were deposited near the North American continental margin, and may not have been separated from the continent by an intervening oceanic trench. It is, therefore, difficult to visualize the Rogue-Galice assemblage as being a far-traveled and exotic component of the Cordillera. As discussed in a later section, a similar conclusion can be drawn for the Logtown Ridge-Mariposa sequence of the western Sierra Nevada, with which the Rogue-Galice sequence has often been correlated.

It is, nevertheless, conceivable that the east-dipping regional thrust fault that now separates footwall rocks (Josephine ophiolite-Rogue-Galice) from higher lithological assemblages (Rattlesnake Creek and Preston Peak) represents a fundamental Late Jurassic plate boundary, particularly since the Josephine and Rattlesnake Creek ophiolites appear to be of considerably different ages (Jurassic and Triassic, respectively). However, in the context of Klamath plutonism it is difficult to view this thrust fault as being other than an intra-arc (intraplate) structure. The thrust postdates strata in the Galice Formation as young as early Kimmeridgian, but it is intruded by plutons of latest Jurassic-earliest Cretaceous ages (ca. 143 to 126 m.y., K-Ar, Snoke, 1977). Plutons of somewhat older age intrude both footwall Rogue-Galice rocks (155 to 145 m.y., K-Ar, Hotz, 1971; Dick, 1973; Young, 1974) and hanging wall rocks of the "western Paleozoic and Triassic" subprovince (155 to 147 m.y., Hotz, 1971). Unfortunately, their age relation to thrust faulting is not known. Since some of the plutons have ages that fall within the Kimmeridgian stage (ca. 157 to 150 m.y., Armstrong, in press) they may be essentially contemporaneous with Rogue-Galice volcanism and, therefore, older than thrusting. More stratigraphic and radiometric age data obviously are needed, but the data reviewed here raise the distinct possibility that Late Jurassic and Early Cretaceous plutons were emplaced respectively before *and* after thrust faulting into rocks of both the western Jurassic and "western Paleozoic and Triassic" subprovinces.

The most reasonable interpretation of these relationships (Fig. 3-5) is that thrusting between the two subprovinces occurred *within* a Late Jurassic-Early Cretaceous magmatic arc, not along a plate boundary. If the widespread Late Jurassic-Early Cretaceous magmatism is related to subduction of oceanic lithosphere beneath the arc, then the convergent plate boundary (trench) lay considerably west of present exposures of Rogue-Galice rocks. Since Late Jurassic and Middle Jurassic plutons are spatially superimposed in the western Klamath Mountains, the possibility exists that all are related to the same east-dipping subduction system.

Collectively, the spatial and temporal relationships between Klamath plutonism and thrust faulting described here indicate that an evolving Middle and Late Jurassic arc complex was internally disrupted and telescoped by Middle (?) to Late Jurassic faulting. This deformation can be tied to the effects of plate convergence between ocean basin and arc, as can of course the widespread Jurassic volcanism and plutonism. This idea, which is amplified in the succeeding section on the Sierra Nevada, is admittedly conservative, but we suggest that it is philosophically at least as satisfying as concepts currently in vogue that three (or more) separate and independently evolving Jurassic volcanic arcs, or their remnants, are now preserved in the narrow, 200-km wide Klamath terrane.[2] Figure 3-7 supports these conclusions. Three separately evolving Jurassic arcs are shown (Fig. 3-7a) without vertical exaggeration prior to their collision and aggregation as postulated by others. Dips of 45° for the three subducting plates were selected to reduce arc widths to geologically conservative values. The arcs pictured include a western Jurassic arc with a Jurassic ophiolitic basement (Josephine), a Rattlesnake Creek arc with an early Mesozoic basement of "Calaveras"-type oceanic rocks, and an eastern Klamath-northeastern Sierra arc with a basement of older Mesozoic continental arc rocks. Figure 3-7b illustrates the present configuration of the thrust-fault-bounded Klamath subprovinces at the same scale for comparison. The space and scale problems implicit in this illustration of a multi-Jurassic arc origin (Fig. 3-7a) for the present Klamath province (Fig. 3-7b) are so extreme that a simpler explanation appears to be necessary.

[2]For comparative purposes, the Cenozoic Aleutian arc is 210 km wide at 176°W longitude (from trench axis to the abyssal floor of the Aleutian Basin).

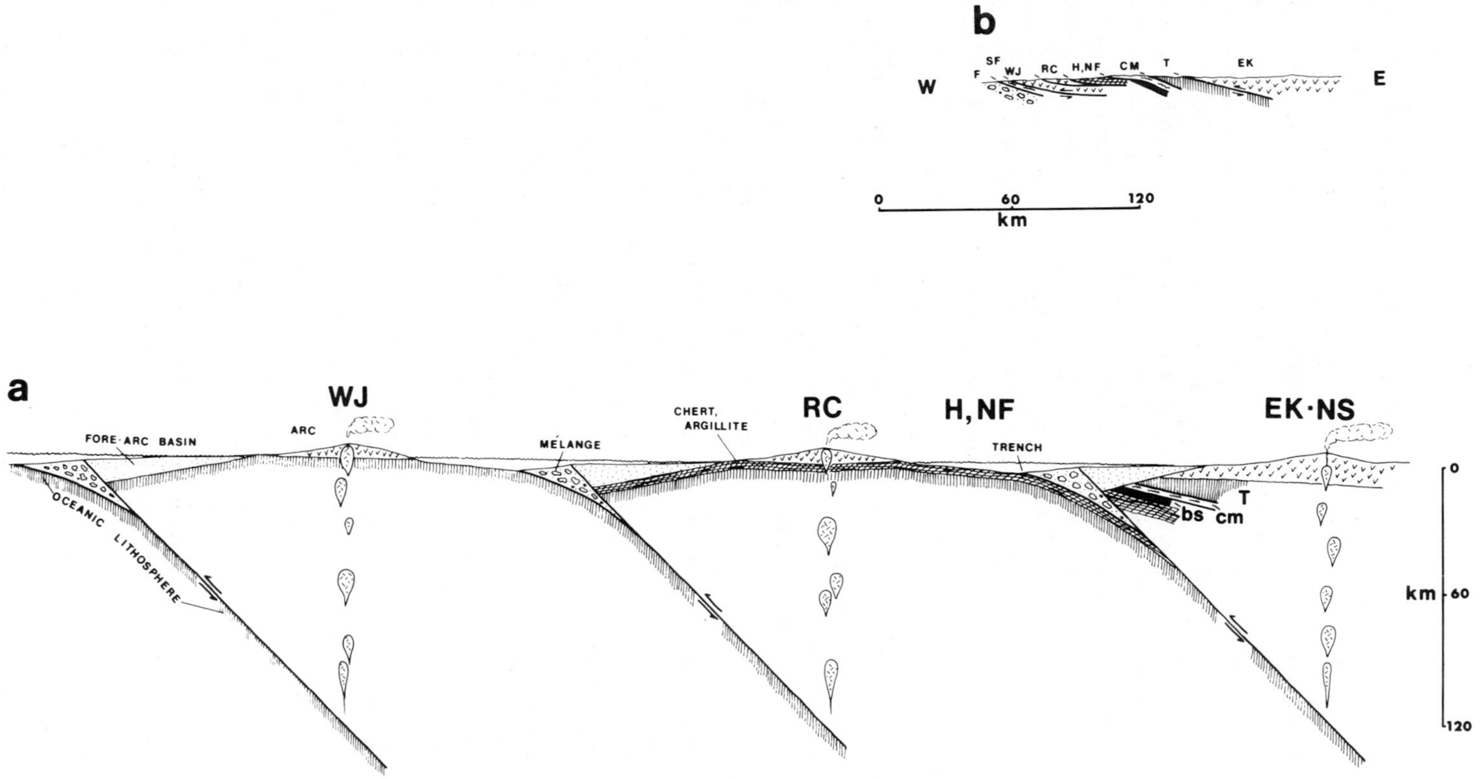

Fig. 3-7. (a) Hypothetical cross section through active, multiple Klamath volcanic arcs in early Late Jurassic time prior to their collision and aggregation (no vertical exaggeration). WJ, western Jurassic island arc; RC, Rattlesnake Creek island arc; H, NF, Hayfork, North Fork oceanic terrane; EK-NS, eastern Klamath-northern Sierra Nevada continental margin arc. Basement for the latter includes: T, Trinity ophiolitic complex (Ordovician?); cm, central metamorphic subprovince (Devonian); bs, blueschist assemblage of type exposed near Fort Jones (Late Triassic). (b) Highly simplified, essentially true scale geologic cross section through central Klamath Mountains, California (plutons omitted). F, Franciscan Complex; SF, South Fork Mountain Schist; other letter designations as in Fig. 3-7a.

Sierra Nevada

Middle Jurassic through Early Cretaceous volcanism and plutonism are well documented in the geologic record of the eastern Sierra and western Nevada. This magmatic activity is representative of an Andean-type arc formed along the leading edge of the North American plate. However, as in the Klamath Mountains, broadly coeval volcanism and plutonism occurred in areas west of what many consider to be the main, continental margin arc. In the case of the Sierra Nevada, these igneous rocks lie west of the Melones fault zone and correlate, in part, with the Rogue-Galice assemblage of the western Jurassic Klamath subprovince. Their plate tectonic setting is controversial. They have recently been interpreted (Schweickert and Cowan, 1975) as exotic components of the Cordilleran Orogen that include from east to west, an ensimatic east-facing island arc, an interarc basin floored by oceanic crust, and a remnant arc split from the main arc to the east. Collision of the east-facing island arc with the west-facing Sierran continental arc is purported to have occurred along the Melones fault, an arc-continent suture of Late Jurassic age. The Bear Mountain fault zone to the west represents a zone of partial collision of the remnant arc with its eastern "parent" arc.

This interpretation is troubling to us for a number of reasons, some of them reminiscent of Klamath Mountain problems discussed previously, but largely independent of them. In the western Sierra Nevada the Smartsville ophiolite is bordered on the east and west by Schweickert and Cowan's Jurassic island arc and remnant arc, respectively. They tentatively interpret the Jurassic ophiolite as the oceanic floor of an interarc basin, formed during rifting of the remnant arc from its parent. The ophiolite, however, has an antiformal structure, and studies by E. Moores and his associates (Bond and others, 1977; Buer, 1977; Xenophontos and Bond, 1978) indicate that flanking Middle and Upper Jurassic volcanic and sedimentary rocks were deposited upon the ophiolite, not separated by it. Another problem with the rifted interarc basin paleogeography of Schweickert and Cowan is that their western (remnant) and eastern (island) arcs are separated near the Consumnes River by a melange belt (Duffield and Sharp, 1975) that contains rocks of "Calaveras" lithology, including limestone blocks bearing a Tethyan Permian fusulinid fauna (Douglass, 1967). This relation is difficult to explain if the hypothesized interarc basin formed by the Jurassic rifting of a Jurassic ensimatic island arc.

The relationships between Mesozoic volcanic and sedimentary rocks in the Sierran Foothills and older oceanic rocks of Calaveras-type are interesting. Saleeby's studies (1977a; Saleeby and others, 1978) between the Kings and Kaweah rivers in the southwestern Sierra Nevada indicate that "Calaveras" rocks accreted to the continent are overlain by continent-derived sedimentary rocks of Late Triassic age and by Early Jurassic volcanic rocks that can be related to a more westerly, east-dipping subduction zone. The location of the Kings-Kaweah area with respect to the Melones fault zone to the north is not definite, but Saleeby believes that it is a southern extension of eastern portions of the Sierran Foothills terrane. If so, Jurassic volcanic rocks and somewhat younger plutons that lie west of the Melones fault were part of the continent in Jurassic time and cannot be exotic with respect to North America.

At the opposite, northern end of the Sierra Nevada, Hietanen (1973) and Creely

(1965) have also concluded that "Calaveras" rocks west of the Melones fault zone form the basement for Mesozoic volcanic rocks. For example, volcanic rocks of Logtown Ridge (= Rogue Formation?) lithology and age overlie "Calaveras" phyllites in the vicinity of Oroville (east of the Smartsville ophiolite), although studies in progress at the University of California, Davis, suggest that the contact may be a low-angle fault (E. Moores, personal communication, 1978). Elsewhere in the Sierran Foothills some lithic clasts in Mariposa (= Galice Formation?) graywackes were derived from a "Calaveras"-type source terrane (Behrman and Parkison, 1978). Others were derived from older Paleozoic units (including the Shoo Fly Formation) that lie east of the Melones fault and were part of continental North America at the time of Mariposa sedimentation (R. Schweickert, personal communication, 1978).

In central parts of the Foothills belt, the relationship described by Duffield and Sharp (1975), that a Calaveras-type "melange" along the Bear Mountains fault zone separates similar Jurassic volcanic terranes, could also be explained by the initial deposition of volcanic rocks atop a previously disrupted "Calaveras" basement. As Schweickert and Cowan (1975) point out, it is extremely significant that no fragments of the adjacent (overlying?) volcanic rocks (Logtown Ridge Formation) occur within the melange. Schweickert (1978, p. 373) has since concluded that Jurassic volcanic rocks west of the Melones fault were deposited on "a structurally complex assortment of slices of oceanic crust, melanges, and sheets of ultramafic rock that evidently were deformed and juxtaposed prior to the birth of the island arc." He also states that chaotic chert-argillite units in the complex basement resemble the "Calaveras Complex." The location of this complex, deformed basement at the time of Jurassic volcanism atop it is of extreme importance to an understanding of Sierran Mesozoic tectonics. Schweickert believes that the basement terrane lay an unknown distance west or southwest of the west coast of North America at the time an island arc was being constructed on it.

Alternatively, we believe that it represents a subduction zone complex that had already been accreted onto the western edge of the continent by Late Triassic or Early Jurassic (?) time. We postulate that Early (?), Middle, and Late Jurassic volcanic and plutonic rocks in the western and eastern Sierra Nevada (as in their Klamath counterparts to the north) belong to one magmatic arc complex that was built across a previously sutured (Middle to Late Triassic) convergent plate boundary. East of the boundary, the Jurassic arc was constructed on an older basement terrane that included the Late Triassic Sierran arc and older lithotectonic elements (from northwest to southeast, the Paleozoic Klamath-Sierran arc, Antler and Sonoma allocthonous units, the miogeoclinal prism, and cratonal basement and cover). Directly west of the boundary the arc was built on accreted late Paleozoic and early Mesozoic oceanic rocks of the "Calaveras" terrane (equivalent to Hayfork-North Fork and Rattlesnake Creek assemblages farther north). Still farther west, its basement was at least locally relatively undisturbed ocean crust of unknown origin—the Smartsville ophiolite (possibly comparable to the Josephine ophiolite of the western Klamath Mountains). Similar, independently arrived at conclusions have recently been drawn for the Sierra Nevada by Behrman and Parkison (1978) and by Saleeby and others (1978).

The arc did not develop statically across the diverse basement terranes just cited.

Marine sedimentation, volcanic activity, and plutonic intrusion were accompanied, as in the Klamath Mountains, by profound internal disruption and imbrication of the evolving arc by strike-slip and thrust faulting. Saleeby's Kings-Kaweah "suture" is certainly the best documented example of the former. Displacement along this zone of transcurrent faulting began roughly 190 m.y. ago and continued until at least the Late Jurassic, as indicated by a 157 m.y. concordant U-Pb zircon age for a synkinematic intrusion pluton emplaced across it (Saleeby, 1977a, 1977). To the north, the Melones fault zone, which may be a component of the Kings-Kaweah "suture," disrupts a 162 m.y. old diorite stock; movement along it appears to have ceased by 140 m.y. ago (Morgan and Stern, 1977).

Relationships between the steep, east-dipping faults of the western Sierra and the low-angle, east-rooting thrust faults of the Klamath Mountains have been explored by Davis (1969). He proposed that these broadly coeval fault systems are linked structurally, and that the thrust faults of the Klamaths must steepen southward into apparently deeper structural levels now exposed in the western Sierra Nevada. The probable right-lateral offset of the "Calaveras Formation" across the Melones fault (Fig. 3-3) and Saleeby's analysis of right slip along the Kings-Kaweah suture indicate that intraplate thrust faulting in the Sierran-Klamath Jurassic arc had a major strike-slip component.

In conclusion, we find no compelling reason to believe that more than one Early (?) to Late Jurassic magmatic arc is represented in the geologic terrane east of the Great Valley of California. The extreme narrowness of separate exotic island and remnant arcs presumed by others to occur in the western Sierra Nevada and western Klamath Mountains can be explained as the consequence of internal imbrication of a single Jurassic arc in response to oblique Pacific-North American plate convergence. The trench and east-dipping subduction zone responsible for arc magmatism lay to the west, but it is doubtful that either has been preserved in the geologic record.

Paul C. Bateman
U. S. Geological Survey
Menlo Park, California 94025

4

GEOLOGIC AND GEOPHYSICAL CONSTRAINTS ON MODELS FOR THE ORIGIN OF THE SIERRA NEVADA BATHOLITH, CALIFORNIA

ABSTRACT

The following established or deduced features of the Sierra Nevada batholith and its country rocks are constraints that must be met by any model of the events that immediately preceded and attended the emplacement of the Sierra Nevada batholith:

1. The source materials for granitoid magmas that solidify to form batholiths are not derived primarily from either oceanic crust or oceanic mantle, although such materials may have played important roles in the generation of granitoid magmas in continental regions.

2. Triassic (?) volcanic rocks, exotic limestone blocks that contain warm-water Tethyan fusulinids, and conglomerates that contain boulders composed of granitoids and rock types like those in the late Paleozoic Calaveras Formation are all present in the strip of Mesozoic strata that lies between the Melones and Bear Mountain fault zones.

3. Two spatially separated volcanogenic sequences were deposited in the western foothills and in the high Sierra region during the same general Mesozoic time span. The volcanic rocks of both sequences range widely in composition, but the sequence in the western metamorphic belt was deposited in a volcanic island-arc environment and has an average composition transitional between andesite and dacite, whereas the Ritter sequence in pendants of the high Sierra Nevada was deposited in a continental subaerial and shallow subaqueous environment and has an average composition of rhyodacite.

4. Thick, complexly folded and faulted Paleozoic and Mesozoic stratigraphic sections on opposite sides of the batholith face inward.

5. Progressive changes across the batholith in the compositions of the granitoids, the country rocks, and the Cenozoic volcanic rocks probably reflect fundamental differences in the composition of the lower crust and possibly of the upper mantle.

6. The parent magmas of the granitoids probably contained abundant crystalline material, and models for producing parent magmas in the mantle by the formation and separation of partial melts devoid of crystals are probably not applicable.

7. A belt of Cretaceous granitoids trending about N 20° W crosses a belt of Jurassic granitoids trending about N 40° W in the central Sierra Nevada. Most of the isotopic ages of the Jurassic granitoids fall between 186 and 155 m.y., and the ages of the Cretaceous granitoids fall between 125 and 88 m.y.

8. Isotopic ages and intrusive relations indicate that the Cretaceous granitoids are progressively younger eastward, but show no pattern for the Jurassic granitoids.

9. Few granitoids with isotopic ages between 155 and 125 m.y. have been identified. The Nevadan orogeny occurred during a part of this interval and appears not to have been coincident with the emplacement of major granitoid sequences.

10. Comagmatic sequences of granitoids are generally elongate in a northwest direction, parallel with the long axis of the batholith, and some are repeated along this direction.

11. No evidence has been found to indicate that the granitoid plutons floated upward as balloonlike forms, leaving behind underlying gneiss, or that they have sharply defined bottoms.

12. The seismic Moho is depressed beneath the batholith and reaches a maximum depth of about 52 km beneath the Sierra Nevada crest.

13. Measured P-wave velocities in the crust are 6.0 km/s at a depth of 10 km, 6.4 km/s at a depth of 20 km, and 6.9 km/s near the base. The relatively slow measured P-wave velocity of 7.9 km/s in the upper mantle may represent the velocity in a slow direction if the mantle rock is tectonite with a fabric parallel to the axis of the batholith.

14. The Bouguer gravity profile slopes eastward from - 80 milligals in the western foothills to about - 225 milligals beneath the high Sierra Nevada, assuming a density of 2.67 g/cm^3. If differences in the specific gravity of the exposed granitoids were taken into account, the magnitude of this negative anomaly would be reduced.

15. The residual magnetic intensity of the granitoids in the western foothills of the Sierra Nevada is extremely low and appears to correlate with a paucity of magnetite.

16. The heat flow from the Sierra Nevada and inferred mantle contribution of 0.4 HFU are anomalously low and suggest loss of heat downward.

INTRODUCTION

Models of emplacement of the Sierra Nevada batholith and attendant events must account for certain geologic and geophysical features or constraints. Some of these features are established facts and can be simply stated, but many are inferred on tenuous grounds, are less obvious, and require extended discussion and interpretation.

I chose to write about constraints rather than to develop a model for two related reasons. First, I published a model sixteen years ago (Bateman, 1962), which was revised in Bateman and Wahrhaftig, 1966, Bateman and Eaton, 1967, and Bateman, 1974 to incorporate new data and concepts. New data will continue to become available and to render any model obsolete in some respects. Second, others have developed or are developing models for the events that occurred in the Sierra Nevada region during the time when the batholith was being emplaced, and a summary of some of the constraints that should be taken into account is likely to serve a more useful purpose than another revision of my model.

The geological studies of the U.S. Geological Survey during the past twenty years have been concentrated largely in a belt across the middle of the batholith between the 37th and 38th parallels of north latitude, although J. G. Moore of the Survey has carried on mapping south of the 37th parallel. Most of the belt between the 37th and 38th parallels has been mapped on a scale of 1:62,500, and much of the following discussion is based on this work.

BATHOLITHS ARE FEATURES OF THE CONTINENTS

I know of no batholith that was emplaced into oceanic crust, except where the oceanic crust is contiguous with, and generally tectonically involved with, continental rocks. Small bodies of quartz diorite, tonalite, and plagiogranite (trondhjemite) occur in volcanic island-arc terranes underlain by oceanic crust, but no batholiths. Apparently, subduction zones involving only oceanic crust and mantle are not overlain by batholiths.

Constraint 1 The source materials for granitoid magmas that solidify to form batholiths are not derived primarily from either oceanic crust or oceanic mantle, although such materials may have played important roles in the generation of granitoid magmas in continental regions.

AGES, COMPOSITIONS, AND STRUCTURES OF THE COUNTRY ROCKS

The Sierra Nevada batholith was emplaced in Mesozoic time into complexly deformed weakly to moderately metamorphosed strata of Paleozoic and Mesozoic age. Precambrian rocks are exposed along the east side of the batholith, but no outcrops of Precambrian rocks have been found west of Owens Valley in the Sierra Nevada (Fig. 4-1).

Paleozoic country rocks occur along the eastern margin of the batholith, in roof pendants, and along the western margin in the western metamorphic belt. The Paleozoic strata in the Inyo and White mountains are sedimentary shelf and miogeosynclinal sedimentary rocks, predominantly carbonate, quartzite, and pelite. Facies changes take place toward the north and west, and in the northern Sierra Nevada volcanic rocks thought to have been emplaced in an island-arc environment are present. The Paleozoic strata in the west side of the central Sierra Nevada between 37° and 38°N latitude (Fig. 4-2) have generally been assigned to the Late Paleozoic Calaveras Formation. In this region, the Calaveras is separated on the west from Mesozoic strata by the Melones fault zone. The Calaveras has been subdivided by Schweickert and others (1977) into four lithologically distinct units that dip steeply east and are believed to be successively younger eastward. From west to east, these are a volcanogenic unit; a schist and phyllite unit that also contains minor chert and marble; a chert (siltstone?) unit that contains some schist, phyllite, and marble; and a quartzite unit that also contains schist, marble, and calc-silicate rocks.

The Mesozoic country rocks include three distinct, spatially separated, groups of strata: a western group that includes the Mariposa Slate and associated volcanic rocks, an eastern group that composes the Ritter Range sequence, and an intermediate group that composes the Kings sequence.

The Kings sequence (Bateman and Clark, 1974) is exposed in a series of roof pendants that lie east of the Calaveras. In the area shown in Fig. 4-2, the rocks of this sequence are metasedimentary–quartzite, marble, schist, calc-silicate hornfels, and biotite-

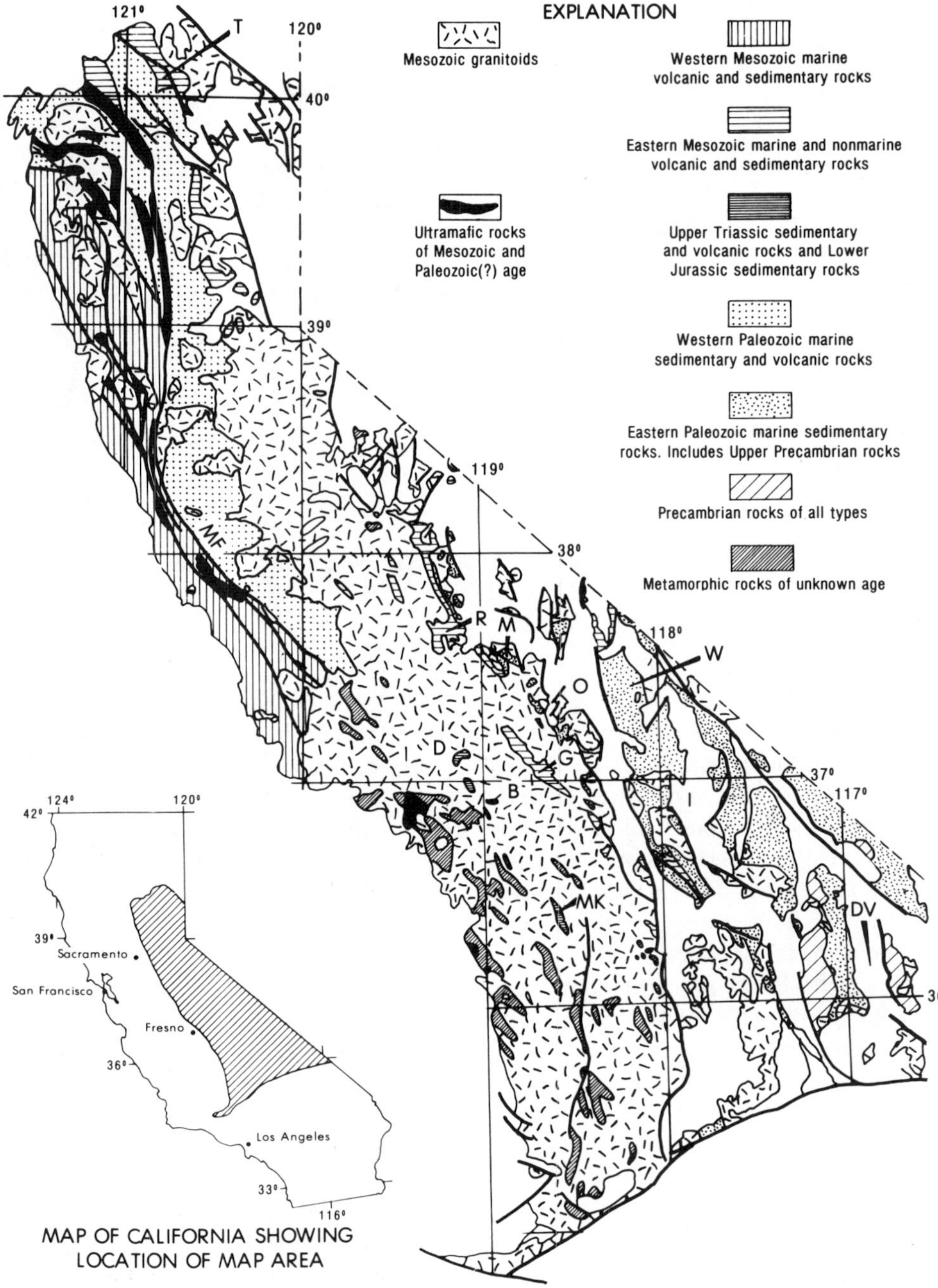

Fig. 4-1. Sierra Nevada and adjacent areas in eastern California showing the Sierra Nevada batholith and its country rocks. Taylorsville, T; Ritter Range roof pendant, R; Mount Morrison roof pendant, M; White Mountains, W; Inyo Mountains, I; Dinkey Creek roof pendant, D; Boyden Cave roof pendant, B; Mineral King roof pendant, MK; Owens Valley, O; Death Valley, DV; Melones Fault, MF.

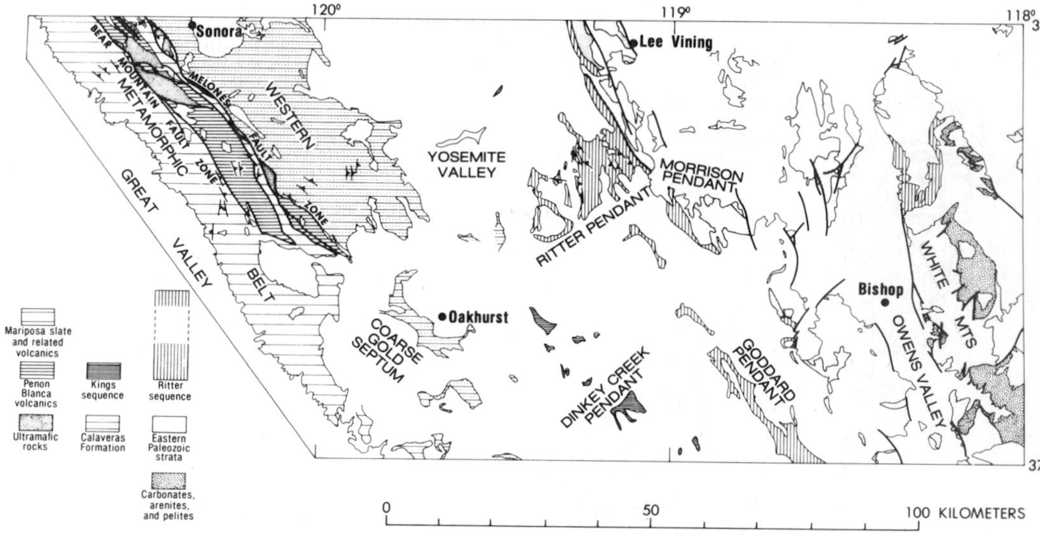

Fig. 4-2. Distribution of country rocks in the central Sierra Nevada and White Mountains. Barbed symbols in western metamorphic belt and Ritter Range roof pendant show top facings of beds (Clark, 1964; Huber and Rinehart, 1965).

andalusite hornfels (Kistler and Bateman, 1966); but farther south, in the Mineral King and Yokohl Valley roof pendants, volcanic rocks are also present (Saleeby and others, 1978). Although the strata are strongly deformed, fossils of Early Jurassic and Late Triassic age have been collected from the Boyden Cave, Mineral King, Yokohl Valley, and Isabella roof pendants, all of which are south of the area shown in Fig. 4-2 (Knopf and Thelen, 1905; Christensen, 1963; Moore and Dodge, 1962; Jones and Moore, 1973; Saleeby and others, 1978). Lithologic similarities, spatial relations, and the probable ages of the strata suggest that the Kings sequence stratigraphically overlies, or is in part correlative with, the upper quartzitic unit of the Calaveras.

The Ritter sequence (Bateman and Clark, 1974) is dominantly volcanogenic and consists of intermediate to felsic volcanic and volcanogenic sedimentary rocks. Fiske and Tobisch (1978) have subdivided the Ritter sequence into a lower and an upper section, which are separated by an unconformity that seems to represent a significant time gap. The lower section is essentially homoclinal and dips steeply west. Common rock types are ash-flow tuffs, flows of andesite and basaltic andesite, tuff breccia, bedded tuff, lapilli tuff, and thin beds of limestone. Crossbeds are common in sedimentary layers and indicate that almost all bedding tops face west. Undoubtedly, the section was deposited partly on land and partly in a shallow subaqueous environment. Although fossils collected from a single locality (Huber and Rinehart, 1965) indicate an Early Jurassic age, isotopic ages suggest the lower section ranges in age from Permian to Middle Jurassic. The present stratigraphic thickness of the lower section is about 5 km, but it has been tectonically thinned from an original thickness approaching 11 km (Fiske and Tobisch, 1978).

The upper section of the Ritter sequence dips gently in most places and rests unconformably on the lower section. It consists of lava flows, tuff, and tuff breccia, all of which were deposited on land and which probably originated in a caldera (Fiske and others, 1977). Isotopic dates indicate a mid-Cretaceous age. The upper section is less strongly deformed than the lower section and has been tectonically thinned much less, from an original thickness of about 4 km to a present thickness of about 3.4 km (Fiske and Tobisch, 1978). Although the Ritter sequence and the Kings sequence overlap in age, the upper section of the Ritter sequence appears to rest locally on, and hypabyssal units to intrude, the Kings sequence.

The western group of Mesozoic strata lies west of the Calaveras Formation in the western metamorphic belt. It is composed largely of slate, phyllite, and a variety of volcanic rocks, including pillow lavas. These strata are divided by the Bear Mountain fault zone into structural blocks. Chemical analyses of composite samples (unpublished data) indicate that, despite a wide range of compositions, the bulk composition of both structural blocks lies between average andesite and average dacite, quite distinct from the Ritter sequence. Graded beds are ubiquitous, and the overall constitution of these strata indicates that they were deposited in an island-arc environment. Sparse fossils indicate that most of these strata are Late Jurassic. However, Early Jurassic U-Pb isotopic dates (Morgan and Stern, 1977) on two small plutons that intrude the Penon Blanco Volcanics in the block between the Melones and Bear Mountain fault zones (Fig. 4-3) indicate that these volcanic rocks are at least as old as Early Jurassic and may be Triassic. In this same block, north of the area shown in Fig. 4-2, is a melange (the "western belt" of the Calaveras Formation of Clark, 1964) in which Tethyan fusulinids of Permian age are present in exotic blocks of limestone (Douglass, 1967). Although these fossils may not indicate the age of the enclosing strata, they do indicate the former presence of Permian source beds that were deposited in a warm-water environment. Jones and others (1977) report that exotic blocks containing Tethyan fusulinids are sporadically distributed as far north as Alaska. They interpret these relations to indicate that Permian strata deposited in warm water moved northward tectonically during the early Mesozoic, shedding fossiliferous remnants en route. In contrast with these exotic materials, conglomerate beds that crop out between the Penon Blanco Volcanics and the Melones fault zone, within the Mariposa Slate, contain chert and granitoid boulders, suggesting a local source.

All the stratified country rocks in the central Sierra Nevada are strongly deformed, and evidence of repeated deformations has been reported from many different areas. The late Devonian and early Mississippian (?) Antler orogeny and the late Paleozoic Sonoma orogeny affected the eastern Paleozoic rocks, and early folds in the Mount Goddard roof pendant suggest deformation during the Triassic or Jurassic before the Late Jurassic Nevadan orogeny. The Late Jurassic Nevadan orogeny is generally presumed to account for the northwest-trending grain in the country rocks, the "Sierran" trend. Other deformations took place as late as the mid-Cretaceous, about 100 m.y. ago. Much of the evidence for them is in the granitoids, but recognition of the mid-Cretaceous age of deformed strata in the Ritter Range roof pendant also indicates Cretaceous deformation (Fiske and others, 1977).

The major structure in the country rocks of the central Sierra Nevada is defined by

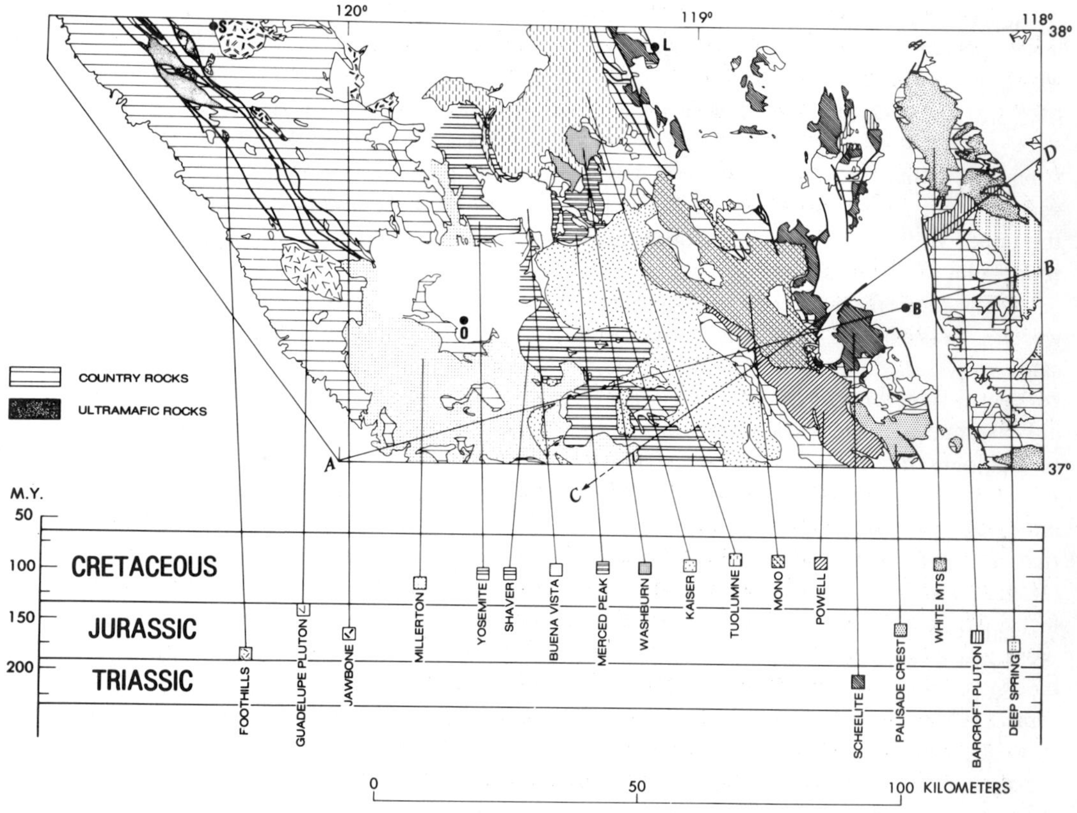

Fig. 4–3. Provisional granitoid sequences in the central Sierra Nevada and their approximate radiometric ages. Ages are based on an evaluation of unpublished Pb-U ages and of published K-Ar ages (Evernden and Kistler, 1970; Crowder and others, 1973; McKee and Nash, 1967; Kistler and others, 1965). Geochemical and geophysical profiles (Figs. 4–4, 4–5) are along line *AB*, except the residual magnetic intensity profile, which is along line *CD*.

the preponderance of east-facing bedding tops in the western metamorphic belt and of westward-younging strata in the Paleozoic strata of the Mount Morrison roof pendant and the Mesozoic strata of the Ritter Range roof pendant. In some earlier reports, I interpreted these inward-facing bedding tops to indicate a complex faulted synclinorium. However, no single stratigraphic unit has been identified in both limbs, and it is possible that the inward-facing tops were caused by some other mechanism than folding.

Following are some specific constraints imposed by the country rocks:

Constraint 2 Triassic (?) volcanic rocks, exotic limestone blocks that contain warm-water Tethyan fusulinids, and conglomerates that contain boulders composed of granitoids and rock types like those in the late Paleozoic Calaveras Formation are all present in the strip of Mesozoic strata that lies between the Melones and Bear Mountain fault zones.

Constraint 3 Two spatially separated volcanogenic sequences were deposited in the western foothills and in the high Sierra region during the same general Mesozoic time span. The volcanic rocks of both sequences range widely in composition, but the sequence in the western metamorphic belt was deposited in a volcanic island-arc environment and has an average composition transitional between andesite and dacite, whereas the Ritter sequence in pendants of the high Sierra Nevada was deposited in a continental subaerial and shallow subaqueous environment and has an average composition of rhyodacite.

Constraint 4 Thick, complexly folded and faulted Paleozoic and Mesozoic stratigraphic sections on opposite sides of the batholith face inward.

COMPOSITIONAL PATTERNS OF THE GRANITOIDS

The batholith is made up of a large number of plutons that range in outcrop area from less than 1 km^2 to 1500 km^2 or more. These plutons are in sharp contact with one another or are separated by septa (screens) of older, generally metamorphic rocks or by later dikes. The plutons can be assembled into a much smaller number of comagmatic sequences (Fig. 4-3), each resulting from a single fusion event (Bateman and Dodge, 1970; Presnall and Bateman, 1973). The concept underlying the grouping of plutons into sequences is that successively emplaced members of a comagmatic sequence solidify from the more fluid core magma from which the preceding pluton crystallized. The simplest comagmatic sequence is a zoned pluton that solidified from its margins inward with falling temperature. Because of crystal fractionation during inward solidification, the pluton is composed of progressively lower temperature mineral assemblages inward. A discontinuity is created during solidification if the core magma moves enough to erode the solidifying margins. If the core magma breaks through the outer coherent carapace repeatedly, the concentric structure may be partly or completely destroyed (Bateman and Chappell, 1979).

In addition to the compositional and textural changes within granitoid sequences, systematic regional compositional changes take place across the batholith. The most conspicuous of these is eastward increase in potassium (Bateman and Dodge, 1970). The granitoids in the western foothills include gabbro, quartz diorite, tonalite, leucogranodiorite, plagiogranite, and sparse granite. These rocks give way eastward to sequences that range from equigranular granodiorite to porphyritic granodiorite and granite to equigranular granite. Still farther east, in the White Mountains, a sequence ranges from monzodiorite to monzonite to quartz monzonite. Crystal fractionation within granitoid sequences results in increasing amounts of both K-feldspar and quartz, so a simple profile showing the amount of K-feldspar reflects crystal fractionation within sequences and regional variation across the batholith. However, plots of C.I.P.W. norms show that the ratio of normative orthoclase to normative quartz increases systemically eastward across the batholith from sequence to sequence (Presnall and Bateman, 1973). Figure 4-4 shows the variation in modal K-feldspar/(K-feldspar plus quartz). Modes are used rather than norms because of a much greater abundance of modal data. This profile represents the variations

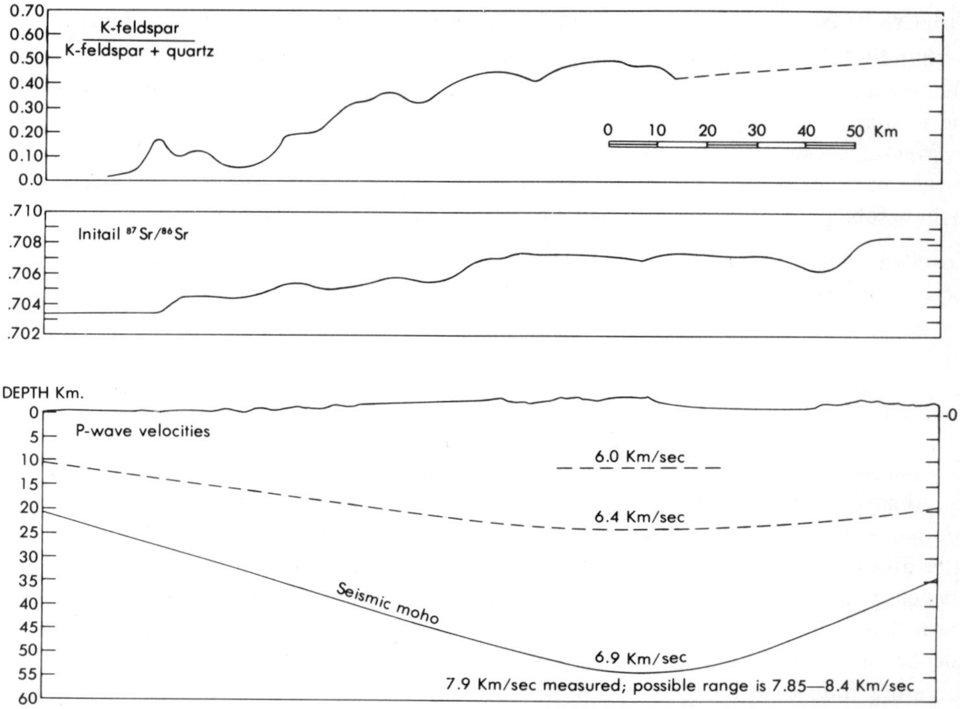

Fig. 4-4. Geochemical and seismic profiles. Profiles are along lines *AB* in Figure 4-3. Geochemical profiles from Bateman (1979) and seismic profile from Bateman and Eaton (1967).

within an 8-mile-wide (13 km) strip that extends from latitude 37°N, longitude 120°W, N 74° E across the batholith through Bishop (Bateman, 1979).

Other compositional changes across the batholith are eastward increases in the amounts of uranium, thorium (Wollenberg and Smith, 1968), beryllium, rubidium (Dodge, 1972a), and in the oxidation ratio (Dodge, 1972b), and a small decrease in the amount of calcium (Bateman and Dodge, 1970). Initial ^{87}Sr-^{86}Sr (Kistler and Peterman, 1973) also increases eastward (Fig. 4-4). These compositional changes across the batholith appear to be independent of the ages of the rocks involved, making it unlikely that they were caused by eastward increasing depth of magma generation along a subduction zone, as has been proposed by Dickinson (1970). Similar variations of K, U, and Th in the country rocks (Wollenberg and Smith, 1970) and eastward increase in K-Na in Cenozoic volcanic rocks (Moore, 1962) support the view that the lateral changes in the compositions of the granitoids across the batholith result from progressive changes with distance from the Pacific Ocean in the composition of the source materials from which the magmas were generated.

Constraint 5 Progressive changes across the batholith in the compositions of the granitoids, the country rocks and the Cenozoic volcanic rocks probably reflect fundamental differences in the composition of the lower crust and possibly of the upper mantle.

NATURE OF PARENT MAGMAS

The nature of parent granitoid magmas can be inferred from textures in the margins of granitoid sequences such as the Tuolumne Intrusive Series (Bateman and Chappell, 1979) and the Mount Givens Granodiorite (Bateman and Nokleberg, 1978). In these sequences, the earliest crystallized, highest temperature mineral assemblages are in the margins and give clues as to the physical constitution of the magma at the earliest stages of crystallization at the present level of exposure.

In both the Tuolumne Intrusive Series and the Mount Givens Granodiorite, the marginal granitoids contain subequant grains of hornblende, biotite, and plagioclase, but only thin intergranular stringers of K-feldspar. These textures are interpreted to indicate that at the time these rocks solidified the magma contained hornblende, biotite, and plagioclase crystals but that K-feldspar crystallized from interstitial melt. In the Mount Givens Granodiorite, quartz is also present as subequant crystals in the marginal rocks, but in the Tuolumne Intrusive Series, quartz is present only in intergranular films, indicating that at the beginning of solidification at the exposed level the Mount Givens magma was also saturated in quartz, whereas the Tuolumne magma was not.

If these interpretations are correct, the earliest magmas on which we have information were mixtures of melt and crystals. This inference is in agreement with temperatures indicated by the metamorphic grade of the contiguous contact-metamorphosed country rocks, which are mostly in the hornblende hornfels facies of contact metamorphism and locally in the pyroxene hornfels facies. Although it is possible that the magmas were devoid of crystals at earlier stages, that seems unlikely in view of the fact that such magmas would have to have been much hotter than the temperatures indicated by the observed grades of contact metamorphism. Surely, at some place a melt of tonalitic composition, if one existed, would have arrived at the present level of exposure and produced a high metamorphic grade. It seems more likely that granitoid magmas originate by partial melting of a volume of rock, most of the volume moving upward under the influence of gravity as soon as enough melt is formed to disaggregate and lubricate the source rock. Denser refractory material would lag behind, and with falling temperature, crystals would settle relatively downward.

Constraint 6 The parent magmas of the granitoids probably contained abundant crystalline material, and models for producing parent magmas in the mantle by the formation and separation of partial melts devoid of crystals are probably not applicable.

AGE PATTERNS OF THE GRANITOIDS

Potassium-argon dating has shown that a belt of Cretaceous granitoids trending about N 20°W crosses a belt of Jurassic granitoids trending about N 40°W in the central Sierra Nevada (Evernden and Kistler, 1970). Between 37° and 38°N latitude, Jurassic granitoids flank the Cretaceous granitoids on both the east and the west and are also present as enclaves within the Cretaceous granitoids. An extensive sequence of Triassic granitoids is also present in the east. U-Pb ages on zircon (unpublished data; Saleeby, 1976) indicate that in the central Sierra Nevada the Triassic granitoids were emplaced about 206 m.y. ago, most of the Jurassic granitoids between 186 and 155 m.y. ago, and the Cretaceous granitoids between 125 and 88 m.y. ago. U-Pb ages (Fig. 4-3) show that the Cretaceous sequences are progressively younger eastward, but show no pattern for the Jurassic granitoid sequences.

Although the isotopic ages now available indicate few ages between 155 and 125 m.y. ago, the fact that ages do fall in this interval leaves room for doubt as to the paucity of plutonic activity. Additional isotopic dating in search of granitoids with isotopic ages between 155 and 125 m.y. is desirable because the Nevadan orogeny falls within this interval, and the time relations between this orogeny and plutonic activity need to be established.

One Jurassic age that falls in this interval of apparently sparse plutonic activity is that of the Guadelupe igneous complex, a dominantly gabbroic (rather than granitic) pluton. The pluton has a U-Pb age of 140 m.y., which is supported by a K-Ar age on biotite of 136 m.y. (Evernden and Kistler, 1970). The pluton intrudes steeply dipping beds of the Mariposa Slate, and, according to Best (1963), probably has not been tilted more than 10° since it was emplaced. The Mariposa Slate contains late Oxfordian and early Kimmeridgian fossils (Clark, 1964). The Jurassic time scale of Van Hinte (1976a) shows the Oxfordian and Kimmeridgian to extend from 149 to 138 m.y. ago. Thus, if both the U-Pb age on the Guadelupe igneous complex and the age assignments of Van Hinte are correct, the Mariposa Slate was both deposited and deformed within the interval of 149 and about 140 m.y. ago. The deformation of the Mariposa is commonly equated with the Late Jurassic Nevadan orogeny, although it is possible that earlier deformations should also be included.

Constraint 7 A belt of Cretaceous granitoids trending about N 20°W crosses a belt of Jurassic granitoids trending about N 40°W in the central Sierra Nevada. Most of the isotopic ages of the Jurassic granitoids fall between 186 and 155 m.y., and the ages of the Cretaceous granitoids fall between 125 and 88 m.y.

Constraint 8 Isotopic ages and intrusive relations indicate that the Cretaceous granitoids are progressively younger eastward, but show no pattern for the Jurassic granitoids.

Constraint 9 Few granitoids with isotopic ages between 155 and 125 m.y. have been identified. The Nevadan orogeny occurred during a part of this interval and appears not to have been coincident with the emplacement of major granitoid sequences.

NORTHWESTWARD TREND OF SEQUENCES

Geologic mapping and isotopic dating show that most sequences are strongly elongate in the direction of the long axis of the batholith. The Millerton sequence appears to be continuous northward with the granodiorite of The Gateway in the Yosemite region, and the Shaver sequence may be continuous with the El Capitan Granite and related granitoids, also in the Yosemite region. Lithology, chemistry, and isotopic ages suggest that the early Late Cretaceous Tuolumne Intrusive Series is comagmatic with the Mono sequence, with the Sonora pluton, north of the area shown in Fig. 4-3, and with a sequence in the Mount Whitney region, south of the area shown in Fig. 4-3.

Constraint 10 Comagmatic sequences of granitoids are generally elongate in a northwest direction, parallel with the long axis of the batholith, and some are repeated along this direction.

BOTTOMS OF PLUTONS

The problem of how plutons terminate downward can be considered here also. Are plutons detached ballonlike forms that floated upward through a gneiss matrix, as proposed by Hamilton and Myers (1967)? Mapping in the Sierra Nevada has failed to reveal even one pluton with inward-dipping contacts to indicate the possibility of downward bottoming at a sharply defined contact. Although reliable studies of compositional variations in a vertical direction have not yet been made, it seems likely that the granitoids grade downward into heavier, more refractory rocks, Gneissic inclusions in the granitoids, such as might be expected if the granitoids were underlain by gneisses, are completely lacking. The most likely composition for deeper rocks is hornblende-bearing garnet-pyroxene rock, a common rock type among inclusions collected by J. P. Lockwood from a basalt pipe near the town of Big Creek, not far from the geographic center of the batholith. This rock has the proper density and P-wave velocity for rocks of the lower crust (J. P. Lockwood, oral communication, 1978).

Constraint 11 No evidence has been found to indicate that the granitoid plutons floated upward as balloonlike forms, leaving behind underlying gneiss, or that they have sharply defined bottoms

P-WAVE VELOCITIES

The geophysical profiles (Bateman, 1979) shown in Figure 4-5 are drawn along the same line as the geochemical and seismic profiles (Fig. 4-4), except for the magnetic profile, which crosses the line of section near its midpoint and is rotated 27° counterclockwise (Fig. 4-3). The seismic profile shows the following P-wave velocities: 6.4 km/s in the upper crust, 6.9 km/s in the lower crust, and 7.9 km/s in the upper mantle. The boundary between the lower crust and mantle is rather sharp. My present belief is that the batholith probably extends downward to the base of the upper crust, becoming more mafic downward because of settled crystalline material, where it grades into lower crust composed largely of hornblende-bearing garnet pyroxene rock left behind by the rising magmas.

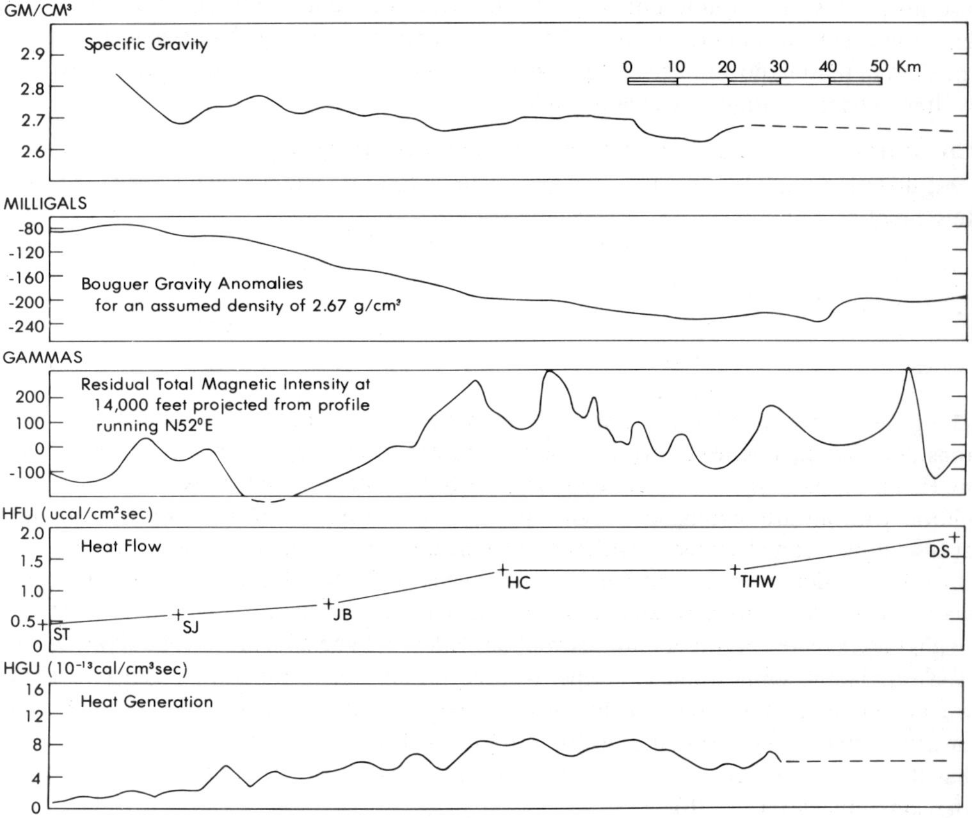

Fig. 4-5. Geophysical profiles. All profiles except residual magnetic intensity are along line *AB* on Figure 4-3. Residual magnetic intensity is along line *CD* on Figure 4-3. Profiles are from Bateman (1979).

The P-wave velocity of the upper mantle was measured by J. P. Eaton (Bateman and Eaton, 1967) to be 7.9 km/s. However, velocity measurements and calculations from petrofabric analysis of tectonite ultramafic nodules (rock types that probably came from the mantle) from the basalt neck near Big Creek suggest the possibility that P-wave velocities actually may vary with direction of measurement between 7.85 to 8.4 km/s (Peselnick and others, 1977). Velocities vary in different directions in ultramafic tectonite primarily because of the preferred orientation of olivine and, secondarily, that of orthopyroxene. The seismic profile was constructed from lines parallel to the long axis of the Sierra Nevada, which possibly corresponds with the slow direction of the regional fabric in the upper mantle. Such a regional seismically anisotropic fabric is known to exist in the upper mantle of the Pacific Ocean basin.

Constraint 12 The seismic moho is depressed beneath the batholith and reaches a maximum depth of about 52 km beneath the Sierra Nevada crest.

Constraint 13 Measured P-wave velocities in the crust are 6.0 km/s at a depth of 10 km, 6.4 km/s at a depth of 20 km, and 6.9 km/s near the base. The relatively slow measured P-wave velocity of 7.9 km/s in the upper mantle may represent the velocity in a slow direction if the mantle rock is tectonite with a fabric parallel to the axis of the batholith.

BOUGUER GRAVITY ANOMALY

The Bouguer gravity anomaly calculated for an assumed density of 2.67 g/cm^3 falls toward the core of the batholith, as would be expected as a result of the relatively light rocks of the batholith (Oliver and Robbins, 1973). However, measured specific gravities show that the assumed constant density is erroneous and that the density of the exposed granitoids actually decreases systematically eastward into the core of the batholith. If this decrease is taken into account, the Bouguer gravity anomaly would be reduced by an amount that would depend on the depth extent of the surface densities.

Constraint 14 The Bouguer gravity profile slopes eastward from -80 milligals in the western foothills to about -225 milligals beneath the high Sierra Nevada, assuming a density of 2.67 g/cm^3. If differences in the specific gravity of the exposed granitoids were taken into account, the magnitude of this negative anomaly would be reduced.

MAGNETIC DATA

The gross changes in residual magnetic intensity (Oliver, 1977) appear to reflect chiefly the abundance of magnetite in the granitoids, but the dip in the eastern part of the profile doubtless reflects the structural and topographic low of Owens Valley, inasmuch as the magnetic data were measured at a constant altitude. The two features of greatest interest are the twin bumps at the west end of the profile, which reflect outcrops of olivine-bearing hornblende gabbro, and the regional magnetic low just east of the bumps. The granitoids that correspond with the magnetic low contain little or no magnetite, a feature Dodge (1972b) expressed in terms of the oxidation ratio and attributed to lower oxygen fugacity contingent on lower water content in the source rocks. Preliminary petrographic studies of these rocks show that they contain rounded blebs of pyrite, suggesting the possibility that sulfur fugacity may have been important in producing the low magnetic values. Sulfur fugacity may indicate a magma source in the mantle or in rocks erupted from the mantle and not exposed to oxidizing conditions.

Constraint 15 The residual magnetic intensity of the granitoids in the western foothills of the Sierra Nevada is extremely low and appears to correlate with a paucity of magnetite.

HEAT GENERATION AND HEAT FLOW

Heat generation in the surface rocks from the radioactive disintegration of isotopes of K.U. and Th increases eastward to the Sierran Crest, then diminishes farther east. Heat flow increases steadily eastward, rising from a little more than 0.4 HFU (heat-flow units) in the west to about 1.8 HFU in the White Mountains (Lachenbruch, 1968; Lachenbruch and others, 1976). All these heat-flow values are lower than the average for continental areas, and 0.4+ HFU is the lowest value yet measured and far below measurements obtained in the ocean basins. A graph in which the heat-flow measurements are plotted against heat generation at the drill sites yields a straight line that intersects zero heat generation at 0.4 HFU. Lachenbruch (1968) has interpreted this amount of heat to represent the mantle contribution beneath the Sierra Nevada, the excess above this amount coming from radioactivity in the crust. This inferred mantle contribution is low as compared to other regions and has been tentatively explained by Lachenbruch as resulting from downward loss of heat from the upper mantle, possibly to a cold subducted slab.

Constraint 16 The heat flow from the Sierra Nevada and inferred mantle contribution of 0.4 HFU are anomalously low and suggest loss of heat downward.

Richard A. Schweickert
Department of Geological Sciences
Lamont-Doherty Geological Observatory
Columbia University,
Palisades, New York 10964

5

TECTONIC EVOLUTION OF THE SIERRA NEVADA RANGE*

*Lamont-Doherty Geological Observatory Contribution No. 2911

ABSTRACT

The distribution and structural grain of metamorphic wallrocks of the Sierra Nevada are a result of the youngest major deformational event in the region, the Late Jurassic Nevadan orogeny. Structures produced during earlier deformational events have in most cases been rotated or transposed into the Nevadan structural grain.

Mesozoic volcanic and volcaniclastic rocks in the central parts of the range occupy the core of the Nevadan synclinorium. They indicate that an early Mesozoic Andean-type arc was isoclinally folded prior to the emplacement of Late Jurassic to Late Cretaceous granitic rocks of the Sierra Nevada batholith in the core of the synclinorium.

In the western limb of the synclinorium, two Paleozoic bedrock complexes are separated by a steeply dipping, pre-Nevadan thrust fault. The lower Paleozoic Shoo Fly Complex, in the upper plate, consists of metamorphosed sedimentary and volcanic rocks that may have rather close counterparts in the lower Paleozoic Roberts Mountains assemblage of north-central Nevada. The Shoo Fly is unconformably overlain by a Devonian to Permian pyroclastic sequence that most likely represents an island-arc accumulation.

The upper Paleozoic (?) Calaveras Complex, in the lower plate, contains metamorphosed volcanic and sedimentary rocks that resemble the upper Paleozoic Havallah sequence in north-central Nevada. Both the Calaveras Complex and the Havallah sequence may have accumulated within a late Paleozoic marginal basin. During Late Permian or Triassic time, the Shoo Fly and overlying pyroclastic sequence were thrust over the upper Paleozoic (?) Calaveras Complex.

Possible equivalents of the Shoo Fly Complex occur in the eastern limb of the Nevadan synclinorium in several areas, but despite their lithologic similarity to rocks of the Roberts Mountains allochthon, they probably occupy a structural position above the Havallah sequence.

In the south-central Sierra Nevada, lower Paleozoic miogeoclinal sedimentary rocks occupy the east limb of the Nevadan synclinorium and also form a major anticlinorium in the White-Inyo Mountains to the east. Their position in the eastern limb strongly implies that metamorphosed miogeoclinal strata in the western limb of the synclinorium in the southwestern Sierra Nevada are lower Paleozoic or Eocambrian.

During Cretaceous time, the Nevadan synclinorium underwent minor dextral strike-slip faulting along the Kern Canyon fault, and minor folds and cleavages developed in many areas.

The most plausible cause of the major deformation that accompanied the Nevadan orogeny was the Late Jurassic collision of an east-facing island arc that now lies west of the Foothills suture in the western Sierra Nevada with the Andean-type continental margin. This island arc developed between Late Triassic and Late Jurassic time on complexly deformed and largely chaotic oceanic basement that probably formed at an oceanic fracture zone.

Only through an understanding of the nature and geometry of the Nevadan structure can the extent and continuity of Paleozoic terranes be deciphered. Structural data and regional relations indicate that the Shoo Fly Complex and pyroclastic sequence occupy the highest Paleozoic thrust plate in the southwestern Cordillera. They rest structurally above the Calaveras Complex and Havallah sequence, which collectively rest structurally upon rocks of the Roberts Mountains allochthon in north-central Nevada. The higher two plates were probably emplaced during the Sonoma orogeny (Late Permian-Early Triassic); the lower was emplaced upon the Cordilleran miogeocline during the Antler orogeny (Late Devonian-Early Mississippian).

INTRODUCTION

The Sierra Nevada range is geologically famous for its extensive exposures of igneous rocks. The Sierra Nevada batholith is perhaps the best-known link in a chain of Mesozoic granitoid batholiths that stretches with only minor interruptions from southeastern Alaska to western Mexico. Compared to the granitoid rocks, the metamorphosed wallrocks that enclose the batholith are somewhat surprisingly rather poorly known, even though they received considerable attention in the early part of this century and have been intensively studied since the late 1950s. In addition, regional relationships between Paleozoic Sierran wallrocks and rocks of adjacent areas in California and Nevada have scarcely been discussed in the literature.

Uncertainty about framework rocks of the Sierra Nevada stems from the facts that the rocks are to a very great extent unfossiliferous and, in addition, are in most cases strongly deformed and metamorphosed. Poor exposure and deep weathering that are characteristic of much of the metamorphic belt in the western foothills further compound the problem.

Because of these drawbacks, no complete summary or account of the geology of Sierra Nevada framework rocks has ever been written, nor is one possible at present. This paper aims at synthesizing the results of research by many geologists on wallrocks of the batholith in the Sierran region during the past two decades, with special emphasis on eugeosynclinal rocks. The coverage of the wallrocks in this paper is not exhaustive. Some areas are treated superficially, and other areas I know firsthand are dwelt upon. There are many controversial issues, most of which cannot be resolved at present. I make no pretense of complete objectivity and state personal opinions in numerous cases.

Known and inferred geologic relations of various terranes are discussed initially, and various interpretations of the geologic evolution then follow. However, I present no comprehensive plate tectonic history for Paleozoic time in this paper, because the Sierra Nevada alone does not provide an adequate basis for such scenarios. Many other regions must be considered. Schweickert and Snyder (Chapter 7, this volume) present Paleozoic plate tectonic models for the Sierra Nevada region that are based on analysis of regional relationships in California and Nevada.

GROSS STRUCTURE OF THE RANGE

In describing the geology of various parts of the range, it is important to emphasize the controls of large-scale Mesozoic folding on the distribution and structural grain of the metamorphic wallrocks. The dominant northwest-trending structural grain of the Sierra Nevada is a product of the youngest *major* deformation, the Late Jurassic Nevadan orogeny. Structures produced during earlier deformational events have in many cases been strongly rotated or transposed into the Nevadan structural grain. Geologists of the U.S. Geological Survey have recognized for many years that the distribution of wallrocks and remnants within and adjacent to the batholith strongly suggest the existence of a major synclinal structure or synclinorium (the Nevadan synclinorium) (Bateman and others, 1963; Clark, 1964; Bateman and Wahrhaftig, 1966; Bateman and Clark, 1974) whose axial region is occupied by the Sierra Nevada batholith.

Saleeby (Chapter 6, this volume) believes that the regional inward facing pattern of stratigraphic units resulted from downward flow of wallrocks compensating the upward magmatic transport of the massive Cretaceous Sierra Nevada batholith. This view seems inconsistent with the facts that the form of the synclinorium clearly *predates* the Cretaceous batholith and that the wallrocks have a regionally uniform mesoscopic structural fabric, which indicates that the batholith was emplaced by passive mechanisms and did not appreciably deform the wallrocks to any degree. It seems more likely that the Nevadan synclinorium formed during a major Late Jurassic regional deformational event in eastern California.

Two other major structural features of the Sierra Nevada range, the Foothills suture (partly marked by the Melones fault zone) and the Calaveras–Shoo Fly thrust fault (Fig. 5-1), are of major importance to an understanding of Sierra Nevada tectonic evolution. The Foothills suture forms the western boundary of and may be contemporary with the Nevadan synclinorium, and the Calaveras–Shoo Fly thrust is an older fault that separates two major tectonostratigraphic units within the western limb of the synclinorium.

Nevadan Synclinorium

Western limb Rocks of the western metamorphic belt that lie east of the Melones fault zone (Figs. 5-1, 5-3, and 5-7) form the steeply dipping, east-facing limb of the Nevadan synclinorium. North of latitude 39°15′N, the lower Paleozoic Shoo Fly Complex, a complexly deformed sequence of metasedimentary and minor metavolcanic rocks, is unconformably overlain on the east by a steeply east-dipping Paleozoic pyroclastic sequence in which stratigraphic units face eastward (see later discussion). The Paleozoic pyroclastic rocks in turn are unconformably overlain by east-dipping Mesozoic pyroclastic rocks throughout their extent and pinch out beneath the Mesozoic rocks near latitude 39°15′N. South of this point the Mesozoic strata rest directly upon the Shoo Fly Complex.

Between latitudes 38°45′N and 37°30′N, the Shoo Fly Complex is structurally underlain by the upper Paleozoic Calaveras Complex, which, despite its extreme structural

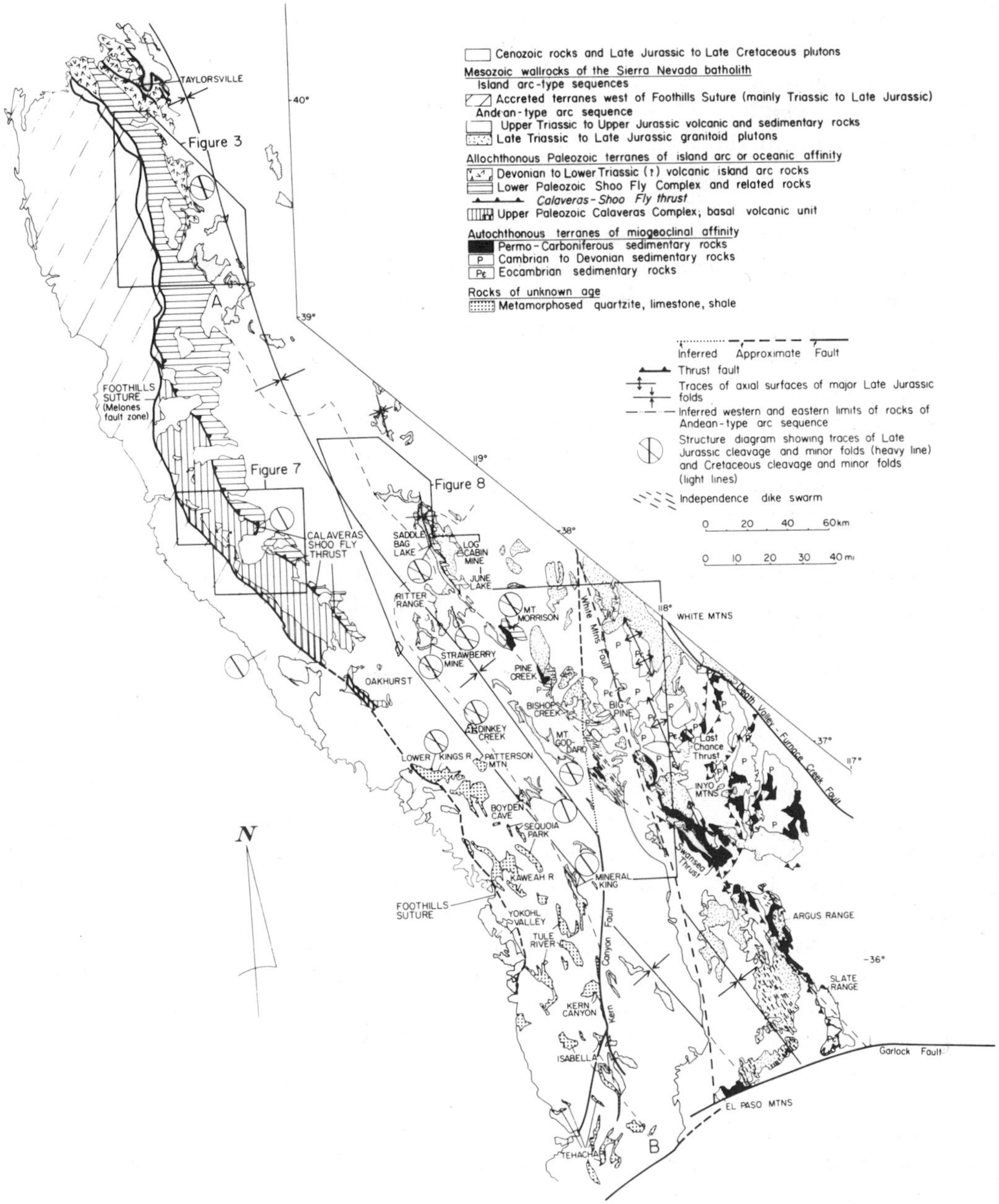

Fig. 5-1. Generalized geologic map of the wallrocks of the Late Jurassic-Late Cretaceous Sierra Nevada batholith. Sources of data: basemap, (Jennings, 1977); thrust faults, Last Chance, (Stewart and others, 1966), Swansea (Dunne and others, 1978), Calaveras-Shoo Fly (Schweickert, unpub. mapping); structural data (Russell and Nokleberg, 1977; Schweickert, unpub. data); Independence dike swarm (Moore and Hopson, 1961).

complexity, shows evidence of an eastward-facing direction (Schweickert and others, 1977). The steeply dipping Calaveras–Shoo Fly thrust forms the boundary between these two complexes (Schweickert, 1977).

From latitude 37° to 40°N the western metamorphic belt east of the Melones fault zone, even with thrust imbrications within it, can be regarded as a nearly homoclinal east-dipping and east-facing western limb of the major Nevadan synclinorium. Because of this relationship, the geologic map of the western metamorphic belt is a fair approximation of a north-south cross section parallel to the axis of the synclinorium (Fig. 5-1). The western Sierra Nevada is unique in providing a nearly continuous structural section across almost the entire Paleozoic Cordilleran orogen.

Unfortunately, Mesozoic metavolcanic rocks have not been preserved at the present level of erosion between about latitudes 38° and 38°45′N, so that the internal structure of the synclinorium is not readily evident through this sector.

From about latitude 37°N to the Garlock fault, controversy exists over the extent and nature of Mesozoic rocks on the western limb. Saleeby and others (1978) argued that all roof pendants in this region contain Jurassic and Triassic rocks, some of which are complexly deformed. If they are correct, the western limb of the synclinorium may lose definition in this region. However, other workers, notably Kistler and Bateman (1966), Kistler and Peterman (1973), W. Nokleberg (personal communication 1978), and the author feel that lower Paleozoic rocks occur in some of the more westerly pendants within the region.

The line AB in Figure 5-1 is drawn at the western limit of occurrences of Mesozoic fossils reported by Saleeby and others (1978). No fossils occur to the west of this line, except along and west of the Foothills suture belt. The Mesozoic fossils in the roof pendants occur generally near the base of east-dipping, east-facing sequences of metavolcanic and metasedimentary rocks in the Boyden Cave, Mineral King, and Isabella pendants (Saleeby and others, 1978). Such strata are flanked to the west by more complexly deformed sequences of quartzite, marble, and pelitic schist that resemble Paleozoic miogeoclinal rocks of the Inyo Mountains. These relations suggest to me that the western limb of the Nevadan synclinorium can thus be traced as far south as the Garlock fault. In the El Paso Mountains (Fig. 5-1), a northwest-striking, northeast-facing succession of Paleozoic rocks (Dibblee, 1967) also apparently occupies a position in the western limb of the synclinorium, although it lies considerably east of rocks in the corresponding structural position in the southern Sierra Nevada. This important structural relation will be discussed more fully in a later section.

Eastern limb North of latitude 38°30′N, Cenozoic high-angle faults that cross the axis of the synclinorium have downdropped Sierran bedrock from view, and therefore the eastern limb is not evident. However, south of about latitude 38°N, metamorphic rocks in roof pendants on the east side of the range clearly define the eastern limb of the Nevadan synclinorium. In addition, farther east, lower Paleozoic and Precambrian rocks of the White-Inyo Mountains form the core of a major anticlinorium (Fig. 5-1 and 5-2) that appears to be structurally and stratigraphically continuous with the east limb of the Nevadan synclinorium.

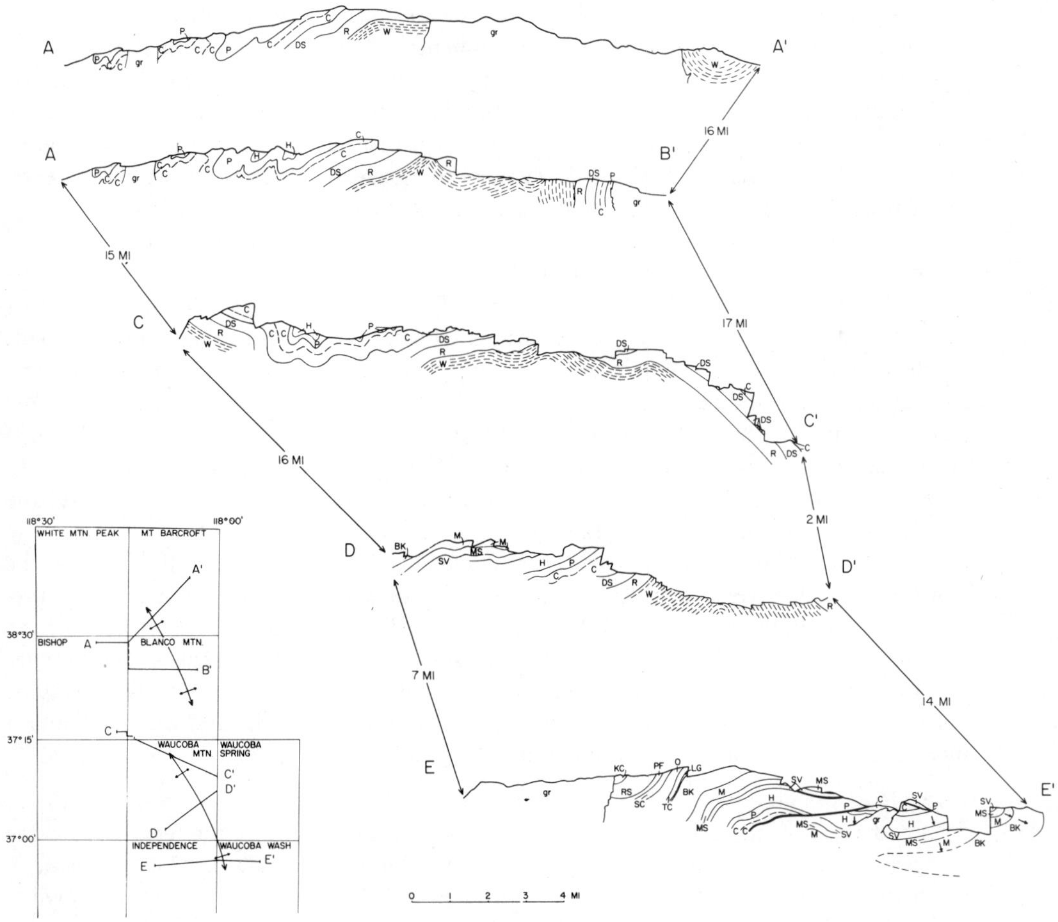

Fig. 5-2. Structure sections in Paleozoic and Precambrian rocks of the White and Inyo Mountains, showing the White Mountains anticlinorium. Prepared by removing displacements on Cenozoic faults shown on cross sections by Bateman, 1965a; Krauskopf, 1971; Nelson, 1966a, b; Ross, 1962, 1967a, b. Units shown: Precambrian (W, Wyman Formation; R, Reed Dolomite; DS, Deep Springs Formation); Cambrian (C, Campito Formation; P, Poleta Formation; H, Harkless Formation: SV, Saline Valley Formation: MS, Mule Spring Limestone; M, Monola Formation; BK, Bonanza King Dolomite; LG, Lead Gulch Formation; TC, Tamarack Canyon Formation); Ordovician (O, Al Rose Formation, Badger Flat Limestone, Barrel Spring Limestone, Johnson Spring Limestone, and Ely Springs Dolomite); Silurian-Devonian (SC, Sunday Canyon Formation); Mississippian (PF, Perdido Formation); Pennsylvanian (RS, Rest Spring Shale); Pennsylvanian-Permian (KC, Keeler Canyon Formation).

In the Ritter Range and Saddlebag Lake pendants (Fig. 5-1), complexly deformed lower and upper Paleozoic rocks are unconformably overlain along their western margin by steeply dipping, west-facing sequences of Mesozoic metavolcanic rocks that closely resemble the east-facing Mesozoic metavolcanic sequences in the western limb north of latitude 38°30′N and south of latitude 37°N.

Southeast of the Ritter Range pendant, west-facing Mesozoic metavolcanic rocks are underlain by west-dipping, west-facing sequences of Permian and Pennsylvanian age. This relationship exists in the Mount Morrison pendant (where the unconformity is apparently faulted; Morgan and Rankin, 1972), the Pine Creek pendant, the Cerro Gordo district in the southern Inyo Mountains, and in the southern Argus and Slate Ranges, indicating that the eastern limb of the synclinorium can be recognized, with several complications (discussed later), as far south as the Garlock fault (latitude 35°30′N).

It should be noted that the stratigraphic data and map relations just discussed accord well with the results of detailed structural mapping over the entire length of the range. Figure 5-1 shows summary stereoplots from all areas of the Sierra Nevada that have yielded data on minor folds and cleavages. The heavy lines show the dominant trends of cleavage and axial-surface traces of minor folds that are believed to be of the same generation as the folding of the synclinorium.[1] In all cases these minor structures show remarkable parallelism to the axial-surface trace of the regional synclinorium. I have observed asymmetric minor folds of this generation throughout the western limb from latitude 37°30′ to 40°N, and the dominant sense of asymmetry everywhere suggests the existence of a synclinal closure to the east.

Consideration of all available stratigraphic and structural data thus forces the conclusion that a very large synclinal structure (or synclinorium) dominates the pre-Cretaceous wallrocks of the entire Sierra Nevada region and is in fact responsible for the predominant N20 to 30W regional structural grain of the wallrocks. This conclusion has important implications for understanding the regional relationships of tectonostratigraphic units of Paleozoic and Mesozoic age throughout eastern California, and will be discussed more fully in the section on structural models.

Age Rocks as young as Callovian age (in the Taylorsville area at the north edge of Figure 5-1) occur within the axial part of the synclinorium and are cut by the regional cleavage that is probably related to the formation of the synclinorium (McMath, 1966). Elsewhere, as on the north fork of the American River (Fig. 5-3) and in the Ritter Range, Boyden Cave, Mineral King, and Isabella pendants (Fig. 5-1), rocks as young as Early Jurassic age are involved in the metamorphism and deformation. Upper Jurassic plutonic rocks cut folded rocks of the synclinorium in the western metamorphic belt near latitude 38°N (Fig. 5-1), thus placing a younger age limit on its formation. These relations indicate that this major fold system developed during the Late Jurassic, and that its formation may have been simultaneous with the deformation of Oxfordian-Kimmeridgian rocks west of the Melones fault zone. The deformational events indicated have long been known as the Nevadan orogeny (Knopf, 1929; Taliaferro, 1942; Bateman and others, 1963; Clark, 1964; Bateman and Wahrhaftig, 1966; Bateman and Clark, 1974).

[1] Nokleberg and Kistler (1980) and Dunne and others (1978) have argued that more than one generation of N30W-trending cleavages and minor folds exist in parts of the region, particularly in what is here regarded as the east limb of the synclinorium. They state that in several areas (not shown on Fig. 5-1) old N30W-trending structures are cut by Late Triassic to Early Jurassic plutons (see section on structural models for discussion).

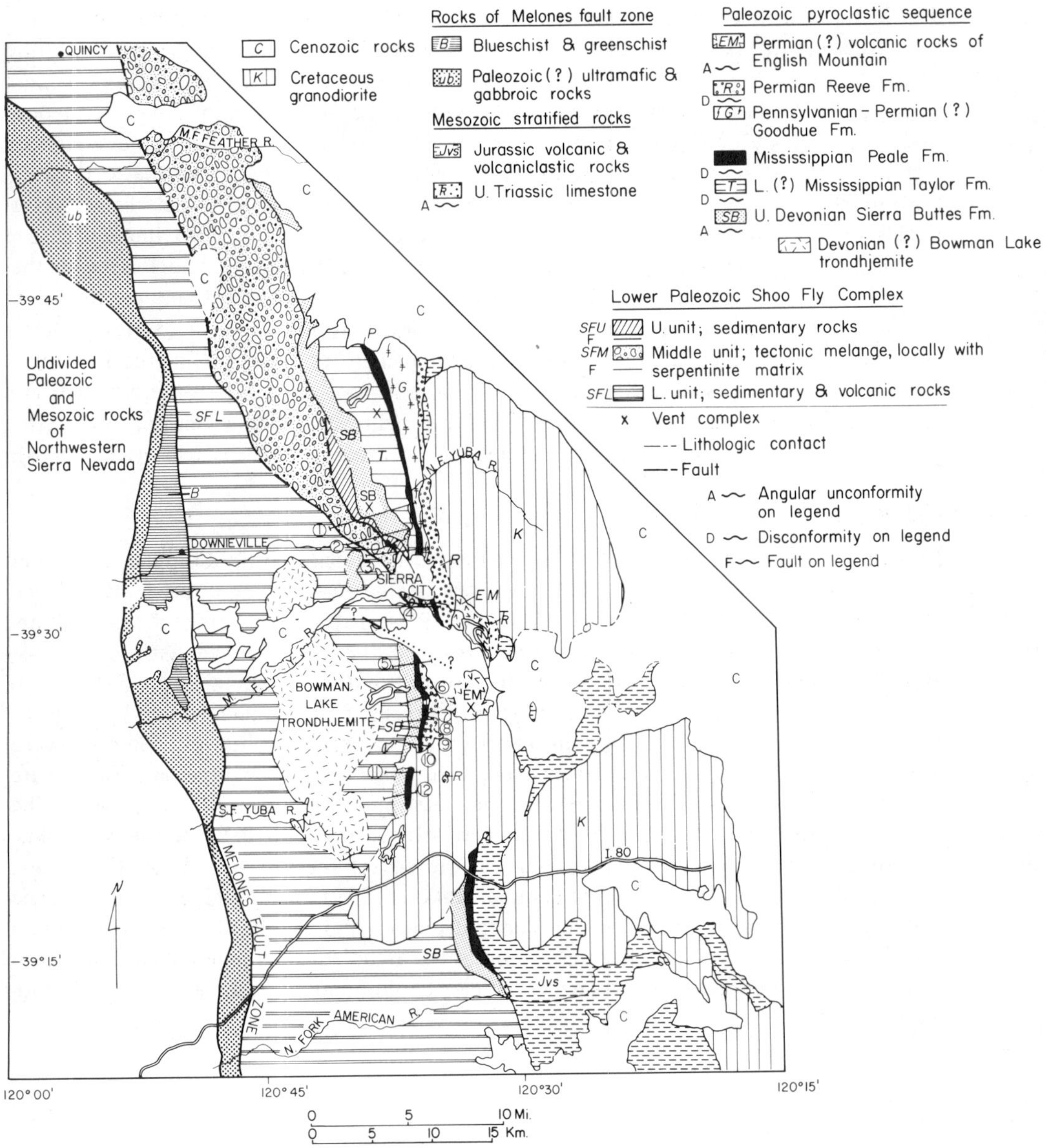

Fig. 5-3. Generalized geologic map of the Sierra Buttes-Bowman Lake region. Mapping by Schweickert and students, 1973-1979. Areas south of I 80 and north of Sierra Buttes from d'Allura and others (1977). Bowman Lake is located at the eastern edge of the Bowman Lake trondhjemite.

Foothills Suture Belt

The Melones fault zone in the western metamorphic belt (Fig. 5-1) marks part of a major suture between terranes that experienced different and unrelated geologic histories prior to the Nevadan orogeny (Davis, 1969; Schweickert, 1976a, b, c, 1978; Schweickert and Cowan, 1975). Mesozoic volcanic sequences west of the Melones fault zone were deposited on complexly deformed oceanic basement; in contrast, Mesozoic volcanic sequences east of the fault zone rest on Paleozoic sedimentary and volcanic rocks that originated both in island arcs and in a Paleozoic miogeocline (Schweickert, 1978). Saleeby (Chapter 6, this volume) considers the Foothills suture to occupy a position approximately 25 km east, near the present trace of the Calaveras-Shoo Fly thrust (discussed later). Between the lower Kings River and the Tule River (latitude 36° and 37°N), I infer that a continuation of the Foothills suture separates a chaotic ophiolitic terrane and an overlying chert-argillite-volcanic sequence (the "western Kings sequence" of Saleeby, Chapter 6, this volume) on the west (Saleeby, 1977a, b) from rocks in roof pendants to the east that are characterized by sequences of marble, quartzite, pelite, and silicic volcanic rocks (the "eastern Kings sequence" of Saleeby, Chapter 6, this volume; Saleeby and others, 1978).

Calaveras-Shoo Fly Thrust Fault

Between latitudes 37°30′ and 38°45′N, the lower Paleozoic Shoo Fly Complex and upper Paleozoic Calaveras Complex are separated by a steeply east-dipping thrust fault (Schweickert, 1977, 1978, and in preparation). Clark (1976) mapped a portion of this fault between latitudes 38°40′ and 39°N, and I have mapped it in detail between latitudes 37°45′ and 38°15′N and in reconnaissance over the rest of its extent shown in Figure 5-1.

Between latitudes 38°45′ and 38°15′N the thrust fault is characterized by a belt of mylonitic rocks that grades upward into upper plate rocks of the Shoo Fly and downward into lower plate rocks of the Calaveras. The thrust fault and both upper and lower plate are tightly folded east of Sonora into a major east-plunging antiform and synform. The Standard pluton (Figs. 5-1 and 5-7) truncates both the mylonite belt and the later folds, and thus places an important constraint on the age of the Calaveras-Shoo Fly thrust. Sharp and Saleeby (1979; personal communication, 1979) have obtained a mid-Jurassic U-Pb age of about 170 m.y. for the Standard pluton, and therefore the Calaveras-Shoo Fly thrust is older than Middle Jurassic. Structural arguments and regional considerations discussed in a later section of this paper indicate a minimum displacement of tens of kilometers on the Calaveras-Shoo Fly thrust.

OUTLINE OF TERRANES EAST OF THE FOOTHILLS SUTURE

Paleozoic Rocks

Shoo Fly Complex The Shoo Fly Formation was originally named by Diller (1892, p. 375) for a sequence of stratified rocks exposed north of latitude 40°N and 8 to 16 km southwest of Taylorsville (Fig. 5-1). Since Diller's (1892, 1895, 1908) original work, various portions of the Shoo Fly terrane have been restudied by a number of workers, including McMath (1966), Clark (1976), and d'Allura and others (1977). Much of the following discussion is based on these reports and on my unpublished work.

Recent investigations in the Shoo Fly terrane indicate a degree of structural and stratigraphic complexity that invites use of the term "Complex" rather than "Formation." At present, the full regional extent of this complex is poorly known, but my detailed and reconnaissance mapping indicate that rocks best assigned to the Shoo Fly extend at least to the Merced River (latitude 37°30′N) (Fig. 5-1), and that rocks of the Saddlebag Lake pendant, recently studied by Brook (1977), have many characteristics of the Shoo Fly (Fig. 5-1). Brook (1977) correlated the rocks at Saddlebag Lake with fossiliferous Ordovician and Silurian strata in the Mount Morrison pendant described by Rinehart and Ross (1964), and for convenience I adopt his correlation. Thus defined, rocks of the Shoo Fly Complex occur on both the west and east limbs of the Nevadan synclinorium, and span a northwest-southeast distance of 340 km, making the complex the most areally extensive Paleozoic tectonostratigraphic unit in California.

The western boundary of the Shoo Fly is marked by the Melones fault zone as far south as latitude 38°45′N (Fig. 5-1), which truncates the base of the complex. As already discussed, the Calaveras-Shoo Fly thrust marks its western boundary from latitude 38°45′ to 37°N.

The most recent detailed investigations of the Shoo Fly suggest that it includes at least three major lithotectonic units (Fig. 5-3) (d'Allura and others, 1977; Schweickert, 1974 and unpublished data), herein referred to as the "lower," "middle," and "upper" units. The contacts between these units appear in most areas to be faults (Standlee, 1977; Schweickert, unpublished data).

The "lower" unit is the most extensive unit within the Shoo Fly, inasmuch as it underlies the entire outcrop belt from Bowman Lake (Fig. 5-3) to the Merced River (Fig. 5-1), and rocks possibly assignable to this unit also occur within the Saddlebag Lake and Mount Morrison pendants (Fig. 5-1). D'Allura and others (1977) described the unit as including subfeldspathic and lithic sandstone, quartz sandstone, and dark green to black shale with subordinate conglomerate.

I have mapped the "lower" unit in reconnaissance from the middle fork Feather River to Lakes Basin, and G. Girty and I have now mapped in detail between the north fork Yuba River and Interstate 80 (Fig. 5-3). Phyllite, chert, and rare thin quartzose sandstone beds predominate north of the north fork Yuba River. South of the north fork Yuba River the "lower" Shoo Fly unit consists of abundant thick quartzose and quartzofeldspathic sandstone beds that form part of a recognizable stratigraphic succession (Girty and Schweickert, 1979). The lowest exposed part of this succession consists of basaltic pillow lava and breccia (Fig. 5-4). These rocks are overlain by up to 200 m of cross-bedded calcarenite and pisolitic limestone, which is succeeded in turn by up to 300 m of rhythmically bedded chert and slate. The sandstone succession conformably overlies the chert-slate unit and is up to 1.3 km thick. It contains numerous olistostromes and abundant evidence of soft-sediment deformation. Still higher units are chert and slate with minor volcanic breccia. This stratigraphic sequence is suggested by abundant facing indicators in all units.

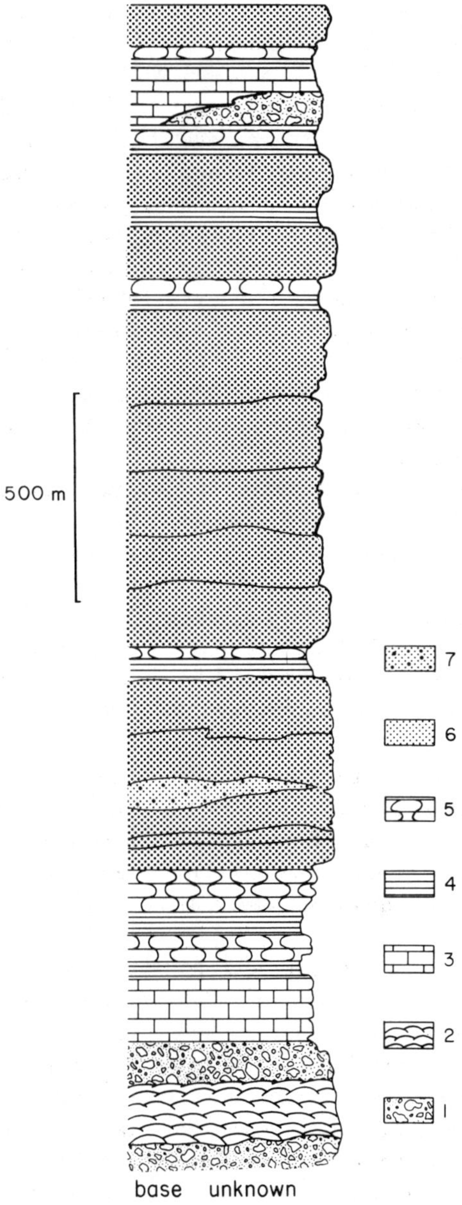

Fig. 5-4. Generalized columnar section of a part of the lower Shoo Fly (SF 1) near Bowman Lake, California. 1, basaltic pillow breccia; 2, basaltic pillow lava; 3, cross-bedded, locally pisolitic calcarenite; 4, slate; 5, bedded chert; 6, quartzose to quartzofeldspathic sandstone; 7, chaotic pebbly mudstone.

The lower part of the sandstone unit near Bowman Lake very closely resembles distinctive lithologies in both the Upper Cambrian Harmony Formation in north-central Nevada and the undated Antelope Mountain quartzite near Yreka, California, in the eastern Klamath belt (Girty and Schweickert, unpublished data; see Fig. 5-5). In addition,

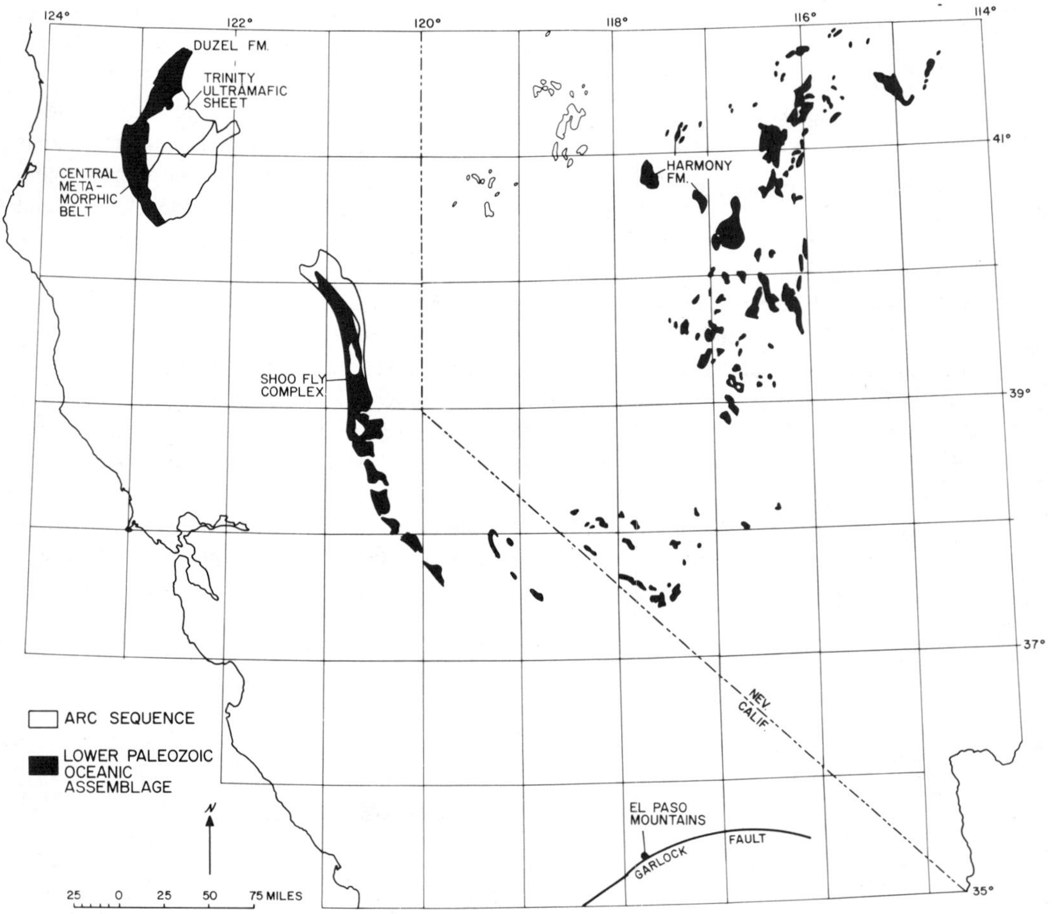

Fig. 5-5. Sketch map of distribution of lower Paleozoic oceanic rocks and upper Paleozoic island-arc rocks in California and Nevada.

the underlying chert-slate unit closely matches the Paradise Valley Chert of north-central Nevada and an unnamed chert unit beneath the Antelope Mountain quartzite in the Klamath Mountains. Other portions of the section, which include phyllite, chert, quartzite, and greenstone, are reminiscent of parts of the Ordovician Valmy Formation in north-central Nevada.

Such marked resemblance of the lithology and sequence of parts of the "lower" Shoo Fly to Cambrian and Ordovician formations in north-central Nevada is suggestive of a lithologic correlation, but such units, by analogy with modern pelagic and oceanic sediments, are to be expected to be time-transgressive, and hence Shoo Fly rocks may range in age from Eocambrian to even Devonian, overlapping the ages of rocks in the Roberts Mountains assemblage.

Durrell and Proctor (1948) noted that the Shoo Fly was folded prior to deposition of a thick sequence of silicic to andesitic volcanic rocks, which are now known to range from Late Devonian (Anderson and others, 1974) to as young as Permian. Mapping near Bowman Lake has produced evidence that the entire sequence of units in the lower Shoo Fly was isoclinally folded about hinge lines presently striking 315° (Girty and Schweickert, 1979) *prior* to both juxtaposition with the "middle" Shoo Fly unit, a tectonic melange, and the eruption of the overlying Upper Devonian Sierra Buttes Formation.

The "middle" Shoo Fly (Fig. 5-3) is a tectonostratigraphic unit consisting of highly sheared shale and sandstone, lenses of limestone, dolomite, chert, volcanics, and small masses of serpentinized ultramafic rocks up to several kilometers long (d'Allura and others, 1977; Schweickert, 1974, and unpublished data). A limestone block in the melange near Taylorsville (Fig. 5-1) has yielded a Late Ordovician fossil (d'Allura and others, 1977), the only fossil known from the Shoo Fly.

The extent of the "middle" Shoo Fly unit is poorly known in the region north of Sierra Buttes, where it was shown by d'Allura and others (1977). I have mapped the melange from Sierra Buttes to a point 6 km south of Sierra City (Fig. 5-3), south of which it

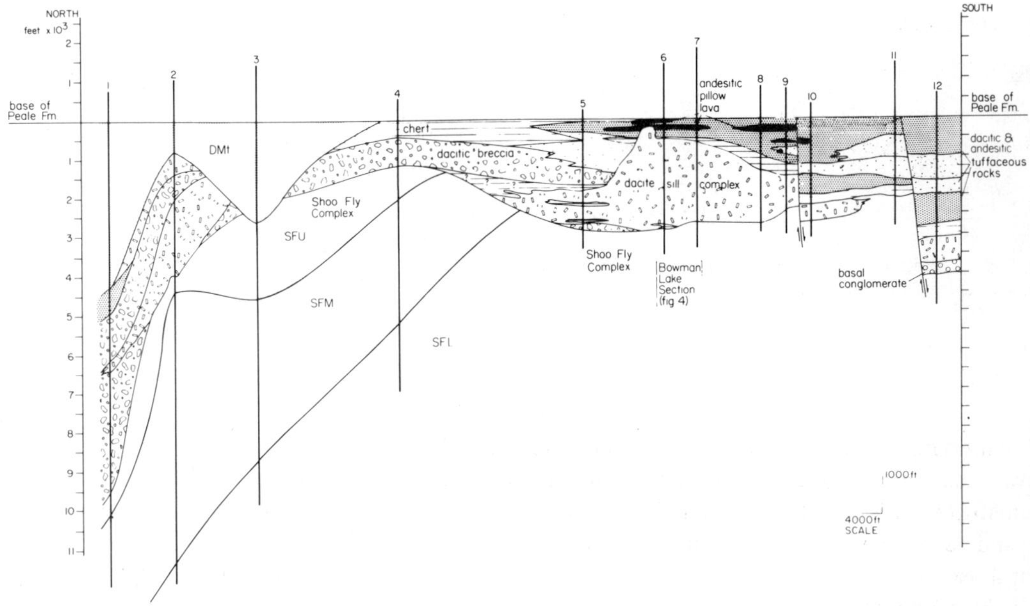

Fig. 5-6. Restored section of lithofacies of Sierra Buttes and Taylor formations, Sierra Buttes-Bowman Lake region. Note how andesitic rocks become more abundant in Sierra Buttes Formation in southern part. See Fig. 5-3 for locations of columns 1 to 12.

is cut out beneath the unconformably overlying Sierra Buttes Formation (Fig. 5-6). Dikes and sills of quartz keratophyre lithologically indistinguishable from the overlying Sierra Buttes Formation also cut the melange.

The "upper" Shoo Fly (Fig. 5-3) consists of black to light gray chert, siliceous argillite, and shale (d'Allura and others, 1977). According to Standlee (1977), the "upper" Shoo Fly also contains greenstone and minor tuff units. D'Allura and others (1977) have suggested that the protolith of the cherts may have been crystal-poor tuff and ash.

Pyroclastic sequence The Paleozoic pyroclastic sequence, as defined by McMath (1966), consists of up to 11 km of pyroclastic and epiclastic deposits that unconformably overlie the Shoo Fly Complex (Figs. 5-1, 5-3, and 5-6). The following descriptions of stratigraphic units are based in part on work by Durrell and d'Allura (1977) and d'Allura and others (1977), and in part on R. Hanson's and my own unpublished mapping.

As noted by d'Allura and others (1977), the pyroclastic sequence consists of two dominantly volcanogenic phases that are separated by an interval of chert and clastic rocks.

The older volcanogenic phase deposited during Late Devonian-Mississippian (?) time consists of the Sierra Buttes and Taylor formations. The Sierra Buttes is 100 to 1250 m thick, and is made up predominantly of fragmental dacitic to rhyolitic rocks, with minor occurrences of andesitic lava and tuff. It contains a few interbeds of clastic rocks and several intervals of black, phosphatic radiolarian chert. Durrell and d'Allura (1977) asserted that the black chert is characteristic of the Elwell Formation, a unit they considered to lie above the Sierra Buttes Formation. However, the chert forms discontinuous lenses throughout the Sierra Buttes Formation (Fig. 5-6) and does not even form a continuous unit at the top. Andesitic rocks that lithologically resemble the overlying Taylor Formation (discussed later) are intercalated within the upper part of the Sierra Buttes Formation in and south of the Bowman Lake area. The southward increase in thickness of these andesitic rocks suggests the contribution from an andesitic volcanic center that lay generally to the south of the present outcrop. D'Allura and others (1977) reported a vent area within the Sierra Buttes Formation on Grizzly Ridge, near Taylorsville (Fig. 5-1); another vent complex occurs in the vicinity of the Sierra Buttes peaks (Figs. 5-3 and 5-4). The upper part of the Sierra Buttes Formation northeast of the Sierra Buttes peaks has yielded a rich Upper Devonian fauna (Anderson and others, 1974).

The Bowman Lake trondhjemite (Fig. 5-3) clearly cuts early folds in the "lower" Shoo Fly and has contact metamorphosed adjacent rocks. It is lithologically distinctive, containing 30 to 40% modal quartz and virtually no K-feldspar. Marginal facies of the pluton consist of quartz-porphyry that is petrographically indistinguishable both from the small dikes and sills that cut the Shoo Fly and the domes and sills of quartz-keratophyre in the Sierra Buttes Formation. Both field and petrographic evidence suggest that the Bowman Lake trondhjemite is coeval and comagmatic with the Sierra Buttes Formation and hence is Late Devonian.

The Taylor Formation, 900 to 3000 m of andesitic tuff and breccia, rests upon the Sierra Buttes Formation. North of Sierra City, the contact is conformable and gradational; however, 2 km south of Sierra City the Taylor pinches out and does not reappear to the south (Fig. 5-6). D'Allura and others (1977) apparently mistook the andesitic rocks within the Sierra Buttes Formation near Bowman Lake for the Taylor Formation, and suggested that the Taylor Formation near Bowman Lake originated from a different source than the thicker parts of the formation north of Sierra City. My data, as noted, indicate that the Bowman Lake area contains andesites that are intercalated within the older Sierra Buttes Formation. No fossils have been recovered from the Taylor Formation.

The Peale Formation separates the older Devonian-Mississippian (?) volcanic phase from a Permian pyroclastic sequence. The Peale disconformably overlies the Taylor in most areas, but where the Taylor is absent, the Peale rests directly on the Sierra Buttes Formation. On the middle fork Yuba River (Fig. 5-3) a 5 to 10° angular discordance exists between the Peale and Sierra Buttes Formations. D'Allura and others (1977) noted that north of the north fork Yuba River the Peale Formation contains a lower member of up to 1 km of silicic pyroclastic rocks. An upper member of rhythmically bedded radiolarian chert, tuffaceous slate, and local intraformational breccia predominates south of the north fork Yuba River. Near Taylorsville, Early Mississippian fossils have been recovered from the upper member of the Peale (d'Allura and others, 1977). Early Permian fossils that d'Allura and others (1977) considered to have come from the upper member of the Peale are from rocks I consider part of an overlying formation (see later discussion).

The Goodhue Formation, up to 2.8 km of olivine basalt, basaltic andesite breccia, and other pyroclastic rocks, conformably overlies the Peale Formation north of the middle fork Yuba River (d'Allura and others, 1977; Fig. 5-3), but extends no farther south than the middle fork.

Overlying the Goodhue is the Reeve Formation, consisting of up to 1.7 km of andesitic breccia, tuff, pillow lava, and subaerial flows. The most distinctive lithology is a plagioclase porphyry with phenocrysts up to 3 cm in length. Near the middle fork Yuba River, the Reeve Formation with the distinctive plagioclase porphyry disconformably overlies the Peale, locally occupying channels up to 30 m deep and containing angular fragments of Peale chert. The Lower Permian fusulinids reported by d'Allura and others (1977, p. 402) from the Peale Formation near Bowman Lake are detrital fragments that occur near the tops of graded chert breccias that are indistinguishable from those in the Reeve Formation 3 km north. This indicates that it is incorrect to attribute to the Peale a period of sedimentation from Early Mississippian to Late Permian time, as suggested by d'Allura and others (1977). Instead, a hiatus involving much or most of the Pennsylvanian Period probably exists between the Peale and the Reeve formations near Bowman Lake. The Goodhue Formation may have been deposited during part of this interval in areas north of the middle fork Yuba River.

Again near the middle fork Yuba River, the Reeve Formation is unconformably overlain by the volcanic sequence of English Mountain (Fig. 5-3), with an angular discordance locally as high as 30°. Pyroclastic rocks of this sequence, shown as Jurassic by d'Allura and others (1977), coarsen rapidly southward to very thick, massive tuff-breccias on English Mountain, which evidently contains a vent complex (Fig. 5-3). This sequence is overlain by probable Upper Triassic limestone north of English Mountain (Fig. 5-5), and therefore I regard this sequence as broadly correlative with the Robinson Formation of McMath (1966) in the Taylorsville area. The Robinson consists of 200 m of andesitic conglomerate breccia, volcanic sandstone, slate, and minor sandstone (McMath, 1966).

North of latitude 40°N, d'Allura and others (1977) mapped pyroclastic deposits overlying the Peale as the Arlington Formation, and included within it Norian (Upper Triassic) limestone from Butt Lake, 20 km west of Taylorsville. They assert that the Arlington contains a conformable transition from Permian pyroclastic rocks to upper Triassic limestone, although Lower and Middle Triassic units have not been identified. If true, the Butt Lake area stands in marked contrast to areas to the south, where Upper Triassic limestone unconformably overlies Paleozoic rocks (Clark and others, 1962; Fig. 5-5). Northeast of English Mountain, limestone that overlies Paleozoic rocks to the west yielded *possible* Triassic fossils, suggesting it is correlative with limestone on the north fork American River. The fossils are currently being studied.

The Paleozoic pyroclastic sequence doubtless represents a volcanic island arc. According to d'Allura and others (1977), the volcanic arc was active from Devonian to Triassic time, but was marked by a long period of volcanic quiescence from Early Mississippian to Early Permian time. Data presented here indicate this statement is inaccurate. A major unconformity separates Upper Triassic limestone from Permian or older rocks in many areas of the southern Cordillera, as earlier recognized by Dott (1971). No unequivocal Lower or Middle Triassic volcanic rocks have been found in the northern Sierra Nevada, suggesting that igneous activity related to the Paleozoic arc terminated prior to the Triassic. Within the Paleozoic pyroclastic sequence, the period of volcanic quiescence marked by the Peale Formation and by a hiatus beneath the Reeve Formation probably represents much of Mississippian and Pennsylvanian time, but did not extend into the Permian.

Calaveras Complex The Calaveras Complex (Fig. 5-1) was defined by Schweickert and others (1977) as the younger of two Paleozoic metamorphic complexes (the older being the Shoo Fly) that lie east of the Melones fault zone. Rocks formerly called "Calaveras Formation" or "Western belt of Calaveras Formation" to the west of the Melones fault zone were specifically excluded from the complex. Subsequent work indicates that the definition of the Calaveras Complex presented by Schweickert and others (1977) needs modification. First, the "quartzite unit," consisting of quartzite, pelitic schist, and augen gneiss, originally included within the Calaveras Complex, is now known to be separated by the Calaveras-Shoo Fly thrust from the chert-argillite rocks typical of the Calaveras (see Fig. 5-7). Such upper plate rocks are now regarded as part of the lower Paleozoic Shoo Fly Complex (see the preceding; Schweickert, 1978, 1979). Second, rocks south of latitude 37°N in the southern Sierra included within the Calaveras Complex by Schweickert and others (1977), and now referred to as "Kings sequence" by Saleeby (Chapter 6, this volume; Saleeby and others, 1978), no longer appear to the author to be related to the Calaveras Complex. The eastern parts, consisting of sequences of quartzite, marble, and with interbedded silicic volcanic rocks in its eastern parts, bear no resemblance to any parts of the Calaveras Complex. The western part of the Kings sequence, consisting of chert-argillite rocks in the Yokohl Valley pendant and the lower Kings and Kaweah River pendants (Saleeby, 1979; Saleeby and others, 1978) that resemble Calaveras Complex chert-argillite rocks, also contains small but

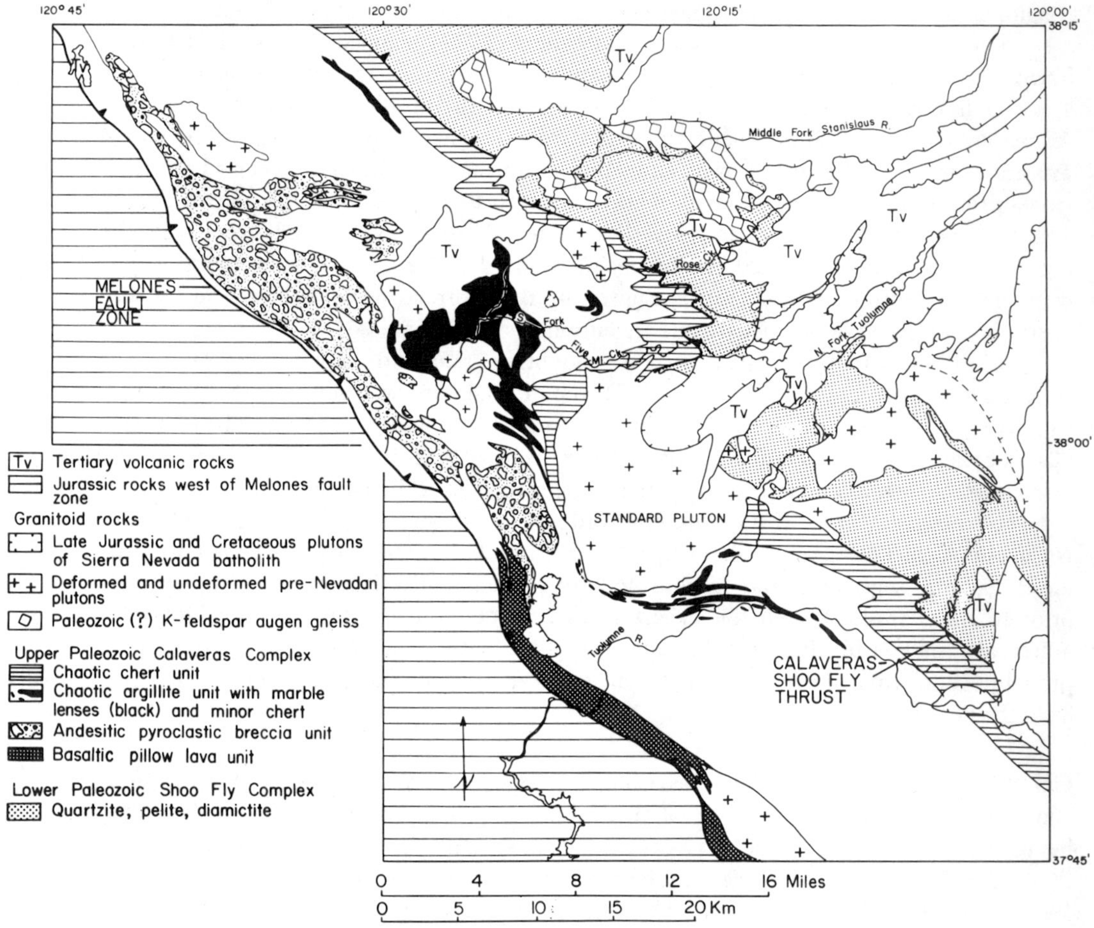

Fig. 5-7. Generalized geologic map of the Calaveras Complex in the central Sierra Nevada. Mapping by Schweickert, 1973–1979, with additions from Clark (1970), Eric and others (1955), and Hart (1959).

important amounts of interbedded augite porphyry basalt and andesite, together with appreciable quantities of epiclastic quartzose sediments. These latter two distinctive lithologies are not represented in the Calaveras Complex, but instead are common in chaotic chert-argillite terranes described by Duffield and Sharp (1975) and Behrman (1978; Behrman and Parkison, 1978) and in Jurassic island-arc sequences west of the Melones fault zone. Significantly, limestone olistoliths bearing Permian Tethyan fusulinid faunas are common both in the Yokohl Valley pendant and in the melange terrane of Duffield and Sharp (1975). Therefore, although both the Calaveras Complex and parts of Saleeby's (Chapter 6, this volume) "western Kings sequence" contain chaotic chert-argillite rocks, they are in detail dissimilar in several important respects. In this paper, the western part of the Kings sequence is shown west of the Foothills suture, together with the similar chaotic rocks west of the Melones fault zone.

Between latitudes 37°30′ and 38°N the Calaveras Complex consists of several mappable units that generally trend northwestward and dip northeastward (Schweickert and others, 1977), reflecting the position of the Calaveras in the western limb of the synclinorium (Fig. 5-7). From lowest to highest, the units are (1) a volcanic-rich unit comprising a lower subunit of basaltic pillow lava and argillite and an upper subunit of basaltic to andesitic tuff-breccia, tuff, and slate; (2) an argillite unit, consisting mostly of chaotic argillite and siltstone and variably sized chert inclusions and lenses of shallow-water limestone in its upper parts; and (3) a chert unit, made up of chert-rich olistrostromes with olistoliths of coherent bedded chert up to 1 km in length. Contacts between these units in all cases appear gradational, and few tectonic breaks have been observed. Thicknesses of the various map units are unknown, in part because the Calaveras has been involved in at least three phases of folding, but also because the argillite and chert units are mainly made up of large numbers of olistostromes in which the only bedded rocks are slabs of chert. Contrary to some recently published statements (e.g., Sharp and Saleeby, 1979), the argillite unit of the Calaveras Complex contains little or no quartzose flysch or coarse clastic detritus. The steep dips and broad outcrop widths of the various units suggest original thicknesses were probably on the order of thousands of meters for each map unit.

Schweickert and others (1977) reported that facing indicators, however few, without exception indicate tops eastward. Numerous lines of evidence indicate that the chaotic parts of the Calaveras accumulated as a result of the sliding and flowage of unlithified and partially liquefied sediments downslope into a basin floored by basaltic volcanic rocks probably representing layer 2 of the oceanic crust (Schweickert and others, 1977).

The Calaveras Complex underwent a complex deformational history involving isoclinal folding and flattening, in part related to thrusting of an extremely thick plate of Shoo Fly rocks over it and, in addition, the rocks experienced several later episodes of folding that produced foliations and cleavages. The mapped trace of the thrust and the form of the chert unit show the effects of a major generation of tight, east-plunging second folds that deform both upper and lower plates and the Calaveras-Shoo Fly thrust. Significantly, these folds are truncated by the Standard pluton, which, as mentioned previously, indicates that the age(s) of folding, thrusting, and hence the Calaveras Complex itself is pre-170 m.y.

The Standard pluton and the second folds are crosscut by an extensive Upper Jurassic swarm of mafic, E–W trending dikes that are subparallel to the F_2 axial surfaces (Schweickert, unpublished data). As the dike swarm is traced westward toward the Melones fault zone, dikes are progressively deformed and locally transposed into a NW-trending foliation that probably developed during Late Jurassic movements along the Melones fault zone. Significantly, these dikes do not occur west of the Melones fault zone. This important structural relationship clearly establishes that the chaotic rocks of the Calaveras Complex were metamorphosed and twice deformed prior to the development of folds and cleavages during the Nevadan orogeny. The fact that the dike swarm does not extend across the Melones fault suggests that it was truncated by the fault.

The "Calaveras" problem Great confusion currently exists in the literature about the extent, age, and tectonic significance of "Calaveras" rocks. In this paper, the name and descriptions of "Calaveras Complex" apply only to the belt of rocks *east* of the Melones fault zone and lying between latitudes 37° and 39°N (Fig. 5-1).

Burchfiel and Davis (Chapter 3, this volume; Davis and others, 1978) and Saleeby (Chapter 6, this volume) have used the name "Calaveras" more broadly and have applied terms like "Calaveras Formation," "Calaveras sequence," or "Calaveras-type rocks" to rocks from the southern Sierra Nevada to the Klamath Mountains and farther north, and including terranes east and west of the Melones fault zone. This procedure is reminiscent of Turner's (1893) original usage, except that Turner meant the name "Calaveras Formation" for all *Paleozoic* sedimentary rocks of the Sierra Nevada, while Burchfiel and Davis and Saleeby apply the name to rocks they regard as Mesozoic. In my opinion, it is unfortunate that some geologists still continue to use the name "Calaveras" so broadly, because terranes with different lithologic assemblages, structural histories, and possibly ages are thus lumped together, and, more importantly, the distinction between known and *inferred* age relationships is completely obscured.

For example, Saleeby (Chapter 6, this volume) states that chert from the "Calaveras sequence" is early Mesozoic in age, and later assumes that the Calaveras Complex (as used here) is of early Mesozoic age. In fact, no chert from the Calaveras Complex has yet yielded identifiable radiolarians. Upper Triassic or Jurassic cherts reported from the "Calaveras Formation" by Irwin and others (1978) are in reality from chert-argillite rocks studied by Hietanen (1973, 1977) to the *west* of the Melones fault near latitudes 38°30′ and 40°N. Because of the occurrence of Mesozoic cherts in this region, Burchfiel and Davis (Chapter 3, this volume) assume that the Calaveras Complex is of early Mesozoic age as well.

Further, both Saleeby and Burchfiel and Davis erroneously imply that the Calaveras Complex has yielded Permian Tethyan fusulinids. In fact, no fusulinid fauna of any kind has definitely been recovered from the Calaveras Complex. *"Fusulina cylindrica"* reported by Turner (1893) has not been verified, and original collections have been misplaced (Clark, 1964). The known occurrences of Permian Tethyan fusulinids in proximity to the Calaveras Complex are from three localities near latitude 38°N that are *west* of the Melones fault zone (Figs. 5-12 and 5-13) in rocks formerly called "western belt of Calaveras Formation" by Clark (1964) and "Sierra foothills melange" by Duffield and Sharp (1975). As noted by Schweickert and Cowan (1975) and Schweickert and others (1977), there exists no evidence that the "western belt" has or had any direct tie with the Calaveras Complex. In fact, this "western belt" contains rocks of very low metamorphic grade that have not experienced the two strong pre-middle Jurassic deformational and metamorphic events represented in the Calaveras Complex (Schweickert, 1979). A fourth occurrence of Permian Tethyan fusulinids is in the Yokohl Valley pendant (Fig. 5-1), which, as Saleeby himself (1979) noted, may correspond tectonically with the Tethyan localities to the north.

Because of their unproven assertions that the Calaveras Complex is Mesozoic and the erroneous idea that it contains Permian Tethyan fusulinids, Burchfiel and Davis and Saleeby all regard the Calaveras Complex as lying to the west of the Foothills suture, which they are obliged to conceal within the Sierra Nevada batholith.

The Calaveras Complex may indeed be Mesozoic, although there is not sufficient age data to prove this. By analogy with terranes containing chert and argillite in the northwestern Sierra Nevada, the Klamath Mountains, and in Oregon, Washington, and British Columbia that have yielded Triassic and Jurassic radiolarian fauna from cherts, the Calaveras Complex could be considered Mesozoic.

However, no cherts from east of the Melones fault zone-Foothills suture in California *or* Nevada have yielded Mesozoic radiolarians. And since the only fossils now known from the Calaveras Complex are Permo-Carboniferous (Schweickert and others, 1977), the complex may instead be Permian or Carboniferous, and may occupy a paleogeographic setting closely analogous to that of the lithologically similar Havallah sequence in north-central Nevada (Schweickert, 1976a; Schweickert and others, 1977). The Havallah sequence contains both Permian and Pennsylvanian fusulinid-bearing limestones and Permian chert (Stewart and others, 1977; D. Jones and W. Snyder, personal communication, 1978, 1979), together with argillite and greenstone. The important message here is that not every chert-argillite sequence in western North America is Mesozoic.

Rather than infer that the Calaveras Complex is Mesozoic even though the only known fossils in it are Paleozoic, I prefer the interpretation that the Calaveras Complex is late Paleozoic and probably correlative with the Havallah sequence. All available structural and lithologic data seem best reconciled with this interpretation.

Roof pendants in the eastern and southern Sierra Nevada South of about latitude 38°10′N, Paleozoic rocks that occur within scattered roof pendants on the eastern side of the range occupy the eastern limb of the Nevadan synclinorium (Bateman and others, 1963; Fig. 5-1). There appear to be two distinct assemblages of lower Paleozoic rocks within these pendants.

In the Saddlebag Lake, Log Cabin Mine, June Lake, and Mount Morrison pendants (Fig. 5-1) presumed and known lower Paleozoic strata are characterized by the presence of bedded chert, argillite, limestone, and quartzite; as pointed out by Rinehart and Ross (1964), these rocks bear a close resemblance to the allochthonous western facies of north-central Nevada referred to here as the Roberts Mountains allochthon. Brook (1977) suggested that the unfossiliferous rocks of the Saddlebag Lake pendant are lithologically and structurally correlative with the fossiliferous Ordovician and Silurian strata of the Mount Morrison pendant 50 km southeast, described by Rinehart and Ross (1964), although Brook and others (1979) have now identified upper Paleozoic rocks in this area as well.

As mentioned earlier, the lower Paleozoic (?) rocks at Saddlebag Lake bear a remarkable lithologic and structural resemblance to the "lower" Shoo Fly, now recognized 50 km west in the west limb of the Nevadan synclinorium. Both the Shoo Fly in the northwestern Sierra Nevada and the lower (?) Paleozoic rocks of the Saddlebag Lake and June Lake pendants are unconformably overlain by Triassic and Jurassic volcanic and sedimentary rocks that lie in the core of the synclinorium. These relations, together with the lithostratigraphic similarities, suggest that the lower Paleozoic rocks in the eastern Sierra Nevada between latitudes 37°30′ and 38°N can be considered part of the Shoo Fly Complex.

The second assemblage of lower Paleozoic rocks, whose full distribution is as yet very poorly known, occurs farther south within the Pine Creek, Bishop Creek, and Big Pine pendants (Fig. 5-1). This assemblage is a westward continuation of the carbonate-shale-quartzite miogeocline characteristic of the White and Inyo Mountains (Fig. 5-1). Moore and Foster (1980) have recently discovered Lower Cambrian archaeocyathids in strata of the Big Pine pendant that are lithologically correlative with the Lower Cambrian lower Poleta and upper Campito formations of the White Mountains. In addition, in the eastern part of the Bishop Creek pendant (Fig. 5-1) they have discovered graptolite-bearing middle Ordovician strata that are correlative with middle Ordovician carbonate-shale sequences in the Inyo Mountains. These findings of Moore and Foster (1980) are of especial significance, for they establish beyond reasonable doubt stratigraphic and gross structural continuity between Great Basin miogeoclinal sequences of the White-Inyo Mountains and roof pendants on the east limb of the Nevadan synclinorium in the eastern Sierra Nevada.

In addition to lower Paleozoic rocks, upper Paleozoic rocks are represented in the eastern roof pendants, but their extent is even less certain than that of lower Paleozoic rocks. Probable Mississippian fossils have recently been reported from chert and marble sequences in the Ritter Range and Saddlebag Lake pendants by Brook and others (1979). Fossiliferous Pennsylvanian and Permian strata in the Mount Morrison pendant (Fig. 5-1), which probably were deposited unconformably upon the lower Paleozoic rocks (Rinehart and Ross, 1964; Russell and Nokleberg, 1977), are of a peculiar silty facies interbedded with rare limestones. Rinehart and Ross (1964) and Bateman (1965a) considered it likely that correlative rocks occur within the Pine Creek pendant. Fossils have not been found in any other of the small masses of metasedimentary rock to the south, but Paleozoic rocks, if present to the south, would most likely be characterized by carbonate-shale facies like their counterparts in the White-Inyo Mountains.

In summary, the Paleozoic strata within roof pendants along the eastern edge of the range are of major importance for three reasons: (1) they form an important structural link in defining the eastern (west-facing) limb of the Nevadan synclinorium; (2) they indicate that allochthonous eugeosynclinal terranes are present in the north, and rocks of the autochthonous Cordilleran miogeocline are present to the south; and (3) their distribution suggests that the apparent boundary between the allochthonous and autochthonous domains occurs to the south of the Mount Morrison pendant and trends at a high angle to the axis of the synclinorium. I infer that one or more thrust faults pass southwesterly between the Mount Morrison and Pine Creek pendants.

These relationships are of major consequence for the interpretation of the rocks in roof pendants to the southwest of the axis of the synclinorium (Fig. 5-1), for they suggest that the rocks in these roof pendants may largely consist of a continuation of the lower Paleozoic and uppermost Precambrian miogeoclinal assemblage of the White-Inyo Mountains and eastern Sierra Nevada, simply repeated on the western limb of the Nevadan synclinorium. Such a view is consistent with the suggestions of Kistler and Bateman (1966), Kistler and Peterman (1973), and Nokleberg and Kistler (1980), but differs strikingly from the views of Saleeby (Chapter 6, this volume; Saleeby and others, 1978), who regard the rocks within roof pendants throughout the entire region as Triassic and Jurassic.

Indeed, the rocks in roof pendants between the Foothills suture and line AB of Figure 5-1 consist of metamorphosed sequences of quartzite, marble, and pelitic schist that locally, as in the Boyden Cave and Lower Kings River pendants, possess much more complex structural fabrics than fossiliferous Mesozoic sequences east of line AB (G. H. Girty, in Moore and others, 1979). This indicates that a structural break, possibly an unconformity, separates the volcanogenic Mesozoic rocks from an older, miogeoclinal sequence that is probably of Paleozoic or Precambrian age.

Mesozoic Rocks

Andean-type arc sequence The following outline is taken in large part from a more more extensive summary by Schweickert (1978). Many original references not cited here can be found in that article.

Metavolcanic and plutonic rocks of an early Mesozoic Andean-type magmatic arc occupy a region up to 150 km wide extending from the eastern Klamath Mountains to the Mojave Desert. Rocks of this assemblage in the Sierra Nevada and the White-Inyo-Argus ranges are shown on Figure 5-1. South of latitude 39°N, metavolcanic and metasedimentary rocks of Late Triassic to Middle Jurassic age lie generally west of exposures of coeval plutonic rocks, and have been invaded by younger plutonic rocks of the Sierra Nevada batholith. Lower Jurassic or Upper Triassic plutonic rocks have not been reported from the Sierra Nevada north of latitude 39°N, but fossiliferous sequences of metavolcanic and metasedimentary rocks extend to the northern tip of the range (Fig. 5-1) and reappear near latitude 41°N in the eastern Klamath Mountains. These volcanogenic sequences of Triassic and Jurassic age contrast markedly and locally interfinger complexly with nonvolcanic sequences to the east in central and southern Nevada and southeastern California.

The occurrences of metavolcanic rocks in the Sierra Nevada are mostly of Early to Middle Jurassic age, although in some areas, as at Mineral King (Saleeby and others, 1978) and the Ritter Range (Fig. 5-1; Fiske and Tobisch, 1978), Triassic metavolcanic rocks are present. The metavolcanic rocks are heterogeneous assemblages of lava, tuff, and breccia of basaltic to rhyolitic composition, locally interbedded with graywacke, siltstone, conglomerate, and limestone. Welded tuff, accretionary lapilli, evaporites, and aeolian sandstone, reported in several areas, suggest that much of the volcanism was subaerial, although the existence of marine fossils and interbedded submarine tuff, turbidites, and limestone, especially in areas north of latitude 38°30′N, indicates marine conditions near or adjacent to volcanic centers.

On the western limb of the Nevadan synclinorium, the metavolcanic rocks unconformably overlie Paleozoic rocks on the north fork American River (Figs. 5-1 and 5-3). A possibly similar unconformity may occur far to the south in the Boyden Cave pendant, as described previously. On the eastern limb, in the Saddlebag Lake, Ritter Range, and Mount Morrison pendants (Fig. 5-1), the same unconformity occurs. In addition, there may be one or more regional unconformities within the volcanic section (Schweickert, 1978).

Radiometric ages of early Mesozoic plutons between latitudes 35°30′ and 39°N show a close correspondence with the periods of volcanic activity indicated by fossils. Triassic ages ranging from 226 to 194 m.y. (Late Triassic to Early Jurassic) have been obtained on numerous plutons between latitudes 36° and 38°N. Plutons of Early and Middle Jurassic (184 to 160 m.y.) ages are more widespread, and occur mainly between latitudes 36° and 39°N (Fig. 5-1) in relatively close association with, but generally east of, Lower and Middle Jurassic metavolcanic rocks. Several Middle Jurassic plutons occur within the Calaveras Complex (Figs. 5-1 and 5-7; Sharp and Saleeby, 1979), 80 to 90 km west of the belt of early Mesozoic plutons in the eastern Sierra Nevada. It is an open question whether these plutons are directly related to the early Mesozoic arc as Saleeby (Chapter 6, this volume) assumes.

The north-northwest trending early Mesozoic Andean-type arc developed across the grain of the northeast-trending Paleozoic Cordilleran orogen and parallel to the truncated edge of the orogen (Schweickert, 1976a, 1978). Differences in the nature of eruptive rocks within the arc can be reconciled with the different character(s) of the basement of the arc, from sialic or continental south of about latitude 37°30′N to crust of increasingly oceanic character to the north. South of about latitude 39°N, volcanic rocks are mainly intermediate to silicic in character, and voluminous ash flows are abundant. Large batholiths are common. Much or most of the volcanic material was erupted subaerially. Volcanic sequences north of latitude 39°N are, in general, slightly more mafic in character, with andesitic to basaltic rocks predominating, and much or most of the material was deposited (and perhaps erupted) in a marine environment.

Within the Sierra Nevada the distribution of the stratified Mesozoic metavolcanic and metasedimentary rocks is clearly controlled by the structure of the Nevadan synclinorium (Fig. 5-1); evidently, the Upper Triassic-mid-Jurassic Andean-type arc sequence of the Sierra Nevada was tightly folded (in some areas isoclinally) prior to the inception of a second, younger Andean-type arc in the region, now represented by the Late Jurassic-Late Cretaceous Sierra Nevada batholith. Tobisch and others (1977) have shown that strains on the order of 100% shortening have occurred in Mesozoic rocks near the axis of the synclinorium. In some areas, early Nevadan thrusting preceded or accompanied the folding. This is true in the western limb north of latitude 40°N (Fig. 5-1), where Paleozoic rocks have been thrust eastward along the Taylorsville thrust over the Mesozoic arc sequence (McMath, 1966; d'Allura and others, 1977). A similar relationship exists in the eastern limb of the synclinorium south of latitude 36°35′N in the southern Inyo Mountains-Argus Range-Slate Range, where upper Paleozoic and locally Precambrian rocks were thrust eastward along the Swansea thrust system, prior to final folding (Dunne and others, 1978).

Lawsonite-bearing blueschist metamorphic rocks have been reported from one area within the Sierra Nevada (Fig. 5-3) along the Foothills suture (Schweickert, 1976c, 1978; Schweickert and others, 1980). K-Ar dating of white micas from the blueschist rocks indicates a minimum metamorphic age of 174 m.y. (Middle Jurassic) (Schweickert and others, 1980). These blueschist rocks, together with their counterparts in the Klamath Mountains, strongly suggest that the Foothills suture represents the site of a subduction zone that was active during the evolution of the early Mesozoic Andean-type arc. This relationship makes the Middle Jurassic plutons in the Calaveras Complex seem anomalous, because they occur within a few kilometers of the Foothills suture. Perhaps these plutons formed by shear melting at shallow levels during subduction.

Cretaceous metavolcanic rocks Metavolcanic rocks of Cretaceous age have been identified by detailed mapping and intensive radiometric dating within the Ritter Range pendant (Fig. 5-1), where Fiske and Tobisch (1978) and Fiske and others (1977) have delineated a portion of the Minarets Caldera (Fig. 5-8). Fiske and Tobisch (1978) showed that 3 to 4 km of pyroclastic rocks and chaotic collapse breccias that have yielded U-Pb and Rb-Sr ages of 98 to 101 m.y. underlie the eastern part of the caldera and rest unconformably upon Lower Cretaceous tuff with a U-Pb age of 127 m.y. The western extent of the caldera fill is not presently known, but the sequence of rocks is intruded on the south by the 90 m.y. old Mount Givens Granodiorite (Fig. 5-8), the largest known pluton within the Sierra Nevada batholith (Bateman and others, 1963). Fiske and Tobisch (1978) noted that the Cretaceous volcanic rocks are less deformed than the fossiliferous Lower Jurassic metavolcanic rocks to the east that underlie them, providing important evidence that the main phase deformation of the Jurassic metavolcanic rocks that produced the form of the Nevadan synclinorium was Late Jurassic.

A probable second occurrence of Cretaceous metavolcanic rocks underlies Mount Dana, within the Mono Craters quadrangle (Fig. 5-8), mapped by Kistler (1966b), and part of the adjacent Tioga Pass quadrangle, where the rocks have been mapped by Brook (1977). These rocks, called the Dana sequence by Kistler (1966a), lack the N20-30W striking folds and cleavages that nearly everywhere else in the Sierra Nevada occur in rocks older than Late Jurassic, and instead possess a single set of folds striking N50-70W (Kistler, 1966a).

The rocks in the Dana sequence are unfossiliferous, and Kistler (1966a) assigned them to the Lower Jurassic on the basis of their lithology. Brook (1977) considered some of the same sequence to be Permo-Triassic, but admitted that, with such an age assignment, their structure is an enigma. I consider these rocks Lower Cretaceous on the basis of structural arguments presented by Russell and Nokleberg (1977). They noted that, in many Sierran roof pendants, second-generation structures ("Nevadan" structures in this paper) trending N20-30W regionally are in many areas crosscut by N50-60W trending third-generation structures that are themselves cut by Upper Cretaceous plutons of the Cathedral Range intrusive epoch. This suggests that rocks bearing solely the N50-60W generation structures are most reasonably assigned to the Lower Cretaceous. In the next section I discuss this problem more fully.

Structural Synthesis

Late Jurassic (Nevadan) folding Figure 5-1 shows the locations of areas for which data bearing on the Mesozoic structural evolution of the Sierra Nevada have been pub-

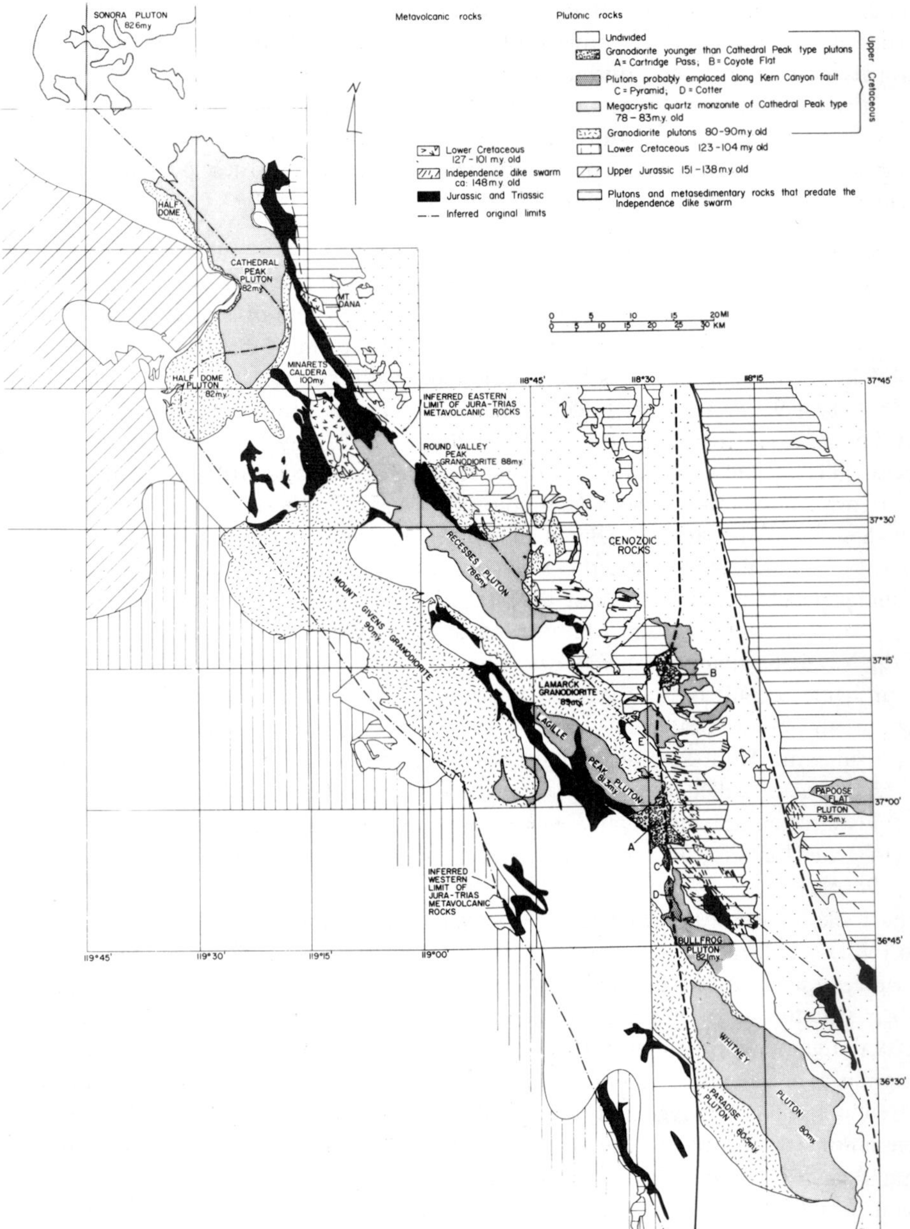

Fig. 5-8. Map of granitic and metamorphic rocks of the east-central Sierra Nevada, showing inferred ancestral trace of Kern Canyon fault. Sources: Bateman, 1965a, b; Bateman and Moore, 1965; Bateman and Wones, 1972; Bateman and others, 1971; Brook, 1977; Chesterman, 1975; Huber, 1968; Huber and Rinehart, 1965; Kistler, 1966a, 1973; Krauskopf, 1971; Lockwood and Lydon, 1975; Matthews and Burnett, 1965; Peck, 1964; Rinehart and Ross, 1957, 1964; Ross, 1962, 1967b; Strand, 1967.

lished. The heavy lines within structural plots depict the trends of axial-surface traces of minor folds that almost certainly developed during the Late Jurassic, because the folds involve strata as young as Callovian near Taylorsville and as young as Pliensbachian (late Early Jurassic) in the Ritter Range pendant, are cut by plutons as old as 90 m.y., and apparently do not involve strata of Early or Late Cretaceous age (as in the Ritter Range pendant).

As stated by numerous authors, the predominant orientation of axial surfaces of these folds is N20-30W, exactly parallel to the structural grain of the Sierra Nevada. It is interesting to note the parallelism of these structures with mapped dikes of the Independence dike swarm (Moore and Hopson, 1961). Several dikes of the Independence dike swarm have recently yielded U-Pb ages of about 148 m.y. (Chen, *in* Dunne and others, 1978). Inasmuch as they are grossly parallel to the axial-surface trace of the synclinorium, they probably were emplaced during a period of relaxation following folding, and their age therefore sets a younger limit on the age of formation of the synclinorium.

Kistler (1966a) and Russell and Nokleberg (1977) have argued that folds of the N20-30W orientation in the Saddlebag Lake pendant are Early to Middle Triassic in age, which, if true, would argue that certain parts of the Sierra Nevada east of the Foothills suture were subjected to deformation along NW trends more than once during the Mesozoic. While this is not an unlikely case, considering the evidence of pre-Early Jurassic Mesozoic deformation in the Inyo Mountains and Argus Range presented by Dunne and others (1978), I think the arguments for pre-Late Jurassic Mesozoic deformation in the Saddlebag Lake and Mount Dana areas can no longer be supported.

Kistler (1966a), as noted, assigned the Dana sequence to the Early Jurassic, and because the N30W-trending folds in question did not involve the Dana sequence, but occurred in the Permian (?) Koip sequence, he argued they were Triassic. If the Dana sequence, with its single set of N50W-trending folds, is Lower Cretaceous, as suggested by regional relationships of this fold system, and by the fact that the Dana sequence rests unconformably on both steeply dipping lower Plaeozoic and Lower (?) Jurassic rocks, and if the metavolcanic rocks in question are in part Late Triassic to mid-Jurassic (as shown on Figs. 5-1, 5-8, and 5-9), as suggested by their lithologic similarity with the fossiliferous Jurassic volcanic rocks of the Ritter Range, then the N30W-trending folds must be Late Jurassic.

Russell and Nokleberg (1977) stated that Permian-Triassic metavolcanic rocks of the Saddlebag Lake pendant with N25W-trending folds are cut by Upper Triassic plutonic rocks of the Lee Vining intrusive epoch of Evernden and Kistler (1970), once again indicating the folds are pre-late Triassic. Careful inspection of Brook's (1977) map of the Saddlebag Lake pendant and Kistler's (1966b) map of the Mono Craters quadrangle shows that only the undated granodiorite of Tioga Lake is in contact with rocks Brook (1977) assigned to the Permo-Triassic. In addition, these "Permo-Triassic" rocks appear to be the northwest-continuation of the Lower Cretaceous (?) Dana sequence. Significantly, Brook (1977) only reported N50W-trending minor folds from them.

In summary, these observations lead to the hypothesis that the N30W-trending folds in most (or all) parts of the central Sierra Nevada are Late Jurassic, and that metavolcanic rocks lacking these folds but containing only a N50-60W trending set of folds are probably Lower to mid-Cretaceous in age. This generalization seems to accord best with all the structural and stratigraphic data presently available.

Cretaceous (syn-batholithic) folding and faulting Figure 5-1 shows the inferred original limits (prior to the emplacement of Cretaceous plutons) of Juratrias metavolcanic rocks in the core of the Late Jurassic Nevadan synclinorium.

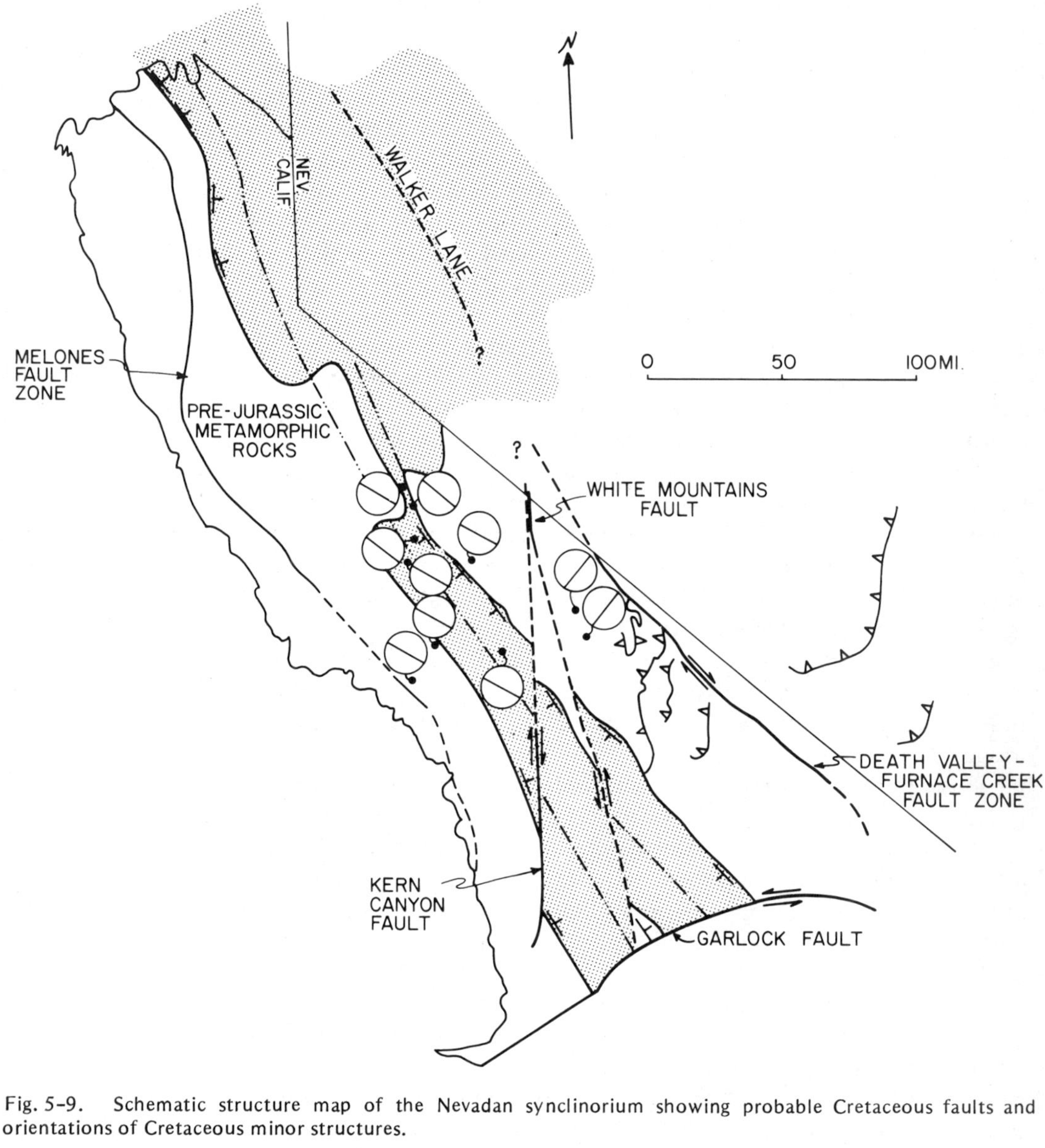

Fig. 5-9. Schematic structure map of the Nevadan synclinorium showing probable Cretaceous faults and orientations of Cretaceous minor structures.

Figure 5-8 depicts this relationship in more detail in part of the eastern Sierra Nevada, and focuses attention on the Kern Canyon fault, along whose trace the southwestern and northeastern limits of Juratrias metavolcanic rocks are offset. However, Moore and du Bray (1978) show that the Kern Canyon fault cannot be mapped as a continuous surface feature north of about latitude 36°40′N, where the fault is lost in the Paradise pluton.

It is important to note that both limbs of the Nevadan synclinorium show an apparent dextral offset along a line corresponding to the projected trace of the Kern Canyon fault. The north limb and an E-W trending set of mafic dikes probably related to the Independence dike swarm (Moore and Hopson, 1961) show comparable offsets of about 19 km. The fact that Late Cretaceous plutons seem to have obliterated the fault north of latitude 36°40′N (Fig. 5-8) indicates that the Kern Canyon fault was active during the Cretaceous, and that most of the apparent offset along the fault occurred prior to about 80 m.y. ago.

Several additional features on Figure 5-8 are highly suggestive that a pre-Late Cretaceous fault exerted varying degrees of control on the emplacement of plutons. The two youngest plutons of the region, the granodiorite of Cartridge Pass (Moore, 1963) and the granodiorite of Coyote Flat (Bateman, 1965a; A and B, Fig. 5-8) occur directly along the projected trace of the fault. Neither shows the characteristic NW-SE elongation of Late Cretaceous plutons in the region. In addition, the Pyramid and Cotter plutons (Moore, 1963; C and D, Fig. 5-8) are extremely narrow and have an anomalous north-south alignment along the projected trace of the fault. The Bullfrog pluton (Moore, 1963) in like manner, although mainly NW-SE trending, has an elongate N-S trending finger that parallels the Pyramid and Cotter plutons (Fig. 5-8). Evernden and Kistler (1970) reported a K-Ar biotite age of 82.1 m.y. on the Bullfrog pluton.

Bateman (1965a) noted a conspicuous plutonic breccia in the Upper Cretaceous Inconsolable Granodiorite (E, Fig. 5-8) near its contact with a pluton of the Cathedral Peak type, and this location is coincident with the projected trace of the Cretaceous (?) Kern Canyon fault.

Late Cretaceous plutons of the Cathedral Peak type, a distinctive megacrystic quartz monzonite, were emplaced between about 83 and 78 m.y.a. (Evernden and Kistler, 1970), mainly within the core of the Nevadan synclinorium. The larger plutons show a marked NW-SE right-stepping en echelon alignment northwest and southeast of the inferred Kern Canyon fault (Fig. 5-8). This suggests that, even though no surface faulting is evident in the plutons, they may have occupied extensional zones related to continued movement on the Kern Canyon fault during the Late Cretaceous (80 to 90 m.y.a.). If so, the plutons must have been incompletely solidified and still able to heal fractures, except along the chilled (?) southwestern margin of the Paradise pluton, where the magma probably crystallized earlier.

Just as a dextral offset of the synclinorium seems required along the Kern Canyon fault, a sinistral offset along a fault beneath the Owens Valley is implied by the configuration of the east limb of the synclinorium as it is traced from the eastern Sierra Nevada to the Inyo Mountains (Figs. 5-1 and 5-8). Dunne and others (1978) have presented evidence that the White Mountains fault zone was active during the Cretaceous, and I infer

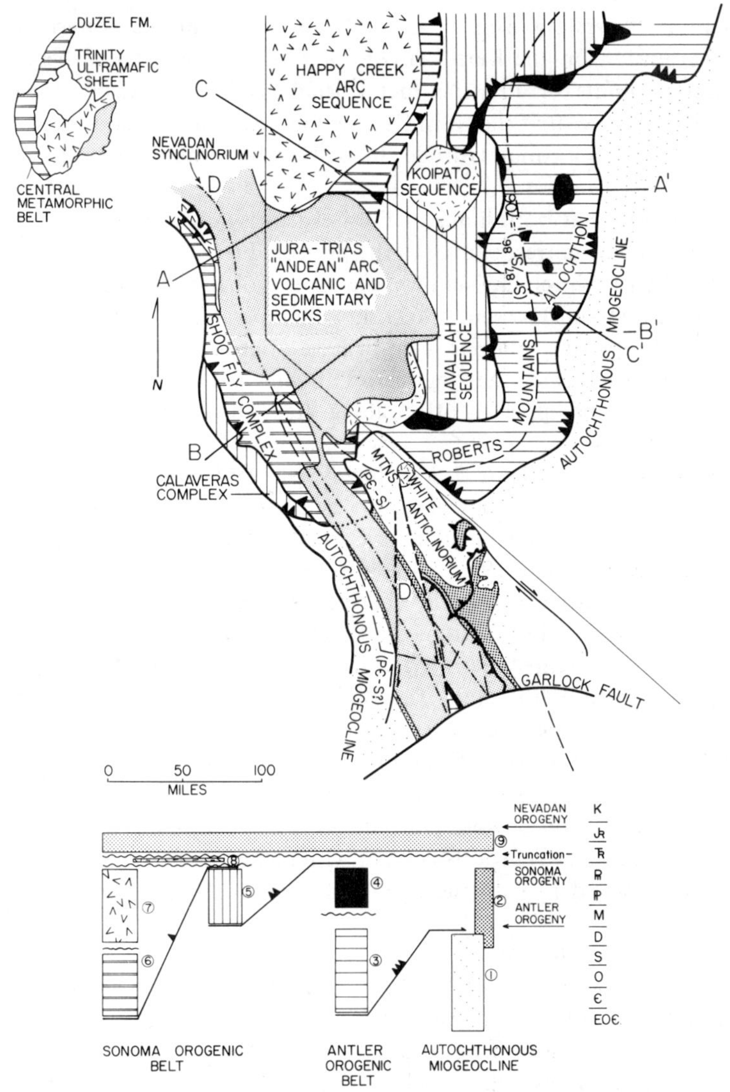

Fig. 5-10. Interpretive structural map of the southern part of the Cordilleran orogen, showing inferred stacking order of major structural units .706 Sr^{87}/Sr^{86} isopleth from Kistler and Peterman (1978).

1. Precambrian to Devonian sedimentary rocks of Cordilleran miogeocline.
2. Devonian to Permian sedimentary rocks of the White-Inyo Mountains.
3. Allochthonous Cambrian to Devonian siliceous and volcanic rocks of the Roberts Mountains allochthon.
4. Pennsylvanian to Permian clastic and carbonate rocks of the "overlap" sequence.
5. Allochthonous Permo-Carboniferous siliceous and volcanic rocks of the Havallah sequence and Calaveras Complex.
6. Allochthonous Cambrian (?) to Devonian (?) siliceous and volcanic rocks of the Shoo Fly Complex, Central Metamorphic belt, Duzel Formation, and Mount Morrison and Saddlebag Lake pendants.
7. Devonian to Permian pyroclastic rocks of island-arc sequence.
8. Permian and Permo-Triassic pyroclastic rocks of Koipato sequence and areas to south.
9. Late Triassic to Late Jurassic pyroclastic rocks of Andean-type arc sequence.

that this was the northern part of a fault now concealed beneath the Owens Valley. A sinistral fault is also required in this approximate position to account for the occurrence in the El Paso Mountains of rocks in the western limb of the Nevadan synclinorium (Figs. 5-1 and 5-10).

Note that the sinistral displacement inferred here for the southern Sierra Nevada with respect to the El Paso Mountains differs from that hypothesized by Burchfiel and Davis (Chapter 3, this volume) for the El Paso Mountains. They infer a large, sinistral displacement of the lower Paleozoic rocks of the El Paso Mountains from a presumed original position over 200 km north where they would have lain nearer the Antler orogenic belt. However, no evidence presently exists for a sinistral fault in the position they require, nor is there compelling evidence that the El Paso Mountains block moved south during a Triassic sinistral truncation event, as suggested by Davis and others (1978). The sinistral megashear of Silver and Anderson (1974), to which Davis and others (1978) related their proposed displacement of the El Paso Mountains, was most likely a Jurassic feature (T. Anderson, personal communication, 1979). In any case, the minor faults discussed here are Cretaceous features.

Is there a relationship between transcurrent faulting in the southeastern Sierra Nevada during the Cretaceous and the widespread (but not ubiquitous) evidence of folding along trends of N50-60W in roof pendants during this time? Figure 5-9 explores this possibility in a summary structural diagram, and reveals that the Cretaceous folds in metamorphic rocks have about the right orientation to be second-order dextral wrench

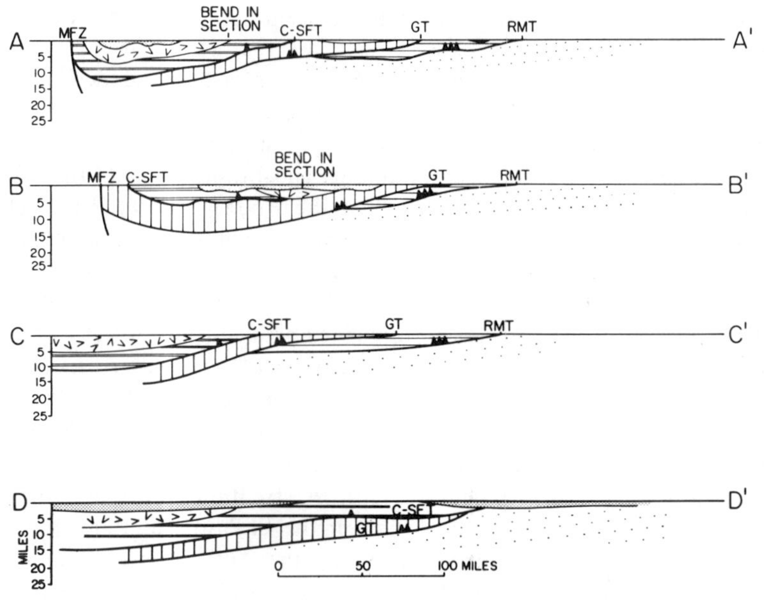

Fig. 5-11. Interpretive cross sections through terranes shown in Fig. 5-10. MFZ, Melones fault zone (Foothills suture); C-SFT, Calaveras-Shoo Fly thrust; GT, Golconda thrust; RMT, Roberts Mountains thrust.

folds related to the evolution of the dextral Kern Canyon fault, although the dihedral angle between the folds and faults is higher than that observed in sedimentary rocks along the San Andreas fault (Moody and Hill, 1957). In addition, undated NE-trending folds that occur in Cambrian and Precambrian strata in the White Mountains (Dunne and others, 1978) have the appropriate orientation to be considered wrench folds related to sinistral Cretaceous displacements on the White Mountains fault. However, A. Sylvester (personal communication, 1979) considers these folds to be pre-Mississippian.

Regional structural framework of the southwestern cordillera Figure 5-9 shows schematically the geometry of Mesozoic structural elements of the Sierra Nevada, as interpreted in this paper. Figures 5-10 and 5-11 show a hypothetical structural map and cross sections of the southwestern cordillera in Nevada and California, derived in large part from analysis of the structural framework of the Sierra Nevada presented here, and supplemented by published interpretations of structural and tectonic elements of Paleozoic age in Nevada.

The Jurassic Nevadan synclinorium in the Sierra Nevada is the single most important structural element of Figures 5-9 and 5-10, for it requires that tectonostratigraphic units to the east of the present Sierra Nevada batholith in Nevada and eastern California be repeated in the western limb. Figure 5-10 graphically indicates a complex overlapping pattern of nine tectonostratigraphic units, as follows:

1. Autochthonous late Precambrian to Devonian sedimentary rocks of the cordilleran miogeocline, extending southwest into the southern Sierra Nevada, where they terminate at the Foothills suture.
2. Autochthonous upper Paleozoic sedimentary sequence (shown only in California), which, near the White Mountains anticlinorium, contains an important hiatus representing parts of Devonian and Silurian time, probably reflecting the Antler orogeny, and the emplacement of the Roberts Mountains allochthon (unit 3) to the northeast. However, there is no evidence to indicate that the Roberts Mountains allochthon ever extended southwest across the White Mountains anticlinorium (Stevens and Ridley, 1974).
3. Allochthonous sheets of Cambrian to Devonian siliceous detrital (or western) facies of the Roberts Mountains allochthon interpreted by most workers as oceanic deposits that were overthrust onto the miogeocline north and east of the White Mountains during the Antler orogeny. An enigmatic occurrence of sedimentary rocks regarded as correlative with the Roberts Mountains allochthon occurs in the El Paso Mountains at the south edge of Figure 5-10 (Poole and others, 1977; Stewart and Poole, 1974) on the western limb of the Nevadan synclinorium. Stratigraphic data from the Inyo Mountains (Stevens and Ridley, 1974) precludes any former structural continuity between these rocks and the Roberts Mountains allochthon. If the El Paso Mountains rocks represent allochthonous sheets emplaced separately during the Antler orogeny, they must have been derived from the west, implying that the mid-Paleozoic North American continental margin was markedly south trending in the region south of the central Sierra Nevada.
4. Pennsylvanian-Permian overlap sequence of carbonate and clastic rocks, deposited

unconformably upon the Roberts Mountains allochthon subsequent to the Antler orogeny.

5. Allochthonous Pennsylvanian-Permian sedimentary and volcanic rocks of oceanic affinity, represented by the Havallah sequence and by the Calaveras Complex, that I infer are continuous beneath the core of the Nevadan synclinorium (Schweickert, 1976a; Schweickert and others, 1977). These terranes were thrust southeasterly along the Golconda thrust onto both the miogeocline (unit 1) and the Roberts Mountains allochthon and overlap sequence (units 3 and 4) during the Sonoma orogeny.

6. Allochthonous Cambrian (?) to Devonian (?) sedimentary and volcanic rocks of oceanic (western) affinity of the Shoo Fly Complex and presumed correlative rocks in the Klamath Mountains, in eastern Sierran roof pendants and their hypothetical extent in northwestern Nevada. I interpret these rocks as containing in part the same lithologic successions as the Roberts Mountains allochthon. In addition, they have a pre-Late Devonian structural fabric that may have evolved during the Antler orogeny. But, significantly, these rocks occupy a structural position *above* the Roberts Mountains allochthon, with upper Paleozoic oceanic rocks of the Havallah sequence and the Calaveras Complex between them (Figs. 5-10 and 5-11B). I interpret the thrusting of these rocks, together with unit 7, the upper Paleozoic volcanic arc sequence, over the miogeocline and the Calaveras-Havallah assemblages along the Calaveras-Shoo Fly thrust and its presumed continuation in northwestern Nevada to have occurred during the Permian-Triassic Sonoma orogeny.

7. Devonian to Permian volcanic and sedimentary rocks of island-arc sequence deposited unconformably upon the Shoo Fly Complex and clearly part of the upper plate transported together with the Shoo Fly during southeastward (?) thrusting along the Calaveras-Shoo Fly thrust. Locally, in the northwestern Sierra Nevada and eastern Klamath Mountains, this sequence contains Triassic rocks, suggesting that in some areas sedimentation and volcanism may have continued uninterrupted during the Sonoma orogeny.

8. Lower Triassic and Upper (?) Permian volcanic and sedimentary rocks that unconformably overlie the allochthonous Havallah sequence and may overlap the Golconda thrust. Also possibly included are volcanic and sedimentary rocks that overlie miogeoclinal strata in the northern White Mountains. This is one of the most enigmatic assemblages of the entire region, and may locally predate and locally postdate the Sonoma orogeny.

9. Upper Triassic to lower Upper Jurassic volcanic and sedimentary rocks of an Andean-type volcanic arc sequence that unconformably overlie parts of nearly all the other tectonostratigraphic units. Present distribution and extent, though strongly influenced by Late Jurassic (Nevadan) folding, suggest north-northwest trending paleogeography of Andean-type arc, parallel to the Foothills suture, and nearly orthogonal to structural trends of Paleozoic terranes. These geometric relationships suggest that a major truncation of northeast-trending Paleozoic tectonostratigraphic units occurred in the vicinity of the Foothills suture during mid-Triassic time, and that Late Triassic to Late Jurassic subduction that produced the Andean-type arc occurred along a trend parallel to the line of truncation (Burchfiel and Davis, 1972; Schweickert, 1967a, 1978; Saleeby and others, 1978).

Schweickert and Snyder (Chapter 7, this volume) present a comprehensive Paleozoic plate tectonic model for the region derived from the structural interpretation shown in Figs. 5-10 and 5-11 and from considerations of stratigraphic and structural data from all "eugeosynclinal" terranes in California and Nevada.

OUTLINE OF TERRANES WEST OF THE FOOTHILLS SUTURE

In this section, it is inappropriate to discuss tectonostratigraphic elements in terms of Paleozoic and Mesozoic terranes, inasmuch as the distribution of Paleozoic rocks is so incompletely known; moreover, where known, Paleozoic rocks appear in most cases to represent isolated blocks floating in Mesozoic melanges. Therefore, I will use another format for considering rocks of this region, one based on the inferred tectonic evolution of the various terranes. Much of this outline is extracted from a more extensive discussion of these terranes by Schweickert (1978).

Island-Arc Terrane

A very extensive terrane of Mesozoic metavolcanic and metasedimentary rocks that are closely associated with remnants of ophiolite complexes and melanges forms the westernmost belt of both the Sierra Nevada and the Klamath Mountains from latitude 36° to 42°N (Fig. 5-12). As discussed later, limestone lenses bearing a Permian Tethyan fauna occur in several parts of this western belt in the Sierra Nevada (Fig. 5-12) and the Klamath Mountains.

Bateman and Clark (1974) and Schweickert and Cowan (1975) subdivided the belt of island-arc rocks in the Sierra Nevada into several structural blocks to simplify description; that procedure is adopted here. Blocks A and B (Fig. 5-12) contain the principal occurrences of arc-generated igneous rocks. They are separated by the Smartville ophiolite north of about 38°30′N and by the complex Bear Mountains fault zone between 37°N and 38°30′N (Fig. 5-12). Block B, mainly a steeply east-dipping homocline, but locally with more complex structure, provides important information about the basement of the arc (Fig. 5-13). Block A has a more complicated folded and faulted structure that is not well understood, but the Smartville ophiolite may be an exposure of its basement.

The stratigraphy of the arc sequence in blocks A and B is fairly simple and consists of a variably thick sequence of pyroclastic rock and lava of basaltic and andesitic composition ranging to in excess of 4 km thick, with locally abundant rhyolitic rocks in block A. The volcanic rocks are concordantly overlain by and also intertongue with pelitic or slaty sedimentary rocks, including appreciable graywacke and conglomerate. Significantly, some conglomerates in this younger sequence may reflect a Sierran source (Bateman and Clark, 1974; Behrman and Parkison, 1978; Schweickert, unpublished data) and imply that the arc terrane was in close proximity to North America when they were deposited. Clark (1964) indicated that in block A a second volcanic sequence overlies and is younger than

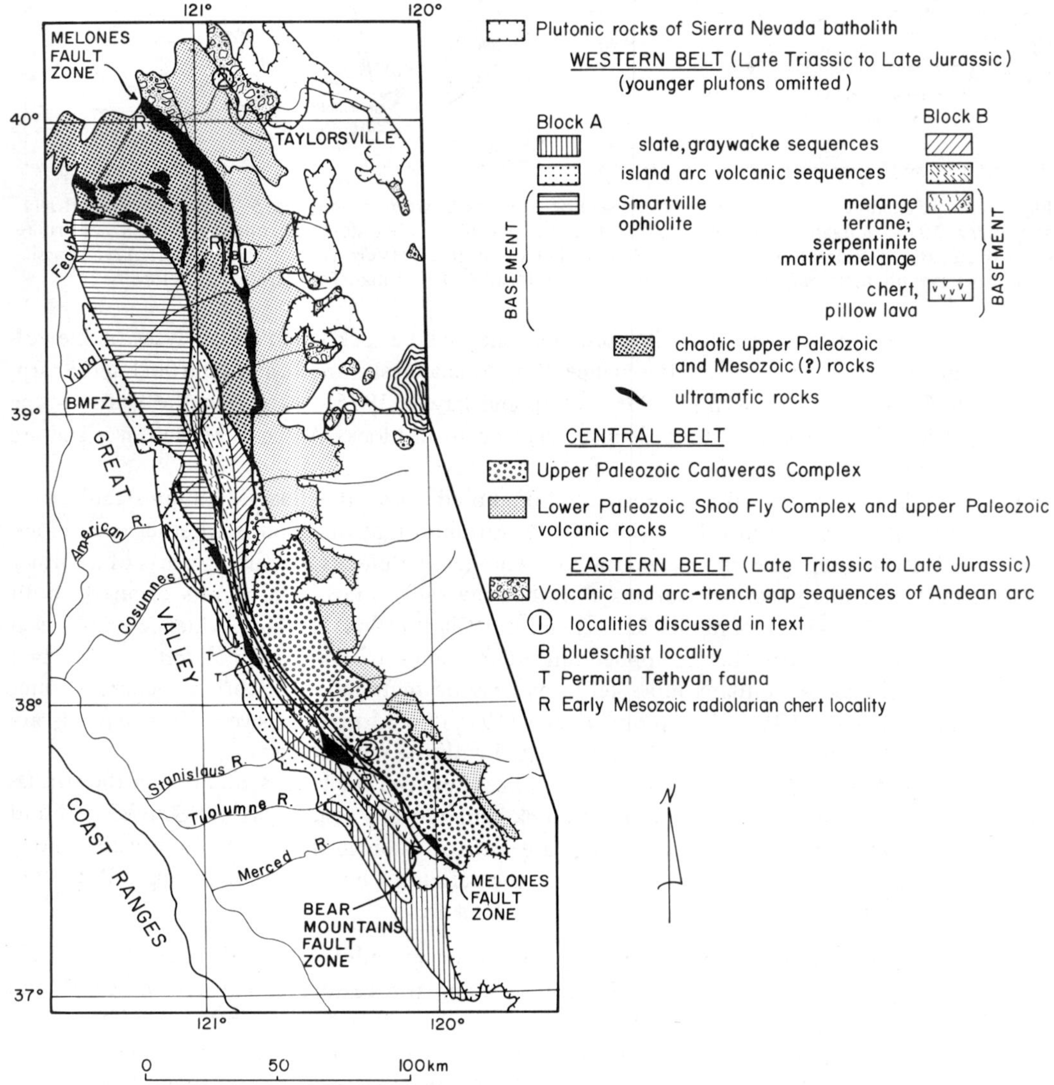

Fig. 5-12. Sketch map of western Sierra Nevada metamorphic belt. Adapted from Schweickert and Cowan (1975) with additions: western belt from Bond and others, 1977; Clark, 1976; Morgan, 1976; Duffield and Sharp, 1975; Hietanen, 1973, 1976; Schweickert, unpub. data; central belt from Schweickert, this paper; eastern belt from d'Allura and others, 1977; Schweickert, this paper.

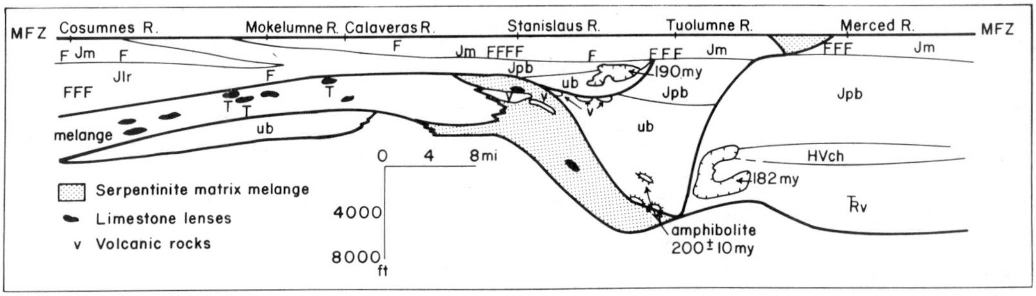

Fig. 5-13. Schematic north-south structure section of block B, western belt, Fig. 5-12. Data from Duffield and Sharp, 1975; Morgan, 1976; Schweickert, unpub. data. MFZ, Melones fault zone; Jm, Mariposa Formation; Jlr, Logtown Ridge Formation; Jpb, Penon Blanco volcanics; HVch, Hunter Valley chert; Trv, Triassic (?) pillow lava; ub, ultramafic rocks; F, Mesozoic fossil locality; T, Permian Tethyan fusulinid locality.

the sedimentary succession. The few published whole-rock chemical analyses of the volcanic rocks indicate the predominance of basaltic andesite and basalt (Duffield and Sharp, 1975). In addition, Kemp (1976; Kemp and Payne, 1975) noted that the Kuroko-type volcanogenic massive sulfide deposits that occur in blocks A and B reflect an island-arc environment.

Blocks A and B are presently juxtaposed along the Bear Mountains fault zone south of the Cosumnes River (Fig. 5-12). An important question is whether each block represents a separate and distinct arc or whether the blocks are simply parts of a former single arc. Many earlier workers felt that the same stratigraphic units occur on both sides of the Bear Mountains fault zone, but Clark (1964), lacking evidence of physical continuity between them, applied different names to stratigraphic units in the two blocks. Block A contains more felsic pyroclastic material and debris is generally finer grained (Clark, 1964). Kemp and Payne (1975) noted that massive sulfide deposits are more abundant in block A and are associated with the felsic rocks.

However, there are no major differences between the blocks, and in fact the stratigraphy is remarkably similar in each block. Therefore, I assume, as did Schweickert and Cowan (1975), that the two blocks are slabs of a single island arc, imbricated along the Bear Mountains fault zone, and that the variations between blocks A and B resulted from facies changes to be expected within island arcs.

Duration of activity in the arc Fossils occur in only one portion of the volcanic sequence, the Logtown Ridge Formation along the Cosumnes River in block B, and they indicate a Callovian age (Duffield and Sharp, 1975). It is tempting to infer, on the basis of remarkably similar lithology, that the volcanic part of the section in other areas to the south is of the same general age. However, Morgan and Stern (1977) obtained a 190 m.y. ^{238}U-^{206}Pb age on a small pluton that apparently intrudes the volcanic sequence south of latitude 38°N (Figs. 5-12 and 5-13), suggesting there that the volcanic sequence is in part Lower Jurassic or older. Plutons as old as Middle Jurassic have been dated near the Kings River (Fig. 5-1) by Saleeby (1977a, b) and the central Klamath Mountains by Lanphere and others (1968). In each of these cases the plutons intrude rocks that I interpret as part of the basement of the arc (see below).

The sedimentary portion of the sequence, especially the Mariposa Formation in block B (Fig. 5-13), contains numerous fossils and for the most part these indicate Oxfordian-Kimmeridgian age (Clark, 1964; Duffield and Sharp, 1975). Volcanism locally persisted until that time, judging from the fact that thick tongues of pyroclastic material locally interfinger with and even overlie the sedimentary sequence (as shown in Figure 5-13; see also Behrman and Parkison, 1978).

In summary, evidence from the central Sierra Nevada indicates that volcanic activity in the island arc may have been initiated in some areas in Late Triassic or earliest Jurassic time, and probably persisted widely until the Callovian, gradually waning in Oxfordian-Kimmeridgian time, when the arc was near enough to North America to receive clastic detritus from the continental margin.

Basement rocks of the arc The Smartville ophiolite, a northward-widening expanse of oceanic crust and upper mantle, separates the arc volcanic rocks of blocks A and B north of the Cosumnes River (Schweickert and Cowan, 1975; Fig. 5-12). The ophiolite evidently forms a broad antiform and is overlain by andesitic pyroclastic breccia of block A on the west (Bond and others, 1977), although a broad shear zone locally forms the boundary between the Smartville ophiolite and block A. Bond and others (1977) and Xenophontos and Bond (1978) have mapped zones of pillow lava, sheeted dikes, and a plutonic sequence including cumulate gabbro in the Smartville ophiolite. The eastern boundary of the ophiolite is a fault that separates it from largely chaotic upper (?) Paleozoic and Mesozoic rocks that lie west of the Melones fault zone (Fig. 5-12; Schweickert and Cowan, 1975).

South of latitude 38°45′N a different, more complex basement terrane intervenes between blocks A and B (Figs. 5-12 and 5-13). Figure 5-13 is a longitudinal section through block B between the Cosumnes River and the Merced River (Fig. 5-12), and shows that from north to south the arc volcanic sequence rests on three different types of basement: (1) melange; (2) an ultramafic sheet; and (3) layers 1 and 2 of an ophiolite sequence.

From latitude 37°50′ to 39°N, volcanic rocks of the Logtown Ridge Formation in block B (Fig. 5-12) are underlain by a melange terrane first described by Duffield and Sharp (1975) and more recently mapped in part by Morgan (1976), Parkison (1976), Behrman (1978), and Schweickert (1978). The melange contains blocks of meta-sedimentary and metavolcanic rocks embedded in a slaty, fine-grained matrix. In addition, blocks of coarsely crystalline quartz-mica schist with rotated earlier structural fabrics occur locally (Schweickert, unpublished data). Duffield and Sharp (1975) noted that some hypabyssal intrusive rocks within the melange have no counterparts in the overlying Logtown Ridge Formation; this suggests that the formation of melange pre-dated the eruption of volcanic rocks of the Logtown Ridge, and that the Logtown Ridge may have been deposited on the melange, even though the present contact between the units is interpreted as a fault (Duffield and Sharp, 1975; Behrman, 1978; Behrman and Parkison, 1978). Limestone lenses containing Permian Tethyan fusulinids have been recovered from three localities in the melange (Fig. 5-13). Between the Stanislaus and Tuolumne Rivers, the melange has a serpentinite matrix (Schweickert, 1978). Locally, in areas to the north, small masses of sheared serpentinite occur as well (Duffield and Sharp, 1975; Behrman, 1978).

Morgan (1976) mapped a depositional contact between the arc volcanic sequence (there known as the Penon Blanco Volcanics) and two larger serpentinized ultramafic bodies between the Stanislaus and Tuolumne rivers (Figs. 5-12 and 5-13). A small pluton that intrudes the contact was dated by ^{238}U-^{206}Pb as 190 m.y. (Morgan and Stern, 1977). Near the base of the larger ultramafic body, Morgan (1976) mapped several tectonic blocks of garnet amphibolite, some with crossite from which he obtained K-Ar amphibole ages of 200±10 m.y. These metamorphic rocks may have formed during original emplacement of the ultramafic body, but according to Morgan and Stern (1977), the ultramafic rocks and their overlying volcanic cover were later sliced into the Oxfordian-Kimmeridgian Mariposa Formation (Fig. 5-13).

South from the Tuolumne River, a thicker part of the Penon Blanco Volcanics rests depositionally on a sequence of red, radiolarian chert known as the Hunter Valley chert that in turn overlies a thick succession of mafic pillow lava (Figs. 5-12 and 5-13) (R. A. Schweickert and N. L. Bogen, unpublished data). Morgan (1976) obtained a ^{238}U-^{206}Pb date of 182 m.y. on a small pluton that evidently cuts the chert (Fig. 5-12); according to D. L. Jones (personal communication, 1978), poorly preserved radiolarians from the chert suggest it is no older than Late Triassic. I interpret the chert and mafic volcanic rocks as layers 1 and 2 of Upper Triassic oceanic crust (Fig. 5-13).

One of the most enigmatic terranes in the Sierra Nevada occurs between latitudes 39° and 40°N immediately west of the northern part of the Foothills suture (Melones fault zone) (Fig. 5-12). As noted earlier, many geologists, including Burchfiel and Davis (Chapter 3, this volume), Saleeby (Chapter 6, this volume), Hietanen (1973), and Irwin and others (1978), use the name "Calaveras Formation" for exposures of chert, argillite, and limestone in this terrane. Although gross lithologic similarities exist between these rocks and the Calaveras Complex, the rocks in the northern Sierra are as yet very poorly known, and there is no direct evidence they are part of the same assemblage as the Calaveras Complex. To avoid confusion, I refer to these rocks as part of the northwestern Sierran terrane.

In the northern and northwestern parts of the terrane, Hietanen (1973, 1976, 1977) mapped a heterogeneous assemblage of metamorphosed chert and mudstone with rare marble lenses that is bounded to the south by structurally complex assemblages of metamorphosed basaltic, andesitic, and rhyolitic volcanic rocks associated with numerous large, lenticular masses of serpentinized peridotite, talc schist, and metagabbro. Hietanen (1973, 1976) noted evidence for at least two phases of folding within the volcanic rocks. The map-scale admixtures of such a diverse assortment of lithologies and reconnaissance observations in several areas led Schweickert and Cowan (1975) to assert that much of this terrane is tectonic melange. Xenophontos and Bond (1978) also noted the existence of melange. In detail, however, the gross structure of the northwestern Sierran terrane is unknown.

The ages of the rocks are highly uncertain as well. Hietanen (1976) correlated certain metavolcanic units with parts of the Paleozoic pyroclastic sequence east of the Foothills suture in the Taylorsville area (Fig. 5-1) and therefore suggested Devonian to

Permian ages for them. She further interpreted the metachert and phyllite to underlie the metavolcanic rocks, requiring a pre-Devonian age for it. However, Irwin and others (1978) reported a Late Triassic radiolarian fauna from the bedded chert mapped by Hietanen (1973). If Hietanen's interpretation that the metavolcanic rocks overlie the chert-phyllite unit is correct, then none of the rocks within the terrane need be older than Late Triassic and, instead, most may be Jurassic. Clark (1976) reported Paleozoic fossils from a limestone lens and from a massive tuff on the south fork Yuba River near latitude 39°15′N (Fig. 5-12). It is presently unclear whether these rocks are simply isolated tectonic blocks or whether they are lithologically or structurally continuous with rocks in the areas mapped by Hietanen (1976).

Clark's (1976) work and Hietanen's (1973, 1976) mapping together with the radiolarians reported by Irwin and others (1978) suggest that strongly deformed Jurassic volcanic and sedimentary rocks in the northwest Sierran terrane rest on structurally complex basement consisting of pelagic sedimentary rocks and ophiolitic rocks. In summary, between latitudes 37° and 40°N (Fig. 5-12), the basement of the island arc is a structurally complex assortment of slices of oceanic crust, melanges, and sheets of ultramafic rock that evidently were deformed and juxtaposed prior to the birth of the island arc, but which have undergone additional tectonic shuffling during the Late Jurassic, both prior to and during the Nevadan orogeny, when the entire terrane was strongly cleaved and folded. Nowhere is there evidence that volcanic rocks of the western island arc (i.e., west of the Melones fault zone) developed on Paleozoic rocks of the North American plate.

Southern Sierra Nevada The complex basement described above is markedly reminiscent of the geology of the Kings-Kaweah ophiolite belt between latitudes 36° and 37°N (Fig. 5-1) described by Saleeby (1977a, 1978a, 1979).

Saleeby (1977b) characterized the structure of the belt as a huge tectonic megabreccia containing slabs of ophiolitic rocks up to 20 km long embedded in a schistose serpentinite matrix. From north to south the size of ophiolite slabs decreases until, in its southern part, the belt is a serpentinite matrix melange. U-Pb ages on plagiogranite suggest the ophiolite is in part of Permo-Carboniferous age (Saleeby, 1977b, 1978a). Plutons with U-Pb ages of 169 and 157 m.y. cut the deformed basement near the Kings River (Fig. 5-1), indicating that breaking and mixing had terminated and that the melange terrane had become the basement of a magmatic arc by mid-Jurassic time (Saleeby, 1975 and 1977b; Saleeby and Sharp, 1977). Saleeby (1977b) has noted that several very small remnants of andesitic pyroclastic rocks apparently were deposited on the melange, and probably are minor erosional remnants of an island arc volcanic sequence like that recognized in blocks A and B of the western Sierra.

Overlying the serpentinite melange is an olistostrome complex containing limestone blocks with a Permian Tethyan fauna (Saleeby, 1977b; Saleeby and others, 1978; Schweickert and others, 1977). I agree with Saleeby's (1979) suggestion that these rocks may be correlative with the melange belt of the western Sierra Nevada that lies west of the Melones fault zone. The Tethyan fauna most likely indicate that the melange terrane is an entity exotic to North America, as suggested earlier by Schweickert and Cowan

(1975) for the melange terrane and associated island-arc rocks north of the Stanislaus River (Fig. 5-12).

Saleeby (1977b) and Saleeby and others (1978) have argued that the deformation and chaotic mixing recorded by the Kings-Kaweah ophiolite belt occurred along a fracture zone that was distant from the North American continental margin, and that the entire fracture zone complex subsequently was accreted to the North American continental margin. The similarity between the Kings-Kaweah ophiolite belt and the exposed basement of the island arc between latitudes 37° and 40°N suggests that the entire western Sierra Nevada island arc may have developed upon structurally complex basement that in part originated along one or more fracture zones.

Tectonic Model

In discussing the tectonic evolution of the island arc in the western Sierra, Schweickert and Cowan (1975) argued that the Coast Range ophiolite, lying some 100 km west of the arc terrane in northern California and forming a discontinuous belt paralleling the arc, provides the prime constraint for deducing the polarity of the island arc.

Radiometric data from a number of localities in the Coast Range ophiolite indicate that it was generated about 150 to 160 m.y.a. (Lanphere, 1971; Hopson and others, 1975b), during the interval that the island arc was active. Saleeby and others (1979) have reported age data indicating that the Smartville ophiolite in the Sierra Nevada and the Josephine ophiolite of the western Klamath Mountains both formed during the same interval (ca. 155 to 162 m.y.a.). The apparent similarity of age of these ophiolites and the fact that they are coeval with the arc activity make it very unlikely that the Coast Range ophiolite formed at a spreading ridge west of and parallel to the arc, with a subduction zone, now concealed beneath the Great Valley, consuming the east flank of the spreading ridge, as suggested by Pessagno (1977a). Furthermore, this interpretation leaves unanswered the questions why the Coast Range ophiolite was accreted bodily to the continental margin during the latest Jurassic to become the basement for the Great Valley Sequence and why Franciscan subduction stepped west of the Coast Range ophiolite.

The alternative view, favored by Schweickert and Cowan (1975), Schweickert (1978), and adopted here, is that the age relations between the Coast Range ophiolite and the Sierran island arc and its coeval Smartville ophiolite mean they are all genetically related to one another. The most obvious tectonic setting in which the production of oceanic crust can be directly linked to the evolution of an island arc is that of a spreading marginal or interarc basin (Karig, 1971, 1972) on the rear side of an island arc. The geometry in this case would imply that the island arc had an eastward-facing polarity and developed above a westward-dipping subduction zone (Schweickert and Cowan, 1975).

Karig (1972) has shown that, on the arc-side of marginal basins, a thick apron of arc-derived pyroclastic material typically mantles the oceanic crust. Fittingly, at several localities discussed by Bailey and others (1970), the Coast Range ophiolite contains a cover of silicic pyroclastic and volcaniclastic debris (Blake and Jones, 1974) that probably represents such an arc-derived apron, as suggested by Karig (1972). These are difficult to reconcile with a west-facing arc. Furthermore, silicic intrusive rocks are more abundant in the Coast Range ophiolite than would normally be expected in mid-ocean ridge ophiolite (Blake and Jones, 1974). In this respect the Coast Range ophiolite is similar to ophiolites that elsewhere have been inferred to have originated within marginal basins (Dalziel and others, 1974; Tarney and others, 1976; Ave'Lallemant, 1976). I conclude that the character of the Coast Range ophiolite and portions of its volcanic cover can best be reconciled with the view that it originated by spreading in a marginal basin west of, and behind, the Sierran island arc. In addition, the Smartville ophiolite may represent an intraarc rhombochasm as suggested by Schweickert (1978). In any case, these arguments lead to the inference that the Sierran island arc and its counterpart in the western Klamath Mountains had an eastward-facing polarity, opposite to that of the partly coeval Andean-type arc developed on the North American continental margin.

During Late Jurassic time, the Sierra Nevada region probably had a paleogeography much like that of the present-day Philippines-Celebes Sea region, where two opposing island arcs, Sangihe and Halmahera, are obliquely colliding. The important difference, however, is that the eastern volcanic arc had developed upon continental crust of North America, while the western arc was an oceanic island arc.

I interpret the Nevadan orogeny to have resulted from the oblique collision of the two arcs (Schweickert, 1978).

During the collisional event, the Andean arc and its Paleozoic basement were tightly folded into the Nevadan synclinorium (Figs. 5-1, 5-10, and 5-11), the rocks of the island arc were tightly folded and thrust faulted, and a regional, penetrative slaty cleavage developed throughout both arc terranes.

As Schweickert and Cowan (1975) suggested, the accretion of the bulky island-arc complex to North America caused the consuming plate margin to retreat to the west of the Coast Range ophiolite, where Franciscan subduction was initiated. The first Sierran magmas related to the Franciscan subduction regime, dated at about 140 to 145 m.y., intruded parts of the deformed, accreted arc terrane.

The fact that the oldest Franciscan blueschists yielded radiometric dates of about 150 m.y. (Coleman and Lanphere, 1971) suggests that Franciscan subduction began slightly before and overlapped in time the final suturing of the two arcs during the Nevadan orogeny. Again, a remarkable parallel exists in the Philippine-Celebes Sea region, where a new subduction zone (the southern Philippine trench) has been initiated east of Mindanao, where the Halmahera and Sangihe arcs are sutured (Hamilton, 1977), even though the collision is still underway in the Molucca strait to the south of Mindanao (see Hamilton, 1977a, Figures 3, 14, 15, and pp. 27-28, and Silver and Moore, 1978).

Saleeby (Chapter 6, this volume) has expressed concern that the western Sierra Nevada lacks the requisite "collisional volume" of crustal rocks to represent a collisional orogen. However, the present crustal structure of the western Sierra Nevada may not necessarily represent the Jurassic or older crustal structure of the region, since phase changes during or after the Nevadan orogeny may have affected the density distribution within the crust. In addition, comparisons with zones of incomplete suturing like the Molucca Sea may not be valid, for much or most of the accretionary material in the collision zone may be uplifted and eroded away by the time suturing is complete. It may additionally be shed longitudinally, parallel to the collision zone, as suggested by Graham and others (1975). Finally, contrary to Saleeby's assertion, the Late Jurassic and Early Cretaceous of California (including the Jurassic parts of the Great Valley Sequence and the Franciscan Complex) contain an exceedingly voluminous record of sedimentary rocks eroded from the Sierran-Klamath orogen, which could indeed account for "missing volume," if any.

Comments on alternative models An elegant and radically different plate tectonic interpretation of the Mesozoic Sierra Nevada has been presented by Saleeby (Chapter 6, this volume; Saleeby and others, 1978). The key difference is that Saleeby interprets the island-arc rocks west of the Foothills suture as having formed essentially *in situ* with respect to North America on fracture zone basement that was juxtaposed against western North America prior to the inception of the island arc. In this view, the Triassic-Jurassic igneous rocks of the eastern and southeastern Sierra, together with the western Jurassic igneous rocks, represent a single complex arc with intra-arc wrench zones.

Major contrasts exist between the oceanic basement of the western part of the arc (west of the Foothills suture) and the continental basement of the eastern parts of the arc. The complex arc interpretation hinges critically upon establishing ties between the two different basement terranes, and between the varied expressions of arc activity in the west and the east.

According to Saleeby (Chapter 6, this volume), "the Kings sequence laps across the Foothills suture," thus tying the western ophiolitic basement to the eastern continental basement by Early Jurassic time. The basis for this assertion is the interpretation of Saleeby and others (1978) that roof pendants of the southern Sierra Nevada (1) contain only Mesozoic rocks, (2) which can be interpreted as a facies model with continentally derived sediment and silicic volcanic rocks on the east interfingering westward with olistostromes containing quartzose sandstone, chert, argillite, and Tethyan limestone olistoliths, and interbedded with flysch and basaltic to andesitic volcanic rocks.

Important points about this interpretation are that (1) the rocks occur in scattered, isolated roof pendants, so continuity of units or assemblages cannot be demonstrated, and (2) within individual roof pendants, structures are exceedingly complex, and as yet even local stratigraphic sequences have not been established, and (3) the gross structure of the Nevadan synclinorium and the proximity of lower Paleozoic miogeoclinal rocks only 40 km to the east in the eastern limb of the synclinorium make it equally (more?) plausible that the unfossiliferous quartzites, marbles, and pelitic schists within the central parts of the "type" "Kings sequence" are actually part of the lower Paleozoic to uppermost Precambrian Cordilleran miogeoclinal succession extending to the western limb of the synclinorium.

Thus, the facies interpretation of Saleeby and others (1978) is in no way established and does not constitute a firm tie between the western oceanic basement and the eastern continental basement, as implied by Saleeby (Chapter 6, this volume).

It has also been argued that stratigraphic sequences west of the Foothills suture contain a major component of "continentally derived" clastic sediments (Behrman, 1978; Behrman and Parkison, 1978; Saleeby, Chapter 6, this volume), again requiring close proximity of the two terranes during or prior to evolution of the western Jurassic arc.

Behrman's (1978) descriptions of sandstones near latitude 38°30′N reveal high abundances of chert and polycrystalline quartz along with quartz and plagioclase, the sum of which is strongly suggestive of recycled orogenic provenances (including subduction complexes) rather than continental provenances (Dickinson and Suczek, in press). Such sediments could indeed have been locally derived from within the belt, since chert and mylonitic granitoids, quartzites, and conglomerates all occur locally, and the original stratigraphic relationships of the clastic sediments are not preserved (Behrman, 1978).

A third relationship that Saleeby (Chapter 6, this volume) argues ties the two terranes together into a single magmatic arc is that the Melones fault zone purportedly lies within the western Jurassic plutonic belt and "thus could not have been a Late Jurassic subduction zone" separating two different arcs.

The fact that Jurassic plutonic rocks occur both east and west of the Melones fault zone does not require that all the plutonic rocks represent parts of the same arc. Furthermore, separating plutons into Jurassic and Cretaceous age groupings carries no fundamental tectonic significance. In fact, Jurassic plutons west of the Melones fault are of two distinct age groupings: (1) scattered, small, 180 to 190 m.y. pre-Nevadan plutons that are genetically related to the thick sequences of extrusive volcanic rocks; (2) post-Nevadan, 145 to 135 m.y. granodiorites and quartz diorites that postdate movements on the Melones and are satellitic to the Late Jurassic-Early Cretaceous Sierra Nevada batholith to the east, and that crosscut the deformed Middle to Late Jurassic volcanic arc sequences.

The Jurassic plutons east of the Melones fault zone are predominantly post-Nevadan discordant plutons ranging in age from about 148 to 140 m.y. Scattered Middle Jurassic plutons (170 m.y. and older) occur as well, but these were forcefully intruded and do not have clear affinity with either the western or the eastern arc sequences. Significantly, no volcanic rocks occur in the region immediately east of the Melones. Furthermore, an east-trending mafic dike swarm (discussed earlier) that is post-170 m.y. and pre-150 m.y. occurs widely east of the Melones fault zone, but does not occur to the west, suggesting that the two terranes were not adjacent to one another and that there were major Late Jurassic displacements on the Melones fault zone.

Thus, the pre-Late Jurassic plutonic rocks do not provide a unique tie between the western and eastern arc sequences. There is no compelling evidence for proximity of the two dissimilar terranes until Oxfordian or Kimmeridgian time, when metamorphic detritus that presumably originated from the east was shed into the Mariposa Formation in the island-arc block.

Saleeby (Chapter 6, this volume) has broadly related the great complexity of Mesozoic structure of the Sierra Nevada to dextral slip and "transpression tectonics" within the complex magmatic arc. However, this model fails to provide adequate explanations for two important features, (1) the Late Jurassic Nevadan synclinorium and the accompanying White Mountains anticlinorium, produced during the Nevadan orogeny, and (2) the existence of 150 to 160 m.y. ophiolitic rocks both within the Sierra Foothills and as a much more extensive terrane within the Coast Ranges to the west.

According to Saleeby (Chapter 6, this volume), the synclinorium formed by marginal sagging of the wallrocks of the batholith, and the Nevadan orogeny was an intraarc transpressive orogeny. If so, it seems most unlikely that the regional homogeneity of Late Jurassic structural fabrics that exists throughout the Sierra (Fig. 5-1) and in areas considerably east of the batholith itself could have developed. In addition, this explanation fails to account for the White Mountains anticlinorium, well east of the main Late Jurassic magmatic locus. More importantly, however, the form of the synclinorium clearly *predates* the most voluminous Late Jurassic and Cretaceous batholiths, which were emplaced mainly by passive mechanisms and did not appreciably deform their wallrocks. Finally, all available evidence suggests that the Nevadan orogeny was an exceedingly short lived and intense deformational event, not the protracted affair envisioned by Saleeby.

The Coast Range ophiolite, together with the Smartville and Josephine ophiolites, are considered to have formed as parts of a major intra-arc rift in mid-Jurassic time by Saleeby (Chapter 6, this volume). Since the Coast Range ophiolite would in this view have to have formed within an Early to Middle Jurassic arc-trench gap, this interpretation requires both the existence of sedimentary rocks in the forearc region that *predate* the rifting and in addition necessitates elements of a pre-Late Jurassic subduction complex within the Franciscan Complex in the Coast Ranges. There is no evidence that either assemblage exists. Thus, the complex arc model cannot at this time be substantiated.

SUMMARY AND OVERVIEW

Despite the fact that many parts of the metamorphic belt of the Sierra Nevada have not yet been studied in detail, a reasonably coherent structural picture is beginning to emerge. As argued in this paper, structural and petrologic data indicate that metamorphosed eugeosynclinal terranes in the western and northern parts of the range may have close counterparts in the eastern Sierra Nevada and in north-central Nevada. Although these terranes may have suffered the effects of one or more Peleozoic orogenies, their present form and structural setting were determined by intense folding during the Late Jurassic Nevadan orogeny. Recognition of the extent and magnitude of the Nevadan deformation in eastern California has yielded important clues both about the nature and extent of Paleozoic terranes and about the nature of the younger Cretaceous deformation of various parts of the range.

Mesozoic geology is now considerably better known as a result of the major efforts of numerous workers in recent years and is proving to be surprisingly complex. As is evident from this and other chapters in this volume, there are many different interpretations currently allowed by the data. In fact, there may never be solutions to many of the problems.

Controversies abound in the Sierra Nevada, and this seems altogether fitting considering the enormous complexity, lack of fossils, and great variety of rock types in the range. Because of such controversy, new studies should be undertaken and a great deal more learned of the geology of this alluring, enigmatic region. For a variety of reasons, the Sierra Nevada may well hold the keys to understanding the evolution of the southern part of the Cordilleran orogen. Our job has just begun.

ACKNOWLEDGMENTS

I am grateful to N. Bogen, G. Bond, D. S. Cowan, W. R. Dickinson, G. H. Girty, R. Hanson, C. Merguerian, J. Saleeby, W. Sharp, W. S. Snyder, O. T. Tobisch, and W. H. Wright, III, for discussions over a period of several years that have contributed a great deal to the ideas and concepts presented here. An earlier version of this manuscript benefited from critical reviews by W. G. Ernst, I. W. D. Dalziel, and W. S. Snyder. Research leading to this paper has been sponsored by NSF grants EAR-76-10979, EAR-78-14779, and EAR-78-23567.

Jason Saleeby
Division of Geological and Planetary Sciences
California Institute of Technology
Pasadena, California 91125

6

OCEAN FLOOR ACCRETION AND VOLCANOPLUTONIC ARC EVOLUTION OF THE MESOZOIC SIERRA NEVADA

ABSTRACT

Recent studies on the history of the Pacific Ocean floor coupled with structural, paleomagnetic, and paleobiogeographic studies of the North American Cordillera indicate that the Pacific floor moved northward and beneath the California continental margin throughout much of Mesozoic time. Northward movement began by Early Jurassic time and possibly as early as Late Triassic time. The northward movement brought equatorial ocean floor into the western Sierran region as an allochthonous ophiolitic basement terrane. Accretion of the oceanic basement was complex, involving both transcurrent faulting and subduction processes. Native basement consisting of Paleozoic borderland sequences and the North American continental shelf sat inboard of the exotic oceanic basement. The contact between the native and exotic terranes was an active break of long duration, herein named the Sierran Foothills suture.

The signature of plate convergence in the Sierran region is a composite volcanoplutonic arc. Voluminous magmatism proceeded semicontinuously from Early Jurassic through Late Cretaceous time. Arc rocks of all ages show compositional variations that reflect the nature of the preexisting basement rocks. Gabbroic-basaltic associations occur primarily within the limits of the oceanic basement, whereas voluminous granitic-rhyolitic assemblages are associated with continental basement. The depositional environments of the volcanic arc rocks varied greatly due to tectonic instability within the arc terrane. These environments consisted of an inherited continental shelf, uplifted and eroded continental and oceanic basement, sedimentary basins receiving epiclastic as well as volcaniclastic turbidites, and newly formed intraarc basins floored by juvenile crust.

The Mesozoic plate juncture history along the California margin was complex, being dominated by oblique subduction with probable interludes of transform faulting. The Sierra Nevada coincided with the axis of volcanoplutonic arc activity from Early Jurassic through Cretaceous time. No consuming plate juncture of Jurassic or younger age can be identified within the Sierran region. Late Triassic to possibly earliest Jurassic consumption occurred in the western foothills during the accretion of the ophiolitic basement terrane. Jurassic consumption occurred west of the Sierran region, but any subduction complex that may have existed was removed by Jurassic and Cretaceous rifting and transcurrent faulting. Cretaceous consumption in the Coast Ranges is marked by the Franciscan Complex. The Sierra Nevada evolved as one composite arc terrane containing numerous internal structural breaks. Deformation of the arc consisted of large-scale dextral wrench faulting with a family of wrench-related structures, oblique rifting, and the development of fold-thrust welts. The result of these processes is a complex igneous and metamorphic terrane consisting of interleaved volcanic arc and older basement fragments that are cut by highly to only slightly deformed plutons of a wide range of ages. Similar structural complexities are evolving today in modern arc terranes with analogous plate-tectonic settings. The processes that operate along a convergent plate juncture that also has a significant strike-slip component are more complex than junctures with normal or near-normal convergence. The unique tectonic style resulting from prolonged oblique convergence is termed transpression after Harland (1971).

Metamorphic country rocks of the Sierra Nevada represent an orogenic collage that was constructed primarily during Mesozoic time. The term collage is adopted from Helwig (1974) and Davis and others (1978), who use it to denote accretionary orogenic terranes characterized by internal discontinuity with extreme lithologic and paleogeographic heterogeneity. The Sierran collage consists of tectonostratigraphic belts that are composed of Mesozoic strata and older pre-Mesozoic basement remnants. The term tectonostratigraphic is used for the belts since they are the vestiges of stratigraphic sequences, but owe their positions and shapes to tectonic processes. The belts are separated by fault and melange zones, and by Mesozoic batholithic rocks (Fig. 6-1). The basement remnants consist of Paleozoic continental shelf and borderland sequences and highly deformed

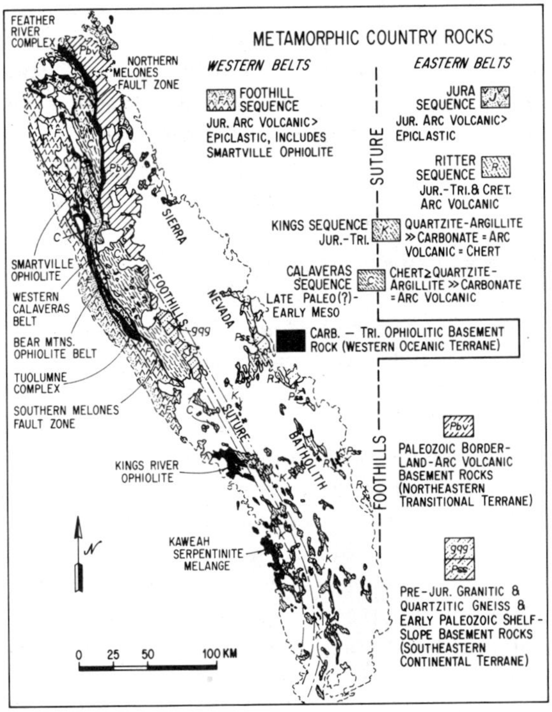

Fig. 6-1. Tectonic map of the Sierra Nevada showing informal rock unit terminology used in text and geographic distribution of units relative to the Foothills suture. References for map distribution and composition of rock units given in Tables 6-1 and 6-2.

Carboniferous to Triassic ophiolite fragments. The early Mesozoic strata consist of hemipelagic, continental clastic, and arc volcanic strata. These are all typical of the "eugeosynclinal facies" of the Cordilleran orogen, which extends as far north as the Arctic slopes of Alaska and in part as far south as western Mexico (Jones and others, 1977; Davis and others, 1978). The ophiolitic and most of the hemipelagic assemblages of the Sierran collage are exotic to North America, whereas the clastic and arc volcanic assemblages were deposited on and adjacent to North America. The emplacement of the exotic assemblages into the Sierran terrane and the construction of the Sierran collage can be interpreted as a result of plate margin processes that operated between North America and the Mesozoic Pacific ocean floor.

The purpose of this paper is to review data on the Mesozoic structure, stratigraphy, and age relations within the Sierran terrane, and to interface these data with both the regional tectonic framework of the North American Cordillera and the kinematic history of the Pacific ocean floor. Of prime importance is the nature of the basement that early Mesozoic strata of the Sierran region were deposited on. The gross structure of this basement holds important clues to the nature of the tectonic framework in which the Sierran collage evolved. After review of the pre-Mesozoic basement geology of the Sierran region, data on the Mesozoic Sierra Nevada and adjoining areas will be presented and incorporated into a general model for the construction of the Sierran collage.

Actualistic models up to this point have emphasized the importance of plate-tectonic accretionary processes operating during the early Mesozoic in the Sierran terrane (Hamilton, 1969, 1978a; Moores, 1970, 1972; Schweickert and Cowan, 1975; Schweickert, 1978). An important assumption in these models is that construction of the collage can be adequately explained by one-dimensional plate interaction models. Such models have nicely explained the gross distribution of rock types or petrotectonic assemblages within a plate-tectonic context. However, they fail to explain many of the important details of the sedimentary, igneous, and metamorphic development of the Sierra, as well as its overall structural configuration. The view set forth in this paper is that one-dimensional plate interaction models can in no way explain the petrogenesis and structural configuration of the Sierran collage, and that tectonic removal processes of longitudinal transport worked in conjunction with, and may have predominated over, tectonic accretionary processes. The resulting pattern of such interplay is an "orogenic collage" of narrow tectonostratigraphic belts, which are small stranded and interleaved vestiges of the once much larger terranes that interacted along an evolving plate juncture.

The complex interplay of tectonic removal and accretionary processes that is apparent in the Sierran terrane is attributed to the northward drift and underflow of Pacific ocean floor relative to North America throughout most of Mesozoic time. This relative plate motion pattern resulted in a complex plate juncture history that involved transcurrent (or transform) faulting and oblique subduction. The combined effect of transcurrent motion and compression has been termed transpression (Harland, 1971). Even though not widely used, this term will be used in this paper because it adequately denotes an important and distinctive type of tectonic regime, and it stresses the importance of two-dimensional plate interaction models.

MESOZOIC PALEOBASEMENT GEOLOGY OF THE SIERRA NEVADA

Jurassic and Upper Triassic strata of the Sierra Nevada were deposited on a complex substrate consisting of deformed crystalline and sedimentary complexes, which constituted a pre-Jurassic paleobasement. The term basement will be used for this substrate, but it must be remembered that much of it no longer exists owing to its obliteration by the Cretaceous batholith. Local remnants of the basement can be observed within the metamorphic country rocks. In addition to these metamorphic remnants, the nature of the basement is also revealed by petrochemical features of the batholith. Some useful generalizations can be made about the basement, one of which is a threefold division of basement terranes. These are (1) western oceanic, (2) northeastern transitional (between oceanic and continental), and (3) southeastern continental. Each of these terranes is distinct in both their metamorphic remnants and their batholithic petrochemical features. The general features of each terrane are given in Table 6–1.

Nature of the Terranes

The western oceanic terrane consists of Carboniferous to Triassic ophiolite fragments. In the southwest Foothills, the Kings-Kaweah ophiolite belt occurs as the western wall of the Cretaceous batholith. Overlying Jura-Trias strata are faulted and folded into the ophiolite belt and occur as scattered roof pendants to the east (Saleeby and others, 1978; Saleeby, 1979). To the north of the Kings-Kaweah belt, oceanic basement rocks occur as fault and melange-bounded slivers of mafic and ultramafic rocks along much of the Melones fault zone (Clark, 1964, 1976; Ehrenberg, 1975; Saleeby and Moores, 1979). A complex belt of metamorphosed ultramafic and mafic rock extends 150 km southward along this zone from the Feather River area. To the southwest of this belt occurs the 90-km-long Bear Mountains ophiolite belt, which represents a nearly complete but highly disrupted ocean-floor sequence. This belt has been faulted up between parallel belts of Jurassic strata (Behrman, 1978). Southeast of the Bear Mountains belt the Tuolumne River peridotite complex sits unconformably beneath Lower Jurassic volcanic arc rocks (Morgan, 1973, 1976). Numerous fault slivers of mafic and ultramafic rock occur in association with hemipelagic and clastic strata of the Calaveras sequence both east of the southern Melones fault and in the northern Sierra (Hietanen, 1973, 1977; Sharp and Saleeby, 1979). These slivers appear to have moved up from the basement into the younger strata. Some of the slivers are ophiolitic melange.

The northeastern transitional terrane consists of the Shoo Fly Complex, a Devonian and possibly older assemblage of continental rise strata containing a zone of Paleozoic melange (d'Allura and others, 1977). The melange zone contains members of the ophiolite suite, suggesting that the Shoo Fly Complex sat on oceanic crust. The Shoo Fly Complex is unconformably overlain by a thick section of upper Paleozoic island-arc deposits (Durrell and d'Allura, 1977). Metamorphosed Paleozoic strata that may be related to the Shoo Fly complex occur in the east-central Sierra roof pendants (Rinehart and Ross,

TABLE 6-1. General Descriptions, Age Constraints, and References for Exposures of Pre-Jurassic Sierra Nevada Basement Rocks

Terrane	*General Character of Protoliths*	*Formal or Informal Name*	*Local Age*	*References*
Northeastern Transitional	Subarkosic-lithic sandstone, quartz sandstone, argillite, conglomerate, chert, and pre-Mesozoic melange zone	Shoo Fly Complex	Ordovician-Devonian	McMath, 1966; Bond and others, 1977; d'Allura and others, 1977; D. L. Jones, pers. commun., 1978
	Quartz keratophyre tuff and breccia, chert, andesite volcaniclastics, keratophyre, basalt, limestone	Northern Sierra upper Paleozoic volcanic sequence	Devonian-Permian	McMath, 1966; d'Allura and others, 1977; Durrell and d'Allura, 1977
Southeastern Continental	Siliceous argillite, carbonaceous limestone, calcareous quartz sandstone	Mt. Dana-Saddlebag Lake roof pendant	Ordovician (?)-Silurian (?)	Kistler, 1966a, b; Brook, 1977
	Siliceous and carbonaceous argillite, quartz sandstone, dirty and carbonaceous limestone chert	Mt. Morrison roof pendant	Ordovician-Silurian, Pennsylvanian-Permian (?)	Rinehart and others, 1959; Rinehart and Ross, 1964; Russell and Nokelberg, 1977
	Limestone, dirty limestone, argillite, quartzite	Bishop Crk. roof pendant	Ordovician-Silurian	Bateman, 1965a; Moore and Foster, 1980
	Oolitic and mottled limestone, mudstone, quartz sandstone	Big Pine roof pendant (Poleta Fm.)	Early Cambrian	Walcott, 1895; Moore and Foster, 1980
Western Oceanic	Harzburgite-dunite tectonite, layers of amphibolite and plagiogranite	Feather River peridotite-gabbro complex	mm: ∿235 m.y. ig: 275–313 m.y.	Ehrenberg, 1975; Weisenberg and Ave'Lallemant, 1977; Saleeby and Moores, 1979
	Harzburgite-dunite tectonite, schistose serpentinite, cumulate dunite, wehrlite, pyroxinite, gabbro, amphibolite, metabasalt, chert, ophicarbonate	Bear Mtns. ophiolite belt	mm: >190 m.y. & ∿300 m.y. ig: 295 & 304 m.y.	Behrman, 1978; Behrman and Saleeby, in press
	Harzburgite-dunite tectonite, cumulate wehrlite, serpentinite matrix melange, with blocks of amphibolite and metachert	Tuolumne ultramafic complex	mm: >200 m.y.	Morgan, 1973, 1976
	Harzburgite-dunite tectonite, amphibolite layers, cumulate wehrlite and gabbro, basaltic dike complex, pillow basalt, chert, zones of serpentinite matrix melange	Kings River ophiolite	mm: >190 m.y. ig: 190–225 m.y.	Saleeby, 1977a, 1978a; Saleeby and Sharp, 1980
	Blocks of serpentinized peridotite, amphibolite, gabbro, diabase, pillow basalt, chert, ophicalcite and silica-carbonate rock in schistose serpentinite and deformed sedimentary serpentinite	Kaweah serpentinite melange	mm: >194 m.y. ig: 270–305 m.y. & 190–225 m.y.	Saleeby, 1977b, 1979; Saleeby and Sharp, 1980

1964). These strata consist of Ordovician-Silurian and Permo-Pennsylvanian strata containing elements of both shelf and rise facies. Further treatment of rocks here assigned to the northeastern basement terrane is given in Schweickert and Snyder (Chapter 7, this volume).

Cambrian and Ordovician shelf facies strata occur as small metamorphic septa in the southeastern Sierra (Moore and Foster, 1980). These septa represent the only known vestiges of the southeastern basement terrane. Paleozoic and possibly older shelf-type rocks of this terrane may also occur along the eastern margin of the Calaveras sequence in the central Sierra region. These rocks consist primarily of highly deformed and recrystallized quartzites. They have been considered the southward extension of the Shoo Fly complex by Schweickert and Snyder (Chapter 7, this volume), but their age is only constrained to pre-Jurassic, and significant elements of the Shoo Fly complex are lacking. Thus a direct correlation must be considered highly speculative. Gneissic granitoids occur in association with the highly deformed quartzites of the central Sierra. At least some of the granitoids are highly deformed plutons of Triassic age, but others are hybrid assemblages containing highly discordant pre-Phanerozoic zircon (Saleeby and Sharp, unpublished data). The hybrid granitoid lenses may represent metamorphosed remnants of old sialic crust or younger deformed Jura-Trias plutons containing significant components of such crustal materials. In either case, continental basement rocks are directly or indirectly expressed along this complex zone. Calaveras sequence rocks located west of the granitic and quartzitic gneiss belt sit on oceanic basement. These Calaveras rocks have traditionally been treated as Paleozoic in age (Clark, 1964; Schweickert and others, 1977). However, as discussed later, these rocks are here considered primarily early Mesozoic in age.

Petrochemical data on the batholith yield valuable insight into the general nature of the pre-Mesozoic Sierran crust. Transverse variations in the bulk composition of the batholith have been documented by Moore (1959), Bateman and Dodge (1970) and Saleeby (1975, 1976). These data show that K_2O increases eastward across the batholith. This variation is pronounced along the western margin of the batholith, where it is joined by a distinct increase in SiO_2. Gabbroic to quartz dioritic rocks predominate along the western margin, whereas granitic rocks predominate to the east. Transverse variations in the isotopic composition of initial Sr and Pb in the southern Sierra Nevada batholithic rocks have also been documented (Kistler and Peterman, 1973; Doe and Delavaux, 1973; Chen, 1977; Chen and Tilton, 1978; Saleeby and Chen, 1978; Saleeby and Sharp, 1980). These data show that the initial isotopic compositions become more radiogenic eastward with a distinct jump in values along the batholith's western edge. Of particular interest is the 1.8 b.y. Pb-Pb isochron derived by Chen and Tilton (1978) in the southeast Sierra, which matches with a major 1.8 b.y. age group of the North American craton (Engel, 1963). These data suggest that the southeastern basement was floored by pre-Phanerozoic crystalline rock. This basement complex is loosely referred to as sialic, although a hybrid assemblage of overall intermediate composition is envisaged. In the southwestern foothills the initial Pb and Sr values are distinctly oceanic in character (Saleeby and Chen, 1978; Saleeby and Sharp, 1980). In the northern Sierra the transverse variations exist, but they are not as pronounced. Of major importance here is a subtle longitudinal variation in the batholith. The highly radiogenic initial Pb and Sr values drop off to the north, and the silicic plutons are generally less potassic. The petrochemical features of the northeastern terrane are thus intermediate between those of the western and southeastern terranes. The batholith can be contoured with respect to initial $^{87}\text{Sr}\backslash^{86}\text{Sr}$ values (Fig. 6-2). The initial Pb data parallel the Sr data. The southeastern terrane lies mainly on the high side of the 0.706 initial Sr isopleth, whereas the western terrane lies on the low side of the 0.704 isopleth.

Geophysical studies of the region show marked changes in Bouguer gravity, crustal density, and heat flow, which coincide with the eastern limit of western terrane basement rocks (Lachenbruch, 1968; Oliver and Robbins, 1975). These data indicate that crystalline rocks west of the Sierran Foothills are generally simatic in character (Cady, 1975). This, in conjunction with the petrochemical data on the batholith and the restriction of ophio-

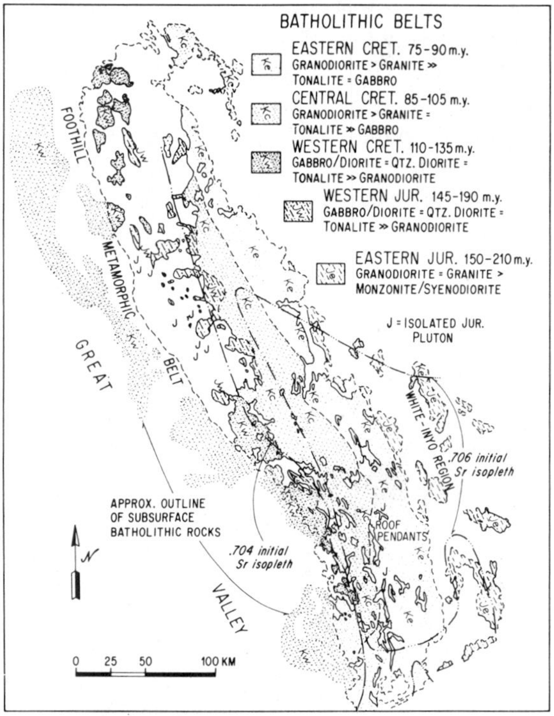

Fig. 6-2. Generalized map showing batholith age-composition belts of the Sierra Nevada region. Time ranges shown for each belt refer to major magmatic pulses, which encompass over 90% of published and unpublished isotopic ages. Ages and Sr isopleths after Evernden and Kistler (1970), Armstrong and Suppe (1973), Kistler and Peterman (1973, written commun., 1978), Chen (1977, pers. commun., 1978), Saleeby and Chen (1978), Saleeby and Sharp (1979, unpub. data), Sharp and Saleeby (1979), and Stern and others (in press). Subsurface batholithic rocks of western Cretaceous belt after May and Hewitt (1948), Cady (1975), Williams and Curtis (1977), and Saleeby and Williams (1978).

litic rocks to the west of the 0.704 Sr isopleth, indicates that a fundamental discontinuity in the earth's crust is situated along the eastern edge of the western terrane.

The present distribution of the Sierran basement terranes is shown in Figure 6-3. The petrochemical and geophysical discontinuity that marks the eastern edge of the western terrane approximates a fossil contact between this terrane and the eastern terranes. The cryptic contact is termed the Foothills suture. The Foothills suture is not visible along most of its length owing to early Mesozoic "superjacent" sedimentation and its obliteration by the Cretaceous batholith. Its approximate trace is defined by the 0.704 isopleth for initial $^{87}Sr\backslash^{86}Sr$ in batholithic rocks. This isopleth runs just east of both the Kings-Kaweah ophiolite belt and along the eastern margin of the Calaveras sequence. In the northern Sierra the suture appears to coincide with the northeastern branch of the Melones fault zone, where the Feather River ophiolite fragment of the western terrane is faulted up between the Shoo Fly Complex of the northeastern terrane and the Calaveras sequence of the western terrane. The trace of the Foothills suture in the central Sierra is a subject of much debate. The view adopted here is that it coincides with the complex deformation zone expressed by the granitic and quartzitic gneiss belt that bounds the eastern margin of the Calaveras sequence. This view best satisfies the compositional patterns seen in the batholith and the ensimatic nature of Calaveras sequence rocks located between the

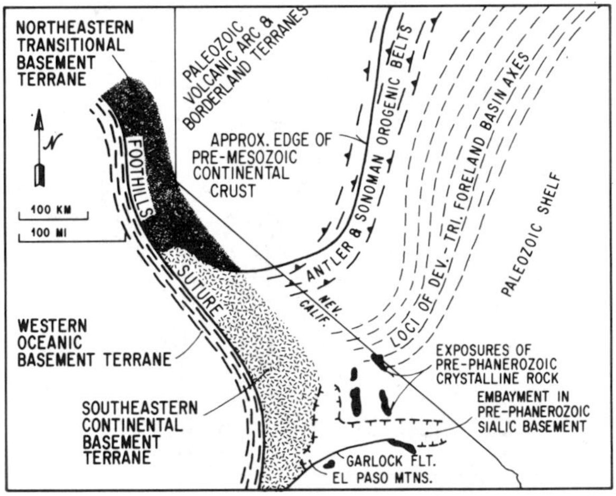

Fig. 6-3. Generalized map showing Sierran pre-Jurassic basement terranes and selected features from the pre-Mesozoic southwest Cordillera. Paleozoic stratigraphic, structural and paleogeographic trends after Burchfiel and Davis (1972), Poole and others (1977), Speed (1977), and Stevens and Ridley (1974). Approximate edge of Paleozoic continent synthesized from these references and R. W. Kistler, written commun. (1978). Pre-Phanerozoic crystalline rocks after Jennings (1977). Embayment in pre-Phanerozoic sialic basement synthesized from Wright and others (1976) and R. W. Kistler (written commun., 1978).

deformation zone and the southern Melones fault zone. The gneissic granitoid lenses with old isotopic imprints mentioned previously, may represent vestiges of pre-Phanerozoic crystalline rock that were mobilized along the Foothills suture. It must be emphasized that the Foothills suture existed in the basement, and that it has been obscured by Jurassic and Cretaceous sedimentary and igneous processes. Jurassic strata were deposited across the suture as it evolved. Thus the structures in these younger sequences only partly reflect the movement history and the position of the suture.

The western oceanic terrane is exotic to North America, with its juxta-position locus being the Foothills suture. As discussed later, its pelagic and hemipelagic strata not only record a complex history of oceanic sedimentation, but they contain limestone blocks with fauna that are exotic to North America (Douglass, 1967; Saleeby and others, 1978). The southeastern continental terrane is native to North America. Stratigraphic correlatives to Paleozoic rocks of eastern California and Nevada have been recognized by Speed and Kistler (1977) and Moore and Foster (1980). The northeastern terrane may be autochthonous or parautochthonous, as suggested by Schweickert and Snyder (Chapter 7, this volume), or totally allochthonous as suggested by Speed (1979).

The contact between the north and southeastern terranes is also a cryptic feature approximated by the 0.706 isopleth, where it turns eastward (in a complex fashion) from the Foothills suture in the central Sierra and continues into Nevada (Figs. 6-2 and 6-3). The 0.704 isopleth continues northward, and thus much of northeastern transitional terrane is contained between the 0.704 and 0.706 isopleths. This, in conjunction with the fact that a thick section of Paleozoic arc deposits formed on continental rise strata of the northeastern terrane, are the criteria for defining it as transitional between continental and oceanic.

Structural Development of the Basement

The Mesozoic tectonics of the Sierra Nevada cannot be divorced from the pre-existing structure of the region. Two fundamental basement contacts are shown in Figure 6-3: the Foothills suture, which is a Mesozoic structure, and the Paleozoic continent edge, which trends through Nevada and into the Sierran region at an angle to the suture. The western limit of the Paleozoic continent edge is shown in the Sierra Nevada as the contact between the two eastern terranes. This contact is deflected northward and then terminated against the Foothills suture.

Throughout much of Paleozoic time the Cordilleran Orogen consisted of a broad eastern shelf built over continental basement and western borderland and volcanic arc assemblages built over oceanic or transitional basement. The eastern basement terranes of the Sierra Nevada are an expression of this paleogeographic regime. Two Paleozoic orogenic pulses are recorded in the southwest Cordillera; both involved eastward thrusting of oceanic assemblages over the continental edge. The first pulse was the Devonian-Mississippian Antler orogeny, and the second was the Permo-Triassic Sonoman orogeny (Roberts, 1968; Speed, 1971a,b, 1979; Burchfiel and Davis, 1972). Deformations related to these orogenies may have extended into the eastern Sierra as suggested by structural studies in eastern Sierran roof pendants (Kistler, 1966b; Russell and Nokleberg, 1977; Brook, 1977).

Paleozoic to Triassic orogenic zones and depositional patterns run through Nevada and project into the Sierra Nevada, where they appear to be terminated and overprinted by northwest Mesozoic trends (Hamilton and Myers, 1966; Burchfiel and Davis, 1972; Jones and others, 1972; Saleeby, 1977b). This structural pattern has been cited as evidence for a major early Mesozoic truncation event of the southwest Cordilleran orogen. Inasmuch as Sonoman trends are thought to be cut by the truncation zone, and since post-Sonoman volcanoplutonic arc activity followed the trend of this zone, the truncation has been considered a post-Sonoman event.

Several models have been proposed to account for the truncation. Burchfiel and Davis (1972) and Schweickert (1976a, 1978) have proposed a continental rifting event. Signs of rifting along the truncation locus, such as sediment-filled extensional basins, alkalic volcanism, aulacogens, and mafic intrusive complexes of the proper age are lacking, so this model cannot be substantiated. Silver and Anderson (1974), Saleeby (1975), Davis and others (1978), and Speed (1978, 1979) have proposed a sinistral transform-type truncation of the orogen. Supporting data include apparent offset sections of Paleozoic strata in southern California and mainland Mexico and disruption of transcontinental pre-Phanerozoic crystalline age belts. Substantiation of this model has run into difficulties since an actual physical break or deformation zone has yet to be identified. It has been alternatively suggested that the Foothills suture represents the main locus of continental truncation, involving dextral movements between the North American plate and the Pacific ocean floor (Saleeby, 1977a,b; Saleeby and others, 1978). In this view the truncation processes not only operated prior to inception of the northwest-trending volcanoplutonic arc, but they continued during the life of the volcanoplutonic arc by intra-arc wrench faulting in response to oblique subduction.

An important point concerning the truncation event(s) that has not been given adequate attention is what the "pre-truncation" configuration of the continental margin may have been, and how this configuration may have affected the truncation processes. Initiation of the southwest Cordilleran shelf began following a late Proterozoic continental rifting event (Stewart, 1972; Burchfiel and Davis, 1972; Dickinson, 1977). Following rifting, shelf deposits were built on a passive continental margin, with the first signs of tectonic unrest marked by the Antler Orogeny. Wright and others (1976) suggest that the late Proterozoic Pahrump Group of the Death Valley region formed in an aulacogen, which was presumably related to the continental rifting event. The Pahrump Group occurs within the limits of the Cordilleran shelf strata. Labotka and Albee (1977) agree that the Pahrump Group was deposited during the rifting event, but they also suggest that the paleogeography was more complex with local source areas and irregular depositional basins. Isotopic data (R. W. Kistler, written communication, 1978) from the southeast end of the Sierra Nevada batholith suggest that plutons of part of that region did not pass through fully developed continental crust. This is shown on Figure 6-2 where an anomalous zone occurs at the southern end of the batholith with initial $^{87}Sr/^{86}Sr$ ratios between 0.704 and 0.706. The anomalous zone connects with the Death Valley "aulacogen," which suggests the existence of a complex zone in and to the southeast of the Sierra Nevada, which lacked pre-Phanerozoic sialic crystalline rock. This zone is shown as an embayment in the pre-Phanerozoic sialic basement on Figure 6-3. It is significant to note that the northernmost exposure of Paleozoic strata thought to have been displaced in a sinistral sense during Triassic time occurs in the El Paso Mountains, which is within the southwestern limits of the embayment. These strata are thought to have been displaced since they have affinities to Paleozoic borderland facies rather than nearby shelf facies (Poole, 1974). However, their spatial relations with the basement variations leave open the possibility that they accumulated within an embayment in the continental margin, and that they are not grossly out of place.

The main component of continental truncation is here believed to have been related to the late Proterozoic rifting event that resulted in the termination of northeast-trending pre-Phanerozoic crystalline age belts. The basement embayment at the southeast end of the Sierra suggests that the rifted continental edge lay nearby the presumed truncation zone. Since the southeastern Sierra Nevada for the most part was underlain by continental-type basement, and vestiges of strata from this basement are correlative with lower Paleozoic shelf strata that occur to the east of the Sierra, the basement embayment may represent the vestiges of a failed arm of the major rift system. Thus the southwest Cordilleran shelf formed on the "ragged" edge of the truncated continent, which probably had both embayments and significant changes in trend. For example, consider the modern western edge of Africa or the southern edge of Australia. This view implies that Triassic truncation of the southwest Cordillera did not involve displacements of large continental masses, but of small slivers or microcontinental masses. These masses may have been necked down and locally detached from North America by the earlier rifting event. The fundamental nature of the presumed Triassic truncation(s) is difficult to resolve owing to the complex tectonics that prevailed along the continental margin throughout Jurassic time. The most important relationship that bears on this problem is that in the Sierra Nevada the oceanic terrane is exotic to North America, and it is juxtaposed against terranes that are visibly and inferentially North American. Thus the approach taken here is that the Foothills suture is the discontinuity that must be emphasized if the history of Mesozoic truncation(s) and inception of volcanoplutonic arc activity are to be understood. The history of the Foothills suture is recorded in the structure and petrology of the early Mesozoic rock assemblages of the Sierra Nevada.

MESOZOIC PETROTECTONIC ASSEMBLAGES OF THE SIERRA NEVADA

Petrotectonic assemblages are defined as rock associations that are diagnostic of specific plate-tectonic regimes (Dickinson, 1972). Specific examples are ophiolite sequences representative of oceanic crust and upper mantle formed along zones of plate divergence, and blueschist-bearing melange belts formed along zones of rapid plate convergence. Caution must be used when interpreting petrotectonic assemblages, however, due to their high mobility along plate junctures. The existence of a given petrotectonic assemblage within an orogenic belt does not imply that the tectonic regime responsible for the development of the assemblage ever existed at that specific location within the orogen. Independent structure-age data must be sought to tie a given petrotectonic assemblage to a well-defined point in time and space within an orogen.

The tectonostratigraphic belts of the Sierran collage consist of different petrotectonic assemblages that originated both within and at some distance from the Sierran terrane. These assemblages are covered in a petrotectonic context since well-defined stratigraphic nomenclatures do not exist on a regional scale, and since the tectonic significance of each assemblage is the point of interest here. Stratified rock sequences from each belt are named informally as shown on Figure 6–1. General descriptions, age constraints, and local formal and informal names of the sequences are given in Table 6–2. Strata that occur to the east of the Foothills suture include the Jura, Ritter, and Kings sequences, all mainly of Jurassic age. The Jura and Ritter sequences are primarily volcanic arc assemblages, whereas the Kings sequence has abundant quartzite, argillite, and carbonate in addition to volcanic arc rocks. The Kings sequence laps across the Foothills suture. Its western facies is much like the Calaveras sequence, which sits west of the suture. In addition to quartzite, argillite, carbonate, and volcanic arc rocks, the Calaveras and western Kings sequences contain chert. Chert from the Calaveras and Kings sequences is early Mesozoic in age, whereas the carbonates range back to Permo-Carboniferous. The Foothills sequence lies entirely west of the suture. This sequence is also of Jurassic age, and it consists of volcanic arc and epiclastic deposits. The petrogenetic histories of the sequences located to the west of the Foothills suture are intimately related to the basement tectonics. Thus pertinent features of the oceanic basement terrane will be discussed along with the western petrotectonic assemblages.

Oceanic Assemblages

Mesozoic oceanic assemblages of the Sierra Nevada consist of the western ophiolitic basement terrane and overlying pelagic and hemipelagic strata of the Calaveras and western Kings sequences. In the discussion that follows the term hemipelagic is used in a broad sense to denote mixtures of chert, argillite, and variable amounts of continent-derived sandstone of probable submarine-fan origin. The sequence terminology is not used in the discussion of these strata since it includes primitive arc-volcanic rocks that were erupted in both the Calaveras and western Kings sequences during the later stages of hemipelagic sedimentation. The eastern Kings sequence is not treated in this discussion since it lacks chert and was deposited east of the Foothills suture. The eastern and western Kings sequences are intergradational along the area of the Foothills suture. The most distinctive member of the sequence that laps across the suture is a quartz-rich flysch unit.

Ocean floor genesis of the ophiolitic basement terrane came in two major pulses, one in Permo-Carboniferous time and the other in Triassic time (Saleeby and Moores, 1979; Saleeby and Sharp, 1980; Saleeby and Behrman, in press). The ophiolitic basement was disrupted and metamorphosed prior to deposition of most of the hemipelagic strata. In the Kings-Kaweah ophiolite belt, Permo-Carboniferous oceanic crust was disrupted by faulting and serpentinite protrusion, and then it was the site of a Triassic pulse of syntectonic ocean-floor genesis. As Triassic magma rose into the zone of ocean-floor disruption, much of it crystallized as gneissic gabbro and greenschist-amphibolite facies tectonics. Triassic-age submarine volcanism and pelagic sedimentation were accompanied by massive flows of sedimentary serpentinite. The diachronous nature of the Kings-Kaweah ophiolite belt, along with its structural and petrologic features has led to the conclusion that it represents an ancient oceanic fracture zone of considerable size. This conclusion is based on comparisons with marine geological data on modern fracture zones (Saleeby, 1977b, 1978a, 1979). The main structural pattern within the Kings-Kaweah belt is coeval northwest penetrative shear and flattening surfaces with down-dip lineations, and northeast faults, fractures, and dikes. This pattern permits a model with the fracture zone trend oriented along the ophiolite belt with a Triassic-age ridge axis trending at a high angle to it.

Most geochronological relations that are well established in the Kings-Kaweah ophiolite belt have emerged from other Foothills ophiolite exposures, and thus they appear applicable for the entire oceanic basement terrane (Table 6-1). These age relations, along with the fact that similar hemipelagic strata occur along the entire terrane, imply that the ophiolitic terrane is the product of one ocean-floor emplacement episode. An important aspect of the Sierran ophiolite geochronology is the existence of a Triassic-age regional deformation and amphibolite facies metamorphic episode. This episode is now recognized and dated at every major exposure of western oceanic basement rock (Ehrenberg, 1975; Morgan, 1976; Weisenberg and Ave'Lallemant, 1977; Behrman, 1978; Saleeby and Moores, 1979; Saleeby and Sharp, 1980). It should be recognized as the major Mesozoic regional deformation-metamorphic episode of the Sierran Foothills, but most workers in Sierran geology still consider structures developed during this episode to be Nevadan or Late Jurassic in age. The main problem has been that most workers have not seen the collated body of geochronological data. Another problem is that Nevadan structures in overlying Jurassic-age strata followed the trends of the Triassic-age structures in their ophiolitic basement. Thus age assignments of structures based on orientation without the aid of geochronology have been quite misleading. The pervasive northwest grain of the Sierran Foothills metamorphic belt was established during the Triassic episode. Later Nevadan deformations followed this trend, but did not reorient structures into this trend. This can be demonstrated at each major occurrence of oceanic basement rock on structural, textural, and geochronological grounds. Later deformations may have steepened the Triassic structures by rigid body rotations, but they did not penetratively transpose them. The Triassic deformation-metamorphic episode is here equated with emplacement of the western oceanic terrane into the continental margin domain. A curious relationship that arises is that the second pulse of igneous petrogenesis in the Kings-Kaweah ophiolite belt was nearly synchronous with the emplacement episode. This suggests a complex geometry of plate junctures during ophiolite emplacement that probably involved transform, spreading, and subducting links. Alternative possible geometries will be presented later after the larger-scale tectonic framework of the Sierran region is discussed.

Oceanic sedimentation progressed throughout the genesis, deformation, and metamorphism of the ophiolitic basement, as shown by the occurrence of pelagic sediments in

TABLE 6-2. General Descriptions, Age Constraints, and References for Mesozoic Metasedimentary and Metavolcanic Sequences of the Sierra Nevada

Informal Name	*Lithologic Character of Protoliths*	*Local Formal or Informal Name*	*Local Age*	*References*
Foothill Sequence	Andesite tuff, breccias, flows, dacite, quartz keratophyre, tuffaceous chert, phyllite, massive basalt	Franklin Cyn., Duffey Dome and Bloomer Hill Fms.	159 m.y.	Hietanen, 1973, 1976, 1977; Bond and others, 1977; Saleeby, unpub. data
	Mafic to silicic hypabyssals, pillow basalt, andesite, dacite, flows, breccias, tuffs	Smartville ophiolite	159–164 m.y.	Bond and others, 1977; Xenophontos and Bond, 1978; Saleeby and Moores, 1979
	Andesite, basaltic andesite, tuff, breccia pillow lava; argillite, lithic sandstone	Logtown Ridge and Mariposa Fms.	Call.-Kim.	Clark, 1964; Behrman and Parkison, 1978
	Basaltic-andesite tuffs, breccias, pillow lavas, dacitic tuffs, flows, argillite, lithic sandstone, argillite	Gopher Ridge-Copper Hill volcanics-Salt Spring slate	Call. (?)-Kim.	Clark, 1964; Behrman and Parkison, 1978
	Basaltic-andesite tuff, breccia, pillow lava; chert	Peñon Blanco volcanics	L. Jurassic	Clark, 1964; Morgan, 1973, 1976; Morgan and Stern, 1977; D. L. Jones, pers. commun., 1978
Jura Sequence	Andesite, dacite breccias, tuffs, flows argillite, lithic sandstone, conglomerate; deposited on Upper Triassic marine strata	Mt. Jura group	L. Jurrasic-Call.	Crickmay, 1933; Imlay, 1961; McMath, 1966
	Basaltic-andesite tuffs, breccias, flows, limestone, conglomerate	Milton Fm.	L. Jurassic	Clark, 1976
	Argillite, andesite tuff, lithic sandstone	Sailor Cyn. Fm.	L-M. Jurassic	Clark and others, 1962; Clark 1976
Ritter Sequence	Andesite, basalt tuff-breccia, flows, lithic sandstone, conglomerate	Mt. Tallac roof pendant	L. Jurassic (?)	Loomis, 1966
	Rhyolite, dacite tuffs, ash flows, lithic sandstone, argillite, conglomerate	Mt. Dana-Saddlebag Lake	No data	Kistler, 1966b; Fiske and Tobisch, pers. commun., 1977
	Rhyolite, dacite, andesite tuffs, ash flows, breccias, basalt flows, limestone	Ritter Range roof pendant	Triassic (?)-M. Jurassic	Rinehart and others, 1959; Jones and Moore, 1973; Fiske Tobisch, 1978
	Rhyolite ash flow	Ritter Range caldera fill	101–98 m.y.	Fiske and Tobisch, 1978
	Rhyolite, dacite, andesite, tuffs, ash flows, argillite, limestone	Mt. Morrison roof pendant	No data	Rinehart and Ross, 1964
	Rhyolite, dacite tuff, ash flows, basalt flow, limestone	Mt. Goddard roof pendant	M. - L. Jurassic	Bateman and Moore, 1965; Saleeby and others, 1978, unpub. data
	Rhyolite, dacite, basalt tuffs and flows	Sierra crest roof		

TABLE 6-2 (continued) General Descriptions, Age Constraints, and References for Mesozoic Metasedimentary and Metavolcanic Sequences of the Sierra Nevada

Informal Name	*Lithologic Character of Protoliths*	*Local Formal or Informal Name*	*Local Age*	*References*
		pendants		1978
Kings Sequence	Quartzite, argillite, limestone, silicic tuff	San Joaquin River pendants	No data	Bateman and Clark, 1974
	Quartzite, argillite, limestone, dacite tuff, basalt	Kings River roof pendants	E. Jurassic	Macdonald, 1941; Kistler and Bateman, 1966; Moore and Marks, 1972; Jones and Moore, 1973; Saleeby and others, 1978
	Quartzite, argillite, limestone, rhyolite, dacite tuffs, ash flows; basalt flows, chert	Kaweah River roof pendants	L. Triassic-M. Jurrassic	Durrell, 1940; Ross, 1958; Christensen, 1963; Saleeby and others, 1978; Saleeby and Sharp, 1979
	Quartzite, argillite, limestone, tuffaceous sandstone	Tule River roof pendants	E. Mesozoic	Saleeby and others, 1978
	Quartzite, argillite, limestone, rhyolite, dacite tuffs, breccias; basalt flows, locally tuffaceous quartz sandstone	Kern River roof pendants	L. Triassic-E. Jurassic	Saleeby and others, 1978
	Quartzite, limestone, argillite, tuffaceous sandstone	Tehachapi roof pendants	No data	Saleeby and others, 1978
Calaveras Sequence	Chert, argillite, quartzose sandstone, limestone lenses	Northern Calaveras Fm.	Triassic	Hietanen, 1973; Clark, 1976; D. L. Jones, pers. commun., 1978
	Quartzite, chert, argillite, siltstone, limestone lenses, mafic volcanics	Calaveras Complex	Permo-Carb-E. Mesozoic	Clark, 1964; Schweickert and others, 1977; Sharp and Saleeby, 1979
	Quartz sandstone, argillite, chert, limestone lenses	Western Calaveras belt	E. Mesozoic	Clark, 1964; Duffield and Sharp, 1975; Behrman, 1978

ophiolitic melange and as an important component in the younger hemipelagic strata. The first signs of terrigenous sediment in the Kings-Kaweah ophiolite belt are argillite interbeds within radiolarian chert depositionally above Triassic pillow lava. Overlying hemipelagic strata of the western Kings sequence escaped the 200 m.y. metamorphic episode. Whether these strata and similar Calaveras strata were deposited in trench, transform axial deep, or forearc basin settings is unclear. Their terrigenous components probably represent outpourings from submarine fans built off the edge of North America. The transition from pelagic to hemipelagic sedimentation roughly coincided with ophiolite.

An outstanding feature of the hemipelagic strata is their widespread soft sediment deformation (Schweickert and others, 1977; Saleeby and others, 1978). Numerous olistostromes are present, some apparently recycled. Clasts and large blocks of shallow-water limestone occur within some olistostromes, and locally ophiolite assemblage blocks and serpentinitic detritus shed from basement exposures also occur within the chaotic deposits. The presence of the olistostromes and the ophiolitic detritus point to the tectonic unrest of their depositional environment. A number of the shallow-water limestone blocks contain Permian fauna. Some of these blocks may be tectonic in origin, but most of them appear to be olistoliths. The Permian fauna are found to be mainly genera of the Fusulinid family Verbeekinidae, which are exotic to Permian North America and most closely resemble fauna of the Tethyan realm (Douglass, 1967; Saleeby and others, 1978).

The Permian age of the fauna from the limestone blocks has been traditionally used for the age assignment of the Calaveras sequence. The recognition of these blocks as exotic to their enclosing hemipelagic material now shows that the Permian ages are actually maximum possible ages for the hemipelagic deposits. In Calaveras sequence rocks east of the southern Melones fault, the larger bodies of Permo-Carboniferous limestone are interlayered with mafic metavolcanic rock. Such associations probably represent the vestiges of ancient seamounts. Nearby, smaller limestone blocks encased in hemipelagic strata may represent slide blocks derived from such seamounts. Similar exotic limestone bodies occur in latest Paleozoic to mainly early Mesozoic pelagic and hemipelagic strata along much of the Cordillera to the north (Monger, 1977; Davis and others, 1978). These limestone bodies are generally less metamorphosed than the Sierran bodies, and can be shown to occur either as isolated blocks within distinctly younger chaotic rocks or as the caps of ancient seamounts that were displaced from equatorial latitudes. It seems likely that the "floating" limestone blocks of the Sierran Foothills and their relatives to the north were derived as slide blocks from seamounts that were being accreted or were en route to the margin of North America. It is thus reasonable that the limestone blocks are consistently older than the enclosing hemipelagic strata. Ongoing radiolarian biostratigraphic studies in the California region and throughout the western Cordillera indicate that Calaveras and similar cherts are primarily early Mesozoic in age and only locally Permian in age (Irwin and others, 1977; D. L. Jones, personal communication, 1978; Seiders and others, 1979). The results of these studies, in conjunction with the fact that hemipelagic rocks of the Calaveras and western Kings sequences appear to have escaped the Triassic amphibolite facies metamorphism that was so widespread in their ophiolitic basement, are the rationale for treating all Calaveras and Kings sequence cherts as early Mesozoic in age.

Oceanic assemblages consisting of chert-argillite with exotic limestone blocks and with ophiolitic basement exposures occur in the western Paleozoic and Triassic belt of the Klamath Mountains (Cox and Pratt, 1973; Irwin, 1973; Ando and others, 1976; Wright, 1979). Davis (1969) first suggested the possible correlation between these rocks and the Calaveras sequence of the northern Sierra. As with the Calaveras, the Paleozoic ages from the Klamaths initially were based on exotic limestone blocks, but new radiolarian biostratigraphic data include only early Mesozoic ages.

The general story told by the oceanic petrotectonic assemblages of the Sierra Nevada is that Permo-Carboniferous to Triassic ocean floor with possible north-south trending fracture zones and limestone-capped seamounts received Permian (?) to mainly early Mesozoic pelagic deposits as it approached the ancient California continental margin. During Triassic time fragments of this ocean-floor assemblage were emplaced into the continental margin domain. At about the same time, continent-derived sediments were added to the oceanic sedimentary sequence as submarine fans and olistostromes. The terrigenous materials were interstratified and mixed with pelagic materials. Penecontemporaneous deformation and mixing appears to have been a widespread phenomena in these hemipelagic deposits. Exotic shallow-water limestone blocks were incorporated into the chaotic deposits in a fashion that is presently unresolved. They may have slid off of seamounts in Paleozoic ocean floor that has accreted farther north in the Cordillera. Or perhaps they were tectonically detached during ocean-floor accretion and then subsequently reworked as the hemipelagic deposits accumulated. These deposits for the most part escaped the regional high-grade metamorphism that nearly pervades the ophiolitic basement. This suggests that sedimentation and tectonic intercalation of the hemipelagic strata followed ophiolite emplacement. The near synchrony of the second pulse of ocean-floor genesis with ocean-floor emplacement calls for a complex geometry of plate junctures. The details of this geometry may be irresolvable owing to later structural and metamorphic overprints.

Melange Assemblages

Melanges occur in several different settings within Sierran country rocks. The Calaveras sequence contains significant domains of melange. Schweickert and others (1977) consider most melanges of the Calaveras exposed east of the southern Melones fault as tectonically deformed olistostromes. However, zones of tectonic melange do occur in numerous places throughout the Calaveras sequence, and some contain admixtures of serpentinite and other ophiolitic basement rock (Duffield and Sharp, 1975; Behrman, 1978; Sharp and Saleeby, 1979). Some of the Calaveras tectonic melanges exposed east of the southern Melones fault were formed under medium- to high-grade metamorphic conditions during the emplacement of syntectonic Jurassic plutons. Both tectonically deformed olistostromes and synplutonic melanges of high metamorphic grade also occur in the Kings sequence (Saleeby and others, 1978).

The Kings-Kaweah ophiolite belt is an ophiolitic melange (Saleeby, 1977a, 1978a, 1979). Internally stratified slabs and monolithologic blocks range from 20 km in length down to pebble size and sit in a deformed serpentinite matrix. Melange mixing occurred both by tectonic processes under high-grade metamorphic to cold, brittle conditions, and by sedimentary processes related to protrusion of ultramafic rock onto the sea floor. Ophiolitic rocks of the Tuolumne complex are partly serpentinite matrix melange, and the Bear Mountains belt is internally disrupted, but not to the extent of being a tectonic melange. The ophiolitic melanges and broken formations of the western oceanic terrane were internally mixed without significant involvement of terrigenous materials. Local mixing of such materials into disrupted ophiolite can usually be shown to be of later fault or sharp in-fold origin. Internal mixing of the Kings-Kaweah belt is thought to have been primarily a fracture-zone process. This is based on the fact that mixing commenced during the igneous petrogenesis of the ophiolite belt, and that oceanic sedimentation followed or progressed during the mixing process. An actualistic explanation for this fact must favor a large oceanic fracture zone over a spreading ridge or an oceanic trench if the pertinent marine geological data are considered. A possible alternative to the fracture-zone model is a ridge migrating along a trench. If such were the case, an extended period of pelagic sedimentation must have passed within the trench environment prior to ophiolite emplacement, or pelagic sedimentation rates were extremely high. A combination of fracture zone and subduction tectonics may have driven the mixing processes in the more northern ophiolitic basement exposures. Behrman (1978) concluded that disruption and metamorphism of the Bear Mountains belt could have occurred by either process. Disruption and mixing of the ophiolitic basement terrane for the most part predated deposition of the Calaveras and Kings sequence hemipelagics, as shown by the paucity of such materials in the ophiolitic melanges and metamorphic tectonite zones.

The Melones fault zone is in part a melange zone. The fault zone ranges between 100 m and several kilometers in width, with individual blocks ranging up to 15 km in length (Clark, 1960, 1964). It contains fragments of the adjacent Foothill and Calaveras sequences, abundant serpentinite and other ophiolitic basement rocks, granitic plutonic rocks, and an array of exotic elements that lack obvious sources. Movement along the Melones zone took place under upper greenschist facies conditions and, as discussed later, involved both strike-slip and reverse components.

Melange assemblages of the Sierra Nevada differ from those of the Franciscan Complex in the California Coast Ranges. Rather than being a vast terrane of imbricated ocean floor and submarine-fan slices showing widespread brittle and ductile shear, the Sierran melanges are either ophiolitic and thus oceanic in origin, deformed olistostromes, the products of synplutonic deformation at elevated temperatures, or local mixtures along relatively narrow fault zones. This points to important differences between the manner in which the Franciscan and Sierran structural frameworks evolved.

Paired and Unpaired Metamorphic Belts

Metamorphic country rocks of the Sierra Nevada along with metamorphites of the Franciscan Complex represent a classic paired metamorphic belt (Hamilton, 1969; Miyashiro, 1961, 1973). The Sierran country rocks display a low-P/T facies series with most rocks in hornblende hornfels, or amphibolite facies, whereas Franciscan metamorphites display a high P/T series. The concept of paired metamorphic belts for the Franciscan-Sierran pair is only applicable for Late Jurassic and Cretaceous time, however. Franciscan metamorphites yield K-Ar ages that range from 150 to 70 m.y. (Coleman and Lanphere, 1971; Suppe and Armstrong, 1972). The subduction regime represented by the Franciscan melanges and metamorphites is related in time and space to Late Jurassic and Cretaceous volcanoplutonic arc activity and associated regional thermal metamorphism of the Sierra Nevada and/or its lateral equivalents. Regional thermal metamorphism in the Sierras gave rise to both strongly schistose and hornfelsic rocks.

Pre-Upper Jurassic volcanoplutonic arc and associated metamorphic rocks of the Sierran region are not paired with a high-P/T belt. However, blocks of lawsonite-bearing blueschist of probable Triassic age occur along a 12-km stretch of the northern Foothills suture (Schweickert, 1976c). This isolated occurrence within the long-active suture does not constitute an extensive belt having genetic linkage to the Sierran arc and its thermal metamorphic derivatives, however. This occurrence suggests that at some point in its history, the suture either had a significant component of subduction along it, or the suture cut into and incorporated a slice of a preexisting subduction complex. Alternatively, Late Jurassic movement along the Melones fault, which branches into the northern Foothills suture, could have carried the blueschist in from an older subduction complex. The northern Sierra blueschist locality most likely represents the southernmost vestige of a discontinuous Middle and Late Triassic blueschist belt that occurs in similar structural positions in the Klamath Mountains (Hotz and others, 1977) and in melanges of eastern Oregon, Northern Washington, and British Columbia (Davis and others, 1978). The main manifestation of this belt in the Sierran region appears to be the Triassic metamorphic tectonites of the oceanic basement terrane.

Metamorphic tectonites of the Sierran ophiolites represent an intermediate P/T facies series, as shown by mineral assemblages and phase chemistry (Behrman, 1978; Saleeby and Sharp, 1980). As mentioned previously, the ophiolitic metamorphics are considered a product of continental margin emplacement. Such a pronounced emplacement event calls for a distinctive change in the plate kinematics along the Triassic California margin. This is thought to represent the onset of east-directed subduction along the trend of the Foothills suture. The apparent gradation from primarily intermediate P/T rocks in the Sierran region to high P/T rocks to the north suggests lateral variation along the early Mesozoic subducting margin. Such variation is also suggested by the near synchrony of Triassic-age ophiolitic magmatism in the southern Sierra with the regional ophiolite emplacement episode and by the existence of structural and petrologic features typical of fracture zones. The evolution of a major transform zone with an adjacent active ridge segment into a subducting juncture is proposed later for the plate juncture history of the Sierran ophiolites. Such a tectonic setting seems most conducive for the development of intermediate P/T facies rocks of the ophiolitic basement terrane.

The metamorphic tectonites of the Sierran ophiolites along with remnants of regional contact metamorphic rocks of the eastern Sierra region constitute a Late Triassic to earliest Jurassic unpaired metamorphic belt. The unpaired nature of this metamorphic regime continued until Late Jurassic time, when pairing with the Franciscan high P/T belt was established. The entire Sierra Nevada was within the volcanoplutonic arc domain following the regional ophiolite emplacement episode. Any 190 to 150 m.y. high P/T metamorphic rocks that may have formed in the interlude between Foothills ophiolite emplacement and the Franciscan regime are now missing. This important point will be discussed later.

Volcanic Arc Assemblages

The Sierra Nevada lies within a broad northwest-trending zone that was affected by numerous pulses of Late Triassic through Cretaceous volcanoplutonic arc activity. The volcanic members of the composite arc are preserved in the metamorphic country rocks. The older Jura-Trias volcanic rocks were preferentially preserved rather than the Cretaceous volcanic rocks. Both Jura-Trias and Cretaceous plutonic rocks are well preserved throughout the Sierra Nevada batholith though. The distribution, age, and petrologic character of the Mesozoic volcanic arc rocks are shown in Fig. 6-1 and Table 6-2.

A close correspondence between the composition of the volcanic rocks and basement exists. Silicic volcanic rocks of the Kings and Ritter sequences were erupted in shallow water and subaerially with numerous ash flows. These occur over the southeastern continental basement (Fiske and others, 1977, 1978; Saleeby and others, 1978). Intermediate and lesser amounts of mafic volcanic rocks of the Jura sequence were erupted mainly as submarine breccias and tuffs over the northeastern transitional basement (McMath, 1966; Rogers and others, 1974; Clark, 1976). Mafic and intermediate with rare silicic volcanic rocks of the Foothills, Calaveras, and western Kings sequences were erupted under marine conditions as pillow lavas, breccias, tuffs, and local domes, and are sited over the western oceanic basement. Thus shallow-water to subaerial ash flows are the main expression of arc volcanism where the arc was rooted in the continent, whereas northward across the Paleozoic continental edge, and westward across the Foothills suture, arc volcanism is expressed primarily by basalt-andesite submarine eruptions. It is apparent that the bulk composition, environment of deposition, and eruptive mode of the Sierran arc rocks were controlled by the nature of the arc basement in a similar fashion as in modern arc systems, such as Tonga-Kermadec to North Island, New Zealand, and the Aleutians (Carmichael and others, 1974).

Not only were the Sierran arc rocks erupted over a variety of basements, but the nature of the depositional interface and the nonvolcanic admixtures also varied considerably. In the western oceanic terrane in the south, Lower to Middle Jurassic mafic to intermediate arc rocks of the western Kings sequence were deposited on a rugged substrate with reworked hemipelagic strata and uplifted ophiolitic melange. These arc rocks were interstratified with volcaniclastic and quartzitic flysch derived both from silicic volcanic centers to the east and the North American craton (Saleeby and others, 1978). Farther north, arc rocks of the Calaveras sequence were interstratified with hemipelagic and continental clastic strata, and similar Lower Jurassic rocks of the Foothills sequence were erupted across the highly deformed Tuolumne peridotite complex (Morgan, 1973, 1976; Sharp and Saleeby, 1979). Middle and Upper Jurassic arc rocks of the Foothills sequence were erupted primarily in the basin and slope facies of a submarine-fan system composed of continental detritus (Behrman and Parkison, 1978). Similar age Foothills arc rocks were also deposited across juvenile crust that formed by intra-arc rifting. This is shown in the Smartville complex, which had its ophiolitic stratigraphy generated in late Middle Jurassic time, and which contains Foothills sequence volcaniclastic and epiclastic rocks within and above its pillow basalt section (Xenophontos and Bond, 1978; McJunkin and others, 1979; Saleeby and Moores, 1979). Smartville complex dike rocks can be traced away from the main ophiolitic block, where they cut older melange consisting of Calaveras and ophiolitic basement rocks.

The arc rocks of the western terrane were thus deposited over a rugged surface in which oceanic basement uplifts were common, submarine-fan deposition was active, and intra-arc rifting was occurring. The eastern terranes were also tectonically active, as shown by uncomformable relations between arc and older basement rocks. Within the northeastern terrane at its northern end, Lower to Upper Jurassic arc rocks of the Jura sequence rest on Upper Triassic marine strata, which in turn rest uncomformably on Permian arc rocks of the basement (McMath, 1966; Bond and others, 1977). Southward there are additional localities where Jura-Trias arc rocks of the Jura and Ritter sequences rest unconformably on Triassic and Paleozoic marine strata (Clark and others, 1962; Brook and others, 1974; Clark, 1976). Farther southeast well within the southeastern basement terrane, the basal contacts of the Jura-Trias Ritter and Kings sequences are not preserved. The arc rocks of this terrane were erupted mainly under shallow marine conditions inherited from the Paleozoic shelf (Fiske and Tobisch, 1978; Saleeby and others, 1978). Here, interbeds of shallow-water limestone are common, and in the Kings sequence significant quantities of craton-derived quartz sandstone are also interstratified with the arc rocks. Within the silicic metavolcanic Ritter Range pendant, a major unconformity separates the highly deformed Jura-Trias section and a Cretaceous caldera fill (Fiske and others, 1977). The latter represents the only remnants of Cretaceous volcanic rock thus far recognized in the Sierra.

Jura-Trias arc rocks similar to those of the Jura and Foothills sequences occur to the north in the Klamath Mountains (Irwin, Chapter 2, and Burchfiel and Davis, Chapter 3, this volume). Triassic and Lower Jurassic arc rocks of the eastern Klamaths were deposited on a basement that is similar to the northeastern Sierran basement terrane (Davis, 1969; Irwin, 1977), whereas Middle and Upper Jurassic arc volcanic rocks of the western Klamaths were deposited on a Tethyan-affinity oceanic basement. Thus a discontinuity similar to the Foothills suture passes through the Klamath Mountains. As in the Sierra Nevada, arc volcanic rocks overlying each basement terrane are interstratified with continental detritus-bearing epiclastic strata. Part of the Upper Jurassic arc volcanic-epiclastic assemblage of the western Klamaths occurs as the upper conformable member of the Josephine ophiolite (Harper, 1979). This relationship is similar to what is observed in the Smartville ophiolite, which is similar in age to the Josephine ophiolite (Saleeby and others, 1979). Within the California Coast Ranges, significant portions of the Middle Jurassic Coast Range ophiolite also have marked similarities to the coeval Smartville complex, which include abundant silicic intrusives and upper stratigraphic intervals of Upper Jurassic andesitic volcaniclastic strata (Hopson and others, 1975a;

Evarts, 1977; Blake and Jones, Chapter 12, this volume). Thus middle to Late Jurassic arc volcanism appears to have been associated with regional scale rifting and ocean-floor generation.

Numerous localities containing Jura-Trias arc volcanic rocks occur east of the Sierra Nevada (Noble, 1962; Merriam, 1963; Crowder and Ross, 1970; Stanley, 1971; Dunne and others, 1978; Speed, 1978). These rocks consist mainly of intermediate to silcic tuffs and breccias that were erupted in a variety of depositional environments and quite commonly interstratified with abundant continent-derived detritus. The continental detritus consists of conglomerates shed off of local tectonic highs and in some locations mature quartz sand derived from the craton. Similar rocks extend southward from the Sierra Nevada into the Mojave region (Bowen, 1954; Hewett, 1956; Grose, 1959; Dunne and others, 1975; Dunne, 1977; C. F. Miller, 1978; Miller and Carr, 1978; Burchfiel and Davis, Chapter 9, this volume). Finally, arc volcanic rocks and interstratified clastic rocks mainly of Jurassic age resting both on older continental and oceanic basements can be traced southward into Baja California (Jones and others, 1976). These rocks are quite likely an extension of the Sierran composite arc.

The general story told by the volcanic arc assemblages is that voluminous arc volcanism commenced in Late Triassic to Early Jurassic time over a broad zone that included all the older Sierran basement terranes, and continued sporadically in time and space until Late Jurassic time. The basement structure was complex, as shown by the variety of rock types underlying the volcanics and the variety in depositional environments. Arc volcanism spread across the continental shelf and slope and onto basin floors. It extended across the Foothills suture, which joined the eastern and western basement terranes. Continent-derived clastic strata were mixed with the volcanics in shallow and deep-water environments on both sides of the suture. As discussed in the next section, the remnants of the roots of the Late Triassic to Late Jurassic arc are preserved in the older portions of the composite Sierra Nevada batholith.

Batholithic Assemblages

The composite Sierra Nevada batholith represents the roots of the volcanic arc rocks discussed previously, and younger arc rocks that have been eroded away, (Hamilton, 1969). The ages of the batholith and volcanic arc rocks both range from latest Triassic to Late Cretaceous. Genetic linkages have been suggested between volcanic rocks and underlying plutons in a number of areas (Schweickert, 1976b; Fiske and others, 1977; Morgan, 1976; Behrman, 1978; Saleeby and Sharp, 1980). Cretaceous batholithic rocks are much more voluminous than the older plutonic rocks, whereas remnants of Cretaceous volcanic rock are rare. Evidently, most of the Cretaceous volcanic rocks have been stripped off by erosion during unroofing of the batholith and its wallrocks, which contain metamorphic remnants of the older volcanic rocks. The discussion of the composite batholith will include plutonic rocks that lie outside the Sierra Nevada range as well as those of the range itself. Such rocks occur to the east of the Sierra Nevada in the region of the White-Inyo Mountains and to the west beneath the Great Valley, as shown by geophysical and basement core data presented by May and Hewitt (1948), Cady (1975), Williams and Curtis (1977), and Saleeby and Williams (1978).

The composite nature of the batholith is displayed by the geochronological data that are summarized in the age belts shown in Figure 6-2. The age belts are modified after Evernden and Kistler (1970), who presented a vast amount of K-Ar data on the batholith. In many cases, zircon U-Pb data provide a better indication of the crystallization age, and in conjunction with K-Ar data on the same rock yield important insights for general interpretations of discordant K–Ar ages (Chen, 1977, personal communication, 1978; Saleeby and Sharp, 1980; Stern and others, in press). The oldest plutons occur in the eastern Jurassic belt. These are primarily Early Jurassic in age, but several dates are latest Triassic. Most plutons of the eastern belt yield Middle and Late Jurassic ages as do most plutons of the western Jurassic belt. A few small mafic intrusions of the western belt are Early Jurassic in age. The eastern and western Jurassic belts were probably once closer together, but they have since been split apart by the Cretaceous batholith. The eastern belt consists primarily of granodiorite and granite with local quartz diorite, but several significant bodies of monzonite and syenodiorite also occur. The western belt consists of gabbro, diorite, trondhjemite, and quartz diorite with local granodiorite, pyroxenite, and wehrlite. Although both belts are predominantly calcalkaline, the western belt along with the Foothills sequence volcanic rocks shows local arc tholeiite affinities (Kemp, 1976; Day, 1977; Xenophontos and Bond, 1978), whereas the eastern belt and adjacent volcanics exhibit some alkalic affinities (Bateman and others, 1963; Dunne and others, 1978; C. F. Miller, 1978). Thus the transverse compositional zonation of the Jurassic volcanoplutonic arc is similar to that reported by Kuno (1966) for modern arc terranes. However, this pattern was probably strongly influenced by the preexisting basement in the Sierras, as shown by the isotopic data on the batholith and the lithologic variations in the wallrocks.

Cretaceous batholithic rocks also show transverse compositional variations. The western Cretaceous belt consists of gabbro, norite, quartz diorite, tonalite, and local granodiorite. These rocks have high-alumina basalt to calc-alkaline affinities (Saleeby and Chen, 1978; Mack and others, 1979; Saleeby and Sharp, 1980). The more mafic members of this suite are nearly identical to cognate cumulate blocks that have been ejected from young island-arc volcanoes (Lewis, 1972; B. D. Marsh, personal communication, 1978). Much of the western Cretaceous belt lies under the Great Valley, as shown on Figure 6-2. The central and eastern Cretaceous belts underlie most of the Sierra Nevada range. These belts consist primarily of granodiorite, tonalite, and granite. The Cretaceous belts display transverse chemical variations that are most strongly shown by the eastward increase in K_2O-SiO_2. The eastern belt contains several extremely large plutons that cover up to hundreds of square kilometers. These batholith-scale plutons occur only over the southeastern basement terrane. They are primarily granodiorite to granite in composition, and they are elongate along the northwest axis of the composite batholith. Such large-scale granitic batholiths are absent in the northeast terrane, where smaller granodioritic to tonalitic plutons prevail. Thus the composite batholith not only shows pronounced transverse compositional variations, but a more subtle longitudinal variation exists along its Late Cretaceous belts. This longitudinal variation is paralleled by compositional differences between the older Ritter and Jura volcanic sequences and by the subtle longitudinal variations in the initial Pb and Sr isotopes.

Genetic linkages between metavolcanic rocks and underlying plutons have been suggested for mainly Jura-Trias arc rocks. This is an effect of the near absence of Cretaceous volcanic rocks. But, in addition to the direct linkage suggested for Cretaceous rocks of the Ritter Range pendant (Fiske and others, 1977), several other points indicate that the Cretaceous batholith was emplaced at shallow levels beneath an active volcanic field. (1) At least two of the batholith-scale granitic plutons (Tuolumne and Whitney) contain phases that vented to the surface (Bateman and Chappell, 1979; J. G. Moore, personal communication, 1978). (2) Evernden and Kistler (1970) have presented a thermal model for Cretaceous plutons that suggests emplacement depths not in excess of 5 km. (3) Throughout most of the Cretaceous age batholith, K-Ar ages are in fairly close agreement with U-Pb zircon ages (Evernden and Kistler, 1970; Chen, 1977; Saleeby and Sharp, 1980), which points to rapid cooling. To rapidly cool such a great expanse of batholithic material requires a shallow emplacement level for nearly the entire batholith. (4) Gabbroic rocks of the western Cretaceous belt appear to have low P phases suggestive of high-level emplacement (Powell, 1978; Saleeby and Sharp, 1980 and unpublished data). Thus, in addition to at least some of the Jurassic plutons of the Sierra being emplaced at shallow levels, it is quite likely that much of the Cretaceous batholith was also emplaced at shallow levels. The surface expression of the composite batholith was probably mafic andesite volcanoes and local silicic domes over the oceanic basement, andesite volcanoes over the transitional basement, and mainly silicic ash flows over the continental basement. There may also have been andesite volcanoes over the continental terrane, perhaps now represented by clusters of smaller, more mafic plutons, but their volume was probably greatly exceeded by ash flows underlain by large-scale plutons. The main volcanic structures of the southeastern terrane were probably cauldrons or volcanotectonic depressions (Williams and McBirney, 1968), considering the shape of the batholith-scale plutons, their shallow emplacement level, and the overall structure of the Sierra.

Magma emplaced into and erupted over the western basement terrane came from the zone of subduction-related melting with little or no continental contamination, as shown by comparisons with modern oceanic arc terranes. Magma emplaced into the northeastern terrane apparently acquired the signature of that terrane in its more overall silicic composition and its more radiogenic initial Sr and Pb. Batholithic rocks of the southeastern terrane acquired an important component of older continental crust, as shown by their silicic composition (Presnall and Bateman, 1973) and highly radiogenic initial Pb and Sr isotopes (Kistler and Peterman, 1973; Chen and Tilton, 1978). The role of deep-level magmatism in the transfer of heat and material to the lower crust of the eastern terranes is recorded in distinct assemblages of mafic igneous and meta-igneous rocks termed "basic forerunners" by Mayo (1941). Preliminary geochronological work along with field relations suggests that the mafic rocks were emplaced at numerous times during the history of the batholith, and not as distinct "forerunners" of the batholith (Moore, 1963; Saleeby and Sharp, 1980 and unpublished data). The mafic rocks occur in several different settings, which point to their genetic significance: (1) distinct intrusive bodies in pendants or septa of country rock, which include gabbro, diorite, anorthosite, pyroxenite, hornblendite, and mafic quartz diorite; (2) deformed and metamorphosed screens of these rock types that occur between silicic plutons; (3) granitized and highly deformed screens between silicic plutons; and (4) deformed blocks and smaller inclusions that have mineralogically equilibrated with enclosing silicic plutons. Geophysical data and the composition of xenoliths present in Cenozoic volcanic rocks of the region indicate that the lower crust of the Sierra Nevada is mafic in composition (Cady, 1975; Oliver, 1977; Bateman, Chapter 4, this volume; P. C. W. Dodge, personal communication, 1978). The "basic forerunner" assemblage and the existence of a lower mafic crust, together with the fact that mafic and intermediate magmatism predominated in the western terrane, which lacked continental crust, point to the importance of mafic magmatism in the development of the entire batholith. Mafic magma derived from depth was emplaced beneath, into, and through the southeastern continental terrane. This magmatism promoted voluminous partial melting of the older crust, which gave rise to large silicic batholiths and ash flow sheets. The present lower mafic crust is probably a combination of underplated mantle-derived metaigneous material and refractory material left from the preexisting crust. Vestiges of such refractory material are common as mafic inclusions or restite within the darker granitoids. These inclusions are notably lacking in high-Al minerals and are enriched in Ca, Mg, and Fe, which is indicative of an igneous or I-type source (White and Chappell, 1976). Granitoids of the western Cretaceous belt that are closely related to gabbroids lack restite, which is consistent with their genesis directly from a mantle source.

The 150 m.y. history of intense Mesozoic magmatism in the Sierra Nevada is manifest in its mature crustal structure. In Fig. 6-4 the gross crustal structure of the Sierra Nevada (Cady, 1975; Oliver 1977) is compared to the structure of the immature Kermadec arc and the mature Japan and Chilean Andes arcs (Karig, 1970; Grow and Bowin, 1975; Uyeda, 1975, 1977). The sequence from Kermadec to the Chilean Andes points to arc maturation as proceeding by the underplating of high-density mafic material, and the growth of an upper crustal intermediate to silicic layer by differentiation of the mafic material and by remobilization of preexisting and/or newly formed crust. Using these three active arcs as standards, the Sierra Nevada compares most closely to the mature end of the spectrum.

Cretaceous magmatism started along the western margin of the batholith and migrated eastward almost continuously for about 50 m.y. (Fig. 6-2). The first magmas were emplaced into the tectonically accreted oceanic basement and its Jurassic arc assemblage. Since the first Cretaceous magmas were mafic to intermediate in composition, and their wallrocks were mainly mafic, the lower crustal mafic layer of the Sierra in effect surfaces along the western Foothills (Fig. 6-4A). Later Cretaceous magmatism involved progressively more preexisting crust during eastward migration from the Foothills region. Similar time-space patterns are common in other composite batholiths and have been explained as a result of progressive trenchward displacement of the older batholithic material during dilation of the arc (Gastil, 1977; Gastil and Rowley, 1978). This dilation results from the upward and outward redistribution of silicic material and the underplating of more mafic material derived from depth. From the age date summarized in Fig. 6-2, it does not appear that Jurassic magmatism followed this pattern, but this may be a result of inadequate sampling. As discussed later, there were several longitudinal structural

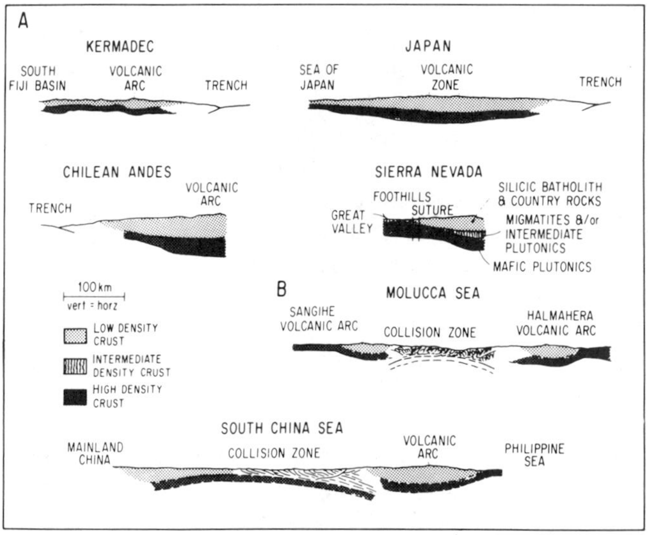

Fig. 6-4. Present-day crustal sections through (A) volcanoplutonic arc terranes, and (B) collision zones. Kermadec arc after Uyeda, (1977); Japan arc, Uyeda (1977); Chilean Andes arc, Grow and Bowin (1975); Sierra Nevada, Cady (1975), Oliver (1977); Molucca Sea collision zone, Silver and Moore (1978); and South China Sea collision zone, Liou and others (1977).

breaks that were active during and after emplacement of the Jurassic belts; accordingly, present-day transverse sections are a poor representation of the Jurassic configuration. There is also reason to believe that the age and extent of the older belts are at present poorly known. There are numerous screens of probable Jurassic-age plutons lodged between plutons of the Cretaceous belts; geochronological data are lacking on such screens. Furthermore, K-Ar data on the older plutons are difficult to interpret, and as of yet few U-Pb data exist. Where U-Pb work has been done, older ages have emerged (Chen, 1977, personal communication, 1978; Sharp and Saleeby, 1979; Saleeby and Sharp, 1980, Stern and others, in press).

The country rock framework of the Sierra Nevada batholith was characterized by significant structural breaks. The breaks not only separated the tectonostratigraphic belts, but they served as passageways for magmatic transport. A similar structural control for batholith emplacement has been suggested for the great batholiths of the Peruvian Andes (Pitcher, 1978), and may be applicable for other segments of the Mesozoic North American batholithic belt. Members of this batholithic belt, which neighbor the Sierran segment, include the southern California and Idaho batholiths and plutonic rocks of the Salinian block, Klamath Mountains, and western Nevada (Hamilton, 1969, 1978a). The batholiths along with coeval volcanic rocks are the signature of Late Triassic to Cretaceous subduction-related arc activity that was superimposed over a highly deformed and actively deforming continental margin.

CONSTRUCTION OF THE SIERRAN COLLAGE

The tectonostratigraphic belts of the Sierran collage contain different variations of the Mesozoic petrotectonic assemblages and basement rocks. The belts represent only small shreds or remnants of the original assemblages and their basement terranes; thus more of the story is missing than is present. In this section the construction mechanisms that brought these remnants together to form the Sierran collage are considered. Of major importance is the time and manner in which the exotic oceanic basement terrane was accreted, and how the accretionary processes related to the evolution of the composite volcanoplutonic arc. Earlier one-dimensional plate interaction models view the Foothills sequence as an exotic arc that was accreted along with the oceanic terrane, whereas later models have viewed the entire volcanoplutonic arc assemblage as having evolved across the previously sutured continental margin.

One-Dimensional Plate Interaction Models

Plate-tectonic models using one-dimensional plate interactions operating normal to the Sierran structural trend have been proposed for the construction of the Sierran collage (Moores, 1970, 1972; Schweickert and Cowan, 1975; Schweickert, 1978; Hamilton, 1969, 1978a). The bases of these models are oppositely polarized subduction-arc systems with an eastern Andean-type arc built over what is here termed the eastern basement, and a western oceanic arc built over the western basement. They envisage Jurassic convergence and ultimately Late Jurassic collision of the arcs by consumption of intervening ocean floor along oppositely dipping subduction zones. The collision event has been equated with the Late Jurassic Nevadan orogeny, and the collision suture has been considered to be the Melones fault zone.

These models meet with a number of problems.

1. Abundant continent-derived clastic material occurs within hemipelagic and volcanic arc strata of the western Foothills; it is difficult to imagine how the continental detritus could have been transported across two trenches to become interstratified with the hemipelagic and "oceanic" arc strata.
2. Post-Nevadan Cretaceous batholithic rocks that were emplaced into the oceanic basement terrane are as "oceanic" or primitive in character as the western Jurassic volcanoplutonic arc rocks; thus the petrogenetic lineage of both the Jurassic and Cretaceous "oceanic" arc rocks can be related to their oceanic basement, rather than a distant oceanic paleogeographic setting.
3. The ophiolitic basement of the "oceanic" arc underwent its continental margin emplacement prior to inception of the arc.
4. The Melones fault zone lies within the western Jurassic plutonic belt and thus could not have been a Late Jurassic subduction zone (Fig. 6-2).
5. The collision volume cannot be accounted for if the Sierran crustal structure is compared with modern collision zones.

Generalized sections of modern arc-continent and arc-arc collisions are shown in Fig. 6-4B. The South China Sea zone represents a collision between an Atlantic-type southeast Asia margin and the Taiwan-Luzon oceanic arc (Liou and others, 1977). This is the only type of modern arc-continent collision known to exist; modern collisions between Andean and oceanic arcs do not exist. The only example of a modern arc-arc collision is in the Molucca sea, where the oppositely polarized Sangihe and Halmahera arcs have met (Silver and Moore, 1978). These represent small oceanic arcs. Compare the crustal structure of the Sierra Nevada in Fig. 6-4A with the structure of the two collision zones in Fig. 6-4B. There is a tremendous deficiency in volume if the Sierra Nevada is to represent two arcs sutured together. The deficiency is greater than what is apparent from Fig. 6-4 since a significant volume of material was added to the Sierra Nevada during the post-Nevadan Cretaceous batholithic activity. Since the "western oceanic arc" and the "eastern Andean arc" are juxtaposed immediately against one another, the frontal arc and subduction complexes of both arcs are missing. From the Sierran crustal structure it can be seen that very little of the missing rock can be accounted for by undertucking. Another alternative is to uplift and erode the missing rock away. This mechanism fails inasmuch as there is no Late Jurassic record of such voluminous erosion and sedimentation, which would be in excess of 10^6 km^3 of rock. Furthermore, the low grade of metamorphism that pervades the Foothills sequence precludes the possibility of those strata having been at the base of an enormous pile of thrust sheets.

The volume deficiency in conjunction with the problems arising from the igneous and sedimentary history of the Foothills necessitates seeking alternative construction mechanisms for the Sierran collage. Such mechanisms must consider two-dimensional plate interactions. Unfortunately, such interactions will in time produce an orogen that defies the application of simple plate-tectonic templates for its analysis (Dewey, 1975). Nevertheless, the genetic implications of the petrotectonic assemblages and the structural configuration of the Sierran collage permit a generalized two-dimensional model constrained by existing data on the relative plate motions between North America and the Mesozoic Pacific ocean floor.

Constraints Imposed by Relative Motion between North America and Mesozoic Pacific Ocean Floor

Inasmuch as the oldest part of Pacific ocean floor is Early Jurassic and is located in the west Pacific (Pitman and others, 1974), detailed reconstructions of early Mesozoic eastern Pacific plate motions are not possible. However, some broad constraints can be placed on the plate motion patterns that affected the early Mesozoic west coast by extrapolation from Middle Mesozoic patterns for which some data are available, and by interfacing these data with what is known about the tectonic development of the greater part of the Cordilleran Orogen.

Plate-tectonic reconstructions of the middle Mesozoic Pacific ocean floor yield a system of spreading centers located to the distant southwest that have generally east-west trends with large north-south fracture zones (Hilde and others, 1977). In addition to a dominant north-south component in spreading from this ridge-fracture zone system, the entire system has apparently migrated 4500 km northward from its middle Mesozoic position (Larson and Chase, 1972). Therefore, since middle Mesozoic time there has been a minimum northward component of 4500 km in north Pacific plate motions. How far back into pre-middle Mesozoic time this pattern extended can only be inferred from the study of on-land oceanic assemblages and the rocks with which they are associated.

The nature of the Cordilleran Orogen changes drastically between northern California and northern Washington. The width of material accreted to the continental margin during Mesozoic time in the north greatly outweighs the material accreted during Mesozoic time in California. Between northern Washington and Alaska, numerous unrelated terranes, several of which are significantly larger than the entire Sierra Nevada, were accreted to the continental margin throughout most of Mesozoic time (Davis and others, 1978; Jones and others, 1978b; Berg and others, 1978). Of these large displaced terranes, the Cache Creek Group lies farthest inboard and was accreted in Late Triassic to Early Jurassic time (Monger, 1977; Davis and others, 1978). The Cache Creek terrane extends for 1300 km between British Columbia and the Yukon Territory. This oceanic assemblage bears shallow-water carbonates with the exotic Verbeekinidae fauna of Mississippian to Permian age. These are tropical reefoidal limestones that were probably deposited on seamounts. Enclosing hemipelagic strata yield radiolaria that range in age from Pennslyvanian to Late Triassic. The Cache Creek Group is thought to represent a fragment of Late Paleozoic equatorial ocean floor that formed within a unique "Tethyan" faunal belt of great lateral extent (Gobbett, 1973; Danner, 1976, written communication, 1977; Monger, 1977, written communication, 1977). Smaller fragments of this group are scattered southward through Oregon and Washington, and are represented in the Sierra Nevada by the Calaveras and western Kings sequences and in the Klamath Mountains by the western Paleozoic and Triassic belts. An important aspect of the Cache Creek terrane and many of its southern equivalents is the association of melange and Late Triassic blueschist blocks. Thus subduction-related accretion of the Cache Creek-type assemblages commenced in Late Triassic time, just as accretion of the western Sierran ophiolitic basement did.

Another displaced terrane of the northwest Cordillera that has direct implications on Mesozoic California is the Alexander terrane of southeastern Alaska. Paleozoic strata of this terrane have strong similiarities to borderland and possibly shelf sequences of the southwest Cordillera, and are known to have undergone northward displacement of about 15° (Jones and others, 1972; M. Jones, and others, 1977). It has been suggested that the Alexander terrane represents a fragment of ancient California. Immediately outboard of the Alexander terrane lies Wrangellia, which consists of voluminous flood basalts and carbonates of Middle and Late Triassic age floored by late Paleozoic island-arc rocks (Jones and others, 1978b). Wrangellia extends at least 2500 km northwestward from the San Juan Islands to central Alaska; it may also extend southward into eastern Oregon. Paleomagnetic data (Packer and Stone, 1974; Hillhouse, 1977) indicate that Wrangellia has undergone significant northward displacements, quite likely from as far as 15° south of the paleoequator. Emplacement ages for Wrangellia and the Alexander terrane are not yet firmly established, but they are bracketed to between Middle Jurassic and Early Cretaceous time (Davis and others, 1978; Jones and others, 1978b).

The three major terranes just discussed, which were accreted to the northwest Cordillera in Mesozoic time, are known to have undergone large northward displacements to reach their present locations. Other major and minor terranes may have also. Structural and petrologic studies of Mesozoic rocks of southwestern Alaska indicate that subduction with a dominate northward component operated there from Late Triassic through Middle Jurassic time, and that intense uplift and unroofing of the arc proceeded until Late Cretaceous time, when northward subduction and arc magmatism resumed (Moore and Connelly, 1979). The relationships in the northern Cordillera are all consistent with the late Mesozoic plate motion patterns recorded on the ocean floor and suggest that northward drift of Pacific ocean floor typified much of Mesozoic time. These general plate motion patterns seem to have had a much different manifestation farther south, as indicated by the following: (1) substantially less mass was accreted to California than in the northern Cordillera; (2) the Sierran and Klamath accreted terranes are oceanic, whereas to the north they are oceanic, island arc, and continental; and (3) California was the locus of continental margin truncation. These observations indicate that the Mesozoic tectonics of the California region were substantially different from those of the northern Cordillera. Tectonic removal processes worked in conjunction with, and may have predominated over, tectonic accretionary processes in California, whereas accretion appears to have predominated in the north. As discussed later, the strong northward component in Pacific ocean-floor motion was probably taken up by transcurrent (transform) faulting and oblique subduction in the California region.

Consequences of Oblique Convergence and Transform Faulting along Continental Margins

Since the advent of modern plate-tectonic theory, much attention has been focused on the recognition of petrotectonic assemblages and their roles in actualistic plate-tectonic models. Petrotectonic assemblages have served as conceptual building blocks for such models. These building blocks consist of rifted continental margins, Andean and oceanic island arcs, marginal and interarc basins, spreading centers, subduction zones, and collisional suture belts. Little attention has been focused on the processes and results of oblique convergence and transform faulting. This is not a result of these processes being insignificant in orogenic areas, but is a result of our approach in modeling ancient orogenic zones. Inasmuch as strike-slip processes do not produce distinctive petrotectonic assemblages, modeling procedures that rely primarily on the genetic implications of the distinctive assemblages will be ignorant of the role of strike-slip processes. One finds, however, that upon studying the modern tectonics of the circum-Pacific, Indian, and Arctic oceans and the Caribbean region, strike-slip components along ancient plate junctures cannot be ignored if our models are to be truly actualistic.

As discussed earlier, there is good reason to believe that the Mesozoic continental margin of California experienced significant components of strike-slip motion in addition to underthrusting. Some attention should thus be given to how this plate motion pattern could have manifested itself within the Sierran region. The best way of doing so is to examine modern and Tertiary orogenic zones where significant strike-slip components can be documented by present-day plate kinematic patterns.

The first-order consequence of oblique subduction is the development of an intra-arc wrench zone; this has been shown to be nearly a one-to-one relationship in modern volcanic arc terranes (Allen, 1962, 1965; Fitch, 1972; Kaizuka and others, 1973; Kaizuka, 1975). The most common expression of the wrench zone is a major transcurrent fault zone, which in most cases runs along the volcanic front. An outstanding example is the Semangko fault zone in Sumatra (Fig. 6-5). The intra-arc transcurrent fault partially or completely decouples the frontal arc and subduction complex from the upper lithospheric plate. The decoupled wedge moves longitudinally with and is underthrust by the converging oceanic plate. The transcurrent zone thus dissipates the strike-slip component of oblique convergence. The normal component of convergence is dissipated by subduction and compression of the wedge and upper plate. Proximity of the major transcurrent zone to the volcanoplutonic arc is attributed to thermal weakening due to arc magmatism. In thermal models of modern arc terranes, Oxburgh and Turcotte (1971) calculate a sharp upward deflection of flat-lying isotherms as they pass from the frontal arc region to the vicinity of the volcanoplutonic front. The sharp deflection yields a

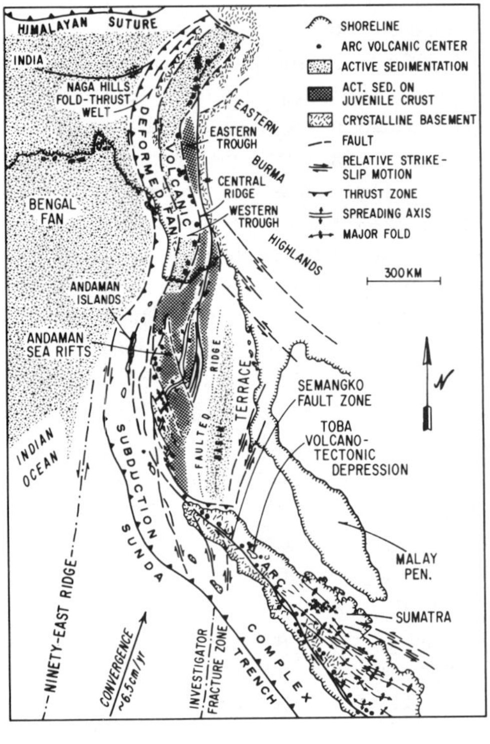

Fig. 6-5. Generalized map showing Late Tertiary to Holocene tectonic features of the Sunda arc region. Data sources: Chhibber (1934), Sigit (1962), Fitch (1972), Sclater and Fisher (1974), Hamilton (1978b), Curray and others (1979), and Karig and others (1979).

zone of high ductility contrast, which is where the main intra-arc break forms. Additional transcurrent faults also occur within and between frontal arc areas and subduction complexes (Fitch, 1970; Karig, 1974; Karig and Moore, 1975a; Karig and others, 1978, 1979; Curray and others, 1979; Karig, 1980). These faults also help dissipate the strike-slip component of oblique subduction.

Transform plate junctures that transect continental margin environments have morphotectonic features that are similar to intra-arc wrench zones except for the absence of an active volcanoplutonic arc. Outstanding examples of such transform junctures are the San Andreas-Queen Charlotte system of western North America (Wilson, 1965) and the Spitsbergen fracture zone and orogenic belt of the Arctic (Lowell, 1972). An important consequence of having several lithospheric plates move on a spherical surface is that major transform junctures in time and space will evolve into and out of oblique convergent and divergent junctures (Dewey, 1975). This is known to be the case in the Cenozoic of California (Atwater, 1970), and is quite likely an important process for much of the circum-Pacific (Hilde and others, 1977).

Second-order consequences of oblique subduction and continental margin transform faulting include several types of morphotectonic features that in serial arrangement are typical of wrench zones. These include extensional basins and fold-thrust welts that form owing to irregularities and terminations in the wrench system, and elongate zones marked by trenches, upthrust and tilted blocks, and fault and fold controlled en echelon ridge-basin systems. All these features can be observed today along the San Andreas transform juncture. In Figure 6-5 a complex wrench system is shown passing through Sumatra, the Andaman Sea, and Burma. This system is inboard of the Sunda trench and has the characteristics of both an intra-arc wrench zone and a transform juncture. The morphotectonic features mentioned previously occur in series along this zone, which not only coincides with the locus of volcanoplutonic arc activity, but is also superposed over Late Tertiary portions of the subduction complex. Note how similar the Andaman Sea rift-transform system is to the Gulf of California-San Andreas system. The same family of morphotectonic features is active in the modern California wrench system, except the subducting trench and volcanoplutonic arc are missing. In the course of evolution of such wrench zones, rock assemblages formed and deformed in one morphotectonic setting are likely to experience the effects of the other settings. In this way, sediment deposited in an extensional basin or young quasi-oceanic crust formed at a rift may shortly thereafter be incorporated into a fold-thrust welt. Conversely, deformed rocks of a fold-thrust welt can be dropped as the floor of a basin and rifted apart with the formation of new crust.

Third-order consequences directly related to wrench tectonics are an array of structural features that in a long-lived wrench zone will lead to a complex pattern of superposed structures. These include synthetic and antithetic strike-slip faults, reverse faults, fold domains, normal faults, and extension fractures (Harding, 1974; Sylvester and Smith, 1976). As deformation progresses, the wrench zone will be segmented into fault slivers by the main wrench fault and its synthetic and antithetic branches. The segmentation facilitates a high degree of mobility within the wrench zone. Slivers or microplates can undergo differential rotations and develop unique structural histories.

These fragments can also be driven out of the wrench zone as thrust sheets with up to tens of kilometers of displacement. Such fold-thrust welts are well displayed in the Transverse Ranges of the San Andreas system and along the Spitsbergen orogenic belt, where the thrust faults root directly into the steeply dipping strike-slip zone (Fig. 6-6a). Thrusting in such zones is driven by mass accumulations at irregularities in the wrench zone, regional compression across the zone, or by thermal-diapiric rise of material within an intra-arc wrench zone. In thermally weakened regions or at depth within a sedimentary basin, rotations can be accompanied by ductile deformation. An important consequence of such ductile rotations is the tightening and transposition of en echelon folds into the trend of the wrench zone. Once transposed, the steep-dipping surfaces are susceptible to buckling about steep axes or to having second-generation wrench folds superposed across them. This process is shown diagramatically in Figure 6-6B, and as discussed later it appears to have been important in the Mesozoic Sierra Nevada.

A significant feature of wrench-zone tectonics is that only a small component of the overall motion is needed to bring basement rocks to high structural levels. Within the Sunda arc, numerous basement exposures occur along the wrench system (Fig. 6-5). These include serpentinite belts and ophiolite remnants, high- and low-P metamorphites, and various granitoids (Chhibber, 1934; Gobbett, 1972; Sigit, 1962; Brookfield, 1977; Bennett, 1978; Jeffrey, 1978; Page, 1978). Similar diversity and involvement of basement rock is typical of the San Andreas system. This is not typical of subduction zones, except in regions starved of terrigenous sediment. The Sunda subduction complex consists of an enormous wedge of accreted sediments with rare slices of crystalline rock (Hamilton, 1973; Curray and Moore, 1974; Karig and others, 1979). Some of these slices appear to be layer 2 of oceanic crust that was imbricated into the accretionary wedge with overlying sediments. However, some of the slices are high-grade metabasites that appear to have migrated upward via strike-slip motion within the subduction complex (Karig, 1980).

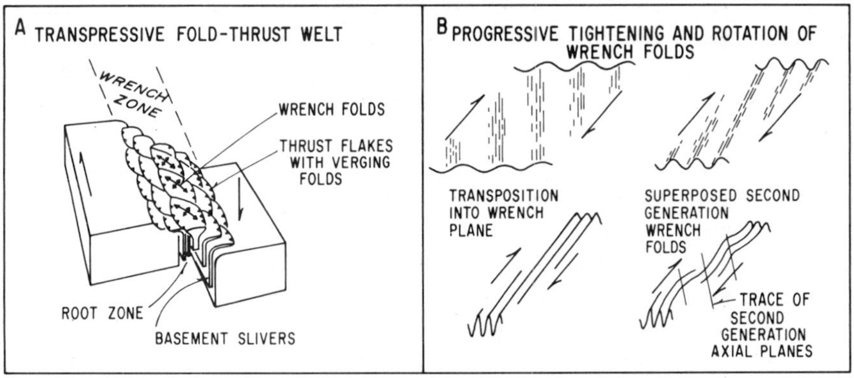

Fig. 6-6. Diagrammatic representation of some folding and thrusting processes typical of transpressive orogens; synthesized from Harland (1971), Lowell (1972), Harding (1974), and Sylvester and Smith (1976). (A) Fold-thrust welt showing fault slivers driven out of root zone as thrust flakes. (B) Basic overprinting pattern in superposed folds in transpression zone of long duration.

Basement exposure via strike-slip transport has also been reported along the Middle America trench, a zone of oblique convergence (Karig and others, 1978). It seems likely that strike-slip motion is of general importance in bringing metamorphosed basement rocks to high structural levels within subduction complexes. It thus seems likely that an important consequence of oblique convergence is basement mobility throughout the entire subduction-arc system.

Mesozoic Structural Framework of the Sierran Region

The main structural features that formed in the Sierran region during Mesozoic time are shown in Fig. 6-7. Most of these structures formed during Late Triassic through Jurassic time. There is little record of the earlier Triassic, possibly because tectonic removal processes dominated, and Cretaceous time was characterized by emplacement of voluminous batholiths. Time constraints and references for the structures of Fig. 6-7 are given in Table 6-3. Also shown in Fig. 6-7 are the Jurassic thrust faults of the Klamath Mountains palinspastically restored to their Early Cretaceous position relative to the Sierra Nevada (Jones and Irwin, 1971; Hamilton, 1978a) and the main Mesozoic structures that formed immediately east of the Sierra in the White-Inyo and neighboring ranges. In Fig. 6-8 a series of diagrammatic cross sections are shown for transverse intervals of the Sierra and adjacent regions. The structural configurations shown in Figs. 6-7 and 6-8 record the effects of dextral transcurrence oriented along the Sierran structural grain and compression across the grain. The term transpression is used to denote the operation of transcurrent and compressional processes in concert. Transpression has been recognized as an important process along young continental margin transform junctures (Harland, 1971; Lowell, 1972; Sylvester and Smith, 1976). The role of transpression along ancient plate edges, particularly within volcanoplutonic arc remnants, has yet to be fully recognized. The modern almost one-to-one correlation between oblique convergence and intra-arc wrench faulting implies that intra-arc transpression was an important process in the past. The Sierran region appears to be an outstanding example; here the thermal and material transport processes of volcanoplutonic arc activity appear to have greatly facilitated the transpressive processes.

All major faults shown on Fig. 6-7 are suspected to have had a significant dextral component. The most fundamental of these is the Foothills suture, which brought exotic ocean floor-bearing equatorial fauna against the truncated edge of the continent in Late Triassic to Early Jurassic time. To what degree this represents transform faulting versus oblique subduction may be irresolvable. The fracture-zone history of the Kings-Kaweah ophiolite belt and possible other ophiolite remnants of the western Sierra, coupled with the truncation of the Antler-Sonoman orogenic belts and the Cordilleran shelf, certainly leaves open the possibility of a transform juncture between oceanic and continental plates. But at some point(s) in time and space, transform motion probably evolved into oblique convergence. Following its history as a plate juncture, the Foothills suture was active as an Early to Middle Jurassic intraarc deformation zone (Saleeby and others, 1978). This

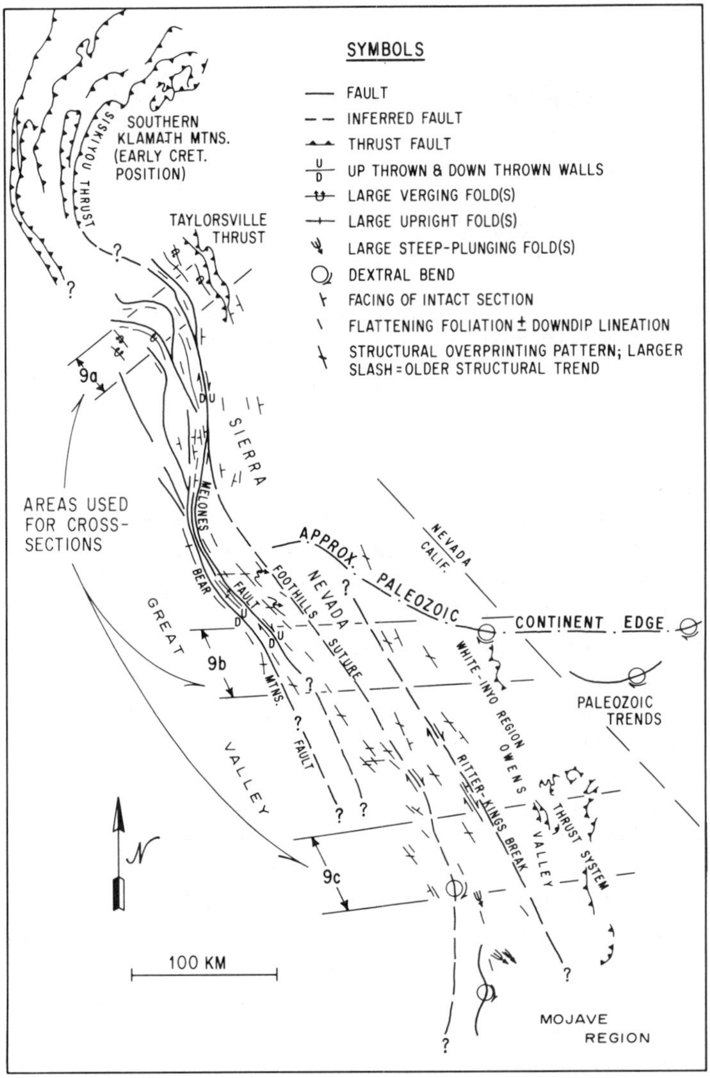

Fig. 6-7. Generalized map showing major Mesozoic structures of Sierran region. See Table 6-3 for age constraints and references. Pre-Late Cretaceous position of southern Klamath Mountains relative to northern Sierra after Jones and Irwin (1971) and Hamilton (1978a).

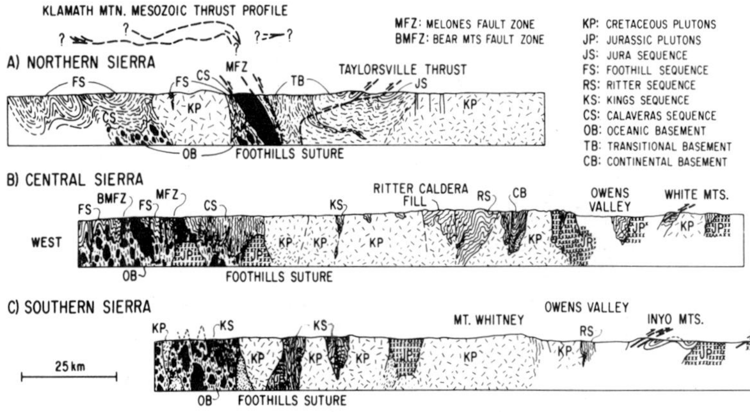

Fig. 6-8. Generalized cross sections through northern, central, and southern Sierra Nevada regions. Map intervals used for cross sections shown in Fig. 6-8. References for sections given in Table 6-3. Southern Klamath Mountains thrust profile after Ando and others (1977) and Hotz and others (1977).

late-stage deformation has obscured the earlier structural relations because it followed the preexisting suture trends.

A hypothetical fault that is believed to have once separated the Ritter and Kings sequences, but has since been obliterated by the batholith, is shown in Figure 6-7. The fault may also have served as an emplacement channel for voluminous plutons of the Cretaceous batholith. The fault brought quartzite and shallow-water carbonates with interbedded silicic tuffs and ash flows of the Kings sequence against great thicknesses of tuffs and ash flows with only local shallow-water carbonates of the coeval Ritter sequence. The Kings sequence probably moved northward from the Mojave region, where similar rocks are presently exposed (Saleeby and others, 1978). A situation similar to the Semangko fault zone of Sumatra (Fig. 6-5) is envisaged for the Ritter-Kings break. In this view, the Ritter sequence and the eastern Jurassic plutonic belt represent the axis of the volcano-plutonic arc. The Ritter-Kings break represents an intra-arc transcurrent fault that brought in the more distal arc facies of the Kings sequence with its dominate nonorogenic lithotypes. Movement on the Ritter-Kings break is bracketed to between Early Jurassic and Late Cretaceous time by the age of the Ritter and Kings sequences and the age of the batholith. Cretaceous movement on the Kern Canyon fault, suggested by Schweickert (Chapter 5, this volume), may have been related to the Ritter-Kings wrench system.

The Foothills fault system is considered another intra-arc dextral wrench system that operated during Late Jurassic time. This system transects the axis of the western Jurassic plutonic belt. It internally disrupts and juxtaposes the Foothills sequence against the Calaveras sequence. The northern Melones branch probably coincides with the Foothills suture. Oceanic basement exposures are common along branches of the system, and its trend appears to have been strongly influenced by the preexisting basement structure (Behrman, 1978; Saleeby, 1978b). Reverse as well as dextral strike-slip is suggested by the geometry of mesoscopic structures (Cloos, 1932; Clark, 1960; Cebull, 1972; Nokleberg,

TABLE 6-3. Sources of Data and Ages for Structures Shown in Figs. 6-7 and 6-8

Structure	*Age*	*References*
Foothills Fault System	Late Jurassic	Clark, 1960
Foothills Suture	Triassic-Late Jurassic	Clark, 1976; Saleeby, 1977a, b
Ritter-Kings break	Jurassic	Jones and Moore, 1973; Saleeby and others, 1978
Taylorsville thrust	Late Jurassic	McMath, 1966; Bond and others, 1977
White-Inyo region thrust system	Middle Triassic-Late Jurassic	Burchfiel and others, 1970; Dunne and others, 1978; Moore, S., in prep.
Southern Klamath thrust faults	Middle to Late Jurassic	Irwin, 1977; Ando and others, 1977; Davis and others, 1978
Eastern Nevada dextral bends	Middle to Late Jurassic	Albers, 1967; Buckley, 1972, 1974
Southern Sierra dextral bends	Jurassic	Saleeby and others, 1978
Northern Sierra large folds	Late Jurassic	Hietanen, 1973, 1976, 1977; Bond and others, 1977
Central Sierra large folds	Jura-Trias and Late Jurassic	Clark, 1964; Schweickert and others, 1977; Sharp and Saleeby, 1979
Southern Sierra large folds	Jurassic	Saleeby and others, 1978
Facing data	–	Clark, 1964, 1976; McMath, 1966; Bateman and Clark, 1974; Saleeby and others, 1978
Flattening foliations ± downdip lineations	Triassic-Cretaceous	Clark, 1964, 1976; Hietanen, 1973, 1976, 1977; Saleeby, 1977b, 1978a; Tobisch and others, 1977; Saleeby and others, 1978; Sharp and Saleeby, 1979
Structural overprinting patterns	Triassic to Cretaceous	Baird, 1962; Kistler, 1966b; Kistler and Bateman, 1966; Russell and Nokleberg, 1977; Brook, 1977; Wright and Schweickert, 1977; Saleeby and others, 1978, unpub. data; P. C. Bateman, pers. commun., 1978; Sharp and Saleeby, 1979

1975; Russell and Cebull, 1977). The true slip cannot be documented inasmuch as the system juxtaposes such contrasting assemblages. A significant reverse component for the Melones branch seems likely, since it may continue northward as a thrust fault in the Klamath Mountains (Figs. 6-7 and 6-8A, B). Reverse slip also seems likely along the Melones branch inasmuch as both Jurassic arc plutonic and volcanic rocks are exposed to the west, whereas mainly plutonic rocks of the same age are exposed to the east. The grade of Jurassic regional metamorphism also increases eastward across the Melones fault, which is consistent with it having an important reverse component. It is possible that the Melones branch was at first low dipping and has since been rotated to its near vertical dip; however, there is no compelling evidence to indicate that this was the case. The east wall of the Bear Mountains branch was the locus of basement uplift (Behrman, 1978), so this branch of the system probably also experienced a significant component of dip-slip movement. The shredding of the petrotectonic assemblages into elongate belts that pinch out and interdigitate is well demonstrated along the Foothills fault system. This is considered a result of primarily strike-slip processes.

A number of oroclinal features all showing a dextral sense of rotation occur in the Sierran region (Fig. 6-7). The term orocline is adopted from Carey (1958), who uses it for a tectonically induced change in the trend of a mountain belt. In the southern Sierra Nevada, steeply dipping rocks of the Kings sequence are deformed into a large asymmetric fold with a wavelength of about 50 km (Saleeby and others, 1978). At the very southern end of the Sierra Nevada the structure of the Kings sequence swings around into northeast trends in a complementary fashion. The age of this regional-scale structure is bracketed by the Early Jurassic age of the Kings sequence and the Cretaceous age of the batholith. Burchfiel and Davis (Chapter 9, this volume), suggest a similar type of bending of the northern Mojave region but of Cretaceous or Early Tertiary age. Dextral sense oroclinal bending of northeast-trending structural and stratigraphic elements of western Nevada (Figs. 6-2 and 6-7) occur just to the east of the eastern Jurassic plutonic belt (Albers, 1967; Buckley, 1972, 1974, written communication, 1978). Cenozoic strike-slip faults have accentuated this pattern, but the large ductile bend is at least in part Middle to Late Jurassic in age, as shown by relations with Jurassic-age plutons. The Antler-Sonoman orogenic zone appears to follow this bend into the eastern Sierra. In a similar fashion, the 0.706 initial Sr isopleth is bent in a dextral sense as it passes from western Nevada into the Sierran terrane (Figs. 6-2 and 6-7). This is thought to represent a pre-Cretaceous oroclinal bend of the northeast-trending pre-Mesozoic continental edge of that region. Within the Sierran terrane, this bend takes on a much sharper curve before ending near the Foothills suture. The sharp part of this bend roughly lines up with the Ritter-Kings break, which suggests that dextral distortion of the southeastern basement terrane was more concentrated along the trace of the hypothetical break. Additional bends in the 0.706 and 0.704 initial Sr isopleths occur near the southern end of the Foothills suture (Figs. 6-2 and 6-7). These bends complement the regional-scale bends present in the southern exposures of the Kings sequence. The bend patterns in the initial Sr isopleths each suggest regional dextral drag along the southeastern continental basement terrane. These and other oroclinal features of the Sierran region are Jurassic to Cretaceous in age. They show the same sense of shear as the fault systems discussed previously, but record a more ductile behavior mode. The oroclinal bends are thought to represent the dextral wrench systems at depth within the thermally weakened volcanoplutonic arc.

The segmentation of the Sierran framework into tectonostratigraphic belts is a product of the fault systems. Within the different belts there are complex sets of structures that in many cases have regional consistency in their overprinting patterns. Kistler and others (1971) have evaluated some of these patterns with respect to the timing and orientation of fold axial surfaces and the timing and orientation of pluton emplacement. Axial surface orientations are on the average 20° west of intrusive trends. Such an orientation in fold structures is consistent with dextral wrench motion parallel to the intrusive loci and fold sets developing en echelon to the wrench zone. In the Kings sequence, steep southeast plunging folds are common in the more highly deformed and metamorphosed areas (Saleeby and others, 1978). The axial surfaces of these folds are generally NNW trending. Locally developed superposed folds and cleavages commonly have W to WNW trends. This pattern is believed to have resulted from the first set of structures having been rotated and flattened into the regional trend (Fig. 6-6B), and then the latter set of axial surfaces and cleavages having developed en echelon across the earlier structures. In the lower Kings River area this pattern has repeated itself twice. The sequence there is Triassic to Early Jurassic NNW trends followed by Middle Jurassic EW trends followed by Late Jurassic (Nevadan) NW trends, and then finally the local development of Cretaceous EW trends (Saleeby, 1978; Saleeby and Sharp, 1980). In the Calaveras sequence exposed east of the southern Melones fault, Middle Jurassic EW trends are superposed on an older NW trend and then in turn cut by Late Jurassic (Nevadan) NNW trends (Baird, 1962; Sharp and Saleeby, 1979). The Late Jurassic NNW trends in the Calaveras are also locally cut by EW trends of probable Cretaceous age. This pattern of Mesozoic deformation is repeated in time and space throughout the Sierra Nevada (Kistler, 1966; Nokleberg, 1970; Russell and Nokleberg, 1977; Brook, 1977; P. C. Bateman, personal communication, 1978). It points to a deformational continuum that started in Triassic time, was most intense in Jurassic time, and waned in Cretaceous time.

In some areas structures that seem at odds with the regional pattern occur. An important example is where conjugate folds and crenulations show a superposed shortening axis within the plane of regional flattening (Tobisch and Fiske, 1976). Another example is the Late Jurassic WNW to NW-trending Independence dike swarm of the eastern Sierra region, which is associated with a small component of sinistral shear deformation in addition to the extension which is nearly within the plane of regional flattening (Moore and Hopson, 1961; Chen, 1977; Dunne and others, 1978). The complexities in the mesoscopic structural patterns of the Sierran region are difficult to reconcile if the deformations are conceived of as compressional episodes directed across the structural grain. Such complex patterns are to be expected in a long-active transpressive regime, however, if the entire family of wrench structures is considered along with differential rotations and inherent space problems that occur along strike. Similar complex structural patterns can be seen in the modern California transform regime and have been suggested for the wrench system of the Sunda arc and neighboring Malay Peninsula (Holcombe, 1977).

In addition to the structural elements that point to a long-lived dextral wrench system, the Sierran metamorphic country rocks show significant transverse shortening and vertical elongation, which occurred along with the wrench deformations. This is shown by penetrative mainly NNW-trending near vertical flattening foliations with common down-dip lineations. These structural elements occur in most metamorphic country rocks and in many plutons of mainly Jurassic but also Cretaceous age. The regional NNW foliations are traditionally viewed as Late Jurassic Nevadan structures. This is certainly the case in the Foothills sequence, which contains deformed Late Jurassic strata whose structures are cut by less or nondeformed Late Jurassic plutons. However, structural and geochronological data show that the regional NNW structural trends in the western oceanic basement are Late Triassic to Early Jurassic in age, and that Early to Middle Jurassic deformation along this same trend affected the Calaveras and western Kings sequences and plutons of the western Jurassic belt (Saleeby, 1977a, b, 1978a, b, 1979; Behrman, 1978).

Furthermore, NNW-trending structures of Middle Triassic to Middle Jurassic age also exist in eastern metamorphic country rocks (Nokleberg and Kistler, 1977; Dunne and Gulliver, 1976; Dunne and others, 1978). There was also an important component of flattening in Cretaceous time in both the Ritter and Kings sequences and at least locally in the Calaveras sequence adjacent to batholithic contacts (Tobisch and others, 1977, personal communication; Saleeby and others, 1978; Saleeby and Sharp, 1980).

To what extent the regional flattening represents overall crustal shortening is difficult to ascertain inasmuch as the batholith evolved during the flattening. In simulated models of crustal diapirism, Ramberg (1967, 1972) shows that during the vertical redistribution of material, intense shortening can develop adjacent to zones of significant dilation. It seems likely that dilation compensated or even overcompensated for shortening if the volume of the batholith and probable underplated material is considered.

The regional flattening foliations and their down-dip elongation lineations represent a significant amount of vertical flow. The surface expression of this flow was probably a system of thrust sheets and folds that diverged from the Sierran root. This is suggested in the cross sections of Figure 6-8, where, from the Klamath Mountains to the southern Sierra Nevada, progressively deeper structural levels are displayed. In the Klamath region, east-dipping thrust sheets of Middle to Late Jurassic age predominate. In the northern Sierra a transition from shallow to steep-dipping structures occurs. In the central and southern Sierran region, steep-dipping structures predominate, particularly in the southern Calaveras and Kings sequences, where down-dip elongation lineations are strongly developed (Baird, 1962; Saleeby and others, 1978; Sharp and Saleeby, 1979). To the east of the central and southern Sierra, Late Triassic to Jurassic west-dipping thrust and reverse faults and east-verging folds occur.

The Sierra Nevada appears to be a root zone from which diverging fold and thrust welts were derived. A strong resemblance exists between the Sierran structural configuration displayed in the cross-sections of Fig. 6-8 and the structural model developed by Lowell (1972) for the Spitzbergen transpressive orogenic belt (Fig. 6-6A). In the Spitzbergen belt, thrusting was a result of transpression along a transform juncture whereby basement slivers and their folded sedimentary cover were driven up and out of the transpressive root zone. A similar mechanism is envisaged for the Sierran root zone, but with the added effect of a much more mobile basement. Not only was the Sierran basement thermally weakened and mobilized by volcanoplutonic arc activity, but the western oceanic terrane was disrupted and highly serpentinized prior to the onset of arc activity. Many of the thin serpentinite "fault-slivers" that occur in the Calaveras and the Foothills sequences were probably emplaced upward as protrusion seams derived from serpentinitic basement rocks (Saleeby, 1979).

The greatest amount of upward transport in the Sierran root zone is represented by batholithic and volcanic rocks. This magmatic transport was probably compensated for by both underplating of mantle-derived magma and by downward flow of wallrocks. Similar downward flow of wallrocks has been simulated in diapirism models by Ramberg (1967, 1972). The regional effect of this process operating during the evolution of the Sierran batholith is the general inward-facing pattern of the metamorphic wallrocks (Figs. 6-7 and 6-8). During the vertical redistribution of materials, the upper levels of the crust deformed both by extensional and compressional mechanisms. East-directed thrusts and folds of the White-Inyo system developed along the eastern margin of the volcanoplutonic arc. This system probably developed along a zone of significant ductility contrast between the arc region and the foreland (Burchfiel and Davis, 1975; Dunne and Gulliver, 1976). At roughly the same time, volcanotectonic depressions or cauldrons were apparently forming over the axis of the arc (Fiske and Tobisch, 1978; Saleeby and others, 1978). Sumatra shows a similar spatial relationship between extensional structures along the arc axis and compressional structures toward the foreland (Fig. 6-5).

Because the central and southern Sierra represent deep levels of the root zone, and inasmuch as the Great Valley to the west offers no basement exposures, it is not possible to observe thrust sheets that may have been shed to the west at these latitudes. The Melones fault zone may represent the root of Late Jurassic west-directed thrusts. Perhaps movement here was also facilitated by magmatic heat and material transfer within the western Jurassic plutonic belt. The northern branch of the Melones fault zone occupies a central position between both east- and west-directed structures of Late Jurassic age (Fig. 6-8A). Along this zone oceanic and transitional basement rocks and Late Jurassic plutons are exposed in the root and core areas of the structures. Here, as in the Klamath Mountains, the basement has not been obliterated by the batholith, so its involvement in thrusting can be observed. Basement preservation is attributed to both higher structural levels being exposed and to the volcanoplutonic arc changing its nature longitudinally with the basement. Clusters of batholith-scale silicic plutons are restricted to central and southern portions of the batholith since those were the only regions where the crustal material needed to produce such plutons existed. In the northern Sierra and Klamath Mountains, smaller individual plutons clearly intrude basement rocks and cut or are cut by the thrust faults.

In the Klamath Mountains, Paleozoic basement rocks have moved up and westward over previously deformed Permo-Triassic ophiolitic and hemipelagic rocks along the Siskiyou thrust (Figs. 6-7 and 6-8). This stage of movement was Middle to Late Jurassic in age, but the Siskiyou break was probably long-lived, with an earlier complex history (Davis and others, 1978). The Permo-Triassic oceanic assemblage in turn is thrust westward over imbricated Middle to Late Jurassic island arc and intra-arc basin fragments (Ando and others, 1976; Snoke, 1977; Davis and others, 1978; Harper, 1979). Imbrication occurred while the young basin floors were still hot, which resulted in dynamothermal metamorphic aureoles along the soles of thrust sheets (Armstrong and Dick, 1974; Klein, 1977; Welsh, 1978). No explanation has been offered as to why the backarc or intra-arc basin(s) should close, however. As pointed out by Davis and others (1978), arc magmatism proceeded throughout and after the imbrication process, so a terminal collision event is most unlikely. The pattern of extensional opening followed by compressional closing is typical in both time and space of an evolving wrench zone. In the Sunda arc the Andaman Sea intra-arc rift environment evolves laterally both to the north and south into fold-thrust complexes of totally different geometries (Fig. 6-5).

In summary, the Sierra Nevada structural framework is considered a transpressive root zone. Major wrench faults appear to have been active through much of Mesozoic time along the composite volcanoplutonic arc. Large-scale drag along the ancient continental margin is a first-order effect of the wrench system. A complex family of structures developed within the evolving system. At high crustal levels or off the active magmatic locus the structures reflect less ductile behavior. Fold and thrust welts were driven up and out from the steeply lineated and regionally flattened root zone. Regional compressive stresses, transcurrent slicing resulting in differential transport and mass accumulations, and thermal-diapiric rise together drove the thrust welts. Where the root zone developed over continental crust, it appears as though much of the basement was exhumed as silicic batholiths and ash flows. Over transitional crust the basement was driven up as thrust blocks and as the cores of large verging folds, only to be locally cut by plutons. Over disrupted oceanic crust, serpentinite was protruded and faulted upward, and hot upper mantle rose close to the surface to yield patches of juvenile crust that were soon detached and imbricated with the neighboring arc rocks and their older basements.

Transpressive Plate Tectonics of Mesozoic California

Mesozoic plate motion patterns of the Pacific call for dextral transcurrence and underthrusting along the California margin. The structural configuration and the petrotectonic assemblages of the Sierran region carry the signature of such processes. The Sunda arc has been used as a modern analogue for the Mesozoic Sierra Nevada. It must be stressed, however, that the Sundra arc is used to demonstrate intra-arc tectonic processes. It is not intended to be an orogenic template that can reproduce the Sierran structural configuration. In considering the userfulness of such a modern analogue, the rates of the modern processes must be considered with respect to the amount of time represented in the ancient orogenic assemblage. Accretion of the Cache Creek terrane in Late Triassic to Early Jurassic time shows that northward displacement of Pacific ocean floor relative to North America started in Triassic time. The Middle Jurassic to Early Cretaceous accretion of Wrangellia shows the profusion of northward displacements during Jurassic time. Paleomagnetic studies on exotic pelagic limestones of the Franciscan Complex show that tremendous northward displacements of Pacific ocean floor continued throughout Cretaceous time (Alvarez and others, 1979). Volcanoplutonic arc activity in the Sierra Nevada was semicontinuous from latest Triassic through Cretaceous time, so the Sierra Nevada evolved in a thermally active state for about 150 m.y. along a transpressive continental margin. In the Sunda arc the Andaman Sea has opened in about 10 m.y. (Curray and others, 1979), and the Semangko fault zone apparently changed its course considerably within the same time span (Holcombe, 1977). Ninety-east ridge was acting as a dextral transform juncture throughout much of Tertiary time (Sclater and Fisher, 1974), but is presently undergoing sinistral transform movements in addition to compressional deformation (Stein and Okal, 1978). It is difficult to imagine the complexities that the Sunda arc could possess in 150 m.y. of evolution. There is no reason to believe that the North American Cordillera evolved in a less complex fashion. For this reason, a detailed plate-tectonic reconstruction of the Sierran region seems unwarranted. Nevertheless, speculations will be presented on the manner in which the exotic oceanic terrane was accreted along the Foothills suture and on the structural and petrologic evolution of the composite volcanoplutonic arc assemblage.

The history of ocean-floor accretion along the Foothills suture must be derived largely from relations within the western ophiolitic basement terrane. Geochronological relations point to disruption and regional metamorphism of Permo-Carboniferous oceanic crust in Late Triassic time, followed immediately by continental margin sedimentation. These relations are taken as the signs of continental margin accretion of the ophiolitic terrane. The synchrony of this event with high P/T metamorphism throughout much of the Cordillera to the north of the Sierra Nevada points to the importance of subduction tectonics in the ophiolite accretionary processes. Much more critical information can be extracted from Sierran ophiolites than simply equating them with subduction, however. As discussed earlier, the Permo-Carboniferous ocean floor experienced a second igneous stage synchronous with ultramafic protrusive activity just prior to or during the early stages of continental margin accretion. These relations, along with the regional setting of the ophiolitic terrane along a truncation suture, lead to the conclusion that a major fracture zone assemblage comprises part or all of the ophiolitic terrane. Evolution of such a transform juncture into a convergent juncture would be most conducive to regional-scale ophiolite accretion. A similar emplacement mechanism has been suggested for a number of other ophiolites (Brookfield, 1977), and it has been suggested that significant links of the western Pacific fringing subduction-arc system were established along major preexisting transform junctures (Hilde and others, 1977). Perhaps the ophiolite tectonics proposed here for the Late Triassic Sierra are typical of the circum-Pacific realm.

Another important consideration is the association of Tethyan-affinity fauna with the oceanic assemblages. Much confusion has arisen in regard to the exotic fusulinids. Hamilton (1978a) assumes that the "Tethyan" affinity of this fauna indicates that they were tectonically transported to North America across the paleo-Pacific ocean from Asia. However, several important facts are ignored in this view: (1) Permo-Carboniferous fusulinids that are truly Asiatic are of the Schwagerinid family, as are those of the North American craton (Yancy, 1975); (2) the "Tethyan" Verbeekinids that do occur in Asia are associated with suture belts and continental blocks that were not completely accreted until well into Mesozoic time (Gobbett, 1973; Burrett, 1974); (3) The "Tethyan" Verbeekinids were a tropical cosmopolitan form, with some genera having spread into west Texas (Gobbett, 1973; W. B. N. Berry, personal communication, 1978; M. K. Nestell, personal communication, 1979). Thus the terms "Tethyan" or Asiatic are quite misleading for the Verbeekinids. A more suitable term may be "panthalassan," referring to the proto-Pacific realm. This is in keeping with the views expressed by W. B. N. Berry, M. K. Nestell, W. R. Danner (1976, written communication, 1977), and J. W. H. Monger (1977, written communication, 1977), who consider the Verbeekinids as a tropical oceanic form that spanned the proto-Pacific and early Tethys. Thus the emplacement of Verbeekinid-bearing oceanic assemblages into the Sierran region marks the northward displacement of equatorial ocean floor into California paleolatitudes.

Two plate tectonic models are presented in Fig. 6-9, which attempt to interface the history of Sierran ocean-floor assemblages with other regional tectonic patterns. Both

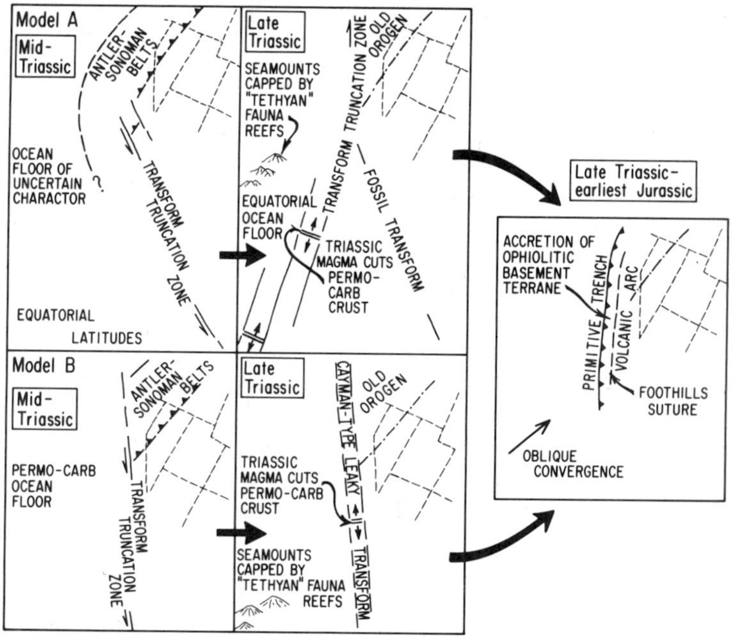

Fig. 6-9. Two models for Triassic truncation of the southwest Cordilleran orogen and the Late Triassic to earliest Jurassic accretion of Sierran ophiolitic basement rocks. Paleolatitudes after Van der Voo and French (1974); palinspastic restoration of northern Basin and Range after Hamilton (1978a). (A) Two-stage truncation with first stage being syn- or immediately post-Sonoman sinistral sense and second stage being post-Sonoman dextral. (B) Single-stage truncation of syn- or immediately post-Sonoman age with sinistral sense. Both models end with accretion of Sierran ophiolites during commencement of subduction along truncation suture.

models end with the same Early Jurassic pattern, which is northeast convergence of proto-Pacific ocean floor beneath the accreted Sierran ophiolites. The northward component of ocean-floor motion is required by the displaced Verbeekinid fauna of the Cache Creek group, as well as those of the Sierran-Klamath region. The commencement of northward ocean-floor motion is unclear. The two models of Fig. 6-9 present alternative time-space relations.

Model A in Fig. 6-9 depicts left-lateral transform truncation of the ancient margin during or immediately after the Sonoman orogeny. This movement pattern is consistent with the tectonic transport patterns exhibited in the Sonoman orogenic belt. Continental basement and Paleozoic strata seemingly offset in a sinistral sense, as discussed in the paleobasement section, owe their present positions to this movement regime. Ocean floor of uncertain character is shown outboard of the Sonoman belt. In Late Triassic time, late

Paleozoic ocean floor with large north-trending fracture zones lay adjacent to the ancient margin. Perhaps this was the same oceanic plate that sat outboard of the Sonoman belt. Transform motion of the Spitsbergen type (Lowell, 1972) is shown cutting into the ancient continental margin across earlier trends and displacing fragments in a dextral sense. The Alexander terrane of southeastern Alaska may represent such a displaced fragment. Active spreading adjacent to or within the fracture zone gave rise to diachronous oceanic igneous assemblages. The large age spread between Permo-Carboniferous and Late Triassic age material requires a large fracture zone(s) like those of the equatorial Atlantic or, alternatively, a ridge jump. In addition to north-trending fracture zones, large seamounts are shown on the Permo-Carboniferous ocean floor with reefoidal buildups of the "Tethyan" fauna. The association of these late Paleozoic fauna with primarily early Mesozoic radiolarian chert is considered a result of the seamounts subsiding during sea-floor transport and being covered by pelagic and hemipelagic deposits. The Calaveras and western Kings sequences are somewhat unique in this regard in that continent-derived sands occur along with more typical hemipelagic deposits. It is possible that these sands were derived from a continental mass that was adrift in the early Mesozoic Pacific, but there is no evidence to indicate this. Most likely the sands were shed from North America onto the ocean floor as it was accreted to the truncated margin.

In Late Triassic to possibly earliest Jurassic time, underthrusting commenced along the transform juncture. The disrupted fracture-zone assemblage and possibly adjacent fragments of standard oceanic crust were accreted to the hanging wall of the primitive subduction zone. The intermediate P/T metamorphic facies rocks that developed in the accreted ophiolitic terrane at this time are considered a result of both obliquity of subduction and the involvement of young, warm crustal material. The Tethyan-affinity limestone blocks were either tectonically dislodged from their substrate during ophiolite emplacement and then at least partly reworked in olistostromes with sandstone, chert, and argillite, or they slid into the trench from bypassing seamounts that were accreted farther north. Shortly thereafter, island-arc volcanism spread over portions of the accreted oceanic material.

Model B utilizes a more simplistic model in terms of continental margin truncation and accretion of the western Sierra ophiolites. The Foothills suture here is equated with the sinistral truncation episode. As with model A, this may be syn- or immediately post-Sonoman. The sinistral truncation is shown as juxtaposing late Paleozoic ocean floor against the truncated continental margin. In Late Triassic time the Foothills suture is shown as a Cayman-type "leaky" transform (Perfit and Heezen, 1978). Thus in this model the diachronous oceanic igneous assemblage was produced parautochthonously. As northeast-directed underthrusting commenced, diachronous ocean floor was accreted along the Foothills suture, and Tethyan-affinity fauna from equatorial latitudes were introduced into the Sierran oceanic assemblages as envisaged in model A.

After ophiolite accretion along the Foothills suture, subduction processes operated to the west of the Sierra Nevada. Thus earliest Jurassic is the latest possible time that subduction occurred in the Sierra Nevada. Jurassic and Cretaceous petrotectonic assemblages of the Sierra are typified by volcanoplutonic arc associations.

The Early Jurassic configuration of the Sierran arc is shown in Fig. 6–10A, which focuses on the central and southern Sierra regions. Volcanic arc activity is shown lapping

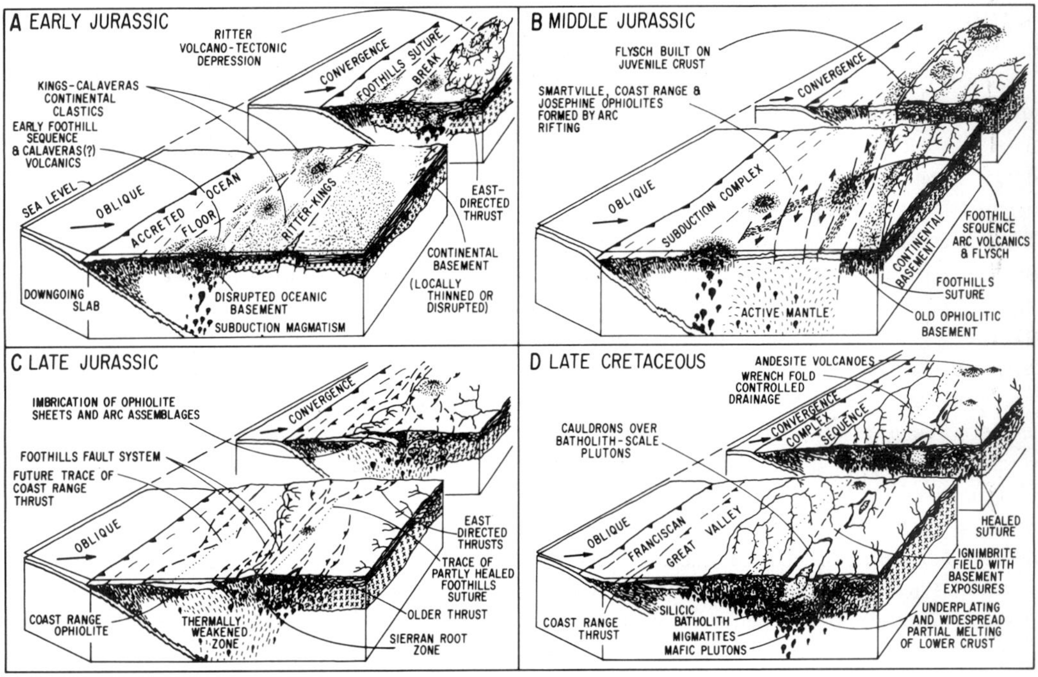

Fig. 6-10. Sequence of block diagrams showing major tectonic evolutionary steps in the development of the Sierran collage and batholith. (A) Early to early Middle Jurassic evolution of Sierran volcanoplutonic arc showing Foothills suture as an active intra-arc structure. (B) Middle to early Late Jurassic intra-arc rifting. (C) Late Jurassic Nevadan orogeny. (D) Late Cretaceous voluminous magmatism and Franciscan subduction.

across the Foothills suture onto the accreted oceanic terrane. Accretionary rocks outboard of the western basement consist of Permo-Carboniferous "Tethyan" and young Jura-Trias oceanic rocks. Continental clastics are shown lapping across the suture and mixing with arc rocks of both continental and oceanic affinity. The traverse variations in the composition of the arc rocks are shown as a function of basement type. Intra-arc structural breaks consist of the Foothills suture and possibly the Ritter-Kings break. East-directed thrusting is shown rooting into the arc axis. Remnants of the subduction complex related to this stage of the Sierran arc were probably stripped off by Middle Jurassic to Cretaceous processes.

A major intra-arc rifting event began in Middle Jurassic time and extended into early Late Jurassic time (Fig. 6-10B). The vestiges of this event are the Smartville, Coast Range, and Josephine ophiolites. The development of a marginal basin similar to the Andaman Sea is envisaged (Fig. 6-5), except epiclastic source terranes are shown as local volcano-tectonic highs rather than a major orogenic mass such as the Himalayas. In Fig. 6-10B, Middle and Late Jurassic Foothills sequence rocks are shown to have been erupted and deposited over both juvenile marginal basin crust (Smartville ophiolite) and older deformed oceanic basement. Similar arc volcanic and epiclastic strata were deposited across parts of the Coast Range and Josephine ophiolites (Jones, 1975; Evarts, 1977; Harper, 1979; Blake and Jones, Chapter 12, this volume). The newly formed marginal basin grew wide enough whereby in some regions only distal tuffs, tuffaceous radiolarian cherts, and in the northern Coast Ranges true pelagic cherts were deposited on pillow lava (Page, 1972; Hopson and Frano, 1977; Blake and Jones, Chapter 12, this volume). As the arc rifted apart the Foothills suture ceased activity as a zone of major intra-arc translation.

Immediately following or in the latter stages of the rifting regime, intense folding and thrusting of the Nevadan orogeny spread across the Sierra Nevada and Klamath Mountain region (Fig. 6–10C). The Nevadan orogeny is here considered an intra-arc transpressive orogeny driven by the forces of oblique subduction and the regional thermal pertubation of the arc and subarc mantle, which was first manifested by arc rifting. Distended arc segments and juvenile crust were transported northward along the Foothills Fault system and possibly other faults that underlie the Great Valley. Flakes of juvenile crust were detached at sub-Moho depths along critical isothermal surfaces and driven away from the root zone as warm thrust sheets. Dynamic contact metamorphism proceeded along the soles of thrust sheets in the fashion envisaged by Armstrong and Dick (1974). Arc magmatism proceeded throughout the thrusting regime, just as it did during the rifting regime.

Obduction of mid-Jurassic ophiolites was an outstanding feature of the Nevadan orogeny. Major differences between Nevadan obduction and the original concept of ocean-floor obduction onto continental edges (Coleman, 1971a) are that the Nevadan ophiolites were imbricated with one another and with fragments of older previously deformed ophiolitic basement, and that imbrication proceeded within the region of active arc magnetism. The westward extent of such intra-arc obduction is not clear. Suppe and Foland (1978) suggest that the older (Nevadan age) high P/T metabasite blocks of the Franciscan Complex were related to obduction of the Coast Range ophiolite. Perhaps these metamorphites are a direct result of the Nevadan orogeny. If such was the case, then there may not have been a fully developed volcanic arc outboard of the Middle Jurassic juvenile crust. Consider the Andaman Sea rift zone, which has retained only a small segment of the Sunda arc (Fig. 6–5). Here a significant segment of the intra-arc juvenile crust is immediately adjacent to the subduction complex. Perhaps the Coast Range ophiolite was thrusted westward from a similar structural setting during the Nevadan orogeny. If so, the line between subduction wedge and intra-arc deformational processes is here arbitrary. This may be a general rule in regions of strong oblique convergence.

A major enigma in both Sierran and Coast Range tectonics is the nature of the events which immediately followed the Late Jurassic Nevadan orogeny. Volcano-plutonic arc activity continued throughout and followed the Nevadan regime, however, there is no direct record of convergence for the time interval between the Nevadan orogeny and Valinginian time (Blake and Jones, 1974). Of course, the paucity of bonified Jurassic subduction assemblages throughout California is also a major enigma. It is reasonable to suspect that pre-Nevadan subduction complexes were for the most part removed from the California region during the rifting episode which gave rise to the Coast Range, Josephine and Smartville ophiolites. However, an additional tectonic removal episode seems necessary to account for the absence of a well-defined Nevadan and immediately post-Nevadan subduction complex. The Franciscan Nevadan-age metamorphic blocks could represent the vestiges of such a complex, however, their ubiquitous but volumetrically insignificant occurrence only adds credence to the concept of a Cretaceous removal event. Whether removal was by subduction wedge wrench faulting, oblique rifting or transform faulting is an important but presently irresolvable question.

The suggestion of a major tectonic removal event of Cretaceous age in the Coast Ranges matches well with the arrival of allochthonous sequences of Coast Range-type rocks in northwestern Washington. These allochthonous sequences consist of mid-Jurassic ophiolite fragments and Franciscan-type melange structurally overlain by Great Valley sequence-type flysch (Whetten and others, 1980; written commun., 1979). These out of place sequences were thrust eastward over older oceanic and island arc assemblages upon their late Cretaceous arrival into northwestern Washington. The California Coast Range region is the most likely source for these sequences.

The late Cretaceous trace of the Coast Range thrust is shown in Figure 6–10d. Most of the material shown outboard of this trace in Late Jurassic time (Fig. 6–10c) is considered to have been removed. The Valinginian timing of Coast Range underthrusting corresponds well with the main magmatic pulse of the western Cretaceous batholithic belt. These events marked the onset of a major component of convergence between the California margin and the Pacific Ocean floor.

The western Cretaceous batholithic belt was emplaced for the most part west of the inactive Foothills suture. As the Cretaceous batholith evolved the locus of magmatism migrated eastward across the trace of the suture. The oceanic affinity of both the western Cretaceous and Jurassic belts is reflected in the high plagioclase to total feldspar ratios in latest Jurassic to Early Cretaceous clastic rocks of the Great Valley sequence which were derived from the Sierran arc (Dickinson and Rich, 1972). In contrast, the continental affinity of the younger central and eastern Cretaceous belts is reflected in lower plagioclase to total feldspar ratios present in the Middle and Late Cretaceous rocks of the Great Valley sequence.

The formation of Late Cretaceous batholith-scale plutons is depicted in Fig. 6–10D, which focuses on the central and southern Sierra regions. Mantle-derived melts are shown underplating the Sierran arc and promoting whole-scale partial melting of the lower continental crust. The surface expression of the eastern batholithic belt is shown as cauldrons filled and flanked by ash flows over the southern Sierra and andesite volcanic centers over the northern Sierra. The final stages of wallrock sagging along the margins of the batholith are shown to involve older Jurassic and Cretaceous plutonic rocks, as well as metamorphic country rocks. The marginal sagging of the wallrocks gave rise to the synclinoral configuration of Bateman and Eaton (1967) and Schweickert (Chapter 5, this volume). This is not a synclinorium in the strict sense, though, since its limbs consist of tectonically interwoven tectonostratigraphic belts rather than continuous stratigraphic units or facies systems. The southern and central segments of the composite batholith occupy the root or axial region of outward-verging fold-thrust systems, part of the two-sided orogen of Burchfiel and Davis (1968). Also shown on Fig. 6–10D are the last-generation wrench folds of the Sierran region and uplifted exposures of older volcanoplutonic arc rocks. The Great Valley is shown as a partly sedimented remnant marginal basin floored primarily by Middle Jurassic ophiolite akin to the Smartville, Coast Range, and Josephine ophiolites.

The tectonic patterns portrayed for the Jurassic and Cretaceous Sierra Nevada in Fig. 6-10 continued to operate during Cenozoic time in California. Migrating loci of volcanoplutonic arc activity (Coney, 1978; Dickinson and Snyder, 1978), the proto-San Andreas fault (Nilsen, 1978), and finally the present plate juncture system are responses of the California margin to ocean-floor consumption and dextral transcurrence whose heritage stems from the Foothills suture.

ACKNOWLEDGMENTS

Conversations and written communications with P. C. Bateman, P. G. Behrman, W. B. N. Berry, C. P. Buckley, B. C. Burchfiel, J. R. Curray, W. R. Danner, G. A. Davis, J. W. Hawkins, D. L. Jones, D. E. Karig, R. W. Kistler, B. D. Marsh, J. W. H. Monger, J. G. Moore, E. M. Moores, R. A. Schweickert, W. D. Sharp, L. T. Silver, and R. C. Speed added greatly to both the data base and formulation of ideas presented in this paper. Part of the geochronological data summarized was gathered under the support of NSF grant EAR 77-08691. Contribution Number 3185, Division of Geological and Planetary Sciences, California Institute of Technology, Pasadena, California, 91125.

Richard A. Schweickert
Department of Geological Sciences
Lamont-Doherty Geological Observatory
Columbia University,
Palisades, New York 10964

Walter S. Snyder
Lamont-Doherty Geological Observatory
Columbia University,
Palisades, New York 10964

7

PALEOZOIC PLATE TECTONICS OF THE SIERRA NEVADA AND ADJACENT REGIONS*

*Lamont-Doherty Geological Observatory Contribution No. 2910

The thick sequences of upper Paleozoic intermediate to silicic volcanic rocks in the northern Sierra Nevada, Klamath Mountains, and northwestern Nevada represent a volcanic island-arc terrane. This arc sequence was deposited unconformably on the western portion of a lower Paleozoic accretionary prism of pelagic sediments, turbidites, and oceanic crust. This accretionary wedge is represented by the Shoo Fly Complex of the Sierra Nevada, the Duzel phyllite and Trinity ultramafic sheet of the Klamath Mountains, and parts of the Roberts Mountains allochthon (i.e., the westernmost exposures of the Harmony and Valmy formations) of north-central Nevada. The eastern portion of the accretionary wedge rode up over the continental margin along the Roberts Mountains thrust during the latest Devonian or earliest Mississippian Antler orogeny. The upper Paleozoic Calaveras Complex of the Sierra Nevada and the Havallah sequence of Nevada are interpreted to be parts of a single tectonostratigraphic unit deposited in a marginal basin that formed between the upper Paleozoic volcanic arc and the continental margin after the Antler orogeny. This unit was thrust onto the continental margin along the Golconda thrust during the latest Permian or earliest Triassic Sonoma orogeny. These structural relationships and broad stratigraphic correlations form the basis for a plate-tectonic model for the Paleozoic evolution of the western Cordillera.

The Cordilleran miogeocline was initiated by rifting about 850 m.y. ago, and by Ordovician time a major oceanic basin had formed by sea-floor spreading. In the Late Ordovician or Silurian, parts of this oceanic basin were subducted westward beneath a new volcanic arc, the Alexander terrane. As subduction continued and the arc migrated eastward toward the continent, a lower Paleozoic accretionary wedge formed. During the Middle to Late Devonian, arc volcanism (represented by rocks in the eastern Klamath Mountains and the northern Sierra Nevada) stepped eastward onto the accretionary wedge. The Antler orogeny marked the culmination of arc-continent collision as the leading edge of the accretionary wedge was thrust over the continent along the Roberts Mountains thrust.

Closely following the Antler orogeny, the arc reversed polarity as a new subduction zone developed on its western edge. Subsequent back-arc spreading created a marginal basin in which sediments of the upper Paleozoic Havallah sequence and Calaveras Complex accumulated. This late Paleozoic rifting stranded that portion of the lower Paleozoic accretionary wedge that had been thrust onto the continental margin (Roberts Mountains allochthon) from equivalent units (Shoo Fly, Duzel), which migrated westward with the arc to form the basement for continued late Paleozoic island-arc volcanism. The rifting also left behind several structual remnants of the Roberts Mountains allochthon within the growing marginal basin; these remnants are now structurally part of the Havallah sequence.

During the Permian, the island arc reversed polarity once again and began to subduct its own marginal basin. By the latest Permian or earliest Triassic, the entire Calaveras-Havallah assemblage was part of an accretionary wedge that was obducted eastward onto the continental margin as the arc collided with the miogeocline during the Sonoma orogeny. Later, probably during the Middle Triassic, the southern part of the Cordilleran orogen was obliquely truncated, severing a portion of the arc complex and the lower Paleozoic accretionary wedge, which migrated together northward to form the present Alexander terrane in southeastern Alaska.

INTRODUCTION

Most recent plate-tectonic syntheses of the southern part of the Cordillera have been developed primarily from regional stratigraphic relationships in north-central Nevada and the Klamath Mountains of northern California. The Paleozoic rocks of the Sierra Nevada, which form an important link between the other widely separated regions, have generally added little to the plate-tectonic syntheses, partly because the rocks are almost totally lacking in fossils, but also because until now there have been no syntheses of their stratigraphic and structural relations. Because of the degree of metamorphism and the lack of fossils in the Sierra Nevada, detailed structural studies have been of major importance to an understanding of the range, and now the structure of the Sierra Nevada is better understood than the structure of the other regions. By contrast, structural analysis has rarely been attempted in the relatively fossiliferous terranes of Nevada, and little is known of detailed structural relations within or between the various allochthonous masses.

Regional tectonic analyses, for completeness, must incorporate both structural and stratigraphic data. Most previous tectonic analyses have drawn primarily on stratigraphic data and have not adequately considered structural relations, and therefore these have suffered severe limitations. In this paper we use available stratigraphic and structural data on Paleozoic rocks in the Sierra Nevada, north-central Nevada, and, to a lesser extent, the Klamath Mountains to constrain and modify plate-tectonic models for the southern part of the Cordillera. We make two key assumptions at the outset: (1) Paleozoic tectonic events in north-central Nevada probably had expression as far away as the Sierra Nevada, an assumption implicit in all previous tectonic interpretations; and (2) Mesozoic thrusting documented in many ranges in Nevada has not seriously modified the geometry of Paleozoic terranes. We present arguments that the structural geometry of Paleozoic terranes of the Sierra Nevada is entirely consistent with the structural style and geometry of Paleozoic terranes in north-central Nevada. These relationships are critically important in determining unique plate-tectonic models for Paleozoic evolution of the Sierra Nevada.

PREVIOUS INTERPRETATIONS OF THE PALEOZOIC TECTONIC HISTORY OF NEVADA

Pioneering studies by members of the U.S. Geological Survey during and after World War II yielded most of our present understanding of the Paleozoic and Mesozoic history of north-central Nevada (Merriam and Anderson, 1942; Ferguson and others, 1951a, b; Ferguson and others, 1952; Muller and others, 1951; Roberts and others, 1958; Silberling and Roberts, 1962; Roberts, 1964; Hotz and Willden, 1964; Gillully and Gates, 1965; Gilluly, 1967; Churkin and Kay, 1967). These geologists recognized evidence that three distinct depositional environments existed during Cambrian to Devonian time: (1) the autochthonous "eastern" assemblage, consisting of miogeoclinal carbonate and shale; (2) the allochthonous "western" or "siliceous" assemblage (called "lower Paleozoic oceanic assemblage," LPOA, in this paper), a eugeoclinal association of shale, chert, greenstone, sandstone, and rare limestone; and (3) an autochthonous and parautochthonous transitional assemblage, consisting of pelitic, siliceous, and carbonate units that were believed to have formed a facies gradational between the shallow-water miogeocline and the deep-water eugeocline. In addition, Kay (1951) clearly recognized evidence that, during the early to mid-Paleozoic, fringing volcanic island arcs existed well offshore from the miogeocline.

Roberts and others (1958) argued that during the Late Devonian to Early Mississippian Antler orogeny, the western assemblage (LPOA; Fig. 7-1) was thrust eastward several tens of kilometers along the sole of a thrust fault (known as the Roberts Mountains thrust) upon the eastern (miogeoclinal) assemblage. Rocks of the transitional assemblage were considered autochthonous, but locally were incorporated into the overriding thrust sheets to become parautochthonous. Figure 7-1 shows the distribution of the various terranes in

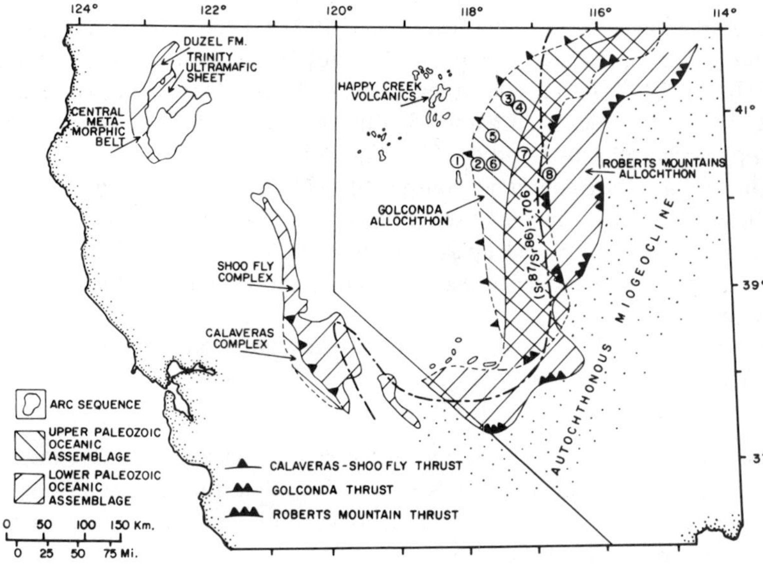

Fig. 7-1. Sketch map of distribution of lower and upper Paleozoic oceanic rocks, upper Paleozoic arc rocks, and Paleozoic miogeoclinal rocks. The ^{87}Sr-^{86}Sr = 0.706 line may approximate the rifted margin of cratonic North America and therefore the westernmost boundary of the miogeocline (Kistler, 1978). Within Nevada, the circled numbers show the locations of (1) Humboldt Range, (2) East Range, (3) Hot Springs Range, (4) Osgood Mountains, (5) Sonoma Range, (6) Tobin Range, (7) Battle Mountain (Galena Range), and (8) Shoshone Range.

Nevada. The "transitional" assemblage is not shown on Fig. 7-1, but would be approximately coextensive with the structurally overlying Roberts Mountains allochthon; in contrast, the "eastern" assemblage is designated as the autochthonous miogeocline.

The Antler orogeny produced a north-northeast trending highland (Antler orogenic belt) that was the source of clastic detritus shed during Mississippian time both to the east (the Eureka-Carlin sequence or Antler flysch) into a rapidly subsiding trough and, starting somewhat later, also to the west into a newly reestablished eugeocline where the Havallah sequence, consisting of chert, shale, sandstone, and greenstone, accumulated. In addition, shallow-water clastic and carbonate rocks were locally deposited within submerged parts of the Antler orogenic belt.

The rocks deposited on the Antler orogenic belt (known as the Antler sequence or overlap assemblage) were subsequently overthrust from the west by the Havallah sequence along a thrust fault (known as the Golconda thrust, Fig. 7-1) during the latest Permian or earliest Triassic Sonoma orogeny. A thick Early Triassic silicic volcanic assemblage (the Koipato sequence) was deposited unconformably on the Havallah sequence, presumably following thrusting accompanying the Sonoma orogeny, according to Silberling and Roberts (1962; Fig. 7-2).

Most recent plate-tectonic models, drawn in large part from the geologic framework just presented, have suggested that the Antler and Sonoma orogenies resulted from collisions between one or more island arcs and the Cordilleran miogeocline (Moores, 1970; Burchfiel and Davis, 1972, 1975; Silberling, 1973; Stewart and Poole, 1974; Poole, 1974; Churkin, 1974a; Schweickert, 1976a; Dickinson, 1977; Speed, 1977). A major Paleozoic island-arc terrane is represented in part by Paleozoic volcanic sequences in the Klamath Mountains (Irwin, 1977; Potter and others, 1977), the northern Sierra Nevada (d'Allura and others, 1977; Schweickert, Chapter 5, this volume), and northwestern Nevada (including Happy Creek volcanics, Speed, 1977; Fig. 7-1).

Following initial arc-continent collision during the Antler orogeny, the island arc withdrew, leaving behind a marginal sea in which the Havallah sequence was deposited. This marginal sea was then closed up as a late Paleozoic arc converged on the continental margin, and the Havallah sequence was expelled as the Golconda allochthon (Fig. 7-1) onto the miogeocline.

As Dickinson (1977) noted, major unresolved questions concerning these plate-tectonic scenarios are as follows:

1. Was the same arc terrane (now visible in the Klamath Mountains, northern Sierra Nevada, and northwestern Nevada) responsible for both collisions, or was a different island-arc complex that is no longer present in the southern Cordillera responsible for the Antler orogeny?
2. What was (were) the facing direction(s) of the arc(s) that led to the Antler and Sonoma orogenies?
3. If the Havallah marginal basin was produced by back-arc spreading after the Antler orogeny, why is there apparently no remnant volcanic arc on the continentward (eastern) side of the former basin?

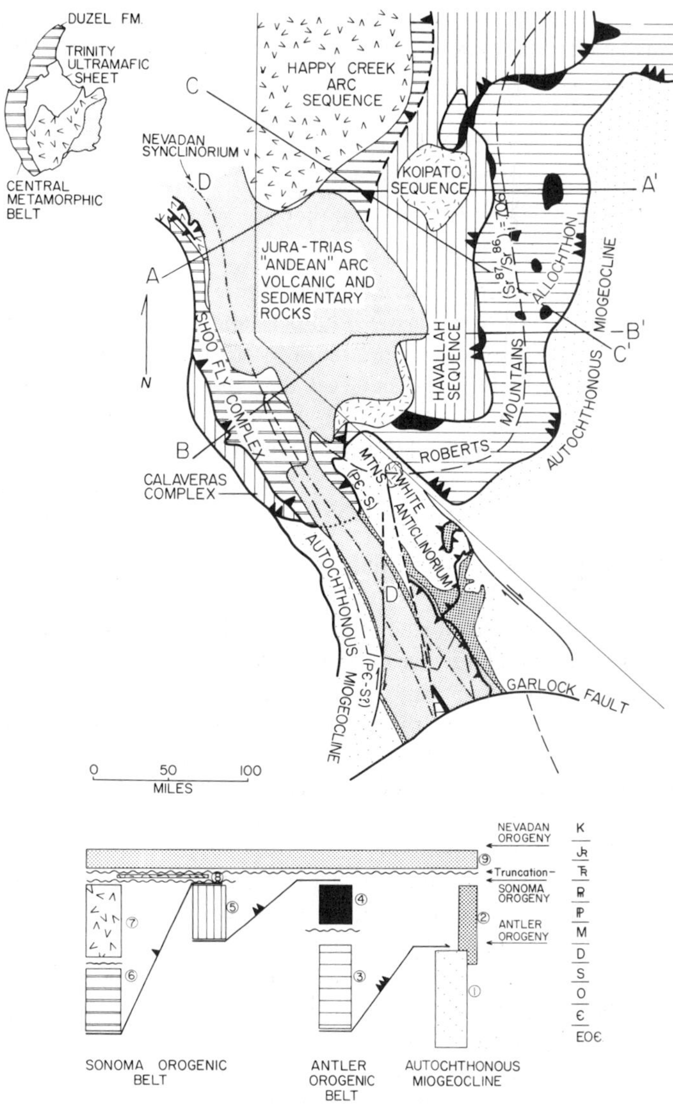

Fig. 7-2. Interpretive structural map of the southern part of the Cordilleran orogen, showing inferred stacking order of major structural units.

1. Precambrian to Devonian sedimentary rocks of Cordilleran miogeocline.
2. Devonian to Permian sedimentary rocks of the White-Inyo Mountains.
3. Allochthonous Cambrian to Devonian siliceous and volcanic rocks of the Roberts Mountains assemblage.
4. Pennsylvanian to Permian clastic and carbonate rocks of the "overlap" sequence.
5. Allochthonous Permo-Carboniferous siliceous and volcanic rocks of the Havallah sequence and Calaveras Complex.
6. Allochthonous Cambrian (?) to Devonian (?) siliceous and volcanic rocks of the Shoo Fly Complex, Central Metamorphic belt, Duzel Formation, and Mount Morrison and Saddlebag Lake pendants, considered to be the rifted equivalent of the Roberts Mountains assemblage (unit 3).
7. Devonian to Permian pyroclastic rocks of island-arc sequence.
8. Permian and Permo-Triassic pyroclastic rocks of Koipato sequence and areas to south.
9. Late Triassic to Late Jurassic pyroclastic rocks of Andean-type arc sequence.

In the next section we discuss the relations of terranes in the Sierra Nevada to terranes in north-central Nevada to arrive at possible resolutions to these questions.

POSSIBLE RELATIONS BETWEEN PALEOZOIC TERRANES OF THE SIERRA NEVADA AND NORTH-CENTRAL NEVADA

Paleozoic rocks that lie east of the Foothills suture in the Sierra Nevada are poor in fossils, but possess certain distinctive lithologic and structural characters that suggest rather close comparisons with the more fossiliferous rocks of Nevada. We cannot prove age equivalence for these terranes, but feel that the correlations presented herein are reasonable considering the lithologic and structural similarity and the present state of understanding of the terranes. Additional work will be necessary to test the suggestions we make here.

Lower Paleozoic Oceanic Assemblage

The Shoo Fly Complex of the northern and central Sierra Nevada includes a wide variety of tectonostratigraphic units, several of which may have rather close counterparts in the Antler orogenic belt of north-central Nevada. As noted by Schweickert (Chapter 5, this volume), part of the lower unit of the Shoo Fly near Bowman Lake is made up of quartzofeldspathic sandstone and pebbly sandstone and a few olistostromes, which together very closely resemble parts of one of the most distinctive stratigraphic units in north-central Nevada, the Upper Cambrian Harmony Formation, which forms the structurally highest unit in the Roberts Mountains allochthon. Recent work by Suczek (1977a, b) has shown that the Harmony originated as a sequence of turbidity flows in an extensive subsea fan complex on the ocean floor an unknown distance west of the Cordilleran miogeocline. The miogeocline directly east of the present exposures of the Harmony Formation apparently lacked exposures of feldspathic source rocks during Late Cambrian time, and thus the sediments are believed to have been derived from the Salmon River arch, 400 km north by present geography (Stewart and Suczek, 1977; see Fig. 7-5A).

The geometry and sedimentology of the feldspathic sandstone bodies in the lower part of the Shoo Fly also indicate deposition within a subsea fan complex (Girty and Schweickert, 1979). The feldspathic Antelope Mountain Quartzite of the eastern Klamath Mountains (Hotz, 1974; Potter and others, 1977) is another unit lithologically similar to Shoo Fly and Harmony sandstones (G. Bond, personal communication, 1978; our unpublished data). The feldspathic sandstones of the Shoo Fly are underlain by a sequence of cherts and shales (Schweickert, Chapter 5, this volume) that may correlate with the Upper Cambrian Paradise Valley Chert, which underlies the Harmony Formation. The feldspathic sandstones and underlying cherts of the Harmony, Shoo Fly, and Antelope Mountain Quartzite may represent parts of a single, large, subsea fan complex deposited on an abyssal plain floored by pelagic siliceous sediments (see Fig. 7-5A). Alternatively, they may represent more than one fan complex. The probable age of the fan complex (or complexes) is no younger than Cambrian, because the only plausible feldspathic source terrane, the Salmon River arch, was buried by Early Ordovician time (Suczek, 1977b).

Another assemblage of rocks in the lower unit of the Shoo Fly consists of phyllite, chert, greenstone, and poorly sorted quartz sandstone with no clear stratigraphic order (Schweickert, Chapter 5, this volume). Our observations suggest that this assemblage may be a close lithologic match for the Ordovician Valmy Formation of north-central Nevada and the Duzel phyllite (Cashman, 1977a, b; Hotz, 1974) in the Klamath Mountains.

The Valmy Formation, according to Roberts (1964), Gilluly (1967), and Gilluly and Gates (1965), consists of conspicuous ledge-forming, poorly sorted quartzite and chert, together with less conspicuous, but significant, amounts of shale and greenstone. No regionally consistent internal stratigraphy within the Valmy has been discovered, and, instead, the larger quartzite bodies and greenstone masses form lenticular bodies of all dimensions. Gilluly (1967) and Gilluly and Gates (1965) suggested that the Valmy has been sliced by numerous thrust faults in order to account for the lack of lithologic continuity.

Both the Valmy and a similar, less quartzitic, more shaly unit to the east, the Vinini Formation, were regarded as early to Middle Ordovician in age on the basis of graptolites from shales and limited faunas from sparse limestones. However, the recent recovery of radiolaria and conodonts from cherts indicates that both units contain appreciable Devonian strata (Stanley and others, 1977; Jones and others, 1978c).

The complex distribution of lithologies in the Valmy and Vinini and the new paleontologic data suggest that in many areas neither "formation" may be a valid rock-stratigraphic unit. Instead, our observations have led us to the conclusion that both probably are in large part tectonostratigraphic units or melanges of various lithologies of various ages. Coherent sequences of chert, shale, and quartz sandstone exist, however. Since most fossils in these units have come from shale and chert, many of the larger, "floating" masses of quartzite and virtually all of the greenstone in the Vinini and Valmy must be regarded as undated, and could reasonably range in age from Eocambrian to Devonian.

Table 7-1 shows our postulated correlations of rock units in the Sierra Nevada,

TABLE 7-1. Postulated Tectonic Correlations of Lower Paleozoic Terranes

Klamath Mountains	*Sierra Nevada*	*North-Central Nevada*
Ordovician Trinity ultramafic sheet	Serpentinite melange in "middle" Shoo Fly or Shoo Fly_2	
Antelope Mountain Quartzite	Feldspathic sandstone in "lower" Shoo Fly or Shoo Fly_1	Upper Cambrian Harmony Formation
Unnamed chert	Chert and shale	Upper Cambrian Paradise Valley Chert
Duzel phyllite and Grouse Ridge Formation	Phyllite, chert, greenstone, and poorly sorted quartzite in "lower" Shoo Fly	Ordovician (?) Valmy Formation and Vinini Formation

north-central Nevada, and the Klamath Mountains. In summary, although neither strict physical nor temporal equivalence can be proven, rocks of the Shoo Fly Complex in the northern and central Sierra Nevada and their counterparts in the Klamath Mountains closely resemble important parts of the allochthonous pre-Mississippian Roberts Mountains allochthon of north-central Nevada.

Upper Paleozoic Terranes

The Calaveras Complex east of the Foothills suture in the central Sierra Nevada contains a very great, but unknown, thickness of chaotic argillite, chert, and limestone that may rest on a substrate of pillow lava and volcaniclastic rocks. See Schweickert (Chapter 5, this volume) and Schweickert and others (1977) for more complete descriptions. The age of the Calaveras Complex is poorly constrained because, apart from several occurrences of crinoid ossicles, only a single occurrence of poorly preserved Pennsylvanian or Permian corals has been found, and this collection is from a limestone olistolith[1] (Schweickert, Chapter 5, this volume; Schweickert and others, 1977). The argillaceous matrix enclosing the limestone could be considerably younger, even Mesozoic, as several authors have recently suggested (Davis and others, 1978; Saleeby and others, 1978; Saleeby, Chapter 6, this volume) by analogy with chert-argillite terranes with Mesozoic radiolaria in the Klamath Mountains, Oregon, and British Columbia. However, we believe that it is dangerous to assume that *all* chert-argillite-greenstone terranes in the southern Cordillera are Mesozoic. One such terrane, the Havallah sequence, contains appreciable amounts of Paleozoic chert (D. L. Jones, personal communication, 1978; Snyder, unpublished data). Schweickert (Chapter 5, this volume) argues that, in the absence of Mesozoic fossils, the Calaveras Complex must still be tentatively presumed to be of late Paleozoic age.

The upper Paleozoic Havallah sequence of north-central Nevada is a widespread assemblage of radiolarian chert, sandstone, siliceous shale, argillite, and mafic pillow lava. The Havallah sequence was originally considered to be Pennsylvanian-Permian based on fusulinids and stratigraphic correlations (Roberts and others, 1958). An older maximum age for the Havallah sequence is suggested by (1) inclusion of several Mississippian units originally mapped as separate formations (Snyder, 1977; Stewart and others, 1977); and (2) recently discovered upper Devonian conodonts from Havallah chert (Snyder, unpublished data). The youngest fauna recovered from the Havallah are Leonardian (late Early Permian) in age.

Lithologically, the Calaveras Complex closely resembles parts of the upper Paleozoic Havallah sequence (Schweickert, 1976a, Chapter 5, this volume). Both terranes characteristically contain extensive exposures of greenstone, chert, and argillite, with lesser

[1] Late Permian Tethyan fusulinids found by Saleeby (in Schweickert and others, 1977) near Yokohl Valley in the southern Sierra Nevada are not from the Calaveras Complex but instead are from a sequence of rocks that forms the basement of a Jurassic island arc (Saleeby and others, 1978) considered exotic by Schweickert (1978) and by Schweickert and Cowan (1975).

amounts of sandstone and limestone. Neither contains much ophiolitic debris. Proportions of the various rocks types vary considerably in each terrane. Speed (1977) objected to correlation of the Calaveras and Havallah because the Calaveras is characteristically chaotic (Schweickert, 1976a, Chapter 5, this volume; Schweickert and others, 1977), whereas he asserted that the Havallah lacks chaotic rocks and is at most a stack of imbricate thrust plates. However, Snyder's (1977 and unpublished data) recent detailed and reconnaissance mapping of the Havallah sequence indicates that chaotic units very similar to those in the Calaveras are much more widespread than previously thought.

Differences in exposure probably account for the discrepancy. In the central Sierra Nevada, stream beds in the more highly metamorphosed and resistant Calaveras rocks yield strips of nearly 100% exposure approximately perpendicular to strike. In Nevada, however, the arid climate together with the lower grade of metamorphism result in generally poorer exposures. However, exceptional exposures in the northern Tobin Range (point 6, Fig. 7-1) reveal that chaotic olistostromes of argillite, chert, and sandstone may underlie the entire range. Furthermore, diamond drill cores from widely spaced localities within the Havallah sequence in other areas suggest that pebbly mudstone or diamictite predominates in the subsurface, despite the fact that widely spaced surficial outcrops (olistoliths?) of the more resistant lithologies suggest an undisturbed stratigraphy. Mapping is underway to determine whether a regional pattern of distribution of chaotic rocks exists in the Havallah.

In general, the Calaveras Complex exhibits a greater degree of deformation and is more strongly metamorphosed than the Havallah sequence. However, MacMillan (1972) recognized at least three phases of deformation in the Havallah sequence in the New Pass Range, 40 km south of the Tobin Range. Two phases occurred prior to the emplacement of the Golconda allochthon (one of which may have been soft-sediment disruption), and one phase accompanied emplacement of the Havallah in the Golconda allochthon. A broad open folding occurred after mid-Jurassic and prior to Oligocene time, possibly during the Late Jurassic Nevadan orogeny. Snyder and H. Brueckner (unpublished data) have recognized three possibly similar structural events in the northern Tobin Range (Point 6, Fig. 7-1).

Significantly, an assemblage of schistose, deformed rocks known as the Leach and Inskip formations (Ferguson and others, 1951a) occurs in the East Range (point 2, Fig. 7-1), west of most exposures of the Havallah sequence but still within the Golconda allochthon. The Inskip Formation, which locally contains Mississippian (?) and Devonian (?) fossils (Roberts and others, 1958; Whitebread, 1978), has not always been consistently distinguished from the Havallah sequence, which it resembles. Recent detailed and reconnaissance mapping in the East Range has shown that the Leach Formation consists of structurally interleaved Valmy and Havallah rocks (Whitebread, 1978; Stewart and Carlson, 1976). These rocks contain evidence of at least three episodes of folding (H. Brueckner and Snyder, unpublished data). The Inskip and Leach formations are polyphase deformed metamorphic rocks that are similar to those in terranes like the Calaveras and Shoo Fly complexes (our unpublished data). Their distribution suggests that the grade of metamorphism and the degree of ductile deformation increased westward within the Golconda allochthon, toward the volcanic arc (Fig. 7-1).

Possible Structural Relations between Sierra Nevada Terranes and Those in North-Central Nevada

Figures 7-2 and 7-3, reproduced from Schweickert (Chapter 5, this volume), are a schematic structure map and cross sections that show diagrammatically the inferred structural relations of Paleozoic terranes in the Sierra Nevada, the Klamath Mountains, and north-central Nevada. The structural relations have been deduced in large part from a consideration of the structural evolution of the Sierra Nevada.

The Calaveras Complex and Havallah sequence are depicted as parts of a continuous tectonostratigraphic unit on both flanks of the Nevadan synclinorium. This upper Paleozoic oceanic assemblage overlies the allochthonous Roberts Mountains allochthon in north-central Nevada along the Golconda thrust and is itself overlain by allochthonous

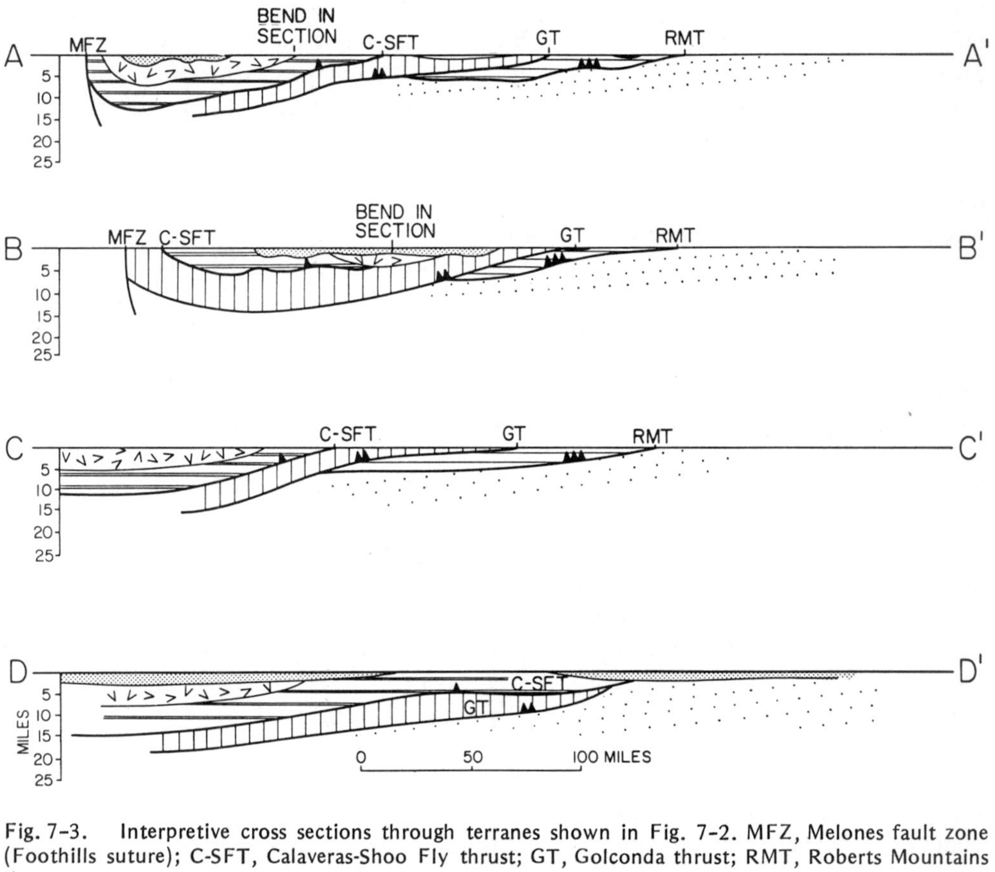

Fig. 7-3. Interpretive cross sections through terranes shown in Fig. 7-2. MFZ, Melones fault zone (Foothills suture); C-SFT, Calaveras-Shoo Fly thrust; GT, Golconda thrust; RMT, Roberts Mountains thrust.

rocks of the Shoo Fly Complex along the Calaveras-Shoo Fly thrust in the Sierra Nevada (a hypothetical continuation of the eastern limb of this thrust is shown in northwestern Nevada (Figs. 7-2 and 7-3). If correct, this structural stack indicates that the Calaveras-Havallah terrane is sandwiched between allochthonous lower Paleozoic oceanic terranes above and below of similar (and possibly the same) provenance and structural evolution.

In Nevada, previous workers have conventionally assigned the Vinini, Valmy, and Harmony formations, the Devonian Slaven Chert, and the Silurian Elder Sandstone of the Roberts Mountains allochthon to a structural position beneath the Havallah sequence of the Golconda allochthon, even where contact relations are unexposed. However, in the East Range (point 2, Fig. 7-1) several lower Paleozoic units, including the Harmony and Leach formations (the latter now considered Valmy Formation by Whitebread, 1978), lie structurally *above* the Havallah sequence along thrust faults that probably predate the Koipato sequence (e.g., pre-Early Triassic), indicating that they reached their structural position during the Sonoma orogeny (Muller, and others, 1951; Wallace, 1977). Although this important relationship has not previously been discussed in the literature, it means that even in north-central Nevada, lower Paleozoic rocks lithologically similar to rocks of the Shoo Fly Complex rest structurally *above* the Havallah sequence in a manner exactly analogous to the structural position of the Shoo Fly above the Calaveras in the Sierra Nevada. In short, might not the Leach and Harmony formations in the East Range be equivalents to the Shoo Fly appearing above the Calaveras-Shoo Fly thrust? Mapping presently being conducted by D. Whitebread in the East Range may help to support or reject this possibility.

Deformed rocks once mapped as Leach Formation (Ferguson and others, 1951a), but now considered as Valmy, also occur along the western edge of the Sonoma Range (point 5, Fig. 7-1), structurally above rocks mapped as Valmy by Gilluly (1967), all of which lie west and northwest of most exposures of the Havallah sequence. This raises the highly speculative possibility that lower Paleozoic rocks of the Harmony and Valmy formations in the East, Sonoma, and perhaps even the Hot Springs ranges (points 2, 5, and 3, respectively, Fig. 7-1) are all part of the Shoo Fly structural plate, and that the hypothetical continuation of the Calaveras-Shoo Fly thrust should be considered to run along the eastern edges of these ranges (east of where it is inferred in Fig. 7-2). Furthermore, in the Humboldt Range (point 1, Fig. 7-1) just west of the exposures of the Inskip and Leach (point 2, Fig. 7-1), the Limerick Greenstone may be a portion of an upper Paleozoic arc complex (Speed, 1977). The Limerick Greenstone could therefore be analogous to the arc complex that was deposited on the Shoo Fly in the Sierra Nevada.

Again, further detailed study will be necessary before these latter possibilities can be seriously entertained, but the structural relations of rocks in the East Range argue that age and lithology of eugeoclinal rocks do not uniquely determine their structural position (e.g., they may lie either beneath or above the Golconda thrust). In summary, we believe that, although many details are speculative at present, plate-tectonic models for the assembly of the various eugeoclinal terranes of Nevada and California must reckon with a structural pattern along the lines of that shown in Figs. 7-2 and 7-3.

PALEOGEOGRAPHY AND PLATE TECTONICS

Pre-Antler Oceanic Basin and Plate Topology

We return now to an important question posed earlier: What was the polarity of the arc responsible for the Antler orogeny? Although it cannot be determined directly, we agree with Dickinson (1977), who argued that, in order to close the extensive Eocambrian to Devonian oceanic basin that existed prior to the Antler orogeny, the arc must have had reversed polarity and faced southeastward. If true, the Roberts Mountains allochthon and possibly the Shoo Fly Complex would have been parts of an immense accretionary prism that developed on the eastward side of the island arc prior to the actual collision. Dickinson's (1977) suggestion that the Roberts Mountains allochthon formed as part of an accretionary wedge carries an important corollary: the stacking order of thrust sheets within the assemblage might well be related to the incorporation of successive packets of strata within the accretionary wedge during precollision subduction; higher structural units probably were accreted earlier and thus were carried farther (hence originated farther west) than lower structural units.

Map relations and structural interpretations in the Shoshone, Galena, and Sonoma ranges (points 8, 7, and 5, Fig. 7-1) indicate that formations within the Roberts Mountains allochthon have a consistent stacking order. In each range the Upper Cambrian Harmony Formation generally occupied the highest structural position after the Antler orogeny (Gilluly and Gates, 1965; Roberts, 1964; Silberling, 1975). The Harmony evidently was thrust over the Ordovician (?) Valmy Formation in these areas, which in turn was thrust over generally younger formations (Devonian Slaven Chert, Silurian Elder Sandstone). Gilluly and Gates (1965) noted that the Vinini Formation lies east of and presumably structurally below the Valmy Formation. If this stacking order consisting of higher sheets with older rocks west of lower sheets with younger rocks formed as a consequence of subduction as the components of the Roberts Mountains allochthon accumulated within an accretionary wedge, then it can be inferred that the Harmony Formation was originally deposited farther west than the bulk of the rocks that make up the Valmy and Vinini formations. This interpretation enables us to make several important deductions about the nature of the oceanic basin that existed prior to the Antler orogeny.

Stewart (1972, 1976; Stewart and Poole, 1974; Stewart and Suczek, 1977) has argued that the Cordilleran miogeocline was initiated during a rifting event that began about 850 m.y. ago, and culminated in the formation of a major oceanic basin by sea-floor spreading during Eocambrian, Cambrian, and Ordovician time (from about 650 to 450 m.y. ago). Symmetrical sea-floor spreading at a midocean ridge would have created an oceanic basin with a crustal age profile as shown in Fig. 7-4. If continuous spreading at an average rate comparable to that of the mid-Atlantic ridge occurred until mid-Ordovician time, an oceanic basin about the width of the present Atlantic could have existed.

The Upper Cambrian Harmony Formation and its presumed equivalents in the Shoo Fly were probably deposited in a major subsea fan on oceanic crust older than

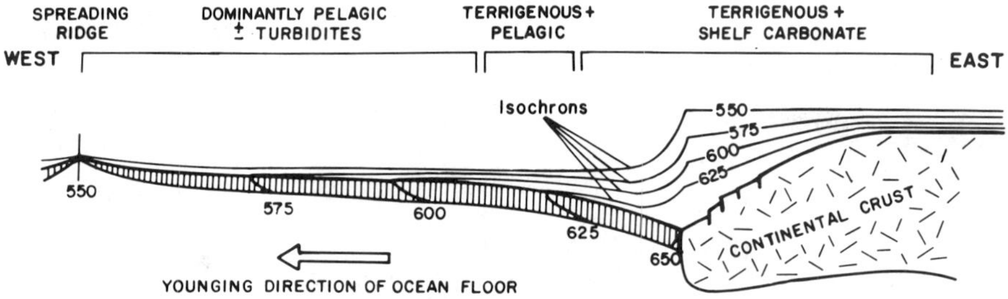

Fig. 7-4. Schematic cross section of eastern half of oceanic basin that probably existed during Middle Cambrian time according to the rifting hypothesis of Stewart and Suczek (1977). Hypothetical isochrons within the miogeoclinal prism and oceanic lithosphere are shown.

Late Cambrian. The underlying Upper Cambrian Paradise Valley Chert in the Hot Springs Range (point 3, Fig. 7-1; Hotz and Willden, 1964) and similar rocks in the Shoo Fly may represent pelagic sediments of layer 2. For the ocean floor to have cooled and subsided enough to have allowed the Harmony subsea fan to develop across it, the oceanic crust beneath the Harmony must have been significantly older than Late Cambrian, perhaps Early Cambrian or Eocambrian in age (Fig. 7-5). If so, the oceanic basin was 100 to 150 m.y. old when the Harmony Formation was deposited upon it, and therefore must have been very extensive (Fig. 7-5B). Furthermore, assuming continued spreading, a midoceanic ridge system must have existed to the west of the depositional site of the Harmony subsea fan (Fig. 7-5C). These considerations make a western source for the Harmony extremely unlikely. The requisite western source, split off from North America during Eocambrian rifting, would have been at least half the width of the oceanic basin away (several thousands of kilometers) and separated from the Harmony depositional site by a formidable topographic barrier, the spreading ridge.

The Valmy and Vinini formations of Nevada probably were deposited closer to the North American craton than the Harmony. They therefore presumably accumulated on the continental slope or on oceanic crust east of, and therefore *older* than, the oceanic crust that underlay the Harmony and Paradise Valley Chert. The question arises, where is the evidence of pre-Ordovician oceanic crust and pelagic sediments that should have originally underlain the Valmy and Vinini sediments as implied by Figs. 7-4 and 7-5? First, much of the greenstone and quartzite within the Valmy and Vinini could in fact be Cambrian or even Eocambrian, inasmuch as fossils have not come directly from these rocks, and uncertainty of stratigraphic relations leaves much of these rocks at present undated. An old age for the quartzite would be consistent with a possible Lower Cambrian cratonic source of coarse quartz sand in the miogeocline (Stewart and Poole, 1974; Fig. 7-5B). In addition, much of the unfossiliferous shale within the Valmy and Vinini could also be Cambrian or older. These shales could lack a representative graptolite fauna because sparse graptolites made their first appearance during the middle Cambrian, and older deep-water shales may have been devoid of fauna that left fossilized remains.

The Vinini could represent lower continental rise deposits, as it contains appreciable Ordovician to Devonian chert and shale, but little greenstone. The Valmy, with its abun-

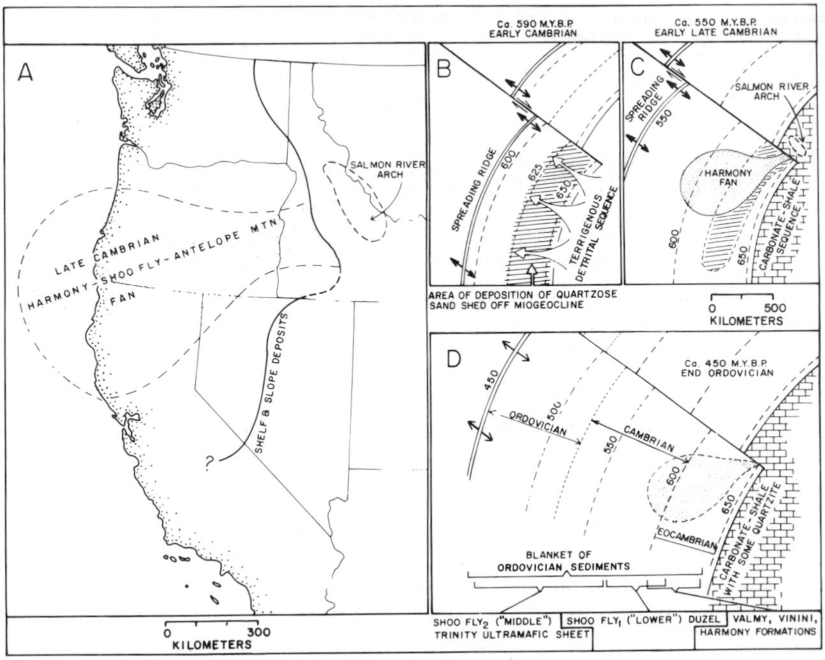

Fig. 7-5. Interpretive paleogeographic maps showing plate topology of the oceanic basin in which the Harmony subsea fan was deposited. Dashed lines in B–D represent hypothetical isochrons of the age of oceanic basement (see Fig. 7-4): (A) Inferred relations between Harmony subsea fan and its presumed source, the Salmon River Arch (after Stewart and Suczek, 1977; Suczek, 1977a). (B) Hypothetical configuration of spreading ridge system and oceanic basin west of miogeocline at the beginning of the Cambrian, showing deposition of quartzites now found in Vinini, Valmy, and "lower" Shoo Fly. (C) Oceanic basin during early Late Cambrian time, showing the Harmony subsea fan deposited across crust that ranged from Eocambrian to Middle Cambrian in age. (D) Oceanic basin at the end of the Ordovician. Deposition of pelagic and hemiterrigenous sediments throughout the ocean basin, while carbonate and shale predominated within the miogeocline.

dant chert, shale, quartzite, and greenstone may represent sediments scraped off an abyssal plain and tectonic slices of oceanic crust and perhaps seamounts. Of course, Silurian and Devonian sediments would have covered the areas of deposition of sediments of the Valmy, Vinini, and Harmony by the time of the Antler orogeny. This fact would account for the intermixing of older and younger rocks within the same structural unit.

The Shoo Fly Complex in the northern and central Sierra Nevada forms the basement on which the Devonian to Permian island-arc sequence was constructed (Schweickert, Chapter 5, this volume). We have suggested that the Shoo Fly is equivalent to parts of the allochthonous, pre-Mississippian Roberts Mountains allochthon (e.g., Harmony and Valmy) of north-central Nevada. The Shoo Fly Complex therefore may have once been part of the lower Paleozoic oceanic assemblage (LPOA) of Nevada that was thrust eastward during the Antler orogeny. An important corollary of this is that the Devonian to Permian island arc that rests upon the Shoo Fly could not have been an entity exotic to North America, but instead must have developed in proximity to a marginal sea (the Havallah basin) that was consumed prior to the Sonoma orogeny.

Figure 7-6 shows that Shoo Fly rocks, together with lower Paleozoic terranes of the Klamath Mountains, can be reconciled with this model to produce an internally consistent view of the sedimentation and structure of the Eocambrian to Ordovician oceanic basin. In fact, Devonian and Silurian metamorphic ages on rocks of the Central Metamorphic belt and Duzel phyllite of the Klamath Mountains (Cashman, 1977a, b) may represent subduction beneath an island arc to the west at about 390 to 345 m.y.

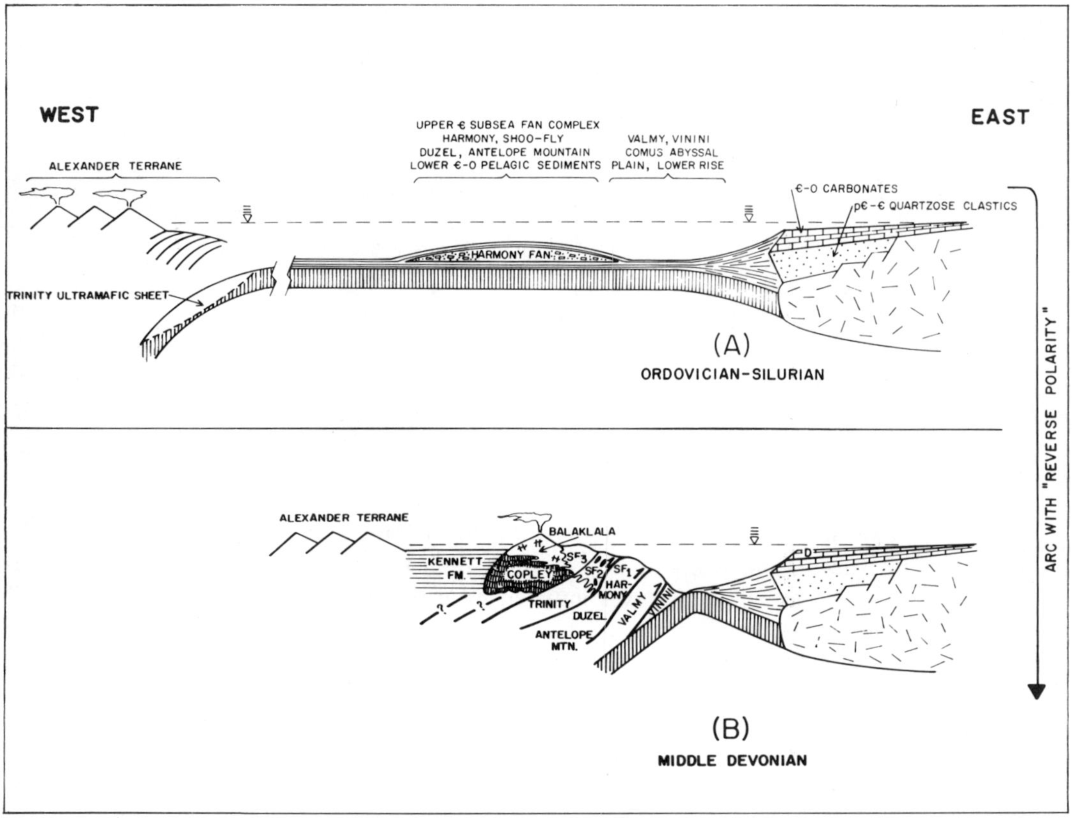

Fig. 7-6. Schematic cross section trending roughly NW-SE showing a plausible Paleozoic plate-tectonic history for the Sierra Nevada, Klamath Mountains, and north-central Nevada, beginning with the early Paleozoic paleogeography depicted in Fig. 7-5. See text for an explanation of the sequence of tectonic events. Abbreviations: AF, Antler flysch; AM, Antelope Mountain Quartzite; B, Balaklala Rhyolite; BA, Baird Fm.; BR, Bragdon Fm.; C, Copley Greenstone; D, Duzel Fm.; DB, Dekkas Andesite, Bully Hill Rhyolite; H, Harmony Fm.; K, Kennett Fm.; M, McCloud Limestone; NRR, Nosoni Fm.; Robinson Fm., Reeve Fm.; OS, overlap sequence; P, Peale Fm.; SB, Sierra Buttes Fm.; SF_1, SF_2, SF_3, Shoo Fly Complex, "lower," "middle," "upper," respectively; T, Taylor Fm.; TU, Trinity ultramafic sheet; VM, Valmy Fm.; VN, Vinini Fm.

Fig. 7-6. cont.

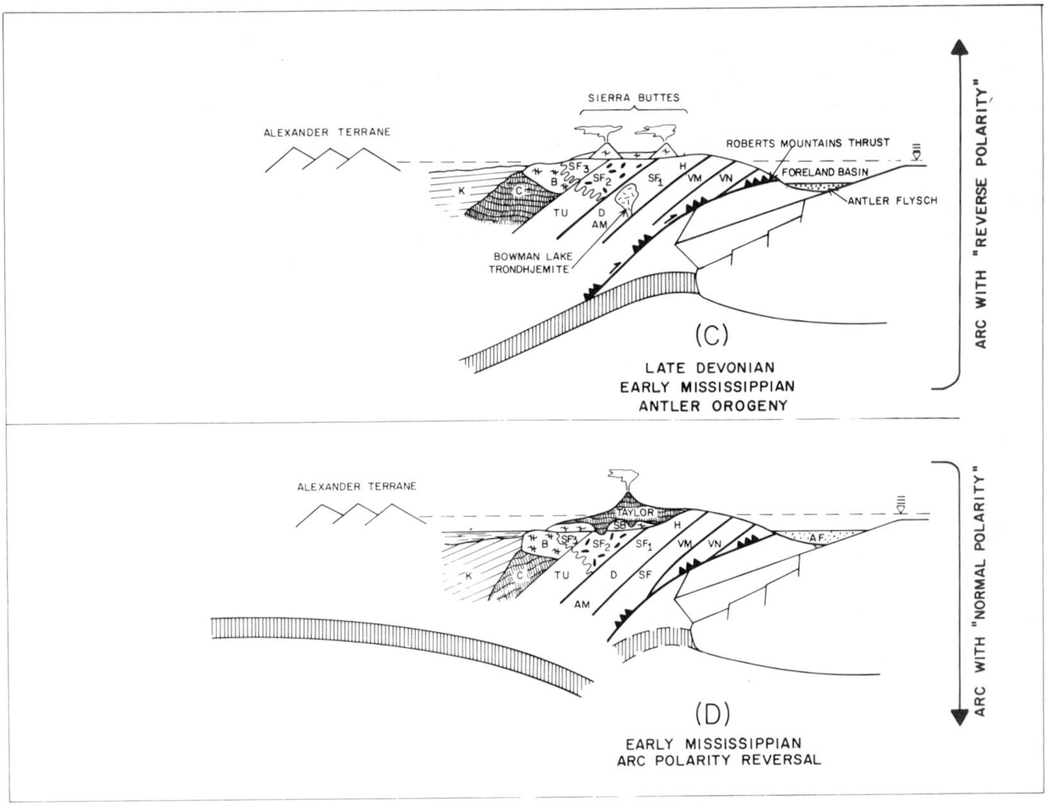

ago. These terranes lie structurally *beneath* the Trinity ultramafic sheet,[2] dated at 480 m.y. (Mattinson and Hopson, 1972), which probably was incorporated into the accretionary wedge as subduction was initiated beneath an early Paleozoic arc that lay still farther to the west.

These considerations require an island arc to have been active in Ordovician or Silurian time as westward subduction of the Klamath and northern Sierran oceanic terranes began. Where is this Ordovician-Silurian arc? Paleomagnetic data of M. Jones and others (1977) have strongly supported the suggestions of D. L. Jones and others (1972) and Schweickert (1976a, 1978) that the Alexander terrane in southeastern Alaska, which contains abundant arclike volcanic rocks of Ordovician, Silurian, and Devonian ages (Churkin and Eberlein, 1977), was at or near the latitude of northern California during Ordovician and Silurian time. Accordingly, we suggest that the preponderance

[2] Although the Duzel Formation has conventionally been considered to lie structurally *above* the Trinity ultramafic sheet, Cashman's (1977b) convincing correlation of Duzel phyllite with rocks of the Central Metamorphic belt strongly implies that the Duzel phyllite was structurally beneath the Trinity during Devonian metamorphism.

Fig. 7-6. cont.

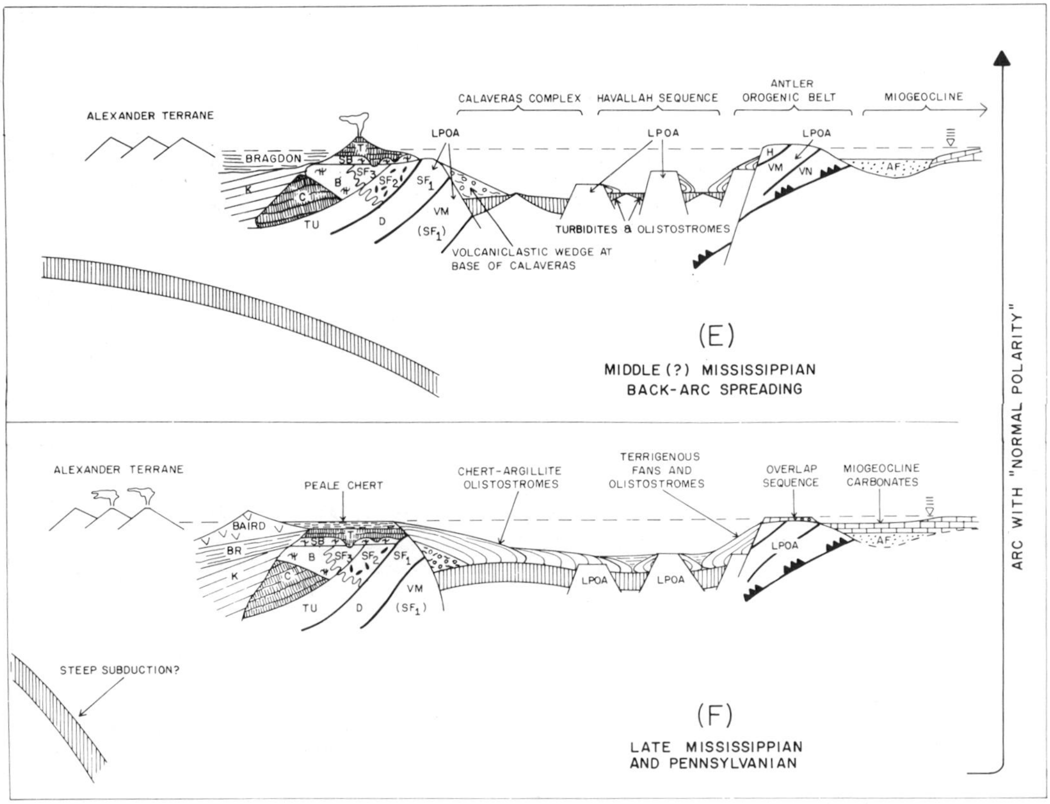

of lower Paleozoic island-arc rocks related to the initiation of subduction of the Roberts Mountains allochthon, Shoo Fly, and eastern Klamath terranes, may now be in the Alexander terrane.

PLATE-TECTONIC EVOLUTION

Figure 7-6 is a cross section trending roughly NW-SE that shows our view of a plausible Paleozoic plate-tectonic history for the Sierra Nevada, Klamath Mountains, and north-central Nevada, beginning with the early Paleozoic paleogeography depicted in Fig. 7-5. Subduction began beneath the Alexander arc in Late Ordovician or Silurian time (Fig. 7-6A) with the incorporation of the Trinity ultramafic sheet into a nascent accretionary prism. During this interval, pelagic and hemiterrigenous sediment (now in part incorporated into the Valmy-Vinini) blanketed the portion of the oceanic basin between the miogeocline and the arc. As subduction continued, the Harmony fan and the Valmy-

Fig. 7-6. cont.

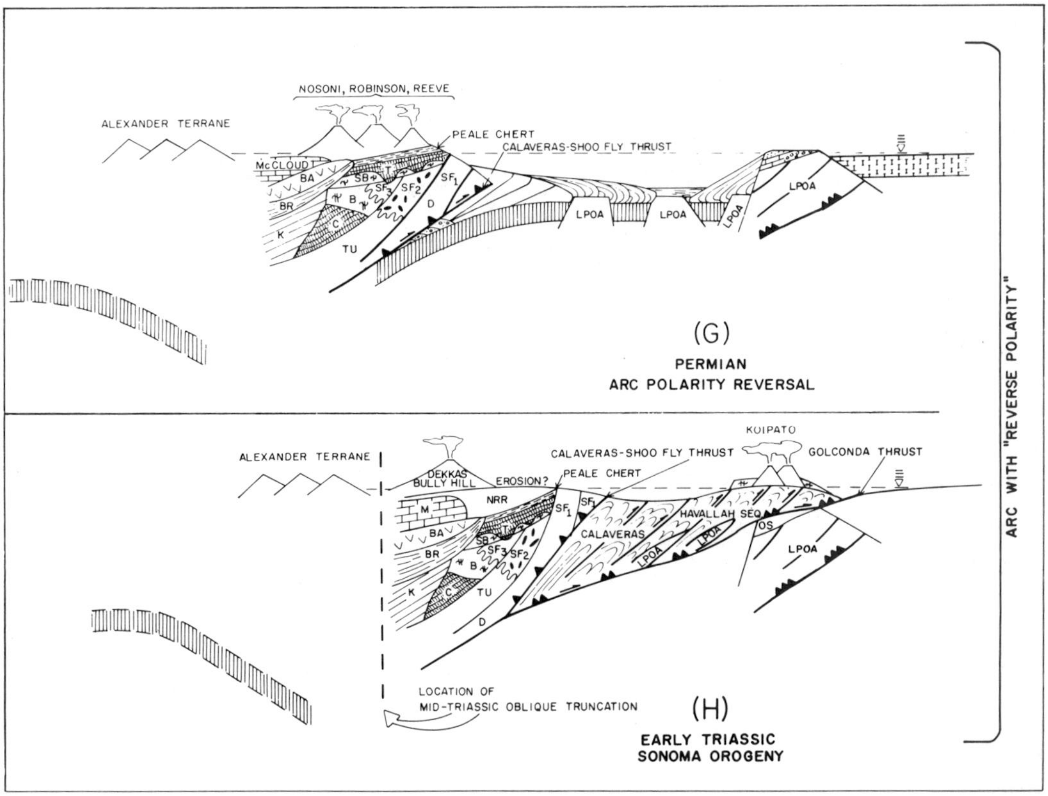

Vinini sediments were tectonically shuffled into a very large accretionary wedge (Fig. 7-6B). This accretionary wedge formed the basement for arc volcanism (represented by the Copley and Balaklala formations in the Klamath Mountains), which had stepped eastward by the Middle Devonian. The Antler orogeny (Fig. 7-6C) marked the culmination of arc-continent collision as the edge of the continent was partially subducted beneath the leading edge of the accretionary wedge which rode over it along the Roberts Mountains thrust. The siliceous volcanic and volcaniclastic rocks of the Sierra Buttes Formation and the Bowman Lake trondhjemite (Schweickert, Chapter 5, this volume) may have formed during the Late Devonian by melting of siliceous Shoo Fly sediments within the accretionary prism.

Closely following the Antler orogeny (Fig. 7-6D), the arc polarity switched to a westward-facing direction and normal island arc andesitic volcanism commenced. The subduction zone initiated back-arc spreading, which created a marginal basin (Fig. 7-6E). Sediments of the Havallah sequence, including clastic debris shed from the remnant Antler highland, accumulated in the eastern portion of this basin, and a thick volcaniclastic wedge (the base of the Calaveras Complex; Schweickert, Chapter 5, this volume) formed on the arc side.[3] The Havallah sequence sediments were deposited both on structural remnants of Roberts Mountains allochthon and on new oceanic crust that formed between these rifted blocks. As the marginal basin grew by back-arc spreading (Fig. 7-6F), olistostromes of chert and argillite with rare lenses of greenstone and shallow-water limestone were shed into the deeper parts of the basin. An apparent period of volcanic quiescence characterized parts of the arc during Mississippian and Pennsylvanian time as the cherts of the Peale Formation were deposited. The Peale may have been the source of many of the chert and argillite olistostromes that were shed into the marginal basin. In some sectors, particularly to the north of that shown in Fig. 7-6, the marginal basin must have been relatively narrow, for Early Permian faunas of two different realms, one typical of the arc, the other typical of the miogeocline (Stewart and others, 1977), occur in limestone olistoliths in the Havallah sequence. The marginal basin thus was composed of some rifted structural remnants of the Roberts Mountains allochthon (early Paleozoic in age), in addition to upper Paleozoic oceanic crust and pelagic and clastic sediments.

During the Permian (Fig. 7-6G), the polarity of the island arc reversed once again, and the arc began to subduct its own marginal basin. Additional olistostromes may have been added during this closure. The Calaveras-Shoo Fly thrust (Fig. 7-6G) marks a deep-seated fault along which the island arc and its lower Paleozoic (Shoo Fly) basement overrode marginal basin rocks of the Calaveras Complex during the initial phase of the Sonoma orogeny.

Eventually, by the end of the Permian or the beginning of the Triassic (Fig. 7-6H) during the Sonoma orogeny, the entire Calaveras-Havallah assemblage with numerous internal imbrications was obducted onto the continental margin as the arc again collided with the miogeocline. Melting within the siliceous sediments of the Havallah sequence may have produced the silicic pyroclastic rocks of the Koipato sequence, a process analogous to that which produced the Sierra Buttes Formation during the earlier Antler orogeny.

Following the Sonoma orogeny, probably during Middle Triassic time, the Cordilleran orogen was obliquely truncated by a new rift that severed the entire Paleozoic island-arc edifice south of Lake Tahoe and the lower Paleozoic parts of the arc farther north. This rifted portion of the arc complex (i.e., the Alexander terrane) presumably included a significant portion of the lower Paleozoic accretionary wedge not found within California or Nevada. This truncation scar became the site of a new subduction zone at the edge of North America in the Late Triassic, which governed the evolution of Mesozoic tectonic events, including the Late Jurassic arc-arc collision that produced the Nevadan synclinorium and related structures (Schweickert, 1976a, 1978, Chapter 5, volume; Schweickert and Cowan, 1975).

[3]We are unable to account for the lack of a remnant arc, which Dickinson (1977) noted is an important problem with the marginal basin interpretation, unless it is presently concealed beneath the leading edge of the Golconda allochthon.

C. A. Nelson
Department of Earth and Space Sciences
University of California
Los Angeles, California, 90024

8

BASIN AND RANGE PROVINCE

ABSTRACT

The geology of the Basin and Range province is the result of a complex history of events related to ancient to recent plate tectonics. Its early sedimentary history developed in response to continental fragmentation along the western margin of North America. Subsequent lithospheric plate convergence was responsible for several episodes of Paleozoic and Mesozoic deformation, arc volcanism, and plutonism. Cenozoic back-arc spreading and right-handed transform wrenching of the western margin of North America were responsible for Basin and Range graben and horst structure, extensive volcanic activity, high heat flow, oroflexural bending, and right-lateral faulting.

INTRODUCTION

The Basin and Range province in California comprises the triangular region lying east of the Sierra Nevada escarpment and north of the Mojave Desert block and the uncertain easterly extension of the Garlock fault marking the northern boundary of the Mojave (Fig. 8-1). The region is but the western edge of the larger structural and physiographic Basin and Range province, which extends eastwardly across Nevada into western Utah and southeastwardly into southern Arizona and New Mexico.

Throughout its extent, the Basin and Range province is characterized by elongate uplifted ranges with a general northward trend and alternating parallel alluvial valleys. The region has been of interest to geologists from the time of the early governmental surveys of the western United States, when speculation began as to the origin of Basin and Range structure and topography. Clarence King and the Fortieth Parallel Survey interpreted the topography as the result of arid-lands erosional processes acting on highly deformed rocks. G. K. Gilbert, on the other hand, based on studies with the Powell Survey, noted the presence of faults marking the boundaries of many of the ranges and reasoned that the ranges had been uplifted along these faults relative to the adjacent basins. This view, modified to some degree by subsequent workers, has stood the test of time. There now appears no question that normal faulting has played a major role in the development of Basin and Range structure.

The present structural subdivisions of California, including the Basin and Range, are of course accidents of late Cenozoic tectonics. Consequently, the internal structures of individual ranges and the regional pre-Cenozoic structure of the province as a whole bear only little resemblance to the present structural pattern, although some parellelism exists between Basin and Range structures, Mesozoic deformational trends, and early Cenozoic igneous patterns. The Cenozoic structures have been superimposed on a wide variety of earlier structures, some of which are as old as late Precambrian.

For the California portion of the Basin and Range we are considering, the Cenozoic structures were superimposed on miogeoclinal and eugeoclinal depositional elements and on the structures produced during the several Paleozoic and Mesozoic deformational events. For the province as a whole, these elements, as well as parts of the western edge of

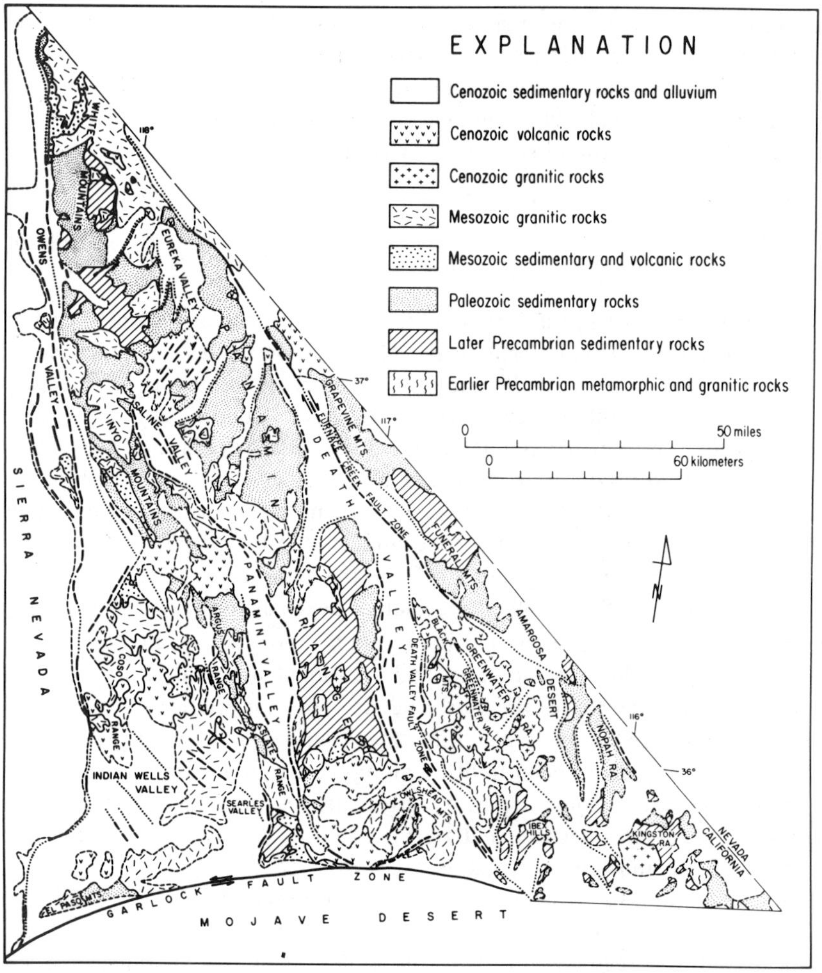

Fig. 8-1. Generalized geologic map of the California portion of the Basin and Range province, showing major time units, rock types, and significant late Cenozoic faults. Compiled from Jennings, 1977; Jennings and others, 1963; Strand, 1967; Streitz and Stinson, 1977.

the continental platform, the craton, in southeastern California and in southern Arizona and New Mexico, and parts of the Columbia Lava Plateau in southeastern Oregon, were disrupted by Cenozoic normal faulting.

PRE-BASIN AND RANGE HISTORY

Precambrian

Within the Basin and Range in California, undoubted Precambrian gneiss and granitic rocks, up to 1700 m.y. old (Silver and others, 1977), are confined to the ranges surrounding Death Valley, principally in the Black Mountains, in the southern part of the Nopah Range, and in the Panamint Range. These rocks are overlain, with profound unconformity, by a thick succession of late Precambrian strata, the Pahrump Group, comprising from the base upward the Crystal Spring, Beck Spring, and Kingston Peak formations (Wright and others, 1974b). The succession is dominated by terrigenous rocks, except for carbonates in the middle Crystal Spring and the dolomite of the Beck Spring. The Kingston Peak Formation contains diamictites, interpreted by some to be of glacial origin, within a generally siltstone succession. Wright and others (1974b) have postulated a broad regional upland, the Nopah-Mojave Upland, developed on the earlier Precambrian gneisses and lying mainly east of the present Panamint Range, and across which an east-west transverse basin, the Amargosa aulacogen, developed. It was in this aulacogen that the bulk of the sediments of the Pahrump Group and a part of the overlying Noonday Dolomite were deposited. The trend of the Amargosa aulacogen, transverse to the essentially northeast trend of the subsequent shelf edge of the continent, and its similarity to the Uinta and Beltian aulacogens to the north suggest that these may represent the earliest evidence of continental fragmentation, representing failed arms of former triple junctions as envisioned by Burke and Dewey (1973). Each is an elongate belt within the western edge of the continental plate, at a high angle to the trend of the continental margin as that trend is marked by the depositional patterns of younger strata.

The Pahrump Group and the overlying Noonday Dolomite are found no farther west than the ranges surrounding Death Valley. To the west, in the White-Inyo Range area, the oldest known strata are younger than the Noonday (Fig. 8-2). Strata equivalent to the Pahrump and Noonday in this western region, including the site of the present Sierra Nevada, are either in structural positions not yet exposed by uplift or have been completely plutonized by Mesozoic igneous activity.

Following aulacogenic deposition and the marginal overlap onto the now separated Mojave and Nopah uplands south and north of the Amargosa aulacogen, respectively, Stewart (1972, 1976) and Stewart and Suczek (1977) have proposed the initiation, as a consequence of continental rifting that took place along the western margin of North America, of the long period of essentially conformable deposition extending across the Precambrian-Cambrian boundary and continuing to mid-Late Devonian time. In places, the base of the conformable succession is at the base of the Kingston Peak Formation and in others at the base of the next overlying unit, the Noonday Dolomite.

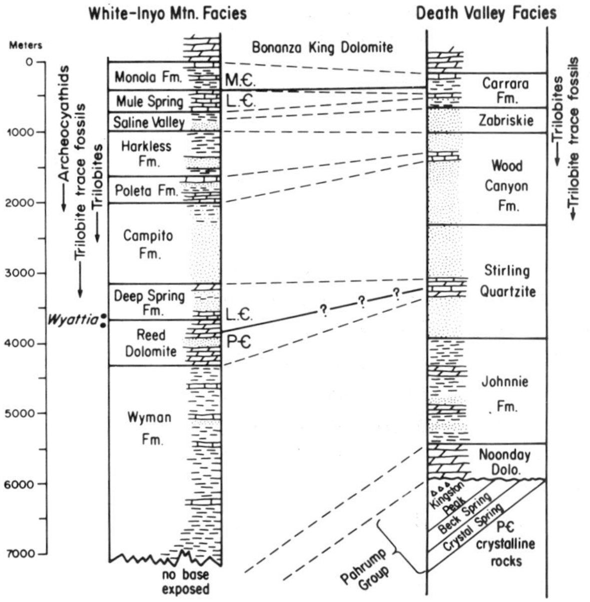

Fig. 8-2. Correlation of White-Inyo Mountains facies and Death Valley facies, showing stratigraphic distribution of various faunal elements and the inferred Precambrian-Cambrian boundary (from Nelson, 1976).

Precambrian-Cambrian

The ranges surrounding Death Valley, and the White-Inyo Range at the western edge of the Basin and Range, display thick successions of shelf-miogeoclinal strata containing the elusive Precambrian-Cambrian boundary. The well-known Death Valley succession (from the base upward, Noonday Dolomite, Johnnie Formation, Stirling Quartzite, Wood Canyon Formation, Zabriskie Quartzite, and Carrara Formation) comprises an essentially terrigenous and poorly fossiliferous Precambrian and Early Cambrian sequence. Its equivalent in the White-Inyo Range is generally more fossiliferous and represents a mixed terrigenous and carbonate sequence (Nelson, 1976). The relations of the White-Inyo facies and the Death Valley facies, the distribution of the several faunas in each, and the position of the Precambrian-Cambrian (?) boundary are illustrated in Fig. 8-2. The strata of the Death Valley and White-Inyo successions can be related to tidal, subtidal, and reefal environments (Moore, 1976; Stewart, 1970).

Paleozoic

The remainder of the Paleozoic succession is dominated by carbonate deposition, from the base of the Bonanza King Dolomite (Middle and Upper Cambrian) to the Upper

Devonian strata, the carbonate dominance broken only by the widespread mid-Ordovician Eureka Quartzite and minor sands in the Devonian carbonates.

This whole conformable succession, from Kingston Peak or Noonday to the Devonian section, is the typical shelf-miogeoclinal succession of the Cordilleran "Geosyncline." In it, each of the sandstone and quartzite units is seen to taper to zero westwardly or northwestwardly; concomitantly, the carbonates and shales thicken. Sedimentary structures such as ripple marks and cross stratification and the general geographic distribution of carbonate-dominated sections and reefal structures also indicate sediment derivation from the craton to the east.

The eugeoclinal equivalents of the preceding successions are not exposed in the Basin and Range in California. They are, however, represented by a succession, more than 9000 m thick, of pelitic hornfels, slate, sandstone, and marble, ranging in age from Early Ordovician to Pennsylvanian and Permian (?), exposed in the Mount Morrison pendant of the eastern Sierra Nevada (Rinehart and others, 1959). Eugeoclinal rocks, from Ordovician to Devonian in age, are extensively exposed farther north in western and central Nevada, where they have been juxtaposed onto miogeoclinal carbonates along the Roberts Mountains and Golconda thrusts during the Antler and Sonoma orogenies, respectively.

In the miogeoclinal area of the Basin and Range, late Paleozoic (Mississippian-Pennsylvanian) rocks of detrital aspect, the Perdido Formation and the Rest Spring Shale of the Inyo Range, represent clastic wedge deposits formed in response to Antler deformation.

The latest Paleozoic, as represented by such units as the Keeler Canyon Formation in the Inyo Range and the Bird Spring Limestone of eastern California, attest to a return to carbonate sedimentation as the dominant style in the time between Antler and Sonoma orogenic pulses.

Pre-Basin and Range Orogenic Events

In the Basin and Range province in general, four major orogenic events have been recognized:

1. Antler orogeny: latest Devonian-earliest Mississippian.
2. Sonoma orogeny: Late Permian-Early Triassic.
3. Nevadan orogeny: mid-Mesozoic.
4. Sevier orogeny: mid-Cretaceous-early Cenozoic.

Each of these orogenies apparently is related to lithospheric plate convergence (Dickinson, 1977) and is represented by either proven or suspected arc volcanism and the emplacement of plutons, as well as deformation and significant thrusting (Burchfiel and Davis, 1972). Each was also followed by the deposition of clastic wedge sequences that now lie to the east of the eastern edges of their respective thrust belts.

In the California Basin and Range area, however, it has not been possible to identify with certainty each of the four orogenies. The Antler orogeny, represented in central

Nevada by a major thrust fault, the Roberts Mountains thrust, which juxtaposes Ordovician to Devonian eugeoclinal strata over Ordovician to Devonian miogeoclinal rocks, is probably represented along the western edge of the province in the Inyo Range (Stevens and Ridley, 1974) by an angular unconformity between Middle Devonian and Upper Devonian (?) to Mississippian beds. A set of northeast-trending folds in the northern Inyo Mountains is interpreted by Sylvester and Babcock (1975) to belong to the same episode of deformation, although no structural or stratigraphic evidence bears directly on their age. The Sonoma orogeny, in which the Golconda thrust of west-central Nevada played an important role in juxtaposing eugeoclinal rocks over miogeoclinal facies strata, is reported by Dunne and others (1978) to be represented in the Inyo Range by a Late Permian-Early Triassic unconformity and possibly by a second generation of folds affecting the early Paleozoic rocks.

Mesozoic

Mesozoic history from the Middle Triassic to the close of the Cretaceous is exceedingly complex. It is represented by scattered occurrences of surficial rocks, both marine and terrestrial sedimentary rocks and volcanics, and by extensive plutonic bodies (Fig. 8-1). Analysis of Mesozoic rocks and structures in the White and Inyo Ranges, the Argus and Slate Ranges to the south, and the Panamint Range west of Death Valley led Dunne and others (1978) to suggest a significant change in the paleogeography of eastern California during early Mesozoic time, from the last phase of marine deposition in the Cordilleran miogeocline in the Late Triassic to the beginnings of terrestrial detrital rocks and volcanics of Middle Triassic to Jurassic age. A similar transition from Early Triassic marine sedimentary rocks and volcanics to terrestrial detritals, evaporites, and volcanogenic rocks of mid-Jurassic age is reported by Speed (1978) in the northwestern Basin and Range. Dunne and others (1978) also recognize three major pulses of deformation in the White, Inyo, Argus, and Slate ranges: Middle Triassic to Early Jurassic, late Middle to early Late Jurassic, and Late Cretaceous.

The Triassic to Early Jurassic episode is represented by fold and thrust fault deformation, including the Last Chance thrust system (Stewart and others, 1966), which, in much of the White-Inyo Range, brings late Precambrian and Cambrian strata above rocks as young as Permian. In fact, most of the range may be underlain by the thrust system. This deformational episode was accompanied by the emplacement of the alkalic plutons (>185 to 167 m.y.) of the White-Inyo region (see Fig. 8-4). These plutons are interpreted by Sylvester and others (1978a) to have originated from primary alkalic magma generated in the upper mantle by partial melting of material rich in alkalies (K_2O and Na_2O). A number of the Sierran-type plutons of Dunne and others (1978), with ages of 180 to 165 m.y., were also emplaced during this episode.

The Middle and Late Jurassic deformational episode (Nevadan) is similarly represented by fold deformation and the Swansea-Coso thrust system (Dunne and others, 1978). The thrust system is present nearly continuously from the southern Inyo Mountains to the Slate Range and is characterized by high-angle faults of only minor slip.

The episode was also the time of emplacement of many of the plutonic bodies (160 to 140 m.y.) of the White-Inyo Range (Crowder and others, 1973), which are regarded as comagmatic with and as statellites of the Sierra Nevada batholith, and of the emplacement of the Independence dike swarm (Moore and Hopson, 1961). The dike swarm, up to 40 km wide, consists largely of mafic dikes up to 1.5 m wide; it extends from the Argus Range northwestward to the White Mountains and beyond the Basin and Range into both the Sierra Nevada and Mojave provinces.

The Late Cretaceous deformational episode inferred by Dunne and others (1978) is represented by minor conjugate strike-slip faults cutting the Papoose Flat pluton in the Inyo Mountains (Nelson and others, 1978; Sylvester and others, 1978b) and the White Mountains fault zone along the west side of the northern White Mountains. The White Mountains fault zone, apparenty a major thrust zone of unknown slip, is bracketed in age by two Late Cretaceous plutons (Crowder and Sheridan, 1972). Several other small plutonic bodies in the White-Inyo Range (Fig. 8-4) fall within this same age range (75 to 90 m.y.).

Early-Middle Cenozoic

That plutonic activity did not end with the close of the Mesozoic is attested to by a number of anomalously young granitic rocks in the ranges west (Panamint Range) and east (Black Mountains, Greenwater Range, Ibex Hills, Kingston Range) of Death Valley (Fig. 8-1). Some have yielded K-Ar ages from 12 to 18 m.y. They all lie east of a sinuous line marking the eastern edge of Mesozoic granitics, suggesting eastward migration during early Cenozoic time of the zone generating granitic magmas.

The stratigraphic record of the early Cenozoic of the Basin and Range is nearly completely lacking. In the California segment, the early Cenozoic record is confined to the occurrence of terrestrial arkose and conglomerate of the Paleocene Goler Formation on the north flank of the El Paso Mountains. Generally, the Basin and Range region was a low upland of westward external drainage and one of very extensive erosion. The earlier formed deformational structures and their accompanying terrains had been severely reduced to a surface probably no higher than a few hundred meters above sea level. The earliest volcanic activity in the Basin and Range is latest Eocene and Oligocene, consists largely of andesites and rhyolites, and is confined to central and western Nevada (McKee, 1971).

Over the period of Oligocene to early Miocene (40 to 20 m.y.) the central portion of the Basin and Range was the site of large-scale caldera eruptions from which were produced very extensive sheets of ignimbrite flows of largely rhyolitic composition (McKee, 1971). These sheets flowed onto the nearly level erosional surface of the upland. This feature and the characteristics of the late Miocene and early Pliocene fossil floras show clearly that the site of the present Sierra Nevada was not an effective rain shadow, and that Basin and Range fragmentation was an event yet to come. Late Pliocene fossil floras, on the other hand, show the beginnings of uplift of the Sierra (Axelrod, 1957, 1962).

The fronts of many of the ranges in the province, including many in California, bear witness to the existence of bounding normal faults. All the classical evidences, such as discordance between internal mountain structure and range geometry, displaced stratigraphic units from the range crests to the foothills of the range front, alignment of springs, fresh scarps, and historic scarp-producing earthquakes, can be found in abundance. There no longer is doubt that most of the range fronts are related to normal faults of substantial throw.

The question of the absolute movements along Basin and Range frontal faults has long intrigued geologists. The paleobotanical evidence shows clearly that the region has experienced absolute uplift. Some of the basins have been elevated from levels of approximately 300 m above sea level in the early Tertiary to their present levels of 1500 m. In general, basins in the central part of the Basin and Range are 600 m higher than basins along the eastern and western margins. On the other hand, some basins have experienced absolute depression. The bottom of the Death Valley trough is filled with more than 2000 m of alluvial fill, and its surface still lies slightly below sea level. In the central part of the Owens Valley, structural relief on the bedrock surface between the Alabama Hills and the main basin to the east is greater than the topographic relief between the Alabama Hills and the crest of the Sierra Nevada at Mount Whitney (Von Huene and others, 1963).

Basin and Range normal faulting has, of course, been accompanied by large-magnitude extension. It has been noted that range fronts in the Basin and Range are commonly concave, suggesting that faults flatten at depth. Assuming an average surface dip of 60°, Hamilton and Myers (1966) calculated that each 2 km of dip-slip requires 1 km of horizontal extension; if the faults flatten downward, the extension approaches the amount of dip-slip. Assuming twenty-five major bounding faults across the province along the fortieth parallel, Hamilton and Myers have suggested extension of from 50 to 100 km.

Davis and Burchfiel (1973), in comparing distances separating correlatable pairs of geologic features across the Garlock fault, have shown that the Basin and Range segment north of the Garlock has experienced a minimum of 60 km more extension than the Mojave block south of the fault.

In addition to the well-documented normal faulting, the Basin and Range province, especially its western portion, is the site of a number of important right-lateral faults of San Andreas style and trend. Significant right-lateral separation in the Owens Valley was suggested by Hill (1954) along the Sierra Nevada fault on which about 6 m of right-lateral slip occurred during the 1872 Owens Valley earthquake. On the basis of the near continuity of trend of the Independence dike swarm across Owens Valley, Moore and Hopson (1961) regard the offset as minimal, no more than a few kilometers; Ross (1962) arrived at a similar conclusion based on correlation of granitic plutons across Owens Valley from the Inyo Range to the eastern Sierra Nevada. However, right slip on the more prominent zones to the east has been demonstrated to be significant. Stewart (1967) postulates

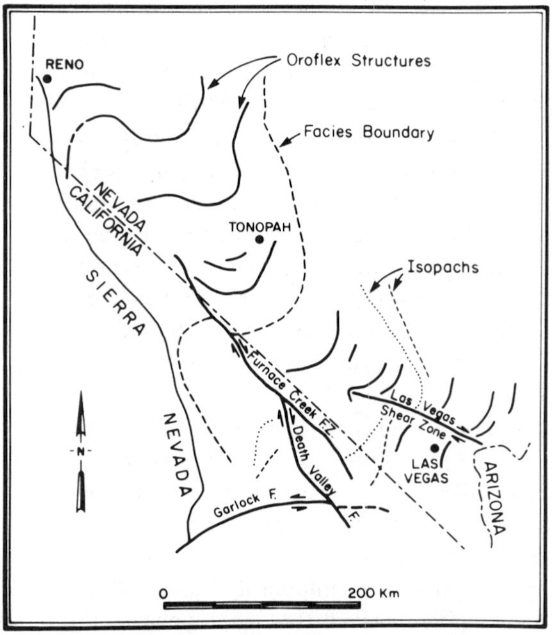

Fig. 8-3. Map of eastern California-western Nevada showing oroflex structures, sedimentary facies (Ordovician carbonates), and isopachs (2000-feet Wood Canyon Formation, dashes; 3000-feet Stirling Quartzite, dots) offset along Furnace Creek-Death Valley fault zones and the Las Vegas shear zone. Compiled from Albers, 1967; Stewart, 1967; Stewart, Albers, and Poole, 1968.

approximately 80 km of offset on the Furnace Creek fault zone of the Death Valley region, based on disruption of facies and thickness trends in late Precambrian and Paleozoic sedimentary formations. Similar data led Stewart to suggest right-lateral offset of about 48 km on the Las Vegas shear zone (Fig. 8-3)

Relation to Plate Tectonics

In 1963, Wise proposed that extensional features of the Basin and Range were the result of major right-handed wrenching of a 480-km-wide zone of western America, as shown by the sigmoidal bend of the Mesozoic plutons of the Idaho batholith-eastern Oregon plutons-Klamath Mountains system-Sierra Nevada complex. Using this model, Wise explained the general north-south trend of normal faults, the lower elevations of the Basin and Range as compared to its adjacent provinces resulting from a thinned crustal slab, and the volcanism of the region.

In 1967, Albers described the belt of sigmoidal bending, 80 km wide by 450 km long, of the ranges and their internal structures, lying generally east of the Mesozoic pluton belt, in the western portion of the Basin and Range. This oroflex structure and the associated right-lateral faulting along such faults as the Furnace Creek zone and the Las Vegas shear zone (Fig. 8-3) indicate a cumulative right-handed horizontal displacement of up to 190 km (Stewart, 1967; Stewart and others, 1968).

To account for the high heat flow in the Basin and Range and the generally thinner crust than in the adjacent provinces, Hamilton and Myers (1968) and Menard (1964), among others, have proposed that the subcontinental extension of the East Pacific Rise, having been overridden by the American plate, lies beneath the Basin and Range province; such a situation would account for the extensional features as well. Current plate-tectonic theory calls for the surface expression of the Rise to have been offset to the northwest from the Gulf of California along the transform system of the San Andreas fault zone, but this does not preclude a mantle thermal high within the Basin and Range as an extension of the Rise.

The timing of Cenozoic volcanic activity in the western Basin and Range has been used by McKee (1971) to develop a simple plate-tectonic model. He suggests that by the Middle Tertiary (40 m.y.), the Farallon plate had been underthrust beneath the American plate to a depth sufficient to cause the eruption of andesitic to rhyolitic volcanics, and that this continued to about 19 m.y. ago, at which time the plate was consumed and volcanic activity was temporarily halted. When the oceanic ridge on the west side of the Farallon plate reached the region beneath the Basin and Range, about 16 m.y. ago, the period of regional extension, Basin and Range faulting, and extensive outpouring of basalt flows began, and has continued to the present..

Stewart (1971, 1978) has provided the most detailed analyses of Basin and Range structure and its relation to plate tectonics. In 1971 he interpreted the structure as a system of horsts, many of which are tilted, and grabens within a rigid slab of crustal material that has been fragmented above a plastically extending substratum. He estimated, based on average width of valleys and assumptions of average dips of bounding faults, as well as on clay-model experiments, that the rigid slab was from 15 to 18 km thick. He visualized the basal part of the slab to have been extended along narrow zones, systematically spaced about 25 to 30 km apart. Each of the major graben-valleys is estimated to represent about 2.5 km of east-west extension, and the total extension for the Basin and Range to be about 50 to 95 km. Based on radiometrically dated volcanic rocks, the extension is thought to have developed during the last 17 m.y., and perhaps most to have taken place during the last 7 to 11 m.y.

Stewart (1978) has recently marshalled arguments in support of the theory of back-arc spreading to account for Basin and Range structures. In this model, it is proposed that extension within the Basin and Range province is related to spreading produced by upwelling from the mantle behind (east of) an active subduction zone. The model explains many other features of the province, such as the Mesozoic and early Cenozoic deformation and plutonism, the widespread mid-Cenozoic volcanism, the low seismic velocity of the upper mantle, high heat flow and thin crust, and regional uplift. The spreading was probably accelerated by a lesser confining pressure following the cessation of subduction along the western margin of the continent and by the development of the right-lateral transform zone.

Within the California segment of the Basin and Range province are a number of areas of special geologic interest. A few features of two of these areas, Death Valley and the White-Inyo Range, will be discussed.

Death Valley

Death Valley is one of the geologic show places of California. Its structural features are excellently exposed in the stark landscape of the eastern desert region. Among its features bearing directly on the problems of Basin and Range geology and plate tectonics are the Amargosa Chaos, the Death Valley turtlebacks, and evidences for very large scale vertical tectonics in the Black Mountains.

The Amargosa Chaos was described by Noble (1941) as a three-phase chaotic breccia with small to very large blocks consisting of Precambrian sedimentary units, Tertiary volcanic rocks, and a combination of those types plus Tertiary granite blocks, lying with structural discordance on Precambrian gneiss. These chaotic masses were originally interpreted as tectonic breccias lying above the Amargosa thrust. More recent work has suggested they are, at least in part, large-scale gravity-slide breccias, having slid from areas uplifted by Tertiary intrusives (Sears, 1953).

The Death Valley turtlebacks are large-scale carapacelike surfaces that plunge northwestwardly from the crest of the Black Mountains toward Death Valley. These structures have been interpreted as the result of compressional folding of the Amargosa thrust surface (Noble, 1941), as surfaces produced by the upwelling of shallow Tertiary intrusives (Sears, 1953), as exhumed undulating erosional surfaces on the Precambrian upon which Cenozoic rocks were deposited (Drewes, 1959), and as large-scale mullion structures on a major gravity fault complex roughly coinciding with the front of the Black Mountains (Wright and others, 1974a). The latter authors postulate deposition of a Cenozoic cover along the fault zone following the beginning of faulting. This cover formed the hanging wall of the fault zone; it slid along the fault zone into the large graben of Death Valley, which had formed as a "pull-apart" structure (Burchfiel and Stewart, 1966) as a consequence of extensional tectonics (Hill and Troxel, 1966). This interpretation of the turtlebacks fits well with the late Cenozoic extensional history of the Basin and Range, whereas the earlier interpretaion of turtlebacks as compressional features is incompatible with such a history. The turtleback mechanism may also provide an explanation for the gravity sliding of the Amargosa Chaos blocks.

The Black Mountains block provides another enigmatic structural feature. In simplified terms, the Black Mountains, lying between the Furnace Creek fault zone on the east and the Death Valley fault on the west (Fig. 8-1), comprise a Precambrian gneissic core on which, with profound unconformity, lie late Cenozoic surficial rocks. Late Precambrian and Paleozoic rocks are nearly completely absent, whereas in the ranges to the east, southeast, and west, sections of late Precambrian and Paleozoic strata are more than 9000 m thick. Removal of this cover, which certainly was deposited on the

Black Mountains Precambrian, requires a truly colossal amount of vertical tectonics since the Mesozoic. Much of the uplift may have been relatively late in the Cenozoic.

White-Inyo Range

Many features of the White-Inyo Range have already been presented. Further discussion of the nature and style of some of the Mesozoic plutons may be of interest, however. In general, the Mesozoic plutons have age spans similar to those of the Sierra Nevada, from about 185 to 75 m.y. (Dunne and others, 1978). A major difference, however, is that in the White-Inyo Range the plutons occur mostly as discrete bodies intruded into relatively unmetamorphosed to only mildly metamorphosed late Precambrian and early Paleozoic rocks as shallower and lower temperature or more rapidly cooling intrusives than those of the Sierra Nevada. Consequently, it has been possible to relate the metasedimentary aureoles of many of the plutons to known stratigraphic sections and to reach thereby some conclusions as to pluton geometries and their emplacement tectonics.

Figure 8-4 is a highly generalized geologic map of a part of the central White-Inyo Range showing the positions of the major plutons of the region and their relations to the major structure of the range. The range is dominated by the N-S trending White Mountains anticline and the NW-SE trending Inyo anticline. This major structure is flanked on the northwest by an overturned syncline and on the east by a pair of folds.

The synclinal fold to the east of the White Mountains anticline is occupied by the plutonic rocks of the Joshua Flat (182 m.y.) and the Eureka Valley (182+ m.y.) monzonites. The probably originally elliptical (in plan) Joshua Flat-Eureka Valley body was invaded and the ellipse breached by the younger Beer Creek quartz monzonite (170 to 180 m.y.). These granitic masses comprise the southern portion of the Inyo batholith (McKee and Nash, 1967). Although the contacts of the plutons with the country rocks are regionally discordant, the southern contact is remarkably concordant; curiously, the body occupies the axial portion of a major steeply plunging syncline.

To the south, and isolated from the Inyo batholith, but along the same trend, the Marble Canyon pluton contains granitic rocks identical to the Joshua Flat monzonite, intrusive into diorite of uncertain but greater age. The Marble Canyon composite pluton occupies the central portion of a synclinal basin, and what is more remarkable is that its eastern contact with Cambrian strata is parallel to sedimentary bedding and dips shallowly (45°) westward toward the pluton, suggesting a funnel-shaped intrusive.

Along or west of the White Mountain-Inyo anticlinal axes are three Cretaceous plutons exhibiting a variety of emplacement mechanisms. At the north, the Sage Hen Flat quartz monzonite (137 m.y.) is a generally circular pluton crosscutting the west flank of the White Mountains anticline. It has produced virtually no deformation at its contacts, and it clearly occupies space formerly held by parts of the sedimentary succession. Even though it contains few xenoliths, it appears to be explainable only as a body that has passively sloped its way upward to its present position.

The Birch Creek pluton (Nelson and Sylvester, 1971), a body of quartz monzonite and granodiorite (81 m.y.), was emplaced along a pregranite fault. Decarbonation of the

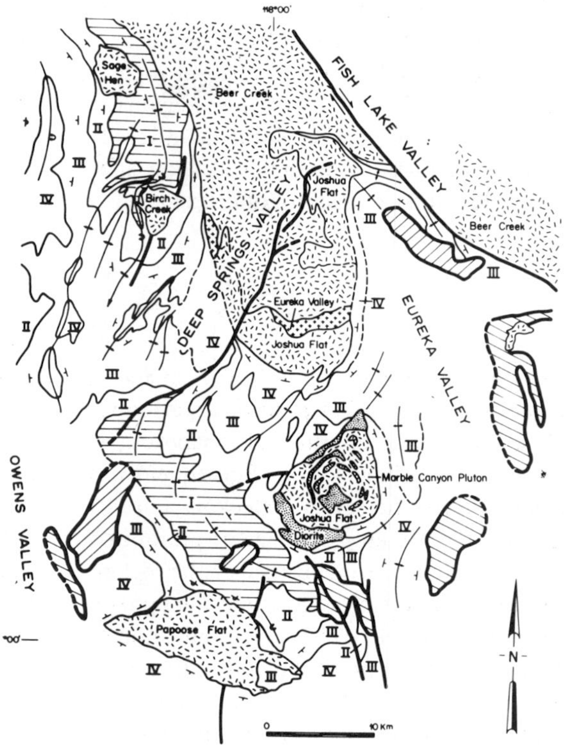

Fig. 8-4. Generalized geologic map of central White-Inyo Range, showing stratigraphic and structural setting of Mesozoic plutons. I, Wyman Formation; II, Reed and Deep Spring formations; III, Campito and Poleta formations; IV, Harkless-Bonanza King formations (see Figure 8-2); NW-SE lined pattern, Mississippian rocks in windows beneath Last Chance thrust; NE-SW lined pattern, thrust blocks within upper plate of Last Chance thrust. Base map from Ross, 1967a.

dolomitic host rocks liberated sufficient CO_2 to induce isothermal crystallization in the upper part of the melt. This resulted in a viscous rind, which, yielding to rising fluid magma from beneath, was forcibly injected northwestwardly, thereby overturning the White Mountains anticline and producing curvilinear traces of strata, crestal axes of subsidiary folds, and prepluton faults, all of which wrap concordantly around the northwest corner of the pluton.

The Papoose Flat pluton (Nelson and others, 1978; Sylvester and others, 1978b) is a body of quartz monzonite (78 m.y.) characterized by its large K-feldspar megacrysts and a highly foliated border zone. It was emplaced within the southwest limb of the southeast-plunging Inyo anticline. The western end of the pluton has drastically disrupted the regional southeast trend of the anticlinal limb by producing a pronounced westward bulge. Bulging has occurred along the northeast contact as well, where the pluton has overturned the stratigraphically lower part of its host rocks. In the western two-thirds of the pluton, foliation parallels bedding in the adjacent metasedimentary rocks; the latter have been attenuated to as little as 10% of their regional thicknesses without loss of stratigraphic identity or continuity. The pluton is inferred to have penetrated discordantly through the lower formations of the succession and to have formed a "blister" beneath the stretched higher units, as a laccolithic structure.

SUMMARY

In summary, it is seen that the varied features of Basin and Range geology can be related to parts of the plate-tectonic process. Depositional patterns, including aulacogenic deposition, are likely the result of early continental separation, which was followed by a long period of shelf and geoclinal sedimentation. Subsequent lithospheric plate convergence was presumably responsible for the several Paleozoic and Mesozoic orogenic episodes, as well as for arc volcanism and the emplacement of plutons so characteristic of the latest Paleozoic, Mesozoic, and early Cenozoic history of the region. Basin and Range structure, high heat flow, thin crust, low seismic velocity in the upper mantle, Cenozoic volcanism, and regional uplift can be related to a simple tectonic model. Upwelling from the mantle behind an active subduction zone produced spreading and its accompanying crustal extension. At the close of subduction of the Pacific (Farallon) Plate and the intersection of the East Pacific Rise with North America, extension was accentuated by right-lateral wrenching of the wide, "soft" margin of western North America (Atwater, 1970), with the development of pull-apart grabens, oroflexural deformation, and right-lateral faulting.

ACKNOWLEDGMENTS

Special thanks are due John H. Stewart of the U.S. Geological Survey for a critical reading of the manuscript and for stimulating discussions of Basin and Range problems, both in the field and office.

B. C. Burchfiel
Department of Earth and Planetary Sciences
Massachusetts Institute of Technology
Cambridge, Massachusetts 02139

Gregory A. Davis
Department of Geological Sciences
University of Southern California
Los Angeles, California 90007

9

MOJAVE DESERT AND ENVIRONS

ABSTRACT

Paleogeographic interpretations of the Mojave region suggest most of the western part of the region was the site of miogeoclinal late Precambrian sedimentation, but cratonic sedimentation during the early Paleozoic and perhaps for the entire Paleozoic. The northwestern quarter of the Mojave contains Paleozoic rocks that are eugeosynclinal in character. We regard them as out of place. These eugeosynclinal rocks may represent slivers of rocks displaced by left-slip faulting during a Permo-Triassic truncation event.

Mesozoic plutonic and volcanic rocks form a northwest-trending Andean-type magmatic arc that is superposed on all previous paleogeographic units. Arc rocks interfinger to the east into largely nonmarine, back-arc sedimentary rocks and interfinger locally with marine sedimentary rocks to the west. Sites of arc magmatism fluctuated irregularly in time; at least one period of arc inactivity is represented by the incursion of nonmarine cratonic rocks of Early Jurassic age into the arc terrane.

Although Paleozoic structures may be present locally, the major structural features of the Mojave region are Mesozoic and Cenozoic in age. Mesozoic structures developed at many different times, and their spatial distributions, while poorly known at present, suggest they fluctuated irregularly in concert with magmatic activity. Episodes of folding and east-directed thrust faulting alternated or were contemporaneous with episodes of strike-slip and normal-slip faulting. Folds and thrust faults typically involve Precambrian crystalline basement rocks.

At least twice during latest Cretaceous and earliest Tertiary time, Franciscan rocks were underthrust to the east at a shallow angle beneath the western edge of the North American plate. These rocks, known as the Pelona, Rand, and Orocopia schists, are now present in windows beneath Precambrian and Mesozoic crystalline rocks. Our interpretation for emplacement of these rocks differs from some recent interpretations, which regard them as forming in an intracontinental ensimatic basin separate from the Franciscan and later being underthrust to the west.

Evidence from regional trends of Paleozoic and Mesozoic paleogeographic and structural elements suggests the rocks of the western Mojave, southern Sierra Nevada, northern Salinian block, and adjacent terranes were oroclinally bent in a clockwise sense. Timing of this proposed right-lateral bend is not clear and may have occurred between latest Cretaceous and mid-Tertiary time.

INTRODUCTION

The Mojave Desert region occupies a large part of southeastern California. It is a wedge-shaped physiographic province that narrows westward (Fig. 9-1) and is bounded by numerous physiographic elements, among them the Tehachapi Mountains-southern Sierra Nevada on the northwest, the Basin and Range province on the north, and the San Gabriel and San Bernardino Mountains on the southwest and south, respectively. The eastern and

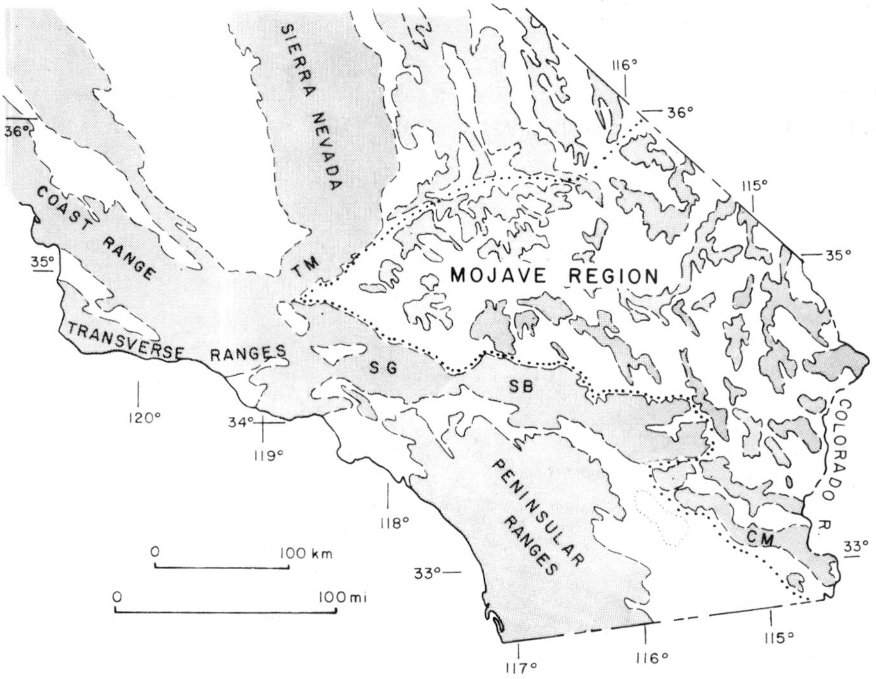

Fig. 9-1. Index map of Southern California showing location of Mojave region. TM, Tehachapi Mountains; SG, San Gabriel Mountains; SB, San Bernardino Mountains; CM, Chocolate Mountains.

southeastern boundaries are more arbitrary, but are drawn for the purposes of this paper along the Colorado River south of Hoover Dam and the northeastern edge of the Salton trough, respectively (Fig. 9-1). The modern physiographic province, however, is a reflection of the most recent geologic history of the area, and its boundaries only rarely follow the more fundamental geological boundaries of pre-late Cenozoic time. For that reason the geology of surrounding regions, particularly the Basin and Range province and the southern Sierra Nevada, must be treated to some extent inasmuch as these areas contain pre-Cenozoic stratigraphic units and structural elements that trend into the Mojave region, where they are less well preserved and exposed.

Approximately 50% of the Mojave Desert is covered by recent alluvium, and of the other 50%, two-thirds is underlain by Mesozoic plutonic rocks and Cenozoic igneous and sedimentary rocks. The remaining outcrops consist of small scattered exposures of Precambrian crystalline rocks, Precambrian and Paleozoic sedimentary rocks, and Mesozoic volcanic and sedimentary rocks. This paucity of pre-Cenozoic stratigraphic units makes reconstruction of pre-Cenozoic geology quite difficult, but in the past few years important discoveries have permitted a significantly more complete geologic synthesis of the Mojave region than was possible only five years ago.

PRECAMBRIAN CRYSTALLINE ROCKS

Precambrian crystalline rocks occur at scattered localities within the Mojave Desert region and more widely in such adjacent areas as Death Valley, western Arizona, and the San

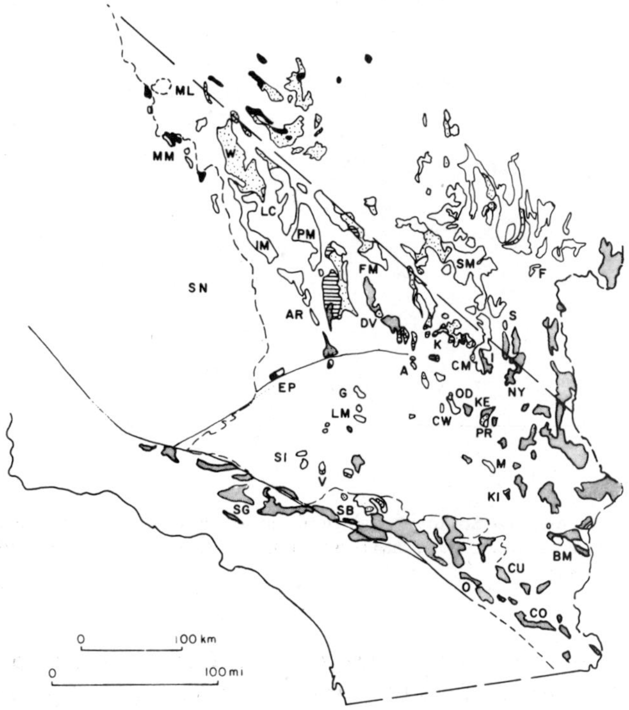

Fig. 9-2. Distribution of Precambrian and Paleozoic rocks in the Mojave region and environs. Black, Ordovician slope and rise sequences; shaded Precambrian crystalline rocks; horizontal lines, late Precambrian Pahrump Group; dots, late Precambrian-Early Cambrian terrigenous sedimentary rocks; white, Paleozoic sedimentary rocks. AR, Argus Range; A, Avawatz Mountains; BM, Big Maria Mountains; CU, Chuckwalla Mountains; CO, Chocolate Mountains; DV, Death Valley; EP, El Paso Mountains; FM, Funeral Mountains; F, Frenchman Mountain; G, Goldstone; I, Ivanpah Mountains; I, Inyo Mountains; KI, Kilbeck Hills; KE, Kelso; K, Kingston Range; LM, Lane Mountains; MM, Mount Morrison; ML, Mono Lake; M, Marble Mountains; NY, New York Mountains; O, Orocopia Mountains; OD, Old Dad Mountain; PM, Panamint Mountains; PR, Providence Mountains; S, Sheep Mountain; SM, Spring Mountains; SI, Shadow Mountains; SG, San Gabriel Mountains; SB, San Bernardino Mountains; SN, Sierra Nevada; V, Victorville; W, White Mountains.

Bernardino and San Gabriel Mountains. Known and suspected outcrops of these rocks are shown in Fig. 9-2. Published data on the Precambrian basement rocks of the Mojave are disappointingly scanty. The most detailed works treat parts of the Clark, Marble, Orocopia, and Chocolate Mountains in the Mojave region proper and, in adjacent areas, the Panamint Range near Death Valley and the San Gabriel Mountains.

Precambrian crystalline rocks in the Panamint Range consist of an older quartzofeldspathic augen gneiss with inclusions of more mafic biotite quartz schist intruded by younger porphyritic granite. Some quartz gneiss with amphibolite and quartz-mica schist layers, quartzite, and pegmatite are also present. Ages on the older metamorphic rocks range from 1.87 to 1.79 b.y., and younger pegmatites have yielded ages of 1.66 to 1.73 (Lanphere and others, 1963; Silver and others, 1977; Wasserburg and others, 1959). Granites in the basement rocks may be about 1.35 b.y. old (Lanphere and others, 1963).

In the Clark Mountains to the southeast, a Precambrian crystalline basement consists of amphibolite-grade metaigneous and metasedimentary rocks, including granitic augen gneiss, biotite-garnet-sillimanite gneiss, hornblende gneiss, and amphibolite (Olson and others, 1954). They are intruded by syenite, granite, and carbonatite. Geochronological work by Lanphere (1964) has indicated that the protoliths of the gneissic rocks were metamorphosed and intruded by pegmatites at about 1.65 b.y. Intrusion of potassium-rich plutons and carbonatites occurred at about 1.40 b.y. Lanphere also reports an age of 1.7 ± 0.065 b.y. on granitic gneiss in the nearby Winters Pass Area.

To the south in the Marble Mountains (Fig. 9-2), Cambrian sedimentary rocks unconformably overlie granodiorite containing pendants of quartzofeldspathic gneiss, marble, and mafic igneous rocks. The basement rocks are intruded by a granite that yields an age of 1.4 b.y. (Lanphere, 1964).

Still farther south, the Chocolate Mountains contain a thick sequence of amphibolite-grade augen gneiss, quartzofeldspathic gneiss, and schist of both igneous and sedimentary origin (Dillon, 1976). Also included in this terrane are metamorphosed mafic and ultramafic rocks and anorthosite. A date of 1.7 b.y. for an intrusive augen gneiss is reported by Dillon. In the nearby Orocopia Mountains the oldest rocks consist of augen gneiss and migmatites, part of the Chuckwalla Complex of Miller (1944), which yield a zircon age of 1.67 ± 15 b.y. from the augen gneiss (Silver, 1971; Crowell, 1975b). Blue-quartz gneiss, which may once have been granulite-grade rocks but now contains amphibolite-facies mineralogy, has been dated at 1.425 b.y. Syenite, anorthosite, and related rocks are also present in the terrane and give ages of 1.22 b.y. (Silver, 1971). Similar rock sequences are present in the San Gabriel Mountains to the northwest (Silver, 1971).

Existing, but geographically scattered data thus indicate that the Precambrian crystalline rocks of the greater Mojave region were deformed and metamorphosed about 1.7 b.y. ago, and were later intruded by granitic rocks at about 1.4 b.y. Syenites and carbonatites were intruded in the Clark Mountains area 1.35 to 1.4 b.y. ago. In the San Gabriel Mountains and formerly adjacent terranes, syenites and anorthosites were intruded at about 1.2 b.y.

LATE PRECAMBRIAN AND PALEOZOIC STRATIFIED ROCKS

Pahrump Group

The oldest unmetamorphosed sedimentary rocks that rest unconformably on Precambrian crystalline rocks in the Mojave region are those of the Pahrump Group. These rocks are present in the Death Valley area (Fig. 9-2) and are locally more than 2 km thick. They have distribution and facies relations that suggest deposition in a fault-controlled trough that trended west to northwest (Wright and others, 1974b). Faults along the northeast side of the trough, called by Wright and others (1974b) the Amargosa Aulacogen, are well documented, but the southern boundary is not well established. The rock sequence consists of a basal sequence (Crystal Spring Formation) of quartzite, siltstone, shale, and dolomite, in many places stromatolitic, overlain by a middle unit of dolomite (Beck Spring Dolomite). A thick upper unit (Kingston Peak Formation) consists of sandstone, conglomerate, and diamictite, which contains debris from the lower two units.

The age of the Pahrump Group is not well constrained, but in the Panamint Range it rests on a basement that contains plutons dated at 1.35 to 1.4 b.y. and is overlain unconformably by upper Precambrian rocks that continue without interruption into fossiliferous Lower Cambrian rocks. The Crystal Spring Formation is intruded in many places by diabase sills and dikes that produce rich "talc" (tremolite) deposits where they are in contact with dolomite. In Arizona, diabase dikes cutting similar rocks have been dated as 1.15 to 1.2 b.y. (Silver and others, 1977). Thus the Pahrump Group is younger than 1.35 to 1.4 b.y. and is older than the latest Precambrian; its lowest formation may be older than 1.2 b.y.

Post-Pahrump Strata

Late Precambrian strata younger than that of the Pahrump Group, and Paleozoic strata are only present in isolated small areas within the Mojave province, but form a more continuous terrain to the northeast in southeastern California and southern Nevada (Fig. 9-2).

The various facies belts of these units have northeast trends across southern Nevada and southeastern California and are shown in a cross section constructed approximately along the California-Nevada state line from near Las Vegas to Mono Lake (Fig. 9-3). Because this section has been disturbed by east-directed Paleozoic and Mesozoic thrust faults, the following discussion is based on a palinspastic reconstruction. The southeasternmost exposures of Paleozoic rocks in the Las Vegas area are cratonal and are characterized by a thin and incomplete sequence of dominantly shallow marine rocks. Northwestward the Paleozoic sequence thickens and becomes more complete, acquires a lower section of latest Precambrian strata, and is miogeoclinal in character. The westernmost exposures of Paleozoic rocks are structurally complex but indicate sedimentation in possible continental slope and rise environments.

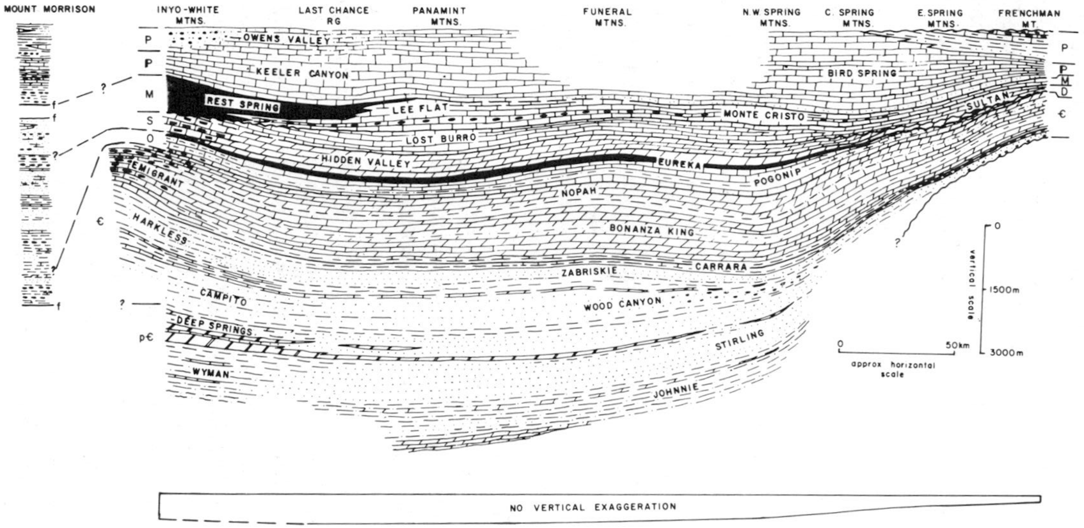

Fig. 9-3. Generalized stratigraphic framework of late Precambrian and Paleozoic rocks on a transect from near Las Vegas, Nevada (Frenchman Mountain), to near Mono Lake, California (Mount Morrison). No Pennsylvanian and Permian rocks crop out in the vicinity of the Funeral Mountains and are thus not shown in this section. Structural relations between the Mount Morrison rocks and thus of the Inyo-White Mountains are unknown, as discussed in the text.

Cratonal Sequence

Rocks belonging to the cratonal sequence crop out at Frenchman Mountain (Fig. 9-3) and Sheep Mountain, Nevada, and in the Ivanpah Mountains of southeastern California. The major components of this sequence include (1) basal terrigenous units of Early and early Middle Cambrian age (Tapeats Sandstone, Bright Angel Shale) overlying Precambrian crystalline rocks; (2) Middle and Upper Cambrian shallow-water carbonate rocks unconformally overlain by stromatoporid-bearing Middle Devonian dolomite (basal Sultan Formation); and (3) Mississippian, Pennsylvanian, and Permian limestone with red beds present in the Early (?) Permian. Total thickness of the Paleozoic cratonal sequence ranges from 2000 to 2500 m.

Miogeoclinal Sequence

Transitions from cratonal to miogeoclinal sequences are obscured by thrust faulting, but appear to occur in three ways: (1) by addition of rock units at the sub-Middle Devonian unconformity; (2) by addition of a thick sequence of upper Precambrian terrigenous rocks below Lower Cambrian rocks; and (3) by thickening of all formations westward (Burchfiel and others, 1974). Transition to the miogeoclinal sequence does not occur uniformly. Formations first thicken slowly, then more rapidly with distance from the craton, but only in some parts of the sequence. Upper and Lower Ordovician rocks first appear below the sub-Middle Devonian unconformity in sequences that on other characteristics would be considered cratonal (Fig. 9-3). The late Precambrian terrigenous rocks appear completely developed at the base of the Wheeler Pass thrust fault in the western Spring Mountains. How these rocks are disposed below the base of the Cambrian or what direction(s) their facies and thicknesses trend cannot be established along the line of cross section.

In the western part of the Spring Mountains the miogeoclinal sequence is complete and consists of more than 3000 m of late Precambrian to early Middle Cambrian terrigenous rocks derived from eastern sources, and 4000 to 5000 m of Middle Cambrian to Permian shallow marine carbonate rocks with only minor terrigenous units (the most important of which is the Middle Ordovician Eureka Quartzite). Farther westward the same general characteristics prevail, although in the western Death Valley-Inyo Range area (Fig. 9-2) different formational designations are used for most of the units. Several important changes take place in this western part of the miogeoclinal sequence. The most important is the presence of Middle Mississippian and Early (?) Pennsylvanian terrigenous rocks in the Perdido Formation and Rest Spring Shale (shown as shaded units in Fig. 9-3). The northwestward transition from carbonate rocks to these terrigenous rocks is abrupt (Stevens and Ridley, 1974). These terrigenous rocks can be considered as the southern continuation of the late Paleozoic clastic wedge shed eastward from the Mississippian Antler orogenic belt that lay north and northwest of the line of section (Pelton, 1966). Other changes in the western part of the miogeocline include the presence of outer-shelf fine-grained clastic rocks at many horizons in the Paleozoic carbonate formations, and the presence of bedded cherts in several early Paleozoic formations.

Changes in rock types present in this western part of the miogeocline can be interpreted as representing changes occurring along the outer edge of the Paleozoic continental shelf. Even though some rock types suggest deposition in deeper water, the rocks were deposited on an older continental basement, as indicated by the dominance of associated shallow-water sedimentary rocks and by high initial ^{87}Sr-^{86}Sr ratios of Mesozoic plutonic rocks that intrude the miogeoclinal sequence (Kistler and Peterman, 1973).

Continental Slope and Rise Sequence

Poorly exposed and little studied rock units consisting of chert, slate, and rare limestone of early Paleozoic age lie north and west of the miogeoclinal sequence of the Inyo-White Mountains area. To the north in western Nevada they are assigned to the highly deformed Ordovician Palmetto Formation, unnamed Devonian rocks (Stanley and others, 1977), and the Cambrian Emigrant Formation. To the west the rocks belong to unnamed Ordovician units in the eastern pendants of the Sierra Nevada (Fig. 9-3). In western central Nevada, rocks of similar facies have been thrust eastward from deep marine environments onto the continental shelf in Late Devonian and Early Mississippian time during the Antler orogeny (Roberts and others, 1958). Highlands created during the Antler orogeny were the source of material shed eastward forming a widespread clastic wedge, part of which is present in the western part of the miogeoclinal cross section. Whether the Palmetto and Emigrant formations and the chert, slate-rich sequences of the eastern Sierra Nevada are part of the allochthonous rocks of the Antler orogenic belt is presently not known. We regard them as having been deformed during the Antler orogeny, but the magnitude of Antler thrusting, if any, affecting these rocks is uncertain. Late Paleozoic rocks in the Mount Morrison pendants of the eastern Sierra Nevada contain Pennsylvanian carbonate and clastic rocks. Although they are in fault contact with lower Paleozoic rocks, they probably represent part of the terrigenous rocks eroded from the Antler highlands.

Trends of Late Precambrian-Paleozoic Sequences

Rock sequences of cratonal character can be identified south of the section (Fig. 9-3) in the New York Mountains (Burchfiel and Davis, 1977), Marble Mountains (Hazzard, and Mason, 1936), Kilbeck Hills (Jones, 1973; Evenson, 1973), and the Big Maria Mountains (Hamilton, 1971). Rock sequences farther west are more difficult to assign to either a cratonal or miogeoclinal sequence because in some areas they have the characteristics of both, depending upon which part of the sequence is examined. Late Precambrian (post-Pahrump) terrigenous units can be identified in the Providence Mountains, Kelso Hills, Old Dad Mountain, and Cowhole Mountains (Fig. 9-2). The units are incomplete in the most southern areas where several unconformities are present, but become more complete northward. In the Avawatz and Clark Mountains and the Kingston Range, the late Precambrian terrigenous units are fully and completely developed. Thus a line separating areas where the cratonal Tapeats Sandstone is present from areas containing older, late Precambrian rocks would pass through the central Spring Mountains, eastern Clark and New York Mountains, and west of the Kilbeck Hills and Big Maria Mountains (Fig. 9-4).

Lower Paleozoic carbonate rocks, however, have a different distribution of cratonal and miogeoclinal facies than the upper Precambrian and Cambrian rocks. An unconformable relation between the stromatoporoid-bearing Middle Devonian of the basal Sultan Formation and the Upper Cambrian Nopah Formation, a typical cratonal characteristic, is present in the Providence Mountains, Clark Mountains, Old Dad Mountain (?), and Cowhole Mountains. Thus these latter areas have transitional or miogeoclinal upper Precambrian sections, but lower Paleozoic rocks (Fig. 9-4) with cratonal characteristics. Because upper Paleozoic rocks in these areas are incompletely exposed or preserved, little can be said regarding their paleogeographic position.

In the western part of the Mojave region only the Victorville area has yielded a recognizable upper Precambrian and lower Paleozoic stratigraphic sequence. Stewart and

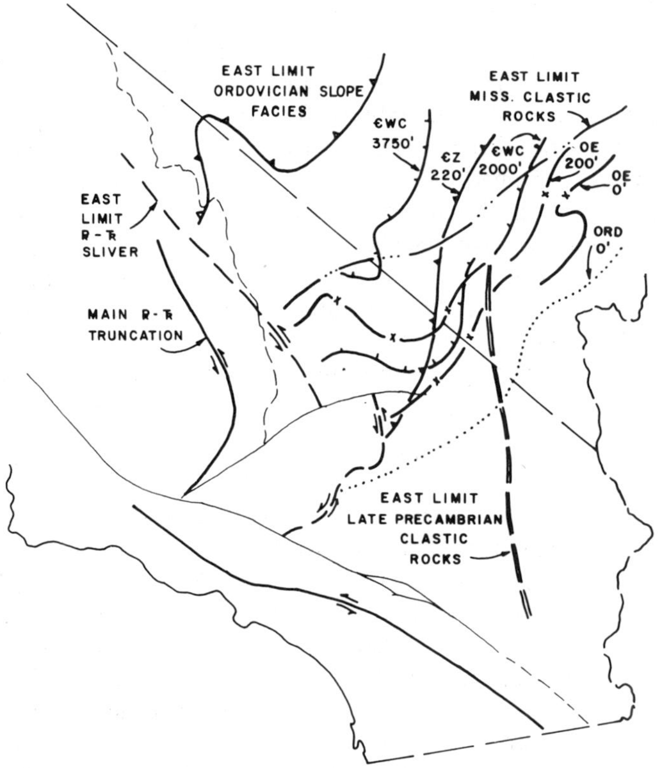

Fig. 9-4. Interpretive map of late Precambrian and Paleozoic facies and isopachous lines and distribution of major paleogeographic elements.

Poole (1975) and Miller (1977) have described an upper Precambrian-Cambrian terrigenous section similar to that found near the northeastern Mojave region. They recognize the Carrara, Zabriskie, and Wood Canyon formations and older quartzites and carbonate rocks that cannot be assigned to formations that normally underlie the Wood Canyon Formation. Conversely, lower Paleozoic rocks are typically cratonal in character since stromatoporoid-bearing Middle Devonian (?) dolomite rests unconformably on Upper Cambrian Nopah Formation (Miller, 1977). Upper Paleozoic rocks are also present, but they are too metamorphosed to subdivide accurately. Thus the section at Victorville includes both miogeoclinal Precambrian rocks and lower Paleozoic sedimentary rocks with cratonal affinity.

A relation similar to that at Victorville is probably present in the eastern San Bernardino Mountains where the basal terrigenous rocks contain the Carrara, Zabriskie, and Wood Canyon formations and older rocks that cannot be assigned to known stratigraphic units (Tyler, 1975; S. Cameron, in progress). Both lower and upper Paleozoic carbonate rocks are present, but metamorphism and structural complications have not yet permitted stratigraphic details of the critical middle part of the Paleozoic sequence to be established.

All these data indicate that the eastern part of the Mojave region contains cratonal Paleozoic rocks (Fig. 9-4), whereas the southwestern Mojave belongs to a transitional area that was the site of miogeoclinal late Precambrian sedimentation, but cratonal early Paleozoic sedimentation. It is clear that the facies trends for these two parts of the stratigraphic record are different (Fig. 9-4). It is also clear that lower Paleozoic miogeoclinal rocks should lie north of a line that passes through the southern Spring Mountains, north of the Cowhole Mountains near Baker, and north of the Victorville area (Fig. 9-2).

Out-of-Place Paleozoic Units, Western Mojave Region

Although facies and isopachous trends for miogeoclinal and cratonal rocks of late Precambrian and Paleozoic age project southwestward across the California-Nevada state line, these rocks do not crop out in projected locations west of the White-Inyo Mountains, the Argus Range and the Avawatz Mountains, and the Victorville area (Figs. 9-2 and 9-4). Only a few small exposures of Paleozoic or probable Paleozoic rocks are present in the western Mojave and southern Sierra Nevada regions, but these rocks are not counterparts of those miogeoclinal or cratonal units that project toward their locations from the northeast. For example, the Garlock Series in the El Paso Mountains (Fig. 9-2) just north of the Garlock fault consists of more than 10,000 m of both lower and upper Paleozoic rocks (Dibblee, 1967). Ordovician and Devonian fossils have been collected from the lowermost 3000 to 4000 m, which includes slate, chert, greenstone, and rare limestones (Poole and others, 1977). These rocks are lithologically similar to the lower Paleozoic rocks of west-central Nevada, which were deposited in a continental slope and rise environment (Roberts and others, 1958). They are overlain unconformably by quartz-rich sandstone and shale that are unfossiliferous, but are believed by Poole and Sandberg (1977) to be part of the Mississippian clastic wedge shed southeastward from the Antler orogenic belt. A still higher sequence of sandstone, shale, chert, and chert-pebble conglomerate contains limestone beds bearing a Permian fusulinid fauna (Dibblee, 1967). Intermediate volcanic rocks within the middle part of the Garlock Series may be wholly or in part intrusive rocks rather than flows and, therefore, not part of the series. The uppermost 4000 to 5000 m of the series consists of slate, limestone, and abundant calc-silicate rocks that have not yielded fossils and may include Mesozoic stratigraphic units. The Garlock Series in the El Paso Mountains terminates southward along the Garlock fault, but outcrops south of the Garlock fault and 60 km to the east at Pilot Knob Valley represent the same sequence of rocks (Smith and Ketner, 1970).

In the Lane Mountain area, in the central western part of the Mojave Province (Fig. 9-2), McCulloh (1952) mapped more than 7000 m of metasedimentary rocks as the Coyote Group. The lower half of this sequence consists of hornfels and slightly schistose rocks whose protolith was a sequence of impure sandstone, siltstone, and shale; associated with these rocks are rare metaconglomerate and limestone. A few hundred meters of andesitic volcanic rocks (Mesozoic?) are present in the middle of the sequence and are overlain by predominantly impure limestones, now calc-silicate rocks. McCulloh (1954) found crinoid debris of probable Paleozoic age in rocks near the middle of the group, and M. Rich (1971, 1977) reports recrystallized fusulinids of possible Pennsylvanian age from marble lenses near the middle of the lower clastic part of the sequence.

Other sequences of impure sandstone, siltstone, chert, and limestone, variously metamorphosed to hornfels, semischist, and marble, are present near Goldstone Lake and in the Shadow Mountains (Fig. 9-2). The rocks in the Shadow Mountains have been mapped recently by Troxel and Gunderson (1970) and Miller (1977). No fossils were found by these workers, although Bowen (1954) reported a possible Pennsylvanian brachiopod from the southern Shadow Mountains. The age of the rocks is thus uncertain. If they are at least in part Paleozoic, they are unlike the nearby pure carbonate Paleozoic rocks to the southeast at Victorville and in the San Bernadino Mountains. Nor do they resemble the miogeoclinal sequences to the northeast, which project toward this area.

Two interpretations have recently been offered for the occurrence of the anomalous or "out-of-place" Paleozoic rock sequences of the western Mojave region. Poole (1974) has suggested that these rocks represent the southward continuations of the middle Paleozoic (Antler) and late Paleozoic (Sonoma) orogenic belts from west-central Nevada. More specifically, Poole and others (1977) interpret the lower Paleozoic rocks in the Garlock Series to lie within the Roberts Mountains allochthon. Following Burchfiel and Davis (1972), Poole (1974) postulates that the allochthon consists of sedimentary and volcanic rocks displaced eastward across the Devonian-Mississippian continental margin. The overlying Mississippian (?) clastic units in the El Paso Mountains are believed to represent autochthonous or parautochthonous flysch deposits derived from the Antler allochthon and partially overlapping it (Poole, 1974; Poole and Sandberg, 1977). Poole and his co-authors visualize continuity of the Antler and Sonoma belts northward from the western Mojave Desert to west-central Nevada, except for the obliterative effects of Mesozoic plutonism. They attribute the irregular trend of the Antler belt between southern California and western Nevada to dextral oroclinal bending of Mesozoic and Cenozoic age.

Dickinson (1977) concludes that the irregular trend of the Antler orogenic belt between the two areas may be the consequence of an initially irregular pattern of rifts and transform faults that outlined the western margin of the North American continent in late Precambrian time. Miogeoclinal rocks were thus deposited along a margin that had protrusions and recesses. Rocks like those of the lower part of the Garlock Series that were initially deposited along the continent-ocean boundary may not, therefore, be appreciably "out of place" with respect to those of the miogeoclinal sequence farther east.

Our interpretation is that the units described previously (Garlock Series, Coyote Group, etc.) are significantly out of place relative to the miogeoclinal sequence to the east. Facies and isopachous lines for the miogeocline all have trends to the southwest or even west. Nowhere do they show indications of turning southward to parallel a different continental margin trend, and there is no reason to doubt that they originally continued westward for some unknown distance. Furthermore, although we agree with some of Poole's stratigraphic correlations, we do not believe that these Paleozoic rocks are the simple southward continuation of the early and late Paleozoic thrust-faulted sequences of west-central Nevada. In their present position just north of the cratonal sequence at Victorville, they would have to have been thrust across the entire width of the Paleozoic miogeocline. Inferred upper Paleozoic rocks at Victorville (Miller, 1977) and upper Paleozoic rocks in the San Bernardino Mountains (Hollenbaugh, 1970; S. Cameron, personal communication, 1978) contain no terrigenous components as might be expected if the Antler or Sonoma highlands had occupied the Mojave locations postulated for them by Poole and Sandberg (1977). Like Poole (1974), we interpret the rocks of the lower Garlock Series to be part of the Antler orogenic belt overlain by an upper Paleozoic clastic wedge. In addition, the lower part of the Coyote Group and at least some of the rocks in the Shadow Mountains are also interpreted as parts of the upper Paleozoic clastic wedge. Lower Paleozoic rocks may also be present in the Lane and Shadow Mountains, but are as yet unrecognized.

We do not, however, believe that the present latitudinal location of these rock assemblages is an expression of the original configuration of the Antler and Sonoma belts. Instead, we hypothesize that these rocks have been moved southward in a major fault sliver or slivers from correlative rocks of west-central Nevada and northeastern California. Attendant faulting is considered to represent the left-slip truncation (Davis and others, 1978) of a part of the North American continental margin during Permo-Triassic time (Hamilton, 1969; Burchfiel and Davis, 1972, 1975). The eastern boundary fault of the displaced continental sliver must now be very irregular (Fig. 9-4), trending southward between the Argus Range and the El Paso Mountains (Fig. 9-2), displaced eastward along the Garlock fault, then curving westward to a position north of the San Bernardino Mountains. The western boundary would mark the main line of Permo-Triassic continental truncation. The inferred sinuous trend of the fragment and its bounding faults is presumably the product of later Mesozoic and Cenozoic deformations, including rotation of the western Mojave terrane, as discussed later. Southward, the Permo-Triassic truncation boundary would pass into northern Mexico, where other out-of-place fragments, such as the upper Precambrian-Cambrian sedimentary rocks at Caborca, may be present. Although not enough is known of the Paleozoic rocks of the western Mojave or northern Mexico, it is possible that the Mojave was part of a westward projection of the North American continent in Paleozoic time, similar to the Peace River arch of Canada (Douglass, 1967) and the Salmon River arch in Idaho (Armstrong, 1975). If so, the area of terrane displaced southward during the Permo-Triassic truncation need not be great.

MESOZOIC ROCKS

Mesozoic rocks of the Mojave region are represented by chiefly plutonic and volcanic rocks that were part of an evolving and shifting magmatic arc of continental margin (Andean) type (Hamilton, 1969; Burchfiel and Davis, 1972, 1975). The full extent of Mesozoic sedimentary rocks within the Mojave region is not yet known, but they are clearly subordinate to coeval igneous rocks. The Mesozoic magmatic arc can be traced northwestward through the Mojave Desert and across all preexisting Paleozoic and Precambrian structural and paleogeographic trends. This change in orientation of tectonic trends is the consequence of convergent plate boundary activity along the Permo-Triassic truncated continental margin.

Mesozoic Back-Arc Sequences

East of the magmatic arc, the Mesozoic stratigraphic sequence begins with Lower Triassic marine sandy limestone of the Moenkopi Formation that rests paraconformably on the Permian Kaibab Limestone. Rocks of the upper part of the Moenkopi are nonmarine red beds. Formations above the Moenkopi are all nonmarine units that form a conformable sequence to the Lower Jurassic Aztec Sandstone. Several different units rest unconformably above the Aztec. Sandstone and conglomerate lie above the Aztec in the Muddy Mountains of southern Nevada; in some places these rocks are Cretaceous (Longwell and others, 1965). In the eastern Spring Mountains, numerous exposures of channel conglomerates of unknown age overlie the Aztec Sandstone (Longwell 1926, 1973; Davis, 1973). An Upper Jurassic-Lower (?) Cretaceous sequence of orogenic conglomerates (Carr, 1977, 1978) rests unconformably on the Upper Triassic Chinle Formation at Goodsprings, Nevada. In the Clark Mountains, the Aztec is overlain by Jurassic volcanic rocks (Burchfiel and Davis, 1971), which represent the transgression of the magmatic arc across cratonal Mesozoic strata. Mesozoic rocks east of the arc are best classified as a back-arc cratonal sequence rather than miogeoclinal.

Magmatic Arc Terrane

The position of the eastern edge of the magmatic arc fluctuated with time, although the entire Mojave region at one time or another lay within the Mesozoic arc. Insufficient data are available to establish the temporal and spatial distribution of arc magmatism or sort out the details of arc evolution. Published age dates are largely K–Ar ages that probably represent cooling or uplift ages and partial or complete resetting of ages from older rocks. Stratigraphic position of many of the Mesozoic volcanic units is poorly known. For these reasons only a generalized and incomplete discussion of the rocks can be presented at this time.

The oldest rocks known within the magmatic arc are plutons of intermediate composition that are present in the western part of the Mojave region and to the west in the now adjacent Transverse Ranges (Fig. 9-5). C. F. Miller (1977a, b) reports an age of 230 ± 10 m.y. from an alkalic monzonitic pluton in the Granite Mountains. A similar pluton in the Victorville area may be as old and has yielded a preliminary ^{40}Ar–^{39}Ar fusion age on a hornblende of 233 ± 14 m.y. (E. Miller, 1977). Farther west in the San Gabriel Mountains, the Mount Lowe granodiorite has yielded ages of 220 ± 10 m.y. (Silver, 1971) and may represent part of the same magmatic episode. Rocks of the same characteristic composition and texture as the Mount Lowe are present in several areas in the southern part of the area, extending as far southeast as the Chocolate Mountains (Dillon, 1976).

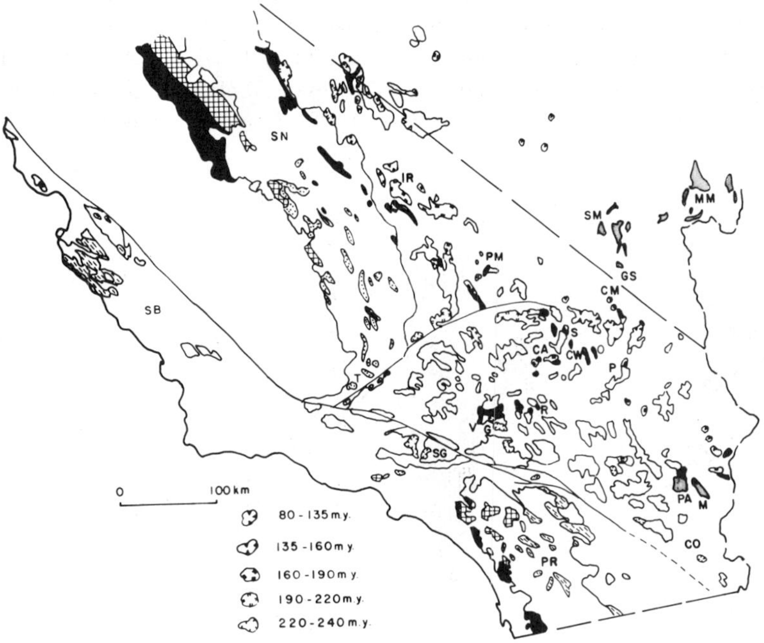

Fig. 9-5. Distribution of Mesozoic rocks. Black, predomininately volcanic or volcanogenic units; white, plutonic rocks, age unspecified except where ages are designated by decorated border; shaded, nonvolcanic sedimentary rocks; squared pattern, early Mesozoic accreted sequences; dot, Kings sequence (Triassic and Jurassic) of Saleeby (1978a, b); dash, metamorphic rocks of unknown affinities. CM, Clark Mountains; CA, Cave Mountain; CW, Cowhole Mountains; CO, Chocolate Mountains; G, Granite Mountains; GS, Goodsprings; IR, Inyo Range; M, McCoy Mountains; MM, Muddy Mountains; O, Old Dad Mountain; P, Providence Mountains; PM, Panamint Mountains; PA, Palen Mountains; PR, Peninsular Ranges; R, Rodman Mountains; S, Soda Mountains; SB, Salinian Block; SG, San Gabriel Mountains; SM, Spring Mountains, SN, Sierra Nevada; V, Victorville.

Marine Lower Triassic rocks are present in the northeastern part of the Mojave region in the Providence Mountains, (Hazzard and Mason, 1936), Old Dad Mountain (?) (Dunne, 1977), Soda Mountains (Grose, 1959), and north of the Garlock fault in the southern Panamint Range (Johnson, 1957) and southern Inyo Mountains (Merriam, 1963).

There are problems in correlating the stratigraphic and geochronologic time scales in this part of the geologic column, but a case could be made that a magmatic arc of Early Triassic age was developed only in the western part of the Mojave and adjacent regions to the south and west, and the eastern part of the area was a shallow marine back-arc area of carbonate changing to nonmarine redbed deposition.

During Middle and Late Triassic time the eastern limit of the arc (Fig. 9-5) clearly shifted to the east along a line from the Inyo Range (Merriam, 1963) through the southern Panamint Range (Johnson, 1957), Soda Mountains (Grose, 1959), and Old Dad Mountain (Dunne, 1977). Southward, volcanic rocks that may belong to this time interval are present in the Providence (Hazzard and Mason, 1936) and Riverside Mountains (Hamilton, 1964a). These volcanic rocks partially overlap age limits for the Mount Lowe pluton, and further refinement is necessary to demonstrate their temporal relations. Plutonic rocks dated at approximately 200 ± 10 m.y. are present along the eastern edge of this volcanic belt and may represent their plutonic equivalents (Fig. 9-6). Structural data discussed later suggest the plutons are somewhat younger than

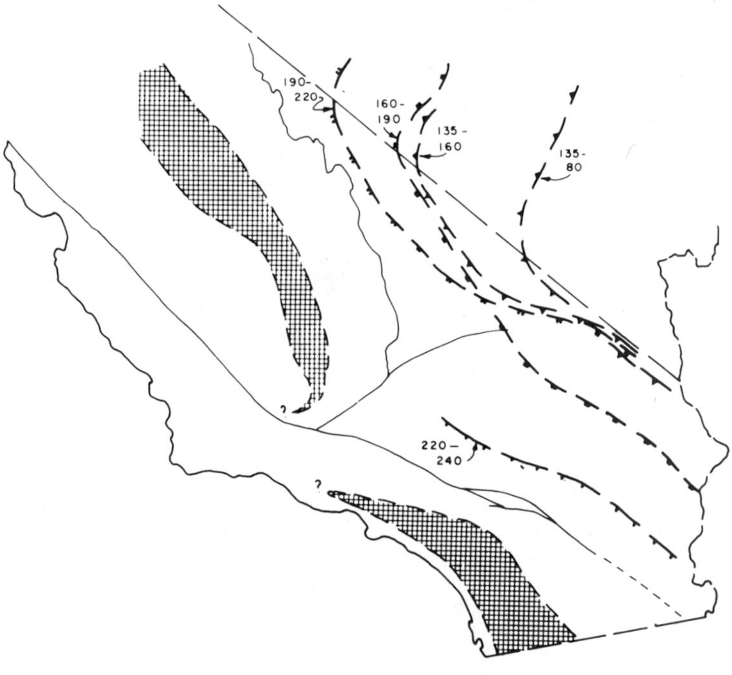

Fig. 9-6. Interpretive map showing eastern limits of plutonic rocks for a given age interval. Crosshatched area is the Traissic-Jurassic accretionary sequence.

the volcanic rocks, as the volcanic rocks predate deformation and the plutons postdate deformation in the same areas.

Stratigraphic data suggest there may have been a pause in magmatic activity within at least the eastern part of the magmatic arc during Early Jurassic time. The Lower Jurassic Aztec Sandstone is present in the eastern and central part of the Mojave region, the site of the Middle and Late Triassic volcanic arc. Aztec Sandstone is present at Old Dad Mountain (Dunne, 1977), Cowhole Mountains (Novitsky and Burchfiel, 1973), Cave Mountain (Burchfiel, work in progress), and in the Rodman Mountains (Miller and Carr, 1978). A southward continuation of these rocks is not known at present, but they may be present in the Palen Formation of the Palen Mountains (Pelka, 1973a, b).

Magmatic activity resumed in the late Early Jurassic and Middle Jurassic time as indicated by age dates of 190 to 160 m.y. (Armstrong and Suppe, 1973; Dunne and others, 1978, Dillon, 1976). These dates are all on plutonic rocks (Fig. 9-5), and although some undated volcanic rocks may belong to this time period, none are established as Middle Jurassic. Dated Middle Jurassic plutons do, however, fall along a trend that lies generally west of the Middle and Late Triassic magmatic rocks of the Mojave region. Their complete distribution must await more detailed geochronologic work.

The eastern limit of Upper Jurassic (160 to 135 m.y.) magmatic rocks is to the east of all previous arc-related rocks (Fig. 9-6). They are present in the Inyo Range (for summary see Dunne and others, 1978) southern Panamint Range (Johnson, 1957), Goodsprings area (Carr, 1977, 1978), and Marble Mountains (Silver, 1966). Rocks farther southwest may belong to this time interval, but the central part of the magmatic arc has had a complex superposition of magmatic events and details must await further study.

Early to early Late Cretaceous (135 to 80 m.y.) magmatic arc activity had its eastern limits near to that of Upper Jurassic arc rocks; they may in fact be part of the same arc. Most of the dated rocks are plutonic, but rare volcanic rocks assigned to this time interval are present in the Muddy Mountains (Fleck, 1970). Some plutons of the central and western Mojave area and adjacent areas yield dates that fall within this Cretaceous time interval (Armstrong and Suppe, 1973). Many of the dates are K–Ar, but inasmuch as in general they represent the youngest plutons in the area, they may be largely valid.

Latest Cretaceous to earliest Tertiary plutons (70 to 60 m.y.) are rare or not present north of the Garlock fault (Cross, 1973; Armstrong and Suppe, 1973) and in the northern Mojave, but are scattered throughout the southern Mojave region and adjacent Arizona. The implications of the change in distribution of post-70 m.y. magmatic activity have been discussed by Armstrong (1974) and Burchfiel and Davis (1975).

Several areas within the magmatic arc contain metasedimentary rocks whose regional affinities are uncertain. These rocks prove that the arc terrane was at times a site of both marine and nonmarine deposition. Near Victorville a sequence, nearly 2 km thick, of sandstone, calcareous sandstone, limestone, mudstone, and conglomerate of the Fairview Valley Formation crops out. Cobbles within a thick upper conglomerate unit contain Permian fusulinids (Bowen, 1954); Dibblee, 1967). E. Miller (1977, 1978) has demonstrated that the Fairview Valley Formation rests with angular unconformity on deformed and metamorphosed Paleozoic rocks that are intruded by monzonitic rocks that may be at least as old as 233 ± 14 m.y. This demonstrates the Fairview Valley Formation is Mesozoic in age, as it is intruded by a Late Cretaceous pluton. It is unconformably overlain by the Sidewinder Volcanics, the age of which can only be established as Mesozoic. The Fairview Valley Formation can be interpreted as a rock unit related to the pre-233 ± 14 m.y. old deformation, which would suggest a Triassic—possibly Early or Middle Triassic—age for it. Some rocks north and northeast of the Victorville area in the Coyote and Starbright groups of McCulloh (1952) and in the Shadow Mountains and near Goldstone Lake have similarities to rocks in the Fairview Valley Formation. It is possible that these rocks may also be Mesozoic in age, although at present there is not positive evidence to prove this suggestion.

In the Tehachapi Mountains in the northwesternmost part of the Mojave Province are several pendants that contain sequences of schist, quartzite, limestone, argillite, and metavolcanic rocks. Two of the easternmost pendants have been recently mapped (Rindosh, 1977; G. A. Davis, work in progress). Preliminary Rb–Sr data on the metavolcanic rocks suggest they are Upper Jurassic (R. J. Fleck, personal communication, 1977). These rock units are similar to those included by Saleeby and others (1978) in their Kings sequence of Triassic and Jurassic age. A pendant at Bean Canyon contains bodies of alpine-type peridotite whose structural position is unclear. The rock units in the pendants strongly suggest they are the southward continuation of the central Sierra Nevada (Saleeby and others, 1978) and lie east and above the Kings-Kaweah suture of Saleeby (1977a, b).

Pelka (1973a, b) has described a thick (7 to 8 km) sequence of impure sandstone, shale, rare conglomerate, and some volcaniclastic rocks in the McCoy and Palen Mountains in the southern part of the Mojave region. Plant fossils indicate a latest Cretaceous to Early Tertiary age for these rocks. Similar rocks are present in adjacent parts of Arizona, for example in the Livingston Hills Formation (Miller and McKee, 1971), and it can be suggested that these rocks are part of a regionally important terrane that may be the youngest and most westerly extension of the Mexican geosyncline (Hayes, 1970). Although Dillon (1976) related these rocks paleogeographically to the Orocopia Schist that crops out to the south, we do not support this interpretation and will discuss the problem later.

Mesozoic Rocks Along the Western Margin of the Magmatic Arc, Southern California

The oldest Mesozoic rocks known to lie west of the Mesozoic arc are Middle Triassic cherts recently dated by D. L. Jones (personal communication, 1978) from rocks assigned to the Bedford Canyon Formation in the Peninsular Ranges. These rocks may represent part of an oceanic sequence that was accreted by subduction to the truncated margin of North America. Data suggest that accreted terrane may continue northward along much of the western margin of North America (Davis and others, 1978). The remainder of the Bedford Canyon Formation consists of structurally disturbed shale and turbidite sandstone containing isolated blocks (olistoliths?) of limestone. The limestone has yielded Middle and Late Jurassic fossils (Bajocian and Callovian; Moran, 1976). The structure, stratigraphy and evolution of the Bedford Canyon Formation is poorly known. It may represent parts of an accretionary prism and forearc basin sequence developed at the western margin of North America and related to early Mesozoic convergent activity that produced the magmatic arc to the east. The formation is overlain, in part unconformably, by andesitic rocks of the Late Jurassic-Early Cretaceous Santiago Peak Volcanics.

Lower to lower Upper Cretaceous plutonic rocks intrude the Mesozoic sequence described previously and indicate a westward migration of the western limits of the arc. Upper Cretaceous sedimentary rocks lie farther to the west in depositional contact on older plutonic, volcanic, and sedimentary rocks of the Peninsular Ranges. Some of these rocks form a coherent sequence of sandstone, shale, and conglomerate that appears to represent a southern counterpart and proximal facies assemblage of the Great Valley sequence of central and northern California. These rocks are part of a forearc basin for the Cretaceous magmatic arc (Jones and others, 1976). Still farther west lies a disrupted terrane of coeval Franciscan rocks that represents the products of subduction eastward beneath the forearc basin.

Thus, during Mesozoic time the Cordilleran belt from Las Vegas, Nevada, to the Pacific (at the latitude of the northern Peninsular Ranges) included all the elements of an Andean-type convergent margin. From east to west these were a back-arc basin of nonvolcanic and largely nonmarine sedimentary rocks, the magmatic arc, a forearc basin, and a subduction complex. Their spatial distribution in Mesozoic time requires a palinspastic reconstruction, part of which is shown in Fig. 9-9. There is an indication of westward migration of arc, forearc, and subduction sequences along this line of section. The eastern limits of the arc changed irregularly (Fig. 9-6). Furthermore, preliminary data suggest the arc was not active continuously, but had times of quiesence or inactivity such as in the Early Jurassic. This may suggest there were distinct arc developments that were partly superposed on one another. Published geochronological data are not sufficient at present to sort out the details.

Salinian Block

Palinspastic reconstruction of southern California prior to Cenozoic strike-slip faulting places the Peninsular Ranges south of most of the Mojave region. More relevant are the Mesozoic rocks of the Salinian block. Although problems regarding a precise positioning of the Salinian block exist (Ross, 1978), we still regard the block as the best candidate for the westward continuation of the Mojave region in Mesozoic time (see discussion later). The early Mesozoic accreted terrane present in the Peninsular Ranges and in the western Sierra Nevada Foothills is not present in the Salinian block, a problem we will discuss later. Upper Cretaceous rocks rests unconformably on an eastern Sierran-type basement and consist of sedimentary sequences in a forearc position, but in a somewhat different geographic setting from the Great Valley rocks farther north (Howell and others, 1977). Upper Jurassic and Lower Cretaceous forearc basin deposits now present farther west rest on oceanic basement but are not easily tied to the Salinian block. (Page, 1970a). Franciscan rocks tectonically underlie these western and older parts of the forearc sequence, thus apparently preserving a trench-forearc terrane for late Mesozoic time west of the Mojave region. Considerable disruption of this terrane has occurred in Mesozoic (?) and Cenozoic time and will be treated later.

STRUCTURE

Pre-Mesozoic Deformations

Precambrian metamorphic rocks in the Mojave region record deformational and metamorphic events that at present are known only in a few isolated areas. The character and extent of these events is unknown and will not be discussed. The late Precambrian Pahrump Series in the Death Valley region has thickness and facies relations that suggest deposition contemporaneously with normal faulting in a sedimentary trough, called the Death Valley Aulacogen by Wright and others (1976). The south flank of the trough has not been determined, and it is possible that it may lie in the northeastern part of the Mojave province. Other than this feature, no late Precambrian deformation is known from the area. Igneous activity, in the form of diabase sills and dikes, intruded the oldest sediments of the fault trough and are responsible for much of the talc mineralization in the area.

Paleozoic deformation (Fig. 9-6) may be recorded in the White-Inyo Mountains where northeast and superposed north-trending folds are interpreted as having been formed during the mid-Paleozoic Antler orogeny and latest Paleozoic-earliest Mesozoic Sonoma orogeny, respectively (Sylvester and Babcock, 1975). Three small klippen of western miogeoclinal Cambrian rocks are present in the White Mountains, contain north-trending folds, and may also have been emplaced in Paleozoic time (Dunne and others, 1978). Dating of the folds and klippen is not firmly established, but if they are Paleozoic or earliest Mesozoic in age, they represent the most marginal effects of either the Antler or Sonoma orogenies in this part of the orogen.

Evidence for mid-Paleozoic deformation is found in the El Paso Mountains, where deformed Devonian and older Paleozoic rocks in the Garlock Series are unconformably overlain by Mississippian (?) and fossiliferous Permian rocks belonging to the middle part of the Garlock Series (Poole and Sandberg, 1977). The deformed lower Paleozoic rocks are probably part of the Antler terrane that was thrust eastward onto the continental shelf in Early Mississippian time and unconformably overlain by the younger Paleozoic overlap assemblage. As discussed previously, we believe these rocks were displaced to their present position as a sliver by left-slip faulting in Permo-Triassic time.

Mesozoic Deformations

The Mojave Desert and areas to the north were the sites of multiple, shifting Mesozoic deformations. We are only just beginning to understand the complexity of this deformational history, much of which in the central and western Mojave areas has been obliterated by late Mesozoic plutonism (the southward extension of the Sierran batholitic terrane). Consequently, most of our information on Mojave Mesozoic tectonic history comes from the eastern half of the region where plutons are more isolated, the stratigraphic section is therefore more complete, and where field studies of the past decade have been concentrated. Because of these comments it is somewhat surprising that the earliest record of Mesozoic deformation in the Mojave region comes from the Victorville area in the southwestern portion of the region, as discussed later.

The earliest Mesozoic deformation in southeastern California is related to the growth of an Andean-type magmatic arc that developed along the previously truncated edge of the North American plate (Hamilton, 1969; Burchfiel and Davis, 1972, 1975). The arc that developed along the truncated continental margin with the onset of eastward oceanic subduction lay parallel to the margin and at a high angle to earlier, northeast-trending paleogeographic elements (Fig. 9-4; Burchfiel and Davis, 1972). In the Mojave region the arc was thus superposed atop a craton composed of (1) Precambrian crystalline basement rocks and overlying upper Precambrian and Paleozoic sedimentary rocks, and (2) to the northwest, a fault sliver or slivers of rocks displaced from farther north in the orogen (including units in the Roberts Mountains allochthon and post-Antler clastic deposits). Thickness and facies trends of the sedimentary rocks and the structural grain of the Precambrian crystalline basement rocks were oblique to the Mesozoic magmatic arc and related structures, demonstrating that mechanical anisotropies in the pre-Mesozoic rocks had little if any influence on Mesozoic structural development.

Latest Paleozoic Through Early Triassic

Recent work by E. L. Miller (1978) has shown that deformation accompanied by metamorphism may be as old as latest Permian (but more likely earliest Triassic) in the Victorville area (Figs. 9-7 and 9-8). In that area, perhaps the site of the oldest deformation related to the Mesozoic continental margin arc, northwest-trending structures are cut by a post-tectonic monzonitic pluton. The pluton is chemically and petrographically distinct, but is similar to monzonitic rocks in the nearby Granite Mountains that have yielded U-Pb (zircon) dates of 230 ± 10 m.y. (C. F. Miller, 1977b). Preliminary ^{40}Ar–^{39}Ar data on hornblende from the monzonite in the Victorville area indicate that it too is early Mesozoic or possibly Late Permian in age (E. L. Miller, 1978). The Fairview Valley Formation, previously regarded by some workers as late Paleozoic, rests unconformably on deformed Paleozoic rocks and the monzonite pluton that intrudes them. It is most likely Mesozoic in age. Clasts in the Fairview Valley Formation consist predominately of carbonate rocks and rare clasts of plutonic and volcanic rocks, quartzite, and granitic

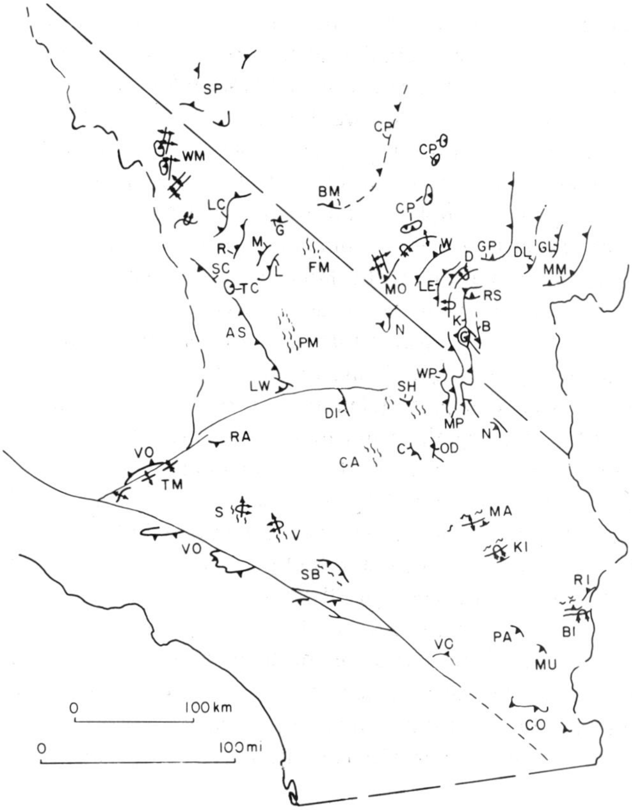

Fig. 9-7. Major Mesozoic structural features of the Mojave region and environs. S-lines are areas of metamorphism. Other symbols for thrust faults and folds are standard. AS, Argus-Stirling thrust; BM, Bare Mountain; BI, Big Maria Mountains; B, Bird Spring thrust; C, Cowhole thrust; CA, Cave Mountain; CO, Chocolate Mountains; CP, CP thrust; D, Deer Creek thrust; DI, Drinkwater Lake; DL, Dry Lake thrust; FM, Funeral Mountains; G, Grapevine thrust; GL, Glendale thrust; GP, Gass Peak thrust; K, Keystone thrust; KI, Kilbeck Hills; L, Lemoine thrust; LC, Last Chance thrust; LW, Layton Well thrust; LE, Lee Canyon thrust; M, Marble Canyon thrust; MA, Marble Mountains; MU, Mule Mountains; MP, Mesquite Pass thrust; MO, Montgomery thrust; MM, Muddy Mountains thrust; N, New York Mountains; NO, Nopah Range; OD, Old Dad Mountain; PA, Palen Mountains; PM, Panamint Mountains; R, Racetrack thrust; RI, Riverside Mountains; RA, Rand Mountains; RS, Red Spring thrust; S, Shadow Mountains; SB, San Bernardino Mountains; SH, Silurian Hills; SP, Silver Peak Range; SC, Swansea-Coso thrust system; TC, Talc City Hills; TM, Tehachapi Mountains; V, Victorville; VO, Vincent-Orocopia thrust; W, Wheeler Pass thrust; WM, White Mountains; WP, Winters Pass thrust.

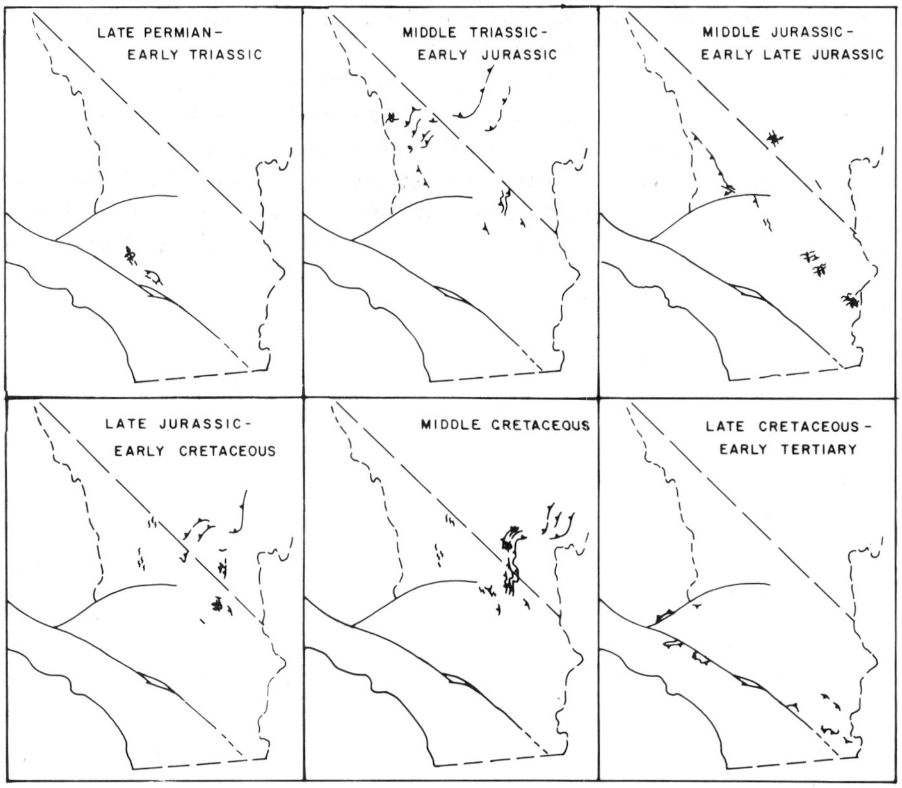

Fig. 9-8. Interpretation of temporal and spatial relations between structures of the Mojave region and environs. Data for age assignments of certain structures are discussed in the text.

and gneissic rocks of probably Precambrian age. Carbonate clasts contain fossils as young as Early Permian (Dibblee, 1967). Thus the deformation affecting the carbonates can be bracketed as post-Early Permian and pre-230 ± 10 m.y. The Fairview Valley Formation most likely represents an orogenic or postorogenic deposit to that deformation, a deformation that may have produced structures large enough to expose the Precambrian crystalline basement.

The extent of this deformation beyond the Victorville area is not known, but structures in the San Bernardino Mountains (Fig. 9-7), first mapped by Dibblee (1964) and locally studied in detail by Tyler (1975), may belong to the same period of deformation. In the San Bernardino Mountains, large eastward overturned folds involving Precambrian crystalline rocks and a Paleozoic cratonal sequence are cut by thrust faults and intruded by a monzonitic pluton dated by Armstrong and Suppe (1973) at 194 ± 3 m.y. (K-Ar, hornblende) and 83 ± 1.2 m.y. (K-Ar, biotite; S. Cameron, personal communication, 1978). The discordant K-Ar age indicates the rocks have been subjected to later reheating and that the hornblende age is a minimum. Thus it is possible that deformation in the San Bernardino Mountains is the southward continuation of the Late Permian-Early Triassic deformation at Victorville.

Silver (1971) has reported a U-Pb age of 220 ± 10 m.y. on the Mount Lowe-Parker Mountain pluton in the central San Gabriel Mountains (Fig. 9-5). A number of adjacent plutons are shown on the 1 : 250,000 San Bernardino sheet (Rogers, 1967) as Permian-Triassic in age and may represent part of this plutonic event. Silver (1971) believes that emplacement of the Mount Lowe-Parker pluton was accompanied by deformation in the surrounding Precambrian crystalline rocks. Crowell's (1962) analysis of displacement

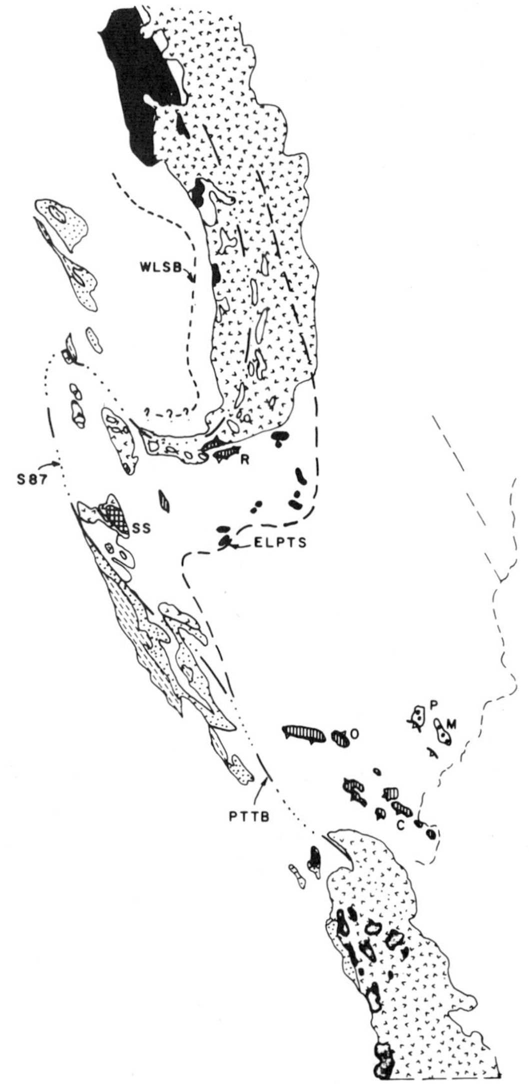

Fig. 9-9. Distribution of certain Mesozoic features and rock units plotted on a reconstructed base for southern California with effects of Miocene and younger displacements on faults removed. V's, westernmost plutonic rocks of the Sierra Nevada, Salinian block, and Peninsular Ranges; black, Paleozoic rocks displaced during Permo-Triassic truncation; vertical lines, Pelona-Orocopia-Rand Schist and correlatives; crosshatched, Schist of Sierra de Salinas (ss); shaded, Triassic-Jurassic accretionary rocks; dots, Great Valley sequence and correlatives; dashes, Franciscan rocks. WLSB, western limit of Sierran basement; S87, ^{87}Sr-$^{86}Sr = 0.706$; PTTB, Permo-Triassic truncation boundary; ELPTS, Eastern limit of Permo-Triassic sliver; C, Chocolate Mountains; M, McCoy Mountains; O, Orocopia Mountains, P, Palen Mountains; R, Rand Mountains; SS, Sierra de Salinas.

on the San Andreas fault places the San Gabriel Mountains originally to the south or southeast of the San Bernardino Mountains (Fig. 9-9). Thus it is not unreasonable to propose that parts of the San Gabriel Mountains are a more deeply exposed southward continuation of the Victorville-San Bernardino Mountains (?) terrane affected by Late Permian-Early Triassic deformation. The validity of this suggestion remains to be established.

Middle Triassic Through Early Jurassic

Younger structures of Middle Triassic through Early Jurassic age are present in the northern Panamint and Inyo ranges, Cowhole Mountains, and Clark Mountains of California (Fig. 9-8). Evidence (Burchfiel and others, 1970; Dunne and others, 1978) indicates east-directed thrusting in the northern Panamint and Inyo ranges (Fig. 9-7) took place in Middle (?) Triassic to Early Jurassic time, prior to the emplacement of the Hunter Mountain Pluton at 183 m.y. (Dunne, 1977). The largest of the thrust faults, the Last Chance thrust, can be extrapolated to the northeast, through Bare Mountain, to the CP thrust of south-central Nevada (Burchfiel and others, 1970). Recent work by Gulliver (1976) suggests parts of this thrust complex can be extended south to the Talc City Hills (Fig. 9-7).

Probable contemporaneous deformation is present in the Cowhole Mountains 250 km to the southeast (Fig. 9-7), where metamorphosed Paleozoic rocks are involved in folding and eastward thrusting (Novitsky and Burchfiel, 1973; Novitsky-Evans, 1978). These structures are overlapped by the Upper Triassic (?) to Lower Jurassic Aztec sandstone. Evidence from adjacent areas suggests the deformation is Middle Triassic.

Two plutons dated at 190 and 200 m.y. (K-Ar, hornblende) cut east-vergent structures in the Clark Mountains (Fig. 9-5; Burchfiel and Davis, 1971). Although the age of the onset of deformation is not well established, these structures are thought to have formed after deposition of the late Early Triassic to early Middle (?) Triassic Moenkopi Formation. This poorly defined belt of deformation lies very near the present eastern margin of the Cordilleran fold and thrust belt in southeastern California, but to the north it occupies a more internal position in the orogen. Relations among the three areas of Middle (?) Triassic through Early Jurassic deformation described previously are not obvious. However, our present interpretation is that the geometry of the geosynclinal wedge controlled in part the trend of the Last Chance CP thrust complex, but as the Triassic magmatic arc developed across geosynclinal trends to the south, stratigraphic controls in the geosynclinal wedge became unimportant and the eastern front of deformation stepped en echelon to the southeast. This hypothesis would suggest that Middle (?) Triassic to Early Jurassic deformation is present over a broad terrane between the Inyo-Panamint ranges and Clark Mountain areas and may extend southward through the eastern Mojave area. The Cowhole Mountains probably represent one such locality in this broad terrane, and the Slate and Argus ranges may represent others (Dunne and others, 1978). Although many possible Middle (?) Triassic to Early Jurassic structures are present, geologic relations do not permit definitive dating, a problem true for most of the structures in the area.

The data suggest that the structures just described are younger than the Late Permian-Early Triassic structures present in the Victorville area. The older ages for crosscutting plutons in that area, if confirmed, indicate that the Victorville structures are older than the Middle (?) Triassic rocks believed to be involved in the Last Chance, Clark, and Cowhole Mountains structures. Middle (?) Triassic through Early Jurassic structures may also be present in the western Mojave region, but to date such structures have not been identified.

Middle Jurassic Through Early Late Jurassic

Structures of Middle through early Late Jurassic age are well dated in only one area. Extending from the southern Inyo Range through the Argus to the Slate Range is the Swansea-Coso thrust system (Dunne and others, 1978), a belt of east- to northeast-directed thrust faults that cut a pluton dated at 156 m.y. and are intruded by a pluton dated at 140 ± m.y. (K-Ar biotite age). The thrusts in turn are cut by northwest-striking faults with left slip that are intruded by dikes dated at 148 m.y. (Dunne and others, 1978). These relations closely bracket the time of thrusting as Middle to earliest Late Jurassic. The thrust faults are clearly younger than the Middle Triassic to Early Jurassic structures of the Inyo Range that they crosscut in the Talc City Hills (Dunne and Gulliver, 1976).

Thrusts of the Swansea-Coso system continue south to the Garlock fault, where their offset equivalent(s) probably continues into the northern Mojave area near Drinkwater Lake (Davis and Burchfiel, 1973). Preliminary studies at Cave Mountain (Fig. 9-7) suggest that metamorphism and deformation were contemporaneous with the intrusion of a pluton (Burchfiel, unpublished) that has yielded a Rb-Sr age of 170 ± 5 m.y. These rocks may extend the belt of Middle and early Late Jurassic deformation into the northern Mojave. No other Mojave structures can be definitely related to this period of deformation, although other possibilities exist.

In the northern Marble Mountains (Fig. 9-7), metamorphosed upper Paleozoic limestone is folded into a recumbent east-trending anticline that has been thrust (?) southward onto unmetamorphosed Precambrian crystalline and lower Paleozoic sedimentary rocks (Burchfiel, unpublished). The thrust (?) contact between the two rock assemblages is intruded by a pluton for which K-Ar ages of 154 and 141 m.y. have been obtained. A Pb-U date on the pluton that cuts the thrust (?) contact is 165 m.y. (Silver, quoted by Bishop, 1963). The deformed Paleozoic marbles are cut by a post-tectonic pluton dated at 148 ± 2 m.y. (K-Ar dates from Armstrong and Suppe, 1973). These data are compatible with either a Middle Jurassic or Middle (?) Triassic-Early Jurassic period of deformation.

One thing evident from the data is that the Middle through early Late Jurassic belt of deformation is not everywhere coincident with the Middle Triassic-Early Jurassic belt of deformation (Fig. 9-8). In the southern Inyo Range the younger belt strikes more northwesterly and crosscuts the older northeast-trending Last Chance and related thrusts. In the Argus and Slate ranges it may be superposed on the older structures, but in the northern Mojave it lies to the west of the older structures. The eastern limit of dated Middle to early Late Jurassic plutons (190 to 160 m.y.) has a comparable spatial distribution to the eastern limit of structures developed during the same time interval.

Latest Jurassic Through Early Cretaceous

Latest Jurassic-Early Cretaceous deformational and metamorphic events are recorded from two areas (Fig. 9-8). In the Ivanpah Mountains (Fig. 9-2) a syntectonic pluton occupies the core of a large eastward overturned fold that is associated with minor thrust faulting (Burchfiel and Davis, 1971). The pluton has been dated at 135 ± 5 m.y. (K-Ar; Sutter, 1968), which places this event near the Jurassic-Cretaceous boundary. These Ivanpah structures lie just east of the younger Keystone thrust fault. Clearly, in this portion of the Cordilleran fold and thrust belt older structures lie east of younger ones. Farther north in the Panamint and Funeral ranges of the Death Valley area (Fig. 9-7), amphibolite-grade metamorphism is dated as ranging from 150 to 100 m.y. (Lanphere and others, 1963). This metamorphism is contemporaneous with Late Jurassic and Early Cretaceous events; its relation to deformational events is uncertain, but speculation concerning its importance is considered later.

Other structures that may be contemporaneous with those of the Ivanpah Mountains are present in the Goodsprings area of southern Nevada (Fig. 9-7). Recent work by Carr (1978) has shown the presence of a thick sequence (more than 300 m) of conglomerates that contain cobbles eroded from Paleozoic rocks. An interbedded tuff near the base of the unit has yielded a biotite K-Ar age of 150 ± 10 m.y., which may make the conglomerates the oldest known orogenic sediments along the eastern margin of the fold and thrust belt in the western United States (cf. Armstrong, 1968). The conglomerates were overridden by the Contact thrust plate, which in turn was cut by later high-angle faults and finally overidden by the Keystone thrust plate. This sequence of events is identical to that established in the Mescal Range and Ivanpah Mountains by Burchfiel and Davis (1971). Correlation of structural events would put the emplacement of the Contact thrust plate at Late Jurassic-Early Cretaceous time. The Red Spring thrust plate farther north, considered by Davis (1973) to be correlative with the Contact plate, and the Bird Spring thrust plate to the east would belong to this event. At the present time, no other structures can be assigned a Late Jurassic-Early Cretaceous age with any certainty.

Early Cretaceous Through Early Late Cretaceous

The last major Mesozoic structural event to affect this region prior to Cenozoic high-angle faulting was Early Cretaceous-early Late Cretaceous in age. In southern Nevada the Keystone-Muddy Mountains thrust system formed after deposition of the Willow Tank

Formation and Baseline Sandstone in the Muddy Mountains that contain ash beds dated at 98.4 and 96.4 m.y. (Fleck, 1970). An upper limit for Keystone thrusting is best established in the Ivanpah Mountains of southeastern California where post-Keystone plutons are 84 to 94 m.y. old (Sutter, 1968). No other data for dating this last phase of southern Cordilleran "foreland" thrust faulting are available.

Although many of the thrust faults in the Clark Mountains can be shown to be crosscut by the same plutons that place an upper limit on the time of emplacement of the Keystone thrust, our structural studies in this area (Burchfiel and Davis, 1971; in preparation) demonstrate that thrust plates above the Keystone plate were involved in multiple deformations, the last of which was probably contemporaneous with the Keystone thrust event.

In the Silurian Hills to the west (Fig. 9-7), Abbott (1971) has mapped a small segment of a thrust that juxtaposes two different terranes; it is only a small segment of the Riggs thrust mapped by Kupfer (1960), most of which is probably due to Tertiary gravity sliding. The fault mapped by Abbott is synmetamorphic and is intruded by syntectonic granite stocks that are partially boudinaged. Dates from the metamorphic and granitic rocks yield ages of 84 to 94 m.y. K-Ar (Sutter, 1968) and may date an event contemporaneous with Keystone-age thrust faulting to the east.

Mesozoic High-Angle Faulting

In the Death Valley and eastern Mojave region are numerous high-angle faults of Mesozoic age. In most places they formed between or during periods of thrust faulting. In the Clark Mountains area (Burchfiel and Davis, 1977) some high-angle faults with variable strike are of Triassic age and have proven dip-slip displacement. Younger high-angle faults elsewhere in the Mojave region have a northwest strike. Some can be dated as Middle Jurassic to early Late Jurassic in the Argus Range (Dunne and others, 1978), Late Jurassic in the southern Spring Mountains (Carr, 1978), and Early Cretaceous in the Ivanpah and southern Spring Mountains (Burchfiel and Davis, 1971; Cameron, 1978). In the Argus Range, evidence suggests that they have a component of left slip, whereas one fault in the southern Spring Mountains has a component of right slip. On some of these faults, field relations suggest strike-slip displacement, but the sense is unknown (Burchfiel and Davis, 1977). The Jurassic and Early Cretaceous high-angle faults may be regionally important and may represent intra-arc strike-slip faults active either during or between thrusting events.

Summary

From the preceding discussion it is clear that east-directed thrusting is the main mode of deformation in the northeastern part of the Mojave region. In the area along the California-Nevada state line (in the northeastern part of the Mojave proper) the thrust faults involve only miogeoclinal Paleozoic or upper Precambrian sedimentary rocks.

Many of the thrusts, however, do not follow specific formations along their surface exposures. Thrust faults such as the Lee Canyon, Deer Creek, and Wheeler Pass thrust faults in the northwestern Spring Mountains cut through preexisting folds and across stratigraphic units along their traces (Burchfiel and others, 1974). The geometry of stratigraphic units in most upper plates suggests that the underlying faults flatten at depth and have a "thin-skinned" type of geometry, although it is unclear how closely the thrusts might follow given stratigraphic horizons at depth. Burchfiel and Davis (1971) have shown that a "thin-skinned" thrust geometry does not, in the Mojave region, require stratigraphic controls and does not preclude the involvement in thrusting of crystalline basement rocks.

In the Mojave and adjacent areas to the south and west, most of the thrust faults and folds involve granitic rocks (Swansea-Coso system) or Precambrian crystalline rocks (San Bernardino Mountains, Layton Well, Winters Pass, and Mesquite Pass thrust faults). In this region the magmatic trends are southeastward across miogeoclinal and cratonal Paleozoic paleogeographic elements. Stratigraphic control for southerly thrust development is lost because of the absence of miogeoclinal sedimentary anisotropy. Instead, ductility contrasts in the crustal rocks related to the Mesozoic magmatic arc become the dominant control on thrust fault localization (Burchfiel and Davis, 1975).

In the southeastern part of the Mojave region, the structural style is not well established. Major east-directed thrust faults can be traced into the craton to the New York Mountains (Burchfiel and Davis, 1977), before disappearing beneath Cenozoic units. The thrust belt re-emerges in the Little Piute Mountains 65 km to the south and extends 45 km southwestward through the Old Woman Mountains to the Kilbeck Hills (Howard and others, 1980). The style of this thrusting is markedly more ductile than that seen to the north in the Clark and New York Mountains. In the Old Woman Mountains the principal allochthon is an east-southeast-vergent nappe, composed of Precambrian crystalline rocks and their now inverted cover of cratonal Paleozoic strata. This cover is metamorphosed up to sillimanite grade, and exhibits both migmatization and, locally, extreme attenuation by flow (Howard and others, 1980). Metasedimentary rocks of Mesozoic age occur in the lower plate of the nappe. Farther south in the Riverside and Big Maria Mountains, Precambrian crystalline rocks, Paleozoic cratonal metasedimentary rocks and Mesozoic metasedimentary and metaigneous rocks are also present in large recumbent folds cut by thrust faults. Folds trend in various directions, but some of the large ones trend generally east and are overturned southward (Hamilton, 1971 and personal communication, 1978). Syntectonic metamorphism in this area reached amphibolite grade. What is clear at present in the southeastern part of the Mojave region is that the Cordilleran orogen continues to the southeast as a zone of deformation and metamorphism. Ages of deformation in this area are poorly constrained. Thrust faults shown on various California state geologic maps in the Colorado River area between Needles and Parker, Arizona, are now known to be low-angle faults of Miocene age and of extensional origin (Davis and others, 1979).

Mesozoic deformation in the Mojave region is unlike that described from other parts of the Cordilleran orogen in the western United States. The earliest deformation thought to be related to the development of an Andean arc is Late Permian to Early Triassic in age and somewhat older than Mesozoic arc-related deformation reported from elsewhere in the orogen. Temporal and spatial patterns of deformation in the eastern Mojave do not reflect the simple eastward migration of deformation with time that is demonstrated for eastern and younger parts of the orogen farther north. In the Mojave region the locus of deformation, or at least the eastern margin of the deformational belt, shifted irregularly (Fig. 9-8). Magmatic activity, or at least the eastern limit of magmatic activity, also appears to have shifted with temporal and spatial patterns similar to those for deformation (Fig. 9-6). We have interpreted the deformation in the eastern marginal fold and thrust belt of the Cordilleran orogen as an expression of yielding within the leading (western) portion of the North American plate, controlled by ductility contrasts related to the Mesozoic magmatic arc (Burchfiel and Davis, 1975). More specifically, we consider the Cordilleran fold and thrust belt to be the consequence of the *westward underthrusting* by the colder, eastern cratonal lithosphere of the thermally weakened leading edge of the North American plate. Such intraplate underthrusting occurred broadly synchronously with the convergence and eastward subduction of Pacific oceanic lithosphere beneath the western edge of the continent. Thus, as the locus of magmatic activity shifted, so should the locus of deformation. In the Mojave region the locus of magmatic activity did shift spatially, presumably in response to changes in plate geometry and interaction along the western edge of North America, although the controlling factors are unknown at present. Other major differences between the Mojave region and Cordilleran areas to the north are the extensive involvement in the Mojave of Precambrian crystalline rocks in thrust faulting and the development within the eastern fold and thrust belt of Mesozoic high-angle faults, probably of both strike- and dip-slip type.

Latest Cretaceous Through Early Cenozoic Deformation and the Orocopia-Pelona Schist Problem

Regional patterns of magmatism and deformation changed pronouncedly in latest Cretaceous time in the western United States. Magmatic activity in the arc terminated about 70 m.y. ago in a segment from southern Idaho to southeastern California (Cross, 1973), as did deformation in the marginal fold and thrust belt east of that sector (Armstrong, 1974; Burchfiel and Davis, 1975). In contrast, magmatic activity continued in southern Arizona and northern Mexico (?) into early Cenozoic time. The major deformation in the Mojave region at this time was related to juxtaposition of the Pelona-Orocopia schist beneath Precambrian crystalline rocks and Mesozoic plutonic rocks along the Vincent-Orocopia thrust system in the western Mojave and adjacent areas. Figure 9-9 shows the distribution of the Pelona-Orocopia schist and closely similar rocks, the schist of Sierra de Salinas in the Salinian block and the Rand Schist of the northern Mojave, on a palinspastic base map for southern California (see Ehlig, this volume, Chapter 10).

A recent summary of the Pelona-Orocopia schist and its relations to adjacent rocks is given by Haxel and Dillon (1978), and only a brief summary will be presented here. The Pelona-Orocopia schist is a metamorphosed sequence of graywackes with some pelite, arkose, and subordinate to rare mafic igneous rocks, limestone, chert, and ultramafic rocks. The parental rocks are of unknown age, but appear to have accumulated in an oceanic environment. They were subsequently metamorphosed under conditions of lower greenschist to lower amphibolite or transitional greenschist-blueschist grade (Ehlig, 1958). The schist is tectonically overlain by Precambrian crystalline and Mesozoic granitic rocks, and the grade of metamorphism locally increases upward in the schist toward the thrust contact. Metamorphic fabrics in the schist and mylonitic fabrics in the retrograded rocks in the upper plate are parallel and indicate that thrusting and metamorphism were contemporaneous. Upper plate rocks for the Vincent thrust carry Mesozoic plutons as young as 80 ± 10 m.y. that are cut by the thrust, but the underlying Pelona-Orocopia schist is not intruded by Mesozoic plutons. The schist of the Sierra de Salinas is intruded by plutons dated at 80 m.y., but its relations to the more southerly Pelona-Orocopia schist are uncertain (see later discussion).

Thrusting of the Pelona-Orocopia schist below Precambrian and Mesozoic metamorphic and igneous rocks probably took place between 50 and 60 m.y. based on ages obtained from mylonites in the thrust zone and from metamorphic rocks below the thrust (Ehlig and others, 1975a; Conrad and Davis, 1977).

Ehlig (1958, 1968, and 1975) and Haxel and Dillon (1978) have presented evidence that the relative motion of the Vincent-Orocopia thrust plate was to the northeast. If correct, the oceanic basin of deposition for the protoliths of the Pelona-Orocopia schist presumably lay *east* of Precambrian crystalline and Mesozoic plutonic rocks now in the Vincent-Orocopia thrust plate and west of the Mojave "mainland." The basin has been hypothesized to be a Late Cretaceous intracontinental but ensimatic back-arc basin formed (1) by crustal dilation above a mantle diapir (Dillon and Haxel, 1975), or (2) by oblique rifting associated with right-lateral faulting along a proto-San Andreas fault system (Haxel and Dillon, 1977). According to these paleogeographic interpretations, the Vincent-Orocopia thrust would have formed as this ensimatic basin was subducted westward beneath continental lithosphere now preserved in the San Gabriel Mountains.

We find the Late Cretaceous development of an ensimatic basin within the Mojave terrane difficult to demonstrate and difficult to accept. There is nothing in the geology of the western Mojave region to suggest a rifting event significant enough to create an ensimatic, intracontinental basin in Late Cretaceous time. No sedimentary rocks of rift or marginal facies to such a basin are known, and no structures or igneous activity associated with rifting have been recognized. Furthermore, the suture left after closure of the hypothesized rift basin would have to trend through the Mojave and into the southern Sierra Nevada, because the present Orocopia, Rand, and "Pelona" schists (the latter north of the Garlock fault) all lie in windows below east- to north-dipping thrust faults (Fig. 9-9). There is no evidence for such a suture, either throughgoing across the entire Mojave region or more locally in areas east of the northern (Rand-"Pelona") schists. We do not regard the thrust fault (or faults) mapped by Pelka (1973a, b) south of the Palen and McCoy Mountains and northeast of the Chocolate-Orocopia Mountains (Fig. 9-9) as a suture representing the line of collision of a western continental mass or fragment with the North American continent.

Alternatively, we return to an idea proposed by Yeats (1968a) that the Pelona-Orocopia schist is correlative with Franciscan rocks. Both terranes have similar rock types of oceanic character, and neither terrane is intruded by Mesozoic igneous rocks. We envision the juxtaposition of Pelona-Orocopia schists beneath older crystalline rocks during eastward underthrusting (subduction) of Franciscan rocks along the continental margin in Late Cretaceous to early Cenozoic time. At this time (ca. 70 m.y.a.) subduction may have occurred along a newly modified zone of very low dip, thus accounting for the cessation of magmatic activity in the Sierran arc and areas to the east (Armstrong, 1974; Burchfiel and Davis, 1975). Some Franciscan rocks were underthrust beneath the western margin of North America for a distance of at least 200 to 250 km. Because granitic rocks of the Cretaceous magmatic arc are cut by the Vincent-Orocopia thrust, the lower part of the arc terrane and the lower part of the continental crust must have also been stripped away during this period of shallow subduction. The shallow underthrusting beneath crust warmed by magmatic activity may have contributed to the inverted metamorphism present in Orocopia-Pelona schists below the thrust. The parental rocks of the geographically separate Rand schist and schist of Sierra de Salinas to the north (Fig. 9-9) may have been underthrust somewhat earlier, prior to the period of very shallow underthrusting and before the final cessation of Sierra plutonism. The timing relations of emplacement of the schists in their northern and southern areas of exposure suggests diachronous emplacement: the northern schist terrane (Rand and schists of Sierra de Salinas) was underthrust prior to 80 m.y., whereas the southern schist terrane (Pelona-Orocopia schist) was underthrust probably between 50 and 60 m.y. The Franciscan rocks provide an available protolith for the schists throughout the Cretaceous and Early Tertiary, whereas an ensimatic basin within the Mojave terrane would require diachronous development.

The present distribution of windows and exposures of the Pelona-Orocopia schist appears to be very irregular even when plotted on a reconstructed base map (Fig. 9-9). As discussed later, the distribution of Pelona-Orocopia schist has been affected by probable oroclinal bending of the southern Sierra Nevada and western Mojave terrane. Viewed in this configuration, the eastward underthrusting of Franciscan rocks prior to oroclinal bending need not have been more than 200 to 250 km.

A major problem for our hypothesis of eastward underthrusting of Franciscan (Pelona) rocks is the evidence cited by others that the Vincent-Orocopia thrust plate moved northeastward relative to underlying rocks. Data supporting this direction of relative motion come principally from two sources. Ehlig (1958, 1975) describes a northeast-facing synform below the thrust in the San Gabriel Mountains, and Haxel and Dillon (1978) apply the Hansen (1967) technique on minor folds below the thrust in the Chocolate Mountains to determine a northeastward thrust plate motion. We believe that the lower plate fold described by Ehlig is equivocal in indicating direction of thrust displacement. It may be truncated upward by the thrust contact rather than having developed as a drag structure below it, and hence may have been part of a larger fold structure the geometry of which is unknown. The Haxel and Dillon analysis is more difficult to discount (and, indeed, may prove to be valid). However, north of the Chocolate Mountains are segments of thrust that carry mylonitized metaigneous rocks northeastward over the Late Cretaceous-Paleocene McCoy Mountains Formation (Palen and McCoy Mountains) and Precambrian granitic rocks over metavolcanic rocks (Mule Mountains; Pelka, 1973b; Fig. 9-9). Perhaps the northeast direction of movement reported for the Vincent-Orocopia thrust by Hazel and Dillon (1978) and for the contemporaneous (?) thrusts north of the Chocolate Mountains represent antithetic faulting above a zone of active, somewhat deeper underthrusting. Haxel (1977) reports that the primary schistosity in the Orocopia schist is parallel to the thrust contact and to mylonitic planar fabric in the basal part of the upper plate rocks. He relates these planar fabric elements to the time of emplacement of the upper plate rocks. However, mesoscopic folds used for determining the northeast direction of movement fold the schistosity and are thus later than the time of emplacement of upper plate rocks. The time between these two events is unknown. It may be that the later folds are related to shallow antithetic thrusting during an episode of eastward underthrusting at a deeper structural level.

At present there is no simple solution to the problem of the origin and emplacement of the Pelona-Orocopia schist. Our interpretation encounters difficulty in explaining the data in support of northeast-directed movement with the terrane in the southern Mojave region and the local intrusion of the Rand Schist and the schist of Sierra de Salinas in northern areas. Other interpretations requiring an ensimatic basin in the Mojave and adjacent regions lack physical evidence for the development of such a basin or for the suture developed during basin closure, which should be present across the central Mojave and into the southern Sierra Nevada.

Cenozoic Structural Bending of the Western Mojave and Adjacent Regions

Structural, paleogeographic, and magmatic belts in the western Mojave and adjacent region show a marked right-lateral deflection, a feature noted by Locke and others (1940) and revived by Howell (1975c). Locke and others (1940) considered only the shape and trend of the Sierran batholith to hypothesize that it had been structurally deformed by right-lateral displacement. Howell (1975c, 1976b) postulated a westward bulge or knee in the area of western southern California, using trends of the Sierran batholith and paleogeographic trends of the Great Valley and Franciscan terranes.

Our reconstruction (Fig. 9-9) of the southern California area at about early Miocene time indicates that all trends in the western Mojave and adjacent terranes show a right-lateral deflection. The following trend lines (Fig. 9-9) are deflected: (1) the eastern boundary of the fault sliver hypothesized to have moved south during Permo-Triassic truncation; (2) the main Permo-Triassic truncation boundary, even though not well constrained; (3) culminations (i.e., windows) in the Rand-Vincent-Orocopia thrust system(s); (4) the western limit of the Sierran batholith; (5) pre-latest Jurassic terranes of the Sierran Foothills and Peninsular Ranges; (6) the ^{87}Sr-^{86}Sr 0.7060 line; and (7) the Great Valley and Franciscan terranes. We believe the evidence supports the hypothesis that a major right-lateral intraplate oroclinal bend has deformed all pre-Eocene units and structures in this region. The eastern limit of the area affected by this bend is constrained somewhat by the Independence dike swarm of Late Jurassic age (Chen, 1977), which is not deflected. A southern limit cannot be defined at present, but the bend may extend at least as far south as Victorville.

Oroclinal bending explains the anomalous structures in the southern Tehachapi Mountains, which consist of isoclinal folds with steep axial surfaces and related foliation that strike east-northeast. When the effects of left-lateral displacement on the Garlock Fault are considered, the Tehachapi rocks are probably a continuation of the Kings sequence of Saleeby and others (1978) from the southern Sierra Nevada. They do not make a simple oroclinal bend but appear to be folded on a smaller scale as well. What is clear is that the ophiolite belt and associated volcanic terrane of the western Sierran Foothills do not continue through the Tehachapi Mountains, but must lie to the northwest (Fig. 9-9).

The reconstruction shown in Fig. 9-9 has the northern part of the Salinian block positioned such that it would be near the hinge of the western part of the oroclinal bend. Ross (1978) has pointed out difficulties in correlating pre-batholithic rocks of the Salinian block with those of the northwestern Mojave. Unfortunately, the Salinian block has had a complex history, and we can only add to this complexity. As Ross (1978) has pointed out, the Salinian block does not contain the accreted early and middle Mesozoic terranes along its western side present in the western Sierra Nevada (Davis and others, 1978) and along the western side of the Peninsular Ranges farther south. These Mesozoic terranes are regional features of the Cordilleran orogen (Davis and others, 1978) and were probably once continuous along its western margin. In the Salinian block, Upper Cretaceous rocks of Great Valley-type rest unconformably on Cretaceous granitic rocks of high initial strontium 87–86 ratio, which are characteristic of the central and eastern part of the Sierran batholith. Immediately west of the Salinian block, the oldest parts of the Great Valley-type rocks rest on an oceanic basement. Their proximity to the Great Valley Upper Cretaceous rocks of the Salinian block suggested to Page (1970a, b) that a considerable breadth of an initially intervening terrane has been removed in some manner. Because Upper Cretaceous sedimentary rocks both to the north and south of the Salinian block rest on rocks that are part of the early and middle Mesozoic accreted terrane (i.e., rocks characteristic of the western Sierra Nevada), regional relations suggest an earlier, perhaps pre-latest Cretaceous modification of the western margin of the Salinian portion of the Sierran block. Internal modification of the Salinian block may have taken place as well.

In conclusion, the rock units of the western part of the Mojave-Salinian block and adjacent terranes are interpreted to have been (1) moved left laterally during Permo-Triassic time in one or multiple fault slivers; (2) augmented by tectonic accretion of rock units along a convergent plate boundary in Early and mid-Mesozoic time; (3) modified tectonically during the Cretaceous to "expose" central and eastern parts of the Sierran batholithic complex to depositional overlap by Upper Cretaceous sediments of proximal facies; (4) underthrust by rocks of Franciscan affinity; (5) perhaps subjected to early Cenozoic strike-slip faulting (Howell, 1975c); (6) oroclinally folded in pre-late Cenozoic time; and (7) disrupted by late Cenozoic right-lateral strike-slip faulting. In light of this complex and incompletely understood history, it is not surprising that the initial paleogeographic location or locations of the Salinian basement rocks are not yet known.

Paleogeographic interpretations of the Mojave region suggest that the eastern three-quarters of the region was the site of cratonal and miogeoclinal late Precambrian and Paleozoic deposition on an older crystalline Precambrian basement. Late Precambrian sedimentary rocks wedge out eastward along a north-south line that lies in the eastern part of the Mojave region. These sedimentary rocks have similarities to those of the better known areas to the north, but rocks in the lower part of the sequence cannot be correlated with described formations farther north. Paleozoic rocks have distinctly different trends. Facies and isopachous lines trend northeast, and most of the rocks in the Mojave region are cratonal in character. Thus much of the region was the site of miogeoclinal deposition in late Precambrian, but cratonal deposition during the Paleozoic. Designation of rock sequences as miogeoclinal and cratonal appears to be strongly time dependent. Known and inferred Paleozoic rocks in the western quarter of the Mojave region are completely unlike correlative rocks to the east and lie athwart their regional trends. We interpret these rocks as being out of place. The rocks are eugeosynclinal in character and may have been emplaced as a sliver or slivers by left-slip faulting during a Permo-Triassic truncation event.

Mesozoic paleogeographic interpretations suggest the Mojave region was the site of one or more northwest-trending magmatic arcs of Andean type that were superposed on all previous structural and paleogeographic elements. The magmatic arc or arcs had a complex and still poorly known spatial and temporal development. The eastern limit of arc igneous activity shifted both to the northeast and southwest at different times. Because of this shift in the locus of arc magmatism, back arc, largely nonmarine and nonvolcanic sedimentary rocks are at times restricted to the very eastern part of the region, whereas at other times they transgress westward over older parts of the arc. West of the arc terrane, Mesozoic sedimentary rocks present a very fragmentary picture, but suggest that largely marine sedimentary rocks interfingered with western parts of the arc. Farther west, beyond the Mojave region proper, are Mesozoic marine rocks that were accreted to western North America during Mesozoic time. These rocks contain evidence of oceanic origin and represent rocks accreted during Mesozoic eastward underthrusting, which also was responsible for the development of the Andean-type magmatic arc. Mesozoic paleogeography is difficult to reconstruct in this westernmost terrane, because this area has been affected by later strike-slip faulting and oroclinal bending.

The major structural features of the Mojave region developed during Mesozoic and Cenozoic time, although vestiges of Paleozoic deformations may be present in the out-of-place rocks in the western Mojave region. The oldest deformations recognized in the Mojave region are latest Permian or Early Triassic in age and are present only locally in the western part of the region. Later Mesozoic deformation, which occurred at several different times, is dominated by east-directed thrusting in the northeastern part of the region. Deformation in the southeastern part of the region is not demonstrably dominated by thrusting, but generally east-trending folds are common. This may represent a significant change in deformational style within the orogen. Spatial and temporal distribution of Mesozoic deformations are similar to those of magmatism. Deformation did not migrate eastward in time, but appears to have shifted both to the northeast and southwest with time. Episodes of both normal-slip and strike-slip faulting are present between or contemporaneous with thrusting and folding.

Mesozoic structures west of the Mojave region are dominated by structures related to accretion of oceanic rocks to the North American continent. At times the west part of the magmatic arc terrane transgressed into the accretionary rock sequence. Evidence is present for one or more episodes of truncation by strike-slip (?) faulting, which modified the accretionary and arc terranes. The most evident result is the absence of the accretionary terrane along the western side of the Salinian block.

At least twice during latest Cretaceous and earliest Tertiary time, Franciscan rocks were underthrust eastward at a shallow angle beneath the western edge of the North American plate. During this shallow underthrusting the lower part of the magmatic arc and continental crust were stripped away. Franciscan-equivalent rocks are now present in tectonic windows in the western Mojave and adjacent regions and are referred to as the Rand, Pelona, and Orocopia schists. Other workers have regarded these rocks as ensimatic (but intracontinental) in origin and emplaced by relatively westward underthrusting. We find the latter interpretation difficult to reconcile with several critical geologic consequences, such as the lack of evidence for ensimatic basin development in the Mojave region and the lack of a required suture in the central Mojave and southern Sierra Nevada. Our interpretation, however, is not without its difficulties.

Evidence from regional trends of Paleozoic and Mesozoic paleogeographic and structural elements suggests that the rocks of the western Mojave, southern Sierra Nevada, northern Salinian block and adjacent terranes were oroclinally bent in a clockwise sense. Timing of this proposed right-lateral bend is not clear and may have occurred between mid-Cretaceous and mid-Tertiary time.

ACKNOWLEDGMENTS

This work has developed from our continuing studies into the tectonics of the Mojave Desert and adjacent regions. Our studies have been generously supported by the Geology and Earth Science programs of the National Science Foundation. We gratefully acknowledge the support of Grants EAR-77-13637 and GA-43311 (Burchfiel) and GA-21401 and GA-43309 (Davis).

P. L. Ehlig
Department of Geology
California State University, Los Angeles
Los Angeles, California 90032

10

ORIGIN AND TECTONIC HISTORY OF THE BASEMENT TERRANE OF THE SAN GABRIEL MOUNTAINS, CENTRAL TRANSVERSE RANGES

ABSTRACT

The San Gabriel Mountains are an east-west-trending range formed from basement rocks uplifted along the southwest side of the San Andreas fault in the central part of the Transverse Range Province of southern California. Uplift began a few million years ago and is still going on as a product of convergence between the North American and Pacific plates along the bend in the San Andreas fault. It was accomplished by reverse faulting along the southern range margin and broad arching across the interior and northern margin, with uplift extending northeastward across the San Andreas fault. The relatively simple pattern and style of uplift are superimposed upon the complex structure of rocks exposed within the range.

Basement rocks exposed within the deeply eroded interior of the range are separated into an upper and lower plate by the Vincent thrust fault. The core of the upper plate originated as part of a Precambrian craton. The oldest rocks consist of amphibolite facies, quartzofeldspathic gneiss, and amphibolite at least 1715 ± 30 m.y. old. These are intruded by 1670 ± 15 m.y. old gneissose granitic rocks and a 1220 ± 10 m.y. old anorthosite-gabbro-syenite complex, adjacent to which is a granulite facies terrane that developed from the older gneiss and granitic rocks. The Precambrian rocks are intruded by a large compositonally zoned pluton of the distinctive, 220 ± 10 m.y. old Lowe Granodiorite and by common types of late Mesozoic granitic rocks. The metamorphosed remnants of a thick sequence of carbonates, quartz arenites, and calcareous and aluminous argillites that occur within migmatitic complexes in the southern and eastern parts of the range probably represent Paleozoic strata deposited on the Precambrian craton and migmatized during emplacement of late Mesozoic granitic rocks.

Exposed rocks within the lower plate of the Vincent thrust consist of Pelona schist. It consists mainly of thin-bedded white mica-quartz-albite schist derived from arkosic graywacke, siltstone, and shale. Greenschist derived from basaltic tuff is common in the upper part of the sequence. Thin beds of quartzite derived from chert are common, particularly in association with greenschist. The depositional age of the Pelona Schist has not been established but is probably Mesozoic, perhaps Late Cretaceous. The schist was probably deposited on oceanic crust within a marine basin that received sediments mainly from a continental source of granitic composition, but with some sediment coming from an offshore volcanic source. The metamorphism of the Pelona Schist occurred about 60 m.y. ago in response to deep burial beneath the Vincent thrust fault. This event probably relates to subduction and involved northeastward thrusting of a microcontinent or peninsula across a marginal basin or narrow gulf.

The basement rocks of the San Gabriel Mountains were adjacent to the Orocopia and Chocolate Mountains in southeastern California about 10 m.y. ago. They owe their present position to 240 km of right slip along the San Andreas fault and 60 km of right slip on the San Gabriel fault.

INTRODUCTION

The geology of California is traditionally treated according to physiographic province because of the obvious relationship between surface physiography and the nature and recent petrotectonic history of the underlying terrane. Until recently, geologists tended to view the physiographic provinces as semipermanent entities that evolved upon a relatively immobile crust through isostatic and other processes primarily controlled by the nature and prior history of the rocks within each province. However, this approach led to conclusions in conflict with observational data and failed to unearth mechanisms to explain the unique evolution of each province. As a result of our newly gained knowledge of plate tectonics, we now realize California is a region of great crustal mobility. The physiographic provinces are petrotectonic domains whose existence is more closely controlled by geographic location relative to plate boundaries and upper mantle processes than by the nature and cumulative history of the rocks within each province.

This is particularly true of the Transverse Range Province of southern California. It is characterized by nearly east-west trending ranges in a belt 500 km long and up to 100 km wide. The physiographic province originated within the past several million years, apparently as the result of deep-seated crustal compression across a left-stepping bend in the right-slip transform boundary between the Pacific and North American plates. Instead of being controlled by the nature or original structure of the underlying basement rocks, the province represents a composite terrane formed from segments of at least four separate basement complexes (Fig. 10-1). Only the San Bernardino Mountain-Mojave Desert basement complex to the northeast of the San Andreas fault appears to have developed in its present location relative to North America. The other three, here referred to as (1) the Franciscan-Catalina, (2) the Santa Monica Mountains-Santa Cruz Island, and (3) the San Gabriel Mountains-Orocopia Mountains basement complexes, originated in separate geographic locations and were moved to their present positions by large-magnitude faulting. The distribution and important features of the four complexes are as follows:

1. The Franciscan-Catalina basement complex probably underlies most of the western third of the province. Where exposed in the Santa Ynez and southern San Rafael Mountains in the northwestern part of the province, it consists of Jurassic-Cretaceous Franciscan assemblage and ultramafic rocks similar to those exposed farther north in the Coast Ranges. The complex is not exposed in the southwestern part of the province, but is well exposed to the south on Catalina Island and has been penetrated in numerous drill holes within the western Los Angeles Basin. The Franciscan assemblage generally is interpreted as trench deposits resting on oceanic crust that experienced subduction and northward displacement along the Pacific-North American plate boundary as described by Blake and Jones (chapter 12) and Page (chapter 13), this volume. The eastern limit of this complex is concealed beneath a Cenozoic sedimentary cover but probably abuts against or passes beneath the San Gabriel Mountains basement complex in the vicinity of the Nacimiento

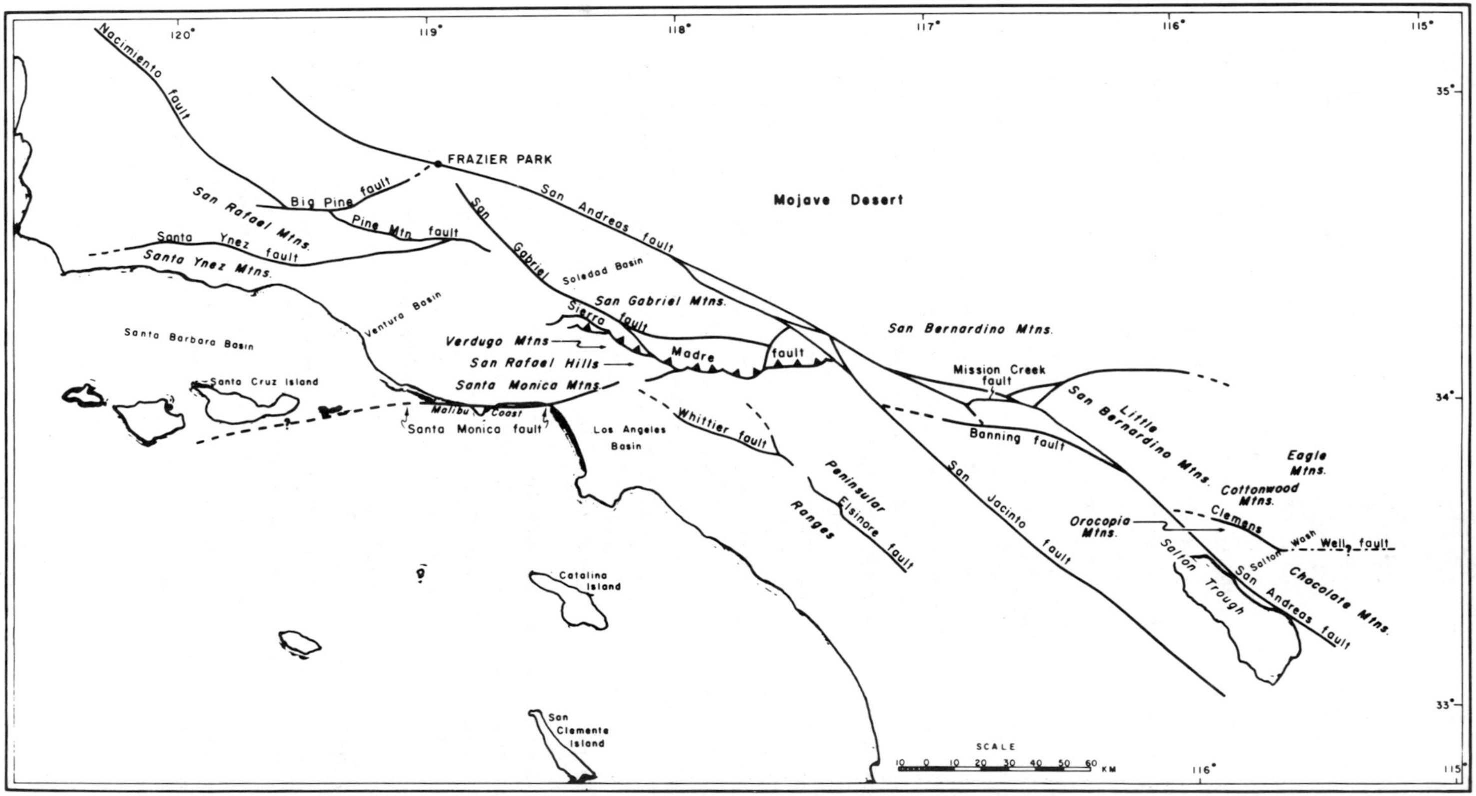

Fig. 10-1. Map of southern California showing location of faults and topographic features referred to in text.

and Pine Mountain faults in the northwestern part of the province. In the southwestern part of the province, the Franciscan-Catalina complex may pass beneath the Santa Monica Mountains and abut against the San Gabriel Mountains complex along a concealed fault located between the Santa Monica and Verdugo Mountains.

2. In the southwestern part of the province, the Santa Monica Mountains contain the Upper Jurassic Santa Monica Formation intruded by Cretaceous granitic rocks, whereas Santa Cruz Island contains Jurassic metavolcanic and related intrusive rocks. The two terranes are dissimilar in lithology but may be part of a single crustal block, with the Santa Monica Formation representing sediments deposited in a forearc or marginal basin and the Santa Cruz basement representing a volcanic arc complex. The Santa Monica Formation has generally been considered correlative with the Bedford Canyon Formation of the northwestern Peninsular Ranges; however, a recent isotopic study by Criscione and others (1978) indicates the two are of different age and have different source terranes. Paleomagnetic measurements by Kamerling and Luyendyk (1977) indicate the Santa Monica Mountains terrane has rotated 75° clockwise and translated northward about 5° of latitude since middle Miocene. Their preliminary work indicates a similar rotation for the Santa Cruz Island terrane. The cause of the rotation and latitudinal displacement has not been established, but one possibility is that the Santa Monica Mountains-Santa Cruz Island basement complex forms a semirigid microplate riding on top of Franciscan-Catalina basement, which has experienced right slip across a broad zone of transform faulting. The southern limit of the Santa Monica Mountains-Santa Cruz Island basement is along, or close to, the Malibu Coast-Santa Monica fault, and the eastern limit lies between the Santa Monica and Verdugo Mountains. The northern limit has not been determined but is probably south of the axis of the Ventura and Santa Barbara basins.

3. The basement complex of the central and southeastern Transverse Ranges, the main topic of this chapter, is greatly affected by faults of several generations. During middle Miocene time this complex extended across southeastern California into the area presently occupied by the northern part of the Salton Trough, where it probably terminated along a transform fault, placing Peninsular Range or Franciscan basement against it on the west. Subsequent displacement along the San Andreas fault and its branches distributed fault-bounded segments of the complex in a semicontinuous belt extending along the west side of the San Andreas fault across the entire Transverse Range Province from the Salton Trough to the southern Coast Ranges. The most extensive exposures west of the San Andreas fault are in the San Gabriel Mountains. East of the San Andreas, the complex is well exposed in the Orocopia Mountains and is discontinuously exposed in adjoining ranges to the east and southeast extending into the southwest corner of Arizona. The complex contains a major thrust fault, referred to as the Vincent thrust in areas west of the San Andreas and as the Orocopia thrust in areas east of the San Andreas. The thrust appears to have originated during the Paleocene as a gently inclined subduction zone. The upper plate contains Precambrian gneisses and plutonic rocks extensively intruded by Mesozoic granitic rocks. In the San Gabriel Mountains and adjacent areas west of the San Andreas fault, part of the upper plate terrane contains Paleozoic (?) metasedimentary rocks, including quartzite, marble, and aluminous rocks, that were extensively converted to gneiss and migmatite during intrusion of Mesozoic granitic rocks. The Vincent-Orocopia

thrust is underlain by a dominantly flyschlike metasedimentary assemblage referred to as Pelona Schist in areas west of the San Andreas fault and as Orocopia Schist in areas east of the San Andreas. The schists were metamorphosed during the Paleocene, evidently in response to deep burial beneath the thrust. Their sedimentary age has not been determined, but is probably no older than Jurassic and is most likely Late Cretaceous inasmuch as they appear to have had a relatively simple history and are not intruded by Mesozoic granitic rocks. The direction of movement on the Vincent-Orocopia thrust has great bearing on the tectonic evolution of southern California. Available evidence suggests that the upper plate moved northeastward across the schist. This would require the upper plate to have originated outboard from the schist, perhaps as a microplate torn from the North American craton farther south, and to have been thrust onto the North American plate margin. This would also require a suture zone to have existed between upper plate rocks exposed in the Orocopia Mountains and the San Bernardino Mountains-Mojave Desert basement complex, which is believed to be part of the North American craton. Such a suture zone may exist along the trend of the Clemens Well fault (Fig. 10-1) but is as yet unproven. An alternate possibility is that the schist was subducted northeastward beneath North America (as advocated by Burchfiel and Davis, Chapter 9, this volume), but major problems are associated with this proposal, as explained later.

4. The San Bernardino Mountains-Mojave Desert basement complex is truncated on the southwest by the San Andreas fault and on the south by a partially conjectured fault (suture zone?) along the trend of the Clemens Well fault. The San Bernardino Mountains and ranges to the east contain Precambrian gneisses and granitic rocks believed to be an essentially in place segment of the North American craton. These rocks are unconformably overlain by remnants of deformed and metamorphosed Paleozoic strata dominated by orthoquartzite and marble. Late Mesozoic granitic rocks are the most abundant component of this complex.

Three things should be apparent from the preceding summary and should be kept in mind while reading this chapter: (1) the crystalline basement beneath the Transverse Ranges is diverse in its nature, history, and origin; (2) our knowledge of the basement complex of the San Gabriel Mountains, the main topic of the chapter, is of little value unless we can determine where these rocks were located relative to other basement complexes during each stage of their evolution; and (3) much remains to be learned about the petrotectonic evolution of the Transverse Ranges.

SAN GABRIEL MOUNTAINS BASEMENT COMPLEX

General Setting

The San Gabriel Mountains, an east-west trending range 100 km long by a maximum of 38 km wide, has been uplifted essentially en masse during the late Cenozoic by reverse faulting along the southern margin and broad arching across the interior and northern margin of the range. Erosion has created rugged topography, exposing basement rocks over all but a few peripheral areas.

A small part of the range lies northeast of the San Andreas fault. It is formed from the upturned margin of the Mojave Desert plate and consists of Mesozoic granitic plutons with intervening migmatite, gneiss, marble, and calc-silicate rocks derived from Paleozoic (?) strata. The genesis of these rocks bears no direct relationship to those southwest of the San Andreas fault and will not be considered in detail.

The main mass of the San Gabriel Mountains lies southwest of the presently active trace of the San Andreas fault and is broken into sub-blocks by the inactive Nadeau-Punchbowl and San Gabriel branches of the San Andreas fault, and by the San Antonio fault (Fig. 10-2). Most of the central and western parts of the range belong to a single sub-block that continues westward across the Soledad Basin and Sierra Pelona to the San Francisquito fault. The Pelona Schist is exposed in the northeastern part of this sub-block and on Sierra Pelona in the northwestern part. Above the schist is a belt of mylonitic and retrograde metamorphic rocks developed along the sole of the upper plate of the Vincent thrust. The upper plate contains a Precambrian gneiss-amphibolite-granite complex intruded by a Precambrian anorthosite-syenite-gabbro complex, all of which are intruded by the Early Triassic Lowe Granodiorite, rhyolitic to basaltic dikes of intermediate Mesozoic age, and Cretaceous granitic rocks. All but the Cretaceous granitic rocks have been metamorphosed and deformed at least once, and some of the oldest rocks have been metamorphosed and deformed as many as four times.

Basement rocks exposed in the south-central part of the range belong to the upper plate of the Vincent thrust and consist of Precambrian rocks intruded by Mesozoic granitic rocks. They are offset from their counterparts to the north by 22 km of right slip on the North Branch of the San Gabriel fault.

The basement terranes south of the South Branch of the San Gabriel fault in the southwestern San Gabriel Mountains and adjacent Verdugo Mountains and San Rafael Hills contain screens of dolomite marble, quartzite, and aluminous and graphitic schists of Paleozoic (?) age immersed in gneiss, migmatite, and Cretaceous granitic rocks. There are no equivalent rocks in the central part of the range, but data presented later indicate that they belong to the upper plate of the Vincent thrust and were brought into their present position by 60 km of right slip on the combined north and south branches of the San Gabriel fault.

A thick sequence of Paleozoic (?) metasedimentary rocks along with gneiss, migmatite, and Cretaceous granitic rocks forms the eastern part of the range to the east of the San Antonio fault. The southern part of this terrane contains an east-west trending mylonite belt derived in part from granulite facies rocks. This terrane appears to have moved into its present position by left slip on the San Antonio fault. No equivalent rocks are exposed within the interior of the range to the west of the San Antonio fault, although they may be related to the terrane south of the South Branch of the San Gabriel fault.

The narrow slice of Pelona Schist and other rocks between the Nadeau-Punchbowl fault and the active branch of the San Andreas fault are offset from their counterparts by 45 km of right slip on the Nadeau-Punchbowl fault, as will be discussed later.

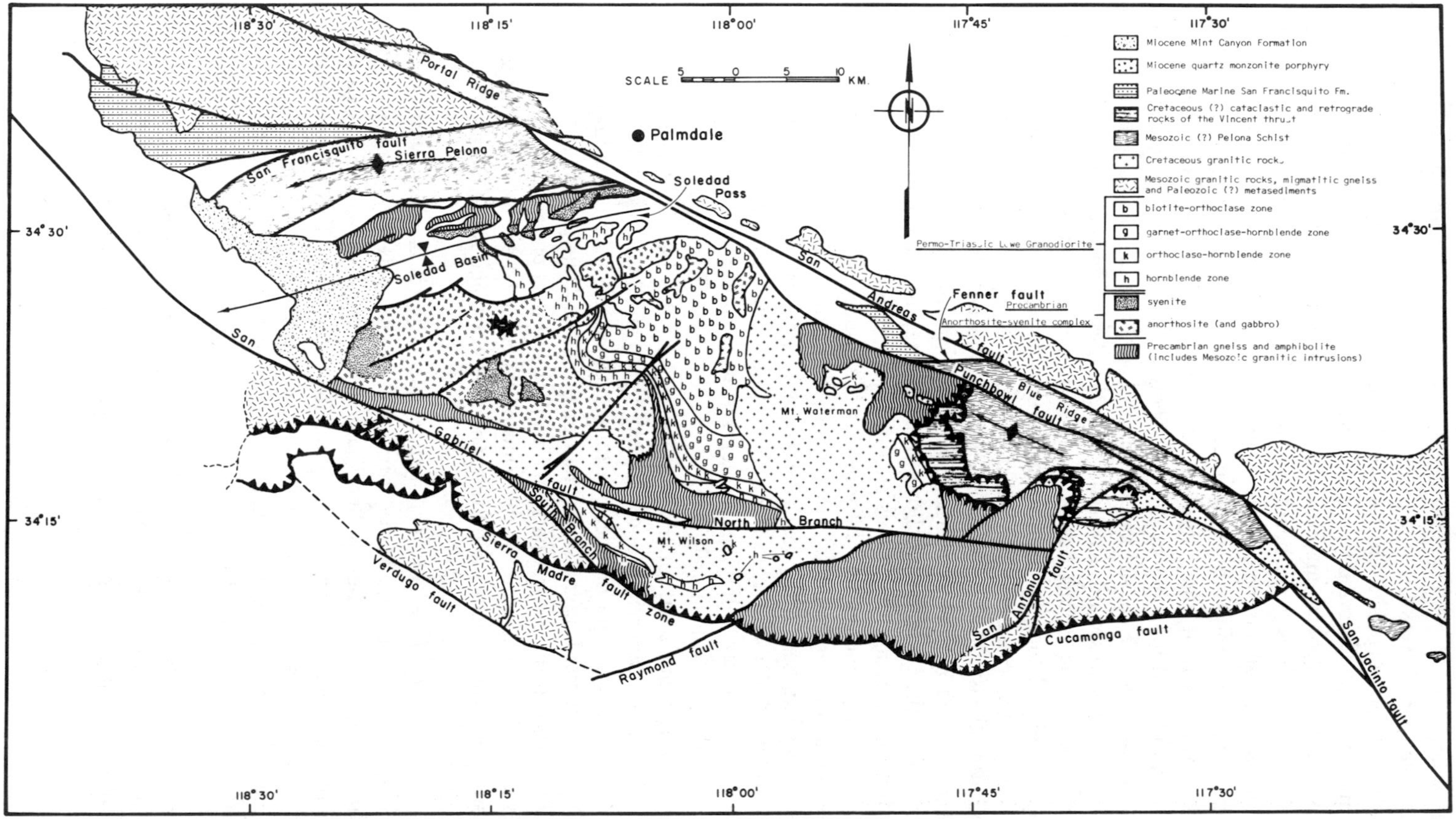

Fig. 10-2. Generalized geologic map of the San Gabriel Mountains and adjoining area.

Precambrian Rocks

These rocks represent a deep-seated segment of Precambrian continental crust, probably derived from the North American craton. They were deformed, metamorphosed, and intruded by granitic plutons during the Mesozoic Era. The oldest recognizable terrane is an upper amphibolite facies complex composed of layered gneiss, amphibolite, migmatite, and granitic rocks. The best preserved parts of this terrane are in the south-central San Gabriel Mountains to the south of the San Gabriel fault and in the north-central Soledad Basin south of Sierra Pelona. These rocks are characterized by a simple mineral assemblage consisting of variable proportions of plagioclase (oligoclase to andesine), quartz, potassium feldspar, biotite, and hornblende. Where well developed, the gneiss consists of alternating light to dark layers, 1 to 20 cm thick, which resemble bedding but lack compositions diagnostic of a sedimentary origin, thus leaving their origin in doubt. The Precambrian granitic rocks are strongly foliated as a result of subsequent metamorphism but have retained their compositional homogeneity. The most distinctive granitic rock is augen gneiss exposed along a ridge that crosses Sierra Highway near Sleepy Valley in the northern Soledad Basin. It contains abundant lenticular pink K-feldspar crystals, up to several centimeters long and typically consisting of a simple Carlsbad twin, in a strongly foliated matrix of quartz, oligoclase, and biotite. This coarse-grained augen gneiss does not occur as such in the San Gabriel Mountains, but a finer-grained variety containing scattered gray-white K-feldspar phenocrysts ½ to 2 cm long, is abundant in the south-central part of the range. Zircons from the augen gneiss of Soledad Basin have been dated as 1670 ± 15 m.y. by the U–Pb isotope method (Silver, 1966), and zircons from foliated quartz monzonite of the south-central San Gabriel Mountains have yielded a $^{207}Pb-^{206}Pb$ age of 1670 ± 20 m.y. (Terry Davis, personal communication, 1978). This is interpreted as the age of emplacement of the granitic plutons and is probably the time when the adjacent Precambrian migmatite and gneiss formed from preexisting units. Zircons from layered gneiss of the Soledad Basin have yielded a U-Pb isotope age of 1715 ± 30 m.y., which Silver (1966) interprets as the approximate age of the original source material.

An anorthosite-gabbro-syenite complex with an exposed area in the western San Gabriel Mountains of about 200 km^2 was emplaced into the older Precambrian terrane about 1220 ± 10 m.y. ago (Silver and others, 1963; Silver, 1971). The structure and petrology of this complex have most recently been described by Carter and Silver (1972) and were previously described by Miller (1931, 1934), Oakeshott (1937, 1954, 1958), Higgs (1954), and Crowell and Walker (1962). Carter and Silver (1972) interpret the main body as an inverted cone-shaped stratiform intrusion, at least 10 km thick and about 15 km in diameter, which differentiated by crystal fractionation to form the rock suite (from bottom to top) andesine anorthosite-leuconorite-norite-jotunite-mangerite-syenite-quartz syenite. The structure has been greatly complicated by Mesozoic and Cenozoic folding and faulting. There has been pervasive alteration of pyroxene to amphibole and microbrecciation of the anorthosite, which most previous workers attribute to deuteric alteration and autodeformation; however, my studies suggest it was caused in large part by Mesozoic amphibolite facies regional metamorphism.

The anorthosite-gabbro-syenite complex is bordered on the south and southeast by gneiss containing a relict granulite facies mineral assemblage. Oakeshott (1958) was the first to distinguish this gneiss from the typical amphibolite facies gneisses in surrounding areas and named it the Mendenhall Gneiss. Mesozoic amphibolite facies metamorphism has largely destroyed the granulite facies minerals, particularly in the southeastern area, but has left distinctive replacement textures. Relict hypersthene, augite, garnet, and alkali feldspar (hairline perthite to antiperthite) have survived locally. Blue to violet quartz is the most common relict mineral. Most of the Mendenhall Gneiss appears to have formed from the 1670 m.y. old amphibolite facies terrane by thermal metamorphism with little associated deformation. This is particularly evident in granulite derived from porphyritic granitic rocks. Granular masses of light bluish-gray mesoperthite form pseudomorphs after K-feldspar phenocrysts. The southeastern part of the Mendenhall Gneiss contains thick sills of metagabbro, which appear to emanate from the anorthosite complex. Some of the thicker metagabbro bodies contain pegmatitic pods of anorthosite. This and the fact that the granulite facies rocks are only known to occur in areas close to the anorthosite complex strongly suggest that the granulite facies metamorphism was produced as a thermal aureole around the 1220 m.y. old anorthosite complex. However, Silver and others (1963) believe the granulite facies metamorphism occurred about 1440 m.y. ago based on discordant U–Pb ages obtained from zircons separated from granulite pegmatite.

In summary, two major Precambrian petrotectonic thermal events are recognizable within rocks of the upper plate of the Vincent thrust. The first, about 1670 m.y. ago, is evidenced by terrane composed of quartzofeldspathic gneiss and migmatite, amphibolite, and granitic rocks; it undoubtedly formed during a major continental orogenic event. No shallowly formed products of this event have been recognized, probably as a result of deep erosion during the late Precambrian. The second event, 1220 m.y. ago, involved emplacement of an anorthosite-gabbro-syenite complex. Our present state of knowledge suggests that, although granulite facies contact metamorphism occurred, this event was not accompanied by regional metamorphism or significant deformation in adjacent rocks. No metamorphic or intrusive events appear to affect this terrane during the billion years between the 1220 m.y. event of the late Precambrian and the emplacement of Lowe Granodiorite 220 m.y. ago in Early Triassic time.

Lowe Granodiorite

The distinctive Lowe Granodiorite forms a compositionally zoned pluton exposed over an area of about 300 km^2 in the San Gabriel Mountains and northeastern Soledad Basin (Fig. 10–2). I interpret the pluton as a floored intrusion, perhaps resembling a large laccolith, which crystallized from the bottom upward forming a differentiation sequence. The pluton subsequently has been tilted northeastward such that its base is exposed along a sharply defined, steeply inclined western margin. The Mendenhall Gneiss and anorthosite in a zone several tens of meters wide directly beneath the base of the Lowe Granodiorite contain intensely deformed, attenuated compositional banding with plunging drag folds showing a left-handed sense of shear. Although subsequent metamorphism and deformation make interpretation tenuous, this zone appears to have been a fault, perhaps a gently to moderately dipping thrust fault, prior to emplacement of the Lowe Granodiorite.

The roof of the pluton is irregular. Southeast of Soledad Pass, the upper part of the Lowe Granodiorite contains large bodies of metapyroxenite and metagabbro interpreted as roof pendants. Farther southeast, to the east of the Cretaceous Mount Waterman pluton, Lowe Granodiorite interfingers with migmatized gneiss in a manner suggesting the gneiss was plastic due to partial melting.

The Lowe Granodiorite is characterized by a high feldspar content, ranging from about 60 to 95%, and a low quartz content, generally averaging about 10%. Its name is misleading in that it varies from hornblende diorite and quartz diorite near the base to albite-rich granite and syenite in the upper part. Megascopic variations in mineralogy permit subdivision of the pluton into zones that subparallel the basal contact. The lowest few meters of the pluton consist of dark gray, medium-grained hornblende diorite. This grades upward into a lighter-colored, coarser-grained diorite to quartz diorite with stout hornblende crystals set in a dominantly sodic andesine matrix. At an average distance of about 800 m above the base, the basal hornblende zone grades over a few tens of meters into the orthoclase-hornblende zone, with the incoming of orthoclase phenocrysts scattered through a matrix of calcic oligoclase and minor quartz. Hornblende phenocrysts are larger and less abundant than in the hornblende facies and, where particularly coarse grained, create a strikingly spotted "dalmatian" textured rock. The garnet-orthoclase-hornblende zone begins roughly 1.5 km above the base of the pluton with the appearance of garnets, both as widely spaced crystals as much as 2 cm across and as small crystals concentrated in seams during crystallization of residual fluids. Orthoclase phenocrysts have their greatest size within this facies, with crystals as much as 10 cm long. The biotite-orthoclase zone, the highest and last to crystallize, occurs over a large area in the northern part of the pluton (see Fig. 10-2). This rock contains up to about 10% biotite in a nearly white matrix. Plagioclase, ranging from oligoclase to albite in composition, is the most abundant mineral. Orthoclase typically occurs as phenocryst, but in some areas occurs instead as part of the granular matrix. Quartz varies from a trace to about 25%. The change from hornblende-bearing to biotite-bearing rocks is abrupt. The two do not coexist as primary minerals in the same rock, although hornblende-bearing layers occur locally in the lower part of the biotite-orthoclase facies.

Much of the Lowe Granodiorite has a primary foliation subparallel to the base of the pluton. Its primary origin is shown by the presence of late-stage aplite and pegmatite segregations and dikelets occurring both parallel to and crosscutting foliation. In addition, there are a few widely spaced layers of nearly pure hornblende, probably of cumulate origin, that parallel the foliation.

The Lowe Granodiorite is affected by metamorphism ranging from upper amphibolite facies near Cretaceous granitic rocks in the central San Gabriel Mountains to lower amphibolite facies in the western part of the range and Soledad Basin. This has enhanced the original foliation and modified igneous textures through granulation and recrystallization. Hornblende is largely replaced by epidote in the northwestern part of the pluton.

The age of the Lowe Granodiorite is 220 ± 10 m.y. as determined by the U–Pb method using zircons (Silver, 1971) and 208 ± 7 m.y. by the Rb–Sr whole-rock method (Joseph and others, 1978). This Early Triassic age is unusual for Cordilleran plutonic rocks, although several plutons of about the same age with similar chemical compositions but different physical appearances occur elsewhere in California (C. F. Miller, 1977 a, b, 1978).

Joseph and others (1978) believe the magma for the Lowe Granodiorite formed by partial melting of subducted oceanic crust, as indicated by a low initial $^{87}Sr-^{86}Sr$ ratio (0.70456 ± 0.00003), a high alkali content ($Na_2O + K_2O \approx 10$ wt. %) and a high Sr content (700 to 1500 ppm). The Lowe Granodiorite may be a product of the initial subduction of Pacific oceanic crust beneath North America.

Amphibolite and Metarhyolite Dikes

Amphibolite dikes, the metamorphosed equivalent of basalt and andesite, are common in Lowe Granodiorite and older rocks but are truncated by Upper Cretaceous granitic intrusions. The dikes are generally several centimeters to a few meters thick with straight parallel walls that typically cut across the structure of the host rock. Some contain angular inclusions of the host rock. Metamorphosed rhyolite dikes containing scattered phenocrysts of quartz in an aphanitic groundmass occur in a belt several kilometers wide extending through the west-central part of the range. The most prominant dikes form gently dipping sheets as much as 15 m thick. Several have been traced for more than 1 km without showing evidence of significant deformation other than possible tilting from an initially steeper orientation. These dikes are truncated by the Mount Waterman pluton.

The straightness of the dikes demonstrates that many rocks achieved their present structural complexity before emplacement of Cretaceous granitic rocks. The aphanitic texture of the metarhyolite dikes, as well as their straight walls, indicates the wall rocks were fairly cool at the time of dike emplacement. Amphibolite dikes in the central part of the range record moderate to extensive postemplacement deformation, including deformation that postdates emplacement of crosscutting Cretaceous aplite and pegmatite dikes. Dikes in the western part of the range show evidence of little postemplacement deformation. None of the dikes has been dated.

Cretaceous Granitic Rocks

All previously described rocks within the upper plate of the Vincent thrust are intruded by plutons composed of common types of medium-grained granitic rocks. None of the plutons is chemically homogeneous. All are composite bodies typically ranging from early-formed melanocratic hornblende quartz diorite to late-formed leucocratic biotite quartz monzonite. These rocks postdate all metamorphism except that associated with the Vincent thrust-Pelona Schist event; however, the youngest regional metamorphic event that affects the host rocks probably occurred synchronous with pluton emplacement and was facilitated by heat from the plutons.

The largest pluton is exposed over an area of about 200 km^2 in the central part of the range and is herein referred to as the Mount Waterman pluton in the area north of the North Branch of the San Gabriel fault and as the Mount Wilson pluton (Wilson Diorite of Miller, 1934) south of the fault. This pluton has a relatively sharp western margin but a diffuse eastern margin bordered by migmatite developed from preexisting gneiss and migmatite. Associated granite aplite and pegmatite dikes are abundant in the terrane east of the pluton but are absent from that west of the pluton. This may reflect the presence of the pluton at shallow depth beneath metamorphic rocks of the eastern area, or it may be the result of postemplacement tilting toward the east or northeast such that the western margin represents the steeply inclined side of the pluton and the eastern margin represents the gently inclined roof.

A series of interconnected plutons trends westward across the western part of the range north of the San Gabriel fault. In most places, these plutons form sharp intrusive contacts with their hosts. The Mendenhall Gneiss has been migmatized locally where it occurs as pendants within the interior and southeast side of these plutons. Irruptive breccias consisting of angular fragments of gneiss and anorthosite-related rocks enclosed in a fine-grained porphyritic matrix occur in several areas along the northern edge of the plutons, indicating rapid cooling of the magma.

These granitic rocks are believed to be Late Cretaceous in age, but only one has been satisfactorily dated. The Josephine Mountain pluton in the west-central part of the range is dated at 80 ± 10 m.y. by the U-Pb method using zircons (Carter and Silver, 1971). Samples from the same pluton have yielded K-Ar ages of 65.8 m.y. on biotite and 70.0 m.y. on hornblende (Evernden and Kistler, 1970), but these are probably cooling ages considerably younger than the emplacement age. Similar K-Ar ages have been obtained from granitic samples elsewhere in the range and are also interpreted as cooling ages.

Paleozoic (?) Metasediments and Associated Terrane

The wide distribution of Precambrian rocks within the main body of the San Gabriel Mountains indicates erosion has stripped away any Paleozoic metasedimentary rocks that might have existed within this part of the upper plate of the Vincent thrust. However, the southwestern part of the range and adjacent Verdugo Mountains, to the south of the San Gabriel fault, contain marble, calc-silicate rocks, quartzite, and aluminous and graphitic schists derived from sediments of probable Paleozoic age. These metasediments occur as lenses and pods within an intensely deformed, upper amphibolite facies complex of gneiss, migmatite, and granitic rocks. This complex has not been dated but shows evidence of only one metamorphism, which appears to have occurred during emplacement of the granitic rocks. Reconstruction of displacement of the San Gabriel fault indicates that this terrane originated directly south of the Mount Wilson-Mount Waterman pluton and that the granitic rocks are offshoots from the pluton. This reconstruction is supported by the occurrence of marble and calc-silicate rock lenses in migmatite in a small area at the south-central range margin to the west of San Gabriel Canyon.

Thus, a Paleozoic cover was probably deposited over the Precambrian terrane in the interior of the San Gabriel Mountains and was subsequently deformed and metamorphosed during emplacement of Cretaceous granitic rocks.

Metasedimentary rocks of probable Paleozoic age are abundant in the eastern San Gabriel Mountains to the east of the San Antonio fault. A stratigraphic section about 2000 m thick has been delineated in this area (Ehlig, 1958), although repetition of section by premetamorphic folding or faulting cannot be ruled out. The original lithology consisted mainly of quartz arenite, dolomite, siltstone, and aluminous shale. The central part of this terrane shows evidence of only one metamorphism, an upper amphibolite facies event marked by the extensive development of gneiss and migmatite around granitic intrusions. The southern part of this terrane contains an east-west trending, northward-dipping mylonite belt that is truncated by the San Antonio fault on the west and the Lytle Creek fault on the east. The mylonite appears to have formed during late stages of the amphibolite facies metamorphism. It is partly derived from and underlain by gneiss and metasedimentary rocks of granulite facies. The petrogenesis of the mylonites and granulites has been investigated by Hsü (1955), but their tectonic significance has not been determined. The granulite facies terrane contains similar types of metasediments as the amphibolite facies terrane farther north, and also contains original mineral assemblages transitional with those of the upper amphibolite facies to the north. Thus, the two appear to be closely related and were probably both formed during the Cretaceous.

The northern part of the terrane passes into a thick zone of mylonitic rocks underlain by Pelona Schist. The mylonites have previously been interpreted as having formed along the Vincent thrust (Ehlig, 1958); however, the geologic relationships, which are complicated by extensive intrusions of Miocene granodiorite porphyry and by Miocene and younger faulting, should be reexamined for alternate interpretations. The apparent absence of Precambrian rocks in this terrane, all the way to the base of the Vincent thrust, seems incompatible with the presence of Precambrian rocks along the west side of the San Antonio fault, as will be discussed later.

Vincent Thrust Fault

From the standpoint of regional tectonics, the Vincent thrust fault is the most important structural feature in the basement complex of the San Gabriel Mountains. It and the Orocopia thrust, its offset equivalent in southeastern California, have placed the Pelona-Orocopia schists under not less than 7500 km^2 of continental basement terrane. The exposed parts of the thrust originated at great depth and owe their present exposure to many kilometers of uplift and erosion. Folding appears to have been the principal mechanism of uplift. Most exposures of the thrust fault are on the flanks of anticlines with schist forming the cores. This is the case in the northeastern San Gabriel Mountains, where the Vincent thrust and underlying Pelona Schist are exposed over an area of about 100 km^2 in a westward-plunging anticline that is truncated on the northeast by the Punchbowl fault. Sierra Pelona, to the west of the range, is a westward-plunging anticline exposing 200 km^2 of schist. The Vincent thrust is discontinuously exposed along the south flank of Sierra Pelona but is not easily recognized owing to deep weathering and complications produced by younger faulting. The San Francisquito fault places unrelated rocks against Pelona Schist along the north side of Sierra Pelona. Mylonitic rocks, which are believed to have formed along the Vincent thrust, are exposed over an area of 2 km^2 in the Mill Canyon anticlinal window within the anorthosite complex in the western San Gabriel Mountains (Fig. 10-2). Mylonitic and retrograde metamorphic rocks that probably formed along the sole of the Vincent thrust are also exposed along the south side of the San Gabriel fault adjacent to the confluence of the North and West Branches of the San Gabriel River in the central part of the range. Of course, if uplift and erosion continue, the Vincent thrust and underlying Pelona Schist will be unroofed over the entire range to the southwest of the San Andreas fault.

The Vincent thrust has a different appearance than most faults with which geologists are familiar. Instead of being marked by a narrow seam of slickensided clay gouge or pulverized rock sandwiched between two well-defined walls, the Vincent thrust is marked by a broad zone of mylonitized and partially recrystallized rocks that commonly appear to form a gradational contact between the two plates. Rocks within the Vincent thrust zone have experienced intense stretching and laminar shear flow parallel to foliation and flattening perpendicular to foliation. The thickness of the thrust zone varies significantly from place to place owing in part to tectonic thickening and thinning during thrusting. In the eastern San Gabriel Mountains the thickness varies from a minimum of about 100 m to a maximum of about 1000 m.

For mapping purposes, the base of the Vincent thrust is placed at the contact between Pelona Schist, which formed by prograde metamorphism of sedimentary rocks, and overlying mylonitic schist, which formed by milling out and retrograde metamorphism of basement rocks from the upper plate. Where well exposed, this contact can be placed within 1 or 2 cm on the basis of differences in the texture and appearance of the rocks on either side of the contact. The contact commonly is marked by a light greenish-gray seam of very fine grained actinolite-chlorite schist. I interpret this to be metamorphosed fault gouge that formed at shallow depth during the early stages of thrusting and then was transported with the lower plate into the zone of metamorphism.

The upper contact of the Vincent thrust zone is more difficult to define than is the base because of the lack of a distinctive change in lithology. A fairly abrupt contact occurs in a few places with mylonite against undisturbed rocks of the upper plate, but in most places there is either a gradual upward decrease in the degree of mylonitization through a thickness of several tens of meters or an irregular upward decrease in the occurrence of narrow zones of mylonite interspersed through relatively undisturbed rock.

The best place to observe the Vincent thrust fault and interpret the nature and history of its movement is within the area extending from The Narrows of the East Fork of the San Gabriel River northwestward to Mount Hawkins in the north-central San Gabriel Mountains. The thrust zone has its maximum observed thickness of slightly over 1000 m within this area. This unusual thickness results from folding and accumulation of rocks within the thrust zone. Movement during the late stages of thrusting was concentrated in ultramylonite along the top of the zone. Earlier-formed mylonitic rocks, including lenses of relatively unmylonitized upper plate rocks, became folded and rumpled as they were dragged along with the lower plate. The bunched-up rocks are greatly displaced from their upper plate source. They include thick layers of mylonitized anorthosite for which there is no known source within this part of the range. Lenses of granulite facies rocks, probably part of the Mendenhall Gneiss, are also present within the thrust zone but absent from the upper plate in this area. A suite of mylonite samples taken across about 500 m of the thrust zone in the vicinity of The Narrows has yielded a Rb-Sr whole-rock isochron of 1715 ± 30 m.y. (Conrad and Davis, 1977). This indicates they were derived from the oldest part of the Precambrian terrane in the upper plate. Although Precambrian gneiss is present above the Vincent thrust in this area, it is intricately intruded by Lowe Granodiorite and Cretaceous granitic rocks and should not yield a Precambrain Rb-Sr whole-rock isochron even if sampling were restricted to the gneiss. Thus, the thrust zone contains rocks that are likely to be tens of kilometers from their upper plate source.

Rocks throughout the thrust zone are characterized by highly developed lineations formed by (1) intersection of multiple foliation surfaces, (2) intersection of compositional bands with foliation surfaces, (3) tiny ridges and furrows on foliation surfaces due to microfolding, (4) quartz rods, including tails along the sides of feldspar porphyroclasts, and (5) preferred orientation of elongate minerals, particularly amphibole, which formed by recrystallization during thrusting. The lineations are subparallel to each other and to fold axes and are therefore interpreted as b-lineations formed perpendicular to the direction of thrust movement.

Tightly appressed folds are common within the thrust zone in most areas. The most readily recognizable folds are less than 1 m in amplitude, but a few have amplitudes exceeding 100 m. The folds are not due entirely to simple drag but have also formed by bunching up of mylonite, much like a soft rug bunches up in front of a sliding object. Some small-scale folds appears to be due to flexural slip within larger folds and have reverse orientations on opposite limbs of the larger folds. Consequently, folds do not provide a consistent sense of shear for interpreting the direction of thrust movement.

The orientation of fold axes and b-lineations clearly indicates that the Vincent thrust moved in a northeast-southwest direction within the eastern San Gabriel Mountains. However, the relative motion of the two plates is of critical importance. Several drag phenomena indicate the upper plate moved toward the northeast. These include (1) a large, asymmetric overturned syncline in Pelona Schist directly beneath the thrust in the vicinity of The Narrows (Fig. 10-3), (2) an asymmetric overturned anticline defined by a layer of anorthosite mylonite in the upper part of the thrust zone southeast of Mount Hawkins, and (3) the prevalence of foliation dipping more steeply toward the southwest within the thrust zone than the overall dip of the thrust zone. All these features reached their present condition during the late stages of thrust movement. In the case of the syncline in Pelona Schist, albite porphyroblasts were nearly their present size before folding began. This is most readily seen in samples from the axial region of the syncline. The interiors of the albite porphyroblasts contain trains of inclusions marking original schistosity. These are rimmed by overgrowths with inclusions parallel to axial plane schistosity that developed after folding ceased. In the case of the anticline in the thrust zone, its asymmetric form and amplitude of about 150 m are shown by a thick layer of

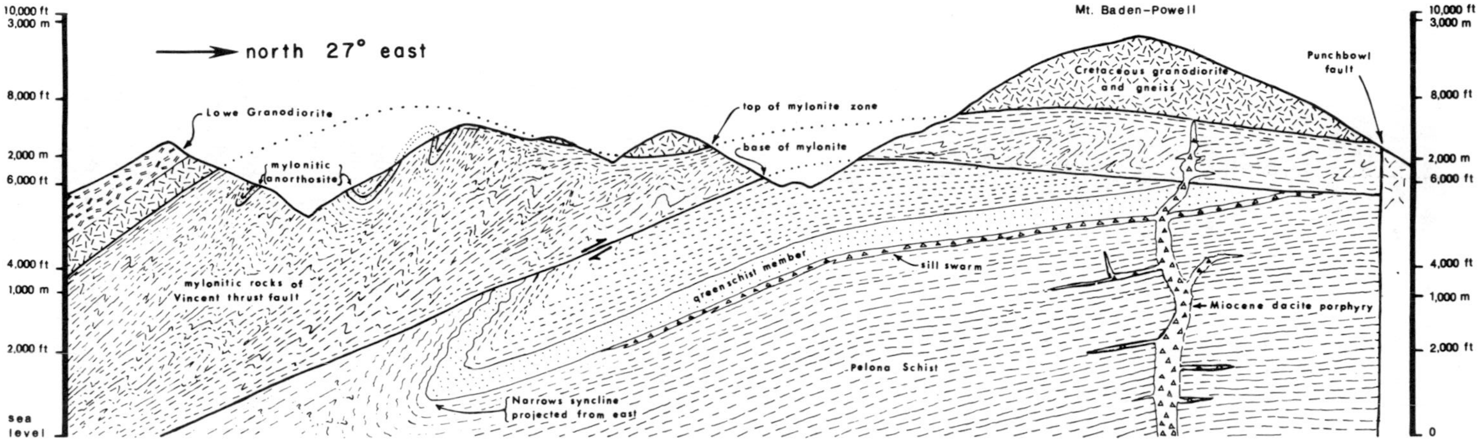

Fig. 10-3. Vertical cross section of the Vincent thrust fault in the central San Gabriel Mountains.

white anorthosite mylonite that crops out on the walls of a steep-sided canyon cut perpendicular to the fold axis. Throughout the area of bunched-up thrust rocks, foliation generally dips more steeply toward the southwest than does the base of the thrust. This appears to result from tight folding with overturning toward the northeast.

The early movement of the thrust is not easily interpreted from structures in the thrust zone because these features have been overprinted by the younger deformation. The ideal method for establishing the direction of movement would involve the matching of rocks in the thrust zone with unique sources in the upper plate. So far, however, anorthosite is the only distinctive rock type with a sufficiently limited known distribution to permit this type of analysis. Mylonitized anorthosite occurs in layers as much as 50 m thick within the thrust zone northeast of Mount Hawkins, as mentioned previously. The only known occurrences of anorthosite in the upper plate within the San Gabriel Mountains are 25 to 55 km to the west and northwest. These are probably not the source of the mylonite because they are located nearly perpendicular to the apparent direction of thrust movement, as indicated by orientation of fold axes and lineations. The most likely source lies east of the San Andreas fault. A possible source terrane containing anorthosite-related rocks, but lacking exposures of anorthosite, crops out in the Little Chuckwalla Mountains. This terrane would have been 40 to 50 km northeast of the anorthosite mylonite prior to creation of the San Andreas fault and would require the upper plate to have moved toward the northeast if it is the source of the mylonite.

The overturning of folds in mylonitic rocks of the Mill Canyon window in the western San Gabriel Mountains indicates the upper plate moved toward the northeast. This is the same as in the eastern part of the range; however, it is also indicative of only the final stages of movement.

Exposures of the Vincent thrust rocks along the south side of Sierra Pelona have yielded few data on the sense of shear owing to the poor quality of exposures. Fold axes and b-lineations in the underlying Pelona Schist generally trend northeastward. The vergence of several large recumbent folds indicates overturn toward the southeast. However, some small folds give the opposite sense of shear. The fact that fold axes are oriented roughly perpendicular to those in the San Gabriel Mountains suggests the two terranes may be rotated relative to each other.

In the Chocolate Mountains to the east of the San Andreas fault, rocks equivalent to those of the Vincent thrust contain asymmetric drag folds, indicating the upper plate moved toward the northeast.

The best available data indicate the Vincent thrust was active 60 m.y. ago during the Paleocene. This is based on a Rb-Sr mineral isochron age of 58.5 ± 4 m.y. (Conrad and Davis, 1977) obtained from a retrograde schist within the lower part of the thrust zone in The Narrows of the East Fork of the San Gabriel River. A prograde assemblage from metachert in the Pelona Schist at the north end of The Narrows has yielded a Rb-Sr mineral isochron of 59.0 ± 1 m.y. (Gary Lass, personal communication, 1977). These dates are probably close to the thermal peak during metamorphism of the Pelona Schist in the San Gabriel Mountains and probably represent an advanced stage of thrusting. (Refer to the discussion of the Pelona Schist for evidence indicating thrust and metamorphism were synchronous.) An ultramylonite from about 500 m above the base of the Vincent thrust on the east side of The Narrows has yielded a K–Ar whole-rock age of 52.7 ± 0.5 m.y. (Ehlig and others, 1975a). This is interpreted as a postthrusting cooling date, as is a K–Ar whole-rock age of 52.4 ± 1.0 m.y. obtained from Pelona Schist 2 km north of The Narrows (Ehlig and others, 1975a). The thrust cuts the Cretaceous granitic rocks in the upper plate, thus placing a maximum limit on its age of about 80 ± 10 m.y., the age of the Mount Josephine pluton (Carter and Silver, 1972). The timing of the Vincent thrust and its regional significance are discussed more fully later.

Pelona Schist

Premetamorphic lithology The Pelona Schist consists of a sedimentary sequence that underwent prograde metamorphism synchronous with deep burial beneath the Vincent thrust. The schist has an exposed thickness of about 3500 m in the eastern San Gabriel Mountains. Here, as elsewhere, its protolith consists dominantly of well-bedded arkosic sandstone, siltstone, and claystone, which have been metamorphosed to a gray-colored white mica-albite-quartz schist, hereafter referred to as grayschist. The gross aspects of bedding are well preserved throughout the section, but metamorphism has destroyed most detailed sedimentary features in the upper part of the section. The lowest 1000 m of exposed section, which consists almost entirely of grayschist, contains well-preserved sedimentary structures, including graded bedding. The original clastic grain boundaries are visible, although flattened, in a few initially coarse grained sandstone beds. Strata within this section tend to be grouped into dark-gray laminated sequences alternating with light-gray medium-bedded sequences. The laminated sequences are derived from interbedded siltstone, claystone, and very fine grained sandstone with most beds less than 2 cm thick. Small-scale pinch-and-swell structures and local discontinuities appear to be original sedimentary features. The medium-bedded sequences are derived from sandstone beds, mostly 10 to 100 cm thick, with siltstone and claystone partings. Claystone rip-up chips occur along the bottoms of some beds, but beds are even and parallel and do not show evidence of channeling. The individual sequences of both types range from a few meters to a maximum of about 100 m thick. These bedding characteristics are suggestive of deposition on the outer fringe of a turbidite fan or on a basin plain. The compositions of the grayschists indicate that the sediments came from a continental source.

In addition to the grayschists, the Pelona Schist of the San Gabriel Mountains contains about 10% greenschist derived from basaltic tuff and 1% or 2% quartzite (with some interlaminated marble) derived from chert. The greenschist occurs mainly in the upper third of the formation and includes one sequence about 60 m thick that is traceable throughout the area. Much of the greenschist is thin bedded with metachert and marble lamellae scattered through the sequence. The metachert is typically laminated and commonly is rich in manganese and iron. Most metachert beds are associated with greenschist, although some in the central part of the formation occur as interbeds in grayschist.

It is important to note that there is almost no intermixing of grayschist and greenschist lithologies, even along contacts between the two. This suggests the two were derived from different areas and transported into the basin by separate currents. The most likely depositional environment appears to be the sporadic encroachment of a pyroclastic-rich turbidite fan derived from an offshore mafic volcanic source onto the outer fringe of a prograded turbidite fan system fed by a continental source. Chert probably accumulated slowly during periods when introduction of volcanic material raised the sea floor above the prograded level of continentally derived sediment. Chert beds not directly associated with greenschist may represent areas cut off from a continental clastic source by turbidite lobes of volcaniclastic sediment or may represent an area temporarily deprived of clastic sediment owing to shifting of turbidite distributaries.

Small serpentinite masses, a few meters across, occur in two widely separated locations within the central part of the Pelona Schist in the San Gabriel Mountains. Their mode of emplacement has not been determined. Small pods of talc-actinolite schist, probably derived from serpentinite, occur locally in strongly deformed schist beneath the Vincent thrust. They may have been introduced by sedimentary processes or perhaps mechanically from an outside source during thrusting.

A section of Pelona Schist about 1000 m thick is exposed on Sierra Pelona. The schist is derived from the same general lithologies as that of the San Gabriel Mountains but is poorly exposed, structurally complex, and more altered by metamorphism. Greenschist is more abundant than in the San Gabriels and is of flow as well as pyroclastic origin. Relict pillow structure appears to be present in a large mass of greenschist that crops out on the west side of Bouquet Canyon, 2 km southwest of Bouquet Reservoir. A highly deformed, laminated metachert-marble member with an original thickness of roughly 25 m forms a mappable unit in several places. It overlies greenschist and probably represents slow biogenic accumulation in an area cut off from clastic sedimentation. Sierra Pelona also contains many lenticular masses of metaserpentinite and talc-actinolite and talc-carbonate rocks derived from serpentinite. The largest body, located on Del Sur Ridge to the west of Bouquet Canyon, is 2 km long and up to 300 m wide. Part of the masses are discordant to adjacent bedding and were probably emplaced diapirically. Others are concordant or consist of small pods within grayschist and are likely to have been distributed by sedimentary processes.

Metamorphism and association with the Vincent thrust The Pelona Schist and its correlative, the Orocopia Schist, have experienced regional greenschist facies metamorphism. Locally, metamorphism has progressed to the lower amphibolite facies in schist close to the Vincent-Orocopia thrust.

The Pelona Schist of the San Gabriel Mountains was metamorphosed at a lower temperature than the schist exposed in most other areas. Mineral assemblages are typical of the lower part of the greenschist facies. Schists derived from interbedded arkosic sandstone, siltstone, and claystone contain the assemblage quartz + albite + white mica ± epidote ± actinolite ± chlorite. Small amounts of sphene, apatite, and cryptocrystalline graphite are invariably present. Allanite, zircon, and sphene occur as relict clastic grains in metasandstone. Microcline is present in a few thick beds of metasandstone, which probably contained clastic grains of potassium feldspar with insufficient admixed clay to convert it to white mica. Stilpnomelane is common in some areas. Scattered crystals of brown biotite locally are present in grayschist close to the Vincent thrust. Schist derived from basaltic tuff contains the assemblage albite + epidote + chlorite + actinolite. Minor amounts of sphene and apatite are ubiquitous, and a small amount of calcite is characteristically present. Magnetite is abundant in some beds. Stilpnomelane is very abundant as a late-forming mineral in several areas. Small crystals of green biotite occur in some greenschist close to the Vincent thrust. Metachert and associated marble have highly variable mineral assemblages depending upon original composition. In addition to the common metamorphic minerals found in the other schists, lamellae within metachert commonly contain high concentrations of one or more of the following minerals: stilpnomelane, soda-amphibole ranging between riebeckite and crossite, biotite, spessartine, piemontite, and specularite. Acmite occurs in a few assemblages with soda-amphibole. Barite and pale-colored tourmaline are minor constituents in some assemblages associated with marble.

The Pelona Schist of Sierra Pelona ranges from intermediate greenschist facies in the structurally lowest schist to lower amphibolite facies in schist close to the Vincent thrust near the east end of Sierra Pelona. The lowest-grade part contains mineral assemblages similar to those in the San Gabriel Mountains except that stilpnomelane has been found in only one outcrop and acmite is fairly common in metachert. The biotite zone is poorly developed. The transition from greenschist to amphibolite facies starts at the garnet isograd and is marked by (1) the conversion of greenschist containing fibrous actinolite to black amphibolite rich in medium-grained aluminous hornblende, and (2) the appearance of numerous small red garnets in some beds of grayschist and amphibolite. The garnet isograd is at a maximum distance of roughly 1000 m structurally below the Vincent thrust at the east end of Sierra Pelona and converges with the thrust toward the west. Garnet-zone rocks have a patchy distribution along the southern edge of Sierra Pelona, with biotite-zone schist in direct contact with the thrust along parts of the contact. The amphibolite facies occurs at the east end of Sierra Pelona within a few hundred meters of the Vincent thrust. It is marked by the development of small, highly twinned oligoclase crystals in amphibolite. The large albite porphyroblasts, which are present at lower metamorphic grades, tend to break down and disappear. Other changes include the widespread appearance of microcline in metasandstone, the replacement of sphene by rutile in much of the amphibolite, and the appearance of zoisite in place of epidote in grayschist and amphibolite poor in ferric iron. The unusual assemblage zoisite-oligoclase-rutile, in combination with other minerals, occurs locally in schist and neomineralized rocks of the adjacent thrust zone.

On the basis of our general knowledge of metamorphic reactions, the Pelona Schist in the San Gabriel Mountains probably formed at temperatures around 350 to 400°C with a very flat thermal gradient perpendicular to the base of the Vincent thrust during the late stages of metamorphism. The schist of Sierra Pelona probably formed at temperatures ranging from roughly 400°C in the lowest-grade schist to a maximum near 550°C in the amphibolite facies schist. The thermal gradient was inverted and was probably uneven, with the highest temperatures adjacent to the sole of the Vincent thrust.

The mineral assemblages are typical of intermediate- to high-pressure facies. On the basis of an extensive microprobe study, Graham and England (1976, p. 143) believe metamorphic pressure was in the range of 6 to 7.5 kbars at Sierra Pelona. This is equivalent to a depth of 20 to 30 km, assuming pressure was produced primarily by the weight of the upper plate of the Vincent thrust. A subduction zone is the most likely environment for transporting sediments to such great depth beneath a thrust fault.

In addition to the previously described associations, a number of features in the San Gabriel Mountains suggest a cause and effect relationship between the Vincent thrust and the metamorphism of the Pelona Schist. These include the following:

1. Metamorphic deformation is more intense and metamorphic grain size is coarser in Pelona Schist close to the thrust than it is in schist far from the thrust.
2. There is continuity between the fabric and structure of the Pelona Schist and that of the overlying Vincent thrust rocks. Pelona Schist has highly developed bedding-plane schistosity that is subparallel to the base of the Vincent thrust and subparallel to foliation within the overlying thrust rocks. Lineations, including those formed by preferred growth of minerals, and fold axes in the Pelona Schist are subparallel to lineations and fold axes in the overlying thrust rocks. The parallelism of lineations terminates at the top of the thrust zone; the overlying upper plate rocks contain an older lineation oriented about 60° to that below.
3. Rocks in the lower part of the thrust zone have experienced extensive retrograde metamorphism to the same mineral assemblages as the underlying Pelona Schist.

The interrelationship between the Pelona Schist and the Vincent thrust fault is visualized as having evolved as follows. The near-surface segment of the Vincent thrust originated as a gently inclined bedding-plane fault within the sedimentary protolith of the Pelona Schist. While the sediments were in a shallow environment, fault movement was probably facilitated by formation of plastic clay gouge (now actinolite-chlorite schist) along the fault surface and by development of superhydrostatic water pressure beneath the fault due to load-induced compaction. As the sediments moved to progressively greater depth, their temperature gradually increased because of residual heat conducted from the sole of the upper plate. Once temperature reached metamorphic conditions, the sediments and clay gouge began to recrystallize to metamorphic assemblages within an anisotropic stress field. The maximum compressive stress was nearly vertical owing to the weight of the upper plate and was essentially perpendicular to the gently inclined fault zone and underlying bedding. This caused bedding-plane schistosity to develop by preferred growth of minerals and by flattening. Shearing strain, amounting to tens of kilometers, was the direct result of convergence of the two semirigid plates and was independent of compressive stress. The tangential nature of the contact between the plates caused shearing strain to be concentrated in a narrow, nearly planar zone along the boundary between the plates. Drag produced a shearing stress parallel to plate motion and added to the compressive stress in that direction. Consequently, the axis of least compressive stress was in the plane of the fault and bedding normal to plate motion. This caused preferred growth of elongate minerals parallel to the b-fabric axis. The fact that shearing movement was concentrated in mylonite within the base of the upper plate rather than in the underlying schist indicates the schist was the stronger of the two. The difference was probably caused by water in the schist, which dissipated frictional heat and promoted the growth of interlocking crystals. The widespread retrograde metamorphism along the lower part of the upper plate was induced by water expelled during progressive metamorphism of the underlying schist. The relatively low metamorphic grade of the schist adjacent to the thrust in the San Gabriel Mountains probably relates to the unusual thickness of the overlying thrust rocks. The bunching up of mylonitic rocks probably insulated the schist from hot rocks within the main body of the upper plate. The local occurrence of amphibolite facies metamorphism along the sole of the thrust might result from shearing away of the sole of the upper plate bringing hotter material in contact with the schist or might result from frictional heating, as suggested by Graham and England (1976).

Pelona Schist-Vincent thrust tectonic development Our present knowledge of the Pelona Schist and Vincent thrust fault is sufficient to reconstruct local details of their evolution with a fair degree of confidence. The intermeshing of their structural history with that of other parts of southern California involves varying degrees of speculation but must be done as an initial step in the unraveling of the pre-Cenozoic evolution of southern California. The initial attempt to fit the Pelona Schist and Vincent thrust into a regional picture was presented by Ehlig (1968). Haxel and Dillon (1978) recently summarized most of what is currently known about the Pelona and Orocopia schists and the Vincent and Chocolate Mountain thrusts and have proposed tentative models for their geologic evolution. My current views are presented next. The reader should also refer to the discussions by Crowell in Chapter 18, this volume, and by Burchfiel and Davis in Chapter 9.

The Pelona and Orocopia schists were probably deposited on oceanic crust within a deep marine basin. Although the basement beneath the schist is not exposed, its oceanic character is inferred from (1) the presence of serpentinite masses within the schist, (2) the presence of schist derived from basaltic volcanics, and (3) the interpretation that the schist was transported to great depth beneath the Vincent thrust without buoying up the upper plate. Most of the Pelona sediment was detritus derived from a continental source terrane of calc-alkaline composition. The source probably included plutonic and/or gneissic rocks, as suggested by the presence of relict clastic grains of allanite, sphene, and zircon. The location of the source terrane is uncertain but was probably toward the east or southeast, as indicated by a decrease in the abundance of interbedded metachert and greenschist in that direction. The source may have been southwestern Arizona or northwestern Sonora. Bedding characteristics and the total absence of conglomerate are suggestive of deposition on the outer fringe of a deep-sea turbidite fan. Basaltic volcanic material (now greenschist) came from a separtate terrane, probably to the west or northwest, as indicated by a regional increase in the abundance of greenschist in that direction. Perhaps the volcanism was associated with the westward-trending spreading center between the Kula and Farallon plates, which should have been in this general area in Late Cretaceous time according to reconstructions by Atwater (1970, Fig. 18). Manganiferous, ferruginous, and calcareous chert were deposited at times when local areas were isolated from the source of detrital sediment.

The Vincent thrust probably began as a low-angle westward-dipping subduction zone while sedimentation was still going on within the deep marine basin. The overriding microplate, consisting of the basement rocks of the upper plate of the Vincent thrust, may have broken from the western edge of North America south of its present position, perhaps along the north side of the spreading center between the Kula and Farallon plates. This microplate could have been carried northward with the Kula plate, then became lodged against a westward projection in the North American plate, and subsequently was bypassed by the spreading center. The microplate then could have been pushed eastward across the Pelona-Orocopia basin by the motion of the Farallon plate. If this model is correct, eastward overthrusting probably continued until the basin was completely overrun and the microplate began to ramp onto the continental margin. The resulting increase in frictional resistance would have stopped the movement along the Vincent thrust and caused initiation of a new subduction zone seaward of the microplate.

This model predicts several relationships that can be tested. Inasmuch as it requires the existence of the microplate derived from outside the local area, the basement terrane of the microplate should have a different history and contain some different rock types than in-place North American basement terrane to the north and east. The two terranes must also have a discrete boundary between them. There does appear to be a change in basement terrane across the San Francisquito-Fenner fault to the west of the San Andreas fault and along an east-trending conjectured fault that joins with the western part of the Clemens Well fault to the east of the San Andreas (see Fig. 10–1). Isotopic studies by Silver (1971) indicate that the upper plate terrane in the San Gabriel and Orocopia Mountains is distinctly different from the basement in the San Bernardino Mountains. However, more data are needed to demonstrate that the differences are regional in extent and to accurately determine the nature and location of the boundary between them if it is indeed real. The model also requires the leading edge of the thrust to occur east of the most easternly schist exposures. Such may be the case at the southern end of the Palen and McCoy Mountains and in the northeastern Little Mule Mountains west of Blythe, California. These areas contain a southwestward-dipping thrust fault underlain by mildly metamorphosed sandstone, conglomerate, and mudstone of the Upper Cretaceous or slightly younger McCoy Mountain Formation. Pelka (1973a, b; personal communication, 1973) suggests that the thrust may correlate with the Vincent-Chocolate Mountains thrust and that the McCoy Mountain Formation may be a near-shore correlative of the Pelona-Orocopia Schist. Strongly deformed sedimentary rocks that may be correlative with the McCoy Mountain Formation crop out in several of the mountain ranges in southwestern Arizona. Thus, the leading edge of the microplate probably passes into southwestern Arizona.

Age of Pelona Schist

The metamorphic age of the Pelona Schist is about 60 m.y. as indicated in the discussion of the age of the Vincent thrust. The depositional age is not known but need not be much older than the metamorphic age. If the Vincent thrust represents a subduction zone, thrusting probably commenced while sedimentation was still going on within a deep marine basin. The leading edge of the thrust might be overriding penecontemporaneous sediments. Assuming the Pelona Schist was deposited on oceanic crust, the oldest sediments are probably only slightly younger than the underlying crust. The underlying crust may have formed along a spreading center between the Kula and Farallon plates and be as young as Late Cretaceous, based on reconstructions by Atwater (1970). As far as is known, the base of the schist and underlying oceanic crust are nowhere exposed.

The Pelona and Orocopia schists are not known to be intruded by Cretaceous granitic rocks, even though such rocks are common within the upper plate of the Vincent thrust and are abundant within the San Bernardino Mountain-Mojave Desert basement complex. A U-Pb age of 80 ±10 m.y. has been obtained from the Josephine Mountain pluton in the San Gabriel Mountains (Carter and Silver, 1971), and an Rb-Sr mineral age of 74.1 ± 1.2 m.y. has been obtained from a pluton in the Little San Bernardino Mountains, 10 km northwest of Orocopia Schist in the Orocopia Mountains (Ehlig and Joseph, 1977). The protolith of the schist may have escaped intrusion by virtue of a younger age, or it may be older but escaped intrusion by virtue of an oceanic location outboard from Cretaceous plutonism.

Miocene Hypabyssal Intrusives

The Pelona Schist, Vincent thrust rocks, and all older rocks in the eastern San Gabriel Mountains have been intruded and locally hornfelsed by stocks of a light-colored, medium-grained granitic porphyry whose composition straddles the border between quartz monzonite and granodiorite. The porphyry contains 5% to 10% biotite as its only dark mineral. Dacite porphyry dikes and sills of the same rock are widely distributed. A swarm of more than thirty sills forms a prominent light-colored band, about 50 m thick, beneath a greenschist member in the Pelona Schist between Mount Baden-Powell and The Narrows. Miller and Morton (1977) have dated these rocks as 14 to 16 m.y. by the K-Ar method.

Shallowly emplaced dikes of andesite, diabase, and, less commonly, olivine basalt and quartz latite are locally common in the southern and east-central parts of the range. These dikes intrude the Miocene porphyry in areas where the relationship can be observed. The dikes have not been dated but are assumed to be related to the middle Miocene Glendora Volcanics along the southern edge of the range (Shelton, 1955). The dikes are locally concentrated along the San Gabriel and San Antonio faults, with dikes locally intruded into fault gouge. Thus, some fault activity occurred prior to dike implacement, although all dikes are truncated by the youngest fault movement.

Paleogeographic Reconstructions

At present, the main mass of the San Gabriel Mountains is a large tectonic slice on the Pacific side of the San Andreas fault. Information obtained from it is of limited value unless we can reconstruct the location and orientation of these rocks relative to other basement terranes at various times in the past. To data, we have been able to establish the magnitude and timing of offsets along the major branches of the San Andreas fault and can reconstruct paleogeography back to about middle Miocene with fair certainty. Reconstructions of paleogeography for the period extending from middle Miocene to when the Vincent thrust was active in Paleocene are sketchy with some problems just now being recognized. Pre-Paleocene reconstructions are highly speculative.

Although the amount of displacement along the San Andreas fault has been the subject of vigorous debate for many years, we now have strong evidence favoring a total right slip of 300 km along it and its branches in both central and southern California. Formations older than about 10 m.y. appear to have the same amount of offset, indicating faulting began during the late Miocene. One of the most precise indicators of the total offset in central California is the correlation of the 23.5 m.y. old Pinnacles and Neenach Volcanic formations (Matthews, 1976). This correlation is important because it places the Salinian basement of the Gabilan Range, upon which the Pinnacles Volcanic Formation rests, against the southwestern edge of the Mojave Desert before faulting began in the late Miocene. It indicates the original trace of the San Andreas fault was close to its present trace in this area and invalidates models, such as those of Suppe (1970) and Anderson (1971), that require the initial offset in southern California to have been along faults farther west. The prefault location of the Pinnacles Volcanic Formation is currently occupied by the north-central Transverse Ranges, including terrane discussed in this chapter. This, along with data presented later, contradicts the views of Baird and others (1974) who believe the Transverse Range Province has existed in nearly the same position and orientation since at least the Cretaceous and perhaps the Precambrian.

Although the segment of the San Andreas fault that extends along the margin of the Mojave Desert has 300 km of right slip with respect to terrane of the Mojave Desert, it has only 240 km of right slip with respect to terrane of the San Gabriel Mountains and Soledad Basin to the south of the fault. The other 60 km of right slip occurred along the San Gabriel fault, which extends southwestward from the San Andreas fault near Frazier Park. The San Gabriel fault experienced progressive right slip during late Miocene and Pliocene, as shown by the offset of sedimentary breccias from their source terranes along the trace of the fault (Crowell, 1952, 1954, 1975c). It formed the main trace of the San Andreas fault during late Miocene at a time when it was located near the northeast margin of the present Salton Trough (Ehlig and others, 1975b). The modern trace of the San Andreas fault developed to the east of the San Gabriel fault during the Pliocene, and subsequently has offset the San Gabriel fault to its present location.

Crowell (1962) and Crowell and Walker (1962) correlated the basement terrane of the San Gabriel Mountains and Soledad Basin with that (1) to the west of the San Gabriel fault near Frazier Park and (2) to the east of the San Andreas fault in the Orocopia Mountains. Their correlation was based on similarities in the types, ages, and histories of rocks present in the three areas. Subsequent studies have strengthened the correlation, including isotopic age dating by Silver (1971), but most have lacked the geometric elements needed to establish net slip. One exception is an analysis of the upper Miocene Mint Canyon Formation and the source of clasts within it (Ehlig and others, 1975b).

The Mint Canyon Formation consists of fluvial and lacustrine sediments that were deposited in a broad westward-trending synclinal trough within the Soledad Basin. Fluvial conglomerate along the trough axis consists mostly of volcanic detritus with east to west paleocurrent indicators. Some clasts are locally derived from volcanics of the Vasquez Formation, but most form an exotic assemblage similar to volcanic rocks that occur in place within the northern Chocolate Mountains. One clast type, a distinctive rapakivi-textured quartz latite porphyry, is petrographically and chemically identical to porphyry that forms dikes at the north end of the Chocolate Mountains directly southeast of Salton Wash (Ehlig and Ehlert, 1972; Ehlig and others, 1975b). The correlation is confirmed by an isotopic study showing that the clasts and dike rocks have identical initial ^{87}Sr–^{86}Sr ratios of 0.7060 ± 0.0001 (Joseph and Davis, 1977). Thus, volcanic conglomerate along the axis of the Mint Canyon trough was derived from terrane directly southeast of Salton Wash on the opposite side of the San Andreas fault. Conglomerate fed in along the south side of the trough axis contains abundant clasts of Lowe Granodiorite, particularly from the biotite-orthoclase zone, and anorthosite whose sources are south and southeast of Soledad Pass. Conglomerate fed in along the north side of the trough axis contains clasts of syenite, Pelona Schist, and other rock types that crop out west of Soledad Pass. This distribution of locally derived materials requires the alluvial channel that transported volcanic detritus from the east side of Salton Wash to have crossed the San Andreas fault near Soledad Pass. Thus, Soledad Pass and Salton Wash were adjacent or in close proximity during the Miocene. This conclusion is supported by the bedrock geology of the two areas. In both areas, Precambrian syenite is faulted against Oligocene-Miocene volcanic rocks that rest unconformably on Lowe Granodiorite. No syenite occurs in the basement terrane to the southeast of the fault in either area, and no Lowe Granodiorite occurs northwest of the fault in either area. Most of the volcanic rocks in Salton Wash are compositionally and texturally different from those in Soledad Pass, but some appear to be the same. The difference between the two suites indicates that most rocks came from different vents and can be accounted for by an unexposed gap between the two sequences. Quaternary sediments conceal bedrock for a distance of 10 km between the San Andreas fault and the most southwesterly exposures of volcanic rocks in Salton Wash. The present topographic lows along Soledad Pass and Salton Wash are probably products of Quaternary erosion along a structural trough of Miocene age. Although a group of volcanic domes aligned along Salton Wash gives the impression of being a young feature and is designated as Pleistocene in age on the geologic map of California (Jennings 1977), they are now known to be part of the Oligocene-Miocene volcanic complex of the Chocolate Mountains (Crowe and others, 1979) and are probably exhumed features. This correlation requires the segment of the San Andreas fault that extends between the two areas to have a total right slip of 240 km and to be no older than late Miocene. This part of the San Andreas fault is probably no older than about 6 m.y. based on the correlation of distinctive granitic clasts in the Pliocene Ridge Basin Group west of Sierra Pelona and the Punchbowl Formation west of the Fenner fault with a source area in the Little San Bernardino and Orocopia Mountains (Ehlert and Ehlig, 1977; Farley and Ehlig, 1977).

Figure 10-4 is a reconstruction showing how the San Gabriel Mountain and Soledad Basin terrane would fit adjacent to that of the Orocopia and northern Chocolate Mountains if displacement were removed from the San Andreas and San Gabriel faults. The present-day outcrop patterns are used in Fig. 10-4 in order to show the existing geometric relationship between exposed rock units. To convert this to a true picture of the geology as it existed about 10 m.y. ago, it would be necessary to undo all subsequent deformation, erosion, and sedimentation. My reconstruction realigns the axes of Soledad Basin and Salton Wash for the previously stated reasons. The Clemens Well fault should realign with the San Francisquito fault. Its failure to do so is probably the result of drag along the San Andreas fault. A conjectured fault is extended eastward from the Clemens Well fault in order to separate the distinctive basement rocks in the upper plate of the Vincent-Orocopia thrust from those of the San Bernardino Mountains and Mojave Desert terrane to the north. An eastward extension of the San Francisquito-Clemens Well fault is also needed to accommodate the great amount of slip indicated by the juxtaposition of Paleocene marine strata against the Pelona Schist, which was undergoing deep-seated metamorphism during the Paleocene. Another conjectured fault is shown to the west of the Coxcomb Mountains. The purpose of this fault is to separate the thrust that overlies the McCoy Mountain Formation from terrane to the west, which is believed to be uninvolved in the thrusting. The need for this fault would disappear if future studies show that the Pelona-Orocopia Schist was deposited in a gulf of spreading origin, similar to the Gulf of California, and that the gulf was subsequently closed by overthrusting from the west.

As shown in Fig. 10-4, exposures of the Pelona-Orocopia Schist lie in a northwest-trending belt less than 30 km wide and about 200 km long. This distribution is the result of anticlinal uplift of the schist along this trend accompanied by erosion of the overlying rocks. Much of the uplift and erosion occurred prior to about 30 m.y. ago, as shown by the presence of extrusive volcanic rocks of this age in close proximity to schist exposures within and to the southeast of the Chocolate Mountains. The earliest documented exposure of schist was about 15 m.y. ago, as shown by the Mint Canyon Formation resting unconformably on the schist at the southwest end of Sierra Pelona and the presence of schist clasts in the Mint Canyon and Caliente formations. The anticline may have been initiated as much as 60 m.y. ago during the final stages of thrusting and developed gradually as erosion removed overburden.

Our knowledge of displacements along faults, such as the San Francisquito faults, which are more than about 10 m.y. old, is insufficient to develop a reliable regional picture of the pre-middle Miocene paleogeography of southern California. It appears likely that the granitic rocks of the Salinian block in the southern Coast Ranges originated in southern California, as suggested by Smith (1977), but unequivocal proof is lacking.

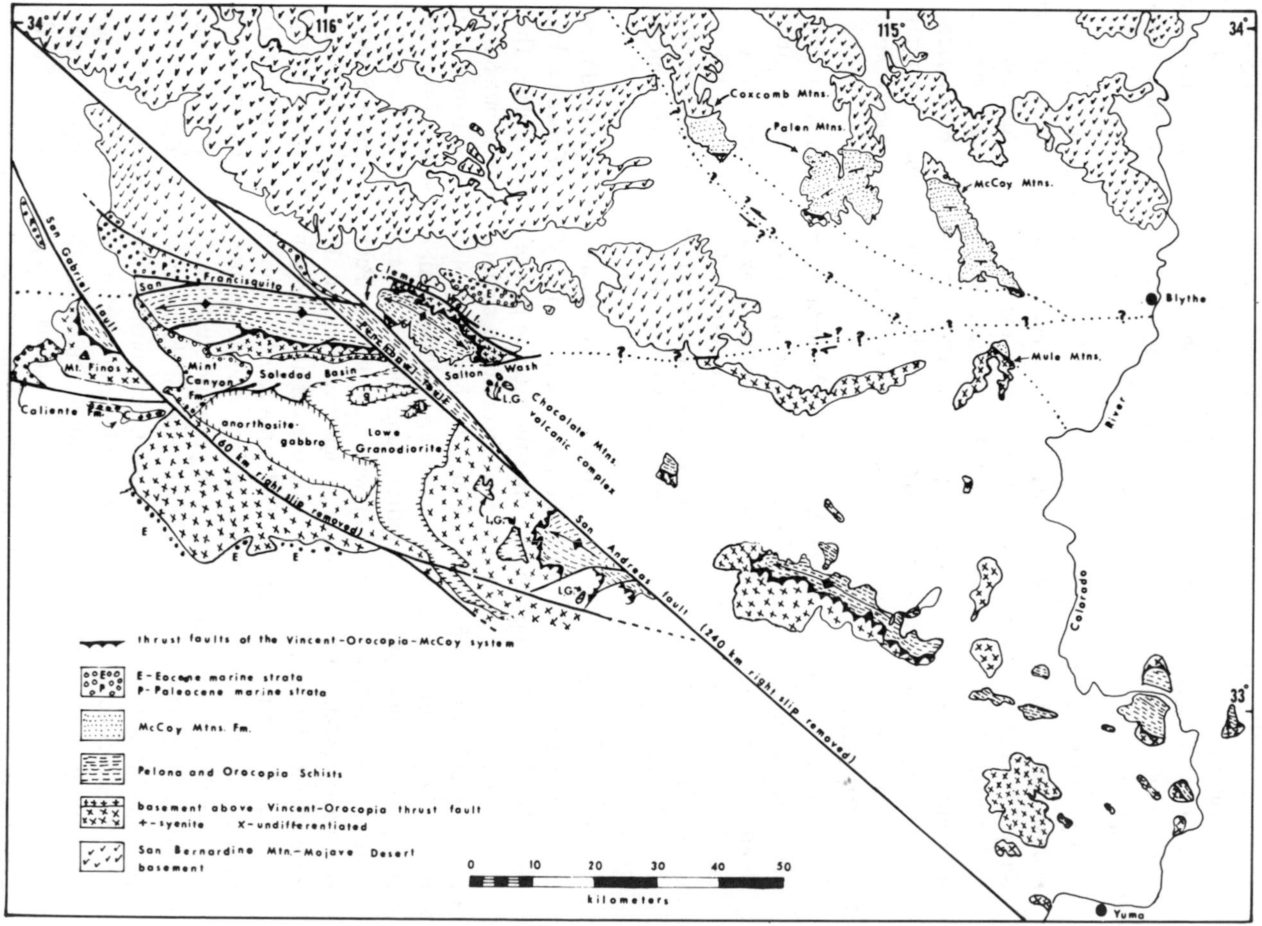

Fig. 10-4. Geologic map showing San Gabriel Mountain basement restored to its middle Miocene position in southeastern California by removing 240 km of right slip on the San Andreas fault and 60 km (22 km on North Branch) of right slip on the San Gabriel fault. Outcrop distribution is that of today modified after the *Geologic Map of California* (Jennings, 1977). See text for explanation of hypothesized faults.

CONCLUSIONS

The development of plate-tectonic concepts and the reconstructions of plate motion have greatly expanded our understanding of petrotectonic events and have yielded models that can be tested and modified through geologic investigation. In this chapter I have attempted to summarize what is known about the basement rocks of the San Gabriel Mountains and how their petrotectonic evolution fits into plate-tectonic events. Among the important conclusions are the following:

1. The Transverse Range Province is a product of late Cenozoic deformation superimposed upon rocks of diverse origin. The immediate cause of the Transverse Ranges is probably interplate convergence along the bend in San Andreas fault. In the case of the San Gabriel Mountains, they are accommodating north-south shortening by reverse faulting along their southern margin and broad arching across their interior and northern margin. The simple geomorphic configuration of the range is superimposed upon the complex structure of the rocks exposed within the range.

2. The basement terrane of the San Gabriel Mountains was attached to that of the Orocopia and Chocolate Mountains in southeastern California as recently as about 10 m.y. ago. It has subsequently been displaced by 240 km of right slip on the San Andreas fault, and part of it has been displaced by 60 km of right slip on the San Gabriel fault. This displacement is associated with the opening of the Gulf of California and transform motion between the Pacific and North American plates.

3. The plutonic and gneissic rocks of the San Gabriel Mountains and related terrane in southeastern California form the upper plate of the Vincent-Orocopia thrust fault. The Pelona-Orocopia Schist forms the exposed part of the lower plate. Thrust faulting and metamorphism of the schist occurred simultaneously about 60 m.y. ago, probably as a subduction event. The best available evidence indicates the upper plate of the thrust moved from the southwest. All exposures of the thrust and schist are in antiforms or uplifted fault blocks. Thus, the leading edge of the thrust is not exposed in contact with recognizable Pelona-Orocopia Schist. A southwestward-dipping thrust fault, which may represent the deeply eroded eastern edge of the Vincent-Orocopia thrust, overlies the mildly metamorphosed McCoy Mountains Formation near the California-Arizona border. A tentative model, pending acquisition of more definitive data, would have the Pelona-Orocopia Schist deposited in a marine basin underlain by oceanic crust that formed along a spreading center between the Kula and Farallon plates or the southwestern edge of the Kula plate and North America during Late Cretaceous. The depositional basin may have been bounded on the west by a microcontinent or may have been an intracontinental embayment, such as the Gulf of California. Closure of the basin by thrust faulting may have occurred when northward migration of the Farallon plate subjected the area to compression in a northeast-southwest direction.

4. The upper plate of the Vincent thrust is part of a continental craton dating back at least 1700 m.y. It appears to have been free from plutonic and metamorphic events following emplacement of an anorthosite-gabbro-syenite complex about 1220 m.y. ago during the Precambrian until emplacement of the Lowe Granodiorite about 220 m.y. ago during Early Triassic time. There was extensive deformation and metamorphism in the time interval between emplacement of the Lowe Granodiorite and emplacement of Cretaceous granitic rocks about 85 m.y. ago. The metamorphosed remnants of a thick sequence of dolomite, quartz arenite, and aluminous and calcareous argillite that occur locally probably represent Paleozoic strata that were deposited on the Precambrian craton. The history and nature of these rocks resemble those of other areas along the western edge of the American craton, but existing knowledge is inadequate to establish their original location.

In spite of the progress made to date, many aspects of the petrotectonic evolution of southern California remain a mystery to be solved by future studies. We need to work backward in time, undoing the younger events and fitting the pieces back together. The picture is clear in most areas back to about 10 m.y. before present, but becomes progressively fuzzier farther back in time, particularly west of the San Andreas fault. As yet, nobody has been able to unravel the structure of the southern Coast Ranges, western Transverse Ranges, or Continental Borderland so as to create a picture of what the region looked like when the East Pacific Rise first encountered the California Coast about 29 m.y. ago.

Gordon Gastil, George Morgan, and Daniel Krummenacher
San Diego State University
San Diego, California, 92182

11

THE TECTONIC HISTORY OF PENINSULAR CALIFORNIA AND ADJACENT MEXICO

ABSTRACT

Tectonism along the southwestern cratonal margin of North America began in late Paleozoic time. Evidence of both convergent and extensional tectonics is found in rocks of Triassic age. During the Jurassic there were two parallel arcs: an oceanic arc off the southwestern edge of the continent and a continental arc well within the old craton. Both subduction planes are believed to have been inclined to the northeast. Convergence along the continental margin moved the oceanic arc toward the continent.

Trench-trench transforms separating the segments of the oceanic arc allowed individual segments to collide with the continent at different times. The boundary between the Peninsular and Transverse ranges is one such ancestral transform, and the Agua Blanca fault zone of Baja California is another. The oceanic arc segment north of the Agua Blanca transform collided with the continental margin during earliest Cretaceous time; the segment to the south began to collide in Albian time. The west coast gravity high of Baja California marks the position of shallow oceanic crust preserved between the western sutured oceanic island arc and the continental apron to the east.

In Late Cretaceous to mid-Cenozoic time the addition of subduction velocities and the crustal dilation resulting from plutonic emplacement caused the magmatic axis to move progressively to the northeast. The spreading of the Basin and Range Province in mid-Cenozoic time caused the edge of the continent to override the Pacific-North American plate boundary. Along that portion of the boundary in which convergence persisted, the trench was displaced to the southwest. Along the portion of the boundary that had become a transform, the plate boundary moved from the continental edge to a position within the continent.

With the southward displacement of the surviving trench, arc volcanism returned to areas that are now located along the Gulf of California. All subduction-related volcanism ceased about 8 m.y. ago. The total transform displacement in the Gulf of California appears to be about 300 km.

INTRODUCTION

The Old Cratonal Boundary

In Fig. 11-1 we show the "old cratonal boundary" drawn west of the San Gabriel Mountains (pre-Miocene position), and the Mojave block (2), along the trace of the San Andreas fault southeast into Sonora, southwest of the Sierra Viejo (13) and west of the Sierra Berruga (17). Farther southeast it is hard to define. To the east of this boundary, older Precambrian metamorphic rocks are overlain by essentially unmetamorphosed late Precambrian and early Paleozoic carbonate-quartzite rocks. To be sure, there are younger Paleozoic and Mesozoic rocks, including large volumes of both volcanic and plutonic rocks east of the boundary, but at the present erosional depth, most Phanerozoic metamorphism is local to the contacts of Mesozoic plutons. West of the boundary, basement rock consists

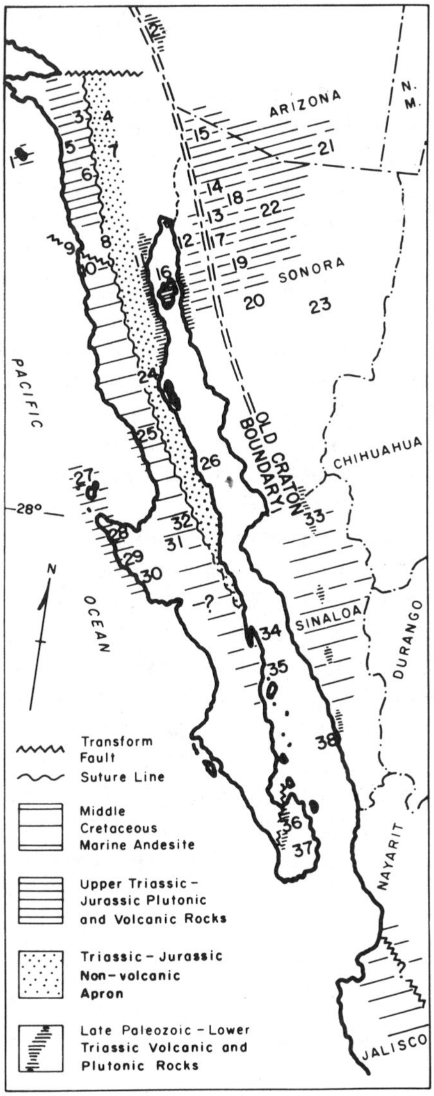

Fig. 11-1. Index map to Mesozoic volcanic-plutonic activity in peninsular California and adjacent areas. The numbers refer to localities cited in the text. The peninsula of Baja California and western Jalisco are located in their pre-15 m.y. positions. Santa Cruz Island (1) is located west of San Diego. Palinspastic adjustments within the peninsula and within the mainland east of the Gulf of California have not been made.

largely of plutons of Mesozoic age separated by belts of regionally metamorphosed host rock, largely of Mesozoic age. Immediately west of the old cratonal boundary are patches of metamorphosed late Precambrian-Cambrian carbonate-quartzite rock. Our old craton boundary, therefore, is gradational. It lies southwest of, and bears no genetic relation to, the "megashear" proposed by Silver and Anderson (1974).

The Crustal Basement of Baja California and the Western Margin of Adjacent Mexico

Except for recognizable "splinters" adjacent to the boundary just described, there are no known rocks of Precambrian or early Paleozoic age west of the craton and no large volumes of enigmatic rocks for which such a designation appears a likely possibility. The two major belts of largely undated host rocks in the peninsula of Baja California and adjacent Sonora (Gastil and others, 1975, 1978) are the predominately sandstone-argillite rocks found in the axial portions of the peninsula and the carbonate + chert + graywacke sequences found in the eastern portion of the peninsula, the midriff islands of the Gulf of California, and the coast of Sonora. These rocks (to be discussed later) appear to be Triassic-Jurassic and late Paleozoic-Triassic, respectively.

There are isolated bodies of rock that may not belong in these categories (see Fig. 9, Gastil and others, 1975, for example), and it may be that there are isolated patches of old crust scattered throughout the region west of the old cratonal boundary. We believe that the volume of such rocks is minor, and that almost all the crust west of the old boundary has formed during the Mesozoic era by the accumulation of volcanic strata on the sea floor, an apron of continent-derived detritus seaboard of the craton margin, and the intrusion of granitic plutons into both of these.

GEOLOGIC HISTORY OF BAJA CALIFORNIA AND ADJACENT AREAS

Late Paleozoic Tectonism?

In the Sierra Pintas of northeastern Baja California (Fig. 11-1, locality 11), McEldowney (1970a, b) describes a sequence of metamorphosed argillite, graywacke, conglomerate, bedded chert, crinoidal limestone, and basalt. Some of the argillite may be tuffaceous, and the conglomerates include volcanic clasts of undetermined composition. Fossils from the limestone have been identified as "definitely late Paleozoic" (oral communication, R. Langenheim, 1978). On the basis of lithology, this association of rocks can be followed down the Gulf of California to Bahia de Los Angeles (26) and to the coast of Sonora (16).

Altamirano (1972) reported the presence of mafic and ultramafic rocks in the metamorphic terrane south of La Paz, Baja California. A hornblende gneiss and a tonalite from this belt yield discordant hornblende ages (K-Ar) of 225 and 325 m.y.

Mullan (1975) attributes a late Paleozoic age to the Rio Fuerte Group of northern Sinaloa. This metamorphosed sequence of predominately fine grained sedimentary rocks includes up to 1500 m of rhyolite and andesite. The age assignment is based upon the fact that the sequence can be traced eastward through nearly continuous exposures to apparently laterally equivalent, unmetamorphosed, fossiliferous Carboniferous limestone (Carillo, 1971; Malpica, 1972). One of the areas of rock in Sinaloa that is similar to the

Rio Fuerte Group (near Mazatlan, locality 38) contains plant remains that set a maximum age of Carboniferous (Rodrigues-Torres, 1972, and personal communication to H. S. Mullan).

Evidence for a continuation of the belt of Permo-Triassic igneous rocks from eastern California (Davis and others, 1978) across Arizona to central Mexico is mentioned under the Triassic Period.

The basalt and mafic-ultramafic rocks of the Sierra Pinta and La Paz localities, associated in the former case with graywacke and chert, suggests a trench or marginal basin environment, whereas the rhyolite-andesite of northern Sinaloa may indicate the presence of late Paleozoic subduction (Fig. 11-3A). The evidence, however, is tentative and very fragmentary. The Mesozoic history, now to be discussed, is schematically portrayed in Fig. 11-2.

The Triassic Period

Exposures of ophiolite, generally believed to represent old oceanic crust, occur in the continental borderland west of Baja California and southern California. In the Vizcaino Peninsula, Robinson (1974) reports a minimum age of 187 ± 3 m.y. (K-Ar, hornblende) from a gabbro in a tectonic melange. At Punta Hipolito, Finch and Abbott (1977) found Triassic radiolarian cherts deposited between basalt pillows.

On Cedros Island, Kilmer (1977) discovered megaclasts of fossiliferous shallow-water limestone and quartzite in a melange of Middle Jurassic age. The fossils are of Middle Permian age (written communication, D. L. Jones, 1977). Limestone might have accumulated on a guyot, or a Bahamas-type platform, but the association with quartzite suggests a continental origin. Both Rangin (1978a) and Gastil and others (1978) suggest that the older ophiolites of the continental borderland formed in Permian to Triassic time by sea-floor spreading near the old cratonal boundary, rafting the limestone and quartzite seaward.

At Punta Hipolito, Finch and Abbott (1977) found a sequence of chert, thin limestone, andesite breccia, tuff, and graywacke, dated by both megafossils and radiolarian as Late Triassic. Both Gastil and others (1978) and Rangin (1978a) picture this as the beginning of a fringing island arc off the west coast of North America.

In the Santa Ana Mountains of Southern California, between the ophiolite and oceanic arc rocks of the Pacific borderland and the old cratonal border and within the terrane now largely occupied by the Peninsular Ranges batholith is a thick section of flysch-type clastic rocks with chert and limestone olistostromes. Part of this sequence has recently been identified as Triassic on the basis of Rb-Sr and radiolarian dating (Criscione and others, 1978; D. L. Jones, oral communication, 1978).

Just south of Romoland, Riverside County, California (4), Michael Murphy, University of California, Riverside, discovered mollusks in a phyllitic slate-quartzite sequence (Lamb, Senior Report, San Diego State University, 1969). The fossils were identified as possibly Triassic by E. C. Allison and G. E. G. Westerman (Gastil and others, 1975). This

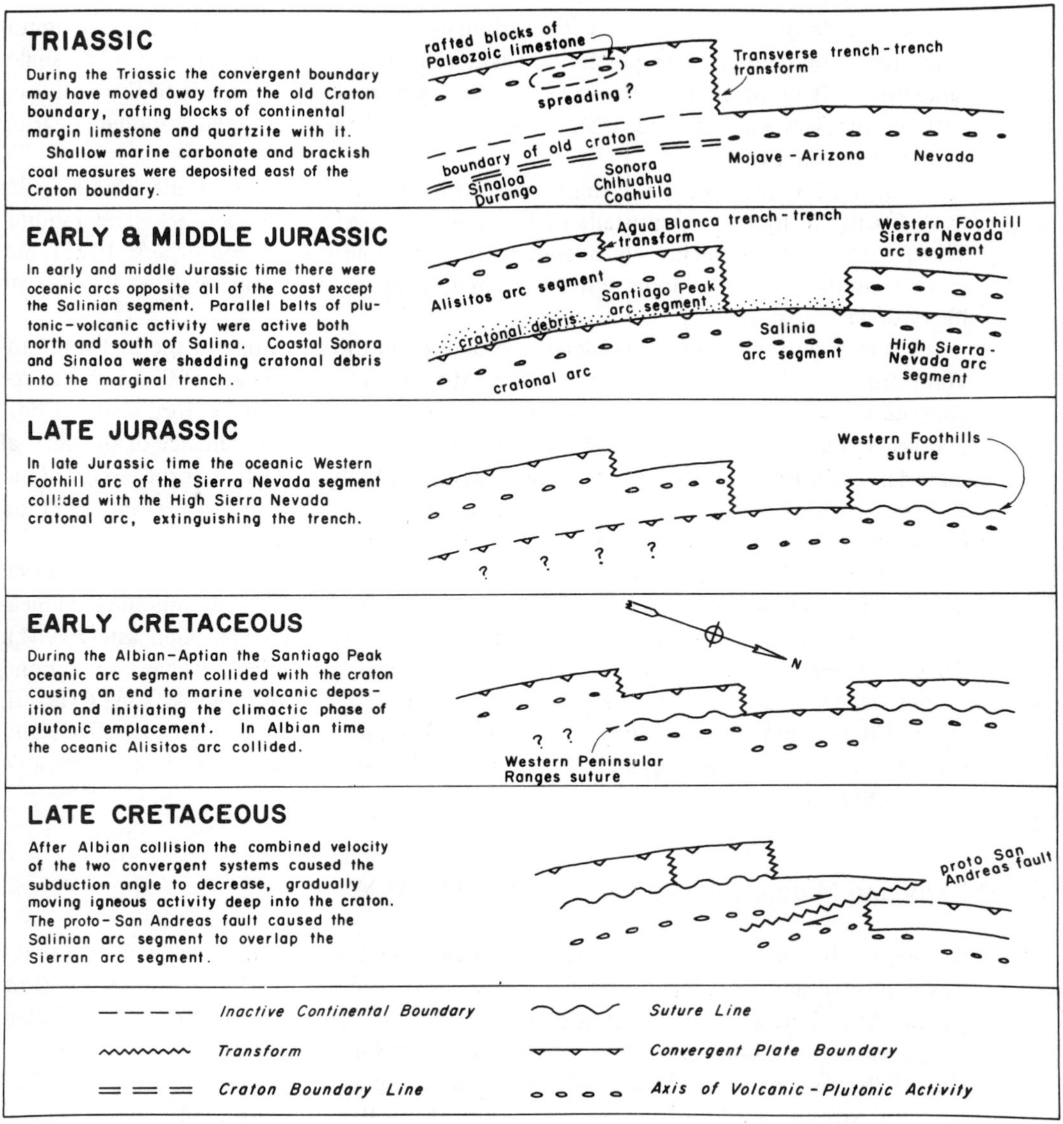

TRIASSIC
During the Triassic the convergent boundary may have moved away from the old Craton boundary, rafting blocks of continental margin limestone and quartzite with it.
Shallow marine carbonate and brackish coal measures were deposited east of the Craton boundary.
rafted blocks of Paleozoic limestone
spreading ?
Transverse trench - trench transform
boundary of old craton
Sinaloa
Durango
Sonora
Chihuahua
Coahuila
Mojave - Arizona
Nevada
EARLY & MIDDLE JURASSIC
In early and middle Jurassic time there were oceanic arcs opposite all of the coast except the Salinian segment. Parallel belts of plutonic-volcanic activity were active both north and south of Salina. Coastal Sonora and Sinaloa were shedding cratonal debris into the marginal trench.
Agua Blanca trench - trench transform
Western Foothill Sierra Nevada arc segment
Alisitos arc segment
Santiago Peak arc segment
cratonal debris
Salinia arc segment
High Sierra - Nevada arc segment
cratonal arc
LATE JURASSIC
In late Jurassic time the oceanic Western Foothill arc of the Sierra Nevada segment collided with the High Sierra Nevada cratonal arc, extinguishing the trench.
Western Foothills suture
EARLY CRETACEOUS
During the Albian-Aptian the Santiago Peak oceanic arc segment collided with the craton causing an end to marine volcanic deposition and initiating the climactic phase of plutonic emplacement. In Albian time the oceanic Alisitos arc collided.
N
Western Peninsular Ranges suture
LATE CRETACEOUS
After Albian collision the combined velocity of the two convergent systems caused the subduction angle to decrease, gradually moving igneous activity deep into the craton. The proto-San Andreas fault caused the Salinian arc segment to overlap the Sierran arc segment.
proto San Andreas fault
Inactive Continental Boundary
Transform
Craton Boundary Line
Suture Line
Convergent Plate Boundary
Axis of Volcanic-Plutonic Activity

Fig. 11-2. Schematic portrayal of major tectonic elements of California and adjacent North America during late Paleozoic and Mesozoic time. We have no control on how far the fringing arcs may have been from the southwestern edge of the continent.

locality has recently been recollected and identified as "probably Triassic" (D. L. Jones, oral communication, 1978).

The Julian Schist (7) of San Diego County, California, is a moderately mature, fine to coarsely clastic sequence with a few carbonate layers. J. P. Smith identified an ammonoid from this formation as possibly Triassic (Hudson, 1922). Similar rocks, the shale-sandstone belt of Gastil and others (1975), are distributed along the axis of the peninsula between the Santiago Peak and Alisitos formations, to the west, and the carbonate-bearing sequences to the east.

Near El Volcan, in the eastern escarpment of the peninsula at the 30th parallel (24), fossils are found in recrystallized limestone associated with graywacke and argillite. These fossils were originally considered to be Paleozoic (Gastil and others, 1975), but more recently (D. V. LaMone, written communication, 1977) are reported to be most likely of Triassic age.

It is important to note that detrital zircons from the sandstone of the Bedford Canyon Formation (Santa Ana Mountains, 3) and the Julian Schist (7) are of Precambrian age (Bushee and others, 1963), and strontium isotope ratios from the Bedford Canyon Formation (Criscione and others, 1978) indicate that the original provenance of the detrital material was a Precambrian terrane. This evidence strongly suggests that the Triassic (and Jurassic) detritus of these formations was derived from erosion of the Precambrian (and Precambrian derived) rocks originating east of the old craton boundary.

Marine Middle Triassic limestone, sandstone, and shale occur at El Antimonio (14), northwestern Sonora (Cooper and others, 1965), and Upper Triassic nonmarine coal measures are extensive in southeastern Sonora (White and Guiza, 1948; Alencaster, 1961). Triassic intermediate to siliceous volcanic and associated plutonic rocks have been found in Nevada and eastern California (Davis and others, 1978), southern Arizona (Drewes, 1976; Cooper and others, 1965), and possibly Durango (Rodrigues-Torres, 1972) and southwestern Coahuila (King and others, 1944). We are not aware of Triassic volcanic rocks in Sonora.

Early and Middle Jurassic Time (190 to 155 M.Y. Ago)

By Middle Jurassic time a belt of intermediate-composition plutonic and volcanic rocks was deposited along the length of what is now the continental borderland. On Santa Cruz Island (1) a diorite has been dated at 150 to 160 m.y. (K-Ar, hornblende; Fred Miller, 1976, reported by Jones and others, 1976) and 162 ± 3 m.y. (U-Pb, zircon; Mattinson and Hill, 1976). This rock may be part of a dismembered ophiolite (Mattinson and Hill, 1976).

In Arroya San Jose (25) on the west coast of Baja California, Minch (1969) and Gastil and others (1975) found a variety of fossils in volcanic-volcaniclastic bedded chert and argillite strata. The paleontologic dating has been contradictory, but the occurrence of *Otapiria cp. O. tailleuri* suggests that the sequence is no younger than Middle Jurassic.

On Isla Cedros (27), Kilmer (1977) has mapped an ophiolite terrane overlain by Bajocian-Callovian shale, sandstone, basalt, and andesite tuff. These strata in turn are overlain by shale-sandstone and megabreccia of late Middle or early Late Jurassic age

(Jones and others, 1976; Kilmer, 1977; the andesite tuff bed has been dated at 159 ± 5 m.y., K-Ar, hornblende).

Troughton (1974) and Rangin (1978b) report extensive meta-andesite in the central part of the Vizcaino Peninsula. These rocks are intruded by tonalite dated at 154 ± 3 m.y. Rangin (1978a) has suggested that the andesitic terrane of the Sierra San Andreas is contemporaneous with the Morro Hermoso and Eugenia formations. This could be partially true, but Barnes (oral communication, 1978) reports that strata of Eugenia age unconformably overlie metamorphosed andesite in the west-central part of the peninsula. It seems probable that the rocks of the San Andreas terrane are at least partially contemporaneous with the similar Callovian to Bajocian sequence of Isla Cedros.

In the Santa Ana Mountains (3) of Orange and Riverside counties, California, parts of the Bedford Canyon Formation are Bajocian and Callovian in age (Silberling and others, 1961; Imlay, 1963, 1964; Moscoso, 1967). This is part of the predominately sandstone-shale, continent-derived sequence mentioned previously.

In western Sonora, immediately east of the craton boundary, shallow marine and lacustrine deposition continued into Early Jurassic time (Cooper and others, 1965; White and Guiza, 1948; Alencaster, 1961). A belt of intermediate-composition Jurassic volcanic and plutonic rocks occurs in central Sonora and southern Arizona. The Sierra Santa Rosa (19) contains both biomicrite-shale and andesite-wacke-conglomerate (Hardy, 1972). In the Cucurpe area a thick volcaniclastic sequence with andesitic interbeds contains a rich Oxfordian fauna (Rangin, 1978b). Merriam (1972) and Rangin and Roldan (1978) believe that much of the volcanic rock in northern and west-central Sonora is probably Jurassic. Rangin (written communication, 1978) reports that the period of volcanism in Sonora extends from Hettangian to Oxfordian, based on ammonites and pelecypods. Anderson and others (1972) report a "conspicuous interval of plutonism from 190 to 160 m.y." (for Sonora and adjacent Arizona). An example is the granitic gneiss at El Capitan (15) dated at 170 ± 3 m.y. (U-Pb, zircon, Anderson and Silver, 1969). Jurassic plutonism in southern Arizona has been described by numerous authors from Gilluly (1956) to Drewes (1976).

To summarize, in Early and Middle Jurassic time, four tectonic belts can be distinguished: (1) to the southwest, a volcanic-plutonic arc closely associated with oceanic crust (the continental borderland); (2) a continent-fringing clastic apron, as in the Triassic (the Peninsular Ranges); (3) nonvolcanic stable shelf deposits (western Sonora); and (4) an ensialic volcanic-plutonic arc (west-central Sonora and adjacent Arizona).

Late Jurassic and Early Cretaceous Time (155 to 120 M.Y. Ago)

Volcanic and volcaniclastic strata, predominantly andesitic, but including pillow basalts and ignimbrites of Late Jurassic to possibly Early Cretaceous age, have been identified at a number of localities from southern California to the western capes of Baja California. In the Santa Ana Mountains (3), the type Santiago Peak Volcanics (Larsen, 1948) consist entirely of volcanic strata. Geologists currently working in the area are not agreed upon the nature of the contact between this sequence and the Triassic-Jurassic Bedford Canyon Formation. The only mineral dates of which we are aware are those reported by Bushee and others (1963), 150 and 155 m.y. ± 10% by lead-alpha zircon. In western San Diego County (5), the name Santiago Peak Volcanics has been applied to intermediate-composition volcanic rocks interbedded with marine volcaniclastic sandstone and argillite. The key fossil is the Portlandian *Buchia piochii* (Fife and others, 1967) in these beds. Jones and others (1976) indicate a range of Tithonian to Hauterivian for this formation.

In the Vizcaino Peninsula, the Eugenia Formation (Robinson, 1974; Minch and others, 1976) and the Morro Hermoso Formation (Rangin, 1978a) contain pillow basalt, ignimbrite, and ash, as well as volcaniclastic sediments containing abundant clasts of intermediate volcanic, ophiolite, granitic, and metamorphic rocks. These formations have been dated by a variety of fossils as Tithonian to Neocomian in age (Robinson, 1974; Jones and others, 1976; Rangin, 1978a). The Eugenia Formation contains volcanic and plutonic clasts, probably of the San Andreas volcanic rocks, and the tonalite + gabbro that intrude the San Andreas volcanic rocks.

Late Jurassic age granitic rocks have been reported from Santa Cruz Island (1) (Hill, 1976, Mattinson and Hill, 1976b), plagiogranite, U-Pb zircon, 141± 3 m.y.; on Isla Cedros (27) (Kilmer, 1977), quartz diorite, K-Ar, 142 ± 13 and 148 ± 6 m.y.; near Asención, Vizcaino Peninsula (Troughton, 1974; Minch and others, 1976), tonalite, K-Ar hornblende 143 ± 3 m.y.; a few kilometers north of Loreto, Baja California, tonalite, K-Ar hornblende, 144 ± 9 m.y. Many andesite dikes cut the Eugenia Formation in both the Vizcaino Peninsula and Isla Cedros. Three dates on these dikes from the northern Vizcaino Peninsula are 128 ± 2 m.y. and 116 ± 5 m.y. (K-Ar, plagioclase), and 125 ± 3 m.y. (K-Ar, whole rock) (Robinson, 1974; Minch and others, 1976).

A gabbro pluton from northwestern Baja California yielded an age of 126 ± 4 m.y., and two gabbros and a tonalite, just to the north in San Diego County, California (6), indicate 129, 135, and 141 m.y. (K-Ar, hornblende; Krummenacher and others, 1975). L. T. Silver (written communication, 1974) has found a similarly old tonalite in northern Baja California (U-Pb, zircon). Silver and others (1975) report a date of 127 ± 5 m.y. (U-Pb, zircon) for hypabyssal volcanic rock near Punta Cabra. Anderson and Silver (1969) report 128 ± 2 m.y. (Pb-U, zircon) for a metarhyolite in the Sierra Bacha, Sonora (12). C. D. Henry (written communication, 1975) reports dates of 129 m.y. for quartz diorite and 130 and 137 m.y. for two gabbros in western Sinaloa (K-Ar, hornblende).

The latest clear-cut evidence of Jurassic volcanism in Sonora is the strata at Cucurpe where "a large volcanic-volcaniclastic sequence with andesitic intercalations contains a rich Oxfordian fauna" (Rangin, 1978a). At Poso Serna, Beauvais and Stump (1976) report "tuffaceous" beds in Oxfordian to Kimmeridgian strata, but the volcanic debris could be entirely reworked from volcanic rocks of an earlier age. No plutonic rocks of Late Jurassic age have been identified in interior Sonora or southern Arizona (T. H. Anderson, P. E. Damon, oral communications, 1978). Drewes (1971) and Hayes (1970) place the dacite-rhyolite Bathtub and Temporal formations of extreme south-central Arizona in the pre-Bisbee Early Cretaceous (Neocomian). Drewes (oral communication, 1978) concedes that these rocks could be latest Jurassic, but prefers the Early Cretaceous designation.

To summarize, in Late Jurassic and Early Cretaceous time a volcanic-plutonic arc, spacially associated with ophiolite rocks, continued to accumulate in the area that is now the continental borderland. By earliest Cretaceous time this oceanic arc was adjacent to the Triassic-Jurassic apron (Figs. 11-2 and 11-3), and the axis of magmatism began to move eastward, involving the area now occupied by the Peninsular Ranges and points as far east as the coast of Sonora.

The continental arc appears to have remained active in south-central Arizona into Neocomian time, but there is no direct evidence of post-Oxfordian Jurassic volcanism in interior Sonora or of Bisbee age volcanism in Arizona.

Mid-Cretaceous Time (120 to 90 M.Y. Ago)

No marine volcanic strata of post-Hauterevian age have been found north of the Agua Blanca fault (9) of northern Baja California. South of the Agua Blanca fault (10), Allison (1964, 1974) described 6000 m (a minimum) of largely marine volcanic-volcaniclastic strata (Alisitos Formation, Santillán and Barrera, 1930). These clastic interbeds are almost devoid of quartz grains. The formation includes rocks ranging in composition from basalt to rhyolite, reef limestones, and volcaniclastic sandstone and shale. It has been identified at the surface as far south as the 28th parallel (32) (Barthelmy, 1975; Rangin, 1978a), and in the subsurface beneath the eastern margin of the Vizcaino desert (31) (A. Guzman, oral communication, 1977).

South of the 28th parallel, the peninsula of Baja California is almost entirely covered by Cenozoic rocks, making the projection shown on Fig. 11-1 very speculative. An area of weakly metamorphosed volcanic-volcaniclastic andesite crops out northwest of Loreto (35). No fossils were found during a few hours of reconnaissance, but an andesite gave an apparent age of 92 ± 2 m.y. (K-Ar, hornblende). This (minimum) age is somewhat younger than ages for rocks of the Alisitos Formation to the north, but is similar to analogous strata to the east in Sinaloa (Bonneau, 1971). In our pre-Miocene reconstruction (Fig. 11-1), Loreto lies opposite central Sinaloa. In northern and central Sinaloa, Bonneau (1971) has described andesite-reef carbonate strata, which he correlates with the Alisitos Formation. His rocks, however, are Albian-Cenomanian, as opposed to Aptian-Albian for the type Alisitos Formation.

The majority of the plutonic rocks of the Peninsular Ranges batholith of southern and Baja California were emplaced between 120 and 90 m.y. ago (Silver and others, 1975). Independent of mineral dates, this igneous activity can be interpreted from the isostatic uplift evident in the stratigraphic record. Great swarms of mid-Cretaceous volcanic dikes cutting the earlier plutons, some of them composed of ignimbrite, testify to the fact that volcanic rock was erupting on the surface during batholithic emplacement. This is apparently true even in areas where no Mesozoic volcanic strata remain today.

It must be kept in mind that, while the batholiths were intruding, the terrane was rising, being eroded, and shedding volcanic and plutonic debris, which accumulated in basins along the continental margin. In the Vizcaino Peninsula (Robinson, 1974; Minch and others, 1976) the "post-batholithic" (molasse) sandstone-shale-conglomerate Valle

Formation contains abundant clasts of Alisitos-type andesite and dacite and granitic rocks typical of the Peninsular Ranges batholith. Fossils in the Valle Formation are as old as Albian, and a 3-m-thick ash bed northwest of Asunción (3) yields an age of 103 ± 2 m.y. (K-Ar, biotite).

Farther east along the coast of Sonora (12), Anderson and Silver (1969) report a date of 103 ± 2 m.y. (U-Pb, zircon) for a quartz diorite, and Gastil and Krummenacher (1977) report discordant dates of 91 ± 2 and 90 ± 3 m.y. for two tonalites from northeastern Isla Tiburón (16) (K-Ar, biotite, hornblende). Farther southeast in Sinaloa, Nayarit, and western Jalisco, both U-Pb and K-Ar dates indicate that the bulk of plutonic emplacement was younger than 95 m.y. (Henry, 1975; Jensky, 1975; Gastil and others, 1976).

Are There Middle Cretaceous Volcanic Rocks in Central and Eastern Sonora?

In their regional syntheses of northwestern Mexico, Gastil and others (1972, 1978) have included a belt of andesitic volcanic rocks of Aptian-Albian age in interior Sonora. References to the existence of andesite and rhyolite associated with marine strata (largely carbonate rocks) of this age are found in such works as King (1939), Salas (1968), and Hayes (1970). However, the Arizona volcanic rocks (Drewes, 1971; oral communication, 1978) have been reevaluated, removing all volcanic strata from the Albian-Aptian (Bisbee Group) interval.

King (1939) describes the Albian Potrero Formation of central and eastern Sonora, Chihuahua, and Sinaloa as consisting of "shale, locally fossiliferous, thin-bedded brown limestone, and contemporaneous flows of andesite." In Table 4, King details forty-two sections that he considered to be of Cretaceous age, twenty-four of which contain volcanic rock, mostly andesite. Three of the andesitic sections are reported to contain Albian fossils. Additional sections are reported to contain marine fossils, but only a Mesozoic or no age assignment was given. However, geologists currently working in Sonora have been unable to confirm the presence of mid-Cretaceous volcanic rocks (written communication, C. Rangin, 1978).

No mid-Cretaceous plutonic or volcanic dates have been reported in southern Arizona or northern Sonora (oral communication from T. E. Anderson and Paul Damon, 1978; Marvin and others, 1973). Accounts of Middle Cretaceous volcanism in Sonora may be erroneous.

To summarize, during Middle Cretaceous time the last of the fringing arc segments became sutured onto the continent (Figs. 11-2 and 11-3), and the interior arc died out. The axis of magmatism from the western arc moved eastward into what is now the Peninsular Ranges, the locus of Jurassic eruptions was eroded down, and an apron of epiclastic (molasse) deposits was laid over the crustal segment now called the continental borderland.

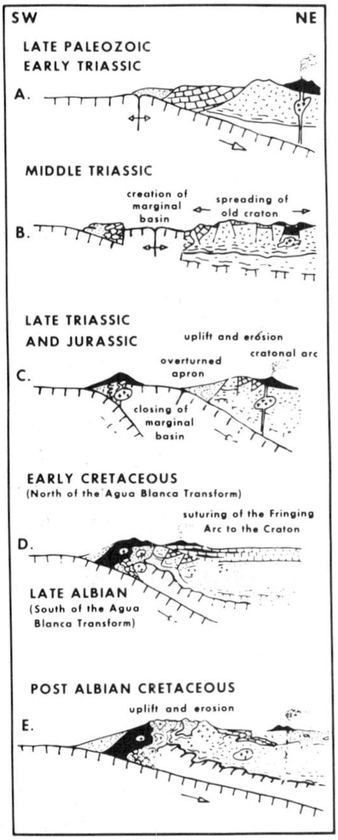

Fig. 11-3. Schematic geologic cross sections northeast across peninsular California and adjacent Mexico: (A) subduction beneath the craton with a spreading center at the continental edge; (B) subduction ceases and a marginal basin opens; (C) there are two northeasterly inclined subduction planes producing a fringing arc to the southwest and a cratonal arc to the northeast; (D) the fringing arc collides with the craton; (E) the axis of magmatism migrates deep into the craton.

Late Cretaceous Time (Post 90 M.Y. Ago)

In Late Cretaceous and early Cenozoic time the axis of batholithic emplacement moved eastward across the old cratonal boundary into southern Arizona, Sonora, Chihuahua, Sinaloa, Nayarit, and western Jalisco (Silver and others, 1975; Henry, 1975; Gastil and others, 1976). Plutonism was accompanied by great extrusions of andesitic to rhyolitic rock (Drewes, 1971; Hayes, 1970; King, 1939; Bonneau, 1971; Jensky, 1975; Gastil and others, 1979a), extending from southern Arizona at least as far south as western Jalisco.

While magmatism continued far to the east, peninsular California was a rugged mountain range shedding coarse, angular, locally derived granitic and metamorphic debris down steep canyons to a fault-line coast, where it was rapidly reworked into deep water (Gastil and Allison, 1966; Gastil and others, 1978). Boulder beds interfinger with marine shale, and bathyl-depth slide deposits formed contemporaneously with deltaic sandstone containing petrified logs and the bones of great dinosaurs (Gastil and others, 1975). Not only the continental apron, but the provenance for the apron lay between the arc and the trench.

Early Cenozoic Time (60 to 35 M.Y. Ago)

The interval of extensive volcanism and plutonic emplacement commonly referred to as "Larimide" extended into Cenozoic time to about 52 m.y.b.p. Between 52 and 35 m.y. we have few areas of known plutonic and volcanic rock. However, recent work between Sinaloa and Chihuahua (Clark and others, 1977, 1978), in Sinaloa (Henry, 1975), and in Nayarit and western Jalisco (Gastil and others, 1979a) suggests that rocks of this interval may be extensive just to the east of the Larimide belt in areas largely covered by Oligocene-Miocene volcanic strata.

By Paleocene time the Cretaceous mountains of the Peninsular Ranges had weathered down to a grus-covered matureland dotted by isolated mountains of more resistant rock. In the Eocene, palms and other tropical vegetation lined the southern California river valleys and conifers covered the uplands (Lowe, 1974). These Eocene rivers brought cobbles and boulders from south-central Arizona and north-central Sonora (Minch, 1969; Merriam 1972; Abbott and Smith, 1978). The volcanic source rocks are Jurassic and possibly Triassic in age, metamorphosed during the emplacement of Larimide plutons (oral communications, R. H. Merriam, P. L. Abbott, 1978).

Oligocene-Miocene Time (35 to 23 M.Y. Ago)

During the interval 35 to 23 m.y.b.p., rhyolite eruptions (many high-silica ash flow tuffs) and a significant proportion of alkaline basalts were deposited over an area extending from Nevada to Central Mexico, and from the eastern coast of Baja California (35) (McFall, 1968; Gastil and others, 1979b) and Isla Tiburón (16) (Gastil and Krummenacher, 1977) on the west to Colorado, New Mexico, and the Big Bend area of west Texas and Coahuila to the east (Barker, 1977). The rocks studied in the southern part of the Sierra Madre Occidental show an eastward increase in alkalinity and progressively younger ages to the west. The extent to which this eruptive interval was accompanied by plutonic emplacement is unknown, but granitic rocks of middle Cenozoic age have been identified at many locations from northwestern Utah to central Mexico. The recent recognition of Cenozoic-age plutonism in metamorphic core complexes through much of the Basin and Range Province suggests the possibility that the Oligo-Miocene volcanic cover is accompanied at depth by a belt of batholithic rocks.

Early Miocene Time (22 to 17 M.Y. Ago)

A gradual eastward migration of the axis of magmatism took place from Jurassic to late Oligocene time. During early Miocene time, this movement reversed, and by 20 m.y.b.p.,

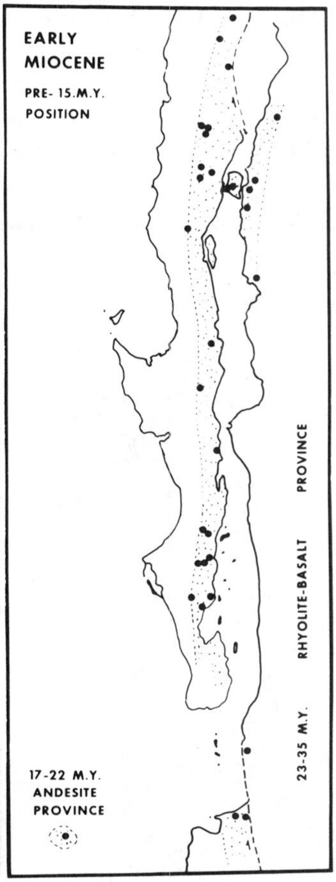

Fig. 11-4. The peninsula of Baja California and western Jalisco are shown in their pre-15 m.y. positions. The black dots show the localities of dated volcanic rocks. The speckled belt is the projected distribution of this volcanic rock province.

volcanism in western Mexico appears to have been limited to a narrow belt of eruptive centers along what is now the eastern edge of Baja California and the western edge of Sonora and Nayarit (Fig. 11-4). In moving toward the Pacific the composition of the volcanic rock changed from rhyolite, with subordinate basalt, to predominantly hornblende andesite. McFall (1968) dated a 20 m.y. old tonalite at Bahia Concepción (34), Henry (1975) has dated a pluton in Sinaloa at 18 m.y., and others have dated plutons of similar age in west-central Mexico. It seems probable that the 22 to 17 m.y. old volcanic rocks are underlain by a narrow belt of granitic rocks analogous to those underlying the Cascades.

It may be significant that the westward migration of the magmatic axis coincided with the dilation of the Basin and Range Province and the movement of the Pacific-North American plate boundary from the continental edge to a position within the continent (creation of the early San Andreas fault).

Middle to Late Miocene Time (16 to 6 M.Y. Ago)

During the interval 16 to 6 m.y.b.p., volcanic rocks erupted from the continental margin eastward to the coasts of Sonora, Sinaloa, and Nayarit (Fig. 11-5). These rocks can be divided into a belt of predominately alkaline basalt near the Pacific coast and a belt of

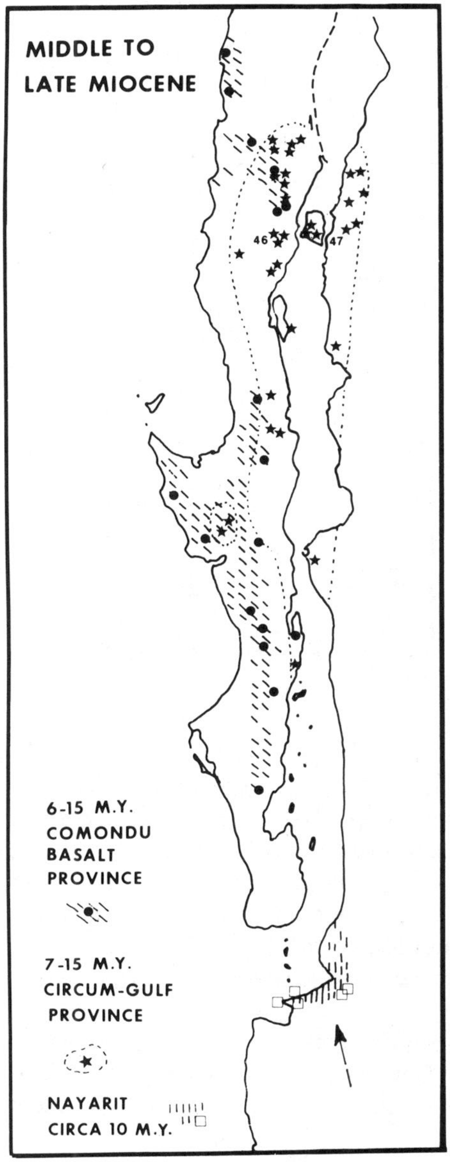

Fig. 11-5. Volcanic rocks with ages ranging from 6 to 15 m.y. are shown by symbols and patterns as indicated in the legend. The dots, stars, and small squares indicate dated localities. Baja California and western Jalisco are shown in a position that they may have occupied about 10 m.y. ago. The arrow, lower right, indicates the hypothesized relative northwest motion of western Jalisco prior to 10 m.y.b.p.

predominately calc-alkaline andesite to rhyolite farther east. In at least three areas, the Sierra Pintas (11), Valle de las Animas (south of 26), and Bahia Concepción (34), volcanic rocks of this interval are thousands of meters thick. Major calc-alkaline volcanism persisted on the Sonora coast until about 10 m.y.b.p. and in Baja California until about 8 m.y.b.p.

Although it is not definite from existing models exactly when subduction should have ceased west of northern Baja California and adjacent Sonora, it has been assumed that it terminated by 10 m.y.b.p. (Atwater, 1970; Pilger and Henyey, 1977). Presumably, volcanism would continue during the interval between the annihilation of subduction and the arrival of the last subducted plate at the depth of volcanic generation. Therefore, 8 m.y. ago would be a reasonable termination age, but it is unclear why the last area of major calc-alkaline volcanism should have been around what is now the northern half of the Gulf of California.

Five Million Years to Present

During this interval, alkaline basalt and associated basaltic andesite were erupted along the western slope of Baja California and somewhat lesser amounts of alkaline basalts at scattered points farther inland (Eberly and Stanley, 1978; Gastil and others, 1979b) (Fig. 11-6). Along the Gulf of California, predominantly dacite-rhyolite has erupted at many points. Some, including the Salton Sea (locality 41, Fig. 11-6), Cerro Prieta (41), and Isla Encantada (42), are clearly related to the transform plate boundary system, whereas other areas such as Puertecitos (43) and the Tres Virgines (45) may somehow be related to the same pattern.

DISCUSSION

A Question of Parallel Arcs

Currently active convergent ocean-continent plate boundaries produce a well-defined line of volcanic-plutonic activity at some distance inland from the older sialic margin (Cascades, Peru, Chile, for examples). During Early and Middle Jurassic time, and possibly continuing into Early Cretaceous time, southern Arizona and central Sonora exhibited a belt of rhyolite-andesite with associated plutonic rocks analogous to modern arcs that are clearly situated within continents. At the same time, to the west, in what is now the continental borderland and the western edge of the peninsula, another belt of andesite and tonalite was building on oceanic crust.

In Figs. 11-2, 11-3, and 11-7 we have attempted to explain these two parallel and contemporary volcanic-plutonic belts as the result of two subduction zones. We cannot as yet suggest how far from the edge of the continent the oceanic arc might have been.

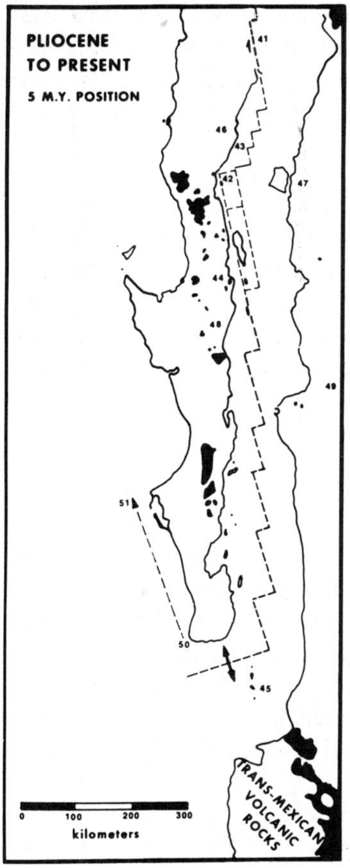

Fig. 11-6. Pliocene to Modern volcanism is shown in black. The dashed line in the Gulf of California represents the spreading centers and connecting transform faults that have been active during the past 4 or 5 m.y. The arrow west of the peninsula represents the motion of Cabo San Lucas (50 to 51) during the past 5 m.y. Numbers relate to citations in the text.

Possible Jurassic-Cretaceous Trench-Trench Transforms

Near its western end, the right-lateral Agua Blanca fault zone (Allen and others, 1960) is complex, including the left-lateral (?) Santo Tomás fault (Krause, 1965). Acosta (1966, 1970) determined that a Santo Tomás fault line valley existed during Late Cretaceous time. The marine volcanic Alisitos Formation, which can be traced 1300 km in Baja California and Sinaloa, ends abruptly at the Agua Blanca fault. The west coast gravity high (Gastil and others, 1975), which extends at least as far south as the 28th parallel, also terminates abruptly at the Agua Blanca fault. We conclude that the Agua Blanca fault marks the location of a fundamental discontinuity of Mesozoic origin.

Another truncation of Peninsular Range elements occurs immediately south of the Transverse Ranges. Neither the Bedford Canyon nor the Santiago Peak Formation are found north of this line. Figure 11-2 includes another discontinuity at the south end of the Sierra Nevada for the purpose of analogy.

We suggest that these truncations originated as trench-trench transforms separating several arc-trench segments, each with a slightly different tectonic history. In this hypothesis the lack of a volcanic belt along the west side of the Salinian block can be explained by the lack of a fringing arc opposite that segment (Fig. 11-2).

Where Is The Suture?

If the Jurassic and Lower Cretaceous rocks of the fringing arc were separated from the continent by a Jurassic subduction zone along the continent margin (Figs. 11-2 and 11-3), underflow along this inclined plate junction would eventually have caused a collision between the rocks of the western arc and the old craton (Fig. 11-3D). If the subduction on the easterly plane ceased before collision, then a belt of oceanic crust should remain between the rocks of the western arc and the old continental margin. Jones and others (1976) believe that the Jurassic volcanic arc of southern and Baja California accumulated on the edge of the continent, but we have rejected this idea inasmuch as it does not account for two parallel and, at least partially, simultaneous arcs. Rangin (1978a) has proposed that there was a belt of back-arc oceanic crust between the Jurassic arc and the continent. This seems like a reasonable model for Permo-Triassic spreading, but again fails to account for the two Jurassic arcs.

Larsen (1948) indicated that the Santiago Peak Volcanics (undated in their type area) depositionally overlie the Triassic-Jurassic Bedford Canyon Formation (3), but recent workers in the area (P. H. Ehlig, oral communication, 1978) report that no definitive contacts have been located. In most localities the two units appear to be separated by faults, and in at least one area, serpentinized ultramafic rock is associated with the contact. Recent workers in the Santa Ana Mountains (Moscoso, 1967; Moran, 1976; Buckley and others, 1975; Criscione and others, 1978) agree that the Bedford Canyon Formation (flysch) is largely, if not entirely, overturned. If overturning has resulted from the underthrusting of a subducting plate, that subduction persisted at least until post-Callovian time and was located adjacent to the overturned section, not approximately 100 km to the west, outboard of the continental borderland (Fig. 11-3C). Thus this structure is evidence for the easterly subduction plane that dipped beneath the continental apron, not for the westerly plane that dipped (eastward) beneath the oceanic arc.

In western San Diego County and northern Baja California we have sought in vain to find the volcanic arc rocks (Santiago Peak and Alisitos formations) in clear-cur contact with the continentally derived Julian schist and sandstone-shale sequences. In a few areas they can be mapped very close to one another, but for the most part their areas of distribution are mutually exclusive. North of Valle Guadalupe (6) (see Fig. 19, Gastil and others, 1975), volcanic strata appear to overlap rocks of the sandstone-shale continental apron, but the former appear to be nonmarine volcanic rocks probably around 100 m.y. in age (on the basis of dated feeder dikes, Krummenacher and others, 1975).

Direct evidence of the suture zone may be found in the pervasively deformed granitic rocks near Lakeside, San Diego County (east of 6), and in the broad zone of gneissic granitic rocks between Valle San Rafael and Valle Trinidad, Baja California (8), including highly deformed gabbro and talc schists. The gravity high beneath the Alisitos Formation, south from the Agua Blanca fault (Gastil and others, 1975), may indicate that from this point south there is a relatively shallow fold of oceanic crust separating the continental apron from the sutured arc. Serpentinite and other ophiolite rocks reported by Fife (written communication, 1968) in the Santa Rosalia quadrangle (northwest of 32) and by Barthelmy (1975) and Rangin (1978a) north of El Arco (32) may be exposures of this ophiolite. Blocks of metamorphosed ultramafic rock were found in the basal Oligocene (?) conglomerate north of Tombobiche (west of 38), and the zone of mafic and ultramafic rocks within the plutonic belt south of La Paz (36) (Altamirano, 1972) may be another remnant of prebatholithic ophiolite rocks pinched into the suture zone.

TECTONIC HISTORY

Evolution of the Arcs: Late Paleozoic to Mid-Cenozoic

In the late Paleozoic, sedimentary and volcanic strata were deposited along or adjacent to the old cratonal margin. The marine basalt suggests an extensional basin. However, Late Permian or Early Triassic andesite-rhyolite volcanism and synchronous plutonism occurred in a belt from British Columbia (Davis and others, 1978; Fig. 11-4) to southern Arizona, and possibly to Durango (Rodrigues-Torres, 1972) and Coahuila (King and others, 1944), suggesting a continental (Andean-type) arc of this age (Fig. 11-3A).

By Late Triassic time, at some distance oceanward from the continent, ocean-floor convergence produced an island arc (Figs. 11-2, 11-3 and 11-7). By Early Jurassic time, two parallel arcs existed, one offshore and a second east of the old cratonal boundary in Sonora and Arizona. During the Triassic and Jurassic periods, the old cratonic margin of southwestern Arizona and western Sonora rose in front of the convergent boundary, shedding a thick wedge of nonvolcanic detritus, perhaps filling the trench in front of the approaching island arc (Fig. 11-3C).

Trench-trench transforms staggered the arrival of the oceanic arc segments against the continent (Fig. 11-2). The collision opposite the Sierra Nevada took place during Late Jurassic time (Jones and others, 1976). Collision of the segment between the Transverse transform and the Agua Blanca transform took place during Early Cretaceous time, and collision south of the Agua Blanca transform did not take place until late Albian time to the north and Cenomanian time farther southeast. Thus, while the collision north of the Agua Blanca fault had caused uplift, terminating marine deposition by earliest Cretaceous time, marine volcanism continued south of the Agua Blanca because that segment of the arc had not yet reached the edge of the continent.

It should be emphasized that collision did not terminate volcanism, but it did result in uplift, which in turn caused subaerial deposition, subject to rapid erosion. Post-Albian volcanism and plutonic intrusion occurred on both sides of the suture.

The collision is believed to have caused a gradual slowing of the eastern subduction plane with a consequent increase of subduction velocity on the western plane. The addition of the two velocities seems to have produced a shallower subduction angle (Luyendyk, 1970), moving the axis of plutonism inland (Gastil and Krummenacher, 1974, 1977; Silver and others, 1975).

By Late Cretaceous time the axis of magmatism had moved to the old craton of southern Arizona and the interior of Sonora (Fig. 11-7). By early Cenozoic time magmatism had passed into Chihuahua and Durango. This continued eastward migration could no longer be attributed to the addition of subduction velocities, but an additional factor causing eastward migration is found in the lateral crustal dilation caused by plutonic

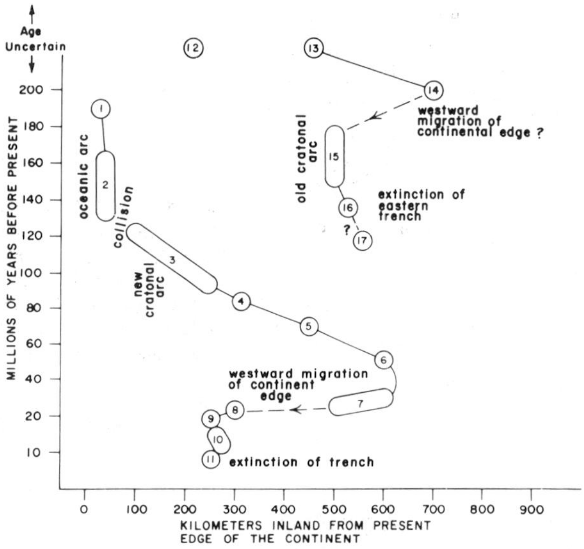

Fig. 11-7. Diagram showing the movement of the axes of magmatism through time relative to the present western edge of North America in the latitudes of Baja California and adjacent Mexico. No adjustments have been made for changes in the width and the edge position of the continent. The numbered localities are as follows: (1) Punta San Hipolito, Baja California; (2) localities extending from Santa Cruz Island, California, to the north to Isla Cedros and the Vizcaino Peninsula of Baja California to the south; (3) peninsular California east of the continental borderland; (4) coastal Sonora; (5) central Sonora; (6) eastern Sonora and Chihuahua; (7) Sierra Madre Occidental; (8) coastal Sonora; (9) eastern Baja California; (10) both sides of the Gulf of California; (11) eastern coast of Baja California; (12) Sierra Pinta, eastern Baja California; (13) El Fuerte area of northern Sinaloa; (14) southeastern Arizona, Coahuila, and Durango; (15) central Sonora; (16) southeastern Arizona; (17) eastern Sonora (?).

emplacement (Fig. 11-8). Geologists are accustomed to thinking of the middle Cenozoic as a time of crustal extension because the surficial rocks of this interval can still be clearly seen. We surmise that the near-surface structures of the Jurassic and Cretaceous batholith belt, now largely eroded away, were also extensional in character.

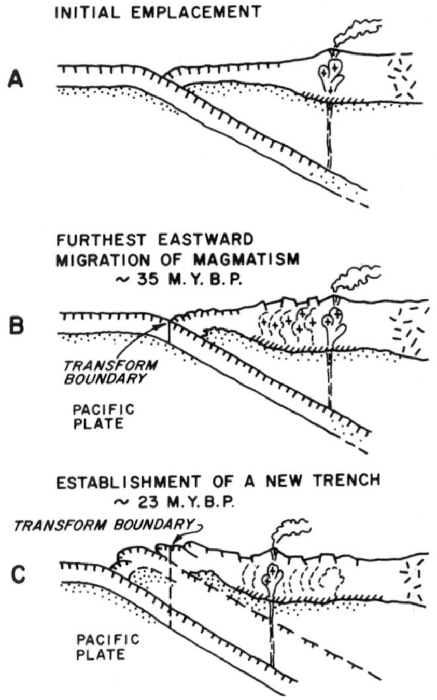

Fig. 11-8. This sketch illustrates the role played by the emplacement of plutons in the sialic crust, causing dilation (B), which ultimately overrides the trench (C), resulting in the creation of a new subduction plane. In this case, where transform motion between plates is taking place along the trench (B), the overriding of the transform (C) resulted in the relative displacement of the transform boundary into the continent.

Reversal in the Migration of Magmatism, Dilation of the Basin and Range Province, and Overriding of the Plate Boundary

During Oligo-Miocene time the axis of magmatism ceased moving east and began moving back toward the west (Fig. 11-7). The Basin and Range Province of southeastern California, southern Arizona, and western Sonora once stood higher than the adjacent Sierra Madre Occidental and Colorado plateaus, probably because the former area absorbed a greater volume of plutonic emplacement. While this highland existed, drainages from central Sonora flowed to the Pacific Ocean across what are now the Peninsular Ranges. Eventually, isostatic uplift caused by plutonic emplacement exceeded the strength of the rock, and this edifice flowed laterally, thrusting the entire western margin of the continent southwestward across the trench (Fig. 11-8C).

The establishment of a new, more westerly subduction plane in early Miocene time produced a new arc of andesitic volcanoes along the area now occupied by the Gulf of

California. Arc-related volcanism ceased about 8 m.y. ago, and subsequent siliceous volcanism appears to be related to the transform motion within the Gulf of California.

The surficial extension of the Basin and Range Province, the reversal in the direction of magmatic migration, and the landward migration of the transform separating the North American and Pacific plates all began near the end of the Oligocene and took place during early and middle Miocene time. Pilger and Henyey (1977) suggest that the approach of the East Pacific Rise to the continent, causing the subduction of progressively warmer lithosphere, resulted in a shallowing of melting and a consequent reduction in the arch-trench gap. However, the westward shift opposite western Mexico (Gastil and others, 1979b) appears to have taken a fairly rapid jump 23 to 20 m.y. ago, with no subsequent western migration of calc-alkaline volcanism. This does not seem consistent with the gradual approach of the rise.

We propose that dilation of the upper lithosphere (Fig. 11-8B) caused the continental rocks to ride out over the North American-Pacific plate margin (Fig. 11-8C) about 23 m.y. ago, with two effects. First, in those segments where subduction persisted, the trench was displaced to the southwest, resulting in a southwestward displacement of the axis of magmatism. Second, the transform motion between the two plates, heretofore at the edge of the continent (Fig. 11-8B), although stationary at depth, gained a surficial location within the North American continent (Fig. 11-8C). A side effect of the shift in the position of the subduction plane was the isostatic uplift of large fault blocks in what is now the continental borderland, causing them to shed ophiolitic detritus continentward during early to middle Miocene time (Stewart, 1976).

The Opening of the Gulf of California

The minimum opening of the Gulf of California would appear to be about 260 km parallel to the transform faults of the Gulf (Larson and others, 1968). The crustal structure east of the Isla Tres Marias (45) might allow a further closing of the Gulf. Thus there has been speculation on how much additional dilation might have occurred during an earlier "proto-Gulf" opening (Karig and Jensky, 1972). There is at least one tie point (pre-15 m.y.) across the Gulf of California that helps to define the Miocene to recent strike-slip offset. This is a conglomerate channel containing clasts of weakly metamorphosed volcanic rock and unmetamorphosed limestone and quartzite containing Early Permian fossils (Gastil and others, 1973). This conglomerate is found on both sides of Valle Noriga, northeast of Bahia Kino (47) and in the Sierra Santa Rosa (46), southwest of San Felipe, Baja California (Figs. 11-5 and 11-6). In Baja California it rests on Mesozoic granitic rocks in what appears to be a southwest-trending channel. The total offset is 300 km measured parallel to the Guaymas lineament. There are also two elements of Mesozoic structure that argue for a transform offset of no more than 300 km. These are the distribution of 110 to 140 m.y. gabbro (Gastil and others, 1976) and the distribution of Aptian-Cenomanian marine andesite (Fig. 11-1). Both of these elements can be clearly followed south to the 28th parallel of Baja California, and both are present in the northernmost part of Sinaloa (Mullan, 1975; Bonneau, 1971). As can be seen from Figs. 11-1 and 11-4 (localities 48 and 49), any additional (more than 300 km) post-Cenomanian movement of the peninsula to the south produces a kink in the trend of these rock belts. Proposed movement of western Jalisco (between Figs. 11-4 and 11-5) makes the aggregate opening at the entrance of the Gulf of California about 400 km.

ACKNOWLEDGMENTS

Most of the work described has been done by graduate and undergraduate students at San Diego State University, many of whom have been cited in the text. Many paleontologic determinations used here have been generously provided by David Jones of the U.S. Geological Survey and others. Our work has been assisted by the National Science Foundation, the National Geographic Society, PEMEX, the Instituto de Geologia, and the Consejo de Recursos Minera. Discussions with many geologists have contributed to the hypothesis described here. These include J. R. Morgan of Gulf Oil Company, Michael J. Walawender of San Diego State University, and Claude Rangin recently at the University of Sonora, now at Universite Pierre et Marie Curie, France.

M. C. Blake, Jr. and David L. Jones
U.S. Geological Survey
Menlo Park, California, 94025

12

THE FRANCISCAN ASSEMBLAGE AND RELATED ROCKS IN NORTHERN CALIFORNIA: A REINTERPRETATION

ABSTRACT

Present plate-tectonic models do not adequately explain the complex geologic relations between the Franciscan assemblage, Great Valley Sequence, Sierran, and Klamath terranes. The presence of several oceanic arc systems of different ages, differences in sandstone petrology and sedimentary environment between the Franciscan assemblage and the Great Valley Sequence, and differences in sedimentological characteristics within coeval parts of the Great Valley Sequence imply a complex paleotectonic setting in which some relations are best explained by juxtaposition of rocks derived from different source areas.

INTRODUCTION

The Franciscan assemblage of the California Coast Ranges consists of graywacke and metagraywacke units that are separated by zones of melange. Because of its extreme structural disorder and the presence of high-pressure metamorphic minerals, the Franciscan assemblage is commonly taken as the type example of an accretionary complex formed by subduction (e.g., Hamilton, 1978a).

The close temporal and spatial relations of Franciscan rocks with rocks of the structurally overlying Great Valley Sequence and the volcanoplutonic rocks that lie to the east in the Klamath Mountains and the Sierra Nevada have led to the widely accepted hypothesis that rocks of the Franciscan assemblage were deposited in an active trench contemporaneously with deposition of the Great Valley Sequence in a forearc basin adjacent to the active Klamath-Sierran arc complex. These three elements comprise the arc-, arc-trench gap, and trench setting (Fig. 12-1) typical of convergent

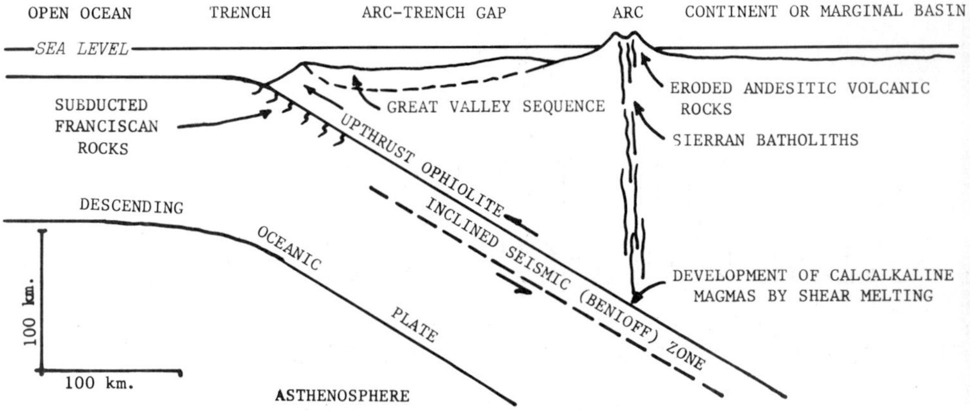

Fig. 12-1. Hypothetical cross section of California showing possible relation of geologic units to proposed plate-tectonic elements seen in present-day island arcs and active continental margins (modified from Dickinson, 1971).

plate boundaries (Dickinson, 1970). A large component of subduction generally has been thought to be eastward, normal to the trend of the continental margin (e.g., Hamilton, 1969).

Although this hypothesis has much merit in a general sense, we believe that it is too simplistic to explain anomalous relations revealed by a growing body of geologic, geochemical, and sedimentologic data obtained from northern California and southwestern Oregon. The purpose of this report is to summarize some of these data and to suggest that a more complex paleotectonic scenario that involves oblique subduction and large-scale transform faulting may be required. Specifically, the present simplistic model fails to explain the following features that characterize the geology of northern California:

1. The presence of oceanic arc systems of Late Jurassic, Cretaceous, and Early Tertiary age west of and genetically unrelated to the Klamath-Sierran arc.
2. Differences in sandstone petrology and subsea fan facies between the Great Valley Sequence and some coeval coherent Franciscan terranes.
3. Differences in sedimentological characteristics, including composition of clasts, within some of the outliers (klippen) of the Great Valley Sequence and correlative strata to the east.
4. Subduction of the Franciscan assemblage and contemporaneous rifting and northwestward movement of the Klamath Mountains are not inconsistent with subduction normal to the continental margin.
5. Quartzofeldspathic Franciscan sandstones that predate and have a different provenance than basal Great Valley strata.

Large-scale northward transport during the Mesozoic of blocks of subcontinental dimensions, such as Wrangellia in Alaska (Jones and others, 1977, 1978b), implies a Pacific marginal tectonic regime dominated by transform faulting until about middle Cretaceous time. This transport regime suggests the possibility that some components of both the Franciscan assemblage and Great Valley Sequence also may be allochthonous and may have originated far to the south of their present position. Paleomagnetic data obtained from Mid-Cretaceous Franciscan pelagic limestone blocks (Alvarez and others, 1979) substantiate this possibility.

SUMMARY OF REGIONAL GEOLOGY

Franciscan rocks in the California Coast Ranges north of San Francisco are subdivided into three major subparallel units (Fig. 12-2) named, from west to east, the Coastal belt, Central belt, and Yolla Bolly belt. Near San Francisco, these three belts are not so easily discriminated, and they may have been juxtaposed by strike-slip faulting and folding related to the Tertiary San Andreas fault system.

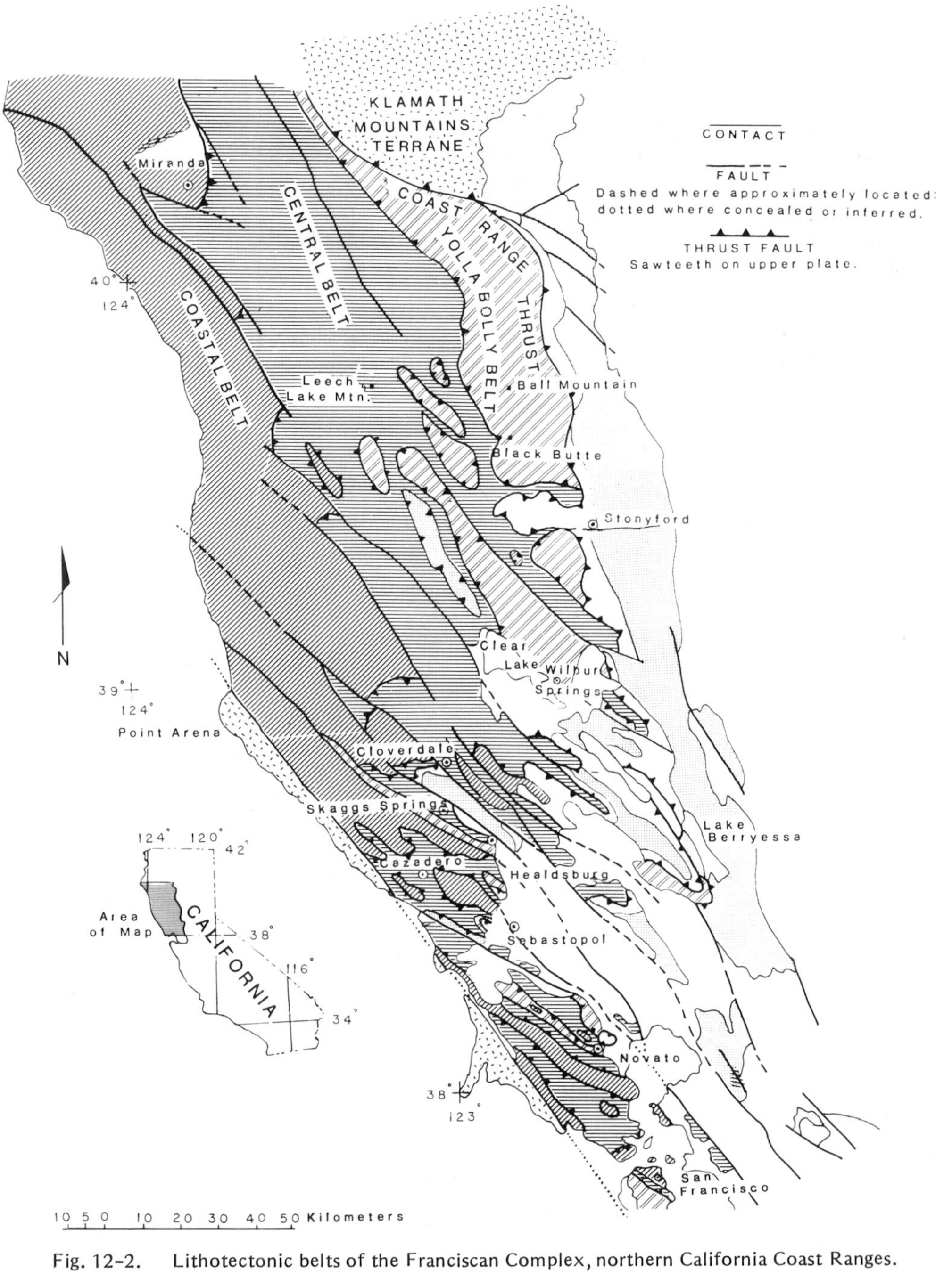

Fig. 12-2. Lithotectonic belts of the Franciscan Complex, northern California Coast Ranges.

Coastal Belt

The westernmost unit of the Franciscan assemblage is the Coastal belt that contains marine fossils of Late Cretaceous, Paleocene, and Eocene age (Blake and Jones, 1974; Evitt and Pierce, 1975). Rocks in the Coastal belt differ from other Franciscan units in that the sandstones are notably more arkosic and contain very little lithic volcanic or chert detritus (Bailey and Irwin, 1959). Although only a few sedimentological studies have been made (e.g., Kleist, 1974; Bachman, 1978), sedimentary structures suggest that these rocks were deposited in deep-sea fan systems.

The metamorphic grade within the Coastal belt appears to be lower than in the other Franciscan belts; laumontite is abundant and prehnite-pumpellyite are only locally present. Newly generated blueschist minerals (lawsonite, glaucophane, jadeitic pyroxene) are unknown. Because widespread Late Tertiary to Holocene faulting related to the San Andreas fault system is superimposed on earlier, subduction-related tectonism, all rocks in the southern part of the map area (Fig. 12-2) are fragmented, and stratigraphically continuous sequences in these youngest Franciscan rocks are unknown.

In the northern part of the belt, recent work by Beutner (1977) shows that the oldest, westernmost beds are Upper Cretaceous (?) to Paleocene deep-water pelagic and hemipelagic rocks that rest on altered basalt. These are overlain by andesitic graywacke, which grades upward and eastward into quartzofeldspathic graywacke. Beutner interprets this graywacke sequence as having been partly derived from an oceanic island arc on the west and partly from the North American continent on the east.

Between Cloverdale and Miranda (Fig. 12-2), the Coastal belt is succeeded on the east by a thoroughly mixed melange that belongs to the Central belt and that contains abundant blocks ("knockers") of serpentinite, high-grade blueschist, and eclogite. Also present within the western part of the melange are blocks and slabs of Coastal belt sandstone. The Coastal belt strata structurally underlie this melange and are isoclinally folded and sheared along a major structure called the Coastal belt thrust by Bailey and McLaughlin (in Jones and others, 1978a).

Central Belt

The Central belt (Berkland and others, 1972) consists largely of melange with numerous small and large bodies of graywacke and metagraywacke (Fig. 12-2), some of which may have been derived from adjacent units, together with slabs and blocks up to a few kilometers long of greenstone, chert, and serpentinite. The matrix of the melange is nearly everywhere sheared dark-gray mudstone with variable amounts of interbedded to structurally intercalated graywacke ("matrix graywacke").

The most common blocks in the melange are greenstone (in many cases pillowed) and radiolarian chert. Also present are the majority of high-grade blueschist knockers known in the northern Coast Ranges. These are most abundant along the western boundary with the Coastal belt, but they also occur along serpentine-marked fault zones in the eastern part of the Central belt.

Fossils are relatively abundant within the melange matrix and are Late Jurassic and Early Cretaceous in age (Blake and Jones, 1974). Fossils of similar age occur within the chert blocks (Pessagno, 1973), which may have been interlayered originally with the clastic rocks or with greenstones and were subsequently disrupted. Several of the graywacke and metagraywacke lenses contain fossils of Late Cretaceous age; in addition, there are blocks of Upper Cretaceous pelagic foraminiferal limestone in the western part of the belt (Blake and Jones, 1974).

As in most of the Franciscan terrane, stratigraphic studies are difficult because of thorough and pervasive mixing. Attempts have been made to subdivide the melange on the basis of composition of blocks (Gucwa, 1975; Maxwell, 1974), but the significance of the noted minor differences in clast populations in terms of source terranes, mechanism of formation, or overall geologic history remains uncertain. The most detailed mapping in this belt has been done in The Geysers area (McLaughlin, 1978), where Franciscan rocks comprise a series of northwest-trending slabs of graywacke, greenstone, and chert, separated by melange or sheared serpentinite. Sedimentary structures within the graywacke slabs are similar to those in the Coastal belt and suggest turbidite deposition in the upper to middle parts of subsea fan systems (R. J. McLaughlin and H. N. Ohlin, unpublished data; McLaughlin and Pessagno, 1978).

The majority of matrix graywackes contain pumpellyite-quartz-albite-chlorite-white mica, without prehnite or lawsonite, as the dominant metamorphic mineral assemblage. Blocks and slabs of higher-grade metagraywacke containing lawsonite and less commonly jadeitic pyroxene are scattered throughout the Central belt (Blake and others, 1967; M. C. Blake, unpublished data; Suppe, 1973).

To the east, the Central belt is overlain structurally by rocks of the Yolla Bolly belt. In many places, the fault zone separating the Central belt and the Yolla Bolly belt is subhorizontal or dips gently to the east (Blake and Jones, 1974, p. 346) and contains numerous small bodies of serpentinite (Brown, 1964; Suppe, 1973).

Yolla Bolly Belt

A discontinuous belt of metaclastic rocks plus minor metachert and metagreenstone is known as the Yolla Bolly belt (Blake and Jones, 1974, 1977). In the Yolla Bolly quadrangle, and to the north, the South Fork Mountain Schist is metamorphosed mudstone and basaltic tuff containing very rare metagraywacke lenses. Metaconglomerate is absent. The metatuff, assigned to the Chinquapin Metabasalt Member by Blake and others (1967), locally contains thin, discontinuous pods of metachert. Fossils are unknown from this unit.

To the west but still within the Yolla Bolly belt, the South Fork Mountain Schist is underlain by quartzofeldspathic graywacke plus minor mudstone, conglomerate, and chert containing fossils at least as young as Early Cretaceous (Valanginian). Also present are thick olistostromes that consist of mudstone intruded by numerous small bodies of quartz keratophyre and less common basalt. The intrusive nature of these greenstones was first recognized in the Leech Lake Mountain-Ball Mountain area (Suppe, 1973) and has since been seen in this belt as far north as southwestern Oregon.

Sedimentary structures imply that the sandstone was laid down by turbidity currents and local submarine slumping in the upper part of a subsea fan. Lithologically, these rocks appear to be very different from the pelites that characterize the South Fork Mountain Schist. They may be similar to bodies of quartzofeldspathic schist that occur to the south of the Yolla Bolly quadrangle in the Leech Lake Mountain-Ball Mountain region (Suppe, 1973) and near Black Butte (Ghent, 1965; Bishop, 1977), which were originally correlated with the South Fork Mountain Schist by Blake and others (1967).

In the Yolla Bolly quadrangle, there exists a complete gradation from little-reconstituted metagraywacke of textural zone 1 through semischist of textural zone 2 to segregated quartzofeldspathic schist of textural zone 3 (Blake and others, 1967; Blake and Jones, 1977; Bishop, 1977). To the south, however, other workers have mapped a thrust fault coinciding with the change from textural zone 2 to zone 3 (Suppe, 1973; Maxwell, 1974). The work of Bishop (1977) and Worrall (1978) suggest that both findings are compatible, and that the original gradation has been disrupted by postmetamorphic faulting. The relations between the quartzofeldspathic schist and the pelitic South Fork Mountain Schist (as redefined by Bishop) remain unclear. They may be separated by an unrecognized fault or they may be facies of one another. That the latter is more plausible is suggested by the total absence of serpentinite along this boundary, and the presence of distinctive metatuffs within both the pelitic schist of the Yolla Bolly area and quartzofeldspathic schist of the Black Butte area.

The age of the South Fork Mountain Schist is also controversial. K-Ar metamorphic ages have been published by Suppe (1973), Suppe and Armstrong (1972), and Maxwell (1974). Apparent ages range from 77 to 143 m.y. The recent work of Lanphere and others (1978) using ^{40}Ar-^{39}Ar and Rb-Sr techniques to both textural zone 2 and 3 metagraywacke and metapelite, and on metabasalt from undoubted South Fork Mountain Schist in the Yolla Bolly quadrangle, suggest that 115 to 120 m.y. is the best current estimate for the age of metamorphism. Paleontologic evidence indicates that some marine sedimentary rocks showing textural zone 2 structures are as young as Valanginian (135 to 130 m.y.) and thus could not have been metamorphosed during Late Jurassic time.

Great Valley Sequence

Structurally overlying the Franciscan assemblage is the Great Valley Sequence composed of well-bedded mudstone, siltstone, sandstone, and conglomerate that range in age from Late Jurassic to latest Cretaceous. The basal strata, which locally include radiolarian chert, lie on the Coast Range ophiolite composed of volcanic rocks, diabase, gabbro, and ultramafic rocks. The fundamental fault contact with the underlying Franciscan rocks is the Coast Range thrust (Bailey and others, 1970). In many places the Coast Range thrust has been strongly deformed by later thrusting, folding, and high-angle faulting, some of which may be related to the San Andreas system (McLaughlin and Pessagno,

1978; McLaughlin and Stanley, 1976) or to Cenozoic uplift of the Coast Ranges (Raymond, 1973b).

Clastic sediments of the Great Valley Sequence were derived from the granitic basement of the Sierran and Klamath regions to the east and north. Compositional changes in the sedimentary rocks through time can be correlated with progressively deeper levels of erosion in the source terrane. The K-feldspar content ranges from zero to very small amounts in the basal part of the sequence to much larger amounts in the middle and upper parts of the sequence (Bailey and Irwin, 1959; Bailey and others, 1964). Recent studies of sandstone composition have documented these changes (Ojakangas, 1968; Gilbert and Dickinson, 1970; Mansfield, 1972; Dickinson and Rich, 1972), and five petrologic units have been recognized based on the amounts and ratios of quartz, feldspar, unstable fragments, and mica (Dickinson and Rich, 1972). The lowest unit, Stony Creek, is low in quartz and mica and has high ratios of plagioclase to total feldspar and volcanic to total lithic grains. The uppermost unit, Rumsey, is quartz-rich with much mica and a low ratio of plagioclase to total feldspar. The intervening intervals have various proportions of these components. As earlier workers have indicated (Dickinson and Rich, 1972, p. 3020), this change from volcanic-rich detritus near the base to granite-derived detritus higher in the sequence records the progressive stripping and erosion of the Sierran-Klamath volcanoplutonic terranes. Not all the volcanic detritus was derived from these sources, however, because locally the basal beds of the Great Valley Sequence consist of reworked fragments from the underlying ophiolite (Blake and Jones, 1974).

Much of the Great Valley Sequence was deposited by turbidity currents in deep water, except for marginal deposits on the north and east that accumulated in shallow water. Paleocurrent studies along the west side of the Great Valley indicate that transport of sediments was primarily toward the south (Ojakangas, 1968; Mansfield, 1972), with minor transport to the west (Colburn, 1970).

Reconstruction of the original geometry of the Great Valley Sequence depositional basin is difficult, particularly for Upper Jurassic and Lower Cretaceous strata, because these rocks are exposed on the west side of the Great Valley on the steeply dipping western limb of a large synclinal fold. On the east side of the Great Valley, only Upper Cretaceous rocks are exposed on the gently dipping eastern limb.

Clues to the lateral facies relations across the basin can be found at the north end of the Sacramento Valley, where dissimilar stratigraphic sequences have been juxtaposed across three major left-lateral fault zones (Jones and others, 1969; Jones and Irwin, 1971; Bailey and Jones, 1973a; Jones and Bailey, 1973; Jones, 1973). Cumulative displacement on these faults of many tens of kilometers have juxtaposed the Klamath Mountains continental crust on the north against oceanic crust at the base of the Great Valley Sequence to the south. These left-lateral faults do not displace Franciscan rocks to the west and are interpreted as tear faults related to the Coast Range thrust (Bailey and others, 1970). Displacement across these three left-lateral faults of at least 90 to 100 km is indicated by an offset Early Cretaceous shoreline (Jones and Irwin, 1971). The present position of these displaced blocks is shown on Fig. 12-3, which is a highly generalized geologic map of the northwestern part of the Sacremento Valley. Because each block has been displaced westward relative to its southern neighbors, a palinspastic

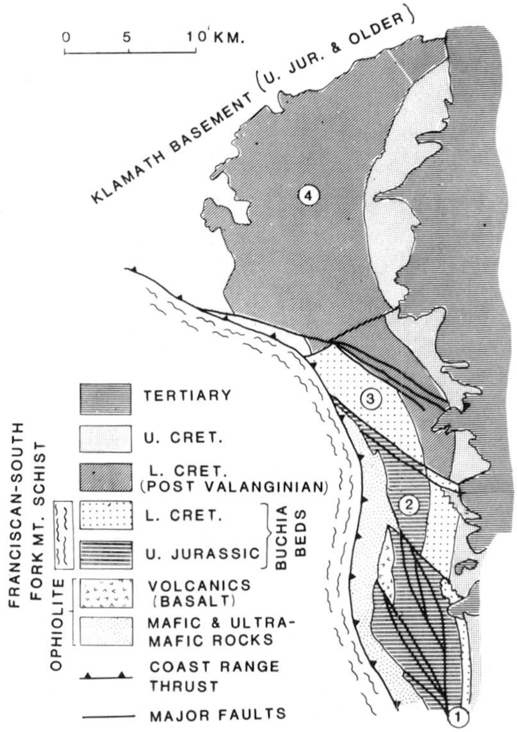

Fig. 12-3. Generalized geologic map of the northwestern part of the Sacramento Valley, California.

reconstruction allows a tentative reconstruction of facies and thickness trends in an east-west direction.

Deposits on the north side of these faults are systematically thicker and contain more sandstone and conglomerate than coeval deposits on the south side. These differences are best seen in the part of the sequence that ranges in age from earliest Cretaceous (Berriasian) to the middle part of the Early Cretaceous (Hauterivian) (Fig. 12-4). In the southern fault block these strata comprise 460 m of dominantly mudstone with minor channels filled with conglomerate and coarse sand. In the central block, more than 3000 m of these beds in present, but the proportion of sandstone is higher. In the northern block, over 4600 m of mudstone, fine- to coarse-grained sandstone, and grit is present. Associated with these rocks is abundant carbonaceous debris, including plant scraps, leaves, stems, twigs, and rare logs. Molluscan megafossils are rare, and most specimens are displaced from their original dwelling site.

The net result of left-lateral movements along these faults was to juxtapose continental crust of the Klamath Mountains on the north against oceanic crust at the base of the Great Valley Sequence on the south. The sequence of events is not readily apparent, however, mainly because some critical relations are deeply buried beneath thick Cenozoic deposits. In 1977 we suggested a complex history of subduction and rifting,

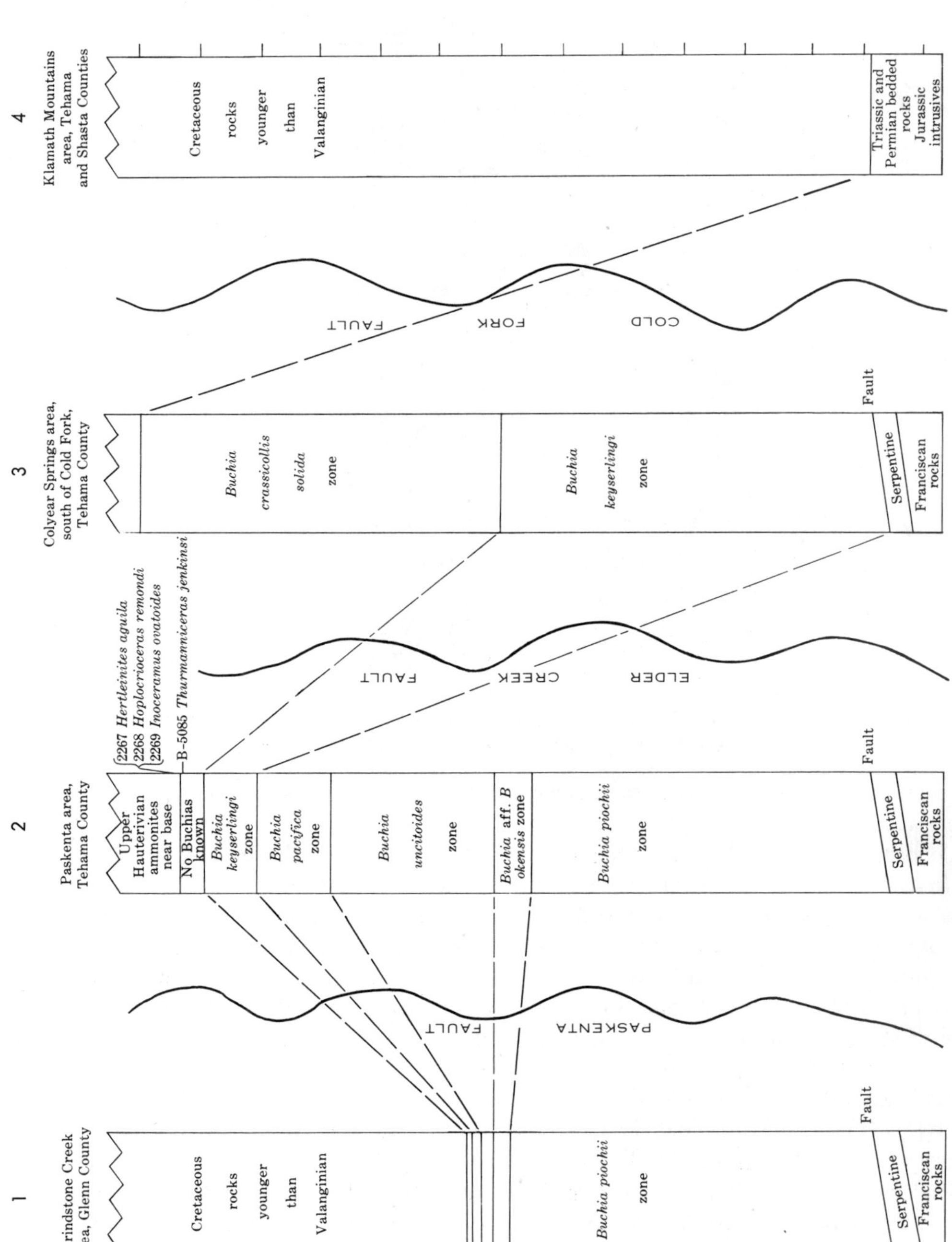

Fig. 12–4. Relationships of Buchia zones across three major fault zones.

beginning with the formation during the Cretaceous of a trench-rift-transform triple junction at the position of the present-day northern end of the Sacramento Valley (Fig. 12-5). Under this scheme, the continental margin broke along a northeast-trending rift; the Klamath plate moved to the northwest, initiating strike-slip faulting along its southwest margin and thrust faulting along its northwestern margin. Distributed shear along the southern margin produced the left-lateral faults along which unlike parts of the Great Valley Sequence were juxtaposed. These faults merge with the Coast Range thrust, and do not affect structurally lower Franciscan rocks. These relations suggest that the rocks being subducted were decoupled from the rocks undergoing rifting (Blake and Jones, 1977).

The Coast Range Thrust

The Coast Range thrust is the fundamental boundary between the deformed and metamorphosed Franciscan assemblage and the overlying Coast Range ophiolite and Great Valley Sequence. In most places along the west side of the Sacramento Valley, the fault trends north-south and is nearly vertical.

Near Lake Berryessa (Fig. 12-2), the Coast Range thrust and overlying ophiolite and Great Valley Sequence have been folded around a tight southeast-plunging antiform. The ophiolite can be traced, albeit discontinuously, for about 70 km to the northwest, to the vicinity of The Geysers (McLaughlin and Pessagno, 1978; McLaughlin, 1977, p. 13, 14). Other outliers of ophiolite and Great Valley Sequence lie even farther to the west (Bailey and others, 1970), and these have been cited as evidence for great horizontal displacement (about 100 km) on the Coast Range thrust.

It has recently been argued (Suppe, 1977; Suppe and Foland, 1978) that the Coast Range thrust is a series of imbricate thrust faults that involve rocks as young as Eocene. Suppe's reasoning is based upon detailed mapping west of Stonyford and Wilbur Springs (Fig. 12-2) and by a complicated correlation with outliers of Great Valley Sequence to the west. In contrast, detailed mapping in the Wilbur Springs quadrangle (Lawton, 1956; E. I. Rich, 1971) indicates that the Franciscan assemblage, the Coast Range thrust, and the lower part of the Great Valley Sequence, including the basal ophiolite, are all folded around the southeast-plunging Wilbur Springs antiform. Upper Cretaceous rocks are not folded around this structure but instead appear to lie unconformably on the earlier deformed strata. This relation suggests that the Franciscan and lower part of the Great Valley Sequence were already juxtaposed by Late Cretaceous time. Further movements, some during the Cenozoic, probably led to the additional complexities elucidated by Suppe (1978).

OBJECTIONS TO EXISTING MODELS

Plate tectonics and the corollary subduction model is a convenient vehicle to explain Franciscan-Great Valley-Sierran relations (Bailey and Blake, 1969; Ernst, 1970; Page, 1976; and many others). Certainly the widespread pervasive shearing seen in melanges

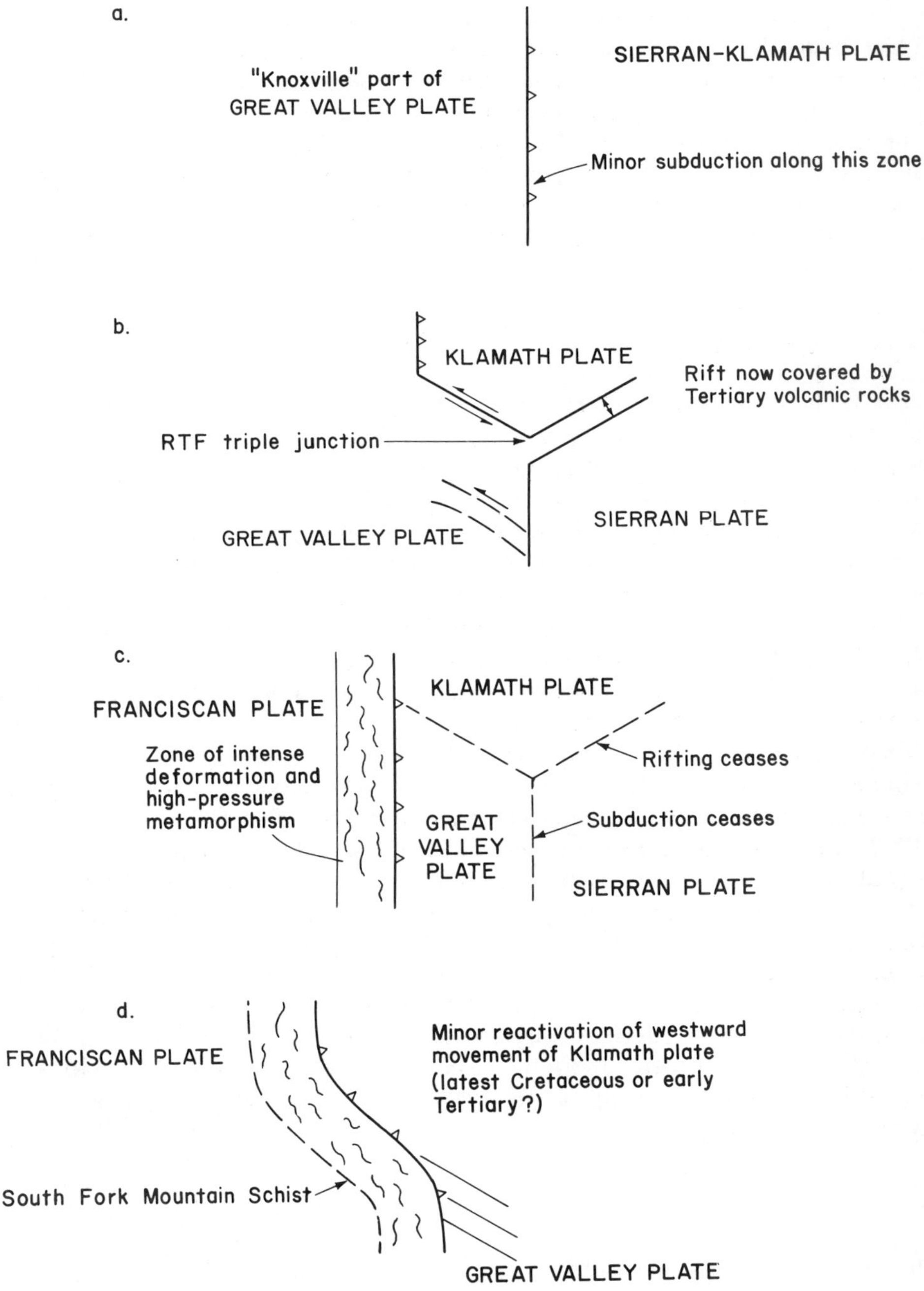

Fig. 12-5. Speculative plate-tectonic history of northern California.

and the high pressures and low temperatures inferred from the metamorphic mineral assemblages are best explained by subduction. Nonetheless, the simple concept of an oceanic plate being subducted along a continental margin with contemporaneous deposition in the trench and arc-trench-gap encounters difficulty when detailed correlations are made between the different parts of this system. Sufficient data to permit such correlations are just now becoming available, and they point to the following anomalous features.

Western Arc Terranes

In 1974, we suggested that an island arc or continental fragment lay between the Franciscan trench and the Great Valley Sequence and that Franciscan graywacke was largely derived from this western source. No existing parts of this hypothetical arc were thought to have survived the combined onslaught of deep erosion and subduction. The presence of Late Jurassic arc-related volcanism west of the Klamath-Sierran arc has now been substantiated in two places, as described later, but no evidence has been found to indicate that the arc terranes provided detritus to the Franciscan.

Within and south of the San Francisco Bay area the Coast Range ophiolite is characterized by very abundant flows, tuffs, and intrusive bodies of keratophyre and quartz keratophyre (Bailey and others, 1970; Evarts, 1977, 1978). Such rocks are absent in the Coast Range ophiolite along the west side of the Sacramento Valley, where the volcanic rocks are mainly high-Ti basalt. Major-element analyses for ten of the keratophyric rocks were presented by Bailey and Blake (1974) and are reproduced here in Table 12-1. As can be seen from these analyses and a normative plot (Fig. 12-6), the rocks appear to be albitized andesite, dacite, and rhyolite, and this conclusion is supported by the occurrence of remnants of zoned plagioclase and rare sanidine.

Preliminary results for trace elements and rare earths in the same rocks (Table 12-2) were presented by Blake and others (1975), but at the time we were not aware of the profound differences that characterize the upper part of the ophiolite to the north and to the south of San Francisco Bay.

The rare-earth patterns (Fig. 12-7), while not indicative of a specific environment, are all very flat and suggest little if any differentiation. These data plus the geologic evidence suggest a two-stage process that involved (1) formation of typical high-Cr basalt at a spreading center such as midocean ridge or marginal basin, and (2) development of an island arc upon this crust by a process of partial melting. The alternative process of differentiation of oceanic basalt to form quartz keratophyre is not supported either by the rare-earth patterns or by qualitative volume estimates, which suggest a quantity of silicic rocks in excess of that expected from the quantity of basalt present.

Additional evidence for non-Sierran andesitic volcanism is seen in the northern part of the Coastal belt near Cape Mendocino (Fig. 12-2), where Beutner (1977) has postulated the presence of an andesitic arc based on sedimentological characteristics of Coastal-belt graywacke.

Support for this postulated arc is provided by Leg 18 of the Deep Sea Drilling

TABLE 12-1. Chemical Analyses of Keratophyre and Quartz Keratophyre from Coast Range Ophiolite

	1	*2*	*3*	*4*	*5*	*6*	*7*	*8*	*9*	*10*
SiO_2	57.5	58.2	65.4	68.04	70.0	72.5	72.6	72.9	74.1	77.08
Al_2O_3	14.5	16.4	14.3	12.09	13.6	14.0	14.4	12.5	12.0	12.43
Fe_2O_3	1.7	3.7	2.9	3.81	2.6	0.77	1.5	2.0	1.8	1.48
FeO	5.3	4.8	3.6	3.21	2.4	3.3	0.60	2.6	1.4	0.55
MgO	3.3	3.2	1.2	1.97	0.73	1.0	0.40	0.35	1.61	0.23
CaO	5.5	6.8	3.3	3.41	2.7	2.5	1.2	2.4	1.5	0.88
Na_2O	5.2	3.3	5.4	5.04	5.2	3.7	5.2	4.7	3.7	6.13
K_2O	0.20	0.54	0.52	–	0.35	0.33	1.1	0.20	1.3	0.15
H_2O^+	2.6	1.1	2.1	1.89	0.93	1.5	1.4	0.86	1.7	0.92
H_2O^-	0.28	0.24	0.39	0.54	0.37	0.08	0.46	0.24	0.45	0.31
TiO_2	0.61	0.77	0.65	0.46	0.62	0.21	0.20	0.30	0.37	0.22
P_2O_5	0.06	0.12	0.12	0.05	0.18	0.06	0.03	0.08	0.08	0.02
MnO	0.08	0.16	0.07	0.10	0.05	0.06	–	0.04	0.05	0.07
CO_2	3.5	–	–	–	–	–	0.02	0.05	–	–
Sum	100.3	99.3	99.9	100.61	99.7	100.0	99.1	99.2	100.0	100.47
Density	2.70				2.70		2.60	2.66		

1. Keratophyre (71-EB-105, Point Sal, Santa Barbara County, Calif. Analysis by L. Artis.
2. Hornblende keratophyre (MB-1B), Del Puerto, Stanislaus County, Calif. Analysis by G. Chloe, P. Elmore, J. Glenn, J. Kelsey, and H. Smith.
3. Quartz keratophyre (B69-8), 1 mile NE of Bradford Mountain, Healdsburg quadrangle, Sonoma County, Calif. Analysis by L. Artis, G. Chloe, P. Elmore, J. Glenn, J. Kelsey, and H. Smith.
4. Quartz keratophyre (DP-2), Del Puerto, Stanislaus County, Calif. (Bailey and others, 1964, p. 55 from Maddock, 1964).
5. Quartz keratophyre (B69-7), 1½ miles ESE of Bradford Mountain, Healdsburg quadrangle, Sonoma County, Calif. Analysis by L. Artis, G. Chloe, P. Elmore, J. Glenn, J. Kelsey, and H. Smith.
6. Quartz keratophyre (CP-6), Quinto Creek, Stanislaus County, Calif. Analysis by G. Chloe, P. Elmore, J. Glenn, J. Kelsey, and H. Smith.
7. Quartz keratophyre (70-B-4), Quinto Creek, Stanislaus County, Calif. Analysis by L. Artis.
8. Quartz keratophyre (B69-16), Fall Creek, Healdsburg quadrangle, Sonoma County, Calif. Analysis by L. Artis, G. Chloe, P. Elmore, J. Glenn, J. Kelsey, and H. Smith.
9. Quartz keratophyre (CP-5), Quinto Creek, Stanislaus County, Calif. Analysis by G. Chloe, P. Elmore, J. Glenn, J. Kelsey, and H. Smith.
10. Quartz keratophyre (DP-1), Del Puerto, Stanislaus County, Calif. (Bailey and others, 1964, p. 55 from Maddock, 1964).

Project, which cored andesite beneath 320 m of Oligocene and younger sedimentary rocks at Site 173, located about 100 km southwest of Cape Mendocino (Kulm and others, 1973). According to MacLeod and Pratt (1973), the volcanic rock from Site 173 is very similar in chemical composition to average andesite and is also similar to basaltic andesite from the Cascade Range of Oregon (see Table 12-3). In trace-element chemistry, the Site 173 rock is also similar to average andesite, although it contains more Ba, Cr, and Y and less Cu and Zr. These data imply the presence of a subduction-related Early Tertiary arc lying within the supposed Franciscan "trench."

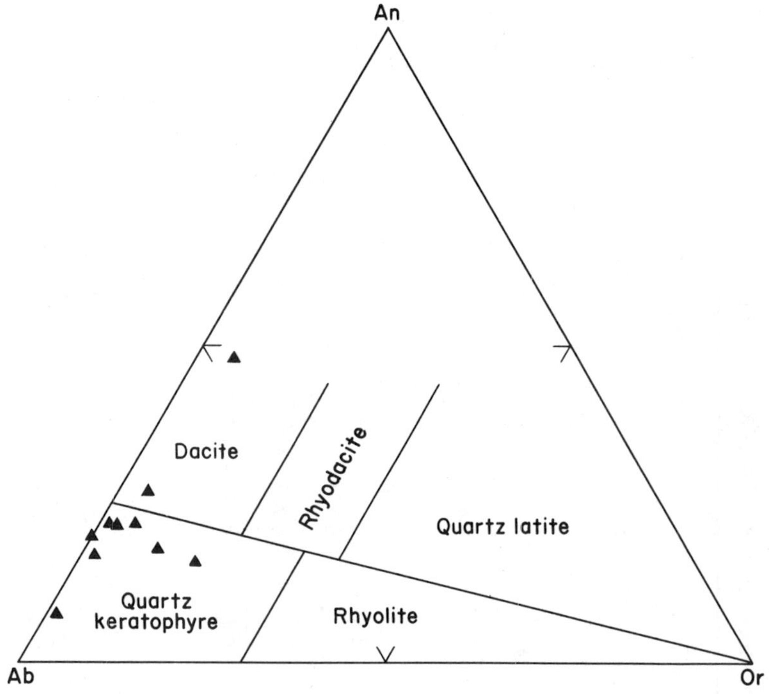

Fig. 12-6. Normative feldspar ratio plot of keratophyres and related rocks from the Coast Range ophiolite (after O'Connor, 1965).

The northernmost occurrence of arc-related volcanic rocks is in southwest Oregon, where the extensive Upper Jurassic Otter Point Formation (Koch, 1966); Dott, 1971; Coleman, 1972) includes andesitic sandstone with intercalated basalt and andesite flows. Similar rocks may occur at Trinidad Head in northern California (R. H. Dott, oral communication, 1978), where keratophyre is abundant and well-bedded tuffaceous sandstone and siltstone are intercalated with thin flows of dacite and andesite. Several recent studies have documented the volcanic nature of the Otter Point Formation and have established that it appears to structurally underlie adjacent coeval Franciscan (Dothan) rocks of differing composition and sedimentological character (Walker, 1976; M. C. Blake, Jr., Francois Roure, and Henry Ohlin, unpublished data). The Otter Point Formation thus appears to record a period of Late Jurassic arc volcanism that originally lay well to the west of the continental margin.

Comparison of Graywackes from the Franciscan and Great Valley Sequence

In 1974, we pointed out that differences in grain size and composition between the Franciscan graywacke of Tithonian to Valanginian age in the Yolla Bolly belt and coeval

TABLE 12-2. Selected Trace Element[a] and Rare-Earth[b] Data of Volcanic Rocks from the Coast Range Ophiolite

	Keratophyre and Quartz Keratophyre								Basalt			
	71-EB-105	*MB-1B*	*B69-8*	*B69-7*	*CP-6*	*70B-4*	*B69-6*	*CP-5*	*B69-4*	*CP-1B*	*B69-14*	*69B-15*
Cr	23	11	2	2	6	2	2	4	430	330	35	88
Ni	16	6	2	2	2	3	2	2	200	76	22	35
Pb	15	15	15	15	31	15	15	63	24	38	15	15
Zr	66	86	100	140	41	89	92	78	39	52	49	54
La	4.81	2.78	2.78	4.81	3.12	6.17	3.61	4.31	2.09	2.33	1.67	1.47
Ce	6.07	8.68	8.10	12.7	5.75	11.6	8.65	10.8	4.81	5.36	4.03	4.38
Md	4.8	7.98	7.18	12.4	4.74	9.87	8.02	–	7.49	–	3.75	–
Sm	1.79	2.28	–	–	1.05	2.81	–	1.84	1.99	–	–	1.03
Ev	0.593	0.664	0.74	1.17	0.399	0.627	1.0	0.710	0.613	0.686	0.510	0.368
Gd	–	3.39	2.25	4.18	1.2	2.23	2.84	2.61	1.99	2.49	1.9	1.28
Tb	–	–	–	0.871	–	–	–	0.564	–	–	–	–
Dy	–	–	2.88	–	1.35	–	–	2.51	2.29	–	–	–
Tm	0.269	0.385	0.317	–	0.262	–	–	0.278	–	0.321	0.279	–
Yb	1.6	2.39	1.87	2.82	1.47	2	2.34	1.96	1.18	1.49	1.26	1.07
Lu	0.297	0.454	0.347	0.526	0.287	0.363	0.444	0.369	0.233	0.285	0.235	0.214

[a] Quantitative spectrographic; C. Heropoulos, analyst. All values in ppr.
[b] INAA; H. T. Millard and R. J. Knight, analysts.

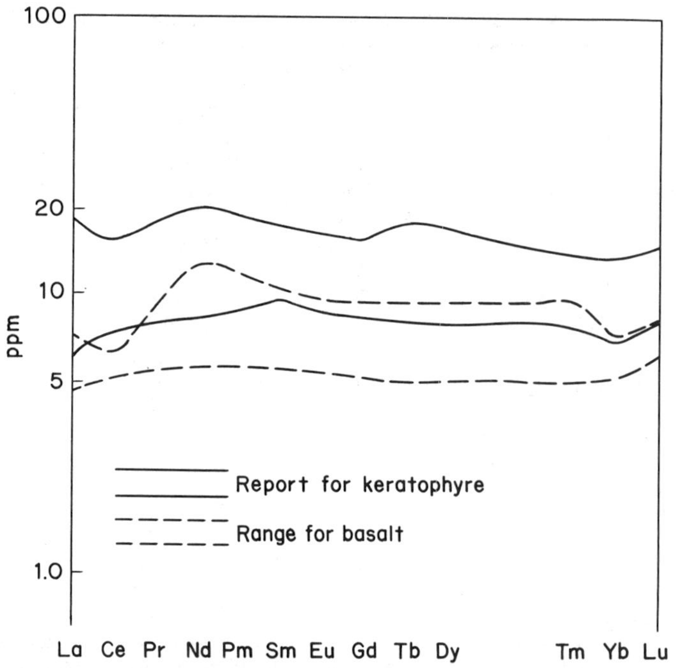

Fig. 12-7. Chondrite-normalized rare-earth patterns for basalt and keratophyres from the southern and western portions of the Coast Range ophiolite.

Great Valley Sequence sandstone make it unlikely that the two suites were derived from the same Sierran-Klamath source terrane. Specifically, Franciscan sandstone is quartzofeldspathic, whereas Great Valley sandstone is volcanogenic. We also presented a palinspastic cross section of the lower part of the Great Valley Sequence that shows a pronounced thinning and shaling out to the west, which would appear to preclude deposition of a second set of proximal, coarse-grained deep-sea fan deposits outboard of the fine-grained distal parts of the Great Valley Sequence fans.

Recent studies by Blake and Wright (1976), Aalto (1976), and Jacobson (1978) document that Franciscan graywacke is typically arkosic to subarkosic in composition, whereas coeval Great Valley Sequence sandstone ranges from volcanic-lithic to subarkosic. While it might be argued that these differences are the result of postdepositional diagenesis or low-grade metamorphism, we see no evidence for this. On the contrary, most Franciscan graywacke contains unaltered volcanic detritus, and only in the completely recrystallized semischist or schist are the framework grains reconstituted.

We have determined modal compositions, shown in Fig. 12-8, for twenty-eight samples of Franciscan sandstone from the north San Francisco Bay region. Most of these samples are from well-dated fossil localities, the youngest being of Campanian age, the oldest of Tithonian age. Also included on Fig. 12-8 are the petrofacies fields for the Great Valley Sequence as presented by Dickinson and Rich (1972) and data for Franciscan

TABLE 12-3. Chemical Analysis of Volcanic Rock from Site 173 (Core 38)

	Chemical Analysis				
	Site 173			*Miocene Basaltic Andesite, Sardine Formation, Cascade Range, Ore., Water Free*[c]	*Basaltic Andesite, Crater Lake, Ore., Water Free*[d]
	Original Analysis[a]	*Water-Free Analysis*	*Average Andesite, Water Free*[b]		
SiO_2	51.3	55.1	54.4	55.1	56.7
Al_2O_3	16.6	17.8	17.8	17.0	18.3
Fe_2O_3	5.0	8.3	9.0	10.4	6.8
FeO	2.7				
MgO	5.0	5.4	4.4	3.5	5.2
CaO	6.6	7.1	7.9	6.8	7.5
Na_2O	3.3	3.5	3.7	4.1	3.7
K_2O	1.1	1.2	1.1	1.0	1.2
H_2O^+	2.1	–	–	–	–
H_2O^-	4.2	–	–	–	–
TiO_2	1.2	1.3	1.3	1.6	0.85
P_2O_5	0.26	0.28	0.28	0.18	0.11
MnO	0.06	0.06	0.15	0.22	0.08
CO	0.05	0.05	–	0.18	–

[a]Analytical methods used are those described in USGS Bulletin 1144A supplemented by atomic absorption.
[b]From Nockolds, 1954.
[c]From Peck and others, 1964, Table 7, col. 8.
[d]From Williams, 1942, anal. 10.
Chemical analysis of volcanic rock from DSDP Site 173 (From MacLeod and Pratt, 1973, p. 938).

graywackes of Tithonian age in the Diablo Range taken from Jacobson (1978). These latter rocks, called the Eylar Mountain sequence by Crawford (1975), are particularly important, as they provide the strongest known contrast with coeval Tithonian rocks of the Great Valley Sequence.

The Diablo Range Tithonian sequence (Eylar Mountain sequence) consists of massive quartzofeldspathic graywacke, minor thin-bedded sandstone and shale, and interbedded radiolarian chert, all metamorphosed to blueschist facies (both jadeite and lawsonite are widespread); this sequence is interleaved with weakly metamorphosed melange. These rocks occur on the east and west sides of the Diablo Range and are underlain by the Garzas melange of Cowan (1974) and the Burnt Hills sequence of Crawford (1975).

Datable fossils are rare in these rocks, but a specimen of *Buchia* cf. *B. fisheriana* of late Tithonian age was found by Crawford in the northwestern part of the Diablo Range (USGS Mesozoic loc. M5794), and Late Jurassic radiolarians are known from a few chert outcrops on the east side of the range. The best material was collected by L. M. Echeverria from the Orestimba quadrangle, where radiolarians probably belonging to Zone 2B were recovered from chert associated with graywacke (L. M. Echeverria,

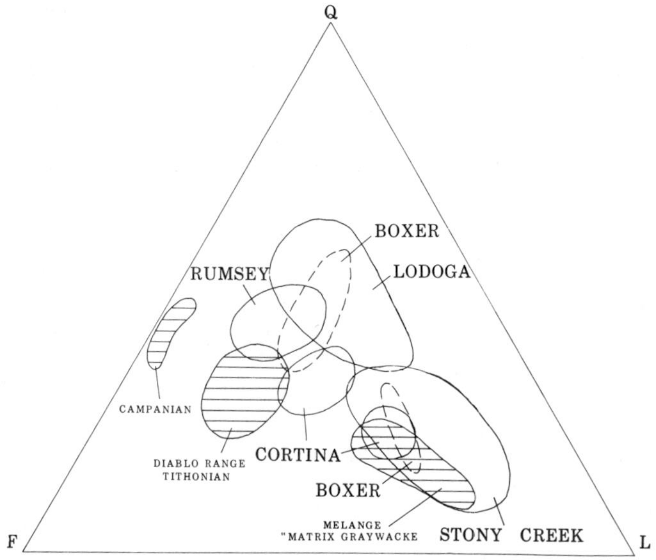

Fig. 12-8. QFL diagram showing composition of graywackes from the Great Valley Sequence (the Stony Creek, Boxer, Cortina, Lodoga, and Rumsey petrofacies of Dickinson and Rich, 1972) and Franciscan assemblage (lined) in the San Francisco Bay area. Subfield labeled "Diablo Range Tithonian" is the Eylar Mountain petrofacies of Jacobson (1978).

D. L. Jones, and E. A. Pessagno, Jr., unpublished data). Elsewhere, silicic tuffs of this age form the upper part of the Coast Range ophiolite (Pessagno, 1977a).

The quartzofeldspathic nature of graywackes (mean composition of approximately $Q_{30}F_{50}L_{20}$) in the Diablo Range Tithonian sequence shown on Fig. 12-6 contrasts markedly with the nearly coeval Stony Creek petrofacies (which range from $Q_{12.5}F_{17.5}L_{70}$ to $Q_{30}F_{27.5}L_{42.5}$). Furthermore, basal clastic rocks of the Great Valley Sequence are dominantly of radiolarian Zone 4 age (Pessagno, 1977a), and thus are younger than the Zone 2B chert-graywacke part of the Eylar Mountain sequence. The only clastic strata of the Great Valley Sequence known to be as old as Zone 2 occur at Paskenta (Jones, 1975; Pessagno, 1977a) in northern California, and these are composed dominantly of mafic and ultramafic detritus derived from the Coast Range ophiolite. It thus appears that Franciscan quartzofeldspathic sandstone was being deposited at the same time that fine-grained siliceous volcanogenic rocks were accumulating on the upper part of the Coast Range ophiolite and before deposition of most of the volcanogenic detritus of the lower part of the Great Valley Sequence. These data clearly imply separate sources for the two clastic sequences and considerable geographic separation for their sites of deposition. In contrast, the melange "matrix graywacke" of the north San Francisco Bay area is quite similar to coeval sandstone of the Stony Creek petrofacies. This similarity was noted earlier (Blake and Jones, 1974) and attributed to tectonosedimentary reworking of the lower Great Valley Sequence and basal ophiolite into the melanges.

Average chemical compositions for 138 samples of Franciscan graywackes from the San Francisco Bay region, including all the samples that were analyzed modally, are given in Table 12-4. The breakdown into the different Franciscan units follows mapping designations of Blake and others (1971; 1974). For comparison, the composition of three sandstones from the basal Great Valley Sequence (Knoxville Formation) is given in Table 12-5. While three samples are obviously not sufficient to characterize

TABLE 12-4. Average Chemical Composition of 138 Franciscan Graywackes and Metagraywackes from the San Francisco Bay Area

	Meta (23)	*Melange* (66)	*Sandstone* (49)	*Average* (138)
SiO_2	72.56	71.29	72.66	71.99
Al_2O_3	13.06	14.11	14.04	13.91
FeO*	4.38	4.57	3.31	4.09
MgO	3.03	2.50	1.78	2.33
CaO	1.79	1.82	2.03	1.89
Na_2O	2.78	3.28	3.59	3.31
K_2O	1.63	1.62	1.93	1.73
TiO_2	0.58	0.62	0.48	0.56
P_2O_5	0.12	0.13	0.13	0.13
MnO	0.07	0.06	0.05	0.06
	100.00	100.00	100.00	100.00

*Total iron as FeO.

TABLE 12-5. Average Chemical Composition of Sandstone from the Lower Part of the Great Valley Sequence, Sacramento Valley, Compared with Samples from Western Outliers

	Avg. Lower Great Valley Sequence (3)	*Avg. Western Outlier* (2)
SiO_2	61.15	73.55
Al_2O_3	15.34	12.93
FeO*	8.00	4.24
MgO	3.83	2.75
CaO	6.61	1.46
Na_2O	2.99	2.49
K_2O	0.74	1.64
TiO_2	0.91	0.66
P_2O_5	0.19	0.16
MnO	0.24	0.12
	100.00	100.00

*Total iron as FeO.

an entire range in composition, work in progress (M. C. Blake and J. Barker, unpublished data) indicates that the three rocks analyzed are typical of the mafic character of a large percentage of the Tithonian to Valanginian rocks in the Great Valley Sequence. Additional differences between the Franciscan and Great Valley rocks are seen in the strontium ^{87}Sr-^{86}Sr) isotope ratios given by Peterman and others (1967). All the analyzed Franciscan graywackes have higher ratios than any of the analyzed Great Valley Sequence sandstones. More data are needed from coeval samples from the two terranes before the significance of these differences can be fully evaluated, but it seems possible that the Franciscan was derived from an older, more radiogenic terrane. However, none of the Franciscan ratios is as high as the mean value obtained from shales of the Middle Jurassic Bedford Canyon Formation by Criscione and others (1978), which had, in part, a Precambrian source terrane (Criscione and others, 1978, p. 393).

Western Outliers of the Great Valley Sequence

Several outliers of the Great Valley Sequence and underlying ophiolite occur north of San Francisco in the vicinity of Healdsburg (Fig. 12-2). The sedimentary rocks in these western outliers differ in composition from those present along the west side. Specifically, sandstone in the lower part of the western outliers contains abundant potassium feldspar, whereas coevel sandstones to the east contain small quantities of K-feldspar.

Chemical analyses of two Lower Cretaceous (Valanginian) sandstones from two outliers North of San Francisco are given in Table 12-5. Note the profound differences in composition between these sandstones and those from the lower part of the Great Valley Sequence to the east. The potassium-feldspar-bearing sandstone is overlain by 2000 to 3000 m of rhyolite-rich cobble to boulder conglomerate that was deposited in a nearshore to inner-fan environment (Blake and Jones, 1974). These rocks are absent in the outliers lying 20 to 30 km farther to the east, which contain thinner units of conglomerate characterized by chert pebbles. It seems unlikely that direct continuity could have existed between these different sequences in their present geographic setting. Because a suitable nearby source terrane is lacking, we suggest that the western outliers are allochthonous and were transported an unknown distance by strike-slip faulting. Perhaps these western outliers were never directly connected with the rocks cropping out along the west side of the Great Valley, and it may be incorrect to connect the Coast Range thrust directly from the eastern to the western outliers.

SUMMARY AND CONCLUSIONS

The widely accepted hypothesis that the Franciscan, Great Valley Sequence, and Sierran and Klamath regions represent a classic subduction-generated triad composed of trench, arc-trench gap, and magmatic arc is probably correct as a basic generalization. However, in its simplest (and most popular) form involving eastward-directed subduction of an oceanic plate, this model fails to explain some local features of northern California geology that indicate a more complicated paleotectonic regime existed. We see evidence for several arcs that suggest a multiplicity of subduction zones; and we see contrasting facies within the Great Valley Sequence and between the Great Valley Sequence and the Franciscan that seemingly are best explained by juxtaposition of rocks derived from different sources and originally deposited far apart.

It seems premature to suggest a more suitable model at this time of rapid data acquisition, although it now is evident that the events that affected the Franciscan and Great Valley Sequence were not isolated but were merely parts of the complex Mesozoic plate interactions that dominated the entire Pacific margin of North America (Jones and others, 1978a). The kinematics of these interactions remain obscure, but paleomagnetic data clearly point to the importance of northward translations and clockwise rotations in relative plate motions between the Pacific realm and North America (e.g., Hillhouse, 1977; Beck, 1976; Packer and Stone, 1974; Stone and Packer, 1977), which resulted in tectonic fragmentation and juxtaposition of unlike terranes (e.g., Jones and others, 1977, 1978b; Saleeby and others, 1978; Davis and others, 1978). Paleomagnetic data suggesting similar large-scale northward translation have also been obtained for Middle Cretaceous Franciscan limestone near Laytonville (Alvarez and others, 1979).

The solution of Franciscan problems now seems to lie in the complex interplay between transform faulting, rifting, and oblique subduction, the resultant northward displacement and juxtaposition of both oceanic and continentally derived sedimentary sequences, and minor changes in plate motion that produced pulses of arc-related magmatism at different times within the oceanic realm as well as on the continental margin. The result is the incredibly complex series of stacked nappes and melanges that form the backbone of the northern California Coast Ranges. How these various elements were assembled, where they came from, and when and how they moved are enigmas that await research attention.

Benjamin M. Page
Department of Geology
Stanford University
Stanford, California 94305

13

THE SOUTHERN COAST RANGES

ABSTRACT

The Southern Coast Ranges include a subduction zone complex (the Franciscan), forearc basin sediments (the Great Valley Sequence), and a magmatic arc (plutonic and metamorphic rocks of the Salinian Block). These assemblages all contain quasi-contemporary Late Mesozoic rocks. The Salinian Block magmatic arc has been displaced hundreds of kilometers from its original position, and is now flanked tectonically on both sides by the Franciscan Complex. The NE boundary of the Block is the San Andreas fault, and the SW boundary is the Sur-Nacimiento fault zone.

The Franciscan Complex consists of melanges and large coherent rock units. Both oceanic and terrigenous materials are represented. Coherent units include bedded Cretaceous sandstone, chert-graywacke sequences, and Upper Jurassic chert-greenstone units. Melanges include blocks of similar materials, plus serpentinite, blueschist, conglomerate, and other rocks, all of which are enveloped in a pervasively sheared argillaceous matrix. Most of the clastic sedimentary rocks appear to have been derived from a Sierra Nevada-type active continental margin. Many of the coherent units and blocks in melanges have undergone blueschist-facies metamorphism resulting from high P/T conditions ascribed to subduction. Although these rocks evidently reached depths of 15 to 30 km, they avoided overheating and somehow returned to the surface. The rise to the surface was probably accomplished in part by the wedging action of tapered slices of "off-scrapings" and deformed slope deposits, but was more largely effected by subduction-driven viscous upward flow of subducted material (Cowan and Silling, 1978), coupled with buoyant rise influenced by a dense "hanging wall." Competent bodies in the subduction complex were dismembered, and the fragments were dispersed and mixed, possibly by the action of olistostromes on the inner trench slope plus recurrent partial subduction. Probably very large strike-slip movements affected the Franciscan during its evolution, but the details and timing are not understood, and the structural features that may have resulted from such events have not yet been recognized in the Southern Coast Ranges.

The Great Valley Sequence and its equivalents mainly consist of stratified terrigenous clastic sediment derived from the Klamath-Sierra Nevada terrane and its former southward continuation. Much of the sediment accumulated in deep-sea fans. The oldest parts of the GVS and its allochthonous counterparts are uppermost Jurassic and Lower Cretaceous and rest on an ophiolite basement beneath which the Franciscan has been thrust. The Franciscan was at first subducted beneath the GVS ophiolite, but probably in the Paleocene it was pushed farther under, en masse, creating the Coast Range thrust.

The out-of-place Salinian Block basement includes metasedimentary rocks that have not been definitely correlated with sequences elsewhere. Wherever these rocks originated, in the late Mesozoic they were situated between the sites of the Sierra Nevada and Peninsular-Baja California ranges, and they were invaded by granitic plutons forming a southward continuation of the axial belt of Cretaceous potassic Sierran intrusives. In the Paleocene (?) the Sierran-Peninsular plutonic belt was intersected obliquely by the ancestral San Andreas fault (essentially on the site of the present SAF) along which the Salinian Block moved 200 km northwestward, probably motivated by oblique plate convergence. During a long hiatus in this motion, Eocene marine sediments were spread across the SAF, and large movements were not resumed until the mid-Miocene (ca. 15 m.y.b.p.). Subduction had ceased at the latitude of the Southern Coast Ranges and was succeeded by a Neogene transform regime. The transport of the Salinian Block and motion along the SAF accelerated about 4.5 m.y.b.p. at the inception of spreading in the Gulf of California. Probably the northernmost part of the Block has moved 80 to 115 km farther than the main part, for a total of about 600 km. The extra motion was accomplished by slip on the San Gregorio-Hozgri fault, which transects the Block at an acute angle.

A large terrane that existed along the SW side of the Salinian Block has been lost, either by mega-strike-slip or by piecemeal subduction, probably in the Paleocene. This impressive but cryptic incident poses one of the major tectonic problems of the Southern Coast Ranges.

Neogene transform tectonics created en echelon compressional basins of deposition, en echelon folds, NW-trending strike-slip faults, and lesser E-W trending thrust faults. Some of these structures are probably still evolving, but this is only well documented for strike-slip faults. Average right-hand relative motion between the Pacific and North American plates appears to be 5.6 cm/yr in a N35W direction in the region of the Southern Coast Ranges (Minster and Jordan, 1978). Of the average annual motion, the SAF and its immediate neighbors may accommodate 3.2 to 3.7 cm/yr (Savage and Burford, 1973), the balance being distributed in a broad zone of deformation and slip.

An important enigma is posed by the Plio-Pleistocene formation and rise of individual ranges and the subsidence of structural valleys. These events took place within the Neogene transform regime, but cannot be readily explained by the familiar kinematics of transform tectonics.

INTRODUCTION

The mild exterior of the Southern Coast Ranges belies a half-hidden record of plate interactions of tremendous overall magnitude. The inferred general sequence and nature of the interactions has been outlined by Dickinson (Chapter 1, this volume), and many of the geological facts presented in this chapter will be seen to relate to his outline.

Previous Work

This review is based largely on the work of others, and the treatment is uneven because of my arbitrary choice of topics and emphasis. A complete survey of data was not attempted; instead, relatively few areas and authors were selected as sources of information. Consequently, I must apologize for having omitted much valuable material and many important contributors. If all had been included, years of work and several volumes of results would have been entailed.

The present understanding of the geology of the Southern Coast Ranges has been built by increments since late in the last century. Early contributions were made by distinguished geologists such as Fairbanks (e.g., 1904) and Lawson (e.g., 1914). The heyday of field work for oil finding brought great progress in deciphering the areal geology, stratigraphy, paleontology, and structural geology, culminating in milestone syntheses by Reed (1933) and Reed and Hollister (1936). Field work was greatly extended by Taliaferro (summarized in Taliaferro, 1943a) of the University of California at Berkeley, and most notably by Dibblee of Richfield Oil Company and (later) the U.S. Geological Survey. Dibblee's remarkable skill and energy have resulted in maps covering most of the Southern Coast Ranges as well as many other parts of California. Hill and Dibblee (1953) showed the likelihood of large-scale strike-slip, a revolutionary idea at that time. Curtis and others (1958) made the startling discovery that granitic rocks of the Salinian Block are younger than much of the Franciscan, which was supposed to rest depositionally on said rocks. This reinforced the probability of large tectonic displacements. Bailey and others (1964) authored a particularly significant and comprehensive treatment of the Franciscan assemblage, a subject that had baffled and intimidated most geologists. The authors included an important summary of the Great Valley Sequence, pointed out the age equivalence of the GVS and Franciscan, and drew far-reaching tectonic conclusions. Dickinson (e.g., 1965, 1966a, 1966b, and many subsequent papers) and Compton (e.g., 1960, 1966a, 1966b) not only contributed their own research, but supervised productive studies by many students. Interest in the Southern Coast Ranges was spurred by the rapid development of plate tectonics in the late 1960s and early 1970s, as it was now realized that the continental margin might have evolved through plate interactions (Hamilton, 1969; Bailey and Blake, 1969; Ernst, 1970; Dickinson, 1970; Page, 1969, 1970a). Meanwhile, the California Division of Mines and Geology under the late Ian Campbell, State Geologist, produced the valuable Olaf P. Jenkins edition of the *Geologic Map of California* (e.g., Jennings, 1958; Jennings and Strand, 1958; Jennings and Burnett, 1961; Rogers, 1966). During the last decade, a new generation of contributors has appeared, as will be seen from the references in this paper.

Geological Highlights of the Southern Coast Ranges

The subprovince under discussion comprises a number of individual ranges and large structural valleys that do not coincide with the distribution of older rocks and structures. Beneath the present landforms and Cenozoic deposits, there are three side-by-side terranes of late Mesozoic rocks, which, if stripped of covering material, would appear to be arranged as in Fig. 13-1. These three terranes are best understood by considering a larger region than the Southern Coast Ranges.

The Franciscan Complex constitutes much of the core of both the Northern and Southern Coast Ranges. Its components were almost certainly both assembled and dismembered by the subduction of oceanic plate(s) beneath the western margin of North America during the Late Jurassic, Cretaceous, and Early Tertiary, probably with superimposed strike slip at times of oblique convergence. As shown in Fig. 13-1, two other late Mesozoic assemblages exist in California, apparently resulting from the same plate

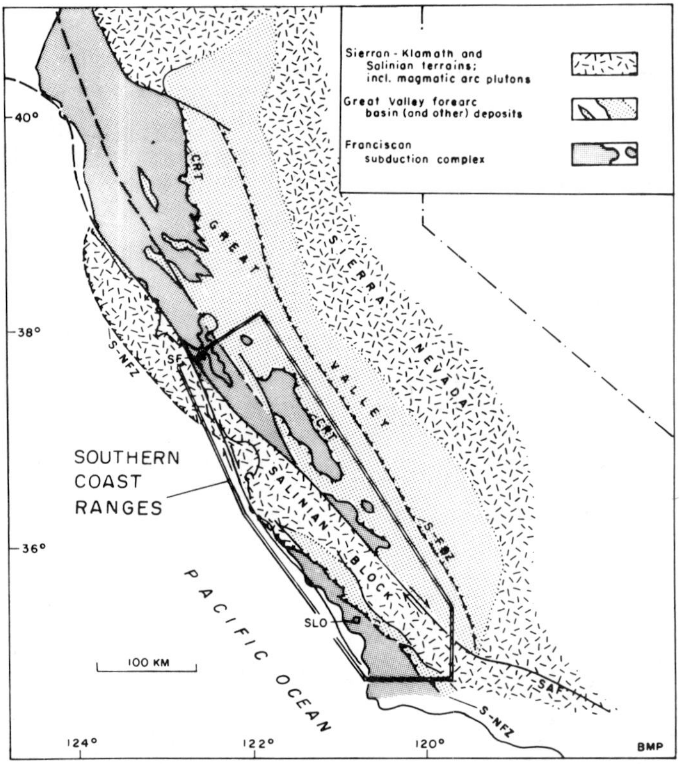

Fig. 13-1. Distribution of principal late Mesozoic petrotectonic assemblages of western California, interpreted in terms of plate-tectonic environments. Parts of all three assemblages are coeval. SF, San Francisco; SLO, San Luis Obispo; SAF, San Andreas fault; S-FBZ, Sierran-Franciscan boundary zone; S-NFZ, Sur-Nacimiento fault zone; CRT, Coast Range thrust. (Modified from Page, 1972, with permission of The Geological Society of America.)

convergence that produced the Franciscan. One of these, the Great Valley Sequence of clastic sedimentary rocks, is interpreted to be largely a forearc basin accumulation. The other, the Sierran-Klamath terrane, includes a magmatic arc assemblage, of which the anomalously positioned Salinian Block is thought to be a fragment. The Franciscan Complex and the Salinian Block will be discussed in considerable detail later, but the Great Valley Sequence will not be fully described here. Blake and Jones (Chapter 12, this volume) give a synopsis of the Great Valley Sequence, and they list references on the subject. Ingersoll (1978) provides an up-to-date interpretation.

The long-continued subduction that played a key role in the evolution of the foregoing assemblages evidently changed markedly in character in the latest Cretaceous and Paleocene, plate consumption becoming more oblique and the descending slab assuming a very low dip. This is suggested as a cause of probable pre-Eocene strike slip along the site of the San Andreas fault, a disputed displacement that is inferred from studies of

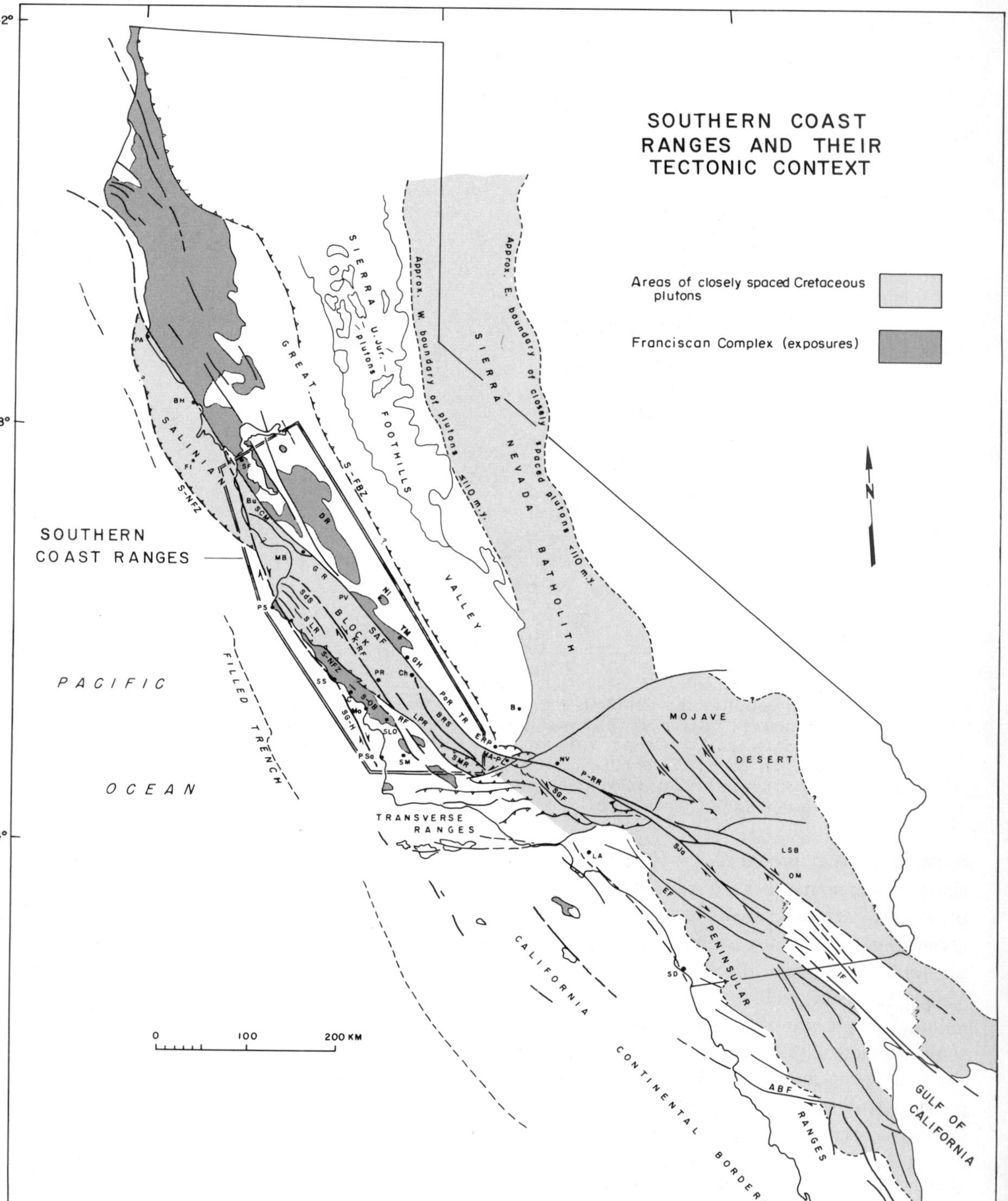
SOUTHERN COAST RANGES AND THEIR TECTONIC CONTEXT
Areas of closely spaced Cretaceous plutons
Franciscan Complex (exposures)
N
42°
38°
34°
124°
120°
116°
0
100
200 KM
SOUTHERN COAST RANGES
PACIFIC
OCEAN
FILLED TRENCH
SALINIAN
S-NFZ
GREAT
VALLEY
S-FBZ
SIERRA
U. Jur. plutons
FOOTHILLS
Approx. W. boundary of plutons <110 m.y.
Approx. E. boundary of closely spaced plutons <110 m.y.
SIERRA NEVADA BATHOLITH
G R BLOCK
SAF
MOJAVE
DESERT
TRANSVERSE RANGES
CALIFORNIA
CONTINENTAL
BORDERLAND
PENINSULAR
RANGES
GULF OF CALIFORNIA
PA
BH
FI
SF
Bu
SCM
DR
MB
PS
SdS
PV
NI
SLR
K-RF
TM
GH
S-NFZ
PR
Ch
SS
S-OB
PoR
TR
B
C
Mo
RF
LPR
BRS
SG-H
SLO
ERP
PSa
SM
SMR
MA-PL
NV
P-RR
SGF
LA
SJa
LSB
OM
EF
SD
IF
ABF

Fig. 13-2. Southern Coast Ranges and their tectonic context.

ABF	Agua Blanca fault	PV	Pinnacles Volcanics
B	Bakersfield	PoR	Point of Rocks Sandstone
BH	Bodega Head	PR	Paso Robles
BRS	Barrett Ridge slice	PSa	Point Sal
Bu	Butano Sandstone	PS	Point Sur
C	Cambria	P-RR	Portal-Ritter Ridge
Ch	Cholame	RF	Rinconada fault
DR	Diablo Range	SAF	San Andreas fault
EF	Elsinore fault	SCM	Santa Cruz Mts.
ERP	Eagle Rest Peak	SD	San Diego
GH	Gold Hill	SF	San Francisco
GR	Gabilan Range	S-FBZ	Sierran-Franciscan boundary zone
IF	Imperial fault	SGF	San Gabriel fault
K-RF	King City-Reliz fault	SG-H	San Gregorio-Hosgri fault zone
L	Logan	SJa	San Jacinto fault
LA	Los Angeles	SJF	San Juan fault
LP	La Panza Range	SLO	San Luis Obispo
LSB	Little San Bernardino Mts.	SLR	Santa Lucia Range
MA-P	Mt. Abel—Mt. Pinos	SM	Santa Maria
MB	Monterey Bay	SMR	Sierra Madre Range
MD	Mount Diablo	S-NFZ	Sur-Nacimiento fault zone
MFZ	Mendocino fracture zone	S-OB	Sur-Obispo Belt
Mo	Morro Bay	SS	San Simeon
NI	New Idria	SdS	Sierra de Salinas
NV	Neenach Volcanics	TM	Table Mt.
OM	Orocopia Mts.	TR	Temblor Range
PA	Point Arena		

Paleogene sediments and from other evidence as well. Continued subduction is indicated by features of Eocene rocks on the San Francisco Peninsula and is clearly manifested in the Coastal Belt of the Franciscan farther north.

A striking unconformity and features indicating changed circumstances of sedimentation have long been recognized at the lower boundary of the Miocene (or, locally, the Oligocene) stratigraphic section in many parts of the Southern Coast Ranges. It is now realized that these features are corollaries of the change from subduction to transform-type motion between plates. In the on-land region under discussion, the development of the Franciscan ceased by the end of the Oligocene, but offshore and especially toward the north, the generation of Franciscan-like complexes must have continued longer.

The accompanying map of the Southern Coast Ranges and their regional context (Fig. 13-2) and the cross section of the Southern Coast Ranges (Fig. 13-3) show the imprint of both subduction (late Mesozoic-Early Tertiary) and translation parallel with the plate boundary, the translation having brought the Salinian Block to its anomalous position with respect to the Franciscan terrane alongside it on the east. Some of the strike slip is ascribed to oblique subduction, but in the Tertiary, subduction was progressively terminated from southeast to northwest and was replaced by transform motion as the Pacific and North American plates came into contact with one another (Atwater, 1970; Dickinson, Chapter 1, this volume).

Many strike-slip faults, several en echelon Neogene basins, numerous en echelon folds, and a few E-W trending thrust faults are interpreted as products of wrench tec-

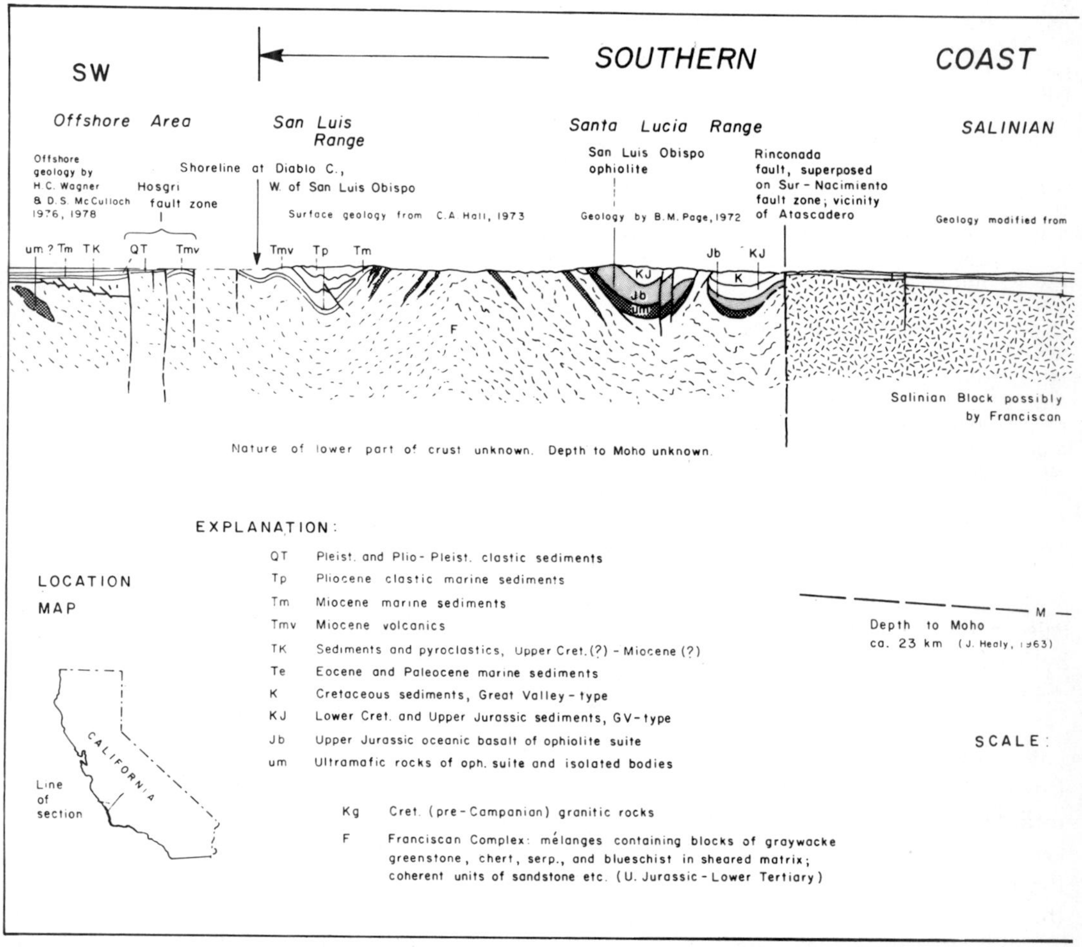

Fig. 13-3. Cross section of Southern Coast Ranges from vicinity of San Luis Obispo to Kettleman Hills. Shows Franciscan subduction complex and coeval Great Valley forearc basin sequence. Salinian Block (atypically simple here) is magmatic arc terrain in anomalous position; it, together with Franciscan terrain to the west

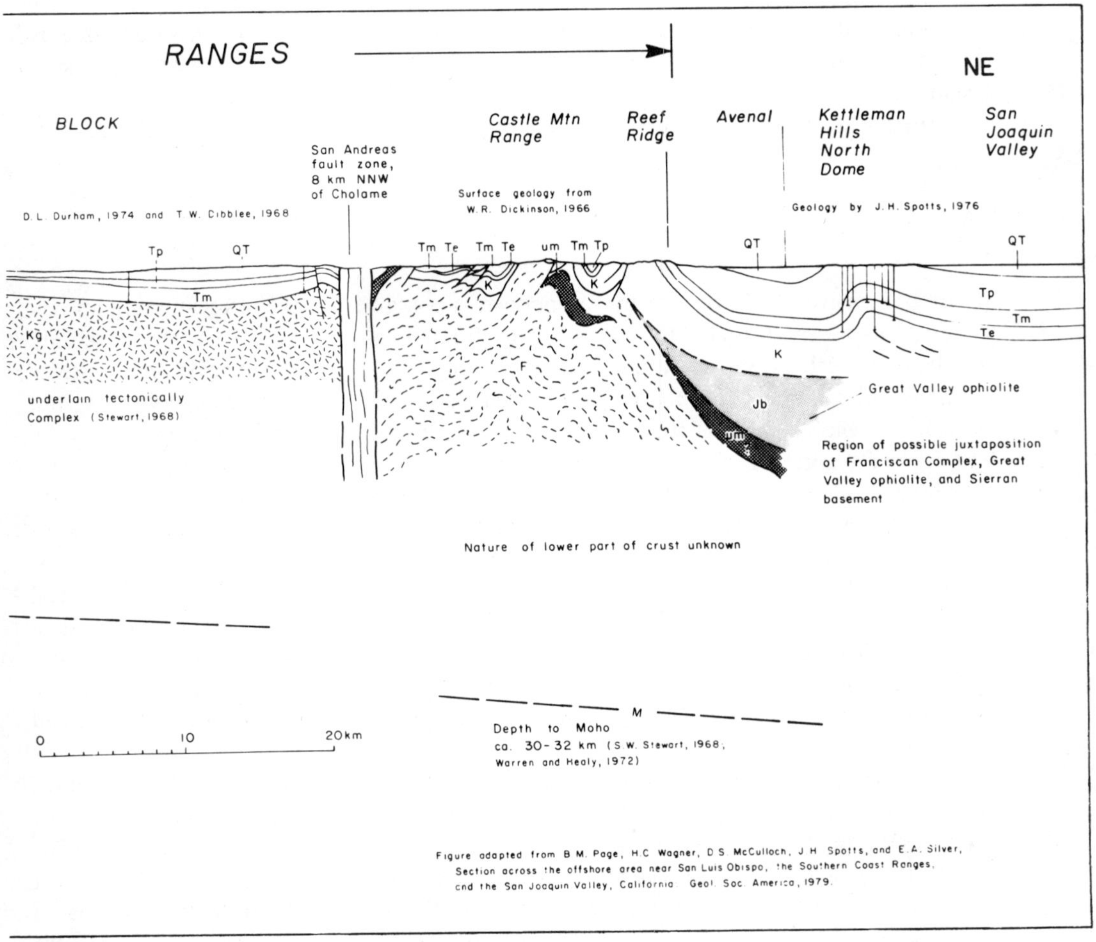

(Sur-Obispo Belt), moved ca. 500 km along San Andreas transform fault. Some Neogene folds shown on section are also result of transform tectonics, but Kettleman Hills probably has a different origin. (Adapted from Page and others, 1979, with permission of the Geological Society of America.)

tonics accompanying the transform regime. One of the outstanding characteristics of the Southern Coast Ranges, as well as some other parts of California, is the folding and faulting of poorly consolidated sedimentary units as young as Pliocene or Pleistocene.

Some of the individual ranges in the subprovince under discussion are flanked by folded or upturned Plio-Pleistocene strata, but the ranges contain complex internal structures and most are not simple antiforms. Some are bounded on at least one side by strike-slip faults, which have accepted vertical motions unrelated to the principal role of the faults.

Faulting and crustal distortion are ongoing today in the Southern Coast Ranges.

THE FRANCISCAN COMPLEX

A comprehensive description of the Franciscan is contained in the classic monograph by Bailey and others (1964); the designation "Franciscan Complex" was proposed by Berkland and others, 1972. In the Southern Coast Ranges, Franciscan areas on both sides of the Salinian Block show most of the features that typify the Complex everywhere, and strikingly resemble some of the terranes in the Northern Coast Ranges discussed by Blake and Jones (Chapter 12, this volume). However, in the Southern Coast Ranges there is no extensive equivalent of the South Fork Mountain Schist or any counterpart of the Coastal Belt except quite possibly offshore. Typical melanges like those of the Northern Coast Ranges commonly appear as chaotic units between (or surrounding) bodies of coherent rock with normal stratification. Most blocks within the melanges are identical with blocks elsewhere in the Franciscan. As in the Northern Coast Ranges, there is a characteristic mixture of terrigenous sediments, oceanic material, and blocks of different ages and degress of metamorphism. Most of the coherent nonmelange units are mainly sandstone, but some are largely chert and some are greenstone with or without chert and sandstone.

To provide a basis for tectonic interpretation, we will briefly describe significant features of two terranes, the Diablo Range and the "Sur-Obispo Belt" (see Fig. 13-2 for locations). The two possess much in common, yet each has some individuality. The Diablo Range is northeast of the San Andreas fault (SAF), whereas the Sur-Obispo Belt is southwest of the SAF and the Sur-Nacimiento fault zone as well, so the two Franciscan terranes have been shifted a great distance relative to each other. This is very different from the Northern Coast Ranges, where a much wider Franciscan tract is exposed, all on one side of the SAF.

Franciscan of the Diablo Range

General structural characteristics The Diablo Range encompasses the widest Franciscan terrane in the Southern Coast Ranges and gives good east-west transects. The main part of the range is antiformal, and the exposed Franciscan core is virtually encircled by rocks of the late Mesozoic Great Valley Sequence (GVS), which is described by Jones and Blake in this volume (Chapter 12). The clastic sediments of the GVS locally rest upon calc-alkaline volcaniclastic rocks underlain by an ophiolitic sequence. Faults separate the GVS (including the ophiolite remnants) from the Franciscan core. In the past there must have been a thrustlike cover of GVS over the entire core; this will be discussed under the heading "Coast Range Thrust."

The Franciscan core consists of melanges and coherent sandstone units. This is shown in Fig. 13–4, which, however, covers less than half the core. The proportion of melanges versus coherent units varies from place to place. Recent mapping in the north part of the core by Cotton (1972) and Crawford (1975, 1976) shows large coherent tracts that tend to dwarf some of the melanges. In describing the structure further, it is difficult to strictly separate observed relations from interpretation. Crawford represents most

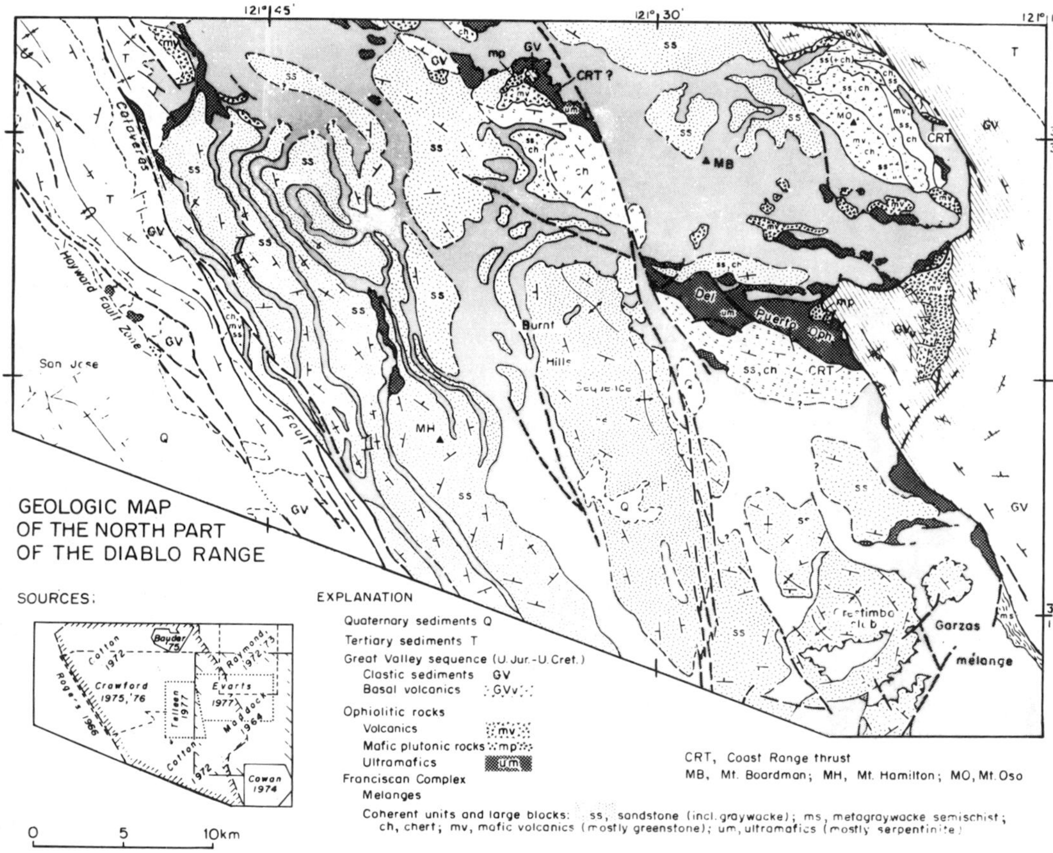

Fig. 13–4. Geologic map of north part of Diablo Range. Franciscan subduction complex occupies most of map area, forming core of an antiform. Franciscan is flanked by coeval Great Valley Sequence, largely forearc basin deposits. Tectonic contact between the two is locally Coast Range Thrust, but more generally consists of Neogene faults. Franciscan greatly simplified; various kinds of melanges are not differentiated here, and different coherent sandstone units are shown by single pattern (Compiled from numerous authors)

of the melanges in the northwest part of the core as narrow corridors in plan and as sheetlike units in cross section. Taking an innovative view, he envisions stacked sequences of alternating coherent sandstone units and intervening melange layers. One of the stacks is the Eylar Mountain sequence, believed to be Upper Jurassic (Tithonian?); another, which is structurally lower, is the Burnt Hills sequence, believed to be Cretaceous (Coniacian?). Sandstone with stratal continuity predominates in both sequences, but the Eylar Mountain sequence contains one ostensibly coherent unit composed of shale, sandstone, chert, and pillow basalt (Crawford, 1975, 1976). In Crawford's judgment, most of the melanges he studied originated as olistostromes, and each melange of this type was covered depositionally by clastic sediment that formed a coherent stratiform unit. He believes, on the basis of sedimentary features indicating tops and bottoms of beds, that one composite stack was folded into a recumbent anticline overturned toward the west. The folding resulted from eventual subduction after the sedimentary pile had been assembled (Crawford, 1975). These ideas are timely and refreshing, although difficult to prove. It is particularly hard to find convincing examples of depositional contacts between coherent units and melanges because of widespread shearing.

The structure envisioned by Cowan (1974) in the east part of the Diablo Range (Fig. 13-9) is entirely different from that just described. Cowan mapped several coherent sandstone (metagraywacke) units separated by tracts of melange, and concluded that all contacts in his area of study are tectonic and that there is no semblance of a sedimentary sequence except within individual coherent units. He says,

> The structural units are grossly sheetlike in external form and are separated from one another by gently to steeply dipping major faults. Unlike low-angle thrusts in imbricated Cordilleran terranes, the faults do not systematically repeat or offset a normal stratigraphic sequence but rather juxtapose rock units that bear no apparent stratigraphic, deformational, or metamorphic relation to one another. The structural units were separately deformed and metamorphosed under a variety of conditions prior to their tectonic juxtaposition during late Mesozoic continental margin subduction.

Somewhat comparable relations have previously been described in the Northern Coast Ranges by Suppe (1972, 1973).

Other persons may tend to visualize the structure of the Diablo Range with some features of Crawford's image and some of Cowan's. In any particular area, the actual complexities and lack of optimum exposures lend themselves to divergent interpretations. However, in some cases, the bedding orientation of coherent units is so discordant with respect to adjacent melanges the contacts must be tectonic, and in some instances inverse age sequences also demand tectonic contacts.

The semiparallel, alternating melange and coherent units are more evident in maps of the northwest part than elsewhere in the core of the Diablo Range, probably because most units elsewhere are not steeply dipping. In the north-central part of the core (Telleen, 1977), in the northeast part (Raymond 1973a), in the vicinity of Panoche Pass (Ernst, 1965), around Pacheco Pass (Ernst and others, 1970), and northwest of the latter (Cowan, 1974), many of the structural units appear to be gently to moderately dipping.

Coherent sandstone units, Diablo Range Most of the coherent units in the core of the Diablo Range are turbidite and kindred sandstones with thin mudstone intercalations. Some contain local conglomerate in the form of channel fillings. The coherent sandstones studied by Crawford (1975, 1976) are predominantly arkosic arenites, which superficially resemble rocks of the Great Valley Sequence and which contain more feldspar and a smaller proportion of lithic particles and matrix than many of the sandstone blocks in nearby Franciscan melanges. Crawford interprets most of the coherent sandstone units in the northwest part of the Diablo Range core as various facies of subsea mid-fan deposits. One such unit 10 km northwest of Mount Hamilton yielded a fossil that was identified by D. L. Jones (cited in Crawford, 1975) as *Buchia fisheriana* of Tithonian (latest Jurassic) age. Nearly all these units have undergone incipient to pronounced blueschist facies metamorphism and are actually metasandstones, but they retain most of their original megascopic textures and structures.

A large slablike (?) body of conspicuously well bedded turbidite sandstone with thin mudstone intercalations lies structurally deep near the axis of the northern third of the Diablo antiform (Fig. 13-4). It is part of Crawford's Burnt Hills sequence, and is apparently more than 20 km long and 6 km wide. The northern part of this unit was studied by Telleen (1977), who shows a tectonic contact with an overlying melange. (The sawteeth in Fig. 13-4 were added by the present author). Telleen interprets the sediments to be channeled subsea mid-fan deposits. Using sedimentary structures, he finds a prevailing southwestward paleocurrent direction. Telleen carefully "unfolded" the folds in determining this direction, but it is conceivable that the sandstone body has rotated somewhat with respect to north, independently vis-a-vis adjacent rocks. This sandstone unit has yielded sparse fossils determined by D. L. Jones (cited in Telleen) to be mid- to Late Cretaceous.

The east side of the Franciscan core of the Diablo Range includes some fairly large coherent units of metagraywacke (e.g., McKee, 1962; Ernst and others, 1970; Cowan, 1974). Cowan's "Orestimba metagraywacke" (Fig. 13-9), which is structurally beneath the extensive Garzas melange, is apparently a somewhat tabular unit more than 7 km long, consisting of massive to thin-bedded quartzofeldspathic graywacke with minor amounts of intercalated siltstone. Some beds show turbidite features. Although metamorphic foliation is generally absent, various parts of the unit contain lawsonite, glaucophane, or jadeitic pyroxene. The radiometric age of metamorphism is 88 to 93 m.y., which means that the sandstone is Cenomanian (middle Cretaceous) or older (Cowan, 1974). Two somewhat smaller sandstone units in the same area are regarded as large blocks or slabs in the melange, although one is at least 4 km long. This points up the problem of deciding how large a block can be before it is placed in some other category. In the area studied by Cowan, most of the small melange sandstone blocks are predominantly metagraywacke with 14% to 40% lithic grains, whereas the Orestimba unit contains only 8% to 14% lithics.

A rather thin-bedded unit of turbidite-type metagraywacke in the east part of the Pacheco Pass area is more than 6 km broad. It is locally veined and slightly foliated, and it contains jadeitic pyroxene; nevertheless, most megascopic sedimentary features are preserved (Ernst and others, 1970).

The petrology of some Franciscan sandstones in the Diablo Range was studied by Soliman (1965) and more recently by Jacobson (1978). Jacobson determined mutual proportions of quartz, feldspar, and lithic detrital grains (QFL), and found that three sandstone units in Crawford's Eylar Mountain sequence are internally consistent and are similar to one another, with mean compositions of approximately $Q_{30}F_{50}L_{20}$ and a mean volcanic-to-total rock fragments ratio of 0.35. He believes the provenance was a volcanically active part of a continental platform, probably the ancestral Sierra Nevada. Jacobson found a different mean composition for sandstone of the principal flysch-type unit of the Burnt Hills sequence, the same unit that Telleen examined. The determined mean was $Q_{38}F_{52}L_{10}$, and the ratio of volcanic to total rock fragments was only 0.07. Some feldspar has unmistakable plutonic characteristics. He correlates the Burnt Hills sandstone with Cowan's Orestimba metagraywacke on the basis of similar petrology. Jacobson thinks the Burnt Hills sandstone was also derived from the Sierra Nevada, but from a (de-roofed) plutonic terrane. He suggests that the sediment was deposited in a trench slope basin, an idea that is currently being proposed by others for sandstones in the Franciscan of the Sur-Obispo Belt, as described later.

Blake and Jones (Chapter 12, this volume) summarize and interpret available petrologic data regarding Franciscan coherent sandstone units, and emphasize contrasts between these rocks and coeval clastic sediments of the Great Valley Sequence. They conclude that these Franciscan sandstones are not distal facies of the Great Valley Sequence, and that some were probably derived from different source terranes.

The various coherent sandstones described previously, and others elsewhere, possess fairly continuous bedding and in small exposures generally appear to have rather simple, orthodox structure. However, geologic maps commonly show rather erratic bedding orientation (e.g., Cotton, 1972). Some sandstone bodies show uneven, nonparallel undulations, some exhibit noncylindroidal folds, and a few (as in the area studied by Crawford, 1975) show steeply dipping beds that are locally overturned. Some (many?) coherent units show disturbance and disruption of bedding in the immediate vicinity of the contact with overlying or underlying melanges. In some localities, there is a semblance of structural harmony among neighboring units, but elsewhere there is discordance. In the Diablo Range core as a whole, there are many different trends. The bedding of a unit north of the Del Puerto ophiolite (Fig. 13–4) strikes east-west, whereas bedding in the metagraywacke east of Pacheco Pass strikes north to northeast (Ernst and others, 1970). In both examples the internal structure is markedly discordant with the nearby contact between the Franciscan and the Great Valley Sequence.

Melanges in the Diablo Range The melanges of this range are much the same as those elsewhere in the Coast Ranges. Some contain blocks of nearly all the well-known Franciscan rock types: graywacke, basaltic greenstone, radiolarian chert, serpentinite, conglomerate, high-grade blueschist, eclogite, amphibolite, and so forth. Others are characterized by the presence of a few types (for example, graywacke and chert) and near absence of the others. The most prevalent rock type in all the melanges is graywacke. This occurs in many varieties; most of the graywacke lacks distinct bedding, but some is clearly stratified; most is lithic graywacke, but some is quartzofeldspathic; most is subtly to visibly metamorphosed; some is veined and foliated. The last-mentioned variety is common in areas where jadeitic pyroxene occurs.

The Garzas melange northwest of Pacheco Pass is described by Cowan (1974), who notes a "distinctive deformational style characterized by small angular phacoids and larger geometrically and dimensionally variable blocks of resistant rock types enclosed in a pervasively sheared, fine-grained matrix." Despite the limited nature of the exposures, he discerns a general east-dipping orientation of the subparallel shear surfaces (Cowan, 1974, bottom of Fig. 2). He finds that sizable graywacke units are progressively sheared toward the margins and tend to structurally grade into the melange. The matrix is predominantly sheared siltstone and mudstone, but "small lenses of fine-grained graywacke are interleaved with the sheared pelitic matrix, and a small part of the dominantly argillaceous matrix may have been derived from cataclasis and comminution of larger graywacke inclusions" (Cowan, 1974, p. 1625).

Franciscan conglomerate Conglomerate constitutes a small but widely distributed component of the Franciscan of the Southern Coast Ranges. It occurs as lenses and channel fillings in graywacke sequences and as blocks (composed entirely of conglomerate) in melanges (e.g., Crawford, 1975; Platt and others, 1976). In the Diablo Range, most such material has undergone varying degrees of blueschist metamorphism, as discussed later, but whether metamorphosed or not, the conglomerates are valuable sources of information concerning the origin and provenance of Franciscan clastic sediments.

The conglomerate that occurs as channel fillings in graywacke sequences largely consists of crowded, rounded pebbles and cobbles that are in mutual contact and show both normal and reversed grading. In contrast, the pebbles in most nonchannel lenses in coherent graywacke are slightly separated from one another in the sandstone matrix, and as a rule there is a higher proportion of subangular clasts. Moore (1977) thinks that the channel conglomerates can be classed as organized conglomerates (A2) of Walker and Mutti (1973), and that they are subsea inner fan channel deposits. She interprets the conglomerate blocks in melanges as fragments of the channel deposits. The subsea fans may have formed in a trench, but the evidence is only circumstantial.

Moore (1977) studied conglomerate from forty-one lenses and blocks in the Diablo Range. The pebbles include volcanics, plutonic igneous rocks, clastic sedimentary rocks, limestone, quartzite, and chert. Considering the compositions, textures, mineralogy, and type of predepositional metamorphism (if any), and considering presently exposed terranes of pre-Tertiary rocks and other possible source areas, Moore concludes that the igneous and metamorphic clasts (which are well rounded) were probably derived from the late Mesozoic continental margin, that is, the Sierran-Klamath terrane. Some angular clasts of sedimentary rocks may have been acquired from the inner slope of a trench. No clasts definitely require a western source.

Igneous rocks, which predominate, are types that could have come from a volcanic-plutonic arc. On the basis of igneous rock types present, Moore divided the conglomerates into three groups. The groups apparently differ in age and represent different stages in the evolution of the Sierran-Klamath arc, in much the same way as proposed by Ojakangas (1968) for sandstones of the Great Valley Sequence. On the basis of their petrology vis-a-vis the Great Valley Sequence, the conglomerates of one group are Tithonian (uppermost Jurassic), those of a second group are lowermost Cretaceous, and those of another group are Turonian (Upper Cretaceous) (Moore, 1977). The conglomerates have not been dated by more direct means.

Detrital glaucophane schist occurs as rare, well-rounded pebbles in the conglomerate of the Diablo Range. The conglomerate itself has undergone blueschist facies metamorphism, but the glaucophane schist was metamorphosed prior to its incorporation in the sediments. The schist pebbles are of higher metamorphic grade than the matrix and have a strong tectonite fabric that is not shared by the surrounding material. Although the schist may possibly have been recycled from older parts of the Franciscan, it is unaccompanied by other typical Franciscan rocks (such as red chert and greenstone) and therefore likely came from earlier blueschist in the Sierra Nevada or Klamath Mountains (Moore and Liou, 1977).

Chert-graywacke associations A sedimentary-tectonic enigma is posed by the occurrence of radiolarian chert interbedded with terrigenous clastic sediment. The coherent jadeitic metagraywacke unit east of Pacheco Pass contains several fairly continuous chert units, mostly 1 to 2 m thick (Ernst and others, 1970). Crawford (1975) describes interbedded chert and sandstone in his lithofacies A north of Mount Hamilton. Perhaps the most striking occurrences are north of Mount Oso (Raymond, 1974), where several coherent formations consist of apparently interbedded metachert and metagraywacke, some of the metachert members being 150 m thick, and some being traceable for more than 1 km. Raymond (1974) discusses possible modern analogs of the sedimentation. Echeverria (1977) reports chert and pillowed greenstone intercalated in a coherent graywacke unit several kilometers long in the Ortigalita Peak quadrangle. Other examples could be cited. In most cases the chert is reddish, occurs in thin rhythmic beds, contains vestiges of Radiolaria, is associated with thin intercalations of red siliceous shale, and is stained with manganese oxide. Although some may be tuffaceous, much of the chert has a pelagic, biogenic aspect.

The alternation of turbidite and chert is difficult to explain. Biogenic chert accumulation would seem to be favored by exclusion of clastic sediment, water depth below the carbonate compensation zone, and a lengthy period of stability. What circumstances could provide a sudden change from clastic influx to a sheltered pelagic environment and a subsequent change back again? One possible setting would be a deep sea floor on the oceanward side of a trench that receives clastic sediment from a continental shelf and slope. Slow subduction might allow the trench to fill completely, and subsequent turbidite flows could spread oceanward. Faster subduction would renew the troughlike topography of the trench, whereupon it would serve as a sediment trap, permitting chert deposition to occur on the outlying sea floor. In any case, the Franciscan sandstone-chert sequences probably originated in the vicinity of a subduction zone, inasmuch as they contain both oceanic and land-derived components, and in all the examples cited here they have undergone blueschist-facies metamorphism, presumably because they were eventually subducted.

Radiolaria contained in a chert-sandstone unit in the Ortigalita Peak quadrangle are reported to be older than the basal sediments of the Great Valley Sequence (D. L. Jones,

1977, oral communication). If this is true, the chert-sandstone unit cannot be a distal facies of the Great Valley sediments, and the provenance of its clastic constituents is unclear. As emphasized by Blake and Jones (Chapter 12, this volume), such mismatches seem to be incompatible with simple convergence normal to the plate boundary. Probably there were large components of relative motion parallel with the continental margin, as a consequence of transform interaction or oblique subduction.

Mafic intrusives in Franciscan subduction complex Another interesting feature of the Diablo Range is the local presence of basalt, diabase, and gabbro intrusives in Franciscan graywacke. These rather unusual bodies are not to be confused with commonplace blocks of greenstone and other mafic rocks in melanges. The mafic material noted here shows unmistakable intrusive features, such as chilled borders, apophyses, baked host rocks, and marginal xenoliths. The host rock is generally coherent graywacke with intercalated shale and, in some areas, chert. Examples of mafic intrusives are found in the vicinity of the western part of the San Luis Reservoir (Ernst and others, 1970) and in the Ortigalita Peak area. The Ortigalita intrusive is a differentiated gabbroic sill more than 8 km long and 200 to 300 m thick, which (together with the enclosing graywacke) has been subjected to blueschist facies metamorphism (Echeverria, 1977). If the Franciscan Complex represents a tectonically stacked pile that once formed an outer arc ridge alongside a trench, the magmatism that produced the mafic intrusives must have occurred near the site of subduction, as in the case of certain anomalously located plutons of the eastern part of the Aleutian arc. Perhaps (1) a spreading ridge "partner" to the trench migrated so close to the latter that one edge of the convective heat source extended beneath the subduction complex, briefly causing melting, or (2) a subducted "leaky" transform fault in the downgoing lithosphere afforded an avenue for rising magma, or (3) as postulated by Marshak and Karig (1977) for other regions, a triple junction involving a spreading ridge at a large angle to the subduction zone provided the locus for magmatism. As pointed out by Ernst (1978, informal communication), the second hypothesis is best in accord with the high P/T ratio recorded in Franciscan rocks of the Diablo Range. A spreading ridge near the subduction zone would necessarily be accompanied by a thin lithosphere and high heat flow.

Franciscan Complex in the Sur-Obispo Belt

The coastal fringe of California that is underlain by the Franciscan Complex and that extends from the vicinity of Point Sur to San Luis Obispo and thence to the Transverse Ranges is here termed the "Sur-Obispo Belt." This belt lies along the southwest side of the Salinian Block and is therefore southwest of the Sur-Nacimiento fault zone (see Fig. 13-2 for location and Fig. 13-3 for a cross section).

Comparison with Franciscan elsewhere Parts of the Salinian Block have been displaced more than 500 km by motion parallel with the continental margin, and the Sur-Obispo Belt has probably been displaced even farther with respect to the original site of accumulation. In view of this large-scale transport, it is remarkable that the rocks of the

Sur-Obispo Belt strongly resemble those of the Diablo Range and Northern Coast Ranges. With few exceptions, the same characteristic rock types are present, and there is the usual association of coherent units and intervening melanges. Structural features, insofar as they have been studied, seem to be comparable. However, there are a few differences that should be mentioned. In the Sur-Obispo Belt, coherent sandstone units are apparently not as prevalent as in the Diablo Range, and there is no large-scale counterpart of the South Fork Mountain Schist of the Northern Coast Ranges, although small blocks of comparable schist and phyllite have been observed. No equivalent of the Coastal Belt of Northern California is visible, but such a terrane may exist offshore.

The narrowness of the on-land part of the Sur-Obispo Belt precludes the recognition of various subdivisions that might be compared with those of the Franciscan Complex in Northern California (e.g., Maxwell, 1974). This disadvantage is aggravated by closely spaced Neogene faults that strike parallel with the edge of the Salinian Block and make it difficult to determine the former distribution of Franciscan units.

Melanges: components and structure Typical Franciscan melanges are present along the coast intermittently from Point Sur (Gilbert, 1973) southeastward to the San Luis Range (Hsü, 1969; Hall, 1973a, b, 1974, 1976; Hall and Prior, 1975) and within the Santa Lucia Range from the Nacimiento Summit road near the 36th parallel southeastward to the Transverse Ranges. Franciscan terranes consisting largely of melanges were studied by Callender (1975) west of Paso Robles and Atascadero and by Brown (1968a, b) at Stanley Mountain. Most of the melanges contain blocks of graywacke, greenstone, chert, and blueschist facies metamorphic equivalents of these. There are occasional blocks of conglomerate, serpentinite, blueschist with tectonite fabric, and other types. All these can be subdivided into several varieties on the basis of texture and composition. For example, there are lithic graywackes, subgraywackes, and metagraywackes ranging in texture and mineral assemblages from textural zone 1 through textural zone 2 of Blake and others (1967). Greenstones include pillow lavas, breccias, and tuffaceous rocks. Cherts vary from red radiolarian varieties to green and gray, and from white microquartzite to crossite-bearing metachert.

Conglomerate of several types occurs as blocks in melanges, and although not volumetrically abundant, it is of considerable interest. The most common type, which is sparsely but widely distributed, is composed largely of rounded pebbles of chert, quartzite, and vein quartz. The chert pebbles are mainly black, gray, or green, unlike typical Franciscan red chert. Recently, Seiders and others (1979) have reported radiolaria from these pebbles, the samples having come from thirteen localities scattered from Cambria to Arroyo Grande. The ages are mainly Triassic; a few pebbles may be Paleozoic and some could be Early Jurassic, but all predate the development of the Franciscan and are not reworked from the latter. The same authors suggest that the source of the pebbles may have been the Calaveras assemblage in the Sierra Nevada. However, the Sur-Obispo Belt has traveled from its original site at least as far as the Salinian Block; furthermore, if large strike slip affected components within the Franciscan (Blake and Jones, 1978, Chapter 12 this volume), the chert pebbles may have a very distant provenance. This would make them all the more significant as possible "tracers." Seiders and others (1979) report that conodonts from pebbles in two of the conglomerates have low color alteration indexes, indicating that the rocks have never experienced high temperatures, a finding quite consistent with the preservation of ancient blueschist in melanges.

A few blocks of conglomerate (much different from the foregoing type) contain material resembling earlier parts of the Franciscan (e.g., Hsü, 1969; Cowan and Page, 1975; Underwood, 1977a, b), apparently indicating recycling.

All in all, the melanges strikingly illustrate the characteristics of the Franciscan elsewhere: mingled oceanic and continent-derived rocks (Fig. 13-5), mixed materials of different ages (e.g., Upper Jurassic chert and mid-Cretaceous graywacke), and juxtaposed rocks of various degrees of metamorphism (e.g., high-grade blueschist alongside pumpellyite-bearing greenstone).

The sea-cliff exposures are particularly important. Hsü (1968, 1969), in his study from Morro Bay to San Simeon and beyond, recognized the diversity of rock types and structural features. The latter include extensional fractures and boudinage in some blocks, in close proximity to shearing features in both the blocks and the matrix. Cowan (1978) restudied exposures near San Simeon, and interpreted successive events, each with distinctive expression. According to Cowan, development of a heterogeneous, nonbedded olistostrome accumulation probably preceded most of the discernible deformation. The most prominent structural feature is a postmixing, NW-striking, NE-dipping crude foliation defined by tectonically flattened inclusions and penetrative shearing in the argillaceous matrix. This has produced imperfect oblate ellipsoidal shapes in most inclusions (even in blueschist). Inclusions of graywacke and greenstone (10 to 20 cm in diameter) that Cowan dug out of the matrix are roughly circular in plan, have mildly scalloped

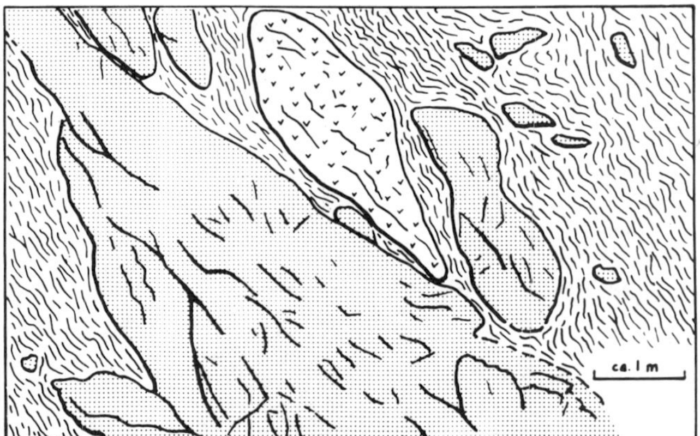

Fig. 13-5. Detail of Franciscan melange, showing solitary greenstone block (V pattern) surrounded by graywacke blocks (stippled). Greenstone presumably came from large oceanic ophiolite mass, whereas the graywacke is terrigenous; hence fragmentation and extreme dispersal and mixing are implied. (Sketched from photograph of exposure in sea cliff near San Simeon. Wiggly dash pattern schematically represents shears in argillaceous matrix.)

surfaces, and taper to sharp distal edges. Pinch-and-swell structures and locally extreme necking and boudinage are associated features. The overall ductile deformation was followed by less conspicuous brittle shearing of the inclusions and corresponding shearing of the matrix, which, however, had already undergone penetrative deformation in the earlier episode. A model which incorporates these chartertistics is illustrated in Fig. 13-6. Finally, Neogene faults locally cut all the aforementioned structures at a high angle.

Intense shearing such as that mentioned is commonplace in nearly all optimal exposures of melanges southwest of the Salinian Block. It is generally expressed by a maze of slip surfaces that curve, diverge, join, and transect one another. In sufficiently large exposures, especially in sea cliffs, the shear surfaces and lensoidal or phacoidal inclusions show an imperfect preferred orientation. Along the sea cliffs the preferred direction of dip is toward the continent (or the Salinian Block), but this cannot be a reliable original orientation because wherever late Cenozoic rocks are present, they record strong Neogene folding and faulting, which must have affected Franciscan melanges everywhere. The Franciscan underlying the San Luis Range and Los Osos Valley is folded into a syncline and an anticline (Fig. 13-3), which must have altered and perhaps reversed the dips of shear surfaces in the melanges, but exposures are inadequate to demonstrate this.

An example of unusually intense and penetrative shearing is at Jade Cove, which is between Point Sur and San Simeon. This was studied by Crippen (1951), who describes nephrite bodies associated with serpentinite, mylonite, and cataclastic schist (phyllonite?). The cataclastic rocks comprise a belt about 3 km long and up to 160 m thick. Apparently the original rocks were graywacke and shale; altered augenlike remnants of the former are visible in the mylonite.

Bedded sandstone slabs, coherent units Well-bedded sandstone units of small to moderate size occur in nearly all Franciscan areas that have been studied southwest of the Salinian Block. Most of these are blocks or slabs within melanges, but they are not the same as the usual blocks of dark graywacke, which commonly lacks distinct, con-

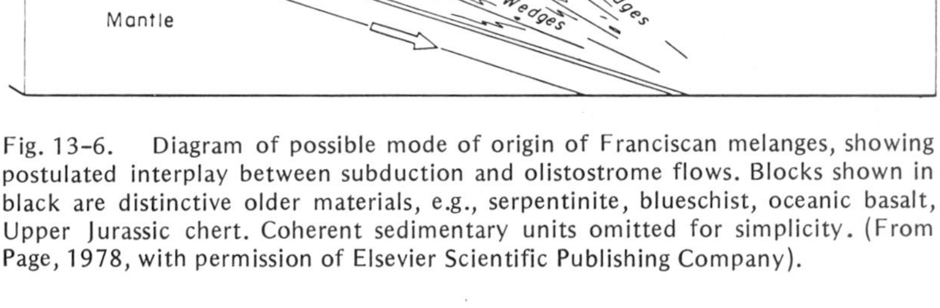

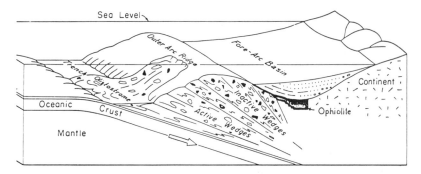

Fig. 13-6. Diagram of possible mode of origin of Franciscan melanges, showing postulated interplay between subduction and olistostrome flows. Blocks shown in black are distinctive older materials, e.g., serpentinite, blueschist, oceanic basalt, Upper Jurassic chert. Coherent sedimentary units omitted for simplicity. (From Page, 1978, with permission of Elsevier Scientific Publishing Company).

tinuous bedding. Most have more K-spar than "typical" Franciscan graywackes (Bailey and others, 1964, plate 2), and they have a smaller proportion of lithic grains vis-a-vis quartz and feldspar. They are chiefly turbidites that strongly resemble those of the Great Valley Sequence. In addition to these bodies of modest proportions, there are large coherent sandstone units such as the Cambria slab that may be either part of the Franciscan or an unrelated superincumbent formation. Depending on the observer, there may or may not be a continuum from small blocks and slabs of well-bedded turbidite in melanges to coherent sandstone masses that are measured in kilometers.

Gilbert (1973) describes a terrane near Big Sur characterized by intersheared well-bedded sandstone-mudstone units and pervasively sheared melange units, each 100 to 1000 m thick. If the units could be mapped individually (which may not be feasible in the Big Sur area), the general aspect might resemble that of the northwest part of the Diablo Range core. The well-bedded sandstone lacks blueschist facies metamorphism, and it contains an average of 6% K-spar. Cretaceous palynomorphs (determined by W. R. Evitt) and a comparison of the petrologic details with those of various intervals of the GVS indicate a Late Cretaceous age (Gilbert, 1973). Underwood (1977b) thinks that one of the coherent slabs in Gilbert's sequence is a trench-slope basin deposit.

Mapping by Hall (1973b) in the San Luis Range southwest of San Luis Obispo shows two coherent sandstone units more than 2 km long that retain stratal continuity and resemble rocks of the GVS. These units are associated with Franciscan volcanic rocks and melanges that tectonically underlie and perhaps surround the coherent bodies; the latter, like the abovementioned example at Big Sur, are regarded as trench-slope basin deposits (Howell and others, 1977).

The sandstones associated with the Franciscan between San Simeon and Morro Bay were included in a stimulating study by Hsü (1969), who divided them into three types of graywacke. Type I is rich in lithic fragments, is devoid of K-spar, and generally lacks distinct bedding. It is of widespread occurrence as blocks in melanges. Type II contains up to 10% K-spar, and type III contains 10% to 20% K-spar. The second and third types occur as modest-sized inclusions in melanges and as "broken formations" large enough to be mapped separately on large-scale maps. Type II graywackes somewhat resemble the lower part of the GVS, but Hsü believes they contain older Franciscan detritus, a debatable point. Type III sandstones are feldspathic graywackes. They are distinctly bedded turbidites resembling the upper part of the GVS, although they reportedly contain small amounts of Franciscan detritus in places.

One of the largest coherent sandstone units in the Sur-Obispo Belt is the Cambria slab (Fig. 13-7), which illustrates some of the problems and differing interpretations engendered by such bodies. The Cambria slab extends for 16 km along the coast southeastward from the town of Cambria (see maps by Jennings, 1958; Hsü, 1969; Hall, 1974; Howell and others, 1977). This unit is largely turbidite-facies sandstone, with some massive channelized sandstone. It is feldspathic arenite, contains 10% to 20% K-spar, has yielded Late Cretaceous dinoflagellates (identified by W. R. Evitt), and superficially resembles sandstones of the GVS. The beds apparently retain stratal continuity; they dip steeply, strike obliquely across the slab, and are truncated at its margins. There are at least three different interpretations of the Cambria slab: (1) it is an outlier of the

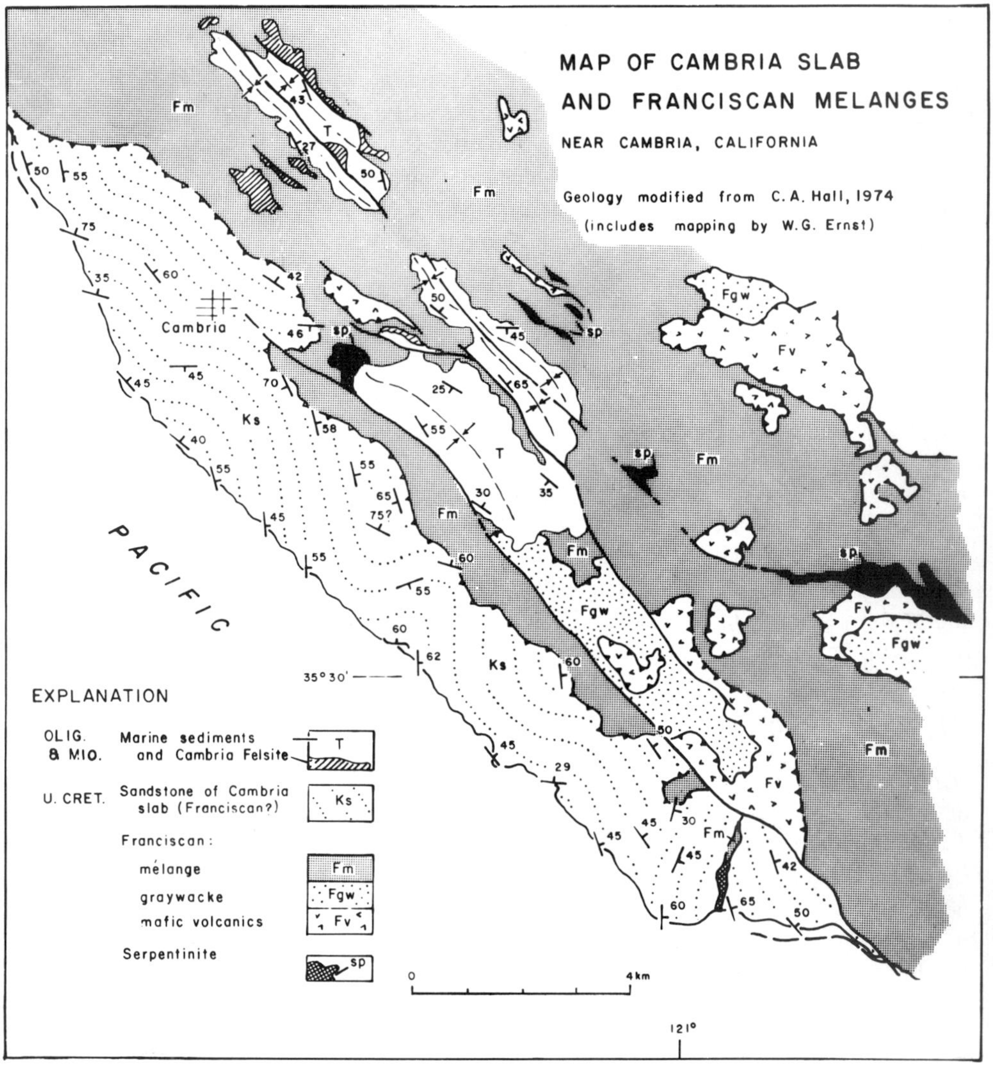

Fig. 13–7. Map of Cambria slab and Franciscan melanges. The slab may be part of the Franciscan; see text for various interpretations. (Adapted from Hall, 1974)

GVS, resting *tectonically upon* the Franciscan, and is not a part of the latter; (2) it is an allochthonous broken formation *within* the Franciscan (Hsü, 1969); (3) it is a local unit *deposited in situ* on a preexisting Franciscan melange, possibly in a basin perched on the continental slope (inner trench slope). The last interpretation is favored by Howell and others (1977) and by Smith and Ingersoll (1978) on the basis of different petrology, structural style, and tectonic situation as compared with correlative GVS strata. Hsü (1969) and the authors just cited report that the sandstone locally contains small amounts of Franciscan detritus, such as blueschist and red and green chert; this would be appropriate for a trench-slope basin deposit. Whatever the origin, the sandstone was caught up in the still-developing melange, which encloses detached pieces of the slab.

Another somewhat comparable sandstone mass more than 6 km long on Pine Top Mountain north of San Simeon poses identical problems. Taliaferro (1944) considered it to be a post-Franciscan formation in depositional contact with the underlying Franciscan, but Hall (1976) shows it as a part of the Franciscan. The present writer finds that it contains considerable K-spar, resembles Upper Cretaceous GVS rocks, and that its lower contact is discordant, tectonic, and thrustlike. The site and mode of origin are still unclear.

A unit of particular interest occurs on Las Tablas Creek west of Paso Robles (Fig. 13-8). It consists of massive sandstone with several zones of sedimentary breccia and conglomerate, some clasts of which are sandstone, shale, intermediate to silicic volcanic rocks, and granitic types. The igneous rocks just mentioned were probably derived from the late Mesozoic magmatic arc of the continental margin. However, other clasts are pillowed basaltic greenstone, chert, serpentinite, and glaucophane schist, all of which are believed to represent recycled earlier Franciscan material (Cowan and Page, 1975). The age of the

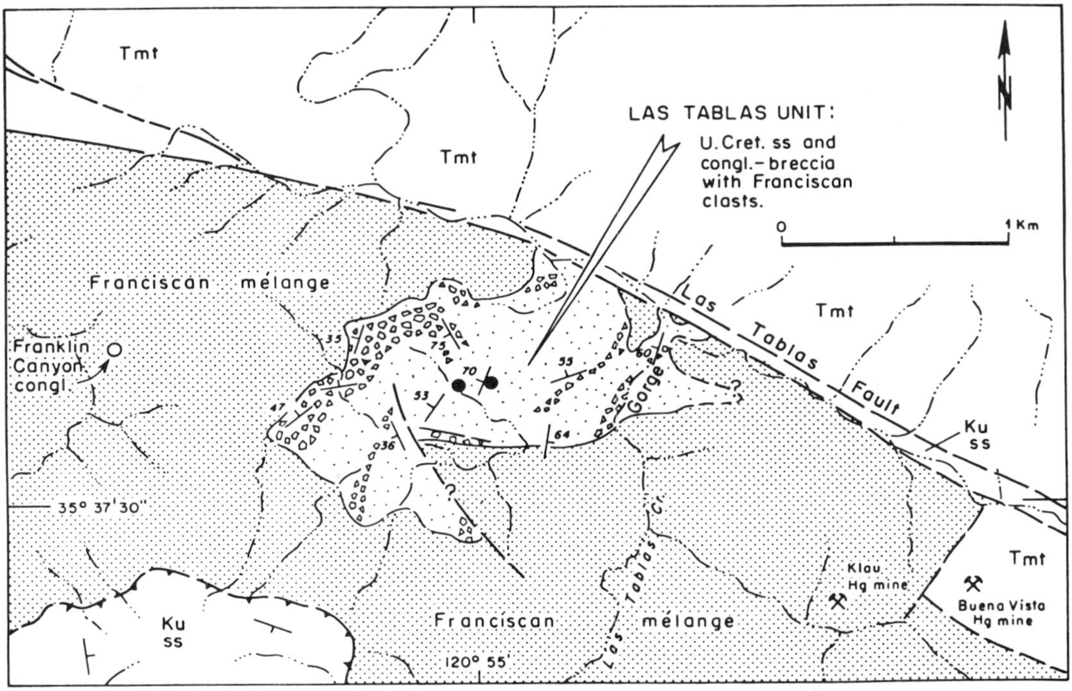

Fig. 13-8. Geologic map of Las Tablas unit. Block of Upper Cretaceous sandstone contains clasts of Franciscan rocks, yet block itself is enclosed in a Franciscan melange. Recycling of Franciscan components is apparently indicated. Solid circles, microfossil localities; Ku ss, Great Valley type Upper Cretaceous sandstone and mudstone; Tmt, Miocene strata, including Monterey Formation. (From Cowan and Page, 1975, with permission of The Geological Society of America).

unit is Late Cretaceous, based on dinoflagellates (Evitt, 1973, written communication). The entire coherent body occupies an area of about 3 km^3 in a melange terrane; it is clearly a large isolated block enveloped in the melange, and it contains pumpellyite throughout, indicating incipient metamorphism possibly caused by partial subduction. Taliaferro (1944) erroneously considered this body to be an erosional remnant of Lower Cretaceous strata resting depositionally on the Franciscan. This illustrates further the divergent interpretations of coherent units in Franciscan terranes. Each occurrence must be studied in detail and must be judged on its own merits.

Conjectures about the Franciscan offshore There is reason to suppose that the Franciscan extends southwestward offshore at least to the Santa Lucia Bank fault, and a Paleogene analog of the Franciscan may extend to the defunct trench at the foot of the continental slope (Fig. 13-2). This is based upon the usual presumption that the Franciscan is essentially a subduction complex lying between the continental plate and the trench. The structure is variously suggested by physical models such as those of Seely and others (1974) and Karig and Sharman (1975), and by the different models of Scholl and others (1977) and Cowan and Silling (1978). According to images featuring offscraping and imbrication, the Franciscan should become progressively younger from the shore southwestward to the continental slope, but this is not inherent in the models of Scholl and others (1977). In nearly all postulated schemes, discrepancies in age progression from northeast to southwest are likely. The offshore portion of the Franciscan presumably consists of melanges and large coherent units (mainly sandstone) like the onshore portions. Some ophiolitic material may be present here and there between the shoreline and the Santa Lucia Bank fault, as suggested by magnetic anomalies, but no such indications are noted seaward between said fault and the filled trench at the foot of the continental slope (Silver, in Page and others, 1979). Judging from the Coastal Belt of the Northern Coast Ranges, melanges may be fewer and blueschist metamorphism may be rare or absent in the westernmost part of the subduction complex, at least near the surface. The eastern part of the offshore Franciscan may contain a large proportion of Upper Cretaceous sandstone, whereas at some distance from the shoreline, Eocene sediments may predominate. (The Coastal Belt of Northern California contains Eocene palynomorphs, as determined by Evitt and Pierce, 1975; also, latest Eocene foraminifera are reported by Bachman, 1978). Beneath the continental slope off the Southern Coast Ranges there could be an Oligocene-early Miocene Franciscanlike assemblage. The age of the youngest components of the complex must decrease from southeastern areas to northwestern areas, as subduction ceased progressively toward the northwest. However, the time differential was partially canceled by the concurrent northwestward movement of the Salinian Block and the somewhat greater translation of the offshore region. Probably the generation of Franciscanlike assemblages terminated at about 28 to 30 m.y.b.p. (late Oligocene) beneath the lower part of the continental slope that presently lies alongside the defunct trench west of San Luis Obispo, but it may have continued into the early or mid-Miocene farther northwest along the same tract. Neogene folding is less pronounced offshore than onshore, so the primary structures of the subsea subduction zone should retain much of their original orientation.

Blueschist Facies Metamorphism in the Southern Coast Ranges

Blueschist facies metamorphism is inherently relevant to structural history. This metamorphism characterizes much of the Franciscan, and the implied high P/T ratio seems to require particular tectonic processes in the development of that complex. In this section we will limit ourselves to observations and conclusions with respect to the Southern Coast Ranges.

The Diablo Range occupies an important place in the study of this type of metamorphism. Maddock (1955, unpublished) recognized lawsonite and jadeitic pyroxene in the Mount Boardman quadrangle. McKee (1962) reported widespread jadeitic pyroxene near Pacheco Pass, and Ernst and co-workers made important studies of metamorphism around Panoche Pass, Pacheco Pass, and in other areas (Ernst, 1965, 1971b; Ernst and others, 1970). Observational, analytical, and theoretical work led Ernst to conclude that rocks containing certain mineral assemblages including lawsonite or jadeitic pyroxene had been subjected to pressures of as much as 3 to 8 kbars, concurrent temperatures of only 150° to 300°C, and that these conditions had been achieved by subduction (Ernst, 1965, 1971a, 1973). Subsequently, the rocks had regained a nearsurface position without being heated (Ernst, op. cit.). Field and laboratory work by others, some of whom are cited in the following, has also contributed to the understanding of blueschist facies metamorphism in the Diablo Range.

In the Sur-Obispo Belt southwest of the Salinian Block, less is known about the overall distribution of metamorphism, but valuable work has been carried out in a few areas (e.g., Gilbert, 1973, 1974; Callender, 1975).

Kinds of blueschist facies rocks The Southern Coast Ranges contain the familiar kinds of Franciscan blueschist facies rocks that are summarized in Table 13-1. All the listed types occur in the Diablo Range (e.g., Ernst, 1965; Ernst and others, 1970), and most occur in other Franciscan terranes in the province. Interestingly, most of these metamorphic rocks are found in the Sur-Obispo Belt southwest of the San Andreas fault as well as in the less displaced Franciscan of the Diablo Range. Admittedly, no eclogite and only a few small blocks of jadeite-bearing metagraywacke have been reported from the Sur-Obispo Belt, to the best of my knowledge. However, the general similarity of blueschist facies rocks on the two sides of the San Andreas fault, parts of which have had 500 to 600 km of displacement, reinforces the belief that these rocks were involved in petrotectonic processes affecting a continuous zone along the California continental margin prior to the birth of the San Andreas. As is well known, the zone extends southward into Baja California.

Distribution of blueschist facies metamorphism There is no continuum of gradual change in metamorphic grade across the Franciscan complex as a whole. Instead, the distribution of rocks of various grades is somewhat erratic and unpredictable, doubtless because tectonic movements have rearranged the metamorphosed units. In the Southern Coast Ranges there is no long, continuous schist or phyllite terrane (such as the South Fork Mountain Schist) along the east boundary of the Franciscan. However, relatively

TABLE 13-1. Typical Blueschist Facies Rocks, Southern Coast Ranges

1. High-grade blueschist facies rocks of basaltic bulk composition.
   Virtually no original minerals or textures are retained.
   1a. *Blueschist, sensu stricto* (= "glaucophane schist").
      Typically contains crossite, white mica, lawsonite, + garnet. Shows pronounced tectonite foliation.
   1b. *Eclogite.*
      Typically contains omphacite, garnet, Na-amphibole, and retrograde minerals.
2. Low-grade blueschist facies *greenstone* (generally metabasalt).
   Contains lawsonite.
   Retains some primary minerals or early "ocean floor" metamorphic minerals, ± pumpellyite.
3. Blueschist facies *meta graywacke* and *metasiltstone.*
   3a. Jadeitic pyroxene-bearing metagraywacke and metasiltstone.
      Contains jadeitic pyroxene, white mica, lawsonite, + glaucophane.
      Commonly retain vestiges of sedimentary structures and clastic textures.
      Faintly to strongly foliated, generally as in textural zone 2 of Blake and others (1967).
   3b. Lawsonite-bearing metagraywacke and metasiltstone lacking jadeitic pyroxene.
      Contain lawsonite, albite, white mica, ± glaucophane.
      Retain sedimentary structures and textures.
      Nonfoliated to distinctly foliated; textural zone 1 to middle textural zone 2 of Blake and others (1967).
4. Blueschist facies *metachert.*
   Some contain fine crossite, ± stilpnomelane, as well as predominant quartz.
   Partially or wholly recrystallized, but bedding may be retained.
5. Blueschist facies *metaconglomerate.*
   Some clasts and/or matrix contain postdepositional lawsonite, ± Na-amphibole, ± jadeitic pyroxene.
   Textures range from virtually unmodified sedimentary to strongly foliated with flattened pebbles.

small units of foliated metasediments occur at intervals here and there in the vicinity of the boundary. Cowan (1974) mapped a narrow body of semischist containing jadeitic pyroxene along the juncture between the Franciscan and the GVS north of Pacheco Pass (Fig. 13-9), but apparently the semischist does not continue very far to the north or south, and to the west it gives way to a melange containing blocks showing various metamorphic grades. Cowan points out the lack of systematic change in metamorphism across his map area as a whole. Coherent metagraywacke bodies studied by him have their own individual metamorphism and directions of metamorphic gradient that are independent of the grade of nearby blocks and partially independent of distance from the tectonic boundary of the Franciscan. Morrell (1978) finds a *downward* increase in metamorphic grade within Cowan's Orestimba metagraywacke unit. Raymond (1973b), working in the northeast part of the core of the Diablo Range, delineates subareas characterized by different metamorphic mineral assemblages, but finds no constant direction of increase or decrease in grade. The individual subareas appear to have a significant degree of internal consistency in the presence or absence of certain critical minerals. Crawford (1975) observes variable blueschist facies metamorphism throughout much of the northwest part of the same Franciscan terrane, and concludes that isograds are not warranted because the inconsistencies are so great.

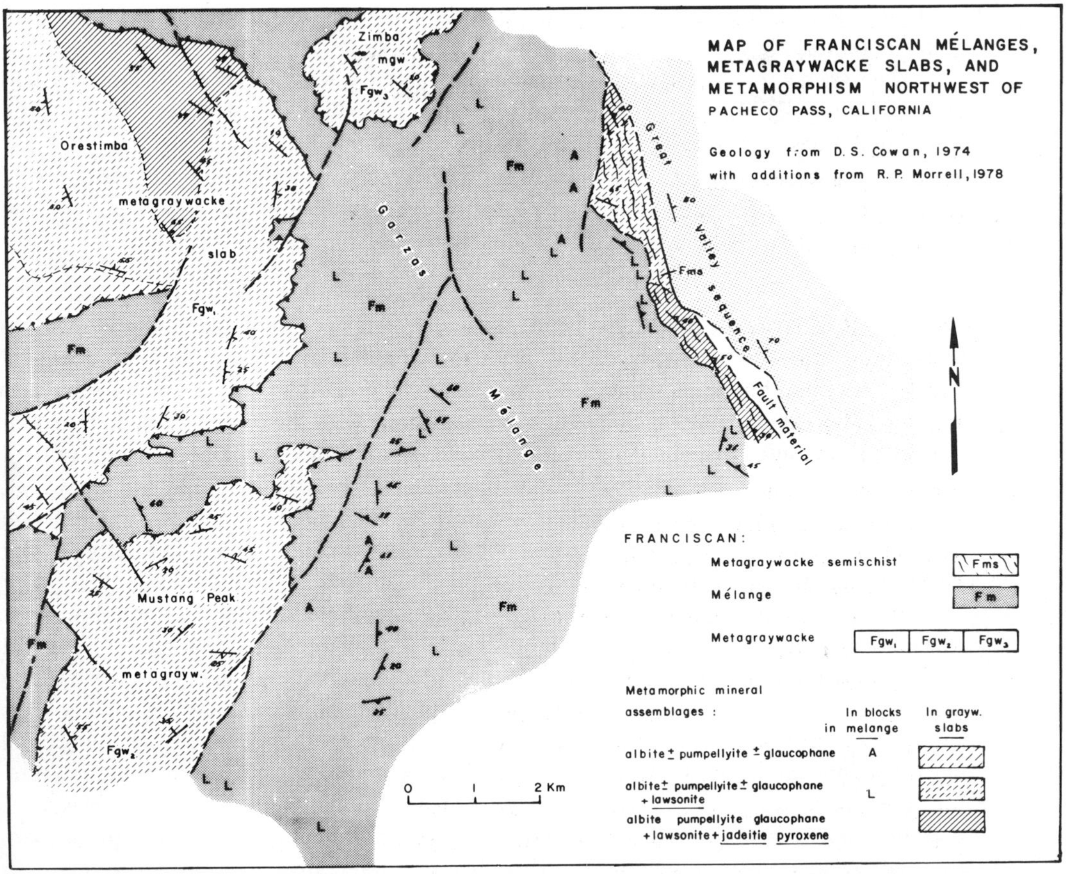

Fig. 13-9. Map showing distribution of Franciscan melanges, metagraywacke slabs, and blueschist facies metamorphism northwest of Pacheco Pass. Small blocks metamorphosed to different grades (denoted by A and L) are mingled in melange. Larger metagraywacke slabs show independent metamorphism, and Orestimba slab shows zonation. (From Cowan, 1974, with additions by Morrell, 1978; permission of The Geological Society of America).

Disruption of previously continuous metamorphic units is indicated by the presence of lawsonite-bearing or jadeitic-pyroxene-bearing rocks in melanges containing blocks of lower-grade material. Fragmentation of earlier units is obvious where blocks of high-grade glaucophane schist with tectonite fabric are enveloped in melanges containing non-foliated greenstone and low-grade metasediments.

On the other hand, the distribution of metamorphism is not altogether random. The boundary that McKee (1962) termed a jadeite isograd near Pacheco Pass seems to be upheld, insofar as it has been tested. Platt and others (1976) found jadeitic pyroxene in metaconglomerates on the appropriate side of the boundary, and found none on the other side. However, the boundary may be a postmetamorphic structural dislocation rather than an isograd (Moore, 1977). In any case, some of the jadeite-bearing rocks are blocks in a melange, the matrix of which shows penetrative shearing but no through-going metamorphic fabric.

The areal concentrations of jadeite-bearing metagraywacke shown by Ernst (1971b) are real, although scattered occurrences of such rocks have been found beyond the designated areas and it would be difficult to draw accurate limits, especially in melanges. Despite all the complexities, if the Diablo Range is visualized as a domical antiform that once had a tectonic "roof" over the presently exposed Franciscan core, many of the known jadeite-bearing rocks appear to be concentrated near the tectonic cover.

Information from metaconglomerates Some years ago, blueschist pebbles in Franciscan conglomerates were generally thought to be Franciscan detritus. However, Bloxam (1960) described blueschist facies metaconglomerates (on Angel Island in San Francisco Bay) in which the clasts had been metamorphosed along with the enclosing rock. Fyfe and Zardini (1967) and Ernst and others (1970) showed that glaucophane-bearing pebbles and other metamorphic clasts in conglomerate near Pacheco Pass were not detrital blueschist, but had acquired blueschist-facies minerals during the metamorphism of the conglomerate as a whole. Subsequently, it has turned out that most Franciscan conglomerates in the Diablo Range are metaconglomerates.

In the Diablo Range, some blocks in melanges are wholly metaconglomerate. Metaconglomerate also occurs as subordinate lenses and channel deposits in coherent metagraywacke units. Probably some of the first-mentioned blocks were derived by disruption of the lenses and channel deposits. In most cases, the pebbles and cobbles show many different compositions and textures. In any one block or channel deposit, all clasts have presumably been subjected to identical metamorphic pressure and temperature, but they have responded in different ways. Where both granite and mafic clasts are present, granitic pebbles may contain lawsonite, which preferentially replaces plagioclase, and minor glaucophane, which merely forms fringes around green amphibole; on the other hand, mafic pebbles may be largely replaced by glaucophane (Platt and others, 1976). As noted by Fyfe and Zardini (1967), some pebbles have a glaucophane-rich rind, suggesting that fluids played an effective role. Lawsonite (and less commonly, glaucophane) occurs in metaconglomerates regardless of the presence or absence of schistosity. In contrast, jadeitic pyroxene occurs only in semischistose metaconglomerates comparable to graywackes of textural zone 2 of Blake and others (1967), indicating that shearing is associated with the circumstances favoring this mineral. However, not all foliated metaconglomerates contain jadeitic pyroxene. Metaconglomerate containing jadeitic pyroxene is generally found only in areas characterized by the presence of this mineral in blocks or coherent units of metagraywacke.

The most detailed study of Franciscan metaconglomerates is by Moore (1977), who considers lithology, provenance, sedimentology, mineral assemblages, element partitioning, and metamorphic conditions and reactions. She finds that the metaconglomerates did not attain full mineralogical or textural equilibrium. However, using a variety of published data and original observation, she concludes that the range of metamorphic temperatures and pressures was 175° to 220°C and 6.7 to 7.5 kbars. The deduced temperature gradient was surprisingly low, 6° to 7° C/km. She finds that the metaconglomerates she examined fall into three distinct groups based on pebble types, postdepositional deformation, and metamorphic mineral assemblages. Interestingly, the groups occupy particular subareas in the Diablo Range. Although the metaconglomerate bodies in a subarea are not physically connected with one another, most being blocks in melanges, they contain similar types of clasts and similar metamorphic minerals and textures, and the separate conglomerate blocks of a subarea all seem to have been metamorphosed under the same range of temperature and pressure. The three subareas differ from one another with respect to these deduced parameters.

As mentioned earlier, Moore and Liou (1977) report rare, well-rounded clasts of preexisting (detrital) blueschist in metaconglomerates of the Diablo Range.

Age and tectonic history of blueschist facies metamorphism Probably the high-grade blueschist blocks in melanges of the Southern Coast Ranges are approximately the same age (i.e., 150 to 155 m.y.) as similar blocks in the Northern Coast Ranges. A blueschist block from the Crevison Peak area of the Diablo Range gave K-Ar ages of 150 m.y. from phengite, 153 m.y. from actinolite, and 120 m.y. from glaucophane (Coleman and Lanphere, 1971, p. 2406). However, most of the coherent metasandstone units and metagraywacke blocks were probably metamorphosed considerably later. In the northwest part of the Diablo Range, one large metasandstone unit locally containing lawsonite and jadeitic pyroxene must have been metamorphosed some time in the Cretaceous, as Crawford (1975) reports that it yielded a Tithonian Buchia (say 141 to 135 m.y. old). Cowan (1974) gives K-Ar dates (mostly whole rock) from ten samples, determined by R. L. Armstrong, from an area northwest of Pacheco Pass, as follows: inclusions from melange, 130 to 131, 116, and 106 m.y.; metagraywacke semischist coherent unit, 111, 110, and 76 m.y.; Orestimba metagraywacke coherent unit, 93, 90, 90, and 88 m.y. Metaconglomerates (also in the Diablo Range) that Moore (1977) considers to be possibly Turonian on the basis of lithology vis-a-vis dated parts of the Great Valley Sequence, must have been metamorphosed after 90 m.y. ago if Moore is correct. On the opposite side of the Salinian Block, metamorphism of three metasandstones in the Lucia area of the Sur-Obispo belt has been dated at 70, 74 and 88 m.y.b.p. by the K-Ar method, using whole-rock material (Suppe and Armstrong, 1972). Although some of these dates are probably inaccurate, it appears that Franciscan rocks in the Southern Coast Ranges were metamorphosed at various times during a span of 60 to 80 m.y. in the Late Jurassic and Cretaceous.

In the Franciscan subduction zone, cool rock must have been carried down, metamorphosed, and somehow returned to the surface without being heated. The high P/T ratio of blueschist facies metamorphism is well established by experimental, analytical, and theoretical physical geochemistry and mineralogy, and it is also supported by other data. Bostick (1974) found that the observed alteration of clastic organic particles (phytoclasts) in fourteen graywacke and metagraywacke samples from the Franciscan of the Diablo Range suggests past confining pressures of about 4.5 to 6.5 kbar and temperatures of only 120° to 150°C. Two of the samples, including the "coolest," contained jadeitic pyroxene. Although hot mantle material initially existed both above and below the subduction zone, the blueschist facies rocks in some cases were probably sandwiched between imbricate wedges of cold sediments. Continued subduction and underthrusting of more and more cold material could have repeatedly shielded the metamorphosed rocks from the normal upward heat flux from the underlying mantle.

At the ground surface the relatively high grade metamorphic rocks (e.g., jadeitized semischist) seem to be localized somewhere near the 'lid" of the subduction zone. This has been explained for the Franciscan as a whole by Platt (1975) as a consequence of postulated dynamics and geometry of a growing subduction complex. The interpretation does not depend upon tectonic overpressure, which, if it alone sufficed, would likely produce blueschist facies metamorphism near ordinary thrust faults as well as in subduction zones. However, penetrative shear, in conjunction with a high P/T environment, would seem to favor the development of blueschist facies tectonites near the "lid" of the system in the early stages of subduction.

As noted by Ernst (oral communication, 1977), it is remarkable that blueschist facies rocks in the Diablo Range could have been metamorphosed in the Mesozoic and could have "kept their cool" for 80 to 100 m.y. or more despite tectonic burial. However, during their sojourn at great depth, these rocks may have been protected from heating as just described, and they may have emerged into the unheated upper part of an outer arc ridge in Cretaceous time. Admittedly, by the Paleocene or earlier, the Franciscan of the Diablo Range was stuffed under a tectonic cover of Great Valley rocks possibly 8 to 12 km thick, and the metamorphic rocks were lodged at a depth where a normal temperature gradient would require warm conditions. However, probably warming was prevented by continued subduction at a deeper level intercepting heat flow from below. Subduction beneath the Diablo Range persisted until about 17 m.y.b.p. in the south and until 6 to 7 m.y.b.p. in the north. The rise of blueschist facies rocks must have been completed prior to these dates, certainly if motivated by subduction mechanisms. Long before a normal thermal gradient could be reestablished, the metamorphosed rocks had reached a safe level.

Origin and Evolution of the Franciscan Complex

The origin of the Franciscan is still obscure, so the following remarks are simply a summary of tentative ideas. For purposes of this resumé, it is necessary to consider some observations made by others in the Northern as well as Southern Coast Ranges.

Evidence for origin in a subduction zone A genetic tie between subduction and the Franciscan is amply established by the collective weight of the following circumstances:

1. The indicative spatial position of the Franciscan with respect to a relict filled trench offshore, apparent forearc basin deposits (GVS), and a plutonic arc (Sierra Nevada).

2. Paleomagnetic and other geophysical evidence for plate convergence in the late Mesozoic when the Franciscan was evolving. One of the compelling conclusions is that ocean-floor spreading was semicontinuous in both the Atlantic and Pacific, and that passive margins persisted in the widening Atlantic. By inference, lithosphere

must have been consumed concurrently in the Pacific. The entire east half of that lithosphere is missing at the latitude of central California. The reconstructions of Larson and Chase (1972) and Larson and Pitman (1972) imply that at least 7000 km (and probably more) of Pacific lithosphere has underthrust western North America since the beginning of Cretaceous time.

3. The quasi-contemporary ages of parts of the Franciscan, GVS, and Sierran plutons.

4. The presence of an upturned ophiolite at the base of the GVS, suggesting that the Franciscan was pushed beneath the crust and mantle of the adjacent plate.

5. The presence of highly sheared melanges in the Franciscan Complex.

6. The presence of high P/T metamorphism in the Franciscan, in part coeval with high T/P metamorphism and plutonism in the Sierra Nevada.

7. The mixture, in the Franciscan, of fragments of oceanic rocks, mantle material, and terrigenous sedimentary rocks, suggesting convergence of oceanic and continental plates.

8. Prevalent soft-sediment disturbance throughout much of the Franciscan. This is appropriate for, although not diagnostic of, a subduction zone environment.

Chronology The time of inception of Franciscan evolution is only tenuously established. It is commonly supposed that the 155 to 135 m.y. old high-grade mafic blueschist, found as fragments in melanges, formed by subduction. Suppe and Foland (1978) propose instead that these rocks originated beneath an obducted slab. In any case, the schist could represent either the initial stage in Franciscan evolution or a pre-Franciscan episode whose products were later recycled. Such fragments are found even in the westernmost part of the Central Belt in the Northern Coast Ranges (Blake and Jones, Chapter 12, this volume), far distant in time and space from their original habitat. Ages of about 150 m.y. approximate the time of the Nevadan orogeny, a critical point in Sierran history that perhaps resulted from a continent-island arc collision (Schweickert and Cowan, 1975; Schweickert, Chapter 5, this volume). Such an event could have caused a westward jump in subduction and thus could have started the Franciscan cycle. Sierran plutonism, which ideally should have mirrored Franciscan subduction, actually subsided for an interval after the Sierran orogeny, but resumed about 135 to 125 m.y.b.p. (Saleeby, Chapter 6, this volume), so by that time the Franciscan was almost certainly being assembled. Voluminous arc plutonism continued in the Sierra during most of the Cretaceous, through 88 m.y.b.p. (Coniacian?), but later magmatic activity moved eastward. Volcanism, plutonism, and crustal deformation affected the Laramide zone roughly 70 to 45 m.y.b.p., presumably because the downgoing slab assumed a very low angle of inclination (Dickinson and Snyder, 1978; Keith, 1978). During this period, much of the Coastal Belt was accreted. The end of subduction must have terminated Franciscan evolution. This event, which is discussed later, coincided with the passage of the Mendocino triple junction northwestward from Mexico to northern California (Atwater, 1970; Dickinson, Chapter 1, this volume). Depending on the latitude, subduction beneath the Southern Coast Ranges apparently ceased between 25 and 6 m.y.b.p., but near Cape Mendocino it continued to the Quaternary, and corresponding neo-Franciscan material must have been assembled concurrently.

The prolonged genesis of the Franciscan Complex is inferred not only from the duration of Sierran arc-magmatism and the width of accreted terrane in northern California, but also by a fragmentary record of successive events in the Complex itself. Various ages of blueschist facies metamorphism have already been cited. In the Goat Mountain area of the Northern Coast Ranges, Suppe and Foland (1978) find relatively early schuppen structure followed by Late Cretaceous thrusting no older than about 90 m.y. After the subsequent advent of the Coast Range thrust, the whole complex was folded. In the Southern Coast Ranges near Paso Robles, a Franciscan melange contains a sedimentary block that apparently incorporates recycled Franciscan detritus (Fig. 13-8), including high-grade blueschist. This implies metamorphism of a protolith at considerable depth, rise of the schist to the surface, fragmentation and sedimentation of the blueschist and other Franciscan rocks in the Late Cretaceous, breakup of the sedimentary accumulation, incorporation of part of it in a melange, and burial (subduction?) sufficient to induce crystallization of pervasive pumpellyite. The length of time represented by this sequence of events could not be trivial.

Possible modern analogs of environment and processes of origin The association of the Franciscan with a deduced paleo-subduction zone implies that conditions well-established in modern subduction zones should apply to the origin of the Complex. Oceanic crust with a burden of pelagic (and other ?) sediment must have been ferried, by ocean-floor spreading, to a trench, as in modern examples. However, most of the pelagic sediment, except for Upper Jurassic-Lower Cretaceous chert, seems to have been carried down and lost (Scholl and Marlow, 1974). The trench probably acquired terrigenous sediments deposited on the pelagic material, and some of this may have been scraped off, as reported in some present-day subduction zones. Excellent documentation of offscraping of sediments is given by White and Klitgord (1976) and White (1977) for the Gulf of Oman, and the same authors note other areas where this process is taking place. Many Franciscan sandstones and argillaceous rocks may have been gathered into the Complex in this manner. Slopes on the landward side of today's trenches are relatively steep (e.g., the Aleutian Trench; Grow, 1973), favoring gravity slides and flows that could cause soft-sediment deformation and disruption of bedding. Large slides of slope material have been reported in the Aleutian Trench by Piper and others (1973), von Huene (1974), and Moore (1977). Soft-sediment deformation as well as disrupted bedding is seen in many Franciscan rocks. Some trench slopes have terraces or basins that collect terrigenous sediment, perhaps explaining the origin of certain independent (?) Franciscan sandstone units such as the Cambria slab. Many modern trench slopes are draped with young in situ slope sediments. Typically, the layers are interpreted (from seismic reflection profiles) to be cut repeatedly by rather small-scale faulting, which might ultimately dismember such sedimentary blankets. By inference, comparable disturbances deeper within an active subduction complex would help to explain the lack of stratal continuity in the Franciscan.

A number of credible subduction-zone models emphasize stacked, imbricated wedges of deformed material (e.g., Seely and others, 1974; Karig and Sharman, 1975). Probably these models do not fit all cases, but in some instances they tend to be substantiated by selected seismic reflection profiles (e.g., Hamilton, 1977a) and by drilling. A drillhole on the inner slope of the Nankai Trough is interpreted to have penetrated uplifted trench deposits that form a recumbent isoclinal fold (J. C. Moore and Karig, 1976). Impressive support for imbricated wedge models can be found in the results of the Deep Sea Drilling Project, Leg 66, near the Middle America Trench (Moore and others, 1979). The following features of the Franciscan Complex largely coincide with the models and with (interpreted) observations of active subduction zones:

1. In a very crude way, the Franciscan of the Northern Coast Ranges appears to "young" oceanward, as the models predict.
2. Bedding and shear surfaces in the Coastal Belt generally dip toward the continent, although folds locally modify this overall orientation. Shear surfaces in the older Sur-Obispo Belt likewise commonly dip landward, but this attitude may not be the original one.
3. In the Coastal Belt, many of the folds have a westward vergence, as the models predict.
4. Slabs of coherent sandstone and intervening melanges, in parts of both the Northern and Southern Coast Ranges, can be interpreted as stacked tectonic units. Where seen, the contacts are shear surfaces or zones.
5. Within the Franciscan there are abrupt juxtapositions of material of different ages, in some cases the younger rocks being beneath the older.

The imbricated accretionary models do not seem to apply everywhere, and indeed it is improbable that Franciscan-like assemblages are generated in all zones of plate convergence. Drillhoes adjacent to the Japan and Mariana trenches (Langseth and others, 1978; von Huene and others, 1978; Hussong and others, 1978) did not find accretionary deposits. Off northern Japan, little-disturbed but fractured continental slope deposits of late Miocene to Holocene age extend to a point only 25 km from the trench axis (von Huene and others, 1978). Some recent seismic reflection studies of the Aleutian Trench slope (von Huene and others, 1979) and a number of other subduction zones have not confirmed the existence of imbrications of the kind that characterize most current models (D. W. Scholl, 1979, written communication). If the imbrications exist, they are probably too far beneath the trench slope to be detected by the methods used, but possibly they are absent, at least in some localities. In some instances, in situ slope deposits, little disturbed to strongly deformed, appear to constitute much of the pile adjoining the trench (D. W. Scholl, M. A. Fisher, and T. R. Bruns, 1978, informal communication). In view of the foregoing and other considerations, including the paucity of pelagic sediments in the Franciscan and its counterparts, Scholl and others (1977) doubt that the offscraping of sediments was a dominant process in the development of coastal mountain belts around the Pacific margin.

Putting aside underwater observations and subduction-zone models, another way of studying the origin of the Franciscan (by presumed analogy) is to examine young exposed complexes, including melanges, on islands near active trenches, as in Barbados (Daviess, 1971) and Nias off the coast of Sumatra (G. F. Moore and Karig, 1976). On Nias, the presence of melanges alternating with normal sedimentary units and the occurrence of exotic blocks are certainly reminiscent of the Franciscan. The association with subduction is manifest. In California, the Coastal Belt offers the best hope of deciphering details of Franciscan origin, as it is the youngest exposed part of the Complex and was near an active trench as recently as the Pliocene or Pleistocene. Bachman (1978) describes the structures and sedimentary features of northern Coastal Belt rocks, and makes a thoughtful interpretation in terms of an accretionary prism and processes known and inferred near modern trenches.

In view of doubts about the importance of offscraping of oceanic material in modern subduction systems, Franciscan sediments should be studied more carefully with respect to evidence for accumulation as "ferried-in" offscrapings and evidence for relatively in situ slope deposition. Either, both, or neither of these options may apply. Chert-graywacke associations and greenstone-chert associations might be regarded as favoring the hypothesis of deposition oceanward from a trench, followed by offscraping, but Blake and Jones (1974) and Scholl and others (1977) think most Franciscan chert may have originally been a capping on ophiolite underlying sedimentary sequences of marginal basin(s) or other sites on the "hanging wall" side of the subduction system. This view is supported by the commonplace limited, relatively old age range of the chert. Franciscan clastic units with sedimentary features of slope, inner fan, or upper channel deposits could indicate deposition on or near a trench slope rather than on the ocean floor beyond a trench. The problem is complicated by the possibility that large-scale strike slip may, in some cases, have transported slope deposits to a subduction site where they were offscraped.

Although the study of active subduction zones and associated young assemblages has improved the understanding of the Franciscan Complex, major questions remain. Even the most fruitful observations have not yet explained the innermost mechanisms of such systems, the origin of Franciscan melange, the manner of incorporation of blueschist facies rocks in melanges, the rise of blueschist-facies rocks to the surface, and the manner in which oceanic crust and mantle fragments entered the system.

Origin of Franciscan melanges The characteristic high-grade blueschist, the serpentinite, most of the greenstone, and the radiolarian chert occurring as blocks and slabs in the Franciscan are probably among the oldest components. The high-grade blueschist is typically about 150 to 155 m.y. old (Coleman and Lanphere, 1971), and hence is late Middle Jurassic or early Late Jurassic. The serpentinized ultramafic rocks, presumably representing mantle material, have not been dated. Greenstone is locally associated with chert resting on it depositionally and is inferred to be late Middle or early Late Jurassic in view of known chert ages. Chert, occurring with or without greenstone, is mainly Late Jurassic (late Kimmeridgian and Tithonian), although Early Cretaceous and even mid-Cretaceous cherts are known (Pessagno, 1977a; McLaughlin and Pessagno, 1978).

Evidently most of these materials could have been acquired by the Franciscan Complex in its early formative stages. Apparently, few such rocks were added during later evolution of the Complex, so some persons have concluded that most of these components were acquired in a "one-time" situation that did not recur except for recycling events. Favorable conditions may have existed only in the first few millions of years of Franciscan subduction history, when the downgoing slab of lithosphere was presented with an adjacent, nearly naked "hanging wall" of oceanic crust and mantle. Conceivably, this "hanging wall" was rasped and plucked, as was the descending oceanic crust, inasmuch as there was not yet a voluminous padding of terrigenous sediment caught between the two plates. Eventually when large quantities of terrigenous sediment occupied the zone of underthrusting, the presence and movements of this material, which was initially water saturated, must have had something to do with the return to the surface of fragmented oceanic lithosphere (with chert) and high-grade blueschist. The returned material may have joined newer scrapings as the subduction complex grew.

Although most of the greenstone-basalt in melanges is probably old and was very likely acquired early, some is probably post-Jurassic. The basalt of Black Mountain south of Palo Alto is presumed to have originated as an elongate submarine extrusive welt, possibly along a leaky transform fault or a spreading system. It is locally overlain by discontinuous bodies of microfossiliferous, early Cenomanian Calera Limestone, believed to have been deposited in perched basins (Wachs and Hein, 1975). The limestone is associated with pillow basalt, is intruded by a diabase sill, and contains an interbedded ash layer. Therefore, the basalt pile is probably only slightly older than the limestone, that is, mid-Cretaceous in age, perhaps 95 to 105 m.y. Basalt in the Coastal Belt of northern California may be even younger. Ways in which post-Jurassic oceanic crust could have been incorporated in the Franciscan include the following: seamounts and also basaltic prominences along leaky transform faults may have been scraped off the downgoing slab at almost any stage of Franciscan development. Parts of sea floor uplifted along faults near the trench may likewise have been scraped off; appropriate faults are found in the acoustic basement near modern trenches. Low-angle thrust faults like those commonly envisioned in the sediments of subduction complexes may have developed in the downgoing slab and may have allowed imbricate wedges of oceanic crust to be emplaced in the pile of deformed scrapings. However, it is not certain which, if any, of the foregoing explanations actually applies to basaltic blocks in the Franciscan.

The old, distinctive rock types (e.g., high-grade blueschist, serpentinite, Late Jurassic greenstone, radiolarian chert) seem to have reappeared again and again during the accretion of the Franciscan Complex and have repeatedly been mingled with younger material. Perhaps submarine slides or flows recurrently carried preexisting melange components down the slope and into the trench, where they were resubducted before appearing again at the surface. This is crudely indicated in Fig. 13-6, which, however, is not intended to show all the other processes that operated. The scheme involves an interplay between subduction and olistostromes. Olistostromes, not necessarily acting as shown in the figure, have been proposed as elements of Franciscan genesis by many persons, including Hsü (1968, 1969, 1974), Blake and Jones (1974, 1978), Suppe (1973), Maxwell (1974), Crawford (1975), and Cowan (1978). Similarities between the Franciscan and the Lichi melange (of Taiwan), which appears to be demonstrably olistostromal have been pointed out by Page (1978). However, if olistostromal units exist in the Franciscan, many of their original features have been obliterated by later deformation.

Regardless of our ignorance concerning mechanisms, repeated deformations are manifest in melanges. The findings of Cowan (1978), already cited in the discussion of the Sur-Obispo Belt, are a good example. Structural features of various ages and environments include soft-sediment structures, boudinage, extensional fractures, pull-aparts, compressional flattening, shearing, cleavage in (some) argillaceous material, foliation and schistosity in metasediments, rotation of blocks, and slickensides on multiple curved surfaces. Some deformation probably occurred in an environment of high fluid pressure, which could have facilitated disruption even at considerable depth and confining pressure.

Rise of blueschist-facies rocks to the surface The rise of blueschist-facies rocks from depths of 15 to 30 km to the surface is difficult to explain. The cause, whatever it may be, probably applies also to the less extreme movements of some other components. A wedging action accompanying subduction, effected by relative movement of tapered, imbricate slices, has been proposed as a cause of uplift of blueschist-facies rocks in accretionary subduction complexes (e.g., Platt, 1975), but this may not adequately account for large vertical components of movement. A different and promising hypothesis is suggested by Cowan and Silling (1978). Their scale-model experiments indicate that the inner regions of mature, active accretionary complexes may behave overall like viscous fluids, and material that has been carried along by the subducting slab may change direction and flow upward well within the interior of the complex. Their model shows distortion and disruption of original layers during the course of these movements. With or without the forced flow proposed by Cowan and Silling, gravitative buoyancy may assist the rise of subducted material. Ernst (1978, informal communication) points out that subducted sedimentary rocks could experience bulk buoyancy despite the increase in density resulting from blueschist facies metamorphism. The density of jadeitized graywacke is only about 2.7 to 2.75 g/cm^3, whereas some of the material of the "hanging wall" of the subduction zone (oceanic basalt, gabbro, peridotite) ranges from 2.8 to 3.4 g/cm^3. Large masses of graywacke may have buoyantly worked their way to shallower levels (not necessarily vertically), perhaps carrying bodies of denser mafic material.

Although all the mechanisms cited may contribute to the rise of blueschist-facies rocks, none fully explains the commonplace isolated occurrences of glaucophane schist, for example, in unmetamorphosed melange matrix. The matrix could not have picked up the high-grade blueschist in its high P/T metamorphic environment without inviting metamorphism of the matrix itself, so there must have been intervening steps leading to the final association of schist and matrix. Regardless of the means by which metamorphic rocks rose to the surface, other processes must account for mixing with unmetamorphosed materials. The lack of continuity of high- and medium-grade metamorphic rocks and the erratic distribution of fragments thereof may be caused by mechanical disruption attending continued subduction, initial dismemberment of slabs during their rise to the surface, and eventual dispersion and mixing caused by olistostromal events at intervals during further evolution of the subduction complex.

Motions parallel with the continental margin, superposed on subduction The ideas and arguments of Blake and Jones (1978, and Chapter 12, this volume) add an important new dimension to the problem of origin of the Franciscan Complex. It now seems that megamovements probably occurred more or less parallel with the continental margin during the evolution of the Franciscan, either by pure strike slip along transform faults or as a result of oblique plate convergence. The latter is exemplified today by the Sumatra-Andaman zone. Very large Mesozoic transport events in the eastern Pacific are inferred from the finding of displaced terranes in Alaska and British Columbia (e.g., Jones and others, 1972, 1977; Davis and others, 1978; Berg and others, 1978). The various times of arrival of these fragments are poorly known, but seem to be between Late Triassic and mid-Cretaceous. Some of the fragments came from the south, as shown by paleomagnetic data. Saleeby (Chapter 6, this volume) believes that the disposition of rock assemblages and structures in the western Sierra are consonant with northward motion of the Pacific realm relative to North America throughout much of Mesozoic time. Evidence cited by Blake and Jones (1978, and Chapter 12, this volume) indicating megatransport of Franciscan rocks is reinforced by other data. Alvarez and others (1979) report that the mid-Cretaceous Laytonville Limestone in the Franciscan Complex of Northern California originated south of the equator. The paleomagnetic and paleontologic data from the limestone mutually lead to this conclusion. Although none of the foregoing findings relates specifically to the Franciscan of the Southern Coast Ranges, there is no reason to doubt that the southern Franciscan shared a generally similar history.

The structural effects of strike slip or oblique movements ("Transpression"; Harland, 1971; Saleeby, Chapter 6, this volume) on the Franciscan in its formative stages are not known, and no features ascribed to such events have been reported in the Franciscan of the Southern Coast Ranges. However, we may expect that some tracts of Franciscan rocks that are now alongside one another originated at different latitudes, albeit in similar subduction zones, and today's Franciscan terranes did not necessarily originate opposite any Great Valley-type rocks that may be nearby at present. If these complications are real, they are analogous to the effects of the more familiar Neogene transform regime. For example, Neogene strike slip must have brought different Franciscan facies into juxtaposition along the offshore Hosgri fault zone (Fig. 13-3) and probably did so along the Santa Lucia Bank fault farther west. Late Cenozoic, including Quarternary, strike-slip movements probably affected the northern part of the Coastal Belt during the most recent stages of its development, as the San Andreas fault is nearby (Bachman, 1978), but apparently specific features resulting from this influence have not been recognized in the Coastal Belt rocks.

Probably strike-slip movements do not produce wide melanges, at least by direct mechanical disruption. If it were otherwise, presumably the San Andreas fault would be bordered by melanges in southern California where Franciscan rocks are lacking along the fault. On the other hand, strike slip might play an important cooperative role. Large-scale submarine strike-slip faulting during the development of the Franciscan Complex could have brought unfamiliar rocks from distant sources, may have provided breccias and slivered rocks for eventual incorporation in melanges, and quite likely could have produced scarps favoring olistostromes.

How many subduction zones and arcs? Resolution of the question of dual or multiple Franciscan subduction zones is beyond the capabilities of this paper, but the existence and importance of the underlying concept should be recognized here. It has been postulated that (older) parts of the Franciscan were formed in one subduction zone and were later joined with Franciscan material that was accumulated independently in a different zone, and that magmatic arc(s) existed oceanward of the Sierran-Klamath arc-trench system (Blake and Jones, 1974, and Chapter 12, this volume). Supporting evidence includes the fact that some Franciscan sediments are older than the basal sediments of the nearby Great Valley Sequence, and some are coarser than, and compositionally different from, coeval GVS sediments to the east, so these particular Franciscan deposits do not seem to represent distal time-equivalents of the GVS. Other evidence includes exposed volcanic arc assemblages oceanward of the Franciscan in Northern California and Oregon, and calc-alkaline volcanic material of unknown provenance at the base of the GVS sediments above the Del Puerto ophiolite (Blake and Jones, 1974). Some of the disparities in textures and ages of Franciscan vis-a-vis GVS sediments might be explained by large-scale strike slip without the existence of more than a single arc-trench system, but the reported volcanic assemblages and volcanic constituents in certain indicative situations seem to require magmatic activity west of the Sierran-Klamath arc. A combination of dual (or multiple) arc-trench systems and large strike-slip displacements cannot be ruled out, and indeed may eventually be confirmed.

Later history of the Franciscan Complex Most Franciscan areas that are now exposed probably remained below the surface until the Neogene, but were not deeply buried. Rare blueschist detritus is reported from the Panoche Formation (Upper Cretaceous) of the Great Valley Sequence near the Diablo Range and from the Eocene Domengine Formation of the same region. This does not mean that an ancestral Diablo Range was uplifted, which would surely have left a more striking record, but it may mean that during subduction to the west an outer arc ridge of Franciscan rocks occasionally contributed small quantities of sediment to the forearc basin. (It is unlikely that the rare blueschist detritus came from old sutures in the Sierra Nevada, as Sierran plutonism had heated most of the prebatholithic rocks.) A more impressive, but localized occurrence of Franciscan debris is in Eocene turbidite-type sandstone on the San Francisco Peninsula west of San Carlos. Sediments interpreted to be submarine debris flows contain angular blocks (1 cm to 2.5 m) of Franciscan rocks, including veined graywacke and red chert. These were presumably derived from an outer-arc ridge adjacent to the forearc (?) depositional basin. The latter mainly received arkosic material from the continent.

Subaerial emergence of some Franciscan areas must have occurred in Oligocene and Miocene time, as noticeable amounts of rounded Franciscan sediment are mingled with granitic detritus in such units as the Lospe, Vaqueros, and Temblor formations. However, there is no reason to believe that the Franciscan source areas remained exposed. No continuing floods of Franciscan debris occurred until late Pliocene and Quaternary time, when certain nonmarine synorogenic sediments (e.g., San Benito Gravels, Livermore Gravels, and Santa Clara Formation) acquired abundant pebbles of graywacke, chert, greenstone, and some blueschist.

Oceanic crust and mantle assemblages were recognized in the Coast Ranges by Bailey and others (1970), who pointed out that some of these are overlain depositionally by sediments of the GVS and are underlain tectonically by the Franciscan Complex, which was "dragged below the rocks of the Great Valley sequence by sea floor spreading" (p. C70). This subject is treated in greater detail by Hopson and others (Chapter 14, this volume).

In addition to classic ophiolite assemblages consisting of several cogenetic units, there are many isolated monolithologic blocks and masses of mafic and ultramafic rock that are also presumably fragments of oceanic crust or mantle material. If we include these in the overall array, the occurrences may be grouped as follows: (1) ophiolite at the base of in situ GVS sediments (eg., the Del Puerto ophiolite of the Diablo Range); (2) tectonic outliers of ophiolite underlying GVS-like sediments of unknown original locations (e.g., Point Sal and Cuesta Ridge ophiolites, Sur-Obispo Belt); (3) blocks and slabs of mafic and ultramafic rock (the latter chiefly serpentinite) in Franciscan melanges; and (4) cold serpentinite intrusions and extrusions localized at diapiric centers (e.g., New Idria diapir in the Diablo Range) or along faults (e.g., Table Mountain, Diablo Range).

A few examples of occurrences, significance, and problems are given next.

Del Puerto The important Del Puerto ophiolite (Maddock, 1964; Evarts, 1977, 1978) is in the northeast part of the Diablo Range (Fig. 13-4). Although in many respects this is the "best" ophiolite exposed at the base of the in situ GVS sedimentary sequence, it is tectonically dismembered and some parts have been lost. The following capsule summary is based on detailed studies by Evarts.

A large mass of peridotite evidently consists of both depleted mantle material (harzburgite) and overlying cumulates of dunite, wehrlite, and podiform chromitites. Interestingly, the penetrative deformation of the depleted mantle continues into the ultramafic cumulates. The middle part of the ophiolite is represented by displaced remnants of a plutonic sequence consisting of ultramafic and gabbroic cumulates, hornblende gabbro, and quartz diorite, all intruded by dikes of diabase, microdiorite, and plagiogranite. The top of the ophiolite, now displaced from the rest by faulting, consists of volcanic flows (keratophyre and basalt) breccias, and hypabyssal intrusions, mostly sills. K-Ar ages of 158 and 160 m.y. were obtained by Lanphere (1971) from dikes cutting ultramafic rocks of the ophiolite.

The Del Puerto ophiolite is overlain depositionally by Upper Jurassic volcaniclastic and tuffaceous sediments, the Lotta Creek Tuff, which differ from the familiar Knoxville-type basal sediments in the GVS elsewhere. (Knoxville-type shale does occur above the Lotta Creek Tuff.) The volcaniclastic sediments consist of interbedded tuffaceous chert and *andesitic* clastic material apparently derived from a nearby active volcanic arc, *not* from erosion of the ophiolite. This fact, together with the absence of a sheeted dike complex, suggests that the ophiolite probably originated in a marginal basin or possibly in an island arc, rather than at a midocean ridge (Evarts, 1977, 1978).

The harzburgite unit is underlain tectonically by the Franciscan Complex, and Evarts reports that antigorite schist locally intervenes at the contact, perhaps marking the

Coast Range thrust of Bailey and others (1970). The ophiolite lacks high P/T metamorphism; mafic dikes cutting cumulates commonly contain the low-pressure assemblage actinolite + calcic plagioclase. On the other hand, rocks of the Franciscan Complex in the vicinity locally contain high P/T minerals, including jadeitic pyroxene in metagraywacke.

The sandstone of the Franciscan metagraywacke originated at a much shallower level than the deep parts of the ophiolite, yet it now shows the effects of higher pressure than the latter. Therefore, it seems likely that components of the Franciscan were subducted, then returned near the surface, were incorporated in the Franciscan Complex as presently assembled, and the Complex was forced en masse under the ophiolite with its cover of GVS sediments. It is doubtful that the underthrusting represented by the Coast Range thrust could have produced the metamorphism now seen in the Franciscan without similarly affecting the ophiolite; hence, the metamorphism predates the thrust.

Point Sal The Point Sal ophiolite, which is in the southern extension of the Sur-Obispo Belt (Fig. 13-2), is advantageously exposed in sea cliffs and, although internally displaced by faults, it is more nearly complete than any other that has been studied in the Coast Range. The ophiolite is evidently an allochthonous fragment tectonically underlain by the Franciscan Complex, and it is overlain by thin chert and clastic Upper Jurassic GVS-type sediments that are not demonstrably attached to any in situ section, but which resemble the "Knoxville" of the classic GVS.

The following information concerning the Point Sal ophiolite is summarized from Hopson and Frano (1977), who have made a meticulous study. Depleted mantle material is virtually absent, but cumulate ultramafic and gabbroic rocks are abundantly represented. Noncumulus gabbro, diorite, and quartz diorite also comprise a large volume, the total plutonic sequence (including cumulates) comprising a thickness of about 1.5 km, the greatest thickness of such material known in the Southern Coast Ranges. This plutonic sequence is succeeded upward by a sill complex with some dikes. Continuing upward, according to Hopson and Frano, there are two successive extrusive units, representing two separate stages of crustal generation. The earlier (and lower) extrusives consist chiefly of albitized, virtually olivine-free basalt flows collectively more than 1 km thick. This pile predates the previously described plutonic unit and sill complex. The sills intruded the early basalt; they and the plutonic unit represent a new batch of magma, which erupted to form a second, olivine-bearing basaltic pile on top of the first.

Hopson and Frano (1977) think that the Point Sal ophiolite originated at a spreading oceanic ridge, basing their opinion on the following evidence: Interpillow pelagic sediment is biogenic, free of terrigenous sediment and arc-derived ash; vesicles in pillow rims are sparse and tiny, indicating deep water; the lavas are mainly tholeiites petrographically resembling oceanic ridge basalts; they are low in TiO_2 and P_2O_5, and are LREE-depleted; the plutonic rocks crystallized from exceptionally hydrous magma, suggesting admixture of sea water admitted by rifting; and the strong internal deformation of cumulates during later magmatic activity points to active tectonism. Hopson and Frano postulate that, some time after the first pile of oceanic basalt was erupted, the spreading center shifted, and the intrusives and second volcanic pile were emplaced.

San Luis Obispo (Cuesta Ridge) Ophiolite The ophiolite of Cuesta Ridge near San

Luis Obispo was reconnoitered by Page (1972) and studied from a petrogenetic-chemical viewpoint by Pike (1974). The ophiolite is inadequately exposed and severely mutilated tectonically, but some aspects, including the structural setting, are instructive.

The ultramafic rocks, which are separated by faults from the rest of the sequence, are largely serpentinized. Judging from serpentinite textures and small remnants of original rocks, apparently the ultramafics include both depleted mantle material (harzburgite) and cumulates (e.g., dunite and wehrlite cut by rodingitized diabase dikes). Gabbro is scantily represented; probably much has been lost by faulting. Pike (1974) thinks a stratiform plutonic complex represented by the ultramafic cumulates and gabbro intruded the contact zone between ultramafic "basement" and the pile of volcanic flows and breccias forming the upper part of the ophiolite. A sill complex (quartz gabbro, diorite, trondhjemite, and plagiogranite) then intruded both the stratiform complex and the basal volcanics. The volcanics are mainly basaltic breccias and pillow lavas, but some keratophyre is present. Mafic rocks have been affected by greenschist-facies metamorphism; no blueschist-facies mineral assemblages have been found. The volcanics are overlain depositionally by a thin but laterally persistent Upper Jurassic radiolarian chert unit, which is followed by GVS-type terrigenous mudstone with turbidite sandstone interbeds. Thus the ophiolite and its sedimentary cover are displaced fragments of a sequence strikingly similar to that of the in situ GVS.

Although it was once presumed that the Cuesta Ridge ophiolite was generated at a spreading oceanic ridge (Page, 1972), it is now uncertain whether it originated at a ridge or in a marginal basin. The biogenic chert is indicative of deep water, and the ocean floor was inaccessible to terrigenous detritus until sea-floor spreading ferried it toward land. Apparently, no active volcanic arc was near the depositional site, judging from the lack of volcaniclastic sediments. These conditions could obtain either at a ridge or in favorable parts of a marginal basin. The lack of a sheeted dike complex may favor a marginal basin origin.

A cross section showing the inferred structural relations of the ophiolite is shown in Fig. 13-10. The Franciscan appears on both sides of the remnant and is assumed to have been thrust beneath it. Furthermore, the ophiolite and its cover of Upper Jurassic-Lower Cretaceous argillaceous sediments have been thrust landward beneath Upper Cretaceous sandstone. The structure may reflect one episode (or several) of underthrusting, possibly related to a late stage of subduction in the Early Tertiary.

Blocks and slabs of mafic and ultramafic rocks in melanges Franciscan melanges contain blocks of basaltic greenstone, some of which are pillowed, and certain melanges contain diabase, which may have been derived from ophiolite dikes and sills. More rarely, blocks of gabbro are present. Ultramafic rocks are common, but are generally serpentinized. A sizable, isolated body of peridotite that is only serpentinized around the margins is exposed at Burro Mountain in the Santa Lucia Range (Burch, 1968; Loney and others, 1971). It is largely harzburgite with pronounced planar tectonite fabric, and it may represent depleted mantle material, but it is cut by many dunite dikes, which make interpretation difficult. Most serpentinite blocks are probably derived from depleted mantle rocks, as they are collectively much more voluminous than gabbroic and felsic plutonic

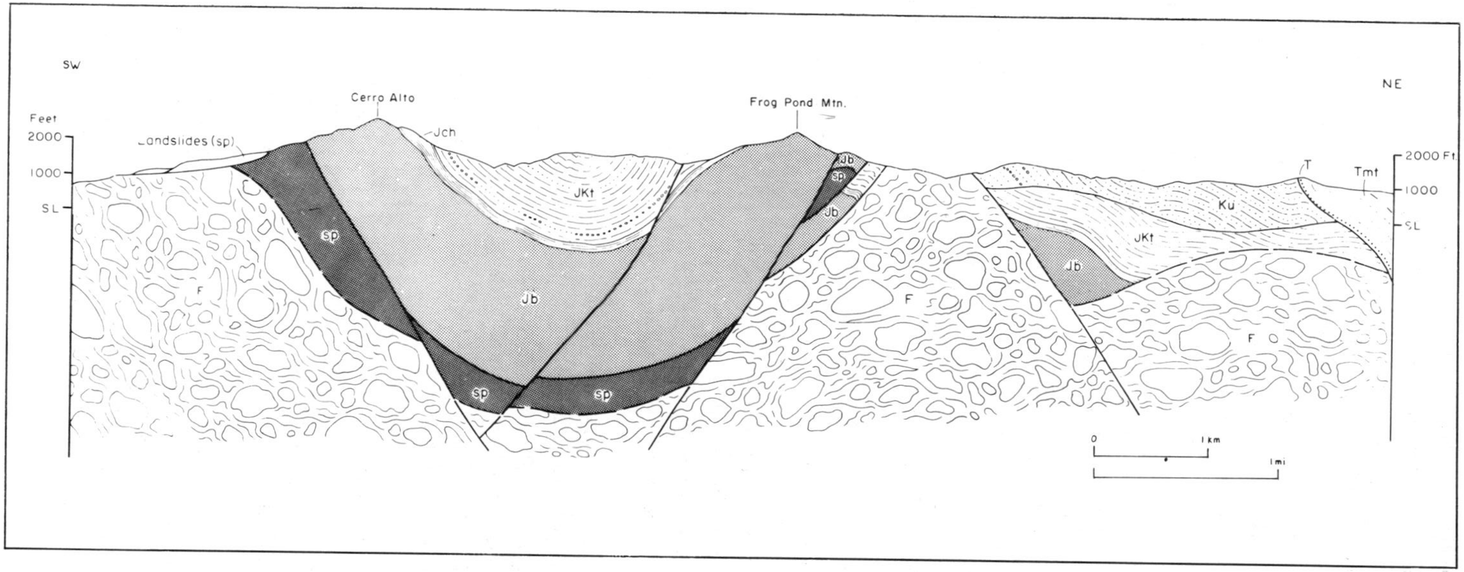

Fig. 13-10. Cross section of Cuesta Ridge ophiolite near San Luis Obispo. F, Franciscan Complex; sp, serpentinite; Jb, Upper Jurassic sill and dike complex and oceanic basalt; Jch, Upper Jurassic (Tithonian) radiolarian chert; JKt, Upper Jurassic-Lower Cretaceous terrigenous sediments (mostly mudstone) of Great Valley type; Ku, Upper Cretaceous (Campanian) sandstone of Great Valley type; T, Tertiary sediments with Oligocene-lower Miocene Vaqueros Formation at base. Thrusts involving Franciscan may be parts of Coast Range thrust or its equivalent. Some thrusts affect Campanian rocks but are overlapped by Vaqueros Formation, far right. (From Page, 1972, with permission of The Geological Society of America).

blocks and, therefore, could hardly represent products of differentiation of gabbroic magma.

The mafic and ultramafic blocks in melanges may have been acquired in any of the following ways: (1) they may have been ripped or rasped from the "lid" of the subduction zone (i.e., from the ophiolite remnants, described previously, at the base of the GVS); (2) they may have been torn from upthrust projections of oceanic crust in the "floor" of the subduction zone (i.e., from the down-going slab); (3) they may have been introduced into melanges as olistostromes fed by submarine or subareal uplifts of sea floor; and (4) they may have been introduced by the intrusion or extrusion of mantle diapirs, which were then disrupted by submarine slides or subduction.

Doubtless much oceanic basement material was indeed removed from the underside of the GVS; locally, the ophiolite that belongs there has disappeared completely. However, it is uncertain whether this is the result of initial shearing off of the oceanic lithosphere at the inception of subduction, of subsequent rasping, or of late-stage underthrusting when the fully developed Franciscan Complex may have been forced "one last time" beneath the edge of the continent en masse. Perhaps all these events took place. Casual examination of geologic maps suggests that, where sizable ophiolite remnants are exposed at the base of the GVS or its equivalents, nearby Franciscan melanges may contain more (and larger) blocks of greenstone and serpentinite than elsewhere (e.g., see Maddock, 1964). However, I know of no systematic attempt to match the petrologic details of isolated blocks with those of nearby large ophiolites, nor do I know of efforts to find isolated pieces of all the components of the ophiolites.

Although the GVS ophiolite is a likely parent of some fragments in melanges, probably not all mafic and ultramafic blocks came from this source. West of Atascadero, a greenstone block retains some attached chert that does not resemble the chert of the nearby Cuesta Ridge (GVS-type) ophiolite. The chert of the block is red and is typical of Franciscan radiolarites, whereas the chert overlying the Cuesta Ridge basalt is brown-black or greenish brown. Many ultramafic blocks are of unknown parentage. Serpentinite blocks containing high-grade blueschist inclusions (Coleman and Lanphere, 1971) could not have been torn directly from the presently exposed GVS ophiolites, which contain no such inclusions and show no signs of blueschist-facies metamorphism.

Cold serpentinite intrusions and extrusions The New Idria serpentinite diapir probably rose from the mantle beneath the Franciscan Complex. Alternatively, it may possibly have been derived from a huge serpentinite mass within a melange, but if so, the source body must have been larger than any ultramafic blocks now exposed. In any case, it is difficult to imagine any way the material could have been acquired from the GVS ophiolite forming the "lid" of the subduction zone. The serpentinite first reached the surface in Middle Miocene time, whereupon it contributed material to the "Big Blue" member of the Temblor Formation (Casey and Dickinson, 1976). The serpentinite that has been squeezed up along faults in a number of localities and that which has flowed out at the surface at Table Mountain (Dickinson, 1966a) may also have originated in the mantle beneath the Franciscan. The Table Mountain extrusion is certainly too voluminous to have stemmed from ordinary ultramafic blocks or slabs in melanges.

IMPORTANT STRUCTURES AND EVENTS DIMLY PERCEIVED

Although the Franciscan is now recognized as a product of long-continued subduction as well as normal rock-forming processes, there is still much doubt as to how and when it arrived at its present tectonic relations with other assemblages.

Coast Range thrust The contact between the Franciscan and the Great Valley Sequence (Fig. 13-1) is apparently everywhere tectonic (Page, 1966), and, because the Franciscan is evidently a subduction complex, the contact may be subduction related (Hamilton, 1969; Ernst, 1970). Bailey and others (1970) showed that, where ophiolite is underlain by Franciscan rocks and overlain by GVS sediments, the major tectonic contact is between the Franciscan and the ophiolite rather than between the ophiolite and the sediments, where it had previously been placed. In the same article, they named the contact the Coast Range thrust (CRT) and suggested that it extends to outliers of GVS that tectonically overlie the Franciscan farther west.

In the Southern Coast Ranges, there is considerable support for the existence of the CRT, but only if regional relations are considered rather than separate localities. The visible boundary between Franciscan and GVS generally consists of Neogene faults that have overprinted the CRT and do not show the underthrusting features one might expect to see. The complications are illustrated by Raymond (1973b) in the Diablo Range, where the Tesla-Ortigalita fault zone of probable Neogene age occupies the site where the CRT might be expected. It is not always easy to distinguish "original" parts of the Franciscan-GVS contact from superposed Neogene segments. Slivers of ophiolite, blueschist, or jadeitic metagraywacke are not diagnostic, as fragments of these rocks have been tectonically strewn along young segments as well as old parts of the contact. A steeply dipping fault could be either a young feature or the CRT, as the latter is locally steepened by folding.

Despite the difficulties just mentioned, the areal geology of the Diablo antiform clearly indicates the presence of something like the CRT. The elliptical core of Franciscan rocks is flanked on nearly all sides by GVS rocks that are in fault contact. A remnant of GVS ophiolite extends across part of the range along a synform in the tectonic roof overlying the Franciscan, showing that the roof was formerly extensive (Fig. 13-4). A small klippe of the ophiolite with superincumbent GVS sediments occurs well within the northern part of the range (Fig. 13-4; Bauder and Liou, 1979). Farther southeast in the Diablo Range the relations are also indicative. Evidently, the GVS with its ophiolitic basement once covered the entire site of the Diablo Range, resting tectonically on the Franciscan Complex.

In the Sur-Obispo Belt southwest of the Salinian Block, the counterpart (or displaced continuation?) of the CRT appears here and there where the Franciscan Complex passes beneath GVS-type rocks. In a number of localities, the tectonic cover consists of Upper Cretaceous turbidite sandstones, plus mudstone and conglomerate, which resemble GVS rocks of the type region (Gilbert and Dickinson, 1970). These rocks form thrust sheets or klippen along the Sur-Nacimiento fault zone (Page, 1970a, Figs. 6, 7, 8). Locally, the Franciscan passes beneath a cover of ophiolitic rocks overlain by Upper Jurassic-Late Cretaceous GVS-type sediments (Fig. 13-10; Page, 1972).

When and how did the Coast Range thrust originate? It may not be the same age everywhere, and indeed it may not be a single unified fault, but let us ignore these complications for the moment. Because the Franciscan is believed to have developed from east to west, the youngest dated part being the Coastal Belt of Northern California, it is natural that the CRT is regarded by some as a feature formed early between the subduction complex and its tectonic "lid" (i.e., the GVS). However, the CRT is probably Early Tertiary as suggested by Page (1966) before the present name was proposed.

In northern California, the CRT seems to have belatedly formed across a subduction complex that was already largely completed. In the Geysers region, early Cenomanian chert occurs in the Franciscan structurally beneath the CRT (McLaughlin and Pessagno, 1978), so the CRT as it now exists in that area must be Late Cretaceous or younger. South of Clear Lake (Swe and Dickinson, 1970) and at Rice Valley (Berkland, 1973), not only the Mesozoic GVS, but also Paleocene strata are thrust over the Franciscan. In the Southern Coast Ranges, the evidence is more equivocal but is nevertheless indicative. Upper Cretaceous rocks of the GVS are emplaced over the Franciscan by the CRT or similar thrusts in the Castle Mountain Range (Dickinson, 1966b), and Campanian (Upper Cretaceous) rocks are cut by imbricate thrusts above the Franciscan in the Sur-Obispo Belt (Page, 1972). If we assume that these scattered observations may all have something to do with the CRT, the latter is most likely Paleocene or Eocene, perhaps varying somewhat from one area to another.

Why should large-scale thrusting occur late in the history of subduction, near the first-formed part of the subduction complex? It would seem that the subduction complex was suddenly pushed en masse farther beneath its cover, although probably still receiving more increments on its seaward underside. W. R. Dickinson (1978, informal communication) suggests that the event may be related to the adoption of a low angle of inclination by the subducting Farallon plate, the same mode of subduction that caused the Laramide orogeny.

Sierran-Franciscan boundary zone The approximate location of this zone is shown in Fig. 13-1. The Sierran basement, which is extremely complicated (see Chapters 3, 5, 6, and 7, this volume), is exposed in the foothills east of the San Joaquin Valley and has been tracked in the subsurface by drilling and seismic reflection to about the centerline of the Valley. On the other side of the Valley, the Franciscan Complex extends eastward beneath the sediments. What is the nature of the subsurface contact or zone of closest approach between the two assemblages? The possibilities include the following: (1) The two complexes may meet at a discrete tectonic contact, either a subduction boundary or a younger thrust system. (2) They may be separated by an intervening assemblage, perhaps oceanic crust and/or mantle.

Whatever form it takes, probably the Sierran-Franciscan boundary zone incorporates a modified downward continuation of the Coast Range thrust. In Northern California, the Franciscan (South Fork Mountain Schist unit) meets the Sierran basement or, more precisely, comparable Klamath basement rocks along a thrustlike, northeast-dipping contact (Blake and others, 1967). West of Redding, the contact swings south and becomes the Coast Range thrust, separating the Franciscan from the GVS rather than Sierran basement rocks, at shallow levels in the crust. However, down-dip and eastward, this fault must pass through the full thickness of oceanic crust beneath the Great Valley sediments. (It could hardly terminate within the sediments or within the mafic crust, inasmuch as its displacement is apparently much too large to allow this.) At greater depth, it probably enters the mantle beneath the Sierran basement, the top of the mantle being placed at 30 km beneath Fresno, for example, by Carder (1973).

If the foregoing speculations are valid, there is no actual contact between the Franciscan and the Sierran basement; instead, oceanic crust and mantle material intervene. The presence of abundant basaltic-gabbroic rock is postulated by Cady (1975) to account for the Great Valley gravity and magnetic high; in fact, Cady's model requires a much greater volume of mafic material than we have implied.

These considerations complicate the visualization of exactly what entities, both visible and unseen, are cut by the San Andreas fault (SAF) at the south end of the San Joaquin Valley. In the Transverse Ranges, granitic and Precambrian gneissic basement rocks are exposed on both sides of the fault. No typical Franciscan or GVS rocks crop out in the vicinity. However, Ross (1970) has drawn attention to rocks of probable ophiolitic affinities (metadiabase, amphibolite, gabbro, pyroxenite, and hornblende quartz diorite-gabbro) at Eagle Rest Peak north of the SAF and west of the westernmost granitic rocks exposed on that side of the fault (see Fig. 13-2 for location of Eagle Rest Peak). If the gabbroic rocks belong to the Sierran basement, the Sierran-Franciscan boundary zone, extending south beneath the San Joaquin Valley, must pass west of Eagle Rest Peak and must be cut by the SAF near that point. On the other hand, if the gabbroic rocks belong to the GVS ophiolite, the CRT lies between the ophiolite and the Franciscan to the west, and the CRT must be cut by the SAF west of Eagle Rest Peak. In either case, the east boundary of the Franciscan must be west of the Peak.

San Andreas fault The SAF (Hill and Dibblee, 1953; Dickinson and others, 1972) slices the Southern Coast Ranges obliquely and is one of the principal features of the province. Although it cannot be fully treated in this chapter, the fault will be taken into account in connection with the emplacement of the Salinian Block and the transport of the Sur-Nacimiento Belt.

As noted previously, the SAF is the northeast boundary of the Salinian Block (Figs. 13-1 through 13-3), so the history of the fault coincides with the history of movement of the Block. The modern role of the SAF as a transform feature began in Miocene time well after the Pacific and North American plates first came into mutual contact (about 29 m.y.b.p.; Atwater, 1970; Atwater and Molnar, 1973). However, there is strong evidence that an ancestral or proto-SAF (Suppe, 1970; Garfunkel, 1973; Nilsen and Clarke, 1975) existed prior to the Eocene. Inferred early movement along the ancestral fault probably resulted from oblique plate convergence; it is identified with the pre-Eocene, pretransform "partway" emplacement of the Salinian Block, which is discussed later. The better known Neogene transform movement along the SAF allowed the Salinian Block to continue its emplacement and reach its present position. The Neogene history of the fault is well documented by Dickinson and others (1972), Turner (1969), Huffman (1972), and Matthews (1973). Other contributors are cited in the foregoing references.

Those who subscribe to a pre-Eocene protomovement along the SAF believe that 100 to 200 km of right-hand slip occurred. Regardless of that debatable issue, there is a general consensus that approximately 305 km of right-hand slip took place along the fault in central California in the Neogene since the beginning of the Miocene, and that most of this accrued in the last 15 m.y. The rate of movement accelerated markedly in the last 4 to 7 m.y. (Dickinson and others, 1972). North of San Francisco, the SAF experienced not only the movements just mentioned, probably totaling more than 450 km, but also as much as 80 to 115 km of additional slip, which was fed into the SAF by the San Gregorio-Hosgri fault system (Graham and Dickinson, 1978a).

Only one distinctive rock suite of limited original extent is believed to be recognized at intervals along the full length of the SAF from the Transverse Ranges to Point Arena. The gabbroic rocks of Eagle Rest Peak (Fig. 13-2) are evidently transected by the fault in the subsurface, and apparently blocks and slivers of that suite have been carried northwestward along the fault for long distances (Ross, 1970). Ross believes that the quartz gabbro at Gold Hill near Cholame and also at Logan on the Pajaro River (see Fig. 13-2 for locations) probably came from Eagle Rest Peak. If this is so, the rocks were transported about 140 km to Gold Hill and 300 km to Logan. Similar rocks occur as clasts in uppermost Cretaceous conglomerate of the Gualala sequence 16 to 45 km south of Point Arena (Ross and others, 1973). If the gabbroic material in all these localities is really from the same source area, it has been strewn along the fault for 560 km. This would make slivering of the Salinian Block irrelevant to the SAF unless it occurred with appropriate timing, location, and orientation. As discussed later, the San Gregorio-Hosgri fault zone (Fig. 13-2) appears to be the only one that satisfies such requirements. It apparently contributed 80 to 115 km of displacement to the northern part of the SAF. The San Gregorio-Hosgri fault originated after the mafic-clast conglomerate had moved 150 to 200 km from its source near Eagle Rest Peak, and the newborn fault joined the SAF just south of the conglomerate. Miraculously, the latter was able to travel a total distance of 560 km, taking advantage of nearly the entire combined displacement of the SAF and San Gregorio-Hosgri zone. The postulated offset of the gabbro also circumscribes the position of any proto-SAF north of the Transverse Ranges, as the protofault must have exactly occupied the site of the present fault, which I think was the actual case. Of course, the foregoing deductions hinge upon the validity of the correlation from Eagle Rest Peak to Logan to Gualala.

Sur-Nacimiento fault zone The Salinian Block is bounded on the southwest by the Sur-Nacimiento fault zone (Figs. 13-1 and 13-2), which may be a continuation of the Sierran-Franciscan boundary zone that has been displaced by the SAF (Page, 1970a). According to this interpretation, the Sierran-Franciscan boundary zone is offset from the Eagle Rest Peak area to the vicinity of Point Arena, whence it passes southward offshore to Point Sur. Near Point Sur, the fault zone intersects the coast and, continuing southeastward, crosses part of the Southern Coast Ranges at an acute angle, reaching the Transverse

Ranges. The fate of the Sur-Nacimiento fault zone in the Transverse Ranges is unknown, but southward it more or less coincides with the Newport-Inglewood structural zone in Southern California and apparently passes near the Vizcaino Peninsula of Baja California. The fault zone cannot be neatly defined, as it consists of many faults of different kinds and different ages. The zone encompasses low-angle thrusts that may be Mesozoic, high-angle Neogene fractures, and strike-slip faults such as the Rinconada.

Probably the Sur-Nacimiento fault zone originated along or near the contact between the Franciscan and the "lid" of the subduction zone, but it has been overprinted and remodeled by large amounts of strike slip and other poorly recognized events. Along this zone, the three major late Mesozoic rock assemblages of the Southern Coast Ranges have been brought into mutual proximity: (1) the Franciscan, with its heterogeneous assortment of terrigenous and oceanic rocks, its blueschist-facies components, and its melanges; (2) the granitic and high-temperature metamorphic rocks comprising the basement complex of the Salinian Block; and (3) GVS-type clastic sediments and fragments of GVS-type basal ophiolite. The tectonic relations of these three assemblages are shown schematically in Fig. 13-11. Evidently, some parts of the three assemblages originated more or less contemporaneously in widely separated environments: the Franciscan as a subduction complex, the Salinian basement as a magmatic arc assemblage, and much of the GVS as a forearc basin accumulation. All these are telescoped together along the Sur-Nacimiento fault zone. It will be noticed that only small remnants of the GVS-type rocks are present along the zone, most of the forearc basin having disappeared. This problem will be discussed later.

Just as in the case of the Coast Range thrust, it is questionable whether or not we see original subduction-related faults in the Sur-Nacimiento fault zone. Near Point Sur, the various faults of the zone not only affect the Franciscan and the Salinian basement

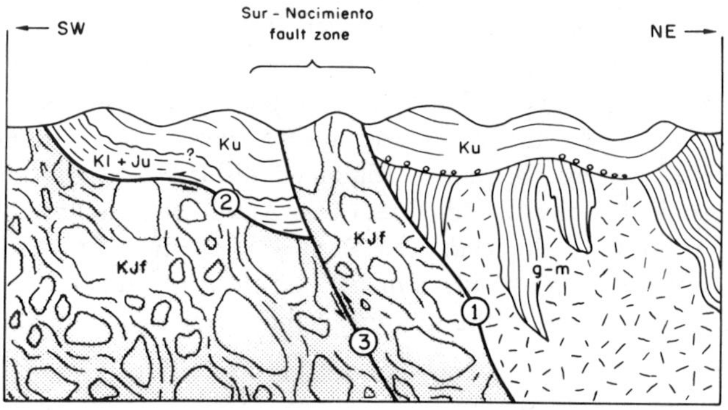

Fig. 13-11. Schematic, idealized cross section of Sur-Nacimiento fault zone and petrotectonic assemblages of late Mesozoic subduction complex, forearc basin, and magmatic arc. KJf, Franciscan Complex; Kl + Ju, Upper Jurassic (Tithonian) and Lower Cretaceous Great Valley-type sediments, chiefly mudstone; Ku, Upper Cretaceous (Campanian) Great Valley-type sediments, chiefly sandstone. (From Page, 1970a, with permission of The Geological Society of America).

rocks, but also cut Neogene formations, including sandstone of probable late Miocene age. The faults that terminate the Franciscan terrane west of Paso Robles and near Atascadero also transect Miocene formations. The southward continuation of the zone, near the Cuyama River, shows a complex Tertiary history (Vedder and Brown, 1968). If any original subduction-related faults remain, perhaps the thrustlike contact inferred to exist between the Franciscan and the Cuesta Ridge ophiolite is one, but it is not actually visible. The low-angle faults between the Franciscan and Upper Cretaceous Great Valley-type rocks west of Paso Robles may be late subduction features. A low-angle fault between Upper Cretaceous sandstone and Lower Cretaceous mudstone, both of Great Valley affinity and both lying tectonically above the Franciscan near Santa Margarita, is overlapped (see east end of section in Fig. 13-10) by Vaqueros Sandstone of late Oligocene-early Miocene age. Therefore, the fault is pre-Neogene and is probably subduction related, although it must be a relatively late product of plate convergence and must have formed far from the trench.

THE OUT-OF-PLACE SALINIAN BLOCK

The geographic source, emplacement, and deep crustal structure of the Salinian Block, together with the disappearance of part of it, comprise a bundle of challenging problems. The Block (Figs. 13-1 and 13-2) is tectonically bounded by the San Andreas fault on the northeast and the Sur-Nacimiento fault zone on the southwest. The San Andreas is understood well enough to throw light on the Neogene part of the story, but there is still doubt as to events prior to the activity of the modern San Andreas. The Block probably consists of more than one entity, as it is subdivided by faults and petrologic differences, so some parts may not be closely related to others. A major puzzle is the difficulty of restoring the Block to some original position where it would match adjoining terranes. A further problem stems from seismic properties of deeper parts of the Salinian area, which raise the possibility that non-Salinian material underlies the Block. Recent articles include an excellent summary by Ross (1978), interpretations of Upper Cretaceous and Paleogene sediments of the Block by Nilsen (1978) and Howell and Vedder (1978), a comprehensive treatment of the tectonic background by Hamilton (1978a), and an analysis of tectonics deduced from sedimentation by Graham (1978). Other important contributions will be cited where appropriate.

Geology of the Salinian Block

Metasediments of the Salinian basement complex The Salinian basement consists largely of granitic plutons and metasedimentary host rocks, the latter being of amphibolite or higher metamorphic grade.

The Santa Lucia Range (see Fig. 13-2 for location and Fig. 13-12 for cross section) has the largest outcrop area of Salinian preintrusive metamorphic rocks. These rocks have been studied by Trask (1926), Compton (1960, 1966a, b), Wiebe (1970), and Ross

(1976a, b, 1977a). Trask named the metamorphic complex in the northwest part of range the Sur Series, but the designation is inappropriate for regional use because of doubtful continuity of the rocks. According to legend, the Sur Series is probably Carboniferous, but the supposed rare crinoids and corals are will-o-the-wisps, some having been collected in the Sierra Nevada and later having been mistakenly ascribed to the Gabilan Range. The actual age (s) could be anything from Precambrian to Early Cretaceous.

The most prevalent metamorphic rocks in the Santa Lucia Range are biotite-feldspar quartzite, quartzofeldspathic granofels, and quartzofeldspathic gneiss, the latter being migmatitic in many localities. In addition, there are lesser amounts of calc-silicate granofels, amphibolite (which could perhaps be metasedimentary), aluminous schist, calcite marble, and metadolomite. Near the coast the metamorphic grade is higher than elsewhere, reaching the granulite facies. Compton (1966a) estimates that of the rocks he studied in the Santa Lucia Range about 70% were originally feldspathic or argillaceous sandstones and siltstones, 10% were argillaceous carbonate rocks, 10% were Ca-poor claystones, 5% nearly pure limestone or dolomite, and 5% very quartzose sandstone. A few thin carbonate or quartzite units are traceable for hundreds or even thousands of meters. Larger mappable units are rare, but Wiebe (1970) found a distinctive graphite- and pyrite-bearing sequence that he traced for more than 30 km. Ross examined many parts of the Salinian Block, including the preintrusive rocks of Point Reyes, the Santa Cruz Mountains, the Gabilan Range, and the Santa Lucia Range. He says, "Field and laboratory studies suggest that the unmetamorphosed parent rocks were predominantly (if not wholly) sedimentary—a thinly bedded sequence of interbedded quartz-rich siltstone and impure sandstone, with lesser amounts of shaly, marly, and calcareous rocks" (Ross, 1977a, p. 371).

Wiebe (1970) thinks the Salinian basement probably matches Paleozoic (and possibly Precambrian) miogeosynclinal rocks of the Mojave Desert on the other side of the San Andreas fault, and he suggests that early northeast-trending folds which he noted in the Santa Lucia metamorphic rocks are among the matching features. However, Ross (1977a) has not found any specific formations or sequences on the other side of the fault that correspond to the most prevalent Salinian metasediments or the protoliths thereof. He has considered, among others, preplutonic rocks of the southern Sierra Nevada, San Emigdio, and Tehachapi Mountains. Most rock sequences of those areas can be ruled out on the basis of such characteristics as abundant volcanics, predominant carbonate formations, large pure quartzite units, thick-bedded coarse clastics, or other distinctive features.

Although most metasediments of the Salinian Block seem to lack obvious equivalents in other terranes, there are two units or assemblages that may indeed correspond with known rocks of distant areas. One of these is schist that forms a belt 6 to 10 km wide in the Sierra de Salinas (the northeast part of the Santa Lucia Range), and which has been studied in detail by Ross (1976b). He describes the rock as a rather homogeneous biotite quartzofeldspathic schist with minor amounts of quartzite, amphibolite, and marble. The schist belt extends beneath the sediments of the Salinas Valley, where it is found in well samples; farther east, part of the belt crops out in the Gabilan Range and part (in the subsurface) approaches within 5 km of the San Andreas fault. Ross (1976b) believes that the schist of Sierra de Salinas may correspond to the schist of Portal-Ritter Ridge 300 km distant on the opposite side of the fault west of Palmdale (see Fig. 13-2 for locations). His detailed petrographic and chemical studies show close similarities between the schists of these two widely separated localities and convincingly indicate that the rocks are metagraywacke. The metagraywacke schist of Portal-Ritter Ridge, like that of Sierra de Salinas, is of amphibolite facies, but in other respects it somewhat resembles the Pelona-Orocopia Schist (likewise metagraywacke) of southern and southeastern California. However, the Pelona-Orocopia Schist is mostly in the greenschist facies, with only local areas of amphibolite facies, and (unlike the Sierra de Salinas schist) it is not intruded by plutons and is probably Mesozoic in age (Ross, 1976b).

A second Salinian metamorphic terrane that resembles rocks known elsewhere is a gneissic complex east of the Red Hills-San Juan fault. The fault is probably an important one (Smith, 1977), as shown by the pronounced difference between the basement rocks on the two sides. Ross (1972b, 1977a) says that the gneissic complex of the Red Hills, Barrett Ridge, and the Mount Abel-Mount Piños areas comprises a sub-block (the Barrett Ridge slice) that petrologically resembles Precambrian complexes of the San Gabriel Mountains in the Transverse Ranges. The three principal outcrop areas differ slightly from one another petrologically, but collectively they are composed of strongly foliated quartzofeldspathic metasedimentary gneiss, lesser amounts of hornblende-bearing gneiss and amphibolite, migmatites and transitional rocks that grade from metamorphic to homogeneous quartz diorite, and granodioritic gneisses. Kistler and others (1973) find an initial Sr isotope ratio of 0.7095 for homogeneous granitic gneiss in the Red Hills, a much higher value than any other measured in the rest of the Salinian Block and comparable to those of plutons in Precambrian terranes of the San Gabriel Mountains. Apparently, these particular basement rocks are common to both the Salinian Block and the Transverse Ranges. Later structures are likewise shared near the provincial boundaries; in fact, the Mount Abel-Mount Piños area, which Ross includes in the Barrett Ridge slice of the Salinian Block, is also a part of the youthful Transverse Ranges if judged by the presence of east-west trending Neogene thrust faults.

Plutons of the Salinian block Most of the Salinian plutons are granodiorite, quartz monzonite (or granite, depending on definitions), and quartz diorite, the first two predominating. Large terranes of these rocks are exposed in the La Panza, Gabilan, and Santa Lucia ranges, where they have been described by Compton (1960, 1966a, b) and Ross (1972a, b, 1974, 1976a, 1978; Ross and McCulloch, 1979). Smaller exposures are scattered all the way to Bodega Head on the coast north of Point Reyes.

The northern part of the Santa Lucia Range is characterized by multiple granitic intrusives, some of which are enveloped concordantly by metasediments of amphibolite facies (e.g., Compton, 1966b; Ross, 1976a). A cross section by Compton is shown here as Fig. 13-12. The various plutons differ somewhat in composition and texture and show sequential rather than simultaneous emplacement (Ross, 1976a; Ross and McCulloch, 1979). The level of exposure appears to be deeper than in the Gabilan Range. Compton (1960) describes charnockitic rocks near the coast, where migmatites consist of wisps of charnockitic tonalite and metamorphic remnants of granulite grade. These rocks perhaps represent a plutonic root zone. Although most of the plutons in the Santa Lucia Range are granitic, there are a number of rather small mafic to ultramafic bodies, mostly mea-

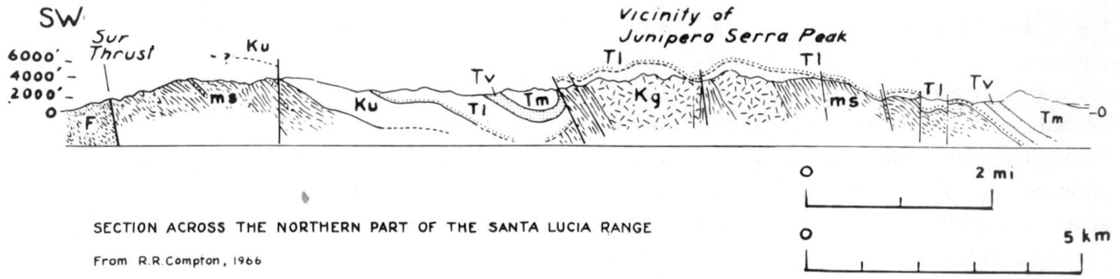

Fig. 13-12. Section across northern part of Santa Lucia Range, showing Plio-Pleistocene folds and high-angle reverse faults. Note involvement of basement massifs in the deformation. Entire section (except part SW of Sur thrust) is in Salinian Block, which is more complex here than in region of Fig. 13-3. F, Franciscan Complex; ms, pre-Cretaceous metasediments (chiefly); Kg, Cretaceous granitic rocks; Ku, Upper Cretaceous clastic sediments; Tl, Paleocene-Eocene-Oligocene sediments; Tv, Oligocene-lower Miocene sediments, including Vaqueros Formation; Tm, Miocene sediments, chiefly Monterey Formation. (From Compton, 1966, with permission of The Geological Society of America).

sured in hundreds of meters. These (especially the ultramafic ones) have sometimes been regarded as "squeeze-ups" along fault zones, but one such body that was recently studied in some detail (Nutt, 1977) shows cumulate textures and features of multiple intrusion. It is thought to have been introduced as a melt at about the same time as the granitic rocks of the region.

The plutonic rocks of the Gabilan and La Panza ranges are largely granodiorite and quartz monzonite (or granite) that is less variegated and apparently less deeply eroded than the plutons of the Santa Lucia Range, and there are relatively limited remnants of metamorphic host rocks.

It is of interest to compare the Salinian granitic rocks with those of the Sierra Nevada. The comprehensive mineralogic and chemical data of Ross (1977b) for the plutons of the Santa Lucia Range seem to me to be comparable with data from some parts of the Sierra. The Peacock index, about 61, is near the boundary between the calc-alkalic and calcic fields. Histograms of the distribution of trace elements show a preponderance of concentrations not very different from average granodiorite, and they are similar to those of the central part of the Sierra Nevada batholith. The $K_2O : SiO_2$ ratios for granitic rocks (excluding tonalite) fall slightly below the ratios shown by Bateman and Dodge (1970) for the High Sierra, but approach those given by these authors for the Shaver sequence. One puzzling difference between Sierran and Salinian terranes noted by Ross (1978) is the lack of metallic mineralization in the vicinity of Salinian plutons.

Many of the Salinian plutons are Cretaceous and might fall within time brackets determined for the Sierra Nevada batholithic complex (Evernden and Kistler, 1970), but their exact ages are controversial. Most K-Ar determinations range from 92 to 70 m.y., and some are impossibly young, younger than unmetamorphosed Campanian sediments (say 76 to 70 m.y.) that unconformably overlie the granitic rocks. A number of ages deduced from Rb-Sr whole-rock determinations (R. W. Kistler, reported by Ross, 1972b) and from U-Pb measurements (Mattinson and others, 1972) fall within the range 117 to 106 m.y. A few ages approximating 80 m.y. are indicated both by K-Ar and Rb-Sr methods for certain plutons (e.g., in the La Panza Range; Compton, 1966a; Ehlig and Joseph,

1977). I conclude that most of the plutons are probably Early Cretaceous (late Barremian, Aptian, and early Albian) and some are Late Cretaceous, perhaps early Santonian.

A quartz monzonite in the La Panza Range is cut by granitic dikes with distinctive "polka-dot" texture. The quartz monzonite and the dikes resemble those in the Little San Bernardino and Orocopia Mountains on the opposite side of the San Andreas fault, according to Ehlig and Joseph (1977), the apparent separation being 330 to 450 km. These authors point to comparable composition, texture, Sr isotope ratios, and age (ca. 80 m.y.). If the match is valid, it is extremely important, being one of the few correlations that bridges the Transverse Ranges.

Deep structure and constitution The Salinian Block has puzzling seismic properties. According to S. W. Stewart (1968), a marked seismic-amplitude attenuation occurs at depths below about 10 km. Stewart conjectures that the cause could be either high temperature or the presence of the Franciscan Complex beneath the sialic basement. The Franciscan might indeed have been subducted under the Salinian basement. At the time the plutons were generated, doubtless the subduction complex was far to the west, and the downgoing slab was 100 to 200 km below the surface in the region under discussion. At that time, the Salinian Block was much farther south. If it moved partway toward its present position in the Early Tertiary, which seems likely, subsequent subduction probably passed beneath it at a low inclination en route to the Eocene-Oligocene magmatic arc east of the Sierra Nevada, and the downgoing slab accompanied by Franciscan material might not have been very deep beneath the presently exposed Salinian basement. Ross and McCulloch (1979) courageously show the Franciscan beneath the Salinian basement rocks of the Santa Lucia and Gabilan ranges, but there is no surface evidence to confirm or refute this interpretation.

Other seismic results are difficult to reconcile with the foregoing. Healy and Peake (1975) carefully derived a preferred velocity model in which V_p for the east part of the Salinian Block near Bear Valley is 6.8 km/s below a depth of 12 km. This velocity may be too high for granite even under considerable confining pressure. It might indicate granulite or charnockitic rocks. It argues against the presence of subducted Franciscan, unless the latter is largely metamorphosed, perhaps with a high jadeitic pyroxene content (see Stewart and Peselnick, 1978). Conceivably, the V_p of 6.8 indicates oceanic crust; if so, the latter would seem to be abnormally thick (perhaps imbricated?), as the Moho is thought to be approximately 23 km deep below the Salinian Block (Healy, 1963). The presence of oceanic crust at a shallow level beneath granitic plutons could perhaps be explained by the relatively recent emplacement of a subduction slab above an earlier one that was responsible for generation of the granitic magma. Could subducted ophiolitic material be reconciled with the observed amplitude attenuation of seismic waves? If so, it would be an alternative to the presence of Franciscan rocks beneath the Salinian Block. So far, the question is unresolved.

Upper Cretaceous-Early Tertiary sedimentary cover The postplutonic sedimentary rocks provide valuable clues to the environment and tectonism of the Salinian Block during latest Cretaceous and Cenozoic time. Summaries by Howell and others (1977), Howell and Vedder (1978), and Nilsen (1978) provide recent information and references to older literature on the Cretaceous and Paleogene sediments. Graham (1978) gives a comprehensive treatment of the Paleogene and Neogene sedimentary-tectonic record of the Santa Lucia Range.

Just before, or during, the early Campanian Stage of the Late Cretaceous, the basement rocks were deeply eroded, exposing plutons at various levels and forming an uneven topographic surface. During the Campanian and thereafter, this surface was largely submerged and covered by marine arkosic sandstone, mudstone, and conglomerate. Probably the sediments were not spread out in continuous, uniform sheets extending across the entire subprovince. Chipping (1972), Graham (1978), and Howell and Vedder (1978) believe that deposition occurred in deep local embayments or basins, and Ruetz (1976) finds evidence for a submarine canyon cut in granitic rock and later filled with sediment.

Howell and Vedder (1978) recognize five Late Cretaceous basins of deposition in the Salinian Block. They are identified by names of places where their sediments are exposed, as follows: (1) Gualala, on the northern California coast; (2) Pigeon Point, on the coast of the Santa Cruz Mountains; (3) Junipero Serra Peak, in the Santa Lucia Range; (4) Nacimiento Lake, also in the Santa Lucia Range, and (5) Pozo district, in the La Panza Range. Four of these localities are in the Southern Coast Ranges. The sediments of all four include subsea fan deposits, suggesting local foundering that might somehow be related to early movements of the Salinian Block.

In parts of the Santa Lucia Range, the basement is directly overlain by an unbroken Campanian to Paleocene sequence consisting of turbidite-facies arkosic sandstone and mudstone, plus much conglomerate. Some of the latter is dominated by granitic and metamorphic cobbles derived mainly from the Salinian basement itself (Howell and Vedder, 1978). In other areas, conglomerates contain large quantities of well-rounded, durable porphyritic-aphanitic volcanic rocks of intermediate to silicic composition, as well as lesser amounts of granitic and metamorphic varieties. The volcanic rocks could be reworked from sediments derived from shallow levels of a magmatic arc assemblage; in any case, they were obtained beyond the present limits of the Salinian Block, which has no known appropriate source. Paleocurrent directions determined by Ruetz (1976) in a Campanian to Paleocene sequence southwest of the Junipero Serra Peak are from the north and northeast. In the Upper Cretaceous rocks near Nacimiento Lake, Howell and Vedder (1978) find paleocurrents from the north and also from the southwest, which supports the interpretation that the coast was deeply embayed.

In the La Panza-Sierra Madre region, the basement rocks are overlain by an enormous continuous section more than 5 km thick, consisting of Campanian to middle Eocene clastic sediments. The sequence is mainly turbidite-type arkosic sandstone and mudstone, plus abundant conglomerate (Chipping, 1972). Conglomerate clasts in the lower part of the sedimentary pile are largely granitic, gneissose, schistose, and quartzitic rocks representing the local Salinian basement complex. The rest of the section includes quantities of porphyritic-aphanitic silicic and intermediate volcanic rocks like those in conglomerates of the Santa Lucia Range, presumably derived from a source beyond the present Salinian Block. Paleocurrent directions determined by Chipping (1972) are mainly from the north, northeast, and east.

Judging from the foregoing sedimentary record, during the latest Cretaceous, Paleocene, and at least part of the Eocene, much of the Salinian Block was receiving detritus from a nearby, perhaps contiguous Sierralike terrane to the east. This is difficult to reconcile with the paleogeography that would result from an early partway emplacement of the Salinian Block (Fig. 13-13), an hypothesis tentatively accepted by the writer on the basis of the data of Nilsen and Clarke (1975) and Nilsen (1978), cited later. The sedimentary record shows that parts of the Salinian Block were submerged much of the time, and the sea floor locally sloped relatively steeply oceanward. The Franciscan Complex was well offshore during most of the time span, and was submerged or covered so that it failed to contribute sediment to the Salinian terrane even where paleocurrents are from the southwest.

Interestingly, Campanian marine transgression and subsequent submergence extended far into the eroded Cretaceous plutonic arc, which initially must have been well inland and at times mountainous. Compared with today's Sierra Nevada, it is as though marine embayments and submarine canyons had formed in an eroded High Sierra terrane. However, the difference in extent and character of latest Cretaceous marine transgression in the Salinian Block vis-a-vis the Sierra Nevada does not preclude the possibility that the two terranes were continuous end to end during the Campanian, as the amount of marine transgression could easily vary along the length of the Cretaceous plutonic belt.

Implications of Eocene and Oligocene rocks Eocene marine deposits have provided insights into probable early displacement of the Salinian Block. Eocene turbidites (Butano Sandstone) of the Santa Cruz Mountains are interpreted as deep-sea fan deposits derived from Salinian granitic basement rocks to the south, and are believed to have been continuous with the Point of Rocks Sandstone in the Temblor Range on the opposite side of the San Andreas fault, indicating a postdeposition offset of about 305 km (Clarke and Nilsen, 1973; Nilsen and Clarke, 1975. See Fig. 13-2 for locations). If one reverses the slip on the SAF to restore the Eocene deep-sea fan, the basement rocks still do not match across the fault. As discussed later, this indicates that the Salinian Block had become displaced (although not to the present extent) prior to the Eocene.

Significantly, Paleocene and Eocene sediments are unknown in the onshore part of the Sur-Obispo Belt north of latitude 34°45′, although they are locally abundant on the neighboring margin of the Salinian Block. Thus the spatial relationship of the two terranes must have changed markedly since some time in the late Eocene.

In general, Oligocene rocks of the Salinian Block are very different from the Eocene sediments. They have important tectonic implications; see, for example, Dibblee (1977). Much of the Salinian Block emerged after the Eocene; marine sediments of Oligocene age (uppermost Refugian and Zemorrian) are not voluminous in the Block except in the Santa Cruz Mountains and northern Santa Lucia Range. Marine sedimentation occurred in shallower, more restricted basins than Eocene sedimentation in the same areas; no Zemorrian subsea fans have been found in the Salinian Block. Exposed nonmarine deposits are relatively extensive, especially in the southeast. Uplift is recorded by sudden change from marine to nonmarine sedimentation in the northern Santa Lucia Range and by conglomeratic sediment there and (more notably) in the Cuyama River region, where an angular unconformity locally occurs at the base of a nonmarine unit. On the basis of relations near the Cuyama River, Vedder and Brown (1968) inferred Oligocene activity along the Sur-Nacimiento fault zone.

Bohannon (1975) made a palinspastic reconstruction in which nonmarine Oligocene sediments (including the Simmler Formation at the southern end of the Salinian Block) are restored to possible original positions. The restoration, which involved "unslipping" the San Andreas fault by 210 km and the San Gabriel fault by 60 km, shows a large region of southern California characterized by sites of nonmarine deposition and intervening elevated areas of basement rock from which sediment was derived. The abundance of conglomerate in some of the deposits emphasizes the marked uplifts of the time. These conditions were presumably related to the cessation of subduction.

After a long interval without marked igneous activity in the Salinian Block, an interval that included the Campanian, Maestrichtian, Paleocene, and Eocene, finally in latest Oligocene-early Miocene time (say 26 to 21 m.y.b.p.) limited igneous intrusion and extrusion occurred both in the Salinian Block and in the adjacent Sur-Obispo Belt. Local igneous rocks of this period include the dacitic intrusive-extrusive centers of the Morro Rock-Islay Hill zone near San Luis Obispo and the Cambria Felsite. These rocks are considered to be about 22 to 26.5 m.y. old (Ernst and Hall, 1974). Rhyolite that is only slightly younger occurs on the northeast side of the Salinian Block at the Pinnacles, and identical rhyolite is found in the Neenach Volcanics on the opposite side of the San Andreas fault in the western Mojave Desert. Both the Pinnacles and Neenach rocks give radiometric ages of about 23.5 m.y. (Huffman, 1970), and their present geographic separation (approximately 300 km) is a highly credible measure of post-early Miocene slip on the SAF. Dacitic volcanics near San Juan Bautista are approximately 21.5 m.y. in age (Turner, 1969), and the Mindego basalts in the Santa Cruz Mountains give ages of 20 to 23.1 m.y. (Turner, 1970).

There have been several attempts to relate this volcanic activity to plate tectonics. Important constraints include the ages of the igneous rocks vis-a-vis known plate-tectonic events, the more or less bimodal compositions of the magmas, the lack of a continuous volcanic chain in the Southern Coast Ranges, an imperfect "younging" from south to north, and the relative nearness of the volcanism to the site of the Paleogene trench.

Ernst and Hall (1974) list several options for the origin of the Cambria Felsite and Morro Rock-Islay Hill rocks. They suggest that the combined effect of subduction (which was about to cease) and the impingement of the East Pacific Rise against the North American plate caused melting at an unusually short distance from the trench. This would relate the volcanism to the Rivera triple junction, of which the East Pacific Rise was an element. The Rivera triple junction may not have been very far from this region at 25 m.y.b.p., as the Cambria-Morro Rock-Islay Hill area was 400 to 500 km southeast of its present position relative to the interior of California. However, the presence of the Rivera triple junction cannot explain the occurrence of late Oligocene volcanism farther to the north, as this triple junction moved in the opposite direction. On the other hand, the passage of the Mendocino triple junction offers a possible explanation for virtually all the late Oligocene-early Miocene volcanism in the Southern Coast Ranges.

Snyder and Dickinson (1979) argue persuasively that the Cambria-Morro Rock-Islay Hill volcanism and other magmatism to the north was likely related to conditions around the migrating Mendocino triple junction, although it did not possess a spreading ridge. They show that the trench between the Farallon and North American plates probably trended more northerly than the direction of relative motion between the Pacific and North American plates, which were progressively coming into mutual contact along a lengthening transform. This geometry suggests that extensional effects and consequent magmatism could be expected to attend the Mendocino triple junction during its northwestward passage. Snyder and Dickinson reconstruct the position of the triple junction at 25 m.y.b.p. and find that it falls opposite the Cambria-Morro Rock-Islay Hill region (restored to its original position). They likewise relate the Pinnacles Volcanics and the Mindego basalt (both in the Salinian Block) to the same extensional effects as the Mendocino triple junction moved northwestward. They also include the Quien Sabe Volcanics, which are east of Hollister and east of the Salinian Block. It seems to the writer that their hypothesis most plausibly explains the sudden outbreak of volcanism in a belt that had long been on the lid of an active subduction system, but which experienced virtually no igneous activity until the cessation of subduction.

The presence of nonmarine Oligocene strata and evidence of igneous activity on both sides of the Sur-Nacimiento fault zone are the earliest indicators that the Sur-Obispo Belt may have arrived at approximately its present position snug against the Salinian Block as it now exists, the postulated intervening terrane having been removed. Apparently, the newly attained (?) proximity of the Sur-Obispo Belt and Salinian Block did not lead to firm coupling, as subsequent relative movement has occurred. No Oligocene formations (indeed, no pre-Quaternary units) extend continuously across the Sur-Nacimiento fault zone, with the possible exception of the limestone at Lime Mountain (Graham, 1978). A further lack of continuity across the zone is shown by differences in sediments on the two sides of the boundary. The nonmarine Oligocene formations of the Salinian block contain granitic and felsic volcanic rocks, but apparently no Franciscan detritus, whereas sediments of comparable age in the Sur-Obispo Belt definitely contain Franciscan as well as other debris. The sediments just mentioned show that the Franciscan was at least locally emergent.

Local uplifts allowed the Upper Cretaceous, Paleocene, Eocene, and Oligocene sediments to be stripped from certain parts of the Salinian basement prior to the Miocene, and the latter epoch witnessed renewed marine sedimentation of markedly different character.

Neogene deposits and tectonics Miocene deposits include shelf sands and sheltered deep-water siliceous shale and porcelanite (e.g., the Monterey Formation) rich in microfossils and organic compounds, and lacking coarse clastic debris. There is little, if any, difference in the general nature of these deposits on the two sides of the Sur-Nacimiento fault zone (e.g., see Vedder and others, 1967), although late Cenozoic movements have broken the continuity of formations across the zone. I conclude that the Salinian Block and the Sur-Obispo Belt were essentially in their present relative positions and shared the same history throughout the Neogene; although both were affected by Neogene folding and faulting, the movements between the two subprovinces probably did not exceed a few kilometers or, at most, a few tens of kilometers.

Middle Miocene sediments in the Santa Lucia Range were deposited in more or less en echelon basins that were probably caused by compressional warping of "wrench-type" origin (Graham, 1978). At about the same time, parts of the Salinian Block and adjoining subprovinces, including the Sur-Obispo Belt and Transverse Ranges, were locally affected by mild to pronounced magmatic activity. Examples of igneous products include the Santa Monica basalt (ca. 14 m.y.), the felsic tuff and diabase intrusions of the Obispo Formation (15 to 17 m.y.), the several basalt flows of the Caliente Range (Relizian to Delmontian according to Vedder, 1975; say 11 to 17 m.y.), and the Page Mill Basalt near Stanford University (14.4 ± 2.4 m.y., Turner, 1970).

The en echelon basins and magmatic events approximately coincide with the resumption and quickening of movement on the San Andreas fault, and hence with the beginning of the main stage of emplacement of the Salinian Block. The en echelon basin formation and volcanism immediately preceded the advent of throughgoing strike-slip faults (other than the San Andreas) within the Salinian Block (Graham, 1978). It therefore seems likely that the deformation and magmatism resulted from wrench-type compressions and extensions representing crustal strain during the eastward transfer of principal transform movement from the area that is now offshore to the present site of the San Andreas fault. Thus the record reflects a broadening of the zone of transform interaction as the edge of the continent gradually became attached to the Pacific plate.

Origin and Emplacement of the Salinian Block

General remarks Data summarized in the foregoing pages provide guidance and constraints for hypotheses about the origin and movement of the Salinian Block and lead to answers of some (but not all) questions on the subject.

We will first deal with the problem of restoring the Salinian Block to its position just after the introduction of the granitic plutons. In doing so, we will be guided by spatial circumstances of the Block and the nature of the plutons themselves. Next, we will consider the original location of the Salinian metasediments, which may or may not have been allochthonous prior to the plutonism; unfortunately, we will reach no conclusions on that subject. Finally, we will list the main stages in the emplacement of the Block, proceeding chronologically.

For present purposes, it seems advisable to forego interpretations that would take into account the very large movements implied by recent paleomagnetic data from the Peninsular Ranges (Teissere and Beck, 1973; Beck, 1976; Beck and Plumley, 1979), the Transverse Ranges (Kamerling and Luyendyk, 1979), Baja California (Pischke and others, 1979), and insecure preliminary findings from the Salinian Block itself. Some of the indicated megamovements are much greater than those of simple palinspastic reconstructions. If real, these large motions would mean that much of the framework of the palinspastic reconstructions was mobile, and the movements deduced herein must be components of larger ones.

Original site of Salinian plutonic belt It is accepted here that the Salinian Block was once situated between the Sierra Nevada and Peninsular Ranges, and that most of its plutons belong to a former southward continuation of the Sierran batholith. This idea was more or less implicit in the classic paper of Hill and Dibblee (1953), who were the first to show very large offset on the San Andreas fault. Although some persons (e.g., G. W. Moore, 1976; Ross, 1978) doubt that the Salinian Block was simply a direct southward extension of the Sierran terrain, the present writer and a number of others (e.g., Hamilton, 1978a; Nilsen, 1978, Fig. 5) still favor this view. Judging from published data, typical Salinian plutons resemble Cretaceous intrusives near the central part of the Sierran batholith in age, overall composition, various chemical ratios, petrologic indexes, and rare-element content, although admittedly there are minor differences.

The southernmost Sierran rocks nearest the San Andreas fault and the Salinian Block might be expected to closely resemble Salinian rocks, but Ross (1978) finds that they differ markedly. Although this is important, it does not preclude a former continuity of the Sierran-Salinian terrane. The southernmost Sierran rocks should match the northernmost Salinian rocks, and the latter (being offshore) have never been seen. Ross (1978) points to a further discrepancy, which is difficult to explain: unlike some plutons of the Sierra, those of the Salinian Block are not accompanied by metalliferous mineralization.

Although most individual Salinian plutons will probably never be matched with specific counterparts on the other side of the San Andreas fault, the polka-dot granitic dikes of Ehlig and Joseph (1977) suggest a possible tie between the intrusive rocks of the La Panza Range and those of the Little San Bernardino and Orocopia Mountains.

The fact that typical Salinian plutons resemble those in the interior of the Sierran magmatic belt instead of those in the Sierran foothills is said to be a problem. It has been pointed out (Gastil and Phillips, 1974) that plutons of the western Peninsular Ranges have Mesozoic volcanic wall rocks and are relatively low in potassium. The Salinian plutons, although on the same side of the San Andreas fault, do not have volcanic wall rocks and are fairly high in potassium. Howell (1975c) attempts to explain the discrepancy by proposing that a westward convexity or "knee" in the batholithic belt was cut by the San Andreas when the Salinian Block was sliced off, so the fault reached into the axial part of the belt. G. W. Moore (1976) shows how the Block could have come from the east side of the Sierran terrane by successive displacements on intersecting strike-slip faults with two different trends. However, neither Howell's or Moore's explanation is necessary as long as any proto-San Andreas fault in the Southern California-Baja region lay well within the plutonic belt, as does the southern part of the modern San Andreas (Figs. 13-2 and 13-13).

Original location of Salinian preplutonic rocks Although the original position of the Salinian plutons does not offer insuperable difficulties if one just considers the plutons per se, when one takes into account their wallrocks, serious problems arise. Most Salinian metamorphic rock sequences are different from most (all?) of the best-known wallrocks and roof pendants of the Sierra Nevada (Ross, 1977a; 1978). No one has made a convincing correlation between Salinian metamorphic rocks and sequences across the San Andreas fault except for the probable match between the schist of Sierra de Salinas and that of Portal-Ritter Ridge (Ross, 1976b). (See the preceding section, Metasediments of the Salinian Basement Complex.)

Eventual correlations cannot yet be ruled out, because there are still many gaps in knowledge. Much (most?) of the Salinian basement is covered by younger rocks or by the ocean and has never been seen. Perhaps the displaced parts of the San Emigdio-Tehachapi terrains are beneath the sea. Parts of the Salinian Block may have been displaced much less or much more than the 300 km and 600 km that are commonly cited as minimum and maximum amounts, so one cannot be sure offhand whether or not there is a correlation of any particular sequence where it should be. Uncertainties about the kinematics of the Transverse Ranges and tectonics of the Mojave Desert add to the difficulty of making palinspastic restorations. In addition, sizable areas in southeastern California are hidden by tectonic cover, young rocks, or sediments. Some exposed Precambrian and Paleozoic terrains are complex, contain poorly known sequences, and may even be far-traveled allochthones (e.g., see Haxel and Dillon, 1978). Lateral changes in sedimentary facies and variations in metamorphic grade could exacerbate the difficulties of correlation. All things considered, it is too early to conclude that the Salinian metamorphic rocks do not match anything on the opposite side of the San Andreas fault.

There is another difficulty in "finding a home" for the preplutonic rocks of the Salinian Block. According to Kistler and Peterman (1973) and Kistler (1978), the spatial distribution of initial ^{87}Sr–^{86}Sr ratios has important tectonic implications. They maintain that the Sr isotope ratios reflect the nature of the lithosphere in which plutons were emplaced. Kistler (1978) proposes that a contour representing initial ^{87}Sr–$^{86}Sr = 0.7060$ in Mesozoic plutonic complexes follows the margin of continental lithosphere. Plutons with 0.7060 (or higher) have been emplaced in a sialic setting; those with a lower ratio reflect a more oceanic lithosphere. The plutons of the Sierra Nevada east of the Melones fault and Kings-Kaweah serpentinite melange have ratios higher than 0.7060, as do the Salinian plutons. However, if the Salinian Block is moved southwest by "unslipping" the San Andreas fault, and after adjustments are made on the Garlock fault, the configuration of the 0.7060 line is still anomalous; the pre-San Andreas Salinian Block appears to lie out of position and is separated from the rest of the continent by a corridor of low ^{87}Sr–^{86}Sr values. Furthermore, the 0.7060 line makes a drastic reentrant in northern Nevada, as though Nevada had lost a sizable piece of sialic real estate. Therefore, Kistler brings the Block from northwest Nevada by a sequence of rifting and left-hand strike-slip translations, commencing with the late Precambrian, continuing intermittently into the Jurassic, and ending prior to the Cretaceous plutonism in the Salinian Block and the Sierra Nevada. This hypothesis is ingenious and is based on fundamental data, but some parts of it may not be required. The portion of the original Salinian terrain with high Sr ratios could have been narrower and could have been farther southeast than Kistler shows it, to make a better pre-San Andreas fit, and the corridor of mafic lithosphere in southern California could be explained by Mesozoic back-arc spreading (Haxel and Dillon, 1978).

The foregoing discussion indicates uncertainties about the site of origin of the Salinian preintrusive rocks, but, unfortunately, it does not lead to any conclusions.

If it should turn out that these rocks did not originate in the region between the Sierra Nevada and the Peninsular Ranges, they probably found themselves in that region in time to be invaded by plutons of the Sierran-Peninsular magmatic belt. In other words, the plutons were not transported with the host rocks from some locality beyond the Sierran-Peninsular axis.

History of emplacement of the Salinian Block The major interpreted events are summarized chronologically in Table 13-2 and Fig. 13-13. Needless to say, there are many uncertainties and some discrepancies. We have purposely omitted the preplutonic history.

The origin and behavior of the Salinian Block is inevitably tied to the history of the San Andreas fault, the Peninsular Ranges, the Baja California Peninsula, and the Gulf of California. One should consider the history of the entire coastal region from Point Arena in Northern California to Cabo Corrientes on the mainland coast of Mexico, a task for which I am not well qualified. Clearly, the tectonic events in the Gulf of California (Hamilton, 1961; Larson and others, 1968; Moore and Buffington, 1968) imply a northwestward movement of the Peninsular Ranges and, consequently, of the Salinian Block. It should be noted that at the mouth of the Gulf the opening that can be ascribed to sea-floor spreading (since about 4.5 m.y.b.p.) is 240 to 260 km, but the total width of the mouth, measured NW-SE between the 1000-m contours, is 450 to 480 km. All of this must be accounted for, somewhere to the northwest. The presence of the little-understood Transverse Ranges athwart the Salinian-Baja California axis is a troublesome complication, and internal slicing and deformation of the Salinian Block add further problems. These matters are briefly addressed in the following.

Discussion of two-stage emplacement Three arguments favoring a two-stage emplacement of the Salinian Block will be discussed in turn.

1. If one imagines that post-Early Miocene 305-km slip on the San Andreas fault (SAF) in central California is reversed, the Eocene as well as Early Miocene rocks of the Salinian Block are restored to their original continuity with counterparts across the fault, but the Block is far out of position with respect to the Sierran-Peninsular axis, so it must have moved prior to the Eocene (Nilsen and Clarke, 1975). I know of no way to counter this argument except to question the correlation of rocks across the SAF. Those who challenge the matching of the Eocene Butano Sandstone with the Point of Rocks Sandstone will experience extreme difficulty in finding alternatives. The amount of pre-Eocene dextral movement of the Block could be as much as 200 km (Nilsen and Clarke, 1975).

2. Eocene deep-sea fan deposits, which Nilsen and Clarke believe were continuous across the SAF prior to Neogene offset, are underlain by Salinian basement rocks on one side of the fault and most certainly by Franciscan rocks on the other side. The Butano Eocene deep-sea fan accumulation in the Santa Cruz Mountains rests on Paleocene strata that are in depositional contact with Salinian granite that is rather widely exposed. The continuation of the same fan accumulation, the Point of Rocks Sandstone, is some 300 km distant on the other side of the SAF in the Temblor Range. Typical

TABLE 13-2. History of Salinian Block, Early Cretaceous to Present

Time	*Principal Events*	*Attendant Circumstances*
Early to mid-Cretaceous (ca. 117–106 m.y.b.p.)	Main granitic intrusive epoch in Sierran-Peninsular site of future Salinian Block	Subduction west of site of Salinian Block before, during, and after plutonism
Late Cretaceous (ca. 80 m.y.b.p.)	Emplacement of youngest plutons?	Subduction west of site of Salinian Block
Latest Cretaceous-Early Tertiary (ca. 75–50 m.y.b.p.)	Deep erosion of plutons, uneven submergence, deep-sea fan sedimentation (in embayments) on site of Salinian Block	Low-angle subduction; Laramide orogeny far to the east (Dickinson and Snyder, 1978; Keith, 1978)
Paleocene (?) (65–55 m.y.b.p.?)	Initial 200-km NW movement of Block along proto-San Andreas fault Somehow Block received sediments from Sierra to E, despite separation West part of Block lost, possibly by subduction rasping or by mega-strike-slip	Oblique plate convergence; rapid subduction of gently dipping slab; strike slip on proto-SAF which, in central Calif., had same location as present SAF
Eocene (53–38 m.y.b.p.)	Salinian Block immobile, largely submerged; deep-sea fans spanned dormant San Andreas (Nilsen and Clarke, 1975)	Subduction in progress West of Block and elsewhere along North American margin
Oligocene (38–23 m.y.b.p.)	Local uplift in Salinian Block; both nonmarine and (limited) marine sedimentation	Subduction in progress; little or no translatory movement along San Andreas
Latest-Oligocene-Early Miocene (26 –21 m.y.b.p.)	Limited basaltic and silicic magmatism in Salinian Block and Sur-Obispo Belt; the two terrains not far apart	Triple junction passed northwestward along margin; subduction ceased ca. 25 m.y.b.p. near San Luis Obispo
Miocene (23–5 m.y.b.p.)	Starting about 15 m.y.b.p., *en echelon* compressional basin development and folding in Salinian Block (Graham, 1978); at about same time, basaltic and silicic magmatism and strike-slip faulting within Block; Salinian Block in motion, having started 15–20 m.y.b.p.; bulk of Block moved 60–100 km in Miocene; slivers moved farther (Fig. 13-14)	Cessation of subduction followed NW migration of triple junction; transform-type tectonics ensued (Atwater, 1970); San Andreas awoke in Middle Miocene (Dickinson and others, 1972); San Gregorio-Hosgri fault zone accommodated movement of major sliver (Graham and Dickinson, 1978a)
Pliocene-Quaternary (5–0 m.y.b.p.)	Accelerated NW movement of Salinian Block (present rate about 3.2 cm/yr; Savage and Burford, 1973); accrued Miocene-Holocene movement of bulk of Block ≃ 305 km; westernmost sliver perhaps accrued 115 km additional (Graham and Dickinson, 1978a)	Opening of modern Gulf of California 4–5 m.y.b.p.; movement of Baja California 200–260 kw NW (Moore and Buffington, 1968; Larson and others, 1968); part of motion absorbed in Transverse Ranges, but most of it shared by Salinian Block

Franciscan rocks are not exposed in the Temblor Range, but their presence in the subsurface may be inferred from the regional distribution of the Franciscan, from the presence of probable ophiolitic rocks at Eagle Rest Peak to the southeast (Ross, 1970), and from local circumstances. The latter include the nearby presence of serpentinite, which in the Coast Ranges is rarely distant from the Franciscan. Serpentinite is exposed in the core of an anticlinorium flanked by Point of Rocks Sandstone. More important is the fact that the Sandstone and underlying sediments rest on a GVS unit that Dibblee (1973, p. 8) believes is the Upper Jurassic-Lower Cretaceous Gravelly Flat Formation. GVS sediments of this age do not (and could not) occur above Salinian basement with its Cretaceous plutons, but such sediments are common in terranes underlain by the Franciscan. All things considered, it is almost certain that the Franciscan is in the subsurface beneath the Point of Rocks Sandstone, whereas the continuation of the latter on the other side of the SAF rests on Salinian basement; hence the Salinian Block had experienced pre-Eocene movement partway toward its present position.

3. A third argument must be considered, although its implications have been modified recently. It is concerned with the *length* of mismatch in basement rocks on either side of the SAF. If Eocene and lower Miocene rocks of central California are restored to former continuity across the fault by reversing 305 km of offset, (see Fig. 13-14), there is still a residual mismatch between the Salinian basement and the Franciscan on the two sides of the fault northwest of the restored Tertiary rocks. The length of the residual mismatch depends on assumptions as to the present location of the northernmost Salinian rocks and of the southernmost Franciscan rocks on the opposite sides of the fault. If we measure from Eagle Rest Peak to Point Arena, the distance is on the order of 595 km. Subtraction of 305 km of Neogene offset determined in central California leaves 290 km of offset unaccounted for, so one might propose this amount for the pre-Eocene partway emplacement of the Salinian Block. However, the San Gregorio-Hosgri fault joins the San Andreas between the point of the 305-km measurement and the northwest anchor point of the 595-km measurement, and slip along it has contributed to the total length of the mismatch (Graham and Dickinson, 1978a. See discussion of "slivering" later). The contribution could be 80 to 115 km toward the unaccounted-for 290 km, leaving a net of 175 to 210 km of displacement that might well represent pre-Eocene partway emplacement of the Salinian Block.

The timing of the postulated early partway emplacement of the Salinian Block is very uncertain. I believe that it was Paleocene (probably between 65 and 55 m.y.b.p.) for reasons mentioned earlier in connection with the proto-San Andreas fault, which was the vehicle for the movement. To briefly recapitulate, thrust faulting affected Campanian rocks on both sides of the present location of the Salinian Block, and evidently Paleocene thrusting occurred at least on the northeast side. No such events are recorded in the Salinian Block, and I conclude that the Block was not there at the time. On the other hand, it was already emplaced partway by the Eocene. Until recently the Paleocene or Maastrichtian were favored because it was supposed that the Kula plate passed along the continental margin during that time interval, inducing strike slip. However, it now seems doubtful that the Kula plate passed the California coast, at least within the last

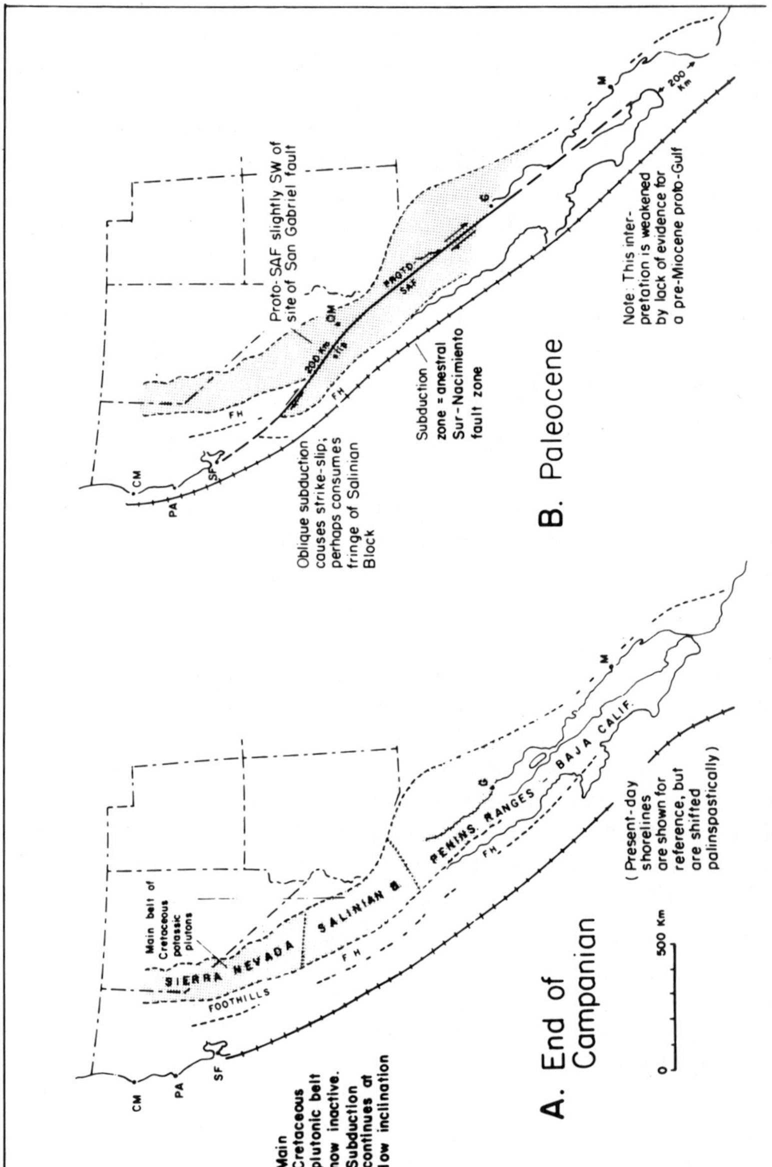
A. End of Campanian
Main Cretaceous plutonic belt now inactive. Subduction continues at low inclination
CM
PA
SF
SIERRA NEVADA
SALINIAN B.
PENINS. RANGES
BAJA CALIF.
FOOTHILLS
FH
G
M
Main belt of Cretaceous potassic plutons
(Present-day shorelines are shown for reference, but are shifted palinspastically)
0
500 Km
B. Paleocene
Oblique subduction causes strike-slip; perhaps consumes fringe of Salinian Block
Subduction zone = anestral Sur-Nacimiento fault zone
Proto-SAF slightly SW of site of San Gabriel fault
PROTO-SAF
200 Km slip
200 Km
Note: This interpretation is weakened by lack of evidence for a pre-Miocene proto-Gulf
CM
PA
SF
FH
G
M

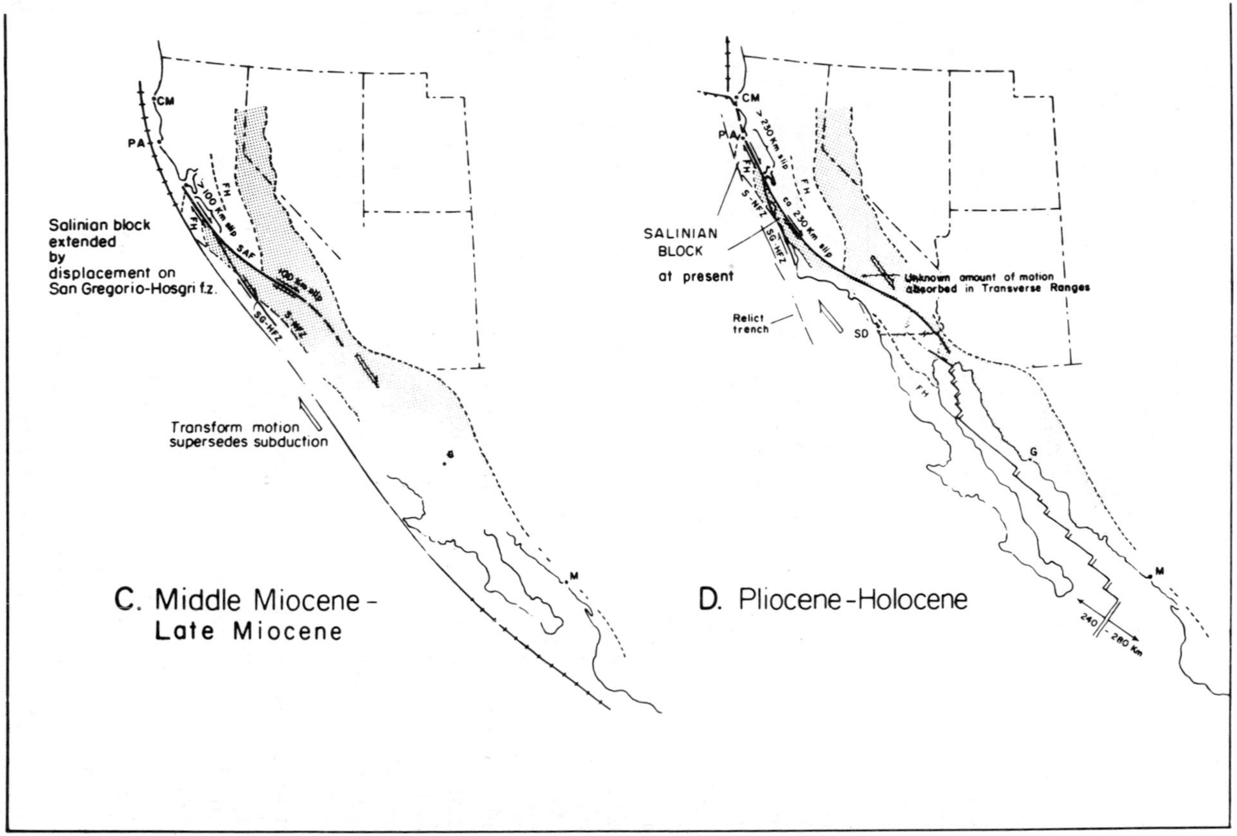

Fig. 13-13. Progressive emplacement of Salinian Block. CM, Cape Mendocino; FH, Sierra Foothills and analogous terranes elsewhere; G, Guaymas; M, Mazatlan; OM, Orocopia Mts.; PA, Point Arena; SAF, San Andreas fault; SD, San Diego; SF, San Francisco; SG-HFZ, San Gregorio-Hosgri fault zone; S-NFZ, Sur-Nacimiento fault zone.

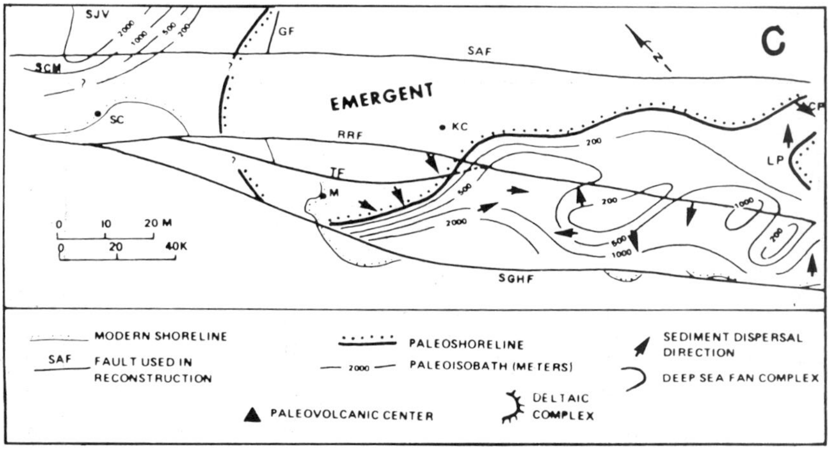

Fig. 13-14. Early middle Miocene basins, Santa Lucia Range, Salinas Valley, and environs. Paleogeography ca. 15 m.y.b.p. showing en echelon basins ascribed to compressional warping associated with transform tectonics. SAF, San Andreas fault; SGHF, San Gregorio-Hosgri fault; RRF, Reliz-Rinconada fault; GF, Garlock fault; SC, Santa Cruz; M, Monterey; KC, King City. (From Graham, 1978, with permission of The American Association of Petroleum Geologists).

80 m.y. (Cooper and others, 1976). An alternative idea, broached earlier, calls upon oblique plate convergence with subduction at a low dip angle. Relative plate motions recently reconstructed by Coney (1978) imply oblique convergence at the time that concerns us. Dickinson and Snyder (1978) and Keith (1978) find evidence for a very gently inclined subducting slab during the Laramide orogeny, roughly 70 to 45 m.y.b.p. Dickinson (informal communication, 1978) thinks the low angle might enable the slab to transmit a strike-slip component to the underside of the continental plate edge. It should be mentioned that Nilsen (1978) proposes that the early partway emplacement of the Salinian Block occurred in the Campanian, based largely on the borderland-type paleogeography of that time, a type of geography associated with instability and deformation.

Loss of part of the Salinian Block One of the great problems of the Southern Coast Ranges is posed by the disappearance of a large terrane from the southwest side of the Salinian Block. Along the Sur-Nacimiento fault zone, the Franciscan subduction complex is locally in contact with the corresponding magmatic arc, the Cretaceous plutons of the Salinian Block (Figs. 13-2, 13-3 and 13-11). Normally, such an arc would lie 100 to 200 km distant from the subduction zone. The entire western flank of the arc assemblage has vanished, as has most of the postulated forearc basin and very possibly part of the Sur-Obispo Belt of Franciscan rocks. The length of the vanished fragment is unknown; it is at least 150 km (the distance from Point Sur to Atascadero), and it could be more than 1000 km if the vanished fragment extended far southward to include part of western Baja California, which seems likely.

How did this sizable fragment disappear? There are two main options: (1) by subduction (Page, 1970a) or (2) by strike slip. The arguments for loss by subduction include the following: Subduction is known to be destructive of the "hanging wall." For example, the ophiolite and other remnants above subduction complexes are never complete sections of the lithosphere, but are generally dismembered fragments. Even low-density rocks can be carried to depths of 10 to 30 km (witness the occurrence of jadeitized matagraywacke), so it is conceivable that some sialic rocks were carried "all the way," never to return. Subduction was indeed occurring at the time that seems most likely for the disappearance of Salinian crust, judging from reconstructions of late Mesozoic and Early Tertiary plate interactions (Atwater, 1970; Coney, 1978). The case for subduction would be won if the Franciscan near the Sur-Nacimiento fault zone contained ripped-off blocks of granite, but there is only one place where this might be proposed, and the evidence does not support the idea. A large body (and several smaller ones) of granitic rock is surrounded by the Franciscan Complex 34 to 45 km southeast of Point Sur. However, Gilbert (1974, Fig. 3) finds that the largest granitic body is a fault block in the traditional sense, and it was probably displaced from the neighboring Salinian basement long after subduction ceased.

The arguments against subduction of part of the Salinian Block are, like most of the arguments in favor of it, largely theoretical. The missing rocks would be chiefly of low to moderate density and would be reluctant passengers in a downgoing transport system. Furthermore, subductive rasping and disposal are inconsistent with the accretion of the Franciscan and with currently favored models of subduction zones (Seely and others, 1974; Karig and Sharman, 1975).

The possibility that mega-strike slip removed part of the Salinian Block is strengthened in principle by the finding of transplated terrains in British Columbia and Alaska that evidently traveled thousands of kilometers from south to north (Jones and others, 1972; Jones and others, 1977; Davis and others, 1978; Berg and others, 1978). The ancestral "homelands" of these terranes are largely unknown, but Saleeby (Chapter 6, this volume) thinks the Cache Creek terrane of British Columbia is represented in the Sierra Nevada by the Calaveras and western Kings sequences, and Schweickert and Snyder (Chapter 7, this volume) suggest that the Alexander terrane of southeast Alaska is a displaced volcanic arc formed by westward subduction in northern California in the Late Ordovician or Silurian. Is it possible that some other fragment, now attached to British Columbia or southeast Alaska, could be the missing part of the Salinian Block? So far, descriptions of the allochthonous fragments do not fit the requirements very convincingly.

The partial destruction of the Salinian Block obviously postdates the youngest truncated plutons (ca. 80 m.y.?). It was probably post-Campanian, judging from structural relations just west of the Block in the vicinity of the Cuesta Ridge ophiolite near San Luis Obispo, where Great Valley-type sediments are preserved as thrust sheets with westward vergence (Fig. 13-10). A sheet of Campanian strata is thrust over an Upper Jurassic-Lower Cretaceous flysch unit that rests depositionally upon an ophiolite suite. The ophiolite is in tectonic contact with an underlying Franciscan melange. By analogy with the relations observed along the west side of the Great Valley, the ophiolite and overlying sediments belong well to the west of the plutonic arc, and the intervening terrane must have been removed during or after the development of the foregoing structural relations. There is no appropriate source for the thrusts at present. Allowing time for the post-Campanian thrusting and subsequent change in tectonic regime, the nearby flank of the Salinian Block was probably destroyed some time after 70 m.y.b.p.

The lost terrane may have vanished in the Paleocene. D. G. Howell (1979, written communication) notes the implications of Campanian to Paleocene (?) strata in the central San Rafael Mountains just southwest of the Salinian Block. In this locality, Vedder and Brown (1968, p. 246) report uppermost Cretaceous (?) nonmarine conglomerates containing first-cycle granitic clasts up to 2 m in size. The nonmarine sediments intertongue with marine strata, which become dominant toward the southwest. These relations indicate derivation of detritus from a provenance like the Salinian Block in the immediate vicinity to the northeast; that is, any previously intervening terrane had already moved away. Perhaps the removal event is represented by a Paleocene depositional hiatus in the San Rafael Mountains (D. G. Howell, 1979, written communication). However, in the La Panza Range adjacent to the zone of removal there is no obvious stratigraphic indication of an extraordinary event. An apparently unbroken marine sedimentary sequence spans an interval of late Campanian to middle Eocene (about 75 to 45 m.y.b.p.). If megamovements were occurring nearby, for some reason they did not affect the sedimentary record. The Paleocene would be an appropriate time for loss of Salinian crust by piecemeal subduction, because initial movements of the Salinian Block during the epoch would have caused the northwest end of the Block to project vulnerably into the subduction zone.

Probably the lost terrane disappeared before the Eocene, because middle Eocene marine strata blanket extensive areas in the southern parts of both the Salinian Block and Sur-Obispo Belt (as pointed out by D. G. Howell, 1979, written communication), and these sequences do not seem to have suffered a truly major displacement at the faulted juncture of the two blocks. Elsewhere in the Southern Coast Ranges, the upper Oligocene-lower Miocene Vaqueros Formation is much the same on the two sides of the zone from which the lost terrane was removed; hence no post-Vaqueros megamovements occurred there.

All in all, the time window for the tectonic "rip-off" seems to fall within 70 to 25 m.y.b.p., and there are reasons for favoring the narrower window of 70 to 45 m.y.b.p. During the period 70 to 45 m.y.b.p., plate convergence is thought to have been oblique and rapid (Coney, 1978), and the downgoing slab had assumed a very low angle (Dickinson and Snyder, 1978; Keith, 1978). Apparently, strike slip, including movement along the proto-San Andreas fault, occurred during this interval, and mega-strike slip is a possibility. However, the low angle of subduction may have facilitated rasping. Whatever happened, it could have taken place either during or after the partway emplacement of the Salinian Block; in fact, the two events could have been brought about by the same set of circumstances.

The postulated passage of the Kula plate parallel with the coast (Atwater, 1970) has been cited as an opportune situation for large-scale strike slip. However, it now seems doubtful that the Kula plate did, in fact, move past the California coast, at least during the last 80 m.y. (Cooper and others, 1976; Coney, 1978). Thus, if part of the Salinian Block traveled parallel with the continental margin in the Paleogene, it probably did so under the influence of oblique subduction rather than transform interaction. If the missing part of the Block cannot be found in British Columbia or southern Alaska, one might wonder if conceivably it could have been subducted endwise at the Aleutian trench. This last fate seems much less likely than piecemeal subduction of the edge of the Salinian Block during some interval in the evolution of the Franciscan Complex.

In summary, the loss of part of the Salinian Block is a great and murky problem. At the moment, I tentatively prefer the subduction explanation, but the matter needs much more study.

Slivering of the Salinian Block The disparity between the total offset along the San Andreas fault in northern California (ca. 550 to 600 km) and the offset of southern California (260 to 300 km) has been explained by postulating a possible proto-San Andreas that bypassed the present SAF in the south and that, by implication, involved early movement of part of the Salinian Block. Suppe (1970) proposed that a proto-SAF extended from the coast of Baja California to a juncture with the modern SAF north of the Transverse Ranges. Johnson and Normark (1974a, b) attempted to explain the same disparity in offsets along the SAF without recourse to a protofault, by means of "slivering" such that parts of the Salinian Block moved farther than the rest along the SAF; the Block was thus elongated by shear. Both schemes are viable in principle, and apparently some elements of both are correct, although the locations and orientations of the postulated faults are subject to adjustment. For example, although the Salinian Block evidently underwent an early movement along a fault that bypassed the present SAF in Southern California, the fault must have been located within the central part of the Peninsular-Baja California plutonic belt. This location is required to explain the presence of potassic Cretaceous plutons in the Salinian Block and the absence of the types of wallrocks that characterize the west side of Baja California.

There is no doubt that the Salinian Block is slivered. Despite its overall narrowness, it is sliced lengthwise into even narrower strips. Some of the internal faults have demonstrable right-hand strike slip. However, there are constraints on slivering as an explanation of elongation of the Salinian Block and unequal offsets along the SAF. For one thing, the only type of shearing that could lengthen the Block along its northeast boundary would be right-hand slip on faults transecting the terrane in a more northerly trend than the axis of the Block itself. In order to affect the apparent slip along the SAF, the sliver faults would have to join the SAF and contribute *their* displacement to *its* displacement. Finally, sliver faults must bypass the region where Tertiary rocks are offset about the same amount as the measured SAF displacement in southern California; in other words, the subsidiary faults must join the SAF north of the Santa Cruz Mountains. These various restrictions mean that faults such as the Rinconada, King City, Huerhuero, the faults within the northern Santa Lucia Range, and those within the Santa Cruz

Mountains could have no substantial effect on the lengthening of the Block along the SAF. What faults, then, could play this role? Virtually the only one that qualifies is the San Gregorio-Hosgri fault zone. It transects the Salinian Block obliquely in a northerly direction, joins the San Andreas near the Golden Gate, and appears to have 80 to 115 km of right-hand slip (E. A. Silver, 1974; Hall, 1975; Graham and Dickinson, 1978a).

Effects of events in the Transverse Ranges The tectonic linkage between the mouth of the Gulf of California and the Salinian Block is broken by the western Transverse Ranges, whose unusual evolution is challenging (Hamilton, 1978a; Hall, Chapter 17, this volume). The Jurassic Santa Monica Slate has moved out of alignment with comparable rocks in the Sierran-Peninsular belt and has rotated about a steep axis, assuming an east-west orientation of foliation (Jones and Irwin, 1975; Jones and others, 1976). In the Santa Ynez Range, Eocene sandstones, and probably Oligocene strata as well, are out of position and show rotated paleocurrent directions. Kamerling and Luyendyk (1979) report 70° clockwise rotation of paleomagnetic poles in middle Miocene (~15 m.y.) basalts of the Santa Monica Mountains, Conejo Hills, and Anacapa Island. These authors and Crouch (1979) have proposed reconstructions of original positions of the rotated terranes. However, the apparent clockwise rotation is in conflict with southerly (oceanward?) paleoslope directions deduced from Eocene sandstones not only in the Santa Ynez Range, which could conceivably comprise a different block, but also in Santa Cruz Island, which could hardly be independent of Anacapa Island and the Santa Monica Mountains. Because of these and other conflicts, the kinematics are uncertain. Nevertheless, large movements and rotations did occur. If the entire thickness of lithosphere was involved, as seems likely, subduction and concurrent rifting and spreading may have accompanied these rather startling events (Hamilton, 1978a).

The post-Eocene date of translation and rotation is too late for any association with the pre-Eocene partway emplacement of the Salinian Block, but it might be associated with the opening of the proto-Gulf of California if this occurred much later than I have proposed here. In any case, the main arrangement of the components of the Transverse Ranges had been accomplished before the opening of the modern Gulf of California starting 4 to 5 m.y.b.p., and the stage was set for the ensuing N-S compressive crunch, which was (and is) partly absorbed by thrusting, folding, and squashing in the Transverse Ranges. This Pliocene-Pleistocene strain alone may have reduced by tens of kilometers the NW movement that would otherwise have been relayed to the Salinian Block. In view of this shock-absorber effect, it is surprising that the post-early Miocene movement (say 305 km) of the bulk of the Salinian Block equals or exceeds the total displacement along the San Andreas fault in southern California (say 260 km), and I am forced to conclude that it is only chance that has caused the two figures to be anything alike. The curvilinear movement of pre-Neogene rocks into their anomalous position in the Transverse Ranges may well have bypassed the present San Andreas fault in southern Califonia and may have nudged the Salinian Block northwesterly without leaving any record along the SAF south of the Transverse Ranges. Somewhat similar bypassing had occurred even earlier, I believe, during the pre-Eocene partway emplacement of the Salinian Block.

It is worth noting that the final (Quaternary) folding and thrusting in the Transverse Ranges postdated all but the last bit of emplacement of the Salinian Block. If it had been otherwise, E-W trending structures would have formed across the Block and would have moved north with it. They would be seen today out of position, north of the Transverse Ranges. This is not the case; only the southernmost structures are deflected by the transverse deformation, and only the southernmost rocks of the Block have been involved in Transverse Ranges thrusting.

TRANSFORM TECTONICS OF THE NEOGENE

The foregoing discussion of the Salinian Block has necessarily spanned a long time interval in which subduction played a prolonged role but was eventually succeeded by a transform regime. We must now go back to examine the transition and will then concentrate on the transform tectonics of the Neogene.

The Atwater model vis-a-vis the Southern Coast Ranges As perceived by Atwater (1970), the NW-SE relative motion between the Pacific plate and North America was first "felt" by the two plates where they initially came into contact off the northern coast of Baja California about 29 m.y.b.p. The ensuing transform-type interaction replaced subduction along the progressively lengthened contact. The latter extended up the coast to the latitude of San Francisco by about 7 m.y.b.p. (Atwater, 1970; Atwater and Molnar, 1973. See Dickinson, Chapter 1, this volume, Figs. 13-12 and 13-14).

The Atwater hypothesis is a milestone in the reconstruction of California continental margin tectonics. It is strongly supported, although not confirmed in every detail, by the following types of geologic evidence:

1. At the present time, a sediment-filled trench lies at the foot of the continental slope west of the Southern Coast Ranges (Fig. 13-2; Page and others, 1979) and apparently elsewhere along parts of the present continental margin. It is no longer accompanied by deep-seated earthquakes or a parallel active volcanic arc, but it is an unmistakable reminder of subduction that continued into Cenozoic time.
2. The Franciscan, which appears to be a subduction complex and which occupies a belt alongside the fossil trench, is locally covered unconformably by Neogene sediments of a "normal" nature, indicating a profound change in dominant processes along the continental margin.
3. Volcanic arc activity in California and Nevada that was associated with Tertiary subduction appears to have been extinguished progressively from south to north in a way that is spatially and chronologically consistent with the Atwater model (Christiansen and Lipman, 1972; Snyder and others, 1976; Dickinson, Chapter 1, this volume).
4. On-land structural evidence for the advent of transform-type tectonics as a replacement of subduction is persuasive and is consistent timewise with the model. The establishment of the transform regime is seen in the history of the San Andreas fault (Dickinson and others, 1972), the middle Miocene initiation of en echelon

folding in the San Joaquin Valley (Harding, 1976), the incidence of north-south thrusting in the Southern Coast Ranges, and the development of Neogene sedimentary basins (e.g., Crowell, 1974a, b, and Chapter 18, this volume; Graham, 1978; Blake and others, 1978).

Time of cessation of subduction in the Southern Coast Ranges In the southern part of the Sur-Obispo Belt southwest of the San Andreas fault, on-land evidence suggests that subduction had stopped by about 25 m.y.b.p. (Page, 1970b). Near San Luis Obispo, the Vaqueros Formation (late Oligocene or early Miocene) unconformably overlies a tectonic contact between the Franciscan Complex and Great Valley-type sedimentary rocks, and east of Cuesta Pass the Vaqueros overlies an apparently subduction related thrust fault between Upper Cretaceous sandstone and Upper Jurassic-Lower Cretaceous shale. Here and there throughout the coastal region between San Luis Obispo and Cambria, the Vaqueros Formation and remnants of somewhat older units (e.g., the Oligocene Lospe Formation) overlap the Franciscan unconformably and do not share the complicated structure of that complex. The Cambria Felsite described by Ernst and Hall (1974) near Cambria unconformably overlies the Franciscan and in turn is overlain by the Lospe and Vaqueros formations; it is well correlated with the Morro Rock-Islay Hill complex of dacitic volcanic necks, plugs, and domes, which are 22.1 to 26.5 m.y. old (Turner and others, 1970). These relationships suggest that subduction had ceased, or at least was not disturbing the Franciscan beneath the Cambria Felsite, by the end of the Zemorrian Stage, which is Oligocene (e.g., see Bandy and Ingle, 1970). If the Cambria area has been shifted 300 km northwest by transform-type movement since the Oligocene, it would have been near the latitude of Los Angeles at that time (say latitude 34°N in terms of present coordinates). The geologic relations just cited correspond well with the inferred arrival (between 29 and 21 m.y.b.p.) of the triple junction at the northwest end of the lengthening transform plate boundary as deduced by Atwater and Molnar (1973, Fig. 1). These conclusions cast doubt on the hypothesis that the Cambria Felsite-Morro Rock magmatism was simply generated by eastward subduction. With greater certainty, it seems that the volcanic components of the Obispo Formation (15.3 to 16.5 m.y.; Turner and others, 1970) are much too young to be direct products of subduction.

Areas that are now due west or due east of San Luis Obispo or Cambria witnessed the cessation of subduction at different times, as they have undergone larger or smaller translatory movement parallel with the plate boundary. If we imagine that North America east of the San Andreas fault has been stationary, and if we accept the Atwater and Molnar (1973) schedule of the migration of the Mendocino triple junction and take into account concurrent strike slip, any area near the continental slope west of San Luis Obispo was formerly perhaps 260 to 475 km to the southeast at a latitude where subduction stopped 26 to 30 m.y.b.p. On the other hand, any area east of San Luis Obispo just beyond the San Andreas fault remained stationary (according to our premise), and subduction was continuing beneath it until about 20 m.y.b.p. Probably subduction continued beneath "stationary" areas east of San Francisco (east of the San Andreas fault) until 6 to 7 m.y.b.p.

Strain accompanying transform motions The partial cover of Neogene rocks in the Southern Coast Ranges clearly shows effects of transform-type deformation that ensued after the long cycle of late Mesozoic-Early Tertiary subduction. The relevant features include strike-slip faults, en echelon basins and folds, appropriately oriented thrusts, and apparently extensional basins. However, some structural and thermal events of the Neogene are not readily understood in terms of transform tectonics despite their time correlation with that regime.

The effects of late Cenozoic transform-type tectonics are distributed in a broad zone that includes much of California and parts of adjacent states and accommodates the relative plate motion (Atwater, 1970). The NW-SE horizontal motion is locally expressed by approximately pure shear and locally by approximately simple shear. (The former involves pronounced transverse shortening, whereas the latter is characterized by differential translation along parallel paths.) In some areas of the Southern Coast Ranges the strain is pervasive and rather evenly distributed, resulting in distortions such as folds. In other places the pattern of pervasive strain is interrupted by discontinuities (faults) along which most of the relative motion is concentrated. The tectonics of horizontal shear are sometimes termed "wrench tectonics" and have recently been discussed by Wilcox and others (1973). Earlier, authors presumed that strike-slip fault movements produced en echelon folds, but it is more logical to regard both the faults and folds as manifestations of regional horizontal shear and to recognize that either may occur without the other.

If horizontal shear were uniformly distributed, a circle inscribed on the ground surface would become an ellipse, and the direction of maximum shortening would be initially at 45° to the direction of shearing. As strain increases, the angle becomes progressively more acute; but in nature it never becomes zero, as there is no way that crustal material could escape altogether between the rotating fold axis and the shear direction. In the case of right-hand shear, the point of the acute angle is aimed to the left as one looks across the area undergoing strain. As a consequence of these "laws," the axes of incipient folds caused by transform tectonics should initially lie at about 45° to the direction of shear (the relative plate motion vector), and the axes should swing around so as to form a progressively more acute angle as the folds become tighter. No folds that lie at angles greater than 45° or that are parallel with the plate motion vector can be explained in a straightforward way by "wrench tectonics." Thrust faults caused by horizontal shear must obey the same rules of orientation as folds. Obviously, in California there are also important time constraints, at least with respect to the commencement of prevalent transform interaction in the Tertiary. In real life, many folds and some thrusts in the Southern Coast Ranges fulfill the predictions of "wrench tectonics" and some do not. Among those that do not, many are of appropriate age but lack the predicted orientation, whereas others show the converse relationships.

En echelon folds In parts of the Southern Coast Ranges, as well as the adjacent margin of the San Joaquin Valley, en echelon folds trend west-northwest at an acute angle with the Pacific-North American plate boundary (Fig. 13-15, from Dibblee, 1976). These have sometimes erroneously been regarded as "drag folds" produced by the San

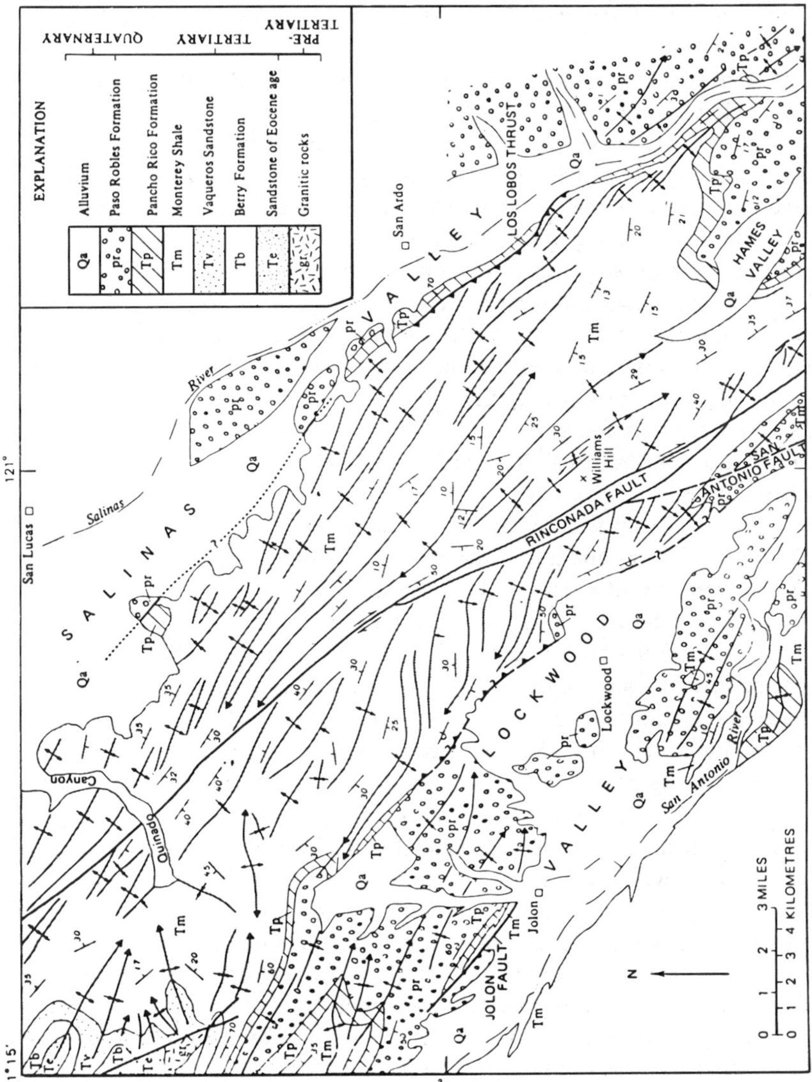

Fig. 13-15. En echelon Plio-Pleistocene folds west of Salinas Valley, and Rinconada fault. Most of the structures shown are expressions of transform tectonics. (By Dibblee, 1976)

Andreas fault, but they are more likely independent symptoms of transform-type relative plate motion. Examples of these folds are seen east of the SAF in the south part of the Diablo Range and vicinity. They are found also west of the SAF in the Santa Cruz Mountains, parts of the Santa Lucia Range (Fig. 13-15), and the area west of Carizzo Plain. The genetic tie between these folds and transform motion is chiefly shown by these circumstances: (1) The orientation of fold axes is appropriate with respect to the orientation of the plate boundary and its right-hand sense of shear. (2) On the whole, the tightest en echelon folds make the most acute angle with the boundary. Moreover, those folds that are arcuate tend to be tightest toward the end that makes the most acute angle. (3) The fold history is compatible with the time of inception and quickening of on-land response to the Cenozoic transform plate interaction.

Harding (1976) makes a strong case for the "wrench" origin of many of the petroleum-bearing folds along the west side of the San Joaquin Valley, showing that their birth and growth were synchronous with the main Neogene activity of the SAF. The fault, which had long been dormant, awoke in middle Miocene time, and many of the folds originated then, as shown by the stratigraphic record. Similar en echelon fold histories are found in parts of the Southern Coast Ranges as well as the San Joaquin Valley (e.g., Graham, 1978). Movement of the SAF and the growth of folds both quickened in the Pliocene and Pleistocene, and new folds appeared at that time. In some areas where Pliocene and early Pleistocene formations are present, if these formations are "unfolded" in one's imagination, the pre-Pliocene anticlines and synclines that remain in the underlying strata are minimal. If the exercise is repeated for Miocene formations, only rarely can a coincident pre-Miocene fold be demonstrated to have existed at the same site, although underlying Paleogene and Cretaceous rocks are generally discordant with respect to the Miocene strata.

Unexplained anticlines and synclines Among the Plio-Pleistocene folds in the Southern Coast Ranges, there are many that lack the orientation predicted by wrench tectonics (Fig. 13-16). For example, a number of folds in the vicinity of the San Andreas fault are parallel with the fault, as are several (most?) folds in the Sur-Obispo Belt (see Hall, 1973b, 1974). The prominent Huasna syncline (Hall and Corbató, 1967; Johnson and Page, 1976), which involves Miocene formations, has an anomalous trend whose origin defies explanation by wrench tectonics, although its slightly sigmoid aspect might be thus explained.

The folds that lie parallel with the plate boundary comprise a rather large family, which is not restricted to the Southern Coast Ranges proper. In the San Joaquin Valley, the Kettleman Hills anticline system is a prominent example, although it may have its own independent origin. Far to the west in the offshore region, a number of open Neogene folds are nearly parallel with the plate boundary.

In summary, although straightforward "wrench tectonics" is a likely explanation of many Neogene folds, especially those with appropriate en echelon arrangement, other influences must have operated during all or part of the same time that transform interaction has been in effect. Perhaps some folds were produced simply by vagaries in plate motion or irregularities in plate margin geometry. The SAF and other long strike-slip

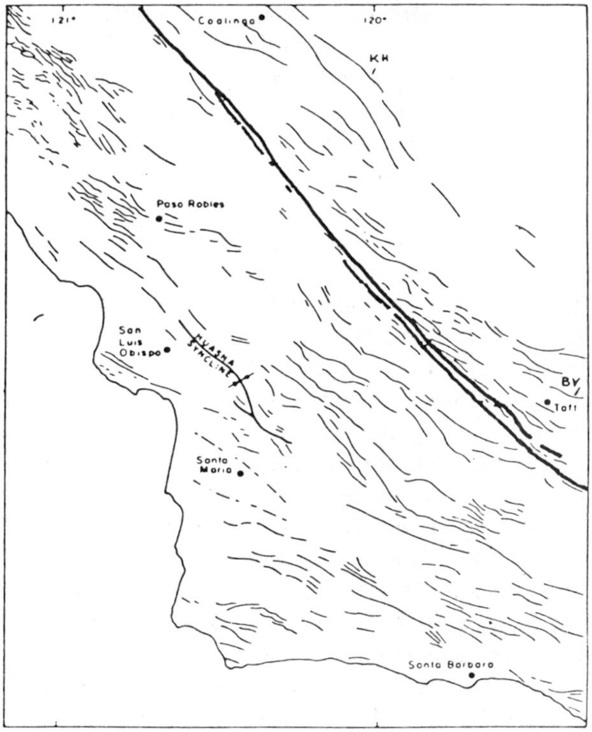

Fig. 13-16. Map of fold axes, Southern Coast Ranges, Transverse Ranges, and southwest part of San Joaquin Valley. The folds are Neogene and originated during the transform regime. Many are oblique with respect to the plate boundary, which is approximated by the San Andreas fault (the only fault shown), and can be explained by wrench tectonics, i.e., horizontal shear. Others, such as those in patterned areas, are quasi-parallel with the plate boundary and are not readily ascribable to wrench tectonics. KH, Kettleman Hills (in San Joaquin Valley); San Andreas fault shown by heavy line. (Redrafted from Johnson and Page, 1976, with permission of Elsevier Scientific Publishing Company).

faults are neither perfectly straight not perfectly oriented with respect to today's plate motion. The present Pacific-North America relative motion vector appears to be about N35W in central California (Minster and Jordan, 1978), but the SAF strikes somewhat more westerly. This must cause abnormal compressive stress across the fault, and it may account for anomalous folds and tectonic ridges in the vicinity.

Neogene thrust and reverse faults Evidently the approximately N-S compression associated with broadly distributed NW-SE shear has produced a few thrust faults of small to moderate size. These trend E-W or ESE-WNW. An example is the Monte Vista vault on the San Francisco Peninsula (Fig. 13-17; Dibblee, 1966, center of map). This interesting structure is nearly horizontal at the surface, but steepens at depth. Along it, the Fran-

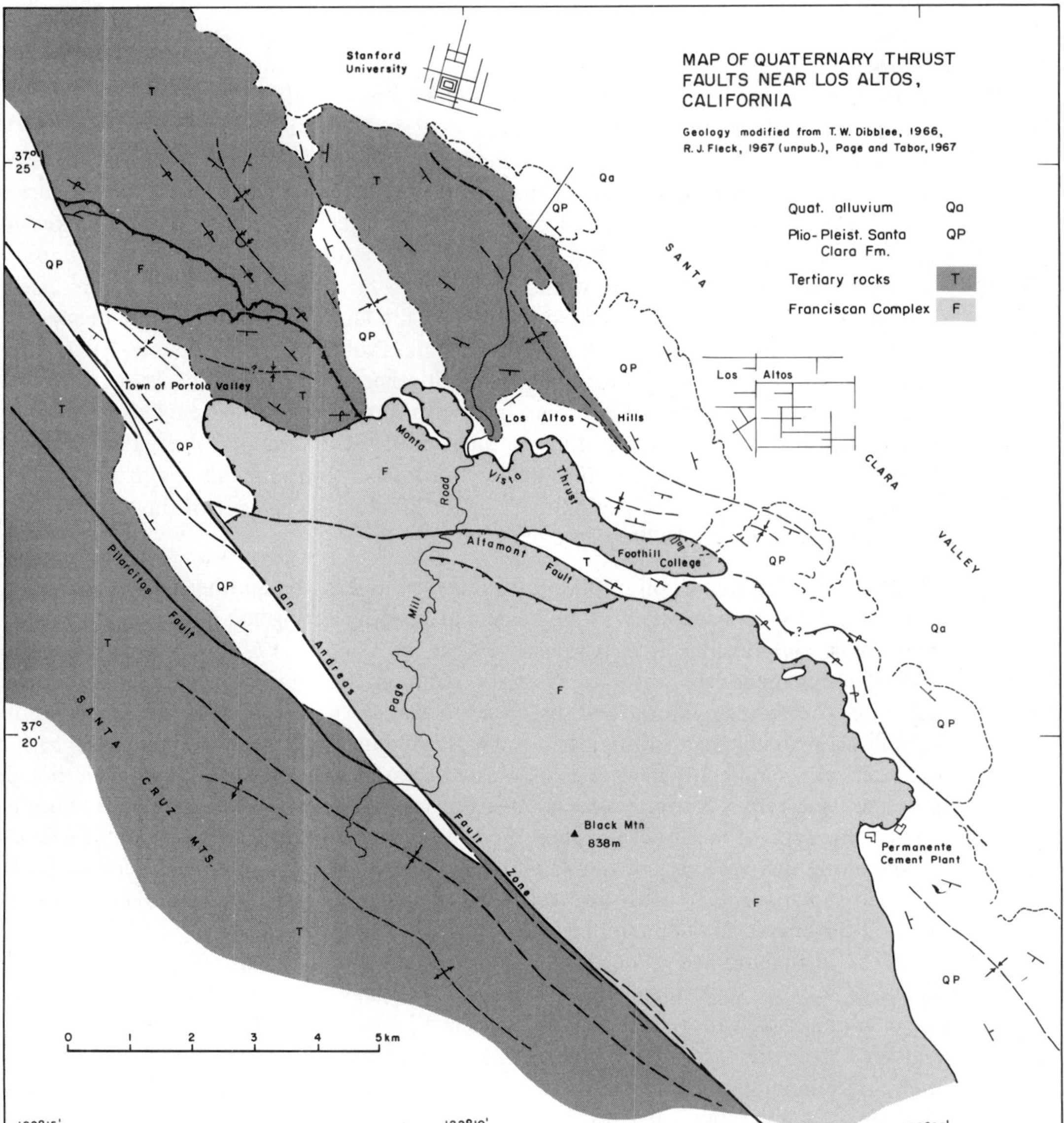

Fig. 13-17. Map of Quaternary thrust faults near Los Altos, California, showing effects of north-south compression associated with transform tectonics. (Compiled and modified from Dibblee, 1966, Fleck, 1967, Page and Tabor, 1967)

ciscan Complex is locally thrust over the Plio-Pleistocene Santa Clara Formation, so there is no doubt that the displacement is Quaternary, yet there is no definitive topographic evidence of activity. Other examples include the Neogene thrusts between Cholame and Coalinga and the small thrusts east of Cholame near Orchard Peak (Marsh, 1960).

Many other Neogene thrusts and reverse faults in the Southern Coast Ranges are not oriented as predicted by "wrench tectonics" and are of unknown origin, like the unexplained folds mentioned in the preceding section. High-angle reverse faults are well known in the northern Santa Lucia Range (Fig. 13-12; Compton, 1966b). They are mechanically puzzling because their dips are steep and their strike approaches the direction of relative plate motion instead of making an oblique angle. Perhaps some strike slip has occurred in conjunction with dip slip, although the direction of maximum compressive stress deduced by Compton would not seem to favor this, and the faults are essentially parallel (in plan) with associated folds. Both the faults and folds are Plio-Pleistocene and therefore formed during the transform regime. Probably the contiguous Pacific and North American plates have not moved past each other in a direction consistently parallel with their mutual boundary. More than likely the motion has varied enough to produce compression and relaxation from time to time across the boundary (Silver, 1975). This is evidenced offshore by Neogene folds and (more rarely) thrusts parallel with the margin, and apparently by brief reactivation(s) of underthrusting at the edge of the "dead" filled trench (Page and others, 1979).

Sedimentary basins In other parts of California, and in the Southern Coast Ranges as well, Neogene sedimentary basins are being explained by rifting and by wrench-type pull-aparts and compressions (e.g., Crowell, 1974a, b; Chapter 17, this volume). Graham (1978) comprehensively analyzed Tertiary sedimentation in the Santa Lucia Range, mainly in the Salinian Block, and discovered a marked influence of Neogene transform tectonics. He found that, beginning in the early middle Miocene, NW-SE trending basins and intervening uplifts began to form in a more or less en echelon pattern, which, although imperfect, suggests N-S compressional effects of transform (wrench) tectonics. The beginning of this type of basin development virtually coincided with the inception of wrench folding along the west part of the San Joaquin Valley (Harding, 1976) and immediately preceded the appearance of throughgoing right-hand strike-slip faults and the commencement of movement on the modern SAF (Graham, 1978).

The just-mentioned en echelon basins are compressional features, oriented in much the same way as subsequent folds. Perhaps surprisingly, middle Miocene basaltic and silicic magmatism, which is commonly associated with extensional tendencies, virtually coincided with the basin development (14 to 17 m.y.b.p.). Inasmuch as the axes of maximum and minimum compressive stress are at right angles, it would be interesting to know whether or not the dikes, feeders, or loci of extrusion show an appropriate N-S or NNE-SSW trend.

Rotations and translations inferred from paleomagnetism One of the most exciting but puzzling results of recent paleomagnetic studies is the apparent rotation and long-distance transport of large and small blocks in otherwise normal-appearing terranes.

Simpson and Cox (1977) find impressive evidence for 50° to 70° clockwise rotation of a block more than 225 km long in the Oregon coast range, and Kamerling and Luyendyk (1979) report 64° to 81° clockwise rotations in part of the Transverse Ranges, a part that they think traveled independently.

Apparently, the Southern Coast Ranges are not exempt from similar discoveries, for Greenhaus and Cox (1979) show that seven late Oligocene igneous bodies in the Morro Rock-Islay Hill complex near San Luis Obispo seem to have rotated 27° to 76° clockwise, the mean being 49°. Greenhaus and Cox think that the rotating entities were 20 km or less in diameter and that they responded independently to a right-hand shear. Thus they turned like ball bearings, as postulated elsewhere by Beck (1976). Because of their small diameter, they are assumed to be decoupled from the bulk of the underlying lithosphere. In addition to adopting the ball-bearing concept, Greenhaus and Cox (1979) suggest that a pull-apart basin may have been forming. A pull-apart could result from the proposed right-hand shear and would provide a favorable setting for uninhibited rotation.

The paleomagnetic results for the Morro Rock-Islay Hill complex are disturbing because there is no compelling geological reason to predict the postulated rotations. There are, however, permissive geological circumstances. The igneous bodies are surrounded by the Franciscan Complex (of which they are not a part), so they are embedded in material that is highly sheared with or without the influence of right-hand relative motion. The boundaries of the rotated entities are not likely to be discovered in the heterogeneous Franciscan terrane. Although many NW-SE faults have been mapped in the area and some are presumed to have undergone strike slip, most elongate rock units in the intervening tracts between faults do not show orientations indicative of rotation. Admittedly, the present structural orientations in the region are strongly influenced by Plio-Pleistocene folding and faulting, which may postdate the rotations. The Morro Rock-Islay Hill igneous bodies lie in a linear zone with an overall trend of about N60W, which is more westerly than most of the faults in the neighborhood. However, the zone could perhaps be subdivided into three or four segments, each with a N40–50W strike that would more closely approximate the prevailing direction of shear on faults with known right-hand slip. The fact that the igneous bodies originated 22 to 26.5 m.y.b.p. probably means that the passage of the Mendocino triple junction was attended by conditions that induced the magmatic activity in the area (Snyder and Dickinson, 1979), and other unusual, transient circumstances of that event may have caused or permitted the rotations.

The above-mentioned paleomagnetic results are not the only ones of relevance to the Southern Coast Ranges. Work in progress at Stanford University and elsewhere tentatively suggests large translations of plutonic and Upper Cretaceous sedimentary rocks affiliated with the Salinian Block. Low paleomagnetic inclinations imply long-distance transport of much greater magnitude than that deduced from ordinary palinspastic reconstructions.

It is too early to completely revise one's concepts of Coast Range tectonics in order to properly incorporate these paleomagnetic findings, but in the near future this may have to be done. In the meantime, it is hoped that tests will be devised that will alleviate lingering doubts about the apparent implications.

PLIOCENE-PLEISTOCENE OROGENY AND RANGE UPLIFTS

It has long been recognized that western and southern California underwent marked crustal activity in the late Pliocene and Pleistocene (e.g., Reed, 1933). The recency of the episode makes it potentially instructive, because the chronologic and physical record is better preserved than would be possible for most older mountain building. Moreover, it is intriguing to observe young, poorly consolidated sedimentary rocks that are strongly folded and faulted. Unfortunately, however, the study of Pliocene-Pleistocene tectonism is not as advanced in the Southern Coast Ranges as it is in the San Joaquin Valley, Transverse Ranges, and Los Angeles region.

The Pliocene-Pleistocene deformation produced or accentuated many of the wrench-type en echelon folds described on preceding pages and was accompanied by activity on strike-slip faults. Clearly, this orogeny was basically an intensification of the continuing deformation associated with transform interaction. It coincided temporally with rifting in the Gulf of California and acceleration of slip on the San Andreas fault. However, it eventually included the delineation and uplift of individual ranges in ways that are difficult to explain in terms of wrench tectonics.

Plio-Pleistocene synorogenic sediments Part of the story of recent events is contained in a number of relatively local nonmarine formations ranging in age from about 4 to 0.5 m.y. There are only very limited marine deposits of this age in the Southern Coast Ranges, which in itself bespeaks of general uplift throughout most of the subprovince. The nonmarine formations under discussion are older than Quaternary terrace deposits and alluvial fans with geomorphic features. Examples are the Santa Clara Formation in the San Francisco Bay region, the Livermore Gravels farther east, the San Benito Gravels south of Hollister, the Paso Robles Formation in the Salinas Valley region, and the Tulare Formation along the west side of the San Joaquin Valley. These sediments are largely deposits of streams and alluvial fans, with minor lacustrine facies, and they commonly rest on Pliocene or Miocene marine strata, locally unconformably. The relations clearly record uplift of large areas above sea level. Within the elevated areas, the parts that were raised highest were deeply eroded, feeding detritus to surrounding aprons of sediment, and the clasts within the Plio-Pleistocene sediments reflect the progressive stripping and eventual deroofing of basement rocks in the source areas. Interestingly, the uplifts did not necessarily coincide with those that immediately followed and produced the present mountain ranges (e.g., Galehouse, 1967). The Plio-Pleistocene formations are locally folded and faulted; in some cases there is some relationship between the structural features and present topography, but in other cases, there is not the slightest relationship, despite the inferred recency of the deformation. Most of the landforms in the Southern Coast Ranges are even younger than the majority of folds in the Plio-Pleistocene formations.

Formation of ranges and structural valleys Among the most important and least understood Neogene structural features of the subprovince are the ranges themselves (e.g., the Santa Cruz, Santa Lucia, Diablo, Gabilan, and Temblor ranges) and the major alluviated depressions (Santa Clara Valley, Salinas Valley, and Carizzo Plain; see Fig. 13-2). These elongate features lie nearly parallel with the San Andreas fault and do not seem to be products of simple wrench tectonics. The ranges are not simple antiforms or fault blocks, but are hybrids and are internally complex. The Diablo Range is indeed antiformal with respect to the remnants of its tectonic cover of Great Valley rocks, but it is crossed obliquely by a couple of synforms and is partially bordered by faults. Most ranges contain internal folds and faults, and there may have been internal movements as well as overall uplift. Part of the Gabilan Range appears to be a southwestward-tilted block with a sloping planar erosion surface and with gently dipping homoclinal strata of the Plio-Pleistocene Paso Robles Formation. The San Andreas fault forms the northeast boundary of the Gabilan Range, but farther north the same fault lies well within the Santa Cruz Mountains, which seem to have formed without regard to its presence. Except for the Gavilan Range, the mountains lack marked asymmetry and give the impression of having been uplifted bodily, although perhaps unevenly. Some are bordered on one or both sides by upturned Plio-Pleistocene strata, and several have imperfect subsummit remnants of erosion surfaces of moderate relief (e.g., the Santa Lucia Range; Howard, 1973).

The Diablo Range provides an instructive history. It is flanked by the marine sediments of the GVS, which, in some areas, are believed to represent all stages of the Upper Cretaceous from Cenomanian through Maastrichtian. The sequence thins south of Panoche Pass, either tectonically or depositionally. If the thinning was caused by circumstances during deposition, as Dibblee thinks is likely, a bathyal submarine ridge may have existed at times near the site of the Diablo Range (Nilsen and Dibblee, 1979), presumably during the Franciscan subduction regime. The GVS sediments give no indication that the Franciscan now comprising the core of the range was ever exposed subaerially during the Late Cretaceous, except for Franciscanlike pebbles reported by Dibblee from the basal beds of the sequence 2 km south of Los Baños Creek, where the Upper Cretaceous Panoche Formation appears to be the lowest part of the GVS (Nilsen and Dibblee, 1979). Elsewhere, GVS conglomerates are notably devoid of Franciscan rocks. Paleocene clastic sediments on the two sides of the range, near Hollister and north of Coalinga, respectively, are subsea fan facies derived from granitic and metamorphic rocks, not from area of the Diablo Range (Nilsen and Dibblee, 1979). Sandstones of the Eocene Domengine Formation are reported to contain grains of glaucophane, probably from the Franciscan Complex, but the sandstones are predominantly arkosic, which argues against the existence of a nearby subaerial exposure of the Franciscan. Overall, the Late Cretaceous and Paleocene record indicates at most relatively mild crustal unrest with no actual uplift of mountains at the site of the Diablo Range.

The first widespread appearance of Franciscan detritus in the vicinity of the present range is in the middle Miocene Temblor Formation. There is compelling evidence that part of the Franciscan core of the Diablo Range had emerged and had been stripped of the tectonic cover of GVS rocks, as the core is locally overlain (east of Hollister) by thin marine Miocene sandstone and limestone, which in turn is covered by the Quien Sabe Volcanics (Leith, 1949). The volcanics have been dated at 7.5 to 10 m.y. by D. C. Powell, according to Snyder and Dickinson, 1979). To the south, the New Idria diapir reached the surface in the middle Miocene, supplying serpentinite to the Big Blue Formation (Casey and Dickinson, 1976). Some areas in or near the site of the Diablo Range were emergent during the Pliocene, as nearby marine and nonmarine sediments of that age contain Franciscan pebbles. Evidently, some crustal stirring was continuing, but without marked folding and faulting.

The principal uplift of the Diablo Range, like that of other mountains in the Southern Coast Ranges, occurred in late Pliocene-Quaternary time. South and east of Hollister the nonmarine San Benito Gravels (Plio-Pleistocene) contain a sedimentary record of the stripping of the Quien Sabe Volcanics from the Diablo Range and the dissection of the Franciscan core, which yielded floods of debris (Griffin, 1967). Eventually, the gravel deposits encroached over the margins of the core. Then the San Benito Gravels themselves were folded and faulted. Elsewhere around the flanks of the range other comparable nonmarine formations (the Tulare Formation, Livermore Gravels, and Santa Clara Formation) played a somewhat similar role.

The events just described were recent, but they did not complete the development of the range. Further uplift occurred, as can be inferred from Fig. 13–18. The figure is a grossly simplified cross section, with exaggerated vertical scale, along a line extending eastward and northeastward from the vicinity of Fremont. Pliocene strata on the flanks of the range are strongly folded and are truncated by faults that lie near the margins of the uplift. Plio-Pleistocene sediments (Santa Clara Formation on the southwest and Tulare Formation on the northeast) are also strongly disturbed and truncated. The range appears to have risen more or less bodily, but unevenly, with a number of internal dislocations. Steep faults, believed to be primarily strike slip in nature, were utilized by the uplift, probably as a matter of mechanical expedience. The structural relief (ca. 3 km) is much greater than the topographic relief (ca. 1 km). Most of the rocks of the range, including Cenozoic material and Franciscan melanges, are not resistant to erosion, so the elevation of the terrane must have occurred recently, and the uplift may still be in progress. There is no broad summit upland that might have formed near sea level, but small subsummit flat areas occur farther south.

A narrow splinter of the Diablo Range lies between the Hayward and Calaveras faults. About 30 km north of San Jose, this narrow elevated tract is transected from one side to the other by an antecedent stream (Alameda Creek), which has cut a deep, winding passageway (Niles Canyon) leading to the alluviated lowland surrounding San Francisco Bay. These relations indicate rapid, recent uplift of this part of the range. The southward continuation of this tract shows other striking geomorphic features leading to the same conclusion, especially between San Jose and Morgan Hill. Alt (1979) finds that repeated first-order leveling along a railroad shows that during a 53-year period the Niles Canyon sector has risen at an average rate of 1.5 to 2.0 mm per year relative to the alluvial plain bordering San Francisco Bay. The same author notes that the plain may have subsided to account for part of the relative movement.

The structural valleys are just as puzzling as the ranges. They are bordered by faults on one or both sides, but the known faults are mainly strike slip and have merely been adapted for vertical adjustments. The valleys are not just grabens, but are probably

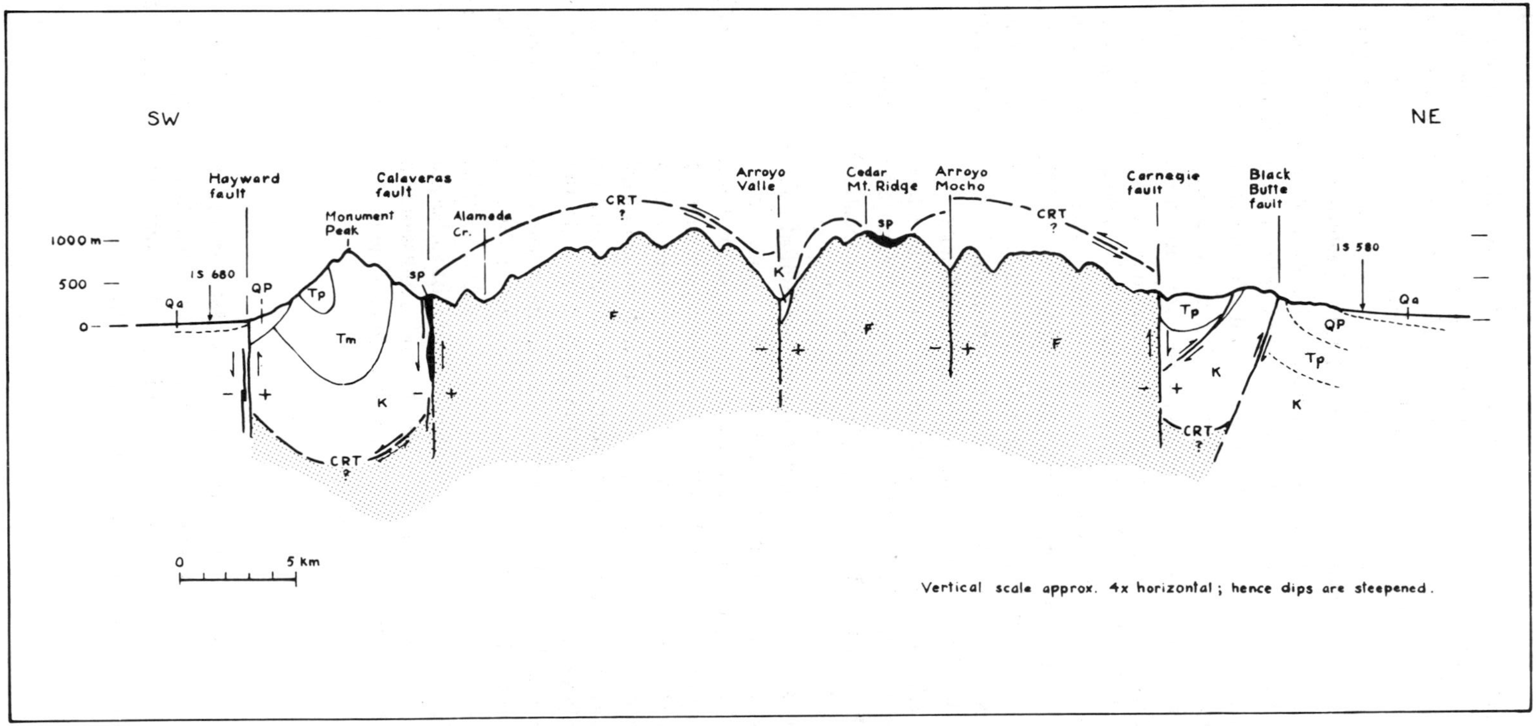

Fig. 13-18. Simplified section across north end of Diablo Range, showing Quaternary uplift and faulting. Vertical scale exaggerated, and complicated structure of Franciscan Complex omitted. CRT, Coast Range thrust; sp, serpentinite; F, Franciscan Complex; K, Cretaceous sediments of Great Valley Sequence; Tm, Miocene sediments; Tp, Pliocene sediments; QP, Plio-Pleistocene sediments; Qa, Quaternary alluvium.

synformal, judging from the dips of bordering formations; they are deeply filled with Late Tertiary and Quaternary sediments. The rocks underlying the low-density sediments of Santa Clara Valley are offset by faults that are longitudinal with respect to the valley trend, as shown by gravity maps with linear belts of steep gradients.

What could have caused major uplifts and structural depressions parallel with the plate boundary? The deformation was largely Quaternary, and took place while "wrench tectonics" were producing widespread effects, some of which are easily identifiable. The Diablo Range has a Franciscan core, and it has been proposed that the Franciscan material bulged up in a huge incipient diapir. But what induced the Franciscan to do this? Was it some thermal influence, a purely mechanical squeeze, gravitative buoyancy, or a combination of these? Why did the Franciscan core wait until the latter part of Cenozoic time before rising toward the surface? Perhaps the presence of Franciscan rocks is not a critical factor at all, as the Gabilan Range rose at about the same time as the Diablo Range, and it contains no Franciscan rocks, at least at the surface. The exposed rocks are Salinian granitic and metamorphic types, plus a partial cover of Tertiary sediments. Admittedly, there is a chance that the Franciscan Complex is present at depth. The Santa Lucia Range contains part of the Salinian Block and part of the Franciscan of the Sur-Obispo Belt, but the uplift of the range has paid no heed to this duality.

Perhaps the ranges rose because of latent buoyancy of sediments subducted in the late Mesozoic and Early Tertiary. If so, several circumstances should be noted. First, there was a lag of some 2 to 20 m.y. (depending on the location) between the cessation of subduction and the marked Plio-Pleistocene folding and uplift. The culminating orogeny correlates chronologically with rifting in the Gulf of California, not with the cessation of subduction, so some aspect of transform tectonics must be of overriding importance. Second, if buoyancy of subducted sediments contributed to uplifts, the rise of light material in rather discrete belts must be explained, inasmuch as structural valleys lie between the ranges. (Could these belts have been determined by imbricate structure and inhomogenous composition of the underlying subduction complex?) Third, if this cause of uplift applies to all the ranges, Franciscan rocks must indeed exist beneath the Salinian Block, but we note that only part of the Block has risen, while the Salinas Valley has subsided.

Regardless of the cause(s) of uplift, the orientation and boundaries of the ranges and structural valleys seem to have been influenced by the presence of Neogene strike-slip faults, inasmuch as the faults locally more or less coincide with range margins. However, the orientation may be partly inherited from the earlier subduction processes, which very likely produced elongate slices, welts, and depressions in the upper part of the crust. These perturbations may have been enhanced by the vertical adjustments of the Neogene.

NEOTECTONICS

One of the fascinations of the Southern Coast Ranges is the fact that tectonic activity is still in progress, as it is in other parts of California. Abyssmal ignorance still clouds the specific sites and characteritics of this activity except in a few localities, but there are many hints (such as seismicity) of crustal unrest.

Present-day regional strain Broadly distributed crustal strain can be deduced, albeit imperfectly, by determining the distortions of areas delineated by classical triangulation, where repeated surveys of the same network show changes in the angles and computed sides of triangles. Unfortunately, classical surveying is subject to errors of a size that precludes firm conclusions except in cases of relatively large strains or high rates. Burford (1965) analyzed two northeast-trending belts of repeated triangulation (1930 to 1932 and 1951) across the Southern Coast Ranges, one belt extending from Monterey into the Diablo Range, and the other from San Luis Obispo into the southern part of the same range. He found apparent linear strains ($\Delta L/L$) that were mostly 1×10^{-4} or less for the period of 19 to 21 years. Maximum contraction appeared to be generally oriented NNE-SSW at an angle of about 75° to the San Andreas fault, more or less normal to many fold axes in the Coast Ranges. Contraction near and within the San Andreas zone was more nearly N-S, and shear strain seemed more pronounced in a band 10 to 30 km wide along the fault zone than elsewhere. The apparent strains deduced by Burford are not quantitatively reliable because of inaccuracies in surveying, but they are qualitatively appropriate for the Neogene structures believed to be the result of transform tectonics.

Present-day strike slip Strike-slip faulting is among the most obvious neotectonic phenomena in the Southern Coast Ranges. Historic surface faulting has been observed both as sudden events accompanied by earthquakes and as creep. The SAF (Allen, Chapter 15, this volume) slipped as much as 10 m at the ground surface in the southern part of this province in 1857, as much as 7 m in the San Francisco region in 1906, and about 0.15 m near Parkfield and Cholame in 1966. The Hayward fault is believed to have ruptured the ground surface in 1936, and definitely did so in 1868, although the slip was probably less than 1 m. Creep is well recorded on parts of the SAF, Hayward, and Calaveras faults (Yamashita and Burford, 1973; Shulz and others, 1976). Typical creep rates on individual faults are 0.02 to 1.0 cm/yr, but rates as high as 3 cm/yr have been recorded.

The relative motion between the Pacific and North American plates is believed to be 5.6 ± 0.3 cm/yr (Minster and Jordan, 1978). Of this amount, probably 3.2 to 3.7 cm/yr is expressed by slip on the SAF and its closest relatives (e.g., the Hayward and Calaveras faults); this is deduced from geodetic measurements (Savage and Burford, 1973) and from detailed geologic evidence for the movement history of the SAF during the past three millennia (Hall and Sieh, 1977). The balance of the relative plate motion, about 1.9 to 2.4 cm/yr, is presumably accommodated mainly by movement on other faults between the foot of the continental slope to the west and central Nevada or even more distant points to the east. The Owens Valley surface faulting of 1872 (with perhaps 3 to 6 m of right-hand strike slip) and Cedar Mountain faulting (possibly 1 m of strike slip) of 1932 in Gabbs Valley, Nevada, testify to the breadth of the transform zone and its activity toward the east.

Toward the west, the San Gregorio-Hosgri fault zone (Fig. 13-2) has also participated in the plate motion. E. A. Silver (1974) found a 90-km right-hand offset in an elongated gravity high and a coinciding granitic ridge in the Salinian Block where crossed by the San Gregorio fault near Monterey Bay. Hall (1975) deduced an 80-km (minimum) right-hand offset for the Hosgri fault, largely based on matching the Upper Jurassic ophiolite and distinctive Oligocene sedimentary breccias (containing ophiolite clasts) at Point Sal with comparable rocks at San Simeon. Graham and Dickinson (1978a, b) propose 115 km of offset for the northern part of the San Gregorio-Hosgri fault zone to explain the displacement of several sedimentary facies and sequences ranging in age from Cretaceous to Miocene, but there is considerable latitude in choosing the favored figure. The fault zone may have adopted its NNW-SSE course when the direction of relative plate motion was somewhat different from the later direction (N22W prior to 10 m.y.b.p., as compared with N30-37W subsequently; Silver, 1975). Although the San Gregorio-Hosgri zone does not seem ideally oriented, its trend of N20-30W is as appropriate for present-day relative plate motion as is the SAF trend of N40-50W in central California. The fault zone is still active, certainly in the north and probably in the south. Seismic epicenters and offsets of the sea floor in Monterey Bay (Greene, 1977) provide local proof of activity offshore. The zone is visible on land at San Simeon, the Big Sur area, and from Monterey Bay to Moss Beach. Near San Simeon and north of Monterey Bay, the fault zone displaces marine terraces along prominent scarps, some of which face east. Weber and Lajoie (1977) studied offsets of sequential marine terraces near Point Año Nuevo and derived a right-slip movement rate of about 0.63 to 1.30 cm/yr for the last 200,000 years.

Other NW-trending faults that are sometimes mentioned in the literature (e.g., San Juan-Chimenas, Huerhuero, Rinconada-King City, Palo Colorado, East Huasna, West Huasna, Oceanic, and so forth) have not been conclusively shown to have Holocene displacements. Although many of them displace Plio-Pleistocene sediments such as the the Paso Robles Formation, several of them pass beneath stream terrace deposits that are not visibly offset. The King City-Reliz fault zone (e.g., Jennings, 1975), which appears to be a northern continuation of the Rinconada fault, is interesting because of the striking morphology along or near the projection of the fault on the west side of the northern Salinas Valley. Faceted spurs and scarps of uncertain origin have attracted attention for many years. Salinas Valley is definitely a structural valley, and the spatially associated King City-Reliz fault zone is real, but a detailed study of the Quaternary geology of the region by Tinsley (1975) shows no continuous zone of faulting younger than the Gloria alluvial fans (probably between 250,000 and 500,000 years old) along the west edge of the valley. Some of the imposing scarps may be products of Pleistocene faulting, but many smaller scarps are of fluvial origin. Fans and fluvial terraces capped by Chualar soils (probably between 25,000 and 75,000 years old) are not known to be offset by faulting at any point along the mountain front. The Reliz fault is crossed by fluvial terraces of the Arroyo Seco; the terraces are unbroken and are capped by Chualar soils.

Needless to say, all the aforementioned faults should be regarded with suspicion except where detailed study has disproved recent activity.

Possible ongoing folding The possibility of present-day folding in the Southern Coast Ranges has not been vigorously tested. I know of no cases of repeated spirit leveling or other accurate surveys that prove historic folding in the province. In other areas where such studies have been made, the rates of uplift are found to be on the order of a few millimeters or fractions of millimeters per year and are likely to escape notice. Although historic folding has not been actually measured in the province, warped Pleistocene terraces are strongly suggestive. One of the most detailed terrace studies to date is that of Bradley and Griggs (1976), who found that wave-cut platforms on the flank of the Santa Cruz Mountains have been arched and progressively tilted seaward. The subdued antiformal configuration of the uplift and its position with respect to the Ben Lomond topographic antiform indicate that Late Tertiary domical uplift has continued into the Quaternary. Because a succession of terraces of different ages has been affected, and because the older platforms are warped more than the younger, the uplift is shown to be progressive, the average rate having been 0.16 to 0.26 mm/yr. One of the warped terraces, the Davenport, is believed to about 100,000 years old on the basis of rather discordant radiometric and amino acid dates. This is young enough to imply that the uplift could still be going on. Marine terraces farther north near Half Moon Bay are also markedly tilted, apparently in accord with their position on an anticline in Tertiary rocks (Lajoie and others, 1972).

Uplift of ranges The large uplifts responsible for individual mountain ranges (discussed previously) are partly or largely Quaternary, and it would be of interest to know whether or not they are still active. Geologic circumstances suggest that slow uplift could possibly be continuing. As pointed out by Christensen (1965), several of the ranges are bordered by upturned Plio-Pleistocene formations that are chiefly nonmarine but that were deposited near sea level. Christensen contoured the present configuration of the surfaces on which these young formations were deposited, and showed uplifts of more than 700 m and subsidences of even greater amplitude. Some of the affected Plio-Pleistocene formations include Pliocene marine or estuarine beds at the very base, but by indirect extrapolations of dated beds in not-too-distant areas, they are believed to contain remarkably young strata in their upper parts. The Tulare Formation (at some distance from the uplifts) contains the tuffaceous Corcoran Clay, which is dated radiometrically at 0.6 m.y. (Janda, 1965), and the Santa Clara Formation near Woodside contains a tuffaceous bed believed to be 0.5 to 1.0 m.y. (Sarna-Wojcicki, 1977). In addition to the upturning of young peripheral strata, the very fact that the ranges persist as topographic highs despite the nondurable character of their rocks and the rapid rate of erosion and landsliding suggests that slow uplift may be continuing. Some of the ranges support elevated remnants of erosion surfaces of moderate relief that are being attacked at the edges by young streams with steep gradients, and this could not have gone on for a very long period or the vestigial topography would have disappeared.

Only a few geodetic studies of present-day uplift have been made in the Southern Coast Ranges. The study by Alt of repeated spirit leveling east of San Francisco has already been mentioned. Alt (1979) not only found an average rate of 1.5 to 2.0 mm/year of relative uplift in Niles Canyon, but also found less certain evidence of comparable uplift in the Altamont Hills east of Livermore. His study is apparently one of the few of its kind in the Southern Coast Ranges. It would be timely to conduct much more research of this sort.

CONCLUDING REMARKS

The Southern Coast Ranges subprovince, like much of California, is providing knowledge of important tectonic events and processes, some of which are of more than local interest. In addition, this region is one of the proving grounds for testing current geological and geophysical ideas. It now seems opportune to emphasize some of the problems that deserve further attention in the near future.

The Franciscan Complex offers potential clues to the disputed inner workings of subduction zones, and there is a continuing need to examine the Complex in innovative ways in order to perceive the mechanisms by which it was assembled. The relatively young Coastal Belt is particularly promising, but the sea cliff exposures near San Simeon have not been exhausted of information. Other aspects of the Franciscan are equally challenging, in view of doubts as to the provenance of constituents, the possibility of more than one arc-trench system, and the likelihood of mega-strike slip during the evolution of the Complex. Close examination and comparison of particular kinds of Franciscan components will inevitably lead to discoveries as important as those of the recent past (e.g., the common age of high-grade blueschist fragments, the widespread distribution of *Buchia*-bearing rocks, the petrologic mismatch between sandstones of the Franciscan and those of coeval Great Valley Sequence rocks). Tentative results of paleomagnetic studies are startling and should lead to further efforts. Franciscan melanges preserve fragments of terranes that otherwise have vanished, affording opportunities for close detective work. The matrix material of melanges has hardly been touched by research and is awaiting scrutiny.

The source and mode of acquisition of mafic and ultramafic rocks in the Franciscan might have become more apparent if details of petrology, chemistry, and age of extrusive and plutonic ophiolite fragments in melanges were compared with the properties of Great Valley-type ophiolites. Also, an effort should be made to prove or disprove progressive detachment of blocks from ophiolite assemblages that are still more or less in place at the base of GVS sediments.

The significance and age of the Coast Range thrust is a lingering problem. Is this structure a feature of early subduction history or, as I think more likely, is it an "afterthought" involving en masse movement of a mature subduction complex?

More difficult is the prime question of the nature of the unseen boundary zone between the Franciscan Complex and the Sierran basement beneath the sediments of the Great Valley. Is it a discrete thrustlike subduction-related contact, or does the Sierran basement grade into the Great Valley ophiolite that separates it from the Franciscan, or is some other option the correct one? The answer will probably require two or more advanced geophysical techniques that can jointly minimize the present ambiguity.

The Salinian Block poses several mysteries. Whence came the preintrusive rocks of the terrane, and how did they find their way to the site of Sierran-type plutonism? How, why, and when did the Block undergo partway emplacement? Perhaps most challenging, because of basic tectonic processes involved, is the puzzling question of how, why, and when did a large slice of the Block disappear. If the process was piecemeal destruction by subduction, this would be important to recognize, but if the vanished slice departed by mega-strike slip, that alternative is equally impressive.

Perhaps related to the last-mentioned problem is the untested hypothesis that the sialic rocks of the Salinian Block are tectonically underlain by the Franciscan Complex. Here, again, we must rely on future geophysical findings. The truth may be difficult to come by, owing to variations in Franciscan compositional, metamorphic, and structural character, together with effects of pressure, temperature, and fluids on the seismic properties. If the Franciscan is found beneath the Salinian basement, we will have a proven frozen subduction system with its "lid" still in place.

The apparent rotations and long-distance translations of individual blocks in the Southern Coast Ranges, as deduced from paleomagnetic data, may open an entirely new field of investigation and interpretation. It is to be hoped that stringent tests can be devised to show whether or not obviously coherent tracts of rock show illusory independent rotations of arbitrarily delineated sectors. Tests would be most telling if they could be applied to young (Pliocene or Pleistocene) rocks having a well-known history, preferably rocks that are strongly folded, but not disrupted.

Several Neogene events and sets of structures, although relatively recent, are almost wholly unexplained. The regional mechanics of Neogene folding parallel with the plate boundary is a puzzle, inasmuch as subduction (which might be invoked) had ceased and the transform regime had supplanted it. Even more cryptic are the causes of the Plio-Pleistocene orogeny, which was timewise related to the opening of the Gulf of California. An outstanding research target is the cause of delineation and uplift of individual ranges late in the history of the subprovince.

Finally, the Southern Coast Ranges as well as other parts of California are a potential laboratory for the study of tectonic processes in action today. We are aware of fault movements, local changes in elevation, horizontal crustal strain, and warped marine terraces, but some of these phenomena have only been studied in a preliminary way. It would be timely to investigate the possibility of present-day folding, continued rise of mountain ranges, and sinking of structural valleys.

ACKNOWLEDGMENTS

I am particularly indebted to the following for information and ideas about the Southern Coast Ranges, both currently and in the recent past: R. G. Coleman, R. R. Compton, D. S. Cowan, K. E. Crawford, T. W. Dibblee, W. R. Dickinson, W. G. Ernst, W. R. Evitt, S. A. Graham, C. A. Hall, D. G. Howell, K. J. Hsü, D. L. Jones, J. G. Liou, R. P. Morrell, T. H. Nilsen, L. A. Raymond, D. C. Ross, D. W. Scholl, and J. G. Vedder.

Clifford A. Hopson
Department of Geological Sciences
University of California
Santa Barbara, California 93106

James M. Mattinson
Department of Geological Sciences
University of California
Santa Barbara, California 93106

Emile A. Pessagno, Jr.
Department of Geosciences
University of Texas at Dallas
Richardson, Texas 75080

14

COAST RANGE OPHIOLITE, WESTERN CALIFORNIA

ABSTRACT

Remnants of the California Coast Range ophiolite (Jurassic) are exposed in the northern Coast Ranges west of the Sacramento Valley, in Sonoma and Lake Counties, in the Diablo Range of central California, in the southern Coast Range west of the Salinian block, and in the submerged continental borderland south of the Transverse Ranges. Twenty-three of the ophiolite localities are described in this chapter, and the sequence of tectonic events at each locality is summarized.

The complete ophiolite sequence in the Coast Ranges, reconstructed from the dismembered remnants, consists of: (1) a basal member of harzburgite tectonite, variably serpentinized; (2) a plutonic member that ranges upward from dunite cumulate through olivine clinopyroxenite and gabbro cumulates to high-level isotropic (noncumulus) uralite gabbro, hornblende diorite and quartz diorite, and minor plagiogranite; (3) sheeted dike or sill complex that separates the plutonic and volcanic members; and (4) an upper member of pillowed and massive submarine lavas with subordinate breccias. Nearly complete sequences are preserved at Point Sal, Del Puerto Canyon, and South Fork of Elder Creek. However, most of the other ophiolite remnants are dismembered by faulting, leaving the sequences tectonically thinned and incomplete. Additional ophiolite units appear in the northern Coast Ranges: (5) polymictic breccias of volcanic and plutonic ophiolite fragments that rest upon the upper surface of the disrupted ophiolite but lie unconformably beneath Upper Jurassic mudstones at the base of the Great Valley Sequence, and (6) ophiolitic melange, consisting of tectonic blocks of the ophiolite (especially basalt and radiolarian chert) enclosed in a matrix of sheared serpentinite derived chiefly from harzburgite. The melange encloses a large slab of stratigraphically intact ophiolite at the South Fork of Elder Creek (Tehama County). The ophiolitic melange lies beneath the ophiolitic breccias and beneath undisrupted Upper Jurassic strata of the basal Great Valley Sequence.

The age of the ophiolite ranges from about 153 to 165 m.y. (Middle Jurassic on the Van Hinte 1976a time scale), based on U/Pb isotopic ages of zircons from plagiogranite and albitite in the plutonic and sheeted intrusive members. The $^{206}Pb/^{238}U$ and $^{207}Pb/^{235}U$ ages are concordant at individual ophiolite remnants (Santa Cruz Island, Point Sal, Cuesta Ridge, Llanada, Del Puerto Canyon, Healdsberg). These remnants represent different age segments of an originally continuous 12-m.y. span of Jurassic oceanic crust. Potassium-argon hornblende ages from the ophiolite remnants at Del Puerto Canyon (Lanphere, 1971) and South Fork of Elder Creek (Lanphere, 1971; Fritz, 1975) are in general agreement with the U/Pb results. A distinctly later magmatic event is the intrusion of plagiogranite approximately 145 m.y. ago into 165 m.y. - old oceanic crust at Santa Cruz Island, shown by U/Pb dating. K/Ar amphibole ages of 140–141 m.y. for gabbroic inclusions in serpentinized ultramafic rocks at Wilbur Springs and Riley Ridge quadrangle (Maxwell, 1974) stem from an alteration event (gabbro→rodingite) that accompanied the disruption, serpentinization, and melanging of the ophiolite. This melanging occurred prior to beginning of deposition of the Great Valley Sequence.

Radiolaria are used to date oceanic sedimentary rocks associated with the ophiolite.

The Radiolarian zonation incorporates a recent revision that extends the range of Zone 1 from Kimmeridgian or early Tithonian (Late Jurassic) back to the base of the Bajocian (early Middle Jurassic). Thin-bedded cherts and tuffaceous cherts that rest on the ophiolite lavas range in age from Oxfordian (upper part of Radiolarian Zone 1) to early Tithonian (Zone 3). The Coast Range ophiolite lies conformably, or in some cases disconformably, beneath these sediments; therefore, the biostratigraphic age of the ophiolite is Oxfordian or pre-Oxfordian (Middle Jurassic or older) in accordance with the radiometric ages. Radiolarian cherts interbedded with high-TiO_2 pillow basalts at Stonyford, west of the Sacramento Valley, yield Radiolaria of Zone 2, Subzone 2B (early Tithonian). The Stonyford lava pile is the remnant of a Late Jurassic seamount, built upon older oceanic crust represented by the Middle Jurassic ophiolite. The earliest terrigenous clastic strata (base of Great Valley Sequence) that rest on the ophiolite or upon the cherts and tuffs just above the ophiolite are middle to late Tithonian (*Buchia Piochii* Zone = Radiolarian Zone 4 in part) south of the latitude of San Francisco, but are early Tithonian at northern Coast Range localities (Radiolarian Zone 3 at Mount St. Helena and Subzone 2A at Paskenta). In the adjacent Franciscan terrane radiolarian cherts that rest directly upon basaltic pillow lava contain Radiolaria belonging to Zones 1-2A, now believed to span from Bajocian or Callovian through early Tithonian (Middle and Late Jurassic). A single still older age (Pliensbachian) comes from a block within melange near Santa Barbara, whose tectonic history is not yet clear. A small percentage of Franciscan radiolarian cherts, including most of those that are interbedded within Franciscan clastic strata (i.e., do not lie directly on pillow lava), are Cretaceous in age (chiefly Valanginian, Subzone 5C; a few are Cenomanian, Subzone 10A).

The petrology and petrogenesis of the ophiolite are discussed in terms of its principal stratigraphic units: the peridotite tectonite member, plutonic member, sheeted intrusive member, and volcanic member. The peridotite member, preserved at seventeen of twenty-three ophiolite localities, is composed of harzburgite, subordinate dunite, and derivative serpentinite. Olivine (Fo_{91-92}) and enstatite (En_{90-92}) are major phases in the harzburgite; chromian spinel is a ubiquitous accessory and minor diopsidic clinopyroxene occurs locally. The peridotite has a weak foliation and relict crystalloblastic texture, indicative of syntectonic recrystallization at high temperature and moderate to low differential stress. This texture is commonly overprinted by mechanical deformation and cataclasis. Small dunite bodies (olivine Fo_{91-92}) form discordant dike-like and irregular masses within the harzburgite. The "trails" of dunite mark the passage of mantle melts that were fractionating olivine. The dunite was subsequently deformed and recrystallized with the host harzburgite. Sparse gabbroic dikes, commonly dismembered, are altered to rodingite where the peridotite host is serpentinized. The peridotite member of the ophiolite is interpreted as the depleted residuum left after removal of partial melt from parent mantle of lherzolitic composition. Serpentinite derived from the peridotite is chiefly chrysotile-lizardite-brucite, indicating formation at low temperatures. The serpentinization occurred during or after emplacement of the ophiolite on land, but some also took place earlier beneath the ocean floor. This is shown at Tehema County ophiolite localities where serpentinite-matrix melanges formed before deposition of basal Great Valley sediments.

Plutonic igneous rocks are exposed at seventeen of twenty-three ophiolite remnants, but a semi-complete plutonic sequence is preserved at only three of these. Cumulates formed by crystal settling onto the floor of a magma chamber form the lower and middle part of the plutonic member, and non-cumulus gabbroic and dioritic rocks that crystallized from residual magma near the top of the chamber comprise the upper part. The most common vertical sequence of cumulus phases is: ol + sp (dunite), cpx + ol ± sp (wehrlite and olivine clinopyroxenite), pl + cpx + ol (olivine-clinopyroxene gabbro). However, a different cumulus sequence occurs at Del Puerto Canyon: ol + sp (dunite), cpx + ol ± sp (wehrlite), cpx + ol ± opx (olivine clinopyroxenite, websterite), pl + cpx + opx (gabbronorite). The cryptic variation curves for olivine, clinopyroxene, and plagioclase in the cumulus sequence at Point Sal show a steep, zigzag trend, indicative of periodic replenishment of fractionating magma by additions of more primitive melt. Replenishment is also indicated by the persistance of olivine throughout most of the cumulus sequence, i.e., the magma was unable to fractionate to the olivine-orthopyroxene reaction point. Such open system fractionation is characteristic of steady-state magma chambers at spreading ocean ridges, which are repeatedly fed by primitive melt as they open at the center. At the Del Puerto Canyon remnant, however, the early disappearance of olivine and trend toward more Fe-rich pyroxenes indicates an early change to closed-system fractionation. This suggests a cessation of spreading. A distinctive structural characteristic of the ophiolite cumulates at most localities is a strong "soft-sediment" style of penecontemporaneous deformation, showing that magmatic crystallization and crystal settling occurred in a tectonically active setting. An oceanic spreading center is the probable site.

The high-level part of the plutonic member consists of non-layered uralitic pyroxene gabbros with hypidiomorphic and diabasic textures, grading to diorite, quartz diorite, and minor plagiogranite. The gabbros include: (1) isotropic "quasi-cumulates" that built upward from the cumulates as the clouds of settling crystals became too dense for the development of layering and planar lamination and (2) "plated" gabbros that simultaneously crystallized downward from the roof. Progressive crystal fractionation of residual magma entrapped between these upward- and downward-growing gabbroic zones produced a sandwich zone of dioritic and quartz dioritic rocks. Siliceous plagiogranite melts were filter-pressed from quartz dioritic crystal mushes and injected into overlying rocks as dike-like intrusions. These closed-system fractionation processes occurred far out at the sides of the continuously opening magma chamber, beyond reach of replenishment from the center. Also, this residual magma became highly enriched in water, producing abundant late magmatic and deuteric hornblende. Where water contents exceeded saturation retrograde boiling occurred, leaving swarms of miarolytic cavities and pegmatitic patches in the high-level rocks. Isotopic studies suggest that seawater penetration has been partly or largely responsible. Hydrothermal processes extended into the subsolidus stage, superimposing metamorphic and metasomation alterations on the high-level igneous rocks. Tectonic disturbances that produced "soft-sediment" deformation of the underlying cumulates caused flowage and intrusion of the high-level crystal mushes, resulting in local zones of flow foliation and schlieren banding.

The sheeted intrusive member, consisting of countless dikes or sills emplaced alongside and within one another, separates high-level plutonic rocks from overlying volcanics at Coast Range ophiolite remnants that are not too badly dismembered. Diabase is the common rock type, followed by microdiorite, microtonalite, epidosite, and plagiogranite. The *sheeted dikes* formed by repeated tensional fracturing, dilation, and intrusion through the roof of the magma chamber, where new melt was conveyed to the surface to feed lava flows. This endless lateral dilation and accretion of the magma chamber roof by dike injection is explained by axial intrusion at an oceanic spreading center. More common than sheeted dikes, however, are *sill complexes* in which the sheeting direction is parallel rather than perpendicular to the volcanic-plutonic contact. The main mass of sills are intrusive into the base of the lavas and truncate their feeder dikes. The high-level plutonic rocks which lie just beneath the sheeted sills are intrusive into the sills and therefore younger. The sills were evidently the vanguard of new masses of magma that rose and spread laterally beneath earlier volcanic crust. This initial injection (as sills) was followed continuously by the main mass of magma which invaded a zone near the base of the sills, inflating it into a new magma chamber. Thus, sill complex forms the roof of the chamber, where it periodically received new additions from the underlying differentiating magma body at successive fractionation stages. A fundamental difference between the sheeted dikes and sheeted sill complex is that the former accreted during continuous opening of an *existing* steady-state magma chamber beneath the spreading ridge axis, whereas the latter marks the early growth stage of a *new* magma chamber. The common occurrence of sheeted sills in the Coast Range ophiolite might be related to: (1) frequent ridge-axis jumping, or (2) *slow* sea-floor spreading, which favors development of small, discontinuous, non-steady state magma chambers (Kusznir and Bott, 1976).

The volcanic member is preserved at eighteen of twenty-three ophiolite localities. This member is very thick (up to 2.5 km), in contrast to the rather thin plutonic and sheeted intrusive members ($<$ 2 km total). The volcanic rocks are deep-sea lavas including pillow lava, massive (ponded) lava, and minor sheet flows, with small scraps of pelagic sediment included. Subordinate fragmental rocks are mainly flow-front rubble and blocky submarine talus breccia. The lavas are chiefly aphyric and microphyric basalt and metasomatically altered basalt (spilite, some keratophyre), with groundmass textures and sparsely microvesicular chilled pillow rinds like those in modern deep-sea lava. Whole rock major-element chemistry and microphenocryst mineralogy show that the lavas were originally tholeiitic basalt; however, pervasive alteration including widespread albitization have substantially changed the mineralogical and chemical composition in many places. Most of the so-called keratophyres are metasomatically altered basalt, *not* lavas with an intermediate (andesitic) original composition. The low abundance of titanium, phosphorous, and rare earth elements, which are relatively immobile during alteration, and the pattern of light rare-earth element depletion, suggest that the basaltic lavas were ocean-ridge tholeiites (MORB). Using FeO*/MgO of least altered aphyric basalts as an index of melt fractionation, the Coast Range ophiolite lavas were fractionated to about the same extent as Mid-Atlantic Ridge lavas. Magma extruded as lavas had the same fractionation range as that which formed the ol cumulates, cpx-ol cumulates, and pl-cpx-ol cumulates in the plutonic section, inferred from olivine compositions in the cumulates and an olivine-melt $K_D = 0.27$. Thus, extrusion tapped only the axial part of the magma chamber, where fractionation was not far advanced. Quartz keratophyre lava, equivalent to quartz diorite and plagiogranite in the plutonic suite, is a local minor member of the ophiolite volcanics. Such highly fractionated lava might have evolved and leaked to the surface from parts of the magma chamber that lay beyond reach of replenishment, or during times when replenishment (and spreading?) abated.

The grade of alteration decreases upward through the volcanic section but it is difficult to distinguish between the partly superimposed effects of: (1) thermal metamorphism above the magma chamber, (2) hydrothermal alteration (metamorphism) from circulation of heated sea water, (3) later sea-floor weathering, and (4) still later burial metamorphism that locally extends up into the lowermost Great Valley Sequence.

The sediment entrapped within the volcanic member provides evidence for the setting in which the ophiolite formed. The sediment scraps within the ophiolite lavas appear to be exclusively pelagic limestone and radiolarian chert, both uncontaminated by admixed terrigenous clastic detritus or arc-derived volcaniclastics or tephra. The pelagic character and very minor amount of these sediments suggest an open-ocean setting that lay beyond reach of any detritus from a continental margin or active volcanic arc.

The Coast Range ophiolite originated at a spreading ocean ridge, indicated by the following collective evidence: (1) morphological features show the volcanic rocks to be deep-sea lavas; (2) pelagic interpillow sediments indicate an open-ocean setting; (3) the lavas, now altered, were chiefly tholeiitic basalt of MORB type; (4) the limited fractionation range of the lavas is best explained by repeated replenishment and magma mixing in open-system magma chambers; (5) the plutonic sequence is roofed only by oceanic volcanic rocks, fed in part by sheeted dikes that indicate recurrent extension of the magma chamber roof; (6) the plutonic sequence shows a cumulus phase succession and cryptic variation indicative of repeated replenishment and magma mixing, expected in a continuously opening ridge-axis magma chamber; (7) the magmatic cumulates show a "soft-sediment" style of penecontemporaneous deformation that fits a tectonically active ridge-axis site; (8) the plutonic sequence (especially the upper part) shows textural, mineralogical, and isotopic evidence for the entry of sea water; (9) the ophiolite is floored by an oceanic-type basement of harzburgite tectonite; (10) the full ophiolite sequence has the thickness and seismic properties of oceanic crust (explained below).

A *slow* spreading ocean-ridge system is inferred from the following features: (1) the plutonic member is very thin, indicating small magma chambers; (2) sill complexes are more common than continuous sheeted dikes, indicating recurrent formation of new ridge-axis magma chambers rather than continuous maintenance of a single steady-state chamber; (3) submarine talus breccias are common within the volcanic member, suggesting a strongly rifted ridge axis with fault-scarp topography; (4) absence of MnFe umbers and Cyprus-type sulfide ore deposits, suggesting a lack of strong ridge-axis hot spring activity. Finally, Middle Jurassic oceanic crust is less than 150 km wide perpendicular to the spreading-axis direction (NNW, see below) but has an age span of 13 m.y. This equates to a half-spreading rate on the order of 1 cm/year, comparable to the Mid-Atlantic Ridge.

Depth to the oceanic crust-mantle boundary (M) within the Coast Range ophiolite is 'too shallow' (3 km or less) compared to depths determined by conventional seismic refraction at sea (4-7 km below the sea floor) if M is assumed to be the contact between gabbroic and ultramafic rocks. However, Nichols and others (in press) argue from measured P- and S-wave velocity profiles at Point Sal, and from synthetic seismograms, that M originally lay deeper, i.e., at a gradational boundary between partly serpentinized and fresh peridotite that existed beneath the deep sea floor. This boundary has been erased from the ophiolite remnants by subsequent serpentinization on land (Margaritz and Taylor, 1976); however, geologic evidence in the northern Coast Ranges shows that extensive serpentinization occurred *beneath the sea floor* in the Late Jurassic, prior to deposition of the Great Valley Sequence. Sea water penetration to depths of 5 km in ophiolites has now been established by isotopic evidence (Spooner and others., 1973; Gregory and Taylor, in press); thus, sea water penetration will extend into the upper peridotite where the plutonic sequence is thin, as in the Coast Range ophiolite. We propose that slow spreading ocean-ridge systems, which have only small magma chambers (Sleep, 1975; Kusznir and Bott, 1976), will commonly experience sea water penetration through the resulting thin plutonic member, producing partly serpentinized peridotite at depths of 3-5 km. The seismic velocity reversal produced by the serpentinized zone (lower part of oceanic layer 3) will be difficult to detect at sea by conventional seismic refraction studies, and will make M appear deeper than it really is (Luyendyk and Nichols, 1977).

The progression of Mesozoic sedimentary deposits within and above the ophiolite bears on its tectonic history. Six stages are recorded: (1) limestone and radiolarian chert entrapped within the ophiolite mark the slow accumulation of pelagic biogenic oozes at the site of ocean crust formation during the Middle Jurassic; (2) A 10-20 m.y. depositional hiatus then followed at several of the ophiolite localities, marking a period of sea-floor erosion or nondeposition below the CCD. (3) Oxfordian to early Tithonian (Late Jurassic) sediments deposited directly upon the ophiolite lavas range from radiolarian chert through tuffaceous chert to waterlaid tuff and coarse volcaniclastic sediments composed of calcalkaline andesitic and dacitic detritus. Viewing the ophiolite remnants collectively, the Oxfordian to early Tithonian sediment blanket formed a tapering wedge, from a thick (700 m) proxymal facies of submarine volcaniclastics including bouldery submarine debris-flow deposits to a thin (25-50 m) distal facies of radiolarian chert with only minor admixed tephra. One step farther takes this lateral progression to the tuff-free manganiferous radiolarian cherts of the same age that cover Franciscan pillow lavas; these represent a still more distal deep-water pelagic facies that accumulated farther seaward on the Middle Jurassic igneous oceanic crust. (4) Polymictic breccias consisting almost wholly of ophiolitic clasts (plus minor chert), rest in depositional contact on tectonically dismembered or melanged ophiolite and chert at northern Coast Range localities. These ophiolitic breccias were submarine talus accumulations, reworked in part by mass flowage, that spread from the foot of large fault scarps. They coincide with the development of serpentinite-matrix melanges at depth, marking a period of large-scale faulting and internal disruption of the deep ocean floor beginning in the Kimmeridgian. (5) Terrigenous strata of the Great Valley Sequence overlap and bury the ophiolitic breccias in the northern Coast Ranges and cover the undisturbed tuffaceous cherts farther south. These mark the encroachment of deep sea fans derived from igneous and metamorphic source terranes to the east and north, beginning in the early Tithonian at localities north of San Francisco and in the middle Tithonian at localities farther south. Interbedded with the lower several hundred meters of the Tithonian land-derived clastics at northern Coast Range localities and at Santa Cruz Island are beds of dark lithic sandstone and microbreccia composed of ophiolitic detritus (including chert), derived locally from uplifted blocks of oceanic basement. (6) Great Valley Sequence terrigenous sediments west of the Sacramento Valley form a thick, uninterrupted progression up through the Cretaceous, but in the central and southern Coast Ranges the Tithonian-Valangian is thinned and most of the Lower Cretaceous is missing beneath a regional unconformity.

The progression of sedimentary environments through which the ophiolite passed from Middle to Late Jurassic is compatible with its formation at an open-ocean spreading ridge, its subsequent transport by sea-floor spreading across a deep oceanic region where pelagic sedimentation below the CCD was nil, its approach to an active volcanic island arc, followed by termination of arc volcanism and an encounter with a tectonically active continental margin. Formation of the ophiolite within an interarc or back-arc basin is ruled out by the absence of volcaniclastic detritus (even fine tephra) at the time of ophiolite formation and for a period lasting up to about 10 m.y. afterward.

The Middle Jurassic spreading-ridge axis (possibly the Farallon-Pacific plate boundary) was oriented approximately parallel to the present Coast Range trend (about NNW), as shown by the lack of an age progression (younging) of the ophiolite remnants either northward or southward along a 1000-km segment of Coast Range terrane (with San Andreas offset restored). The ocean ridge lay several hundred kilometers west of the Late Jurassic (Tithonian) continental margin, calculated by using the 20–25 m.y. time span from ophiolite formation to terrigenous sediment encroachment, and by assuming slow sea-floor spreading rates. The oceanic lithosphere created at the ridge spread eastward toward a consuming plate boundary (trench) that lay just west of the Nevadan volcanic island arc, active from Middle to Late Jurassic (Callovian to Kimmeridgian or early Tithonian). The trench prevented terrigenous or volcaniclastic detritus (except air-borne tephra) from reaching the oceanic plate until the Late Jurassic Nevadan orogeny, when subduction ceased at that site and the trench was filled and bridged. Intraplate magmatic activity in the oceanic realm during this period (early Tithonian) included growth of the basaltic Stonyford seamount and minor siliceous intrusions farther south (Diablo Range and Santa Cruz Island).

Eastward subduction of the oceanic (Farallon?) plate became reestablished farther west, in the Coast Range realm, but not until the Early Cretaceous when Franciscan melanging began. The interval from Kimmeridgian to Valanginian time was marked by a series of westward-stepping "subduction false starts," which disrupted the oceanic crust and created great fault scarps on the ocean floor. This wave of "Nevadan" deformation passed first through the northern Coast Range terrane, where submarine fault scarps, talus accumulations and debris flows, and ophiolitic (serpentinite-matrix) melanges formed prior to the arrival of terrigenous sediments in the early Tithonian. Uplift of oceanic crustal blocks, locally accompanied by serpentinite diapirism (Moisseyev, 1970; Maxwell, 1977) continued there into the Early Cretaceous, shown by ophiolitic and serpentinitic detrital intercalations within the lower Great Valley Sequence. The same wave of deformation passed through the central and southern Coast Range terrane but occurred later, mainly in the Early Cretaceous, after undisrupted oceanic crust was covered by the terrigenous Great Valley sediments. This deformation is recorded by the regional unconformity beneath the Upper Cretaceous in the Great Valley Sequence and by great thinning of the Tithonian-Valanginian strata beneath the unconformity. These resulted from repeated Early Cretaceous uplift of oceanic basement blocks and sliding off of the overlying submerged, newly deposited soft sediments. Submarine sliding was directed mainly westward, toward the developing Franciscan trench.

Structural dismemberment of the Coast Range ophiolite involves several stages of high angle dip-slip (and strike-slip?) faulting, which thin and cut out elements of the ophiolite stratigraphy. Repetition of the ophiolite stratigraphy by low-angle thrusts is not observed. Dismemberment began in the Middle Jurassic at a rifted ocean ridge where the ophiolite originated, and continued through the Late Jurassic and Early Cretaceous during the deformations described above. Ophiolite dismemberment seen along the Franciscan-Great Valley contact, however, was accomplished mainly by Cenozoic high-angle faulting (rotated later to lower angles) which brought deeply buried Franciscan high-pressure metamorphic rocks up against the unmetamorphosed Great Valley Sequence (Ernst, 1970; Maxwell, 1974). We find no conclusive evidence for large-scale thrust faulting that involves the Coast Range ophiolite.

INTRODUCTION

The California Coast Range ophiolite is an assemblage of serpentinized ultramafic rocks, gabbros, minor diorite and plagiogranite, diabasic to microdioritic dike and sill complexes, and submarine volcanic rocks that occurs at the base of the Upper Jurassic to Upper Cretaceous Great Valley Sequence in the California Coast Ranges. Radiolarian cherts, tuffaceous cherts, or water-laid tuffs with radiolarite interbeds lie in depositional contact upon the submarine lavas where faults do not intervene, and these strata pass conformably upward into Tithonian mudstones and flysh comprising the basal terrigenous clastic rocks of the Great Valley Sequence.

Rocks of the Coast Range ophiolite assemblage tend to occur in the following stratigraphic order (top to bottom) at localities where the sequence is most nearly complete: lavas, sheeted dike or sill complex, noncumulus gabbro ± diorite and plagiogranite, cumulus gabbro, cumulus ultramafic rocks (olivine clinopyroxenite, wehrlite, dunite) that are strongly serpentinized and cut by basic dikes, and a basal zone of serpentinized harzburgite tectonites. The ophiolite is tectonically dismembered, however, so that at most localities parts of the sequence are missing or thinned. Also, at some localities in the northern Coast Ranges the ophiolite is internally disrupted and occurs as serpentinite-matrix melange. The base of the ophiolite, generally marked by intensely sheared serpentinite, is structurally underlain by rocks of the Franciscan Complex in most parts of the Coast Ranges. Cenozoic high-angle faulting has commonly offset this contact however, so that one finds ophiolite faulted against Tertiary strata, or Franciscan Complex faulted against Great Valley Sequence with no intervening ophiolite exposed.

The unity of these submarine lavas, plutonic rocks, and serpentinized ultramafics went unrecognized over the years, and their relationships to the Franciscan and the Great Valley Sequence were given diverse interpretations (discussed in Bailey and others, 1964). A major advance in Coast Range geology came in 1970, when these rocks were recognized as an ophiolite sequence and interpreted as remnants of Jurassic oceanic crust and mantle emplaced on land (Bailey and others, 1970; see also Bezore, 1969). Subsequent workers have widely accepted Bailey, Blake, and Jones's proposal that Late Jurassic strata of the Great Valley Sequence were deposited upon oceanic crust represented by the ophiolite, and that rocks of the Franciscan Complex were subsequently thrust beneath them (Page, 1972; Blake and Jones, 1974; Jones and others 1976; Evarts, 1977; but see Maxwell, 1974, for a different interpretation).

This chapter reviews the field occurrence, petrology, and age of the ophiolite, based on much new field work since 1970, and presents a tectonic interpretation.

Occurrence

Scattered remnants of the ophiolite at the base of the Upper Jurassic Great Valley strata are found along the Coast Ranges from the Paskenta area on the north to Santa Barbara County on the south, a distance of more than 700 km (Fig. 14-1). Those remnants in the southern Coast Ranges are separated from the others by the San Andreas fault. Restoration of 600 km of inferred displacement along the San Andreas fault system extends the original north-south span of the ophiolite to approximately 1300 km. Moreover, remnants of Jurassic ophiolite beneath Tithonian strata equivalent to basal Great Valley Sequence are known from Vizcaino Peninsula and Cedros Island in Baja California (Jones and others 1976), from southwestern Oregon (Jones and Imlay, 1973), from the Cascade Mountains in south-central Washington (Hopson and others, 1973, and unpublished), and in the San Juan Islands of Puget Sound. The California Coast Range ophiolite is thus a segment of a much larger system of Jurassic ophiolite remnants that extends for more than 3000 km along the western margin of North America.

A summary of the Coast Range ophiolite occurrences is presented before attempting a synthesis and tectonic interpretation. The geologic relations and petrology of twenty-three of the known ophiolite remnants, including their Jurassic sedimentary cover, are described, and the sequence of events recorded at each locality is summarized. We begin this review of ophiolite field occurrences at the south and move northward, the direction of increasing structural complexity. This is followed by sections dealing with isotopic ages of the ophiolite and with the radiolarian biostratigraphy; many new data are presented in both these sections. A section on petrology and petrogenesis summarizes the petrologic characteristics of the ophiolite, the magmatic processes that are recorded, and the origin and tectonic setting of the igneous sequence. A section on Jurassic sedimentary

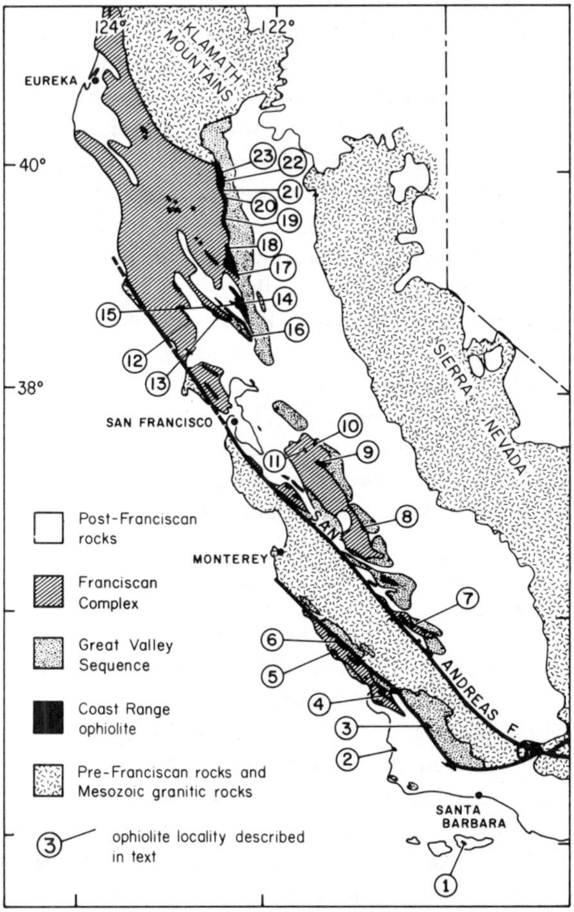

Fig. 14-1. Simplified geologic map of western California, showing the location of Coast Range ophiolite localities described in the text.

rocks associated with the ophiolite briefly summarizes the main characteristics of these sediments and their probable tectonic environments. The final section on the tectonic history of the Coast Range ophiolite is heavily dependent on all the preceding sections.

FIELD OCCURRENCE OF COAST RANGE OPHIOLITE REMNANTS

This section systematically describes the field relations, stratigraphy, and petrology of each of the twenty-three ophiolite remnants, beginning at the south. These data will form the basis for our synthesis of the petrogenesis of the ophiolite and its tectonic history, presented later.

ERRATA SHEET FOR

The Geotectonic Development of California
(W.G. Ernst, Editor)

The following figure should have appeared on page 429 as Figure 14-2:

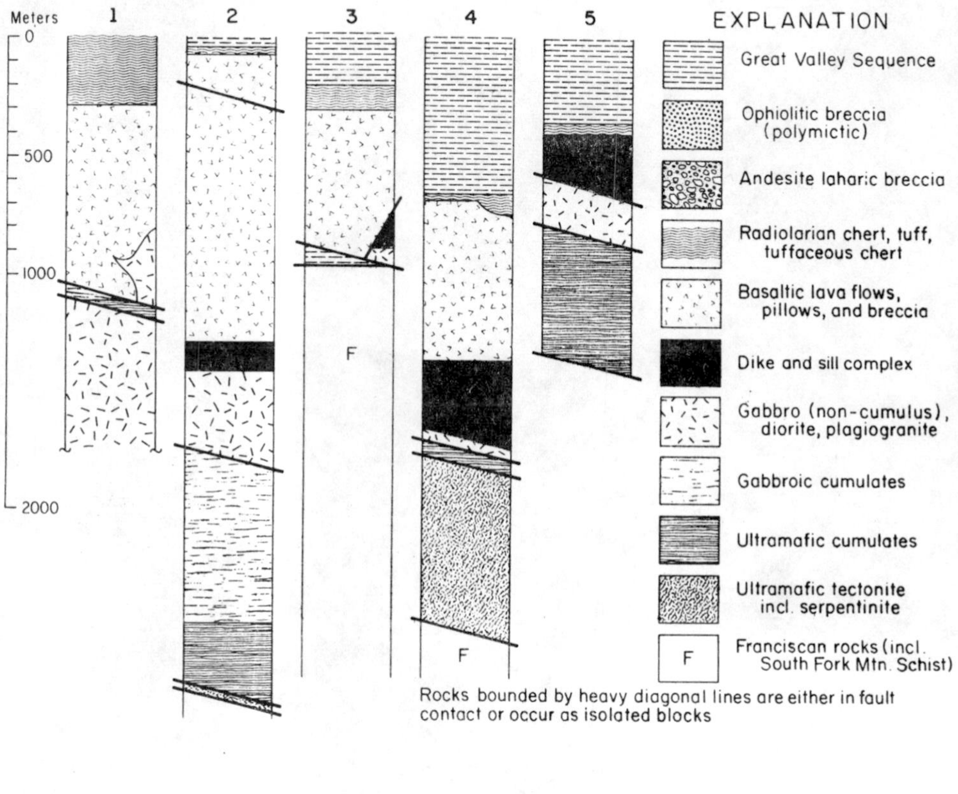

Fig. 14-2. Columnar sections showing details of the ophiolite remnants in the southern Coast Ranges (Nacimiento block), at localities 1 to 5 shown in Fig. 14-1.

1. Santa Cruz Island Ophiolite Remnant, Santa Barbara County

Jurassic basement rocks are exposed in a large (4 by 20 km) erosional window through Tertiary strata on Santa Cruz Island, off the Santa Barbara coast. These rocks have recently been interpreted as an oceanic crustal remnant and correlated with the Jurassic ophiolite of the Coast Ranges based on their similar lithologic and chemical characteristics and age (Hill, 1976; Mattinson and Hill, 1976). Structural complications, poor exposures, and the lack of a well-defined ophiolitic stratigraphic sequence, however, are obstacles to a clear understanding of the petrogenesis and original tectonic setting of the Santa Cruz Island rocks. Moreover, a greenschist-facies dynamic metamorphism has imposed schistose fabrics on the volcanic and overlying sedimentary rocks, an event not recorded at ophiolite localities in the Coast Ranges farther north. It now appears likely, from paleomagnetic evidence, that the Santa Cruz Island terrane has been tectonically rafted to its present position (and rotated >90° clockwise) from a place of origin that originally lay much farther south off Baja California (Kamerling and Luyendyk, 1979). Thus, the metamorphism that renders this southernmost ophiolite remnant unique may perhaps reflect differences in the tectonic history far to the south.

Three mappable units (Fig. 14-2, 1) are present in the Santa Cruz Island basement remnant (Hill, 1976): (1) massive to schistose metavolcanic rocks and metasedimentary schists, all in the greenschist facies (Santa Cruz Island Schist); (2) a small plagiogranite pluton (Alamos Leucotonalite) that intrudes the schist; and (3) a plutonic assemblage (Willows Plutonic Complex) comprised chiefly of hornblende diorite and quartz diorite, with less abundant gabbro, ultramafic rocks, mafic to intermediate dike rocks, and

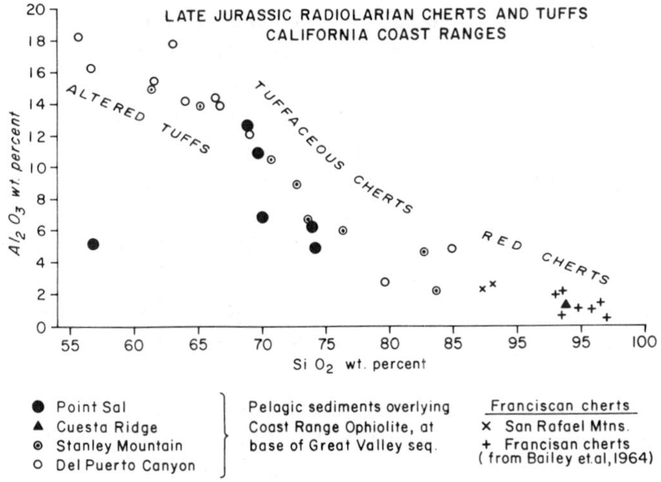

Fig. 14-2. Columnar sections showing details of the ophiolite remnants in the southern Coast Ranges (Nacimiento block), at localities 1 to 5 shown in Fig. 14-1.

minor plagiogranite. The following brief summaries are taken mainly from Hill's (1976) work.

The metavolcanic rocks are massive to semischistose keratophyre, quartz keratophyre, and spilite, derived mainly from massive lavas. The stratigraphic order appears to be keratophyre and spilite low in the sequence, keratophyre and quartz keratophyre toward the middle, and metasedimentary rocks (slaty to phyllitic metatuff and volcanic arenite, metachert, metapelite, and minor metagabbroic sandstone) at the top. Mafic to silicic dike rocks are locally abundant in the lower part of the sequence.

The Willows Plutonic Complex, composed mainly of hornblende diorite and hornblende quartz diorite, with small plagiogranite bodies and abundant crosscutting mafic to silicic dikes is evidently the differentiated upper part of a larger plutonic mass that has cumulus gabbros and ultramafic rocks at deeper levels. Olivine-clinopyroxene cumulates (ol clinopyroxenite, feldspathic and hornblende-bearing ol clinopyroxenite, and wehrlite), locally interlayered with gabbroic cumulates (pl-cpx-ol, pl-cpx, pl cumulates), occur as fault slivers along the North Valley Anchorage fault, which separates the plutonic (Willows) and metavolcanic (Santa Cruz Island Schist) terranes. Clinopyroxenite and hornblendite (cumulus cpx, ol, and possibly hb) and two-pyroxene gabbro (pl-cpx-opx cumulate) plus high-level noncumulus gabbros (hb-cpx gabbro, hb gabbro, hb-qz gabbro) also crop out at other scattered localities, but they are poorly exposed and their structural relations uncertain. Hill (1976) interprets the Willows diorite and quartz-bearing plutonic rocks as late differentiates of magma(s) that left more mafic material at deeper levels than those now exposed. Slices from the deeper gabbroic and ultramafic zones have subsequently been carried up, mainly along faults, to their present positions within the dioritic to quartz dioritic upper zone.

The Willows paragenetic sequence, during magmatic crystallization, was (1) cpx (2) pl + cpx + ol, (3) hb (or opx) + pl + cpx, and (4) qz + hb + pl. Subsequent metamor-metamorphism did not substantially alter the igneous fabric of the Willows plutonic rocks, but produced the subsolidus phases antigorite, tremolite-actinolite, talc, chlorite, epidote, sericite, quartz, and minor albite, which are sporadically developed.

The Alamos pluton is a small (3 sq. km) intrusive mass of plagiogranite within the metavolcanic rocks of the Santa Cruz Island Schist. It has been metamorphosed with the enclosing volcanic rocks, as shown by its schistose margins and apophyses and by the albitization of its primary zoned plagioclase. Small bodies of similar plagiogranite cut parts of the Willows Plutonic Complex and may be offshoots from the Alamos magma, but correlation across the fault that separates the plutonic and metavolcanic-metasedimentary terranes makes this uncertain. Hill (1976) concludes that the Alamos pluton represents the latest stage of differentiation of a magma deficient in K_2O, which invaded older oceanic crust.

Correlation of the Santa Cruz Island basement assemblage with the Coast Range ophiolite is based on the following points: (1) the Willows plutonic assemblage closely resembles the uppermost parts of the ophiolite plutonic sequence as displayed at Point Sal and other Coast Range localities: (2) the Willows gabbros and ultramafic rocks, though displaced stratigraphically, resemble the middle and lower parts of the Coast Range ophiolite plutonic sequence; (3) the 162 m.y. U-Pb isotopic age of the Willows

Plutonic Complex is within the narrow age range of the Coast Range ophiolite (Mattinson and Hill, 1976; also this chapter); (4) the protoliths of the Santa Cruz Island Schist show a general resemblance to the spilitic-keratophyric lavas of the Coast Range ophiolite, and to the tuffaceous cherts and Lotta Creek submarine tuffs that overlie the ophiolite at localities in the Diablo Range (locs. 7 to 10, this chapter). The unique "gabbroic sandstone" of the upper Santa Cruz Island Schist even finds its counterpart in the ophiolitic sandstones of the lowermost part of the Great Valley Sequence, where it overlies the ophiolite at northern Coast Range localities (locs. 18, 22, this chapter). Thus, we concur with Hill (1976) that the Santa Cruz Island basement terrane most likely represents the Jurassic ophiolite and its sedimentary (largely volcaniclastic) cover, now metamorphosed. The crosscutting Alamos plagiogranite pluton (140 m.y., Mattinson and Hill, 1976), is substantially younger than the ophiolitic assemblage, but is matched by minor intrusions of similar composition (quartz keratophyre) and age (Tithonian), which cut the ophiolite and its sedimentary cover at localities in the Diablo Range (locs. 7, 9, this chapter).

Jones and others (1976) have instead correlated the Santa Cruz Island Schist and Alamos pluton with the Late Jurassic (Nevadan) volcanic arc rocks (Logtown Ridge-Mariposa) of the western Sierran-Klamath belt, perhaps because of their schistosity. We disagree because (1) the Santa Cruz Island metavolcanic and metasedimentary rocks differ significantly in composition and petrography from the western Sierran Upper Jurassic rocks; (2) albite granite plutons of Alamos type are unknown in the western Sierran terrane; (3) the Sierran Upper Jurassic volcanics were erupted upon an ophiolitic basement more than 200 m.y. old (Saleeby, Chapter 6, this volume), whereas the Santa Cruz Island metavolcanics are underlain by ophiolitic basement with a much younger "Coast Range" age.

In conclusion, the Santa Cruz Island basement assemblage records the following events: (1) formation of igneous oceanic crust, about 162 m.y. ago (Mattinson and Hill, 1976); (2) deposition of tuffaceous sediments, chert, and ophiolitic sandstone upon the ophiolite lavas; (3) intrusion of plagiogranite about 140 m.y. ago (Mattinson and Hill, 1976); and (4) a subsequent greenschist facies metamorphism under tectonic conditions that produced schistosity in the incompetent rocks, especially the sediments.

2. Point Sal Ophiolite Remnant, Santa Barbara County

One of the best exposed and stratigraphically most complete ophiolite remnants crops out along the southern California coast at Point Sal and Vandenberg Air Force Base in northwestern Santa Barbara County (Fairbanks, 1896). The exposed igneous sequence, approximately 3 km thick, is concordantly overlain by Upper Jurassic marine sedimentary rocks and is cut off at the base by the late Cenozoic Lions Head fault, which throws Miocene Monterey Formation against serpentinized dunite. Tertiary strata and Quaternary deposits blanket the surrounding area for many miles to the north, east, and south, isolating the Point Sal mass from other remnants of the Coast Range ophiolite and the Franciscan Complex.

A composite stratigraphic section through the ophiolite at Point Sal, combining three partial sections (Figs. 14-2, 2 and 14-3) consists of the following units: (1) pillowed and massive submarine lavas, consisting of (a) an upper group of mainly olivine basalts with pervasive low-grade alteration and minor interpillow coccolithic limestone, and (b) a lower group of spilitic cpx-pl microphyric and aphyric basalts and keratophyre, with low- to high-grade greenschist facies alteration and local pockets of red radiolarian chert; (2) dioritic and gabbroic rocks, grading progressively downward from (a) hornblende quartz diorite and diorite to (b) hornblende gabbro and uralitic clinopyroxene gabbro, (c) olivine-clinopyroxene gabbro, and (d) olivine-clinopyroxene gabbro with troctolite and minor anorthosite; (3) layered ultramafic rocks, grading downward from (a) chiefly olivine clinopyroxenite to (b) serpentinized dunite with subordinate ol-clinopyroxenite and wehrlite to (c) serpentinized dunite. Swarms of basaltic, diabasic, microdioritic, and epidositic sills and dikes form a sheeted sill complex (zone 1c) between the volcanic and upper plutonic rocks; also, swarms of similar dikes cut the overlying volcanics (zone 1b) and underlying diorite and gabbro (zone 2a–b). A different swarm of low-angle dikes, mainly noritic microgabbros and feldspathic wehrlites, cut plutonic zones 2c–3c. The Lions Head fault cuts the sequence off just above the level where a basal zone of peridotite (harzburgite) might be expected. Slivers of serpentinized harzburgite occur within the Lions Head fault zone, however, suggesting that alpine-type peridotite was originally present beneath the exposed sequence.

The igneous sequence here is thicker and stratigraphically more complete than anywhere else in the Coast Ranges and comes closest to representing an unbroken ophiolite sequence above the harzburgite level. Thus, it forms a convenient yardstick for comparison with the other, less complete ophiolite remnants. The plutonic sequence ranges from ol-sp cumulates at the base up through cpx-ol cumulates and pl-cpx-ol cumulates, then through noncumulus gabbro, diorite, and quartz diorite that differentiated along the liquid line of descent. It represents a nearly continuous record of crystallization and differentiation by crystal settling in a suboceanic magma chamber. The dike and sill complex that separates the plutonic and volcanic rocks is well exposed and unique in the development of subhorizontal sheeted sills, rather than vertical sheeted dikes. The volcanic sequence above is exceptionally thick (>1.3 km) and bears clear-cut evidence of eruption in an open-ocean setting, at water depths ranging from slightly above to below the CCD, where pure biogenic pelagic oozes were the only sediments accumulating (Hopson and Frano, 1977).

Above the ophiolite a thin sequence (>25 m) of greenish-gray tuffaceous radiolarian chert lies in depositional contact on the upper basalt (ophiolite zone 1a). Overlying the chert are mudstones with thin sandstone interbeds (distal turbidites), representing the base of the Great Valley Sequence. The cherts range in age from Oxfordian through early Tithonian, based on their radiolarian faunas, and the mudstones above are middle to late Tithonian based on their *Buchia* fauna (see Biostratigraphy, this chapter). These strata represent pelagic and then distal terrigenous sedimentation on top of the igneous sequence during the latest Jurassic. The Oligocene Lospe Formation cuts unconformably across the tilted and eroded Jurassic sequence.

In conclusion, the Point Sal ophiolite records the following events: (1) formation of

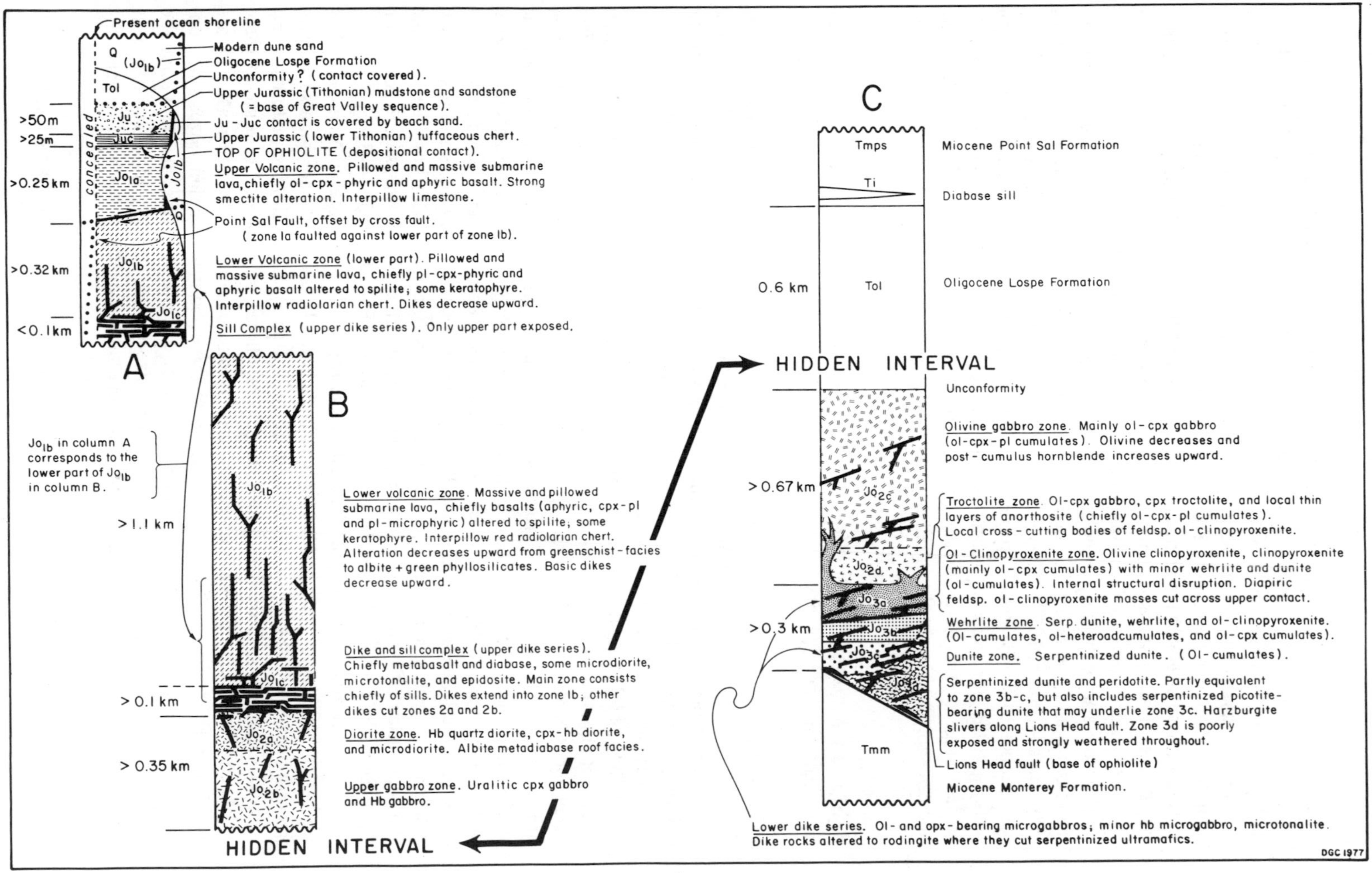

Fig. 14-3. Columnar sections of Point Sal ophiolite showing different stratigraphic levels: (A) coastal strip extending north from Point Sal; (B) Point Sal Ridge, north and northwest of Point Sal Beach State Park; (C) Lions Head-Point 1 ospe area. Reproduced by permission of Oregon Department of Geology and Mineral Industries.

igneous oceanic crust about 160 m.y. ago (see Isotopic Ages, this chapter); (2) accumulation of a thin blanket of tuff-bearing radiolarian ooze upon the pillow lavas, probably at depths below the CCD, during Oxfordian to early Tithonian stages of the Late Jurassic; and (3) encroachment of the distal edge of a deep-sea fan of clastic detritus (flysch sediments) beginning in the middle Tithonian. The igneous oceanic crust formed in a deep-sea setting beyond reach of terrigenous or volcaniclastic sediment, and it may have formed in two stages: an early stage in which mainly fractionated pyroxene basalts were erupted and red radiolarian chert was accumulating, and a later stage when more primitive olivine basalts were erupted while pelagic calcareous ooze accumulated. Oceanic ridge-crest jumping (Sclater and others, 1971; Menard, 1978) has been suggested as a possible explanation (Hopson and Frano, 1977). The apparent gap in ages between the igneous rocks (161 m.y. corresponds to the Bathonian stage of the Middle Jurassic on the Van Hinte, 1976a, time scale) and the beginning of pelagic sediment accumulation on top of igneous crust (Oxfordian stage of the Late Jurassic) suggest that a major depositional hiatus (submarine unconformity) occurs at the basalt-chert contact.

3. Stanley Mountain Ophiolite Remnant, San Luis Obispo County

Fifteen miles northeast of Santa Maria, southern San Luis Obispo County, a remnant of Jurassic ophiolite overlying Franciscan melange is exposed in a window through the cover of Tertiary strata (Brown, 1968a, b; Bailey and others, 1970). Only the upper part of the ophiolite is preserved here, but the sequence of pelagic and terrigenous strata that rests upon it is particularly well displayed (Fig. 14-2, 3).

The ophiolite remnant consists of pillowed and massive submarine lavas and minor tuff, cropping out in a belt 10 km long and up to 3 km wide between the Cuyama River gorge and Alamo Creek. The lavas are chiefly spilite and keratophyre (called andesite by Brown, 1968b). Most of these rocks were originally aphyric and cpx-pl microphyric basalts like those at Point Sal (zone 1b), and here too they have been albitized and altered to secondary mineral assemblages of the zeolite and greenschist facies. The lower part of the volcanic pile, exposed in Cuyama River gorge, is penetrated by dikes and juxtaposed (faulted?) against a small remnant of intrusive complex consisting of hornblende-augite diorite, albitic-uralitic diabase, and minor plagiogranite dikes. These intrusive rocks resemble those at the contact of zones 1c and 2a at Point Sal (Figs. 14-2, 3, 14-3).

Lying conformably on pillow basalt at the top of the ophiolite at Alamo Creek is more than 100 m of thin-bedded radiolarian chert. These cherts are reddish brown, greenish-brown and dark gray due to admixture of fine montmorillonitic material. Faint outlines of shards and small flattened pumice lapilli can be discerned in some of these rocks, and very tiny microlites of plagioclase, pyroxene, hornblende, and magnetite are found at some horizons. These rocks evidently came from radiolarian oozes that contained a tuffaceous component. Moreover, several discrete tuff beds occur within the lower 15 m of the sequence. These include two size-graded, strongly zeolitized tuff beds 2 m and 6 m thick, which grade from coarse crystal-lithic tuff at the base to very fine "flinty" tuff at the top. Thin beds of limestone replacing chert occur at 20, 21, and 29 m above the base of the sedimentary sequence, and limy nodules occur locally within the chert at higher horizons. The age of the chert sequence, based on its radiolarian fauna, spans from Oxfordian through early Tithonian stages of the Late Jurassic (see Biostratigraphy, this chapter). Above the chert is a graywacke-dark shale unit 100 to 200 m thick with dark-green tuff in the lower part (Brown, 1968b), and above this are Valanginian (and older?) mudstones, sandstones, and conglomerate lenses comprising the Jollo Formation (Brown, 1968b; Hall and Corbató, 1967). Upper Cretaceous sandstones of the Carrie Creek Formation cut unconformably across the ophiolite and its cover of Upper Jurassic to Lower Cretaceous strata.

The ophiolite remnant is in fault contact with underlying Franciscan melange, which carries tectonic blocks of metagraywacke, metaconglomerate, metachert, basaltic greenstone, glaucophane schist, amphibolite, and serpentinized ultramafic rocks in a sheared argillaceous matrix (Brown, 1968b). A tectonic slice of "Knoxville" Shale with interbedded arkosic graywacke and conglomerate lenses intervenes between the Franciscan melange and the ophiolite remnant along part of this contact. The middle Tithonian index fossil *Buchia piochii* occurs at several localities within this unit (Easton and Imlay, 1955), which indicates that its original stratigraphic position was above the lower Tithonian chert but below the Valanginian Jollo Formation.

Events recorded at the Stanley Mountain ophiolite remnant include (1) formation of igneous oceanic crust (mainly submarine lavas); (2) accumulation of impure radiolarian ooze during Oxfordian through the early Tithonian; and (3) burial of the igneous crust and its pelagic sedimentary cover by encroaching terrigenous clastic strata (distal flysch sediments) beginning in the middle Tithonian (see Biostratigraphy, this chapter). Volcanic ash is interbedded and mixed with the lower radiolarian cherts, while impurities in the upper cherts appear to be argillaceous.

4. Cuesta Ridge Ophiolite Remnant, San Luis Obispo County

A synclinal remnant of dismembered ophiolite crops out along a belt 25 km long that extends northwestward from Cuesta Pass near San Luis Obispo (Strand, 1959). The southern two-thirds of this belt is mapped and described by Page (1972), and its petrology and geochemistry have been studied by Pike (1974, 1976). The ophiolite here is conformably overlain by Upper Jurassic to Lower Cretaceous sedimentary strata, and it is structurally underlain by melange of the Franciscan Complex. The ophiolite-Franciscan contact, thought to be the Coast Range thrust (Page, 1972), is mainly obscured by landslides and soil cover.

The ophiolite here is strongly dismembered, especially along the eastern limb of the syncline where it is represented mainly by poorly exposed volcanic rocks and fault slices of serpentinite. The least dismembered ophiolite sequence crops out along Cuesta Ridge (the western synclinal limb) between U.S. 101 and state route 41. Here Page (1972) shows a lower unit of harzburgite more than 1 km thick cut by altered diabasic dikes, and an upper unit more than 1.2 km thick consisting of basaltic and keratophyric pillow lavas and breccias, with countless mafic to silicic dikes and sills in its lower part. Near the head of Stenner Creek the volcanics pass downward with increasing dikes and sills into a zone of 100% sheeted intrusives more than 400 m thick. This consists of subparallel sills of uralitic pyroxene diabase, hornblende microdiorite, microtonalite, and plagiogranite, with late-stage crosscutting dikes of olivine-microphyric basalt. The sheeting direction parallels the upper surface of the ophiolite (i.e., the volcanic-chert contact). Thus, the sheeted intrusives are a sill complex (Pike, 1974; personal observation), remarkably similar to zone 1c at Point Sal. Screens of gabbro appear between the sills in the lowest exposed part of the Stenner Creek sill complex, where it is faulted against ultramafic rocks. Pike (1974) calls these gabbros cumulates, but those examined by us are high-level noncumulus uralitic gabbros like those near the top of zone 2a at Point Sal. Beneath this gabbro, across a fault, is about 50 m of serpentinized wehrlite (ol-heteroadcumulate), lherzolite (ol-cpx orthocumulate with postcumulus opx), and olivine clinopyroxenite (ol-cpx-adcumulate) (Pike, 1974; personal observation). These ultramafic cumulates are faulted against sheared serpentinite.

Summing up, the Stenner Creek ophiolite section resembles zones 1b, 1c, 2a, and 3b at Point Sal. The faults that truncate the top and bottom of the ol-cpx cumulates may have cut out gabbroic and ultramafic cumulates equivalent to zones 2c, 2d, 3a, and 3c at Point Sal, which are missing here. Fifteen km to the north, however, serpentinized dunite (ol-chr cumulate) cut by altered microgabbroic dikes, closely resembling zone 3c at Point Sal, is exposed in roadcuts along route 41. In Fig. 14-2, 4 all these units are placed in their inferred original stratigraphic order. Elsewhere along the ophiolite, near Cerro Alto Peak, Pike (1974) describes brecciated sill complex, amphibolite, and gabbro intruding serpentinite, but poor exposures leave structural and petrologic relations uncertain.

Thin-bedded gray to dark-brown radiolarian chert 5 to 130 m thick rests in depositional contact in the pillowed and massive volcanic rocks along the crest of Cuesta Ridge (Page, 1972; personal observation). This sediment was mainly a biogenic siliceous ooze, with abundant radiolaria and a relatively low content of admixed tuffaceous constituents (microlites and glass-derived clay). Conformably overlying the chert is the Toro Formation (Fairbanks, 1904; Page, 1972), consisting of 660 m of marine shale with thin rhythmic interbeds of feldspathic graywacke and local beds of pebble conglomerate (Page, 1972; Seiders and others, 1979). The chert is early Tithonian (radiolarian zones 2B–3), and the overlying Toro Formation ranges from middle Tithonian to Valanginian, based on its *Buchia* fauna (Page, 1972). Upper Cretaceous (Campanian) arkosic sandstone, shale, and conglomerate discordantly overlie the Toro and underlying units with a thrust contact (Page, 1972).

Events recorded at Cuesta Ridge include (1) formation of igneous oceanic crust and upper mantle about 153 m.y. ago (see Isotopic Ages, this chapter); (2) accumulation of radiolarian chert on the pillow lavas during the early Tithonian (see Biostratigraphy, this chapter); (3) deposition of distal terrigenous flysch sediments upon the chert during middle Tithonian to Valanginian time; and (4) fault emplacement of basal ophiolite units against Franciscan melanges.

5. San Simeon Ophiolite Remnant, San Luis Obispo County

An ophiolite fragment northwest of San Simeon is mapped and described by Hall (1975, 1976). He postulates that the ophiolite and its cover of Jurassic to Miocene strata were originally connected to the similar sequence near Point Sal, but are now displaced 80 km by late Cenozoic right-lateral movement along the San Simeon-Hosgri fault system (Hall, 1975).

Elongate slices of the ophiolite occur in a fault-bounded belt that trends northwest from the lowlands east of Piedras Blancas Point to the coast at Ragged Point. The ophiolite abuts Franciscan Complex on the east along the San Simeon fault and Franciscan melange and Jurassic-Cretaceous graywacke on the west along the Arroyo Del Oso fault. Both are late Cenozoic high-angle faults. Reconstruction of the ophiolite sequence, which is displaced internally by additional faulting, reveals a lower zone of serpentinized dunite, wehrlite, and olivine clinopyroxenite, all cut by rodingitized gabbroic dikes, a middle zone of gabbro, and an upper zone of sheeted dikes. (Fig. 14–2, 5). The ultramafic zone is at least 500 m thick but cut off by faults at the bottom and top; the rocks are ol and ol-cpx cumulates, internally much deformed. The gabbro, also fault bounded at both base and top, is no more than 200 m thick where exposed along the coastline at Breaker Point. This rock is chiefly nonlayered, heterogeneous-textured opx-cpx and hb-cpx gabbro that resembles zone 2b high-level gabbro at Point Sal. Layered (cumulus) olivine gabbros are missing, probably faulted out. Magnificent exposures of sheeted dike (or sill?) complex, faulted against high-level gabbro, crop out along the coastline 2 to 3 km northwest of Piedras Blancas Point. The sheeted complex consists of 100% dikes, chiefly uralitic pyroxene diabase and basalt, much altered and partly brecciated.

The top of the ophiolite is poorly exposed inland from the coast. Volcanic rocks, equivalent to zone 1 at Point Sal, are missing. The sheeted dike complex is overlain by 46 m of greenish-gray tuffaceous radiolarian chert, and above that lies Upper Jurassic shale with thin interbeds of sandstone and local beds of black chert-pebble conglomerate (Hall, 1976). It is not clear whether the missing volcanics at the top of the ophiolite were cut out by post-Jurassic faulting or by tectonic disruption or erosion of oceanic crust prior to deposition of the Jurassic sediments.

6. Marmolejo Creek Ophiolite Remnant, San Luis Obispo County

A poorly exposed, fault-bounded belt of ophiolitic rocks 16 km long and 0.1 to 2.5 km wide crops out within the Franciscan terrane east and northeast of San Simeon (Strand, 1959; Hall, 1976), on strike with the Cuesta Ridge (San Luis Obispo) ophiolite remnant. Near Marmolejo Flats the ophiolite is sandwiched along thrust faults between Franciscan melange below and Upper Jurassic-Lower Cretaceous flysch of the Toro Formation (Great Valley Sequence equivalent) above (Hall, 1976). Elsewhere along this belt the ophiolite abuts Franciscan melange along a steeply dipping thrust fault and along later high-angle faults. The narrow belt of ophiolite south of Marmolejo Flats is mainly serpentinite. Farther north, where the belt widens, Hall (1976) reports predominantly microdiorite dikes and sills, diorite, serpentinite and altered pyroxenite, and some volcanics. A more detailed picture of the ophiolite stratigraphy and petrology here must await further study.

7. Llanada Ophiolite Remnant, San Benito County

The Llanada ophiolite, exposed along Bitterwater Canyon in the southern Diablo Range, 7 km southeast of Panoche Pass, contains a strongly dismembered, poorly exposed plutonic sequence, but a remarkably thick, continuous volcanic sequence that is conformably overlain by radiolarian chert and basal terrigenous clastic strata of the Great Valley Sequence (Fig. 14-4, 7). This well-preserved upper part of the ophiolite provides an important record of ocean-floor volcanism and volcanic, pelagic, and terrigenous sedimentation during Late Jurassic time.

The Bitterwater Canyon area is geologically mapped and described by Enos (1965). Bailey and others (1970) first identified the igneous assemblage here as a remnant of

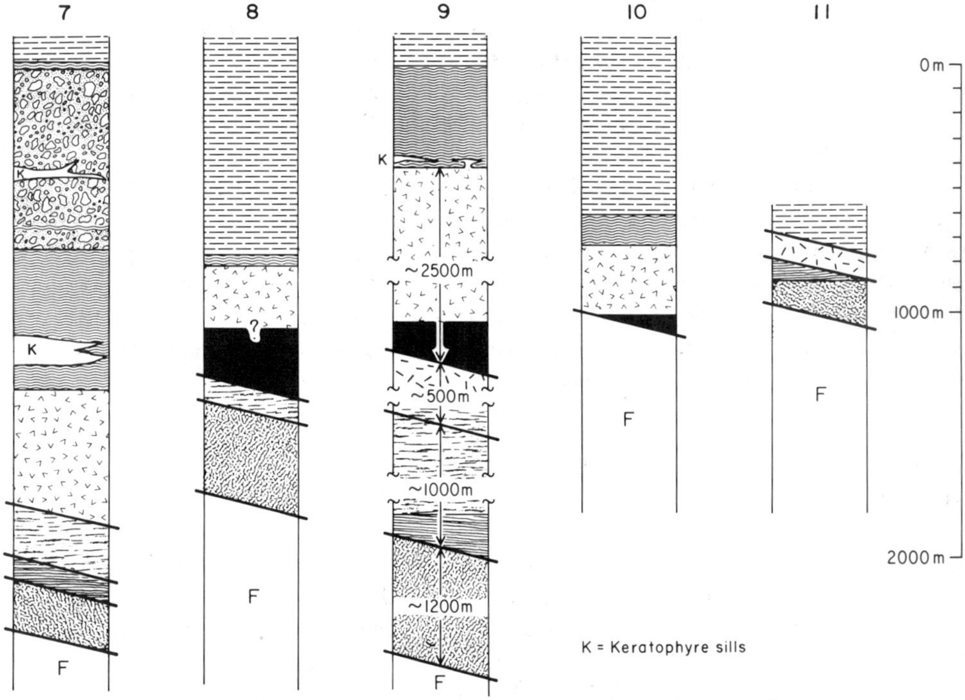

Fig. 14-4. Columnar sections showing details of the ophiolite remnants in the Diablo Range (central California), at localities 7 to 11 shown in Fig. 14-1. See Fig. 14-2 for explanation.

oceanic crust. The present description comes from reconnaissance studies by Hopson and Pessagno; a more detailed study by Nancy Emerson (M.A. thesis, University of California, Santa Barbara, 1980) was recently completed.

The ophiolite remnant and its conformable cover of Tithonian to Valanginian marine volcaniclastic and terrigenous strata (Lotta Creek Tuff and basal Great Valley Sequence) are overlain by Upper Cretaceous strata of the Panoche Group. Franciscan Complex lies structurally beneath the ophiolite and is separated from it by faults.

The ophiolitic plutonic assemblage crops out in three poorly exposed, fault-bounded remnants in the hillside southeast of Bitterwater Canyon. This assemblage consists of uralitic cpx gabbro, ol-cpx gabbro and mela-gabbro, feldspathic wehrlite, and silica-carbonate rock probably derived from serpentinized dunite or harzburgite. The gabbros and wehrlite have well-developed cumulus textures, and the gabbros have cumulus planar lamination. This plutonic assemblage probably represents a lower part of the ophiolite cumulate sequence, underlain by a small fragment of dunite or harzburgite (now silica-carbonate rock); however, no stratigraphic order is evident in the scattered exposures. Wehrlite is faulted against the volcanic sequence, and high-level noncumulus gabbro, dike complex, and basal volcanics are missing. Thus, the middle members of a typical ophiolite sequence are not represented owing to structural dismemberment and burial by Upper Cretaceous sediments.

The volcanic sequence is exceptionally thick (more than 1750 m; base not exposed) and includes both oceanic basaltic lavas and arc-derived, subaqueously deposited andesitic and silicic pyroclastic rocks (Fig. 14-4, 7). This sequence is readily separated into three distinctive parts: (1) a lower zone of basaltic pillow lavas, (2) a middle zone of thin-bedded siliceous tuffs, interstratified locally with basaltic breccia, and (3) an upper zone comprised of increasingly coarse (cobbly to bouldery), massively bedded laharic breccias and tuff-breccias of chiefly andesitic composition.

The lower basaltic member, exposed along the lower part of Bitterwater Canyon, is remarkable for the continuous sequence of well-preserved pillow lavas, more than 500 m thick (the base is not exposed). These are typical deep-sea basaltic lavas, with chilled glassy (now chloritic), nonvesicular pillow rims, distinctive skeletal quench textures (Bryan, 1972), and red radiolarian chert and light-gray limestone as local interpillow sediments. These lavas were evidently erupted onto the deep ocean floor at depths near the CCD ($<$3000 m in the Late Jurassic) in a tectonic setting in which pelagic biogenic ooze was the only sediment accumulating. One massive lava flow with poorly developed columnar structure in its lower part occurs near the top of the pillowed sequence; this resembles massive submarine lavas observed by one of us (CAH) in the Mid-Atlantic Ridge rift valley during submersible diving. The lavas, now spilitic, were originally aphyric basalts and basalts with phyric or microphyric ol, pl-ol, pl-cpx-ol, and pl-cpx. Plagioclase has been altered to albite, olivine and some pyroxene to calcite and/or chlorite, and the glassy to variolitic or aphanitic groundmass is altered to the assemblage albite-chlorite (and/or smectite)-iron oxide-leucoxene-calcite, ± pumpellyite, clays, and secondary silica. Amygdule fillings are chiefly calcite and chlorite.

The top of this continuous basaltic sequence is taken as the top of the ophiolite. Not only does it mark the end of a period in which submarine tholeiitic basalt built up more or less continuously to a great thickness, but it corresponds also to a striking change in sedimentation, that is, the change from biogenic pelagic ooze accumulation *within* the basaltic sequence (interpillow limestone and red radiolarian chert) to voluminous pyroclastic sedimentation above it.

The middle part of the volcanic section, about 485 m thick, consists of a well-bedded sequence of fine-grained vitric tuffs, coarse pumice-lapilli tuffs, and minor fine tuff-breccia, interbedded with three thin units of pillow lava and one of basaltic breccia. The tuffs occur in thin-laminated to massively bedded units several centimeters to tens of meters thick. The lowest tuff unit is strongly graded, from coarsely pumiceous at the base to aphanitic vitric ash at the top. Many of these tuffs, now altered to albite-quartz-green phyllosilicate assemblages, were plagioclase-quartz phyric, suggesting dacitic or more siliceous original melt compositions. Some other tuffs with only plagioclase (now albite) and mafic (now chloritic) microphenocrysts may have originally been andesitic. Poorly preserved radiolaria in some of the very fine ash layers indicate submarine deposition. A massive quartz keratophyre unit more than 30 m thick is a sill (Nancy Emerson, personal communication, 1979). Thus, this tuffaceous middle part of the volcanic sequence, with a keratophyric intrusive within its lower part, corresponds to the Lotta Creek Tuff at ophiolite localities farther north in the Diablo Range. Here, however, the tuffaceous sequence is continuous into a coarser upper member, not seen farther north.

The upper member of the volcanic section, about 760 m thick, shows an upward coarsening progression from thin-bedded tuffs, tuff-breccias, and volcanic conglomerate to thick, massive units of bouldery andesitic laharic breccia. Bedded tuff-breccias (andesitic to siliceous, lithic to pumiceous) with local thin tuffaceous interbeds predominate in the lower 180 m; relict radiolaria are found in one aphanitic vitric tuff bed. A massive volcanic conglomerate unit about 70 m thick, with siliceous volcanic cobbles, also occurs in this lower interval. Higher in the section thick-bedded to massive units of laharic breccia and andesitic conglomerate without internal stratification predominate; thin interbeds of tuff-breccia occur only locally. The uppermost 250 m of the upper volcanic member consists mainly of very thick, massive units of bouldery laharic breccia, with almost no trace of bedding. The rocks called laharic breccias in the upper volcanic member consist of heterolithologic assemblages of subangular to rounded blocks, cobbles, and boulders that are unsupported in a dark ashy matrix. The predominant clast type are dark-gray boulders of porphyritic andesite, with phenocrystic plagioclase, augite, and hypersthene, but light-gray dacite and dark aphyric siliceous volcanic clasts are locally conspicuous. This unit represents an upward-coarsening sequence of siliceous to andesitic volcaniclastic detritus deposited rapidly in a submarine environment. A quartz keratophyre sill intrudes the lower part of this section (Nancy Emerson, personal communication, 1979).

Next stratigraphically comes a thin, discontinuous interval of radiolarian chert and tuffaceous chert that separates the upper volcanic breccias from the overlying, Tithonian mudstone-sandy turbidite unit, which represents the base of the Great Valley Sequence. A continuously exposed section along the steep north side of Bitterwater Canyon (near corner of secs. 26, 27, 34, 35) reveals the following concordant depositional sequence, from the base upward: massive laharic breccia, 2 m of pale green ashy tuff, 1 m of pale gray tuffaceous chert with radiolaria near the top, 0.6 m of terrigenous mudstone, 0.6 m of micrograywacke, overlain by more mudstone. Six hundred meters farther east (gully at NE quarter of SW quarter, sec. 26) along this contact the Great Valley basal mudstones rest directly upon the laharic breccia, but 5 m beneath this contact a thin (15 cm) bed of gray radiolarian chert separates breccia beds. Thus, a period of rapid emplacement of andesitic mudflows onto the sea floor was interrupted near the end, and then followed, by brief quiescent periods when radiolarian ooze accumulated, just prior to the encroachment of terrigenous muds and turbidite sands.

Summarizing the main events deduced from the Llanada Jurassic sequence, we have (1) the formation of new oceanic crust in a tectonic setting where calcareous and radiolarian ooze were accumulating; (2) the submarine deposition of airborne tephra (chiefly pumice and fine vitric ash) and tuffaceous radiolarite; (3) continued volcaniclastic sedimentation with the accumulation of progressively thicker, coarser-grained units on the sea floor, culminating in repeated large-scale andesitic submarine mudflows; (4) the intrusion of quartz keratophyre sills; (5) a brief period of pelagic radiolarian ooze deposition, which overlapped and outlasted the closing stage of volcanic mudflow emplacement; and (6) the deposition of terrigenous muds and distal turbidite sands, marking the encroachment of an orogenic clastic wedge out over the deep sea floor. These events spanned the interval from the Bathonian through the Tithonian stages of the Middle and Late Jurassic, as will be shown later.

8. Quinto Creek Ophiolite Remnant, Stanislaus County

The Quinto Creek ophiolite fragment crops out in a belt about 5 km long on the north canyon wall of Quinto Creek, northeastern Diablo Range about 13 km north of Pacheco Pass. The brief description of Bailey and others (1970), who first recognized the igneous sequence here as an oceanic crustal remnant, provides the only modern account. Hopson and Pessagno subsequently visited the area briefly, made a reconnaissance map (Hopson, unpublished), and collected material for petrographic study and age determinations.

The igneous sequence is thin and dismembered by faulting (Fig. 14-4, 8). The exposed remnants include submarine lavas (ca. 250 m thick), sheeted dike complex (>200 m), gabbro (50 to 100 m ?), and peridotite (>300 m). The peridotite is faulted against metagraywackes and melange of the Franciscan Complex, which lies structurally beneath the ophiolite. Above the ophiolite, resting with depositional contact upon the volcanic member, is about 20 to 30 m of thin-bedded tuffaceous radiolarian chert and aphanitic crystal-vitric tuff. Conformably overlying the chert-tuff unit (correlative with the Lotta Creek Tuff Member of the Del Puerto Canyon area) is up to 100 m of dark shale and terrigenous siltstone, overlain in turn by a thick unit of cobble conglomerate. The shale is Tithonian to Valanginian in age (Schilling, 1962) and represents the base of the Great Valley Sequence. The conglomerate is Cenomanian in age and is the local basal member of the Panoche Group (Schilling, 1962).

The volcanic member of the ophiolite consists chiefly of pillow lavas and smaller amounts of massive lava. Nonvesicular chilled rims on the pillows suggest a deep-sea origin, and the presence of interpillow limestone (at two widely separated localities) also points to an oceanic setting. Olivine-phyric basalt and some pl-cpx microphyric basalt are the major rock types. Clinopyroxene is fresh in these rocks, but plagioclase is pseudomorphed by albite and the olivine by carbonate. A large lenticular mass of keratophyre crops out at Hawk Rock; this composite body includes polymictic lithic tuff and metasomatically albitized pl-microphyric lava.

The sheeted dike complex is poorly exposed but can be seen to intrude the basal pillow lavas at one locality (Woodchopper Gulch). The dikes dip vertically and strike N10–60E, approximately perpendicular to the strike of the volcanic unit and its overlying tuffaceous chert beds. The dike rocks are chiefly keratophyre, spilite, microdiorite, and diabase. Also part of the sheeted complex are two large dikes (7 and 10 m wide) of plagiogranite porphyry, exposed along the road. Dikes cutting hornblende gabbro south of Woodchopper Gulch may represent the basal part of the sheeted complex. If so, the sheeted zone is remarkably thin (only about 200 m).

Gabbro is very poorly exposed and limited in extent. Uralitic clinopyroxene gabbro and anorthositic gabbro (plagioclase cumulates with poikilitic cpx) crop out on the hillside southwest of Hawk Rock, where they may be faulted against the sheeted complex. Quartz gabbro, leucogabbro, hornblende gabbro, and hornblendite are reported by Bailey and others (1970, loc. 6) from this area.

Peridotite, exposed farther west in the Bald Eagle Mine area, is the structurally lowest member of the ophiolite. The peridotite contact trends obliquely across the gabbro and volcanic units, suggesting that the peridotite forms a separate fault slice. Specimens collected by us are remarkably fresh harzburgite with a coarsely granular tectonitic fabric. Some dunite is also reported here (Bailey and others, 1970).

Events recorded at the Quinto Creek locality include (1) the formation of an igneous ophiolitic sequence; (2) deposition of thin-bedded tuffaceous radiolarian ooze and volcanic ash; (3) accumulation of dark shales and terrigenous siltstone during Tithonian to Valanginian time; (4) deposition of marine cobble conglomerate during the Cenomanian, following a period of nondeposition or erosion that spanned much of the Early Cretaceous; (5) fault emplacement of the igneous sequence against the Franciscan. The igneous submarine sequence formed where only calcareous ooze was accumulating, probably in an open-ocean setting. The subsequent succession of sedimentary deposits records pelagic sedimentation near a region of active explosive volcanism, followed by the encroachment of terrigenous clastics.

9. Del Puerto Canyon Ophiolite Remnant, Stanislaus County

The Del Puerto (or Red Mountain) ophiolite in the northern Diablo Range is one of the more complete and also most thoroughly studied of the oceanic crustal remnants at the base of the Great Valley Sequence (Bodenlos, 1950; Maddock, 1964; Himmelberg and Coleman, 1968; Saad, 1969; Lanphere, 1971; Bailey and Blake, 1974; Evarts, 1977). Evarts (1977) has provided an exceptionally comprehensive account of the field relationships and petrology of this mass, and the present brief summary draws chiefly on his work.

The ophiolite here is structurally dismembered and occurs in four fault-bounded slices that include (1) a peridotite member, consisting of harzburgite tectonite overlain by partially serpentinized massive dunite and lesser wehrlite; (2) a lower plutonic member, consisting of feldspathic peridotite overlain by pyroxene gabbronorite; (3) an upper plutonic member composed of pyroxene gabbro (partly uralitic) and hornblende gabbro, overlain by hornblende quartz diorite; and (4) a volcanic member, with a basal sill complex that grades upward into spilitic and keratophyric lavas, breccias, and dikes and sills that decrease upward (Fig. 14-7, 9). The total stratigraphic thickness of these four fault-bounded slices is approximately 5 km, including 3.5 to 4 km of igneous rocks above the harzburgite (Evarts, 1977, Fig. 3). The volcanic rocks are conformably overlain by tuffaceous radiolarian chert and bedded tuffs up to 400 m thick. This unit, called the Lotta Creek Tuff by Maddock (1964) and the basal member of the Great Valley Sequence by Evarts (1977), grades abruptly upward into nontuffaceous mudstones and terrigenous turbidite sandstones of the Great Valley Sequence proper. These latter strata, with latest Jurassic (Tithonian) megafossils near their base (Maddock, 1964), reach a total thickness of nearly 10 km in the Del Puerto Canyon area (Bishop, 1970). The ophiolite remnant is tectonically underlain by jadeitic metagraywacke-chert melange with a pervasively sheared argillaceous matrix—a part of the Franciscan Complex that comprises the core of the Diablo Range. The harzburgite-Franciscan contact, marked by sheared antigorite serpentinite, may represent a rare remnant of the Coast Range thrust that originally brought the ophiolite and the Franciscan Complex into contact (Evarts, 1977). The other, high-angle faults that juxtapose Franciscan against higher members of the ophiolite and the Great Valley Sequence are later and probably related to late Cenozoic post-thrusting diapiric uplift of the Franciscan core of the Diablo Range (Raymond, 1973b).

The lower part of the peridotite member consists of partially serpentinized alpine-type harzburgite with small scattered tabular to irregular bodies of dunite. The harzburgite has a weak foliation and a granular tectonitic microfabric that points to syntectonic recrystallization under high temperature, low strain-rate conditions. This fabric and the refractory composition of the harzburgite leads Evarts (1977) to conclude that it represents a remnant of upper mantle that was depleted by partial melting and extraction of basaltic melt. The dunite-wehrlite unit that overlies the harzburgite also shows microstructural evidence of plastic deformation of olivine, but it contains distinctive cusp-shaped clinopyroxenes that evidently crystallized from entrapped interstitial melt. Local podiform chromitites within these rocks show cumulus textures. Thus, Evarts interprets the dunite-wehrlite unit as ol ± sp ± cpx cumulates that underwent solid-state penetrative flow along with the underlying mantle harzburgite, prior to complete solidification of the entrapped postcumulus melt.

The combined plutonic members (western and eastern fault blocks) preserve at least 1 km of peridotite and gabbro cumulates, overlain by about 200 m of noncumulus hornblende quartz diorites (Evarts, 1977, Fig. 3). The peridotites are chiefly ol + sp ± cpx (± rare opx) cumulates with postcumulus plagioclase and pyroxene. The overlying gabbros are pl + cpx + opx cumulates without olivine. Hornblende and titanomagnetite are postcumulus phases, and the hornblende partly replaces pyroxenes. The settling order deduced from the cumulus sequence is ol + sp, ol + sp + cpx ± opx, pl + cpx + opx (± mag?). The cumulates grade up into hb-cpx gabbros with interstitial late-magmatic green hornblende, and into noncumulus hornblende quartz diorites with banded and schlieren structure thought to be due to flowage. Abundant small dikes of diabase, microdiorite, and plagioclase cut the plutonic sequence. The cumulus dunite, peridotites, gabbros and noncumulus diorite represent the fractionation products of basaltic magma that crystallized and differentiated in place in a shallow magma chamber. The crosscutting dikes may represent residual melts that were filter-pressed from the cumulates.

The exceptionally thick (2½ km) volcanic member consists of submarine lava flows, breccias, and dike (and sill) rocks that grade downward into an intrusive sill complex (Evarts, 1977). Massive lavas exceed pillowed flows and breccias bulk large in the sequence, leading Evarts to conclude that the volcanics were erupted into shallow water. The chilled outer rims of the pillows, however, are nearly nonvesicular (amygdules are microscopic and very sparse), a hallmark of pillows formed in very deep water (Moore, 1970; Moore and Schilling, 1973). This feature, plus the local occurrence of minor interpillow limestone and clast-supported nature of the breccias, suggests to us that the volcanic sequence represents deep-sea lavas with associated pillow and talus breccias.

The volcanic rocks are spilites, keratophyres, and quartz keratophyre, the later rocks occurring mainly as late-stage sills. The spilites are derived from aphyric and pl-cpx microphyric basalts, but phenocrystic and microphenocrystic olivine (now pseudomorphed by calcite or smectite), including some with tiny chromian-spinel inclusions (Evarts, 1977, and personal observation), occur in some flows. The keratophyres are regarded as originally andesitic by Evarts (1977), but as metasomatically altered (albitized) oceanic basalts by Hopson (see Hopson and Frano, 1977). The quartz keratophyres with phenocrystic β-quartz clearly formed from siliceous melts, probably the same late-stage derivative magma that solidified to form plagiogranite in the underlying plutonic sequence (Evarts, 1977). It is noteworthy that quartz keratophyric sills intrude up into the lower part of the overlying tuffaceous sedimentary strata, showing that the closing stage of ophiolite magmatism overlapped with the volcaniclastic sedimentation (Evarts, 1977), as it did in the Llanada ophiolite.

The volcanic rocks are altered to greenschist facies assemblages in the lower part of this section, and to a lower-grade albite-pumpellyite-green phyllosilicate ± prehnite assemblage in the uppermost 1 km (Evarts, 1977). Plagioclase is pervasively altered to albite throughout.

The entire igneous sequence, including the siliceous plagiogranites and quartz keratophyres, are regarded as the fractionation products of low-K tholeiitic magma (Evarts, 1977).

The volcanic member of the ophiolite is overlain by up to 400 m of interbedded tuffaceous radiolarian chert, MnO-stained jaspery chert, flinty vitric tuff, and silty to sandy crystal-lithic and crystal tuff ("volcaniclastic sandstone"). These beds, called the Lotta Creek Tuff Member by Maddock (1964), were deposited directly upon the ophiolite lavas, but later faulting has modified this contact in most places (Evarts, 1977). The section is well exposed along the ridge near the head of Falls Creek, where an upward coarsening sequence is revealed. Here the lower 70 m is chiefly white to tan porcelaneous tuff beds and gray tuffaceous radiolarian chert, with a thin olivine basalt flow 3 m above the base. Thin interbeds of silty tuff appear above 70 m and sandy tuff beds above 90 m. Interbedded sandy tuff, cherty tuff, and dark gray tuffaceous radiolarian chert then extend up to about the 300 m level, where the tuffaceous sequence passes conformably upward into dark mudstones (with terrigenous sandy interbeds) of Knoxville type. The sandy tuff beds of the Lotta Creek member are moderately well sorted, clast-supported crystal and crystal-lithic tuffs. Rare graded beds and current structures suggest turbidity current deposition, but the more common well-sorted nongraded beds probably reflect the periodic settling of crystals from massive infalls of airborne tephra. The tuffaceous radiolarite interbeds record the slow accumulation of pelagic radiolarian ooze and suspended fine ash. The absence of limy interbeds suggests deposition below the CCD. Lithic components of the sandy tuffs are andesitic and dacitic, and the crystal fraction consists chiefly of euhedral plagioclase, hornblende, pyroxenes (fresh augite and altered relicts of hypersthene), and magnetite. These are the common crystal components in andesitic and dacitic tephras from modern volcanoes in the Cascade Mountains (Kittleman, 1973; Mullineaux and others, 1975) and other calc-alkaline volcanic chains.

Thus, the formation of oceanic crust (ophiolite sequence) in this area was followed (and slightly overlapped) by deep-sea pelagic sedimentation interspersed with periodic influx of tephra from a calc-alkaline volcanic source. This terminated in the latest Jurassic with the encroachment of voluminous terrigenous muds and distal turbidites ("Knoxville sedimentation").

10. Lone Tree Creek Ophiolite Remnant, San Joaquin County

Slices up to 300 m thick of altered volcanic rock, gabbro, and serpentinized ultramafic rocks crop out 13 km north of Del Puerto Creek in the upper drainages of Lone Tree and Hospital creeks (Lone Tree 7½′ quadrangle), northeastern Diablo Range (Raymond, 1970, 1973b). These rocks, representing dismembered remnants of the Coast Range ophiolite, are juxtaposed against metamorphosed Franciscan melange terrane along the high-angle Tesla-Ortigalita fault system, and are depositionally overlain by Upper Jurassic cherty-tuffaceous strata (Lotta Creek Tuff) and Cretaceous (Albian to Maestrichtian) terrigenous clastic strata of the Great Valley Sequence (Raymond, 1970, 1973b) (Fig. 14-4, 10).

The ophiolitic volcanic rocks are described as keratophyre, quartz keratophyre, and minor basalt by Raymond (1970). Marshall Maddock, who guided us in a brief visit to the area, shows that part of the volcanic sequence includes sheeted intrusive complex, which contains the more siliceous rocks. For example, a remnant of sheeted complex exposed along upper Lone Tree Creek consists of diabasic and keratophyric dikes, hosted locally by plagiogranite porphyry. The sheeted complex here appears to have a sill-like, rather than a dikelike, orientation according to Maddock (personal communication, 1975). No one has studied the gabbroic and ultramafic rocks in the Lone Tree-Hospital Creek area and few details are known.

The lower part of the sedimentary succession that overlies the ophiolite is exposed along the ridgecrest between Lone Tree and Hospital Creeks (sec. 33, T4S, R5E). Here

Upper Jurassic thin-bedded cherty and tuffaceous strata more than 90 m thick (Lotta Creek Formation of Raymond, 1970) lie concordantly upon spilitic and keratophyric volcanic rocks, and are disconformably overlain by dark shales of Albian age, which represent the base of the Great Valley Sequence in this area. The lower 60 m consists mainly of thin-bedded, tuffaceous radiolarian chert, finely laminated radiolarian-bearing vitric tuff and aphanitic crystal-vitric tuff, and subordinate thin interbeds of silt- to sand-sized crystal tuff. Very fine pumice-lapilli occur in some layers. From 67 to 92 m above the base of the sequence, the fine tuffaceous layers are interbedded with silt-sized volcanic arenites and quartz-bearing silty graywackes, some in thin (centimeter scale) graded beds. This appears to represent a gradation upward into the terrigenous Great Valley strata. The dark brownish-gray Albian shales begin at 114 m above a 20-m nonexposed interval that probably contains the Tithonian-Albian disconformity.

Thus, igneous oceanic crust in this area first received thin accumulations of radiolarian ooze mixed with very fine (dust to silt-sized) volcanic ash, probably representing airborne andesitic or dacitic tephra. A distant arc-type volcanic source seems indicated.

11. Cedar Mountain Ophiolite Remnant, Alameda County

This small, dismembered ophiolite remnant lies 15 km northwest of the Del Puerto (Red Mountain) ophiolite, at the northern end of the Diablo Range. The following brief description is abstracted from Bauder and Liou (1979); we have not personally visited this area nor studied its rocks.

A northwest-trending mass of serpentinized harzburgite more than 5 km long comprises about 70% of the ophiolite; the remainder includes dunite, pyroxenite, gabbro, diorite, and plagiogranite. Harzburgite is locally separated from overlying hornblende gabbro by sheared dunite and pyroxenite. The gabbros, diorites, and plagiogranite are faulted against underlying harzburgite along the western side of the ophiolite. Felsic dikelets locally cut more mafic rocks, as might be expected of late-magmatic differentiates. No remnants of volcanic rocks or sheeted complex belonging to this ophiolite are recognized.

The ophiolite tectonically overlies a coherent Franciscan unit of greenstone and chert along much of its length, and it abuts these same rocks along a high-angle fault on its western side. Overlying the ophiolite but apparently separated from it by fault contacts are Great Valley sandstones and shales. Locally the sandstones contain megafossils of probable Valangian and Cenomanian age. The first sediments to be deposited on top of the ophiolite are not found here.

Sonoma-Lake County Region

Small remnants of dismembered ophiolite lie at the base of Great Valley Sequence outliers amid Franciscan terrane in the northern Coast Ranges south of Clear Lake (Healdsburg, Mount St. Helena, Harbin Springs, Geyser Peak localities). The ophiolite sequences here are incomplete and structurally thinned by faulting, but the igneous rock assemblages are similar to those in the central and southern Coast Ranges. A new element, not seen farther south, is the occurrence of epiclastic diabasic-basaltic breccias, which commonly separate the igneous ophiolitic sequences from overlying sedimentary strata at the base of the Great Valley Sequence. These breccias also extend east to the Lake Berryessa area and north to Paskenta. Thus, they are a regional feature and mark a significant difference in the tectonic history of the northern group of ophiolites from those farther south.

It should be noted that major right-lateral strike-slip faulting has moved some of the Sonoma-Lake County ophiolite remnants from original positions that lay farther to the southeast (McLaughlin, personal communication, 1975).

12. Healdsburg Ophiolite Remnants, Sonoma County

Basaltic greenstone-diabase-gabbro-serpentinized peridotite sequences mapped by Gealy (1951) in the Healdsburg quadrangle as part of the Franciscan Group were first recognized as ophiolite remnants at the base of the Great Valley Sequence by Bailey and others (1970, loc. 4). The larger ophiolite remnant occurs at Bradford Mountain, 3 to 15 km northwest of Healdsburg, and a smaller, strongly faulted remnant crops out at Porter Creek, 6 miles south of Healdsburg (Gealy, 1951, Plate 1). The main elements of an ophiolite assemblage appear to be collectively represented in these areas (Fig. 14-5, 12), but poor exposures plus dismemberment of the sequences by faulting render this one of the less satisfactory areas for detailed study.

The gabbroic and ultramafic rocks in the Bradford Mountain and Porter Creek areas were interpreted by Gealy (1951) as thick sills intruded into the top of the Franciscan Group. The lack of any intrusive contact relations, however, and the upward progression above the Franciscan from ultramafic rocks to gabbros to diabasic dikes to volcanics accords better with the concept that the igneous sequence represents an oceanic crustal remnant that has been faulted over the Franciscan (Bailey and others, 1970).

The ultramafic rocks, up to about 250 m thick, are massive to strongly sheared serpentinite chiefly derived from peridotite. Harzburgite was the main protolith, judging from the brief description by Gealy (1951, p. 16). Pyroxenite occurs at the base of the gabbro in the remnants at Bradford Mountain (Dorman Creek) and Porter Creek. The gabbro and diabase (which includes dike complex) are up to 600 m thick at Bradford Mountain. Here olivine gabbro comprises the lower part of the gabbroic section (near Dorman Creek), whereas augite gabbro (± small amounts of orthopyroxene) occurs above (Gealy, 1951, p. 14). At Falls Creek (northwest end of Bradford Mountain remnant) the "gabbro unit" (Gealy, 1951, Plate 1) consists of massive hb-cpx diorite, hb plagiogranite and granophyric albite granite, plus abundant dikes of diabase and microdiorite. This corresponds to the uppermost part of the ophiolitic plutonic sequence and perhaps a transition into dike or sill complex. Elsewhere in the area diabasic dike complex is included in Gealy's overlying greenstone map unit.

Volcanic rocks up to 300 m thick overlie the gabbroic and dike sequences in the

Bradford Mountain and Porter Creek areas (Gealy, 1951, Plate 1). These are chiefly spilitic and keratophyric lavas with small amounts of breccia. Pillow structure is recognized at only one locality (Gealy, 1951), but poor exposure of this unit makes it impossible to estimate the proportion of pillowed to massive lava originally present. A small suite of the lavas studied in thin section by us suggests that they were chiefly aphyric basalts and microphyric (pl-ol, pl-cpx) basalts, now extensively altered to albite, chlorite, epidote, sphene, calcite, and locally pumpellyite. At Falls Creek a keratophyre unit with tiny microphenocrysts of albite and quartz appears to be the extrusive equivalent of plagiogranite, which occurs in this section.

Dark-gray mudstone, siltstone, and minor sandstone, representing the base of the Great Valley Sequence (Knoxville Formation) overlies the ophiolitic volcanic rocks in the Bradford Mountain and Porter Creek areas. Megafossil assemblages reported by Gealy (1951) from just north of Bradford Mountain include *Buchia* species that range from Tithonian to Valanginian (latest Jurassic to Early Cretaceous). These strata are disconformably overlain by Upper Cretaceous beds. The top of the ophiolite is faulted against the Knoxville mudstone at Falls Creek, and the tuffaceous radiolarian chert expected between them is missing. Poor exposures of tuffaceous chert crop out at the base (?) of the Knoxville shales near Grape Creek, but the underlying volcanics are not exposed. Gealy (1951) reports radiolarian shales (tuffs?) from the small Knoxville remnant that caps the ophiolite at Porter Creek, but their stratigraphic level is uncertain. Thus, the nature of the first sediments to be laid down upon the igneous oceanic crust here (Healdsburg remnants) remains to be clearly determined.

13. Mount St. Helena Ophiolite Remnant, Lake and Sonoma Counties

Small, fault-bounded remnants of the Coast Range ophiolite crop out along the northwest margin of the Pliocene Mount St. Helena volcanic field, on the Lake County-Sonoma County line. The igneous complex here was mapped and first called an ophiolite by Bezore (1969, 1971), who was also one of the first to recognize that ophiolite in the Coast Ranges represents remnants of Jurassic oceanic crust. The area was subsequently remapped by McLaughlin (1975) as part of his regional mapping of the Geysers geothermal area. Our columnar section (Fig. 14-5, 13) is based on McLaughlin's map with modifications resulting from our own work in the north half of section 20. The area is poorly exposed and some contacts uncertain owing to deep weathering and heavy brush cover.

Bezore (1969) divided the igneous ophiolitic complex into four stratigraphic units: (1) basal peridotite 2000 ft thick, comprising serpentinized harzburgite with dikes of orthopyroxenite and olivine gabbro, (2) dunite serpentinite 1200 ft thick with dikes of olivine and hornblende gabbro, (3) massive olivine-free gabbro 500 ft thick, capped by (4) 1000 ft of diabase breccia. Our observations in the same area, however, lead to a somewhat different reconstruction of the ophiolite stratigraphy (Fig. 14-5, 13).

The thick basal unit is serpentinite derived mainly from harzburgite, as indicated

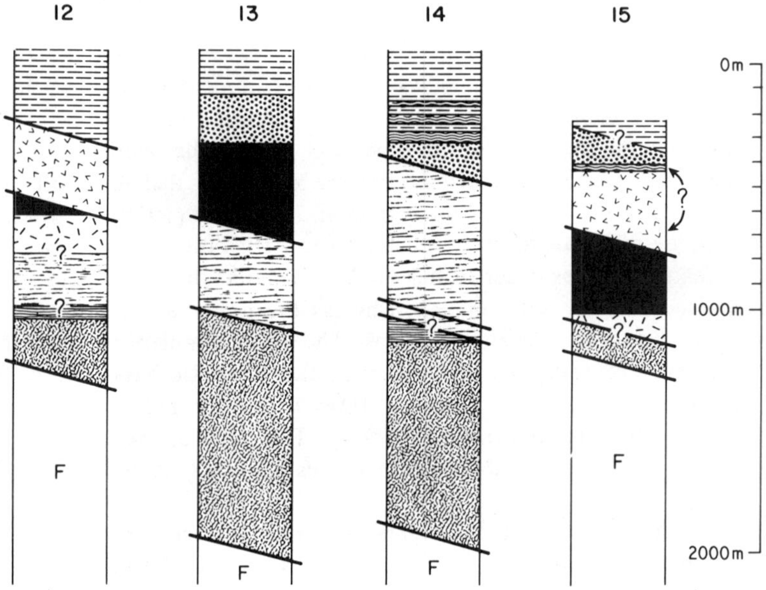

Fig. 14-5. Columnar sections showing details of the ophiolite remnants in the Sonoma-Lake County region, at localities 12 to 15 shown in Fig. 14-1. See Fig. 14-2 for explanation.

by Bezore. It consists chiefly of strongly sheared and slickensided chrysotile-lizardite serpentinite that encloses massive serpentinized harzburgite remnants that still preserve an original high-temperature tectonic fabric. Sparse dikes of pyroxene and uralite microgabbro cut through the serpentinite. The margins of the thicker dikes are altered to rodingite, and the dikes less than a meter thick are converted almost entirely to rodingite. Thin dikelike layers of serpentinized orthopyroxenite occur sparsely.

Pyroxene gabbro is juxtaposed against serpentinized peridotite on the county-line ridge north of Tanbark Canyon and in three smaller areas on the north flank of this ridge in sections 16 and 17 (McLaughlin, 1975). The gabbro-serpentinite contact is abrupt where clearly exposed in a road cut. The gabbro-serpentinite contact is probably a fault everywhere in this area, because (1) it is abrupt rather than gradational, and (2) it juxtaposes cumulus gabbro against serpentinized harzburgite without intervening remnants of olivine and clinopyroxene cumulates that normally separate gabbro from harzburgite in unbroken sequences of the Coast Range ophiolite. Bezore's transitional "mixed zone" of serpentinized dunite with abundant gabbro dikes was not verified by our own investigation of the same area.

The gabbro is best exposed on the south-facing slope due north of the Trout Farm (section 20). Here it is ol-cpx gabbro and cpx gabbro with a well-defined planar lamination and cumulate texture. Plagioclase-pyroxene ratio layering is well developed locally, including millimeter- to centimeter-scale layers of pure plagioclase (anorthosite). These gabbroic rocks are clearly cumulates. The gabbros in the three smaller northern areas (secs. 16, 17) appear to be similar. Apparently missing from the Mount St. Helena area are the distinctive noncumulus hornblende gabbro and diorite that typically mark the upper part of the gabbro sequence in more complete sections of the Coast Range ophiolite.

Bordering the main gabbro unit (section 20) on the southeast is a belt of fine-grained mafic rocks mapped as diabase breccia by Bezore, diabase-basalt by McLaughlin (1975), and diabase breccia by McLaughlin and Pessagno (1978). Bezore (1971) points out that diabase dikes predominate in the lower part of the unit, whereas diabase breccia with only scattered dikes occur higher up. We find that the lower 350 m of the diabase unit is sheeted dike or sill complex, whereas the upper 200 m is polymictic diabase breccias that lack any crosscutting dikes. The dike complex consists of fine-grained uralitic microgabbro (diabase), microdiorite, and aphanitic basaltic rocks; silicic dikes are absent or rare. Dike rocks comprise 100% of the sheeted complex; septa of earlier gabbro, lava, or breccia appear to be lacking. The dikes dip vertically and trend NNE, approximately parallel to the gabbro contact. This suggests that the dike unit may actually be a sill complex.

The polymictic diabase breccia consists of angular fragments and blocks of different kinds of diabase, microdiorite, dark aphanitic volcanic or dike rock, tightly packed and held together by red-weathering microbreccia of the same composition and by iron oxide. No gabbro or ultramafic fragments are observed. Thus, the clast assemblage in the breccia was quite likely derived from the adjacent dike complex. The clast angularity and complete lack of size sorting or bedding argue against any prolonged transport, and the lack of silty-sandy matrix suggests that no fine-grained sedimentary detritus was available. We interpret this breccia as a submarine talus deposit.

Lava and radiolarian chert are missing from this ophiolite remnant. Bezore (1971, p. 23) describes a depositonal contact of diabase conglomerate (reworked breccia) and mudstone, representing the base of the Great Valley Sequence, upon the diabase breccia (top of the ophiolite) along Mill Stream (sec. 20). Radiolarians in calcareous mudstone approximately 76 m stratigraphically above the diabase breccia are of early Tithonian age (McLaughlin and Pessagno, 1978).

Thus, the small ophiolite remnant at Mount St. Helena is strongly dismembered, with faults bounding most of the litholigic units. Some elements of a complete ophiolite sequence are missing, because they were faulted out. Cumulate gabbro is faulted against serpentinized harzburgite, eliminating pyroxenite and dunite cumulates that normally occur in between. A sheeted dike or sill complex is faulted directly against cumulate gabbro and serpentinite, cutting out intervening noncumulus hornblende gabbro and diorite. Finally, the entire volcanic section is missing, so that diabase talus breccia, overlain by basal Great Valley Sequence sediments, rests on the truncated dike complex.

Bezore (1971) suggests that subaerial erosion removed large parts of the ophiolite sequence, including the volcanic unit, prior to the beginning of deposition of the Great Valley Sequence. Alternatively, we believe that the ophiolite was tectonically dismembered and the volcanic unit removed locally by faulting on the ocean floor in pre-middle Tithonian time. The diabase breccias support this latter interpretation: they evidently represent thick piles of talus breccia that accumulated at the foot of submarine escarpments, which exposed sheeted dike complex. This suggests that large-scale faulting had locally exposed deeper layers of oceanic crust, prior to the beginning of deposition of mudstones and distal turbidites of the basal Great Valley Sequence.

14. Harbin Springs Ophiolite Remnant, Lake County

The ophiolite here lies only 10 km north of the Mount St. Helena remnant and resembles it closely, although the two are separated by the northwest-trending Collayami fault system of possible large strike-slip displacement. Geological mapping that includes the Harbin Springs area has recently been done by Swe and Dickinson (1970) and by Goff and McLaughlin (1976). McLaughlin was the first to recognize the existence of an ophiolite sequence here and to map its stratigraphy (McLaughlin, 1976; Goff and McLaughlin, 1976; McLaughlin and Pessagno, 1978). The brief account here is from McLaughlin's work, supplemented by our own field observations and thin section study of rocks collected by us on a field trip that he led.

McLaughlin distinguishes three ophiolite members: an upper diabase breccia unit, a middle gabbro unit, and a lower ultramafic unit, whose combined stratigraphic thickness is about 2 km (McLaughlin and Pessagno, 1978, Fig. 4) (Fig. 14-5, 14). The diabase breccia is overlain by tuffaceous strata and mudstones representing the base of the Great Valley Sequence, and the basal serpentinite is in fault contact with structurally underlying metagraywacke and melange units of the Franciscan Complex. This fault is regarded as part of the Coast Range thrust (McLaughlin and Pessagno, 1978, Fig. 3).

The basal ultramafic unit is pervasively shared serpentinite derived from harzburgite and dunite. The serpentinite adjacent to the base of the gabbro on Harbin Springs ridge is chiefly after dunite of probable cumulus origin. This is suggested by the mesh-textured serpentine, which appears to pseudomorph an adcumulus olivine fabric with peri-euhedral accessory chromite, and by thin chromatite layers. However, large bastites that pseudomorph enstatite megacrysts (xenocrysts?) occur very sparsely in this rock, and they greatly increase in abundance about 100 m below the gabbro contact. This suggests a gradual transition from cumulus dunite with relict refractory enstatite downward into harzburgite tectonite, but the matter deserves more study. Mafic dikes up to 3 m wide cut the serpentinite locally. These were microgabbros now altered mainly to fibrous amphibole, clinozoisite, and prehnite (rodingite).

Up to 600 m of gabbro occur above the serpentinite, across a fault contact (Goff and McLaughlin, 1976; McLaughlin and Pessagno, 1978, Fig. 4). Available samples suggest that cpx gabbro and ol-cpx gabbro are most abundant, and that these are of cumulus origin. Igneous planar lamination and adcumulus textures are well developed, and ratio layering is locally present. Cpx-pl and ol-cpx-pl are the cumulus phases. Sparse clots of fibrous amphibole in the cpx gabbro may pseudomorph orthopyroxene, but no actual opx relics are preserved in our specimens.

Dike rocks are locally abundant within the gabbro, and these increase in abundance upward, according to Goff and McLaughlin (1976). Uralitic diabase, hornblende microdiorite, and fine-grained plagiogranite are the common types. Bronzite-augite gabbro dike rock with abundant interstitial magnetite occurs near the base of the exposed gabbro section. All these dikes cut gabbro and are not part of a sheeted complex.

Missing from the Harbin Springs sequence are ol-cpx cumulates (clinopyroxenites, wehrlites) that normally underlie cumulate gabbro in the Coast Range ophiolite; presumably these are faulted out. Also apparently missing are the noncumulus amphibole-rich gabbro and diorite that normally overlie the gabbro cumulates, and also a sheeted dike or sill complex. These missing units are also probably faulted out. Thus, cumulus gabbro now lies directly against diabase breccia along a fault contact (Goff and McLaughlin, 1976).

Diabase breccia at least 100 m thick is the next highest stratigraphic unit exposed. The breccia is polymictic, consisting of angular clasts of a wide variety of diabasic and microdioritic dike rocks, diverse basaltic and spilitic volcanic rocks, and rare fragments of red chert. The breccia fragments are size unsorted, and no bedding can be recognized at outcrop scale. Many of the clasts are in the size range of gravel, but blocks up to 1 m are common. Progressively smaller clasts fill in between the larger ones, and all fragments are cemented by iron oxide. An epiclastic origin is evident.

More than 300 m of fine-grained clastic sedimentary strata of Late Jurassic and Early Cretaceous (Tithonian-Valangiian) age, representing the base of the Great Valley Sequence, lie above the diabase breccias in this area (Goff and McLaughlin, 1976; McLaughlin and Pessagno, 1978). Resting directly upon the breeccia is approximately 50 m of fine-grained to aphanitic bedded tuff. Above this are three more tuffaceous intervals approximately 45, 30, and 9 m thick, each separated by dark-green terrigenous mudstone intervals approximately 12, 18, and 21 m thick. Thus, the tuff intervals become thinner and the mudstone intervals thicker going up in the section. A monotonous sequence of mudstones with sparse thin interbeds of siltstone and sandstone overlies the highest tuff unit. Rare, thin beds of basalt-diabase breccia with clasts up to gravel size occur in the mudstone beneath the upper tuff unit, and some of the dark sandstone interbeds ('basaltic sandstones") are actually basalt-diabase microbreccias. These local "basaltic" beds seem best interpreted as local debris aprons that accumulated adjacent to submarine fault scarps, which exposed the breccias or upper part of the ophiolite.

The tuff beds are mostly flinty aphanitic rocks composed of altered glass shards and dust. Tiny, sparse microlites of plagioclase, plus minor pyroxene and rare quartz, are also present. Some beds also contain tiny pumice lapilli, replaced by smectite. The flinty vitric tuffs locally alternate with the pumiceous tuffs on a scale of 1 to 5 m. The massive internal character of some of these beds suggests rapid accumulation in still water, probably from massive infalls of airborne tephra. Some of the tuffaceous strata, however, show abundant features indicative of current action and sorting, such as graded bedding, small-scale cross lamination, and the alternation of well-winnowed sand-sized crystal tuffs with aphanitic tuffs on a scale of millimeters and centimeters. Such features are most abundant in the basal tuff unit.

Thus, the Harbin Springs ophiolite is strongly dismembered by faulting, and the strata above it record (1) massive accumulations of submarine fault-scarp breccia from the upper part of the ophiolite; (2) subsequent voluminous deposition of very fine tephra from a distant volcanic source; and (3) the encroachment of a submarine fan of terrigenous muds and fine sands, which interfingered with and then overwhelmed the tuffaceous deposits. High relief on the sea floor, presumably submarine fault-scarp topography, was a local source of ophiolitic detritus in the basal Great Valley Sequence muds. A final critical point is the occurrence of rare fragments of red chert in the diabase-basalt breccias. These could only have come from interpillow radiolarian cherts within the original volcanic sequence, which suggests that the ophiolite formed in a deep-sea environment.

15. Geyser Peak Ophiolite Remnant, Sonoma County

Serpentinized peridotite, gabbro, a diabasic sill complex, and basaltic lavas form an ophiolitic sequence that underlies Geyser Peak and Black Mountain, 20 km northwest of Mount St. Helena. The mapping and description of this remnant is mainly the work of McLaughlin (1974; McLaughlin and Pessagno, 1978), who guided us in a brief visit to the upper part of the section. The ophiolite remnant, slightly over 1 km thick, is separated from the underlying Franciscan Complex by the Coast Range thrust, and it is depositionally overlain by a small remnant of diabase breccia and Great Valley Sequence sedimentary rocks (Fig. 14-5, 15).

As yet no detailed description is available for this ophiolite. Sheared and serpentinized peridotite and dunite approximately 100 m thick at the base of the sequence are overlain by a thinner zone of uralitized gabbro (McLaughlin and Pessagno, 1978, Figs. 3 and 4). A diabasic sill complex, possibly 200 to 300 m thick, crops out above the gabbro on the higher parts of Geyser Peak. Abutting the diabasic complex on the east, across a steep fault, is a volcanic unit approximately 200 to 300 m thick, composed chiefly of basaltic pillow lavas and pillow breccias (McLaughlin and Pessagno, 1978). Locally capping the basalt unit is a small remnant of diabase-basalt breccia, and above this a thin mudstone remnant of possible Early Cretaceous (?) age, representing the base of the Great Valley Sequence (McLaughlin and Pessagno, 1978; McLaughlin, 1974).

An important additional unit is thin-bedded vitric tuff and tuffaceous radiolarian chert on the order of 10 m thick that lies within or upon the volcanics. These strata are poorly exposed along the fire trail on the east ridge of Black Mountain, near its summit. McLaughlin contends that these tuffaceous-cherty strata are interbedded *within* the lower part of the basaltic sequence because of their map position (personal communication, 1975; McLaughlin and Pessagno, 1978). One of us (CAH), however, argues that these beds, which crop out along the ridge crest in brushy, poorly exposed terrane, could just as well be resting on *top* of the volcanics, as they do nearly everywhere else in the Coast Range ophiolite remnants. The significance of the matter lies in the dating of the ophiolite, for the cherts contain well-preserved zone 1 radiolaria (see Biostratigraphy section).

16. Fir Creek Ophiolite Remnant, Napa County

The following ophiolitic sequence, overlying Franciscan metagraywackes and blueschist-bearing melange, is described by Moiseyev (1970) at the base of a synclinal infold of Great Valley Sequence west of Lake Berryessa: (1) a lower unit of serpentinite, separated from Franciscan rocks by a subvertical fault; (2) gabbro, interlayered with leucogabbro, pyroxenite, and olivine pyroxenite near its contact with the underlying serpentinite, and grading upward into diabase of decreasing grain size; (3) an upper unit of basalt, overlain by a breccia with mafic igneous and siltstone fragments cemented by mudstone. Graded beds within the breccia unit suggest a sedimentary origin. Conformably overlying the mafic breccia are shale and siltstone representing the base of the Great Valley Sequence. The ophiolitic sequence appears to be relatively complete, but the combined thickness of gabbro, diabase, and basalt is only about 300 to 400 m maximum, and they are cut out by faulting along strike (Moiseyev, 1970, Figs. 5, 6). Thus the ophiolite remnant here is probably incomplete and structurally thinned by faulting. We have not visited the area personally and cannot supply further details.

Sacramento Valley Ophiolite Belt

A continuous belt of serpentinite separates the Franciscan Complex from the Great Valley Sequence along the mountain front that borders the west side of the Sacramento Valley (Fig. 14-1). This belt, which includes the Wilbur Springs, Stonyford, Thomes Creek, Paskenta, and South Fork of Elder Creek localities, represents the Coast Range ophiolite, but an orderly ophiolite sequence is missing except at two local places (Wilbur Springs and South Fork of Elder Creek). Two reasons are: (1) pre-middle Tithonian disruption of the oceanic crust-mantle sequence converted it into serpentinite-matrix melange along the northern part of this belt, and (2) Cenozoic (?) high-angle faulting has juxtaposed the basal serpentinized peridotite member of the ophiolite directly against the Great Valley Sequence, cutting out the gabbroic and volcanic members in most places. This latter effect is best displayed at Elk Creek, where only 30 m of sheared serpentinite separates Franciscan (South Fork Mountain Schist) from Great Valley Sequence Tithonian mudstone.

17. Wilbur Springs Ophiolite Remnant, Colusa County

This dismembered, poorly exposed ophiolite remnant occurs in the northern Coast Ranges between Clear Lake and Sacramento Valley. Brief descriptions of the ophiolite sequence are provided by Taliaferro (1943b), Moiseyev (1966, 1968, 1970), and Bailey and others (1970, loc. 3). One of us (EAP) visited the area briefly for the collection of fossils, but we have done no geologic or petrographic work here.

This ophiolite sequence is represented by serpentinite, gabbro, diabase, and basaltic pillow lava with intercalated red chert. An anticlinal fold of serpentinite, forming the basal member of the ophiolite, is cored (underlain) by Franciscan Complex "composed chiefly of slates, phyllites, and sheared graywackes which show no sign of continuity and exhibit boudinage effects" (Moiseyev, 1968). The basal serpentinite member is in large part pervasively sheared, but includes massive blocks with prominent bastite, probably indicating a harzburgite protolith. The limited occurrence of gabbroic rocks is described as follows by Moiseyev (1968): "a few boulders or slivers of gabbro, diabase, and greenstone are also present, especially near the upper contacts of the [serpentinite] intrusion." Subsequently, Moiseyev (1970) describes this part of the section as "a thick serpentinite unit containing large masses of gabbro and diabase, especially near its upper contact," overlain by "a unit of submarine basaltic flows containing gabbro and diabase interbeds [sills?] in the lower portion . . . ." The poor exposures in this area greatly hamper any reconstruction of the stratigraphic and structural relations of the gabbroic rocks.

The serpentinite and gabbros are locally overlain by basaltic lavas up to 600 m thick, associated with radiolarian chert and tuff (Taliaferro, 1943b; Moiseyev, 1968, 1970, Fig. 3). The lavas commonly have pillow structure, and sedimentary intercalations are scarce except for radiolarian chert (Bailey and others, 1970). Higher in the section the volcanic rocks are overlain by tuff, shale, and graywacke (Bailey and others, 1970), representing the base of a very thick section of Great Valley Sequence. Tithonian fossils are found in the lower 1500 to 2100 m of the Great Valley Sequence, and Early Cretaceous fossils are found above that level (Moiseyev, 1970, p. 1723 and Fig. 3).

A significant feature of the Wilbur Springs section is the presence of mappable beds of detrital (sedimentary) serpentinite in the lower part of the Great Valley Sequence. Moiseyev (1970, Fig. 3) mapped two such beds within the Tithonian part of the section and a distinctive third bed (Abbott Mine serpentinite) within the Lower Cretaceous. He concludes that the detrital serpentinites had their source in mobilized basement serpentinites that were locally extruded onto the deep sea floor in response to tectonic activity.

The main belt of serpentinized peridotite at the base of the ophiolite sequence was once regarded as an intrusion between the Franciscan and Great Valley assemblages (Moiseyev, 1968), but Moiseyev later concluded that the "non-detrital serpentinites and overlying mafic rocks represent the basement upon which the Great Valley sequence was laid." "The ultramafic-mafic assemblage may represent a remnant of oceanic crust or an ophiolite extrusion" (Moiseyev, 1970). Bailey and others (1970) also interpret the Wilbur Springs ophiolitic sequence as an oceanic crustal remnant. We concur, but conclude that the ophiolite here is strongly dismembered by faulting, so that much of the sequence is missing. Thus, the basal serpentinite is juxtaposed directly against the Tithonian shales and sandstones (Great Valley Sequence) in most parts of this district, and the mafic igneous rocks intervene only locally in the small area northwest of Wilbur Springs (Moiseyev, 1970, Fig. 3). Even here the ophiolite sequence, described here, seems thin and incomplete.

18. Stonyford Ophiolite Remnant, Colusa and Glenn Counties

A broad belt of serpentinized peridotite separates the Great Valley Sequence from the Franciscan terrane along the western side of the Sacramento Valley between Wilbur Springs and Paskenta (Fig. 14-1), a distance of 96 km. The ultramafic belt represents the Coast Range ophiolite, but higher members of the ophiolite sequence are cut out by faults that juxtapose the ultramafic directly against Tithonian Great Valley strata all along the belt, except locally near Stonyford. Here basaltic pillow lavas and radiolarian chert occur at the base of the Great Valley Sequence (Brown, 1964; Bailey and others, 1970; Raney, 1974). Similar basalts also form a thick (>1000 m) outlying mass that is faulted over the adjacent Franciscan terrane (Brown, 1964).

The Stonyford sequence, based on the mapping of Brown, Raney, and also our own field work, proceeds upward as follows: (1) serpentinized peridotite, faulted against Franciscan rocks (South Fork Mountain Schist) on the west (Bailey and others, 1970; Raney, 1974) and against basalt on the east (Brown, 1964); (2) a very thick pillow basalt unit (2 to 3 km, see Brown, 1964), with locally interstratified radiolarian chert; (3) several hundred meters of dark mudstone with interbeds of basaltic ("tuffaceous")

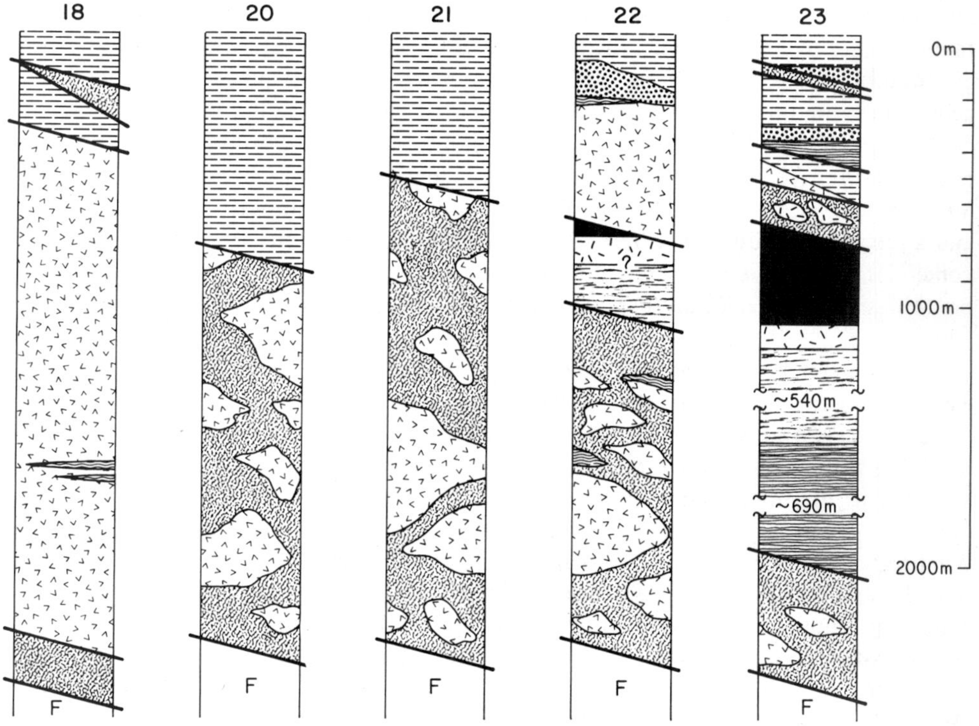

Fig. 14-6. Columnar sections showing details of the ophiolite remnants along the mountain front west of the Sacramento Valley, at localities 18 to 23 shown in Fig. 14-1. See Fig. 14-2 for explanation.

siltstone and sandstone; (4) a thin, fault-bounded slice of serpentinite (sections at Stony Creek and Little Stony Creek); and (5) orderly Great Valley Sequence, beginning with 1500 m of Tithonian flysch (terrigenous mudstone with silty-sandy interbeds) and including distinctive ophiolitic lithic wacke intervals (Fig. 14-6, 18).

The serpentinized peridotite unit is mainly foliated harzburgite, remarkably fresh in places and preserving a well-developed high-temperature granular tectonite fabric (Nicolas and others, 1979). Locally (at Black Diamond Ridge) this peridotite contains enough accessory clinopyroxene to plot in the edge of the lherzolite composition field.

The basaltic unit consists of abundant pillow lava, plus pillow breccia, massive lava, thin beds of basaltic aquagene tuff, and local diabase sills. Petrographically the lavas are mostly aphyric, olivine-poor clinopyroxene basalts and spilites with distinctive purple-tinged Ti-rich clinopyroxenes. These lavas are also distinctive chemically, with TiO_2 values of 1.9% to 3.0% (average 2.4%) compared to 0.5% to 1.8% (average 0.84%) TiO_2 for basaltic and spilitic lavas and diabase from all other remnants of the Coast Range ophiolite (Bailey and Blake, 1974, Table 4). Such high TiO_2 values and titaniferous clinopyroxenes are common to ocean-island basalts, but lie beyond the composition range of ocean-ridge basalts (Bryan and others, 1976).

Reddish-brown manganiferous (nontuffaceous) radiolarian ribbon chert is locally interbedded within the pillow lavas. Along Stony Creek near the campground (Sec. 35), chert units 15 and 30 m thick are interstratified within the basalt. The age of these cherts is early Tithonian, based on the presence of zone 2B radiolaria (see Biostratigraphy, this chapter).

Radiolarian cherts or tuffs are not found, however, above the lavas at the base of the Great Valley Sequence. The basalt here is overlain by middle Tithonian (*Buchia piochii* bearing) strata of two distinct types: (1) terrigenous flysch, composed of dark mudstone with thin silty to sandy interbeds, and (2) dark-colored, thin to massively bedded sandstone. The first type, called tuffaceous siltstone by Brown (1964), has a mixed clastic grain assemblage somewhat high in volcanogenic components (Stony Creek petrofacies of Dickinson and Rich, 1972) that is normal to the lower part of the Great Valley Sequence and not to be confused with water-laid tephra of Lotta Creek type. The second sediment type, called basaltic sandstone and basaltic tuff by Brown (1964), are polymictic lithic wackes composed of the following clast types (listed in order of decreasing abundance): (1) spilitic and keratophyric volcanics, (2) chert, including red manganiferous and tuffaceous radiolarites, (3) slaty argillite, metachert, and siltstone, (4) serpentinite, and (5) diabase, epidosite, microdiorite, and limestone. Clinopyroxene, albite, and chromian spinel (minor) are conspicuous single crystal grains in this sandstone, but quartz grains are rare. We conclude that these dark wacke sandstones are neither tuffaceous nor basaltic, but represent detritus mainly eroded from ophiolitic and possibly proto-Franciscan sources. Basalt clasts with purplish clinopyroxenes (Stonyford basalts) are missing or rare.

In summary, the Stonyford section differs significantly from the ophiolite remnants farther south. Some important conclusions are as follows:

1. The Stonyford volcanic pile is much thicker (2 to 3 km), younger (early

Tithonian), and of a different composition (TiO_2-rich basalt) than the other Coast Range ophiolite volcanic units. The Stonyford volcanics (including the St. John Mountain-Snow Mountain outliers) are probably the remnants of a large off-axis submarine volcano (here named the Stonyford seamount) that rose above the older (pre-Tithonian) oceanic crust (Coast Range ophiolite).

2. Harzburgite tectonite is the only remnant of pre-Tithonian oceanic crust and mantle preserved at Stonyford. Higher members of the ophiolite sequence, including ultramafic cumulates, gabbro, sheeted complex, low-TiO_2 ocean-floor lavas and their pelagic sediment cover are faulted out, as they are everywhere all along the ophiolitic belt west of Sacramento Valley. All that remains, faulted against the peridotite, is the Stonyford seamount, whose top is surrounded and covered by Tithonian clastic sediments of the Great Valley Sequence.

3. Great Valley sediments came from two sources: a terrigenous source, identified as the Sierran-Klamath terrane, mainly the Klamath Mountains (Ojakangas, 1968; Dickinson and Rich, 1972), and an ophiolitic source, probably nearby uplifted segments of the Jurassic sea floor.

19. Elk Creek Ophiolite Remnant, Glenn County

The ophiolite belt in Glenn County (Stonyford to Grindstone Creek) consists of a narrow (0 to 250 m) belt of serpentinite, bounded on both sides by subvertical faults (Jennings and Strand, 1960). A typical section is well exposed where this belt is crossed by the south Fork of Elk Creek. Here the "ophiolite" is represented by only 20 to 40 m of strongly sheared and pulverized serpentinite. The eastern contact is a vertical fault (Stony Creek fault) that truncated the bedding of steeply dipping Tithonian mudstones belonging to the lower Great Valley Sequence. The western contact is another vertical fault that sets the serpentinite against dark slaty argillite and metagraywacke of the South Fork Mountain Schist (Franciscan Complex). This fault contact is sharp, lacks imbrication, and does not noticeably disturb the adjacent bedded metasedimentary sequence. The question of whether this fault represents a regional thrust (Coast Range thrust) with many tens of kilometers displacement (e.g., Bailey and others, 1970) or is a Cenozoic high-angle fault with only a few kilometers displacement (Raney, 1973, 1974; Maxwell, 1974) remains controversial.

Tehama County Ophiolite Melange Belt

The ophiolite belt greatly widens in northern Glenn County and Tehama County and fundamentally changes its character to serpentinite-matrix melange. The sections at Bennett Creek, Thomes Creek, and Paskenta, described next, all lie within this melange. Farther north, the South Fork of Elder Creek section preserves a stratigraphically intact ophiolite sequence within a large (2 by 6 km) native slab enclosed by the melange. This south-to-north progression along the ophiolite belt from serpentinite-matrix melange to melange that encloses intact ophiolite megaslabs is strikingly similar to relationships along the older Kings-Kaweah ophiolite belt in the southern Sierra Nevada Foothills (Saleeby, 1978a, b, 1979, and Chapter 6, this volume).

20. Bennett Creek Ophiolite Remnant, Tehama County

The ophiolite belt is well exposed along the Covelo road at Bennett Creek, 11 km southwest of Paskenta. The belt is 1.7 km wide at this point and composed entirely of ophiolitic melange. Steeply dipping faults bound the melange on both sides, separating it from Tithonian strata of the Great Valley Sequence on the east and from metagraywacke and phyllite of the South Fork Mountain Schist (Franciscan) on the west.

The ophiolite melange consists of tectonic blocks of pillow lava, massive lava (greenstone), rare blocks of radiolarian chert, and remnants (native blocks) of serpentinized periodotite enclosed in a matrix of serpentinite. The blocks of volcanic rock range from less than 1 m up to 500 m in diameter, with blocks 100 to 200 m being commonplace. The larger blocks form resistant knobs and hillocks amid the hummocky melange terrane. Volcanic blocks comprise approximately one-half to two-thirds of the melange surface area, but the proportion of serpentinite is locally much higher.

The melange matrix consists of intensely sheared, slickensided sepentinite. Shear surfaces dip steeply and trend subparallel to the ophiolite belt, but in detail these surfaces conform to the enclosed tectonic blocks. Commonly, the serpentinite cuts up through some of the larger blocks in dikelike sheets; here the serpentinite has clearly intruded or been plastically squeezed in along faults or shear zones. Locally, two or more subparallel serpentinite intrusions isolate slices of volcanic rock between them, and these slices become progressively disrupted into boudinlike blocks. Finally, zones of serpentinite 5 to 10 m wide contain trains of disrupted volcanic blocks or only isolated blocks. It is evident that a once-intact volcanic formation has been broken by faulting and shearing, and then pervasively invaded and internally disrupted by tectonically mobile serpentinite, creating a melange.

The serpentinite is composed of chrysotile-lizardite serpentine and brucite, indicative of formation at low temperature (Coleman, 1971b). It originated mainly from harzburgite, for serpentinized harzburgite is the ubiquitous type of native melange block.

Pillow lava is abundant among the volcanic melange blocks. One large block contains 7 m of thin-bedded radiolarian chert resting in depositional contact upon pillows. Petrographically, the lavas are spilites derived chiefly from aphyric and plagioclase-microphyric basalt. Primary igneous textures, including variolitic quench textures, are well preserved, but alteration has produced mainly secondary assemblages of albite-chlorite-epidote-sphene-calcite ± quartz. Clinopyroxene is the main surviving volcanic phase. These volcanics match the volcanic member of the Coast Range ophiolite at the other localities, where disruption and melanging are described later.

The eastern contact of the ophiolitic melange belt against the Great Valley Sequence is covered by landslide debris, but the western contact against the South Fork Mountain Schist is cleanly exposed in two roadcuts at Bennett Creek. This contact is a steeply dipping fault with blocky serpentinite against bedded metagraywacke and phyllite. Only a few centimeters of gouge lie along the fault plane, and the rocks to either side are not notably more deformed than they are much farther away. It is easier to accept this as a high-angle fault that has brought up the South Fork Mountain Schist from a deeper level than as a regional thrust fault (Coast Range thrust) involving tens of kilometers of displacement (Irwin, 1964; Bailey and others, 1970).

21. Thomes Creek Ophiolite Remnant, Tehama County

Serpentinite-matrix melange extends continuously to the canyon of Thomes Creek, 3 km north of Bennett Creek, and comprises 100% of the ophiolite belt. The character of the melange remains similar here, but exposures are not as good owing to landsliding from the canyon walls. The fault contacts against the Great Valley Sequence on the east and South Fork Mountain Schist on the west, however, are cleanly exposed. The eastern fault contact (Stony Creek fault) truncates bedding in the Tithonian mudstone-sandstone flysch; within 50 m of this contact the mudstone loses its stratification and becomes somewhat slaty. The western fault contact (Coast Range thrust ?) is perfectly exposed in the stream-polished walls of a gorge. Here the fault dips steeply to the northeast and sets schistose metagraywacke sharply against massive basaltic greenstone (part of a large melange block). The fault plane cuts across the metagraywacke schistosity and clearly postdates it. The fault plane contains only about 1 cm of sheared gouge filled with calcite, and the rocks on either side seem little disturbed. It is difficult to imagine that this is the "Coast Range thrust," which purportedly has brought the ophiolite and overlying Great Valley Sequence into juxtaposition with Franciscan sediments from a distant, perhaps different, depositional realm (Bailey and others, 1970; Blake and Jones, 1974).

The ophiolitic melange contains mainly tectonic blocks of pillowed and massive basaltic greenstone (spilite), commonly up to several hundred meters long, and much smaller native blocks of serpentinized harzburgite, which grade into sheared serpentinite that forms the melange matrix. One immense volcanic block (slab), through which Thomes Creek is incised to form The Gorge, measures 1200 by 500 by >200 m. Basaltic breccia and radiolarian chert are very rare as melange blocks, and only one small gabbroic block was found. Another rare but notable feature is an exotic block of gneissic garnet amphibolite more than 30 m in diameter, which occurs within the melange near its eastern contact. The mineral assemblage of this amphibolite is mainly green hornblende and garnet, plus minor diopside, saussurite pseudomorphing plagioclase, sphene, and secondary albite. Whether this garnet amphibolite was brought up along the Stony Creek fault or was transported within the invasive serpentinite is hard to say.

22. Paskenta Ophiolite Remnant, Tehama County

The ophiolite belt broadens to more than 5 km wide along the mountain front west and northwest of Paskenta, where it displays particularly important stratigraphic and structural features. Recent studies of the ophiolite and adjacent rocks in the Paskenta area include the work of Jones (1975; Jones and others, 1969), Bailey and others (1970), Bailey and Blake (1974), Fritz (1974, 1975), and Maxwell (1974). The following description, however, comes mainly from our own reconnaissance mapping (CAH).

On the west, the ophiolite belt is emplaced against South Fork Mountain Schist (Franciscan terrane) along a near-vertical fault (Coast Range thrust?) marked by sheared serpentinite. On the east, the ophiolite is bounded by the Great Valley Sequence; lower Tithonian mudstones at the base of this sequence rest in depositional contact upon the ophiolite. Within the belt there are four main structural units, which are, from west to east (bottom to top), (1) serpentinite-matrix melange, continuous from the melanges at Bennett and Thomes creeks, (2) a stratigraphically intact slab of ophiolite that includes ultramafic and gabbroic cumulates, high-level gabbro and diorite, and a sheeted dike or sill complex, (3) serpentinized peridotite that locally includes plutonic melange blocks, and (4) pillowed and massive basaltic lava, overlain by polymictic ophiolitic breccias. Basal mudstones of the Great Valley Sequence overlie the ophiolitic breccias with a depositional contact. Two sections across this belt will be described: (1) a section west of Paskenta along Toomes Camp road, which exposes mainly units 1 and 4, and (2) a section farther north (South Fork of Elder Creek), which crosses mainly units 2 and 3, plus fault slices of unit 4 that are imbricated into the base of the Great Valley Sequence.

West of Paskenta the lowest (western) structural unit of the ophiolite consists of serpentinite and serpentinite-matrix melange containing tectonic blocks of basaltic lava plus minor radiolarian chert. The highway crossing this unit passes through alternating zones of barren serpentinite, serpentinite filled with volcanic melange blocks, and zones of intact volcanic rock up to 1 km across, which are actually parts of very large melange blocks. For example, the volcanic zone at Round Mountain belongs to a melange block 2.5 km long and 1 km wide.

The volcanic rocks are chiefly pillowed and massive lavas derived from aphyric and pl-cpx microphyric basalt, now altered to spilite. The volcanic rock is capped by ribbon cherts in several of the melange blocks. Radiolaria from one of these cherts indicates a pre-Tithonian Late Jurassic age (see Biostratigraphy, this chaper). Melanging clearly postdates the chert deposition.

The serpentinite zones consist of pervasively sheared and slickensided serpentinite, containing phaccoidal blocks of massive bastite-bearing serpentinite derived from harzburgite. Shear surfaces are subvertical and trend northward, parallel to the ophiolite belt. The zones of barren serpentinite grade laterally into serpentinite-matrix melange.

The transition from the large intact masses of volcanic rock into serpentinite-matrix melange shows relationships that reveal the melanging mechanism. Beginning in the volcanic unit, massive, pillowed lavas are locally broken by narrow fault zones that are occupied by sheared, pulverized serpentinite. As the adjacent melange is approached, the serpentinitic fault slices within the volcanics become wider and more abundant, until they gradually isolate masses of volcanic rock between them. This grades into melange as the proportion of serpentinite (host) to volcanic rock (isolated masses and blocks) increases. Finally, melange grades into barren serpentinite as the volcanic blocks become fewer and then disappear altogether. Thus, a complete transition occurs, from volcanic rock through serpentinite-matrix melange to serpentinite. This melanging evidently results from (1) faulting and tectonic disruption of the volcanic unit, (2) the intrusion of incompetent, plastically flowing serpentinite into the fault zones and disrupted brittle masses of volcanic rock, and (3) further disruption and mixing by penetrative shearing and plastic flowage. These melanges (Maxwell, 1974) are the product of strong tectonism, similar to the serpentinite-matrix ophiolitic melanges of the southwestern Sierra Nevada (Saleeby, 1979).

East of the melange unit the highway passes through a narrow slice of gabbro cut by dikes, the extreme southern tip of the intact ophiolite slab of Elder Creek. This gabbro and overlying sheeted dikes widen to the north, at Eagle Peak.

Structurally overlying the gabbro and sheeted dike zones and truncating them obliquely is a belt of volcanic rocks up to 600 m thick, capped by ophiolitic breccias that are locally more than 200 m thick. The basal contact of the volcanic unit, which cuts across melange, gabbro, and sheeted dikes, is a major fault zone, possibly an extension of the Stony Creek fault. Only massive and thoroughly sheared lavas are exposed along the highway section, but pillow lavas are abundant east of Eagle Peak and elsewhere along the belt. These are spilitic lavas derived from aphyric and microphyric (pl, cpx, ol) basalts, like those in the melange unit.

Red radiolarian chert overlies the basalt at Crowfoot Point, intervening locally between the basalt and overlying breccia. Some beds within the chert are tuffaceous, with microlites and tiny pumiceous and lithic fragments among the radiolaria. Evidently, infalls of tephra periodically interrupted the accumulation of radiolarian ooze.

Polymictic breccia approximately one hundred meters thick forms the top of the ophiolite at Crowfoot Point, just beneath basal strata of the Great Valley Sequence. The breccias consist of unsorted angular fragments of different kinds of basic volcanic rock, fine-grained dike rocks, gabbro, anorthosite, plagiogranite, and pyroxenite; however, dark volcanic rock fragments predominate. The fragments are mostly small, but blocks up to 1 m or more in diameter are not uncommon. The fragments form a tightly packed, clast-supported framework, with a matrix of microbreccia cemented by iron oxide. No bedding is apparent. This breccia is very similar to the ophiolitic breccias that overlie the ophiolite at the Sonoma-Lake County localities. It is interpreted as a submarine talus or scree deposit that developed adjacent to active fault scarps on the deep ocean floor.

The Upper Jurassic mudstone at the base of the Great Valley Sequence rests in depositional contact upon the breccias along the road at Crowfoot Point. For several tens of meters above this contact the mudstone contains blocks of the breccia, some of them 10 to 20 m in diameter. Also, several beds of dark ophiolitic microbreccia are interstratified with the mudstone. These relations suggest that the mudstone accumulated around submarine fault-scarp topography, which shed large blocks as well as finer clastic detritus into the muds and turbidite sands that were being deposited from an external source.

Strong, superimposed deformation further complicates these relations at the upper contact of the ophiolite. Near Crowfoot Point the polymictic breccia is intensely sheared and interspersed with gouge zones and with fault slices of mudstone (from above) and serpentinite (from below). Above the breccia (top of the ophiolite) the Upper Jurassic strata are steeply overturned and strongly deformed internally, and fault slivers from the ophiolite are imbricated into the mudstone.

In summary, the sequence of events recorded at Paskenta include (1) formation of igneous oceanic crust, (2) deposition of radiolarian chert with sporadic infalls of tephra, on top of the deep-sea basalt, (3) large-scale faulting and strong internal disruption of the oceanic crust, preserving intact crust locally but resulting in the widespread development of serpentinite-matrix melanges within the volcanic upper crust, and also forming ophiolitic talus breccias that spread out from submarine fault scarps, (4) deposition of terrigenous Great Valley strata, which filled in around fault-scarp topography but postdated most of the melanging, and (5) large-scale faulting, which emplaced the disrupted oceanic crust (ophiolite) and its terrigenous sedimentary cover (Great Valley Sequence) against metamorphosed Franciscan terrane (South Fork Mountain Schist). Step 1 occurred in the Middle Jurassic, steps 2 and 3 in the early Late Jurassic (Oxfordian-Kimmeridgian), and step 4 began toward the end of the Late Jurassic (early Tithonian) (see Isotopic Ages and Biostratigraphy sections, this chapter). Step 5 was probably a Cenozoic event, but the evidence is lacking at Paskenta.

23. South Fork of Elder Creek Ophiolite Remnant, Tehama County

The best preserved plutonic and sheeted sequence to be found among the ophiolite remnants in the northern Coast Ranges occurs in the toe of Riley Ridge, above where the South Fork of Elder Creek branches. Here it forms the middle part of the Elder Creek ophiolite section, which also has other important features. The full section consists of a lower structural unit of serpentinite, a fault-bounded middle unit consisting of dunite, wehrlite, clinopyroxenite, layered gabbro, diorite, and sheeted dike or sill complex, partly repeated by faulting, and an upper structural unit composed of tectonic blocks or slices of various plutonic rocks separated by faults and zones of sheared serpentinite. Above this upper unit, two additional thin fault slices of the ophiolite are imbricated into the lower part of the Great Valley Sequence, east of the main ophiolite belt. Both these slices preserve the original depositional contact where Tithonian mudstone rests upon polymictic breccias at the top of the ophiolite.

The ophiolite is bounded on the west by the Early Cretaceous South Fork Mountain Schist (phyllite, lawsonite metagraywacke, albite-epidote blueschist) along a straight vertical fault (Coast Range thrust of Bailey and others, 1970; Coast Range fault of Fritz, 1975), whose interpretation remains controversial.

The lowest (western) structural unit of the ophiolite is strongly sheared serpentinite derived mainly from harzburgite. This fault-bounded unit narrows and wedges out to the north.

The middle structural unit consists of stratiform ultramafic and gabbroic rocks that dip steeply to the east and are overlain by diorite and sheeted dike (or sill?) complex. The sequence from the base upward is serpentinized dunite and wehrlite (240 m), clinopyroxenite, olivine clinopyroxenite, and wehrlite (450 m), layered ol-cpx gabbro with minor opx-cpx gabbro, anorthositic gabbro, and clinopyroxenite (540 m), massive hb-cpx gabbro and diorite (0 to 150 m), and sheeted complex consisting of hundreds of dikes (sills?) of diabase, microdiorite, and granophyre ($>$400 m). The upper part of the sheeted complex is cut off by a fault that brings up more layered gabbro, diorite, and sheeted complex in a structurally higher block on the east. The ultramafic rocks and layered gabbros are igneous cumulates: they display adcumulus and mesocumulus textures, planar lamination, and locally well developed ratio layering, phase layering, and grain-size layering. The crystal settling order (crystallization order?) in the cumulus sequence is ol + sp (chromite), cpx + ol, pl + cpx + ol, opx + pl + cpx, and pl + cpx + ol. The upper, noncumulus gabbro and diorite crystallized plagioclase and clinopyroxene, followed by hornblende, magnetite, and quartz.

The upper structural unit begins with 450 m of strongly sheared serpentinite, containing native blocks of harzburgite and diabasic rodingite. Higher up section are poorly exposed plutonic breccia, gabbro, clinopyroxenite, and zones of sheared serpentinite; probably this comprises a serpentinite-matrix melange with plutonic melange blocks. This upper structural unit widens to the north, to where serpentinite comprises the entire width of the ophiolite belt (Bailey and Jones, 1973b).

A distinctive plagiogranite-rich boulder breccia occurs along the South Fork of Elder Creek as a fault slice (?) within the upper structural unit. In addition to plagiogranite, the clast assemblage of the breccia contains gabbro, diabase, and microdiorite. The boulder- to pebble-sized clasts are angular, unsorted, and held together by microbreccia of the same composition. This breccia is unmistakably sedimentary. It probably originated as a talus deposit at the foot of a sea-floor fault scarp that exposed an upper plutonic level of the oceanic crust.

The top of the ophiolite is preserved in two fault slices that are imbricated into the Great Valley Sequence, 200 and 600 m above its base. The lower slice, accurately described by Bailey and others (1970, p. C72), exposes basal mudstone of the Great Valley Sequence resting in depositional contact upon 30 m of ophiolitic breccia. The breccia is composed of unsorted angular clasts of mafic volcanic and dike rocks, diorite, gabbro, and pyroxenite, cemented by microbreccia and iron oxide. The breccia rests in depositional contact with a 70-m intact sequence of cumulus clinopyroxenite, wehrlite, and serpentinized dunite, faulted off at the base against mudstone. A gabbroic dike that intrudes the pyroxenite but is truncated by the overlying breccia has a K-Ar hornblende age of 151 ± 5 m.y. (Lanphere, 1971). Despite its structural displacement, this ophiolite slice preserves an important relationship, the depositional contacts between deep-seated clinopyroxenite, surficial ophiolitic breccia, and Tithonian terrigenous flysch. Evidently, ultramafic cumulates from deep within the oceanic crust were uplifted and directly exposed on the sea floor. Subsequently, they were covered, first by coarse fragmental debris coming from deep plutonic as well as shallow exposures of oceanic crust, and later by land-derived muds and tubidite sands. Large-scale faulting of the deep sea floor, before the onset of Great Valley Sequence sedimentation, is clearly indicated.

In summary, the following sequence of events is deduced from the South Fork of Elder Creek section: (1) formation of igneous oceanic crust and upper mantle; (2) disruption of the oceanic crust-mantle sequence by large-scale faulting, and also by melanging that involved the pervasive invasion of hydrated mantle material (serpentinite) into the plutonic and volcanic levels of the oceanic crust; (3) continued faulting, bringing non-matching segments of intact and melanged oceanic crust into juxtaposition; (4) deposition of terrigenous muds, distal turbidite sands, and gravels over the faulted and disrupted oceanic crust, beginning in the latest Jurassic; (5) large-scale faulting that brought the Great Valley Sequence and its deformed oceanic crustal basement into juxtaposition with the South Fork Mountain Schist, after the latter had undergone low-grade blueschist-facies metamorphism during the Early Cretaceous (Fritz, 1975; Lanphere and others, 1975).

Additional post-Early Cretaceous deformations include the high-angle faulting that offsets and locally imbricates the eastern side of the ophiolite belt (Stony Creek fault zone extended), and the northwest-trending sinistral shearing within the lower Great Valley Sequence (e.g., Paskenta fault zone).

The stratigraphic thickness of the original oceanic crustal sequence down to the "petrologic Moho," obtained by adding the Paskenta volcanic section to the Elder Creek plutonic and dike section, is more than 2.5 km (Fig. 14-6, 22,23). This represents a minimum thickness, since the base of the ultramafic cumulates, the top of the sheeted complex, and the base of the volcanics are each faulted out.

ISOTOPIC AGES OF THE COAST RANGE OPHIOLITE

Introduction

Geochronologic work on the Coast Range ophiolite has been rather limited. Undoubtedly, this has been due in large part to the mafic, ultramafic, and altered nature of most of the rocks in the ophiolite suite, which render them unsuitable for isotopic age determinations. However, dike rocks that cut various levels of the ophiolite and plagiogranite that occurs locally at the differentiated top of the plutonic member provide material suitable for dating. Specifically, hornblende gabbro dike rocks have been dated by K-Ar methods, and plagiogranites yield zircons that have been dated by U-Pb methods, as summarized next.

Previous Work

Lanphere (1971) reported the first isotopic ages for the California Coast Range ophiolite. He measured K-Ar ages of 160 ± 8 m.y. (for hornblende from a gabbro dike cutting cumulus dunite) and 158 ± 5 m.y. (for amphibole from an isolated block of feldspathic periodotite) at Del Puerto Canyon (loc. 9), and 151 ± 5 m. y. (for hornblende from a gabbro dike cutting clinopyroxenite cumulate) at the South Fork of Elder Creek (loc. 23). Deborah Fritz reports a K-Ar hornblende age of 163 ± 5 m.y. for hornblende gabbro within the ophiolite, also near the South Fork of Elder Creek (1975, Table 3 and Fig. 21). Three other K-Ar amphibole ages of 140 to 141 m.y. were obtained by McDowell and others (reported in Maxwell, 1974, and Fritz, 1974, 1975) for hornblende gabbro inclusions (dismembered dikes?) occurring in serpentinite at the base of the ophiolite remnants at Wilbur Springs (loc. 17) and in the Riley Ridge quadrangle north of the South Fork of Elder Creek.

Isotopic U-Pb ages of zircons from the Coast Range ophiolite at Point Sal (loc. 2) have been reported by Hopson and others (1973, 1975a, b) and at Santa Cruz Island (loc. 1) by Mattinson and Hill (1976). The ages from Point Sal and Santa Cruz Island (Willows Plutonic Complex) as originally reported were identical within limits of error at 162 m.y. Mattinson and Hill (1976) also report an age of about 141 m.y. for the Alamos pluton, a plagiogranite body of uncertain origin that intrudes the Willows Complex and overlying metavolcanic rocks.

Present Study

New U-Pb zircon ages are reported here from four additional Coast Range ophiolite localities: Cuesta Ridge (loc. 4), Llanada (loc. 7), Del Puerto Canyon (loc. 9), and Fall Creek near Healdsburg (loc. 12). In each case the zircons were separated from plagiogranite (or albitite) that we believe to be late-stage differentiates of the plutonic member of the ophiolite (see also Coleman and Peterman, 1975). The new zircon data are part of a continuing study of the geochronology of the Coast Range ophiolite by Mattinson.

The isotope data are presented in Table 14-1 and plotted in Fig. 14-7. For completeness, previously published data for Point Sal and Santa Cruz Island are included. Two of the Santa Cruz Island ages have been recalculated as a result of a tracer recalibration. They are slightly higher than, but still agree within the stated limits of uncertainty with, the preliminary ages reported in Mattinson and Hill (1976). One new determination for a fine-grained zircon fraction from the Willows Complex is also reported in Table 14-1.

The occurrence of the dated rocks in Table 14-1, briefly, is as follows. The older ages from Santa Cruz Island (loc. 1) come from a leucocratic phase (aplitic segregation) of the hornblende quartz diorite that composes part of the Willows Plutonic Complex. The younger Santa Cruz Island age comes from one of the plagiogranite dikes that intrudes Willows Complex, identified as apophyses from the Alamos plagiogranite pluton. The Point Sal (loc. 2) age comes from an albite granite segregation within a dioritic dike that occurs near the top of the cumulus gabbro, zone 2c. This dike is regarded as fractionated intercumulus liquid that was filterpressed and expelled from the gabbro cumulate. The Cuesta Ridge (loc. 4) age comes from a thick (ca. 10 m) plagiogranite sill that occurs within the sheeted sill complex at Stenner Creek. The Llanada (loc. 7) age comes from a

TABLE 14-1. Isotopic Data for Zircons from the Coast Range Ophiolite

Sample[a]		Concentrations (in ppm)[b] ^{238}U	$^{206*}Pb$	Isotopic Composition[c] $^{206}Pb/^{208}Pb$	$^{206}Pb/^{207}Pb$	$^{206}Pb/^{204}Pb$	Ages (in m.y.)[d] $^{206*}Pb/^{238}U$	$^{207*}Pb/^{235}U$	$^{207*}Pb/^{206*}Pb$
Healdsburg	c	650	14.40	1.797	15.35	930	163 ± 2	163 ± 2	166 ± 15
Red Mountain	c	302	6.41	8.589	19.29	5965	156 ± 2	157 ± 2	166 ± 10
Del Puerto	f	336	7.07	7.610	17.49	1768	155 ± 2	154 ± 2	142 ± 15
Llanada	c	61.5	1.362	4.139	14.00	674	163 ± 2	164 ± 2	180 ± 20
	f	59.1	1.318	3.830	13.60	617	164 ± 2	165 ± 2	186 ± 20
San Luis Obispo	b	734	15.21	–[e]	18.84	3881	152 ± 3	153 ± 3	162 ± 20
Point Sal	nm	624	13.52	9.010	19.22	5830	161 ± 2	162 ± 2	180 ± 10
	m	850	18.49	8.512	19.22	5822	160 ± 2	161 ± 2	172 ± 10
Santa Cruz Island W	c	140.9	3.16	7.726	18.33	3121	165 ± 2	167 ± 2	184 ± 10
	f	152.9	3.37	2.421	5.896	121.7	162 ± 2	161 ± 2	152 ± 60
Santa Cruz Island A	c	282	5.52	6.801	16.27	1296	144 ± 2	148 ± 2	201 ± 15

[a]c, coarse-grained fraction; f, fine-grained fraction; b, bulk sample; nm, nonmagnetic fraction; m, slightly magnetic fraction. W, Willows Complex; A, Alamos pluton.
[b]Zircons were analyzed by methods similar to those described by Krogh (1973). Radiogenic Pb indicated by asterisk.
[c]Observed isotopic compositions corrected for mass fractionation via replicate analyses of NBS reference Pb standards SRM 981, 982, and 983.
[d]Constants used in age calculations: $\lambda^{238}U = 1.5513 \times 10^{-10}$ yr^{-1}, $\lambda^{235}U = 9.8485 \times 10^{-10}$ yr^{-1} (Jaffey and others, 1971); ^{238}U–^{235}U (atom ratio) = 137.88.
[e]Isotopic composition recalculated from spiked run; no separate isotopic composition run.

hornblende-bearing albitite dikelet within the cumulus gabbro. The dikelet is regarded as the albitized derivative of fractionated intercumulus liquid that was filterpressed from the gabbro. The Del Puerto Canyon (loc. 9) zircon ages come from a meter-thick dike of plagiogranite porphyry (with quartz phenocrysts) that cuts across sheared, foliated hornblende diorite in the upper part of the plutonic member. This sharply crosscutting dike is undeformed and appears to postdate both the flow structure and subsequent shearing of the dioritic host rock. The Healdsburg (loc. 12) age comes from a large (ca. 10 m) dike or sill of plagiogranite that occurs wtihin a dike (or sill) complex at the top of the plutonic member.

With the exception of the data for the Alamos pluton (Santa Cruz Island), all the ages are concordant, or nearly so. Moreover, the ages from different fractions of zircon from the same sample are identical within the limits of uncertainty of the analyses. This is strong evidence that the ^{206}Pb-^{238}U ages are essentially magmatic emplacement ages. Thus the data indicate formation of the exposed part of the Coast Range ophiolite during a rather limited interval of time, corresponding to the Bathonian and Callovian stages of Middle Jurassic on the Van Hinte (1976a) time scale.

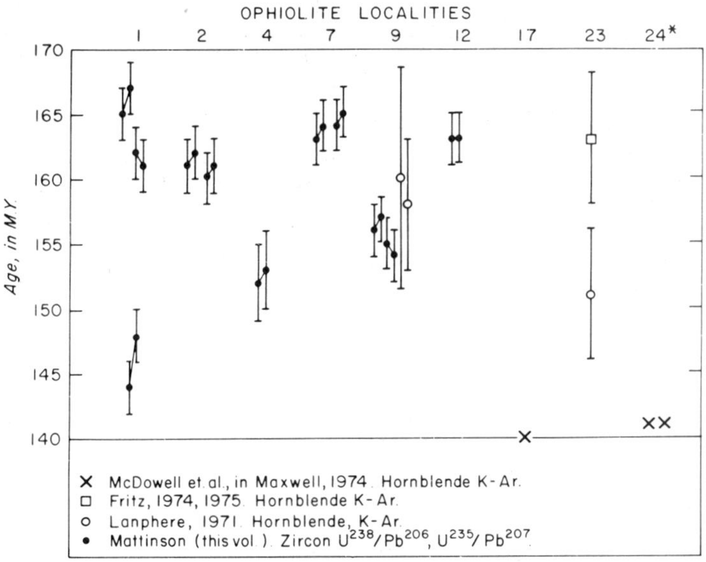

Fig. 14-7. Plot showing radiometric mineral age measurements for the Coast Range ophiolite. Numbered localities 1 to 23 are those shown in Fig. 14-1. Locality 24 lies just northwest of locality 23 in the Riley Ridge quadrangle.

The limited age range for the time of ophiolite formation is rather remarkable, considering the great length of the ophiolite belt (i.e., the samples listed in Table 14-1 span some 1300 km if offset along the San Andreas and related faults is restored). In particular, the ophiolite remnants at Healdsburg, Llanada, Point Sal, and Santa Cruz Island (Willows Complex), and also the older K-Ar age at South Fork of Elder Creek, fall within the very narrow interval of 161 to 165 ± 2 m.y. Point Sal may be 2 to 3 m.y. younger than the other three, but this cannot be demonstrated with the present levels of analytical uncertainty. The Del Puerto and Cuesta Ridge remnants are distinctly younger than Point Sal. Taken all together, we regard this array of ophiolite remnants as representing different age segments of once-continuous oceanic crust that spanned an interval of about 12 m.y.

The interpretation of the Del Puerto (loc. 9) ages is uncertain. The quartz porphyry dike with the concordant zircon ages appears to be somewhat younger than its dioritic host rock (see previous discussion), and it is younger also than the K-Ar ages (Fig. 14-7) from much deeper in the stratigraphic section. This dike might be related to the compositionally similar quartz keratophyre sills that invade the tuffaceous sedimentary cover of the Del Puerto ophiolite (Evarts, 1977). If so, the ophiolite itself could be older, perhaps nearer the age of the Llanada remnant, as expected on geologic grounds. More age dating here is needed.

The Alamos plagiogranite, which invades the ophiolite of Santa Cruz Island (loc. 1) at approximately 145 m.y. (Fig. 14-7), is evidently a later intrusion into older oceanic crust. Jones and others (1976) compare this rock with trondhjemitic plutons in the Sierran Foothills and Klamath Mountains, and infer that it represents island-arc magma. We find, however, that it bears little petrographic or chemical resemblance to the Sierran-Klamath trondhjemites (e.g., Compton, 1955; Davis and others, 1965; Hotz, 1971; Hietanen, 1973), but corresponds closely to the oceanic plagiogranites characteristic of ophiolites (Coleman and Peterman, 1975). The explanation of Mattinson and Hill (1976), that Alamos pluton represents a later increment of the oceanic crust, seems more attractive at present.

The 140 to 141 m.y. K-Ar amphibole ages from Wilbur Springs and Riley Ridge quadrangle (Fig. 14-1, locs. 17, 24) were obtained from gabbroic inclusions in serpentinite at the base of the ophiolite (Maxwell, 1974; Fritz, 1975). We have no firsthand knowledge about these samples, but gabbroic dikes and inclusions in Coast Range serpentinites are invariably altered (partly or wholly) to rodingite, with uralitic amphibole replacing pyroxene. This alteration accompanies the serpentinization of the ultramafic host rock (Coleman, 1967) and may be significantly later than the age of the ophiolite. There is strong evidence in the Paskenta-South Fork of Elder Creek area for extensive serpentinization and serpentinite-matrix melanging of the lower oceanic crust or upper mantle (see loc. 22 description) during the latest part of radiolarian zone 1 or earliest part of Subzone 2A time (i.e., the Kimmeridgian stage of the Late Jurassic; see Biostratigraphy section, loc. 22). This strong pulse of sea-floor serpentinization during the Kimmeridgian provides the likeliest explanation for the K-Ar ages at 140 to 141 m.y.

To summarize, isotopic age determinations indicate that ophiolite remnants throughout the Coast Range belt and extending into the southern California borderland (Santa Cruz Island) formed during a 12 m.y. span of time in the Middle Jurassic. Renewed magmatic activity (plagiogranite intrusion) is recorded at the southern end of the belt in the Late Jurassic (ca. 144 m.y.b.p.) and may correlate with the minor quartz keratophyre diking that occurred during the early Tithonian in the central part of the belt (see descriptions of localities 7 and 9). Additional geologic evidence for two separate stages of ocean crust formation comes from the Point Sal remnant (Hopson and Frano, 1977), but only one stage has been dated. Radiometric and geologic evidence for extensive serpentinization and tectonic disruption of oceanic crust at about 140 m.y. comes from the northern end of the Coast Range ophiolite belt.

Additional zircon work planned and in progress should help resolve some of the existing uncertainties and problems and contribute to further understanding of the exact mode of origin and evolution of the Coast Range ophiolite belt.

BIOSTRATIGRAPHY

The development of the hydrofluoric acid technique (Pessagno and Newport, 1972) has made it possible to extract matrix-free Radiolaria from radiolarian cherts. The use of this method together with the ongoing development of a detailed system of radiolarian zonation for the Mesozoic (Pessagno, 1976, 1977a, b; Pessagno and Blome, in progress) has opened the door to the study of complex Mesozoic orogens such as those of the California Coast Ranges.

The system of zonation proposed by Pessagno focused primarily on the lower Tithonian (Upper Jurassic) to upper Maestrichtian (Upper Cretaceous), stratal interval. This zonal scheme was developed through the study of the stratigraphic distribution of Radiolaria occurring in the Great Valley Sequence and in underlying pelagic strata (mostly radiolarian chert) associated with the Coast Range ophiolite. Radiolarian zonal units were cross-correlated with biostratigraphic data or zonal schemes supplied by other fossil groups (e.g., ammonites, inoceramids, planktonic foraminifera, Buchias). However, until recently, virtually no biostratigraphic control from the more extensively studied invertebrate fossil groups existed below the lower part of radiolarian zone 2 (Fig. 14-8).

Pessagno (1976, 1977a) believed that zones 1 and 0 were Kimmeridgian to early Tithonian in age. However, work now in progress in eastern Oregon (Suplee-Izee area) and in the Queen Charlotte Islands, B.C., indicates that the base of zone 1 corresponds approximately to the base of the Bajocian and the Middle Jurassic. In both of these areas our radiolarian assemblages have been cross-correlated with data supplied by the ammonites.

The base of zone 1 was originally defined to correspond to the biohorizon offered by the first occurrence of the Parvicingulidae Pessagno and *Parvicingula* Pessagno. In eastern Oregon, *Parvicingula* ranges down into the lower middle Bajocian; it probably ranges down to the base of the Bajocian. To date we have not observed *Parvicingula* in Pliensbachian samples from either eastern Oregon or the Queen Charlotte Islands, and we have no well-dated Toarcian samples.

In future reports, zone 1 will encompass strata of early Bajocian (Toarcian?) to Kimmeridgian age. It is apparent that even in the Bajocian to lower Callovian interval, zone 1 can be subdivided into at least five subzones. Studies of radiolarian faunas and associated megafossils from the upper Callovian to Kimmeridgian interval are still underway. It should be noted that zone 0 of the previous zonal scheme has been abandoned.

Tithonian, as used in this report, is equated with the Russian Volgian. The base of the Tithonian using this definition is defined by the first occurrence of *Gravesia.* By this definition, the *Buchia rugosa* and *B. mosquensis* zones of western North America are placed in the Tithonian; the *Buchia concentrica* zone would include all the Kimmeridgian and most of the upper Oxfordian.

With this background, we turn now to the radiolarian biostratigraphy of the cherty and tuffaceous pelagic and hemipelagic strata that rest upon or occur within the Coast Range ophiolite beneath the Great Valley Sequence.

Point Sal, Santa Barbara County (Ophiolite Locality 2)

At Point Sal, 23 m of green, greenish gray, and black tuffaceous radiolarian chert with very minor lenticular masses of light gray pelagic limestone rests with concordant depositional contact on basalt at the top of the ophiolite. The contact of the tuffaceous chert

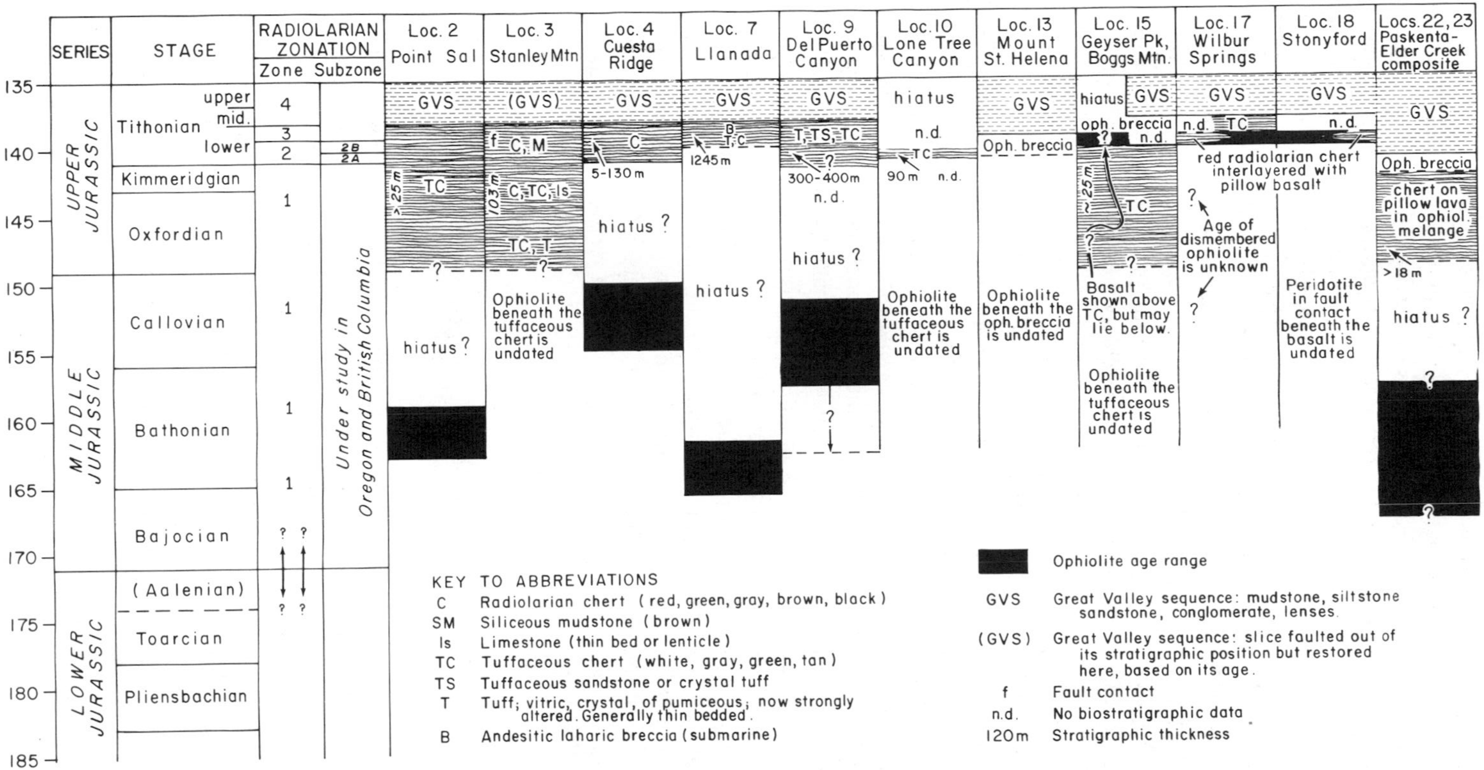

Fig. 14-8. Jurassic stratigraphic sequences plotted against the geologic time scale of Van Hinte, 1976a, and a revised zonation for the Radiolaria.

succession with the overlying Great Valley Sequence is covered at Point Sal, but this contact is conformable along Point Sal Ridge and in nearby Corallitos Canyon.

Radiolaria extracted from the cherts and pelagic limestone nodules indicate that the interval from (0.9 m to approximately 6 m above the contact with the basalt is assignable to the upper part of zone 1 (Oxfordian?-Kimmeridgian). Certain species such as *Hsuum obispoense* Pessagno, *Eucyrtidium* (?) *pyctum* Riedel and Sanfilippo, and *Acanthocircus variabilis* (Squinabol) have not been observed in strata of Callovian age in eastern Oregon. Previously, the lower part of this interval was assigned to zone 0 (Pessagno, 1977a), but this zone has now been abandoned and incorporated in the upper part of zone 1.

The interval from 10.6 to 11.5 m above the ophiolite contact is assignable to zone 2, subzone 2A (lower Tithonian). The boundary between zones 1 and 2 occurs somewhere between 6.4 and 10.3 m above the ophiolite. Strata from 13.5 to 15.8 m above the base of the section are assignable to zone 2, subzone 2B (lower Tithonian). Thus, the boundary between subzones 2A and 2B occurs in the interval between 11.8 and 13 m. Samples from 17.6 to 31 m above the contact are assignable to zone 3 (lower Tithonian). The boundary between zones 2 and 3 therefore occurs in the interval from 16 to 17 m. No well-preserved radiolaria were recovered from the upper 1.8 m of the chert unit.

Buchia piochii (Gabb) and *B. fischeriana* (d'Orbigny) have been noted at several localities at the base of the Great Valley Sequence in this area. Hence, it is certain that the base of the Great Valley Sequence is assignable to the middle or upper Tithonian. The *B. piochii* zone corresponds to all but the uppermost part of radiolarian zone 4.

Detailed faunal lists for the Point Sal section were presented in Pessagno (1977a). It should be noted that some of the species occurring in zone 2 (e.g., "*Triloncbe*" *ordinaria* Pessagno and *Tripocyclia jonesi* Pessagno) are longer ranging than indicated by Pessagno (1977a) and, in fact, range down as far as the lower Callovian in eastern Oregon. Nevertheless, a sufficient number of species (e.g., *Obesacapsula morroensis* Pessagno, *Mirifusus guadalupensis* Pessagno, *Hsuum cuestaense* Pessagno, and *Emiluvia hopsoni* Pessagno) make their first appearance at the base of zone 2 and, hence, make the base of zone 2 readily discernible.

Stanley Mountain (Alamo Canyon), San Luis Obispo County (Ophiolite Locality 3)

Preliminary studies of the radiolarian assemblage of this section were made by Pessagno (1977a). Subsequently, the succession was measured and collected in detail by Hopson and Pessagno. Detailed biostratigraphic analyses are currently being undertaken by Pessagno and Blome. Preliminary results are indicated below.

At Stanley Mountain, 103 m of dark gray, green, black, and reddish-brown tuffaceous chert with minor lenses and nodules of light gray pelagic limestone and two thick beds of tuff rest conformably (?) on weathered pillow basalt. Dark gray cherts predominate in the lower 60.6 m, whereas reddish-brown cherts and siliceous mudstones predominate above this horizon.

The interval from the base of the section up to 16 m contains spars, often poorly preserved Radiolaria characterized by *Eucytidium* (?) *ptyctum* Riedel and Sanfilippo, *Hsuum maxwelli* Pessagno, *Parvicingula* sp., and *Archaeodictyomitra rigida* Pessagno. This assemblage may be assignable to upper Zone 1 or to Zone 2 (Oxfordian? to lower Tithonian). From 17 to 60 m above the contact the radiolarian is assignable to Zone 2, Subzone 2A (lower Tithonian). A subzone 2B (lower Tithonian) fauna was recovered from 62 to 82 m. Zone 3 (lower Tithonian) Radiolaria occur from 83 to 92 m. A Zone 4 radiolarian assemblage (middle to upper Tithonian) was recovered at 97 m. The boundary between Zones 3 and 4 occurs between about 92 and 95 m. It should be noted that the base of Zone 4 is redefined to correspond to the first occurrence of *Obesacapsula rotunda* (Hinde); its top corresponds to the final occurrence of *Parvicingula* (?) *altissima* (Rüst).

Above the cherty pelagic strata, possibly in fault contact, are dark gray mudstone and graywacke 100 to 200 m thick with dark green tuff near the base (Brown, 1968b), overlain by Valanginian (and older?) mudstones, sandstones, and conglomerate lenses of the Jollo Formation (Hall and Corbato, 1967; Brown, 1968b). Tithonian fossils have not been found in the strata that occur above the cherts, but *Buchia piochii* (middle to upper Tithonian) occurs at several localities within a fault slice of dark mudstone with thin graywacke interbeds that is tectonically interposed between the ophiolite volcanics and the Franciscan Complex. Thus it seems likely that chert deposition ending in the early or middle Tithonian was succeeded by flysch deposition beginning in the middle Tithonian and extending into the Valanginian.

Cuesta Ridge, San Luis Obispo County (Ophiolite Locality 4)

At Cuesta Ridge, 5 to 130 m of dark gray, black, and greenish-gray, partly tuffaceous radiolarian cherts rests conformably upon pillow lavas at the top of the ophiolite (Page, 1972; Pessagno, 1977a). The chert sequence here was not sampled in great detail, but the faunal assemblage recovered is sufficient to bracket the age of the pelagic succession.

A reanalysis of the data presented by Pessagno (1977a) as well as a reexamination of the residues was undertaken to determine the age of the base of this section. Because of the presence of *Obesacapsula morroensis* Pessagno (not recorded in Pessagno, 1977a) and *Praeconocaryomma immodica* (Pessagno and Poisson, in press) (= P. magnimamma Rüst of Pessagno, 1977a) the base of this section is still assigned to zone 2, subzone 2A (Lower Tithonian). *Obesacapsula morroensis* makes its first appearance at the base of zone 2, subzone 2A, whereas *P. immodica* makes its final appearance at the top of subzone 2A. Several samples from the upper part of the chert succession are assignable to zone 3 (lower Tithonian).

Page (1972, p. 966) noted the presence of *Buchia piochii* (Gabb) in the overlying Great Valley Sequence in this area. The *Buchia piochii* zone (middle to upper Tithonian) is equivalent to the lower eight-tenths of radiolarian zone 4).

Llanada (Bitterwater Canyon), San Benito County (Ophiolite Locality 7)

At Bitterwater Canyon more than 500 m of basaltic pillow lavas comprise the upper part of the ophiolite. Interpillow pockets of nontuffaceous red chert, low in this section, contain Radiolaria too badly recrystallized to be age diagnostic. An exceptionally thick sequence (1245 m) of upward-coarsening volcaniclastic beds lies depositionally on top of the ophiolite pillow lavas and conformably underlies basal mudstones and sandstones of the Great Valley Sequence. The lower 485 m of this volcaniclastic unit consists of thin-bedded vitric tuff, coarser pumiceous and lithic tuffs, and very minor tuffaceous radiolarian chert. The pillow lava flow and massive quartz keratophyre still that occur within the lower part of this interval probably do not belong to the ophiolite but represent a later period of sea-floor volcanism and intrusion. The upper part of the volcaniclastic unit is an upward-coarsening sequence of bedded andesitic tuff-breccia, volcanic conglomerate, and massive units of bouldery andesitic laharic breccia approximately 760 m thick. Up to 3 m of pale green tuff and gray tuffaceous radiolarian chert conformably overlie these breccias and grade upward into basal terrigenous mudstone and sandstone of the Great Valley Sequence.

Biostratigraphic data from the section at best are meager. Greenish-gray to black cherty tuff resting on pillow lava at the upper contact of the ophiolite (i.e., the "waterfall" lava flow near the base of the tuffaceous sequence) contains a radiolarian assemblage assignable to zone 2, subzone 2B (lower Tithonian). Green tuffaceous chert directly beneath the Great Valley Sequence contains new species of *Parvicingula* which first appear at the base of Zone 4 at Stanley Mountain. The basal mudstone, limestone, and sandstone of the Great Valley Sequence contain poorly preserved calcified Radiolaria and *Buchia piochii* of middle to upper Tithonian age (Enos, 1963, 1965).

Del Puerto Canyon, Stanislaus County (Ophiolite Locality 9)

In the Del Puerto Canyon area, up to 400 m of silty to sandy crystal-lithic tuff, flinty vitric tuff, tuffaceous radiolarian chert, and MnO-stained jaspery chert (Lotta Creek Tuff Member of Maddock, 1964) overlie submarine lavas of the ophiolite (Maddock, 1964; Evarts, 1977). Many of the dark gray, aphanitic tuffaceous-cherty beds contain abundant radiolaria, but their internal structure is poorly preserved (many are casts) and we were unsuccessful in separating them from their tuffaceous matrix. Using thin sections, *Parvicingula* species that probably lie within the range of zones 1 to 5 (Bajocian to upper Valanginian) are recognized in tuffaceous chert from high in the Lotta Creek sequence. Another thin section, from strata at the top of the Lotta Creek Tuff sequence (head of Falls Creek), reveals *Parvicingula* species and *Mirifusus* (?) that probably fall within the range of zones 2 to 5 (lower Tithonian to upper Valanginian). The Lotta Creek strata are conformably overlain by mudstones of the lower Great Valley Sequence that contain *Aucella* (= *Buchia*) *piochii* of middle to upper Tithonian age (Maddock, 1964). This places a middle Tithonian upper limit on the age of the Lotta Creek Tuff.

These thin-section identifications of poorly preserved radiolaria have no precise value, but indicate that the Lotta Creek Tuff of the Del Puerto area is in the age range of radiolarian cherts and tuffs that overlie the ophiolite elsewhere in the Diablo and Coast Ranges (Fig. 14-8).

Hospital Canyon-Lone Tree Canyon, San Joaquin County (Ophiolite Locality 10)

In this area more than 90 m of dark gray to green tuffaceous chert and thin-bedded tuff (Lotta Creek Formation of Raymond, 1970) rest in depositional contact upon submarine lavas of the ophiolite and lie nonconformably beneath the Great Valley Sequence. This section was sampled in detail by Pessagno, Maddock, and Hopson. Abundant radiolaria are evident in thin section in most of the tuffaceous chert beds, but identifiable radiolaria could be separated from only two samples. Our data indicates that both these samples, from 4.8 and 10.9 m above the base, contain radiolaria assignable to Zone 2, (undifferentiated) (lower Tithonian). Great Valley strata (Adobe Flat Shale) that concordantly overlie the Lotta Creek Tuff and cherts in this section are upper Lower Cretaceous (Albian) (Raymond, 1970).

Mount St. Helena, Sonoma County (Ophiolite Locality 13)

No radiolarian chert overlies the ophiolite at Mount St. Helena. Instead, epiclastic ophiolitic breccias rest upon sheeted dike or sill complex, and mudstone flysch of the Great Valley Sequence overlies the breccia. Here, McLaughlin and Pessagno (1978, p. 720) found zone 3 (lower Tithonian) radiolaria in the lower part of the Great Valley strata, 76 m above the ophiolitic breccia.

Geyser Peak, Sonoma County (Ophiolite Locality 15)

Dark gray, black, and greenish-gray tuffaceous radiolarian ribbon chert and tuff approximately 10 to 20 m thick, cropping out on the brushy east ridge of Black Mountain, lie upon or within basalts of the Geyser Peak ophiolite remnant. R. J. McLaughlin concludes from his mapping of the area that these cherts are interbedded within the lower part of the basaltic sequence (McLaughlin and Pessagno, 1978, Fig. 4). However, the field relations in this poorly exposed, structurally complex area are not easy to interpret, and it is possible that the cherts *overlie* the basalts and are faulted into their present position. Modern studies at oceanic spreading centers find that the volcanic layer is built up rapidly, and thick intercalated layers of pelagic sediment (which accumulate very slowly) are not abundant. It therefore seems unlikely that the tuffaceous chert sequence at Geyser Peak, which is tens of meters thick, occurs *within* the basaltic section. Moreover, the tuffaceous cherts *overlie* the ophiolite basalts everywhere else in the Coast Ranges, although at Llanada and Del Puerto Canyon keratophyric lavas and sills believed to be younger than the ophiolite occur within the lower part of the tuff-chert sequence. In all cases where radiolarian chert is clearly seen to occur *within* the ophiolite volcanics, either as interbeds (Wilbur Springs, Stonyford) or interpillow pockets (Point Sal, Llanada), the chert is red, manganiferous, and nontuffaceous.

The Geyser Peak chert contains a radiolarian assemblage (Pessagno, 1977a) assignable to the upper part of zone 1 (Oxfordian?, Kimmeridgian) and to zone 2, subzone 2A (lower Tithonian). A small remnant of lower Great Valley strata, which locally overlies epiclastic ophiolitic breccias in the Geyser Peak area, contains Dinoflagellate fragments identified as Early Cretaceous (?) by W. R. Evitt (McLaughlin and Pessagno, 1978, locality 16 of Fig. 2 and Table 2). Fourteen kilometers farther north, in the Boggs Mountain area, basal Great Valley Sequence contains *Buchia piochii* of middle to upper Tithonian age (McLaughlin and Pessagno, 1978, locality 17 of Fig. 2 and Table 2). Swe and Dickinson (1970, Fig. 2) also assign a Tithonian age to the lowest part of the Great Valley Sequence in this area.

Wilbur Springs, Colusa County (Ophiolite Locality 17)

The ophiolite here is dismembered by faulting and poorly exposed. Well-preserved radiolaria were recovered by Pessagno (1977a) from red, manganiferous chert (nontuffaceous) interbedded with amygdaloidal basalt. Another chert mass, which occurs as an isolated outcrop apart from the basalt, consists of 14 m of black, dark gray, and green tuffaceous ribbon chert. The black chert grades downward into black tuffaceous mudstones with minor sandstone (possibly crystal tuff). The mudstones are in fault contact with gabbro (Pessagno, 1977a, p. 62). Green and gray tuffaceous cherts of the upper part of the section are faulted against mudstone. Stratigraphic relations here are clearly equivocal owing to dismemberment of the sequence, but the gradation from tuffaceous cherts into mudstone suggests a closer association with earliest Great Valley Sequence sedimentation than with ophiolite volcanism. Tentatively, the tuffaceous cherts are regarded as overlying the ophiolite.

The red chert has yielded radiolaria assignable to zone 2, subzone 2B. Samples from a measured section of the tuffaceous chert range from zone 2, subzone 2B (lower Tithonian) at the base to zone 3 (lower Tithonian) at the top (Pessagno, 1977a, p. 72). The lower part of the Great Valley Sequence in this area contains *Buchia piochii* (middle to upper Tithonian). However, we have not observed strata assignable to the *B. piochii* zone in depositional contact with either the ophiolite or its associated radiolarian cherts.

Stonyford, Colusa, and Glenn Counties (Ophiolite Locality 18)

Red, manganiferous (nontuffaceous) radiolarian chert is interbedded with pillow basalts along Stony Creek, about 5 km west of Stonyford. This section of Ti-rich submarine basalts, 2 to 3 km thick, is interpreted on petrologic and stratigraphic grounds as part of a large seamount that was built on top of earlier oceanic crust represented by the Coast Range ophiolite. Large-scale faulting (Stony Creek fault) has cut out all parts of the ophiolite sequence here except the basal peridotite, which is set directly against the Stonyford seamount basalt in some places and against the lower part of the Great Valley Sequence in others.

Radiolaria in the cherts are mostly badly recrystallized. Where radiolaria are well preserved, they could not be extracted from the chert by the HF technique. Pessagno (1977a) was able, however, to identify well-preserved specimens on chert chips etched with HF. The radiolarian assemblage is assignable to zone 2, subzone 2B (Pessagno, 1977a). Great Valley Sequence strata that overlie the basalt-chert sequence contain *Buchia piochii* (Brown, 1964). Thus, the Stonyford basaltic seamount was built during the early Tithonian and first covered by Great Valley Sequence muds and sands in the middle Tithonian. The ophiolite (peridotite) here is evidently older.

Paskenta, Tehama County (Ophiolite Locality 22)

Radiolarian chert, in part somewhat tuffaceous, occurs in two structural units of the ophiolite along the Toomes Camp road, west of Paskenta: (1) chert melange blocks within the serpentinite-matrix melange, and (2) a discontinuous horizon of red chert that directly overlies the upper basalt unit and underlies the epiclastic ophiolitic breccia at top of the ophiolite (Hopson unpublished map; also Fritz, 1975). The radiolaria are mostly recrystallized in both these occurrences, and the well-preserved specimens proved difficult to extract. The following species belong to a radiolarian assemblage that comes from light-green ribbon chert (PK 14-2) interbedded with red, gray, and black ribbon cherts more than 18 m thick that are in contact with (resting upon?) basaltic lava, all within a large block in the serpentinite melange (NW quarter of SE quarter, sec. 28): *Parvicingula* sp. C. of Pessagno (1977a), *Hsuum maxwelli* Pessagno s.s., *Acanthocircus variabilis* (Squinabol), and *Tripocyclia jonesi* Pessagno. More recent data than that presented by Pessagno (1977a) indicate that *P.* sp. C does not occur above zone 1. However, it ranges downward in zone 1 to the base of the Callovian in eastern Oregon. *Tripocyclia jonesi* and *H. maxwelli* are likewise long-ranging species that first appear in the Callovian and extend as high as lower Tithonian (zone 3 and subzone 2B, respectively); *A. variabilis* has not been observed in zone 1 material below the Oxfordian-Kimmeridgian. Hence, the evidence at hand suggests that this sample is assignable to the upper part of zone 1 (Oxfordian-Kimmeridgian).

As noted by Pessagno (1977a, p. 66), Jones (1975, p. 330, and personal communication) found *Buchia rugosa* (Fischer) as well as Buchias that appear transitional between *B. rugosa* and *B. concentrica* in finely laminated, medium gray mudstone above the ophiolite at Paskenta. According to Jones, approximately 1500 m of clastic sediments occur between the top of the ophiolite and strata assignable to the *Buchia piochii* zone (= radiolarian zone 4, in part). A radiolarian assemblage assignable to zone 2, subzone 2A (lower Tithonian) was recovered from the *B. rugosa* horizon (sample *JP–1* from Jones; see Pessagno, 1977a). These represent both the oldest megafossil and microfossil dates on the lower part of the Great Valley Sequence (terrigenous clastics) in the California Coast Ranges.

Summary

1. Radiolarian cherts, tuffaceous cherts, and rare nodules of pelagic (coccolithic) limestone that occur within the tuffaceous chert range in age from Oxfordian (upper zone 1) to early Tithonian (zone 3).

2. The Coast Range ophiolite lies conformably or disconformably beneath the tuffaceous cherts, in most cases, and is therefore older (Oxfordian or pre-Oxfordian). Interpillow sediments within the ophiolite lavas are red manganiferous cherts (nontuffaceous) and/or gray pelagic limestone (nontuffaceous), with Radiolaria too badly recrystallized to be age diagnostic. A possible exception is at Geyser Peak, where equivocal structural relations and poor exposures leave it uncertain whether the tuffaceous chert sequence lies upon or within the ophiolite volcanics.

3. Quartz keratophyre sills and thin basaltic pillow lava that occurs within the lower part of the Lotta Creek Tuff sequence at Llanada and Del Puerto Canyon are probably early Tithonian, based on their occurrence within tuffaceous beds of this age (radiolarian zone 2, subzone 2B) at Llanada. This represents a late Jurassic period of minor sea-floor volcanism that is significantly younger than the underlying middle Jurassic ophiolite (Fig. 14-8).

4. High-TiO_2 pillow lavas of the Stonyford seamount are also early Tithonian in age, based upon their interbedded red manganiferous radiolarian cherts (zone 2, subzone 2B). This seamount was built upon older oceanic crust and mantle (Coast Range ophiolite, not dated at Stonyford).

5. The base of the Great Valley Sequence south of the latitude of San Francisco is middle to late Tithonian in age (*B. piochii* zone = radiolarian zone 4, in part).

6. The base of the Great Valley Sequence north of San Francisco is early Tithonian in age (zone 3 at Mount St. Helena and zone 2, subzone 2A at Paskenta).

These age relationships are summarized in Fig. 14-8.

Notes on the Age of Franciscan Radiolarian Cherts and Pelagic Limestone

Current age status of Franciscan pelagic cherts and limestones are briefly summarized as follows:

1. Most Franciscan radiolarian cherts, including all cases but one in which the cherts rest with depositional contact upon basaltic submarine lava, yield Radiolaria of zones 1 to 2A (Pessagno, 1977a; McLaughlin and Pessagno, 1978). Pessagno's earlier conclusion that these represent a limited age span, from upper Kimmeridgian to lower Tithonian (see McLaughlin and Pessagno, 1978, Fig. 2, zonation column), must now be amended in light of his new finding that zone 1 extends back to the Bajocian (lower Middle Jurassic), as discussed previously. Most Franciscan radiolarian chert deposition, therefore, particularly those cherts laid down directly upon basaltic oceanic crust, evidently occurred

during a broad span of time within the Middle to Late (but not latest) Jurassic (i.e., from Callovian or Bathonian through the early Tithonian).

2. The single exception comes from the San Rafael Mountains of southern California (Happy Canyon area, Santa Barbara County, NSF 955-960 of Pessagno, 1977a), where Early Jurassic (Pliensbachian) red ribbon chert rests upon pillow basalt. The Pliensbachian radiolarian fauna at this locality is both rich and diversified and correlates well with faunas from the Maude Formation (lower Pliensbachian) of the Queen Charlotte Islands, B.C., and the Nicely Formation (upper Pliensbachian) of eastern Oregon. Thus, this chert appears to be upper Pliensbachian. No other cherts of early Jurassic age have been encountered in the Franciscan thus far, but red ribbon cherts with similar Early Jurassic faunas overlie Late Triassic ophiolite of the North Fork and Rattlesnake Creek terranes of the southern Klamath Mountains (Irwin and others, 1977). It is important to note, however, that the Franciscan of the San Rafael Mountains is chaotic melange terrane, and the Early Jurassic chert and the pillow lava that it rests upon are both part of one large melange block within graywacke-matrix melange. Thus, it may represent an exotic block, transported tectonically or within an olistostrome from older terrane (North Fork or Rattlesnake Creek equivalent) to the east.

3. A small percentage of Franciscan radiolarian cherts, including all those that rest upon a sedimentary substrate, are Cretaceous in age. Most of these are Early Cretaceous (late Valanginian, zone 5, subzone 5C), but a few are Cenomanian (zone 10, subzone 10A; Pessagno, 1977a; Mclaughlin and Pessagno, 1978).

4. Pelagic limestone within the Franciscan Complex (i.e., Calera Limestone and Laytonville Limestone) are Late Cretaceous (early Cenomanian to Turonian, and perhaps Coniacian (Bailey and others, 1964; Wachs and Hein, 1975; Gucwa, 1975).

PETROLOGY AND PETROGENESIS

This section summarizes the petrologic characteristics of the main rock groups within the Coast Range ophiolite and interprets its origin and tectonic setting. The rocks are grouped as members, taken in their normal stratigraphic order beginning at the base: peridotite tectonite member, plutonic member, sheeted intrusive member, and volcanic member. Included also are the basal sediments deposited upon the upper surface of the ophiolite, which provide information about its tectonic setting. Conclusions about the origin of the ophiolite come particularly from field and petrographic evidence, but draw also on the studies of its major-element chemistry (Bailey and Blake, 1974), rare-earth and trace-element geochemistry (Menzies and others, 1977a, b), oxygen isotope geochemistry (Wenner and Taylor, 1971), strontium isotope geochemistry (Davis and Lass, 1976; Lass and Davis, 1979), seismic velocity structure (Nichols, 1977), and paleomagnetism and magnetic stratigraphy (Kempner, 1977). Our conclusions also lean heavily on the detailed petrologic studies from a few of the better-preserved sections of the ophiolite, particularly those at Point Sal (loc. 2, Hopson and Frano, 1977), Cuesta Ridge (loc. 4, Pike, 1974), and Del Puerto Canyon (Evarts, 1977).

Peridotite Tectonite Member

Alpine-type peridotite of harzburgite subtype (Jackson and Thayer, 1972), partly to completely serpentinized, is the lowest stratigraphic member of the Coast Range ophiolite. It is preserved at seventeen of the twenty-three ophiolite localities (i.e., loc. 4, 6 to 9, 11 to 14, and 16 to 23); faulting at the other localities has cut off the ophiolite sequence above the peridotite level. The harzburgite is voluminous and moderately well preserved at only about five of these localities (locs. 4, 8, 9, 11, 18), and is well studied at only one of them (loc. 9, Del Puerto-Red Mountain, Himmelberg and Coleman, 1968; Evarts, 1977). Elsewhere these remnants are small, almost wholly serpentinized, and in some cases strongly sheared and/or melanged. It might be noted that the peridotite at Burro Mountain in the Franciscan Complex, which is the freshest, best perserved, and most thoroughly studied alpine-type harzburgite in the Coast Ranges (Loney and others, 1971; Coleman and Keith, 1971), lies along strike and nearly connects with the Marmolejo Creek ophiolite remnant (loc. 6); however, a direct relationship, if any, remains to be proven.

The Coast Range ophiolite peridotite member is a metamorphic tectonite, more or less typical of the alpine-type harzburgites that form the basal member of ophiolites everywhere (e.g., Coleman, 1971a, 1977; Nicolas and Jackson, 1972; Dick, 1977; Sinton, 1979; Nicolas and others, 1979; Boudier and Coleman, in press). Structurally, it shows a foliation and compositional layering (ol-opx ratio layering) that ranges from locally strong to more generally weak, with gradations into massive harzburgite. Textures are typically coarse-granular (crystalloblastic) to coarse-porphyroblastic, thought to be indicative of syntectonic recrystallization and recovery at high temperatures and moderate to low differential stress during upper mantle plastic flowage (Nicolas and others, 1979). Olivine (60% to 90%) and enstatite (10% to 36%) are the major phases, and chromian spinel is a ubiquitous minor accessory. A few percent of diopsidic clinopyroxene is also commonly present, locally in sufficient abundance to plot slightly into the lherzolite field (e.g., loc. 18). Whole-rock chemistry for the harzburgites, mainly from the Del Puerto locality (loc. 9), is tabulated and summarized by Bailey and Blake (1974). Mineral composition data are available only for the Del Puerto locality, where Evarts (1977) reports the following 100 Mg/Mg + Fe*[1] values: olivine, 90.7 to 91.8, orthopyroxene, 90.6 to 91.8; clinopyroxene, 93.7 to 95.0 (with 0.4 to 1.4 weight percent Cr_2O_3). The chromian spinels show Cr/Cr + Al = 0.33 to 0.73 and Mg/Mg + Fe^{2+} = 0.42 to 0.68, with Mg/Mg + Fe decreasing as Cr/Cr + Al increases, characteristic of alpine-type harzburgites elsewhere (Irvine, 1965; Malpas and Strong, 1975; Dick, 1977).

Small bodies of dunite that form concordant layers, discordant dikelike sheets, and highly irregular masses are scattered throughout the harzburgite. These have crystalloblastic tectonic fabrics that indicate comtemporaneous, high-temperature deformation and recrystallization with the enclosing harzburgite. The origin of such dunites has been debated, and a magmatic derivation is favored (Loney and others, 1971). We believe that these trails of dunite mark the passage of mantle melts that were crystallizing and fractionating out olivine, and then were subsequently deformed and dismembered by

[1]Asterisk indicates total iron.

plastic flowage of the peridotite host (Hopson and Pallister, 1979a, b). Small gabbroic dikes occur also within the harzburgite; these are very sparse but widespread. Commonly, the gabbroic dikes are dismembered into strings of disconnected fragments that resemble xenoliths, and they are invariably altered to "rodingite" mineral assemblages wherever the harzburgite host is extensively serpentinized (Coleman, 1967).

Bailey and Blake (1974) conclude that harzburgite in the Coast Range ophiolite is the depleted residuum left after removal of a partial melt from parent mantle material of pyrolite or lherzolite composition (Green and Ringwood, 1967; Coleman, 1971a; Nicolas and Jackson, 1972; Ringwood, 1975). Evarts (1977) reaches a similar conclusion, based on his detailed study of the Del Puerto ophiolite remnant (loc. 9). We concur with this interpretation.

Serpentine derived from the Coast Range ophiolite harzburgite (bastite serpentinite) and dunite (mesh-textured serpentinite) is nearly all of the chrysotile-lizardite variety accompanied by brucite, indicating formation at relatively low temperature (Coleman, 1971b). Antigorite serpentinite is reported only from the Del Puerto-Red Mountain peridotite (loc. 9), where it occurs locally as remnants of an earlier, higher-temperature serpentinization (Evarts, 1977). Much or perhaps most of the serpentinization occurred during or after emplacement of the ophiolite onto land. This is shown by the strongly serpentinized rims of peridotite bodies where they are faulted against terrigenous formations, and by isotopic δD-$\delta^{18}O$ ratios, which indicate interaction with meteoric water rather than with sea water (Wenner and Taylor, 1971). Some serpentinization is evidently still going on in the weathering environment (Barnes and O'Neil, 1969). Some other serpentinization, however, probably took place earlier in an oceanic setting. This conclusion is supported by oxygen and hydrogen isotope studies of Franciscan ultramafic rocks and the San Luis Obispo (Cuesta Ridge, loc. 4) ophiolite (Magaritz and Taylor, 1976). Also, direct field evidence for low-temperature serpentinization of mantle peridotite beneath the deep ocean floor comes from the northern Coast Range localities (loc. 22; also 20, 21). Here stratigraphic relations show that harzburgite was serpentinized and injected into tectonically disrupted upper oceanic crustal rocks to form serpentinite-matrix melange *before* deposition of the basal Great Valley deep-marine flysch sediments.

Plutonic Member

Plutonic igneous rocks are exposed in seventeen of the twenty-three ophiolite remnants, but the full plutonic sequence is preserved at only three of these. Faulting has cut out parts or all of the plutonic sequence at the other localities. At no place does peridotite pass up into sheeted intrusive complex through an *unbroken* sequence that lacks plutonic rocks.

The plutonic member consists of a lower group of cumulate ultramafic and gabbroic rocks and an upper group of noncumulus gabbros, diorite, and minor quartz-bearing rocks, where the full sequence is preserved. Also important within the plutonic sequence are dike rocks, which cut across both the cumulate and high-level zone and are locally very abundant.

Cumulate sequence Cumulus rocks formed by crystal settling provide evidence for the existence of a magma chamber, and the sequence of settled phases indicates the crystallization sequence of the magma (Irvine, 1970). The crystallization sequence in turn provides critical information about the initial composition of the magma and the course of its fractionation. The evidence for a cumulus origin in plutonic rocks of the Coast Range ophiolite includes layering (phase layering, grain-size layering, mineral ratio layering), layer grading, planar lamination (i.e., planar orientation of platy or elongate crystals parallel to layering), cumulus textures, and cryptic variation (i.e., vertical variation in the composition of cumulus phases) going up through the sequence.

The cumulus phase assemblages observed in different remnants of the Coast Range ophiolite are given in Table 14-2. Relatively unbroken cumulate sequences are preserved at only three of the ophiolite remnants (locs. 2, 9, 23), and only at one of them (loc. 2, Point Sal) are exposures sufficiently continuous that the vertical sequences of cryptic and phase variation have been established (Fig. 14-9). The fault-bounded cumulate remnants at the other localities nevertheless provide evidence that magma chambers existed, and help to establish the magmatic fractionation patterns for the Coast Range ophiolite as a whole.

The cumulus sequence begins with olivine-chromian spinel cumulates (dunites) at the base, followed by clinopyroxene-olivine cumulates (wehrlite, ol-clinopyroxenite,

TABLE 14-2. Cumulus Phase Assemblages

Location[a]	*Cumulus Phase Assemblages*[b]
1. SCI	/cpx + ol/pl + cpx + ol/hb + pl + cpx + ol/opx + pl + cpx
2. PS	ol + sp, cpx + ol ± sp, pl + cpx + ol, pl + cpx
4. CR	ol + sp/cpx + ol/
5. SS	ol + sp, cpx + ol/
7. L	/cpx + ol/pl + cpx + ol, pl + cpx/
8. QC	/pl + cpx/
9. DPC	ol + sp, cpx + ol + sp/cpx + ol ± opx, pl + opx + cpx
11. CM	(dunite, pyroxenite, gabbro)
12. Hb	(pyroxenite, gabbro)
13. MSH	/pl + cpx ± ol/
14. HS	ol + sp ± opx/pl + cpx ± ol/
15. GP	(peridotite, gabbro)
16. FC	(serpentinite, ol-pyroxenite, pyroxenite, gabbro)
17. WS	(serpentinite, gabbro)
22. Pk	(serpentinite, gabbro)
23. EC	ol + sp, cpx + ol ± sp, pl + cpx + ol, opx + pl + cpx, pl + cpx + ol/

[a]Locations refer to Fig. 14-1.
[b]Cumulus phase assemblages are listed in order of occurrence from lowest exposed part of the cumulate sequence upward. Commas separate assemblages observed to be in continuous stratigraphic sequence. Diagonal lines separate assemblages or groups of assemblages that are bounded by faults. ol, olivine; sp, spinel; cpx, clinopyroxene; opx, orthopyroxene; pl, plagioclase; hb, hornblende.

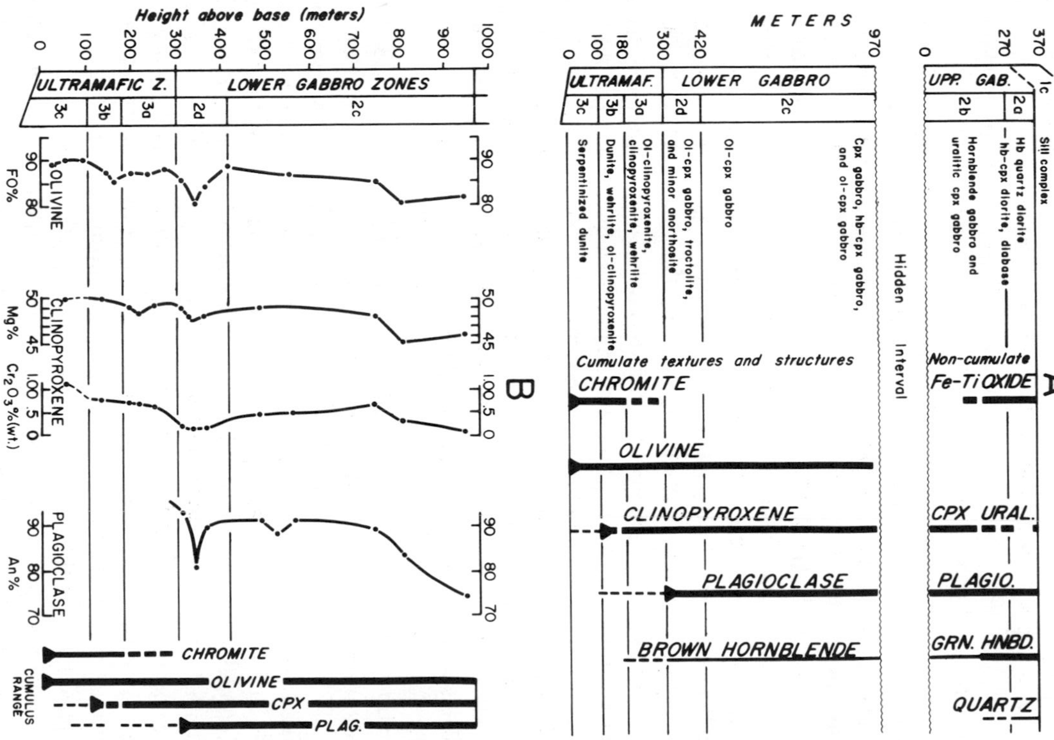

Fig. 14-9. (A) Phase layering, ophiolite zones 2a to 3c. Heavy lines show the stratigraphic distribution of cumulus phases in zones 2c to 3c and early crystallizing phases in the noncumulate rocks of zones 2a to b. Thin lines signify postcumulus phases in zones 2c to 3c and late-crystallizing phases in zones 2a to b. Broken lines signify sporadic occurrence of phases. (B) Cryptic variation (chemically graded layering) in cumulate rocks of ophiolite zones 2c to 3c (lower plutonic section). Stratigraphic range of cumulus phases shown on the right. Reproduced by permission of Oregon Department of Geology and Mineral Industries.

ol-poor clinopyroxenite) where the basal cumulates are preserved (locs. 2, 4, 5, 9, 23). This sequence from ol-sp to cpx-ol cumulates probably also holds for ophiolite remnants where the cumulates are less well preserved (locs. 1, 7, 14). Cumulus plagioclase appears next, forming gabbroic cumulates (locs. 1, 2, 7 to 9, 13, 14, 23, and probably 12). The gabbroic cumulates at most localities have olivine, clinopyroxene, and plagioclase as the the cumulus phases (ol-cpx-pl cumulates with cpx-pl and pl cumulate interlayers). Orthopyroxene-clinopyroxene-plagioclase cumulates (gabbronorites) occur locally within two of the ophiolite remnants (locs. 1, 23) and comprise all the gabbroic cumulates at the Del Puerto Canyon remnant (loc. 9; Evarts, 1977). Hornblende occurs widely as a postcumulus phase throughout the pyroxenites and gabbros, but only at one place (on Santa Cruz Island, loc. 1) are rocks with cumulus hornblende known to us.

The significance of the Coast Range ophiolite cumulus sequence appears to be as follows. The initial magma had a relatively primitive composition, with olivine (Fo_{90}) and chromian spinel as the early crystallizing liquidus phases. Crystal fractionation of this melt as it entered the magma chamber resulted first in olivine cumulates, then olivine-clinopyroxene cumulates, and finally olivine-clinopyroxene-plagioclase cumulates as the fractionating melt reached the three-phase cotectic. Whether the melt reached the three-phase cotectic via the ol-cpx cotectic (as suggested by the abundant ol-cpx cumulates) or via the ol-pl cotectic (which is compatible with the cumulate stratigraphy if plagioclase settling was delayed by its low density) remains an unsolved question. After reaching the three-phase cotectic, fractionation under *closed* system conditions would move the melt to the olivine reaction point (ol-opx-cpx-pl invariant point) and then down the opx-cpx-pl cotectic, forming gabbronorite cumulates (Irvine, 1970). This appears to have happened at Del Puerto Canyon (loc. 9; Evarts, 1977). At Point Sal, however, the thick sequence of ol-cpx-pl cumulates and lack of cumulus opx shows that the magma was held on the three-phase cotectic despite continuous crystal fractionation. This can only mean *open-system* fractionation with recurrent replenishment by primitive melt (O'Hara, 1977), a condition mainly achieved at spreading ocean-ridge magma chambers (Hopson and Pallister, 1978, 1979a, b). Continuous replenishment of an open-system magma chamber is further indicated by the steep, zigzag cryptic variation patterns for olivine and plagioclase (Pallister and Hopson, 1979), as shown at Point Sal (Fig. 14-9), and also by the "step back" from orthopyroxene to Mg-olivine as a cumulus phase at high stratigraphic levels in the cumulate sequence, shown at the South Fork of Elder Creek (loc. 23). In fact, the common occurrence of olivine gabbro (ol-cpx-pl) cumulates and relative scarcity of gabbronorite (opx-cpx-pl) cumulates in remnants of the Coast Range ophiolite may be an indication that open-system fractionation was the normal pattern. This has important tectonic implications, mentioned later.

A final important feature of the cumulates (especially the basal cumulates) within the plutonic member of the Coast Range ophiolite remnants is the abundant evidence for strong penecontemporaneous deformation under hypersolidus conditions. This includes chaotic styles of folding, disruption and dismemberment of cumulus layers, internal unconformities, and diapiric intrusion of cumulates (resembling sandstone diking), all with negligible effect on the primary igneous textures. The striking resemblance of the cumulate deformation to soft-sediment deformation points to tectonic disturbance while magmatic crystal sedimentation was in progress and while buried layers still retained intercumulus liquid (Hopson, 1975, 1976). An environment of strong, *recurrent* tectonic deformation, such as might be expected at an active plate boundary, is suggested. The fact that this type of "soft-sediment" deformation is a characteristic of basal cumulates in other ophiolites such as Troodos (George, 1978), Vourinos (Ewing, 1976), Samail (Hopson and Pallister, 1979b), and Bay of Islands (Malpas and Talkington, 1979) points to oceanic spreading centers as the most likely tectonic setting.

High-level plutonic sequence The upper part of the Coast Range ophiolite plutonic sequence is fairly well preserved at Santa Cruz Island, Point Sal, Del Puerto Canyon, Healdsburg (Bradford Mountain), and South Fork of Elder Creek (locs. 1, 2, 9, 12, 23). Smaller, incomplete remnants of the upper plutonics occur also at several other localities (locs. 3 to 6, 8, 10, 11, 22).

The high-level plutonic sequence typically consists of noncumulus gabbroic and dioritic rocks, commonly accompanied by small bodies or dikes of plagiogranite. The gabbros are structurally, texturally, and mineralogically distinctive from the cumulus gabbros, which occur beneath them stratigraphically. They lack cumulus layering and planar lamination and are structurally and texturally isotropic. The texture is hypidiomorphic rather than cumulus, reflecting continuous crystallization and solidification of melt *in place*. Plagioclase crystals show strong progressive zoning, typically through the range An_{75-20}, reflecting continuous growth from melt as it cooled through a large temperature range. The high-level gabbros are typically rich in hornblende and/or uralitic amphibole, minerals generally lacking in the cumulates except as minor postcumulus phases. Most of the high-level gabbros crystallized plagioclase and proxene (cpx ± opx) early; then hornblende crystallized as a late magmatic overgrowth around pyroxene and also formed uralitic replacements. Thus, the melts from which these rocks crystallized were becoming enriched in water. Titanomagnetite is also an important late-crystallizing phase, swelling in abundance from <0.1% in the cumulates to 5 to 10% in some of the high-level gabbros. This reflects a high fO_2 and strong enrichment in iron in the late-stage, high-level gabbro magma. This gabbro grades into green-hornblende diorite, which commonly contains relict clinopyroxene and a few percent of interstitial quartz. The diorite grades into green-hornblende quartz diorite, which may be structurally isotropic or may have flow-foliated textures and schlieren banding (Hill, 1976; Evarts, 1977). Small segregations or dikes of aplitic plagiogranite (granophyric albite granite) occur within the quartz diorite, representing the extreme liquid end product of crystal fractionation.

The stratigraphic sequence of the high-level plutonic rocks is best preserved at Point Sal (loc. 2). Here cumulus gabbro is overlain by noncumulus pyroxene gabbro, which grades upward through magnetite-rich uralite gabbro and hornblende diorite into local zones of quartz diorite. Above this is a thin upper zone of thermally and metasomatically metamorphosed diabasic gabbro, immediately underlying the sheeted intrusive member (sill complex) (Hopson and Frano, 1977). Here at Point Sal the dioritic and quartz-bearing differentiates were evidently filterpressed upward from the crystallizing high-level residual gabbroic magma, but were trapped along a sandwich horizon beneath the zone of diabasic gabbro that was crystallizing and being plated downward from the roof of the magma chamber. At other Coast Range ophiolite localities the differentiated siliceous magma evidently migrated out of the sandwich horizon, forming small plagiogranite intrusions that cut through the uppermost gabbro and into the sheeted complex.

The differentiating magma that formed the plutonic sequence at Coast Range ophiolite localities was following a tholeiitic iron-enrichment trend, reflected by both the whole-rock and mineral chemistry (Bailey and Blake, 1974; Pike, 1974; Hopson and others, 1975a, b; Hopson and Frano, 1977; Evarts, 1977). Iron enrichment is only moderately shown in the cumulates, where there was open-system fractionation (i.e., periodic replenishment), but it becomes very strong in the high-level plutonic sequence, which clearly crystallized and fractionated as a closed system (i.e., no more replenishment). This upward change within the plutonic sequence from open-system to closed-system fractionation seems best explained by crystallization within a continuously opening magma chamber beneath an oceanic spreading center, as modeled by Hopson and Pallister (1979a; Pallister and Hopson, 1979). Here open-system fractionation occurred within the center of the magma chamber, beneath the spreading axis, where new primitive melt was repeatedly introduced from below. Closed-system fractionation gradually took over toward the top of the growing pile of cumulates as spreading carried the two halves of the chamber apart, and replenishment could no longer reach the residual magma left far out to the side.

The Coast Range ophiolite magmas appear to have had a relatively high water content. Some indication of this comes from as low in the plutonic sequence as the pyroxenite and olivine gabbro cumulates, where small amounts of postcumulus hornblende crystallized from the entrapped intercumulus melt. But water enrichment was mainly localized near the top of the magma chambers, as shown by the drastic increase in magmatic and uralitic hornblende in the upper high-level gabbros and diorites. Here water locally reached extreme concentrations in the magma, as shown by amphibole-rich pegmatitic patches within the upper gabbro and by miarolitic cavities filled by finely fibrous green amphibole. The latter feature shows that the magma near the top of the chamber became oversaturated in water and underwent retrograde boiling during a late stage of crystallization and that the resulting gas cavities were subsequently filled by vapor-phase amphibole (Hopson and Frano, 1977). Fracture-controlled uralitization and related forms of high-temperature hydrothermal alteration are also common in the high-level plutonic rocks, showing that aqueous fluids continued to migrate through the newly solidified plutonic crust during the subsolidus stage of cooling. The Sr-isotopic studies of Davis and Lass (1976; Lass and Davis, 1979) indicate that penetration of sea water was largely responsible.

Dikes within the plutonic sequence The cumulate and high-level zones within the plutonic member are both riddled by cross-cutting dikes of several origins and relative ages. The following types are identified within the *cumulate* zone, especially near its base:

1. Pyroxenite and wehrlite dikes, formed by the intrusion of hot, mobile cumulus mushes, analogous to sandstone dikes injected within deformed soft sediments. These dikes root in the ultramafic cumulates, but they commonly cut up into the overlying

gabbro cumulates. The ultramafic dikes commonly contain feldspathic segregations, which evidently crystallized from intercumulus melt that was squeezed out during mush intrusion.

2. Small gabbroic dikes, formed by filterpressing of residual intercumulus melt from the cumulates. These dikes also are rooted in the cumulus sequence, and their composition is always more fractionated than those of the cumulus layers they cut. These dike rocks are usually norite, hornblende gabbro, or diorite.

3. Olivine gabbro dikes that cut across the harzburgite tectonite and extend up into the basal cumulates. These dike rocks have high Mg/Mg + Fe* values and anorthositic plagioclase, reflecting a relatively primitive melt composition. They evidently represent residual (fractionated) mantle melts that were squeezed up along tectonic fractures from somewhat deeper levels within the peridotite.

4. Diabasic-textured dikes of gabbroic and dioritic (even granophyric) composition, which cut across all other features. Their origin is obscure. They might represent differentiated melts from a spreading-axis magma chamber that were injected laterally into newly solidified oceanic crust that was spreading away at the sides, or they might be melts injecting older oceanic crust from a new magmatic source.

Dikes that cut through the upper part of the high-level (noncumulus) plutonic sequence generally have fine-grained diabasic to granophyric textures and fractionated compositions that range from pyroxene and hornblende gabbro through diorite and tonalite to plagiogranite. Some of these obviously rose from the top of the magma chamber, along fractures through the "plated" gabbro that solidified downward from the roof. These dikes tapped the magma chamber at various stages of its differentiation, accounting for the wide range of dike compositions (Hopson and Frano, 1977, Fig. 9 and Table 6). Some of these dikes rise completely through the upper plutonics to join the sheeted intrusive member as late additions, and some may even reach the surface to feed late, differentiated lava flows (presumably this is the source of some of the keratophyre and quartz keratophyre flows near the top of the volcanic sequence). Commonly, these dikes that cut through the top of the plutonic sequence are highly altered inward from their margins, even to such extreme products as epidosite. This alteration is thought to result from the action of gases generated by retrograde boiling at the top of the magma chamber; these gases evidently were channeled through newly injected dikes as they streamed through the roof.

Sheeted Intrusive Member

A sheeted intrusive complex separates the volcanic and plutonic members of the Coast Range ophiolite where their junction is preserved unbroken (locs. 2, 4). At other localities the high-level plutonic rocks merge upward into dense dike swarms or sheeted complex (locs. 1, 5, 12, 22, 23), or the volcanic member passes downward into dike or sill complex (locs. 3, 8 to 10), but in each case faulting truncates the sequence. It appears that a sheeted complex, or a transition into sheeted complex, is present wherever the top of the plutonic sequence, or base of the volcanic sequence, is found. Stated differently, we know of no unbroken sequence from volcanic to plutonic rocks in the ophiolite where an intervening sheeted complex is lacking. Thus, sheeted intrusives seem to link the volcanic and plutonic zones physically and probably also genetically, and their development is an integral part of the formation of oceanic crust now preserved in the Coast Ranges.

The term "sheeted intrusive" as used here refers to countless dikes or sills emplaced side by side, chilled against or inside one another, with little or no intervening volcanic or plutonic host rock. Sheeted *dike* complexes, well known from the large ophiolite masses in Cyprus, Oman, Newfoundland, and elsewhere, have become a recognized part of normal ophiolite stratigraphy (Coleman, 1977). The dikes (sheeting direction) stand perpendicular to the volcanic and plutonic units and clearly formed by repeated tensional fracturing, dilation, and intrusion through the roof of the magma chamber, where melt was conveyed to the surface to feed lava flows. This endless lateral dilation and accretion of the magma chamber roof by dike intrusion seems best explained by axial intrusion at oceanic spreading centers, with the sheeting direction marking the trend of the spreading axis.

A curious feature of the Coast Range ophiolite is the occurrence of sheeted complex in which the sheeting direction is parallel, rather than perpendicular, to the ophiolite stratigraphy, that is sill complexes. Sill complexes occur at Point Sal, Cuesta Ridge, Del Puerto Canyon, and possibly at Geyser Peak (locs. 2, 4, 9, 15), whereas sheeted dikes are definitely known only from Quinto Creek (loc. 8). The direction of sheeting at the other localities (locs. 1, 3, 5, 6, 10, 12, 13, 22, 23) remains uncertain, either because too little of the original stratigraphy is preserved or because not enough work has been done.

The sheeted sill complexes have been studied in detail only at Point Sal (loc. 2; Hopson and Frano, 1977) and Cuesta Ridge (loc. 4; Pike, 1974). The main features, briefly summarized, are as follows: (1) The main body of sheeted sills (Fig. 14-10, stage 2) is diabasic and basaltic with strong deuteric and hydrothermal alteration. The sills are intrusive into the lower part of the overlying lavas, and they truncate steeply dipping basaltic dikes that fed the lavas (Fig. 14-10, stage 1). The uppermost plutonic rocks, which lie immediately beneath the sheeted sills, are *intrusive into the base of the sill complex* and are therefore younger. These uppermost plutonic rocks are the "plated" diabasic gabbros and diorites that crystallized downward from the sheeted sill complex, which forms their immediate roof. Younger dike and sill rocks (Fig. 14-10, stage 3) cut up across the upper plutonic rocks into the earlier sill complex, where they commonly branch laterally to form additional sills. These late-stage dikes and sills, which locally are quite abundant and substantially thicken the sheeted complex, range from diabase through microdiorite and microtonalite to plagiogranophyre and quartz albitite. The more silicic dikes generally cut the more basic ones, and the entire group (stage 3 of Fig. 14-10) was evidently intruded from the underlying magma chamber (now the plutonic sequence) during progressive stages of its differentiation (Hopson and Frano, 1977, Tables 5, 6, Fig. 9). Finally, thin olivine basalt dikes (Fig. 14-10, stage 4) cut across everything else.

A key to understanding the sheeted sill complexes is their age relations to the

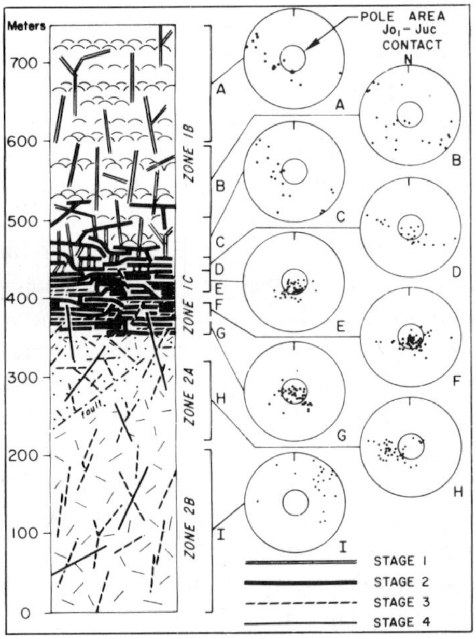

Fig. 14-10. Point Sal columnar diagram and stereographic plots, showing dike and sill attitudes and age relations for the Upper Dike Series. Letters relate each stereoplot to the part of the column and to the geographic location (Fig. 14-3, lettered areas) where the measurements were made. Plotted points are the poles to dikes and sills. The stereoplots have been rotated to bring the upper surface of the ophiolite (Jo_1-Juc contact) back to a horizontal position. The small circle within each plot denotes a ±20° uncertainty in the Jo_1-Juc pole position. Stages of dike intrusion (1, oldest; 4, youngest) determined from field relations. Reproduced by permission of Oregon Department of Geology and Mineral Industries.

rocks that lie immediately above and below. Normally, sills are floored and roofed by older host rocks. In the ophiolite the older roof rocks (i.e., the lower volcanics) are present, but the original floor rocks have been displaced by plutonic rocks (high-level and cumulate sequences) that are still *younger* than the main body of sheeted sills. Thus, the age sequence from oldest to youngest is (1) lower volcanic rocks with their feeder dikes, plus a substrate (plutonic complex or peridotite?) that is no longer directly attached, (2) sill complex, emplaced between the volcanic rocks and their substrate, and (3) plutonic rocks, intruded between the sill complex and the substrate. The plutonic sequence crystallized and differentiated in place beneath the sill complex, sending younger, differentiated dikes and sills up into it, and perhaps also feeding the higher lava flows. The sill complex (stage 2) may thus be regarded as the intrusive vanguard of the main mass of magma that formed the plutonic sequence (stage 3).

These ophiolitic (oceanic) sheeted sill complexes, with younger co-genetic plutonic masses immediately beneath them, corespond closely to thick sill complexes that cap intrusive epizonal plutons on land, particularly in the Cascade Mountains. For example, the Tatoosh pluton near Mount Rainier grew by first sending out enormous swarms of sills, followed by the main mass of magma that inflated a chamber just beneath them, which crystallized to form the main pluton (Fiske and others, 1963). The ophiolitic sill complex at Point Sal and Cuesta Ridge, and probably at other Coast Range ophiolite localities, appear to have developed in much the same way.

We conclude that the sheeted sill complexes mark the initial stages of development of *new* magma chambers that grew beneath already solidified volcanic oceanic crust. As these magma chambers crystallized and differentiated, they sent more dikes and sills into their roof and also fed lava flows at the surface (Hopson and Frano, 1977).

This discussion of sheeted sills should not obscure the fact that vertical dikes are also plentiful, and that "normal" sheeted dike complexes, like those in other ophiolites, are also developed. Vertical dikes (some of which doubtless fed lavas) precede, postdate, and in part are contemporaneous with the sheeted sills at Point Sal (Fig. 14-10) and probably the other localities. Sheeted dike complex occurs at Quinto Creek (loc. 8), probably at San Simeon (loc. 5), and quite likely at some of the other localities where remnants of sheeted diabase or dense dike swarms occur near the volcanic-plutonic contact but are poorly exposed or poorly preserved (locs. 1, 3, 6, 10, 12, 13, 22, 23).

We believe that the development of vertical dike swarms and sheeted dike complex supports other evidence (to be summarized) that points to formation of the ophiolite at an ocean spreading center (or centers). The sheeted sill complexes mark the initial growth stage of small, discontinuous magma chambers that periodically inflated and then crystallized along the spreading axis. This harmonizes with the widely accepted idea that the lavas at slow-spreading ocean ridges are fed from small, discontinuous chambers rather than from a single large steady-state chamber (Hall and Robinson, 1979). Whether dike or sill complexes form, as these new chambers grow, would depend on the balance between spreading rate and the inflow rate of new magma. Thus, if new magma welled in more rapidly than tensional fissures were created (slow-spreading situation), the magma might tend to spread laterally along horizontal zones of weakness, as demonstrated by model experiments (Ramberg, 1970) and by shallow intrusions on land.

Other examples of sill complexes that occur along the junction between volcanic and plutonic rocks in ophiolites are reported from the Caledonian Leka ophiolite in Norway (Prestvik, 1979) and from Tethyan ophiolite in the Ergani district of southwestern Turkey (Bamba, 1974).

Volcanic Member

Volcanic rocks of the upper member of the ophiolite are widely preserved (locs. 1 to 4, 6 to 12, 15 to 17, 20 to 22), but nowhere is there a full, unbroken sequence. The ophiolite volcanics are missing due to faulting at San Simeon, Stonyford, and Elk Creek (locs. 5, 18, 19); however, at Stonyford there is a thick pile of younger (early Tithonian) high-Ti pillow lavas that represent the remains of a post-ophiolite seamount (see loc. 18 description). The ophiolite volcanics are missing also at Mount St. Helena, Harbin Springs, and at the top of the South Fork of Elder Creek section (locs. 13, 14, 23), where they were faulted off before deposition of the ophiolite breccias. The ophiolite lavas occur only as disrupted masses and native melange blocks in ophiolite (serpentinite-matrix) melange at northern Coast Range localities 20, 21, and 23.

A remarkable feature of the volcanic member is its great stratigraphic thickness, where well preserved. For example, the volcanics are more than 2.5 km thick at Del Puerto Canyon (loc. 9; Evarts, 1977), more than 1.4 km thick at Point Sal (loc. 2; Hopson and Frano, 1977), and probably more than 1 km thick at Santa Cruz Island (loc. 1; Hill, 1976). Impressive partial sections, truncated by faults, are also present at several other localities (loc. 3 >500 m; loc. 7 >550 m; loc. 15, 500 m; loc. 22, 600 m). In contrast, the plutonic and sheeted intrusive members are relatively thin. Thus, thick volcanic and thin plutonic-hypabyssal sequences appear to be a distinctive facet of the California Coast Range ophiolite.

Field and petrographic evidence indicates that the bulk of the ophiolite volcanic rocks are deep-sea lavas, chiefly of basaltic composition. Lavas greatly predominate over fragmental rocks, and pyroclastic products are virtually unknown. The lavas occur as pillowed flows, massive flows, and occasionally as thin sheet flows. The various pillow forms that characterize deep-sea basalts (Ballard and Moore, 1977) can be found, especially bulbous and elongate (tubular) pillows. Vesicles (amygdules) are very sparse and tiny in the chilled pillow rims (outermost 1 cm), indicating eruption into deep water (Moore, 1965, 1970; Moore and Schilling, 1973). The occurrence of massive, ponded lava flows and thin sheet flows has recently been documented by submersible diving along the rifted axes of the Galapagos Rise (Ballard and others, in press), East Pacific Rise (Cyamex Scientific Team, 1978), and Mid-Atlantic Ridge (Ballard and others, 1979), and similar features are found in the Coast Range ophiolite. For example, massive (ponded) lavas in the ophiolite, including some flows with poorly developed columnar jointing, are duplicated in wall sections of the Mid-Atlantic Ridge rift valley (Hopson, personal observation from *DSRV Alvin*). Fragmental volcanic rocks, found locally in the ophiolite, are chiefly blocky or rubbly monolithologic basaltic breccias. The blocky breccias resemble talus deposits that occur along submarine fault scarps, and the rubbly breccias resemble flow-front rubble within submarine lava flows; both types of deposits are common in the Mid-Atlantic Ridge rift valley (Ballard and Moore, 1977; Hopson, personal observation). Also, deep-sea drilling in the North Atlantic has penetrated abundant basaltic breccias at some sites (Hall and Robinson, 1979). Thus, it is not necessary to invoke a shallow-water origin (Evarts, 1977) for such breccias in the Coast Range ophiolite. Aquagene tuff and finely fragmental (including bedded) hyaloclastites, which are chiefly restricted to relatively shallow-water environments (Williams and McBirney, 1979, Chapter 12), are rare in the Coast Range ophiolite.

Sediments that occur sparsely *within* the ophiolite volcanics provide additional evidence for a deep submarine, open-ocean origin. These sediments are almost exclusively radiolarian chert and/or pelagic limestone (locs. 2, 3, 7, 8, 17, 18, 21, 22), occurring within pockets between pillows or as small gobs and stringers where the soft ooze was injected up into overriding lava flows. These baked and recrystallized sediments in the lavas were evidently derived from coccolithic and radiolarian oozes, as shown by their very poorly preserved microfauna. They correspond to the thin coatings of biogenic ooze that accumulate during repose periods between eruptions in modern ocean-ridge settings (e.g., Ballard and Moore, 1977). Deposition occurred at ocean ridge-crest depths (see later discussion of sediments) and in settings that lay beyond the reach of terrigenous sedimentation or airborne tephra from island-arc volcanism. This is deduced from the absence of admixed clastic detritus or volcanic tephra components (microlites, shards, pumice) within the interpillow limestone and chert. Possible exceptions are some tuffaceous beds that crop out locally within the basaltic volcanic sections at Llanada and Geyser Peak, but these are probably faulted in rather than interbedded (see loc. 7, 15 descriptions).

Primary petrographic features of the basic lavas resemble those of oceanic basalts, although most of the lavas are now extensively altered (metamorphosed) to secondary mineral assemblages. Phyric (porphyritic) and aphric lavas are both present, and quench textures characteristic of deep-sea basalts (Bryan, 1972) are beautifully preserved in pillow-rim samples. Aphyric and sparsely microphyric lavas predominate. The phyric lavas have as phenocryst and microphenocryst assemblages ol, pl, ol + pl, and ol + pl + cpx, corresponding to the common types of phyric lavas from the Mid-Atlantic Ridge (Hekinian and others, 1976; Bryan and others, 1976; Bryan and Moore, 1977; Hall and Robinson, 1979). Pl-cpx lavas are also common among the Coast Range ophiolite lavas; this type is widespread though not abundant in DSDP basalts from the Pacific, Atlantic, and Indian oceans. Ol-cpx phyric lavas occur among the upper lavas at Point Sal (loc. 2); this rather unusual phenocryst assemblage is probably accumulative rather than primary (Hopson and Frano, 1977). In summary, most of the Coast Range ophiolite lavas are essentially identical to deep-sea basalts in their relict primary textures and phenocryst mineralogy.

The whole-rock major-element chemistry of the volcanic rocks has been comprehensively studied by Bailey and Blake (1974), with further input by Pike (1974), Hill (1976), and Hopson and Frano (1977). The reader is referred to these articles for the data and discussions. The strong alteration of most of these rocks has significantly changed their chemistry, which can lead to erroneous interpretations. For example, pervasive albitization has significantly increased Na_2O and lowered CaO. This tends to move the composition from the tholeiite to the alkali basalt field on alkali versus silica plots (Bailey and Blake, 1974, Fig. 2), and toward the alkali corner on AFM plots (Bailey and Blake, 1974; Pike, 1974; Hopson and others, 1975a). The latter effect tends to produce a pseudo calc-alkaline trend from a tholeiitic rock suite. The effects of rock alteration (metamorphism) on the whole-rock chemistry has been considered by Pike (1974) and Hopson and Frano (1977), who conclude that the lavas originally comprised a tholeiitic suite. The moderately strong iron-enrichment trend, comparable to that of Atlantic abyssal tholeiites, is shown in Fig. 14-11.

Minor-element chemistry indicates that the ophiolite volcanics (but excluding Stonyford high-Ti basalts) correspond to ocean-ridge tholeiite. One indication comes from TiO_2-P_2O_5 plots, where the ophiolite lavas fall with the ocean-ridge basalts (Bailey and Blake, 1974; Hopson and Frano, 1977). Analyses of thirty minor elements plus eleven rare-earth elements from ophiolite lavas at several Coast Range localities, plotted on diagrams used to distinguish basalts of various geologic settings, led Blake and others (1975) to conclude that "most of the Coast Range (ophiolite) basalts, and the keratophyres and quartz keratophyres as well, fall in the fields of oceanic ridge basalts."

The most detailed study of minor and rare-earth elements in the ophiolite is that of Menzies and others (1977a; see also Menzies and others 1977b), conducted at Point Sal (loc. 2). They conclude that "the Point Sal lavas have REE characteristics similar to oceanic tholeiites." A key point here is that these lavas have light rare-earth element (LREE) *depleted* patterns, which are characteristic of ocean-ridge tholeiites and distinguish them

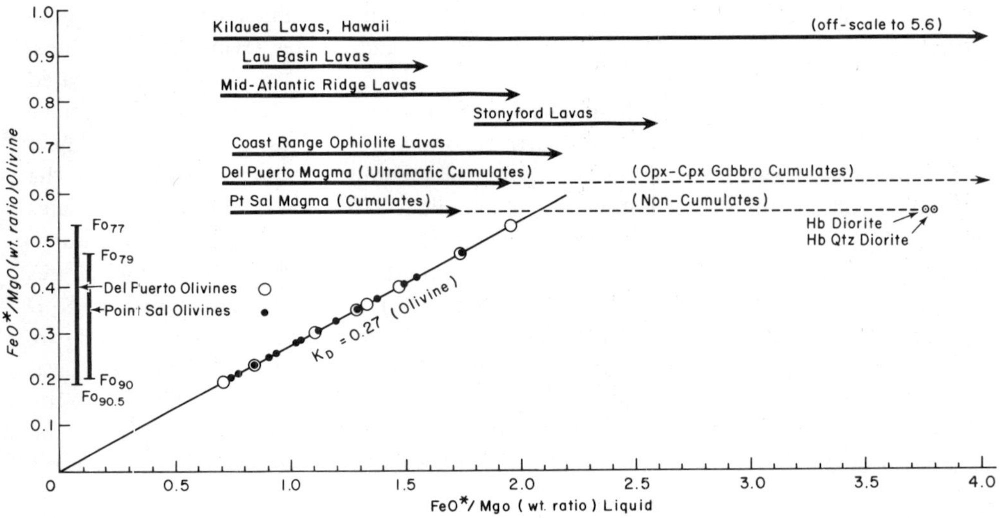

Fig. 14-11. Plot of FeO*-MgO (wt. %) of olivines within cumulate rocks of the Point Sal and Del Puerto ophiolite remnants against FeO*-MgO (wt. %) of the liquids from which they crystallized, employing a partition coefficient for olivine-liquid = 2.7 (Rhodes and others, in press; Bryan, 1979). Shown also for comparison are the range of lava compositions for the Coast Range ophiolite (taken from various sources), the Stonyford seamount lavas (taken from Bailey and Blake, 1974), the Lau interarc basin lavas (Hawkins, 1977), the Kilauea lavas, and the range of basaltic glass compositions for the Famous and Amar rift valleys on the Mid-Atlantic Ridge (Bryan, 1979).

from other types of oceanic tholeiites (Bryan and others, 1976). Some Coast Range ophiolite lavas that occur high in the stratigraphic sequence have rather flat REE profiles (Menzies and others, 1977a; see also Blake and Jones, Chapter 12, this volume). Menzies and others (1977a) suggest here that LREE-depleted profiles were flattened when light rare-earth elements (particularly La, Ce) were mobilized owing to interaction with sea water during sea-floor weathering.

Volcanic rocks described as keratophyre and quartz keratophyre are widely reported from the Coast Range ophiolite (Bailey and others, 1970; Page, 1972; Bailey and Blake, 1974; Hopson and others, 1975a; Hill, 1976; Evarts, 1977). These rocks are characterized by intermediate to high SiO_2 contents, high Na_2O, low CaO, and very low K_2O (except rocks rich in secondary smectite), and their primary plagioclase is extensively to wholly albitized. They have been interpreted as "albitized andesite, dacite, and rhyolite" (e.g., Bailey and Blake, 1974), and it has been suggested that they are the remnants of an island arc built upon oceanic crust that is represented by other rocks of the ophiolite (Blake and Jones, Chapter 12, this volume). We disagree with this interpretation, as explained next.

The keratophyric rocks fall into two groups: those with no quartz phenocrysts and intermediate silica (SiO_2 ~ 55% to 65%), and those with quartz phenocrysts and high silica (70% to 77%). Those of the first group were *not* originally andesites and dacites but mainly basalts, now strongly albitized (Pike, 1974; Hopson and Frano, 1977). They have basaltic, not andesitic, textures. They lack hypersthene and hornblende (or their pseudomorphs), which are ubiquitous in arc-type andesites and dacites (Williams and others, 1958). They have little or no groundmass quartz, which would be abundantly released during low-grade alteration (metamorphism) of arc-type andesite and dacite (average andesite has 11% to 17% normative quartz according to Chayes, 1969, which is held mostly in siliceous glass or groundmass tridymite-cristobalite). Responsible instead for the high silica content of these keratophyres is abundant secondary albite (albite contains 68.7% SiO_2) that pseudomorphs primary plagioclase and preserves the basaltic groundmass texture. Another indicator is the low K_2O content of the Coast Range keratophyres (generally less than 0.5% and commonly less than 0.2%), which contrasts with 1.6% K_2O for calc-alkaline andesites (Chayes, 1969) and 2% to 3% for dacites. Finally, andesitic arc terranes have a high ratio of pyroclastic rocks (including volcaniclastic sediments) to lavas (Garcia, 1978), whereas the keratophyric rocks in the Coast Range ophiolite are exclusively lavas and dike rocks. We conclude that the lavas in the Coast Range ophiolite, including most of the keratophyres, were originally submarine basalts. Primary andesitic lavas were absent or rare.

The second type of keratophyre associated with the ophiolite is that with quartz and albite phenocrysts and abundant quartz in an igneous-textured groundmass. These quartz keratophyres doubtless solidified directly from highly differentiated siliceous melts. We do not believe that they are albitized rhyolites (Bailey and Blake, 1974; Blake and Jones, Chapter 12, this volume), however, because of their very low K_2O (mostly below 0.5%), lack of LREE enrichment, and lack of sanidine or biotite phenocrysts (or pseudomorphs). The composition of these quartz keratophyres is virtually identical to end-member plagiogranites (albite granite) in the ophiolite, and we concur with Evarts (1977) that both of these rock types probably came from the same K-poor, highly siliceous residual melts. Coleman and Peterman (1975) conclude that low-K plagiogranite magmas are generated in oceanic settings.

It has been suggested that some of the Coast Range ophiolite lavas, particularly the olivine-clinopyroxene phyric basalts that occur locally at Point Sal, are related to the high magnesian pillow lavas of western Pacific island arc regimes, specifically the Bonin Islands, Mariana Islands, and SE Papua New Guinea. However, we see major differences. For example, these distinctive Mg-rich arc-type lavas (boninites, marianites) have enstatite phenocrysts and exhibit the unusual crystallization sequence olivine → enstatite → pigeonite → augite owing to high MgO and SiO_2 in the melt and very low plagioclase components (see Sharaskin and Dobretsov, 1979; Hickey and Frey, 1979). The Point Sal lavas, however, never carry orthopyroxene nor do they exhibit the marianite crystallization sequence, and their plagioclase content is much higher. Moreover, the boninites and marianites show LREE enrichment (Hickey and Frey, 1979), in contrast to the Point Sal lavas. Detailed comparisons show many other chemical and mineralogical differences. Most telling, the Point Sal lavas were *not* erupted in close proximity to a volcanic island arc, as shown by the sparseness and nonvolcanogenic character of the intercalated sediments (Hopson and Frano, 1977). Present evidence indicates that arc-type volcanic rocks, including boninites and marianites, are not represented in the Coast Range ophiolite volcanic member.

Distinct from the ophiolite lavas is the enormous pile of submarine lavas that lie at the base of the Great Valley Sequence at Stonyford (loc. 18). These lavas are younger (lower Tithonian) than the ophiolite (Middle Jurassic) and faulted against it, and their chemistry allies them with oceanic island basalts. This is shown by their high TiO_2 (2 to 3%) and high P_2O_5 (0.25 to 0.44%) content (Bailey and Blake, 1974), and by their high REE level (20 times chondrite; Blake and others, 1975). Thus, the chemistry supports the conclusion from field and age data (loc. 18 description) that the Stonyford lavas are the remnants of a large, Late Jurassic submarine volcano (seamount) built upon Middle Jurassic ocean crust (ophiolite).

Most of the Coast Range ophiolite volcanic rocks are strongly altered, as already indicated. Rocks low in the volcanic sections are generally changed to greenschist-facies mineral assemblages (chlorite + albite + epidote + actinolite + sphene ± calcite ± quartz), while the stratigraphically higher volcanics are commonly altered to lower-grade assemblages composed chiefly of albite and green phyllosilicate (Fe-rich chlorite) plus or minus small amounts of pumpellyite, calcite, sphene, hematite, quartz, and clay. Zeolites are rare. The uppermost basaltic lavas at Point Sal retain their primary mineralogy except for pervasive alteration of olivine and groundmass glass to green smectite; this has been ascribed to sea-floor weathering (Hopson and Frano, 1977). More commonly, however, pervasive albitization and choritization extend all the way to the top of the ophiolite volcanics and in some cases high up into the overlying Great Valley Sequence (Dickinson and others, 1969). There has been no systematic study of alteration (metamorphism) in the ophiolite volcanics, but it is apparent that the alteration patterns will be complex in space and time. The lower volcanics and sheeted intrusives have been metamorphosed and metasomatically altered by heat and fluids rising from the underlying magma body (plutonic sequence). Higher in the section it will be difficult to separate the superimposed effects of (1) thermal metamorphism, which decreases upward, (2) pervasive or fracture-controlled alteration throughout the pile, which may have resulted from the circulation of heated sea water shortly after the lavas accumulated, (3) later sea-floor weathering, which dies out downward, and (4) much later burial metamorphism beneath the Great Valley Sequence. The latter achieved laumontite and locally prehnite-pumpellyite grade in the base of the Great Valley Sequence, according to Bailey and Jones (1973b).

Petrogenesis

Bailey and others (1970) first suggested that the ophiolitic sequence of igneous rocks at the base of the Great Valley Sequence is a remnant of Jurassic oceanic lithosphere. Their conclusion has been strengthened by subsequent investigations and now seems well established. The main points are these:

1. The volcanic rocks, now altered to spilite and keratophyre, correspond to ocean-ridge basalt in their eruptive forms, primary textures, and phenocryst mineralogy, and in distinctive facets of the major- and minor-element chemistry;
2. Radiolarian chert and pelagic limestone intercalated within the volcanics indicate a deep-sea setting;

3. The volcanics were fed through sheeted intrusive (dike or sill) complexes;

4. The plutonic sequence is roofed only by oceanic volcanic rocks (with basal sheeted complex), and is floored only by peridotite (harzburgite) tectonite;

5. The harzburgite tectonite corresponds to oceanic mantle, and it (including derivative serpentinite) is the only basement rock type;

6. The plutonic assemblage forms a sequence of ultramafic and gabbroic cumulates overlain by noncumulus gabbro and dioritic rocks, corresponding to a complete differentiation sequence from tholeiitic olivine basalt magma;

7. The least-altered plutonic rocks have Sr-isotopic compositions (near 0.7027) of oceanic tholeiite;

8. Petrographic and isotopic evidence indicates that sea water penetrated the upper and lower crustal rocks and entered the magma chamber, affecting late magmatic crystallization and hydrothermal alteration;

9. A strong "soft-sediment style" of hypersolidus deformation in the layered cumulates points to magmatic crystallization in a setting of continuous tectonic activity, such as a spreading center;

10. The ophiolite lithologic assemblage matches rocks drilled and dredged from the deep ocean floor;

11. The seismic properties and thickness of the ophiolite sequence correspond to oceanic crust and upper mantle, as shown in the next section.

These and other features, taken together, indicate an oceanic crustal and upper mantle origin for the Coast Range ophiolite.

Current controversy hinges mainly around two questions. First, does the Coast Range ophiolite represent "normal" oceanic crust formed at a spreading ocean ridge (accreting plate boundary), or did it form in a spreading back-arc basin? Second, are the ophiolite remnants composed solely of oceanic crustal and mantle rocks, or have there been later magmatic additions, perhaps in new tectonic settings? Specifically, have new island-arc type magmas invaded the Coast Range oceanic crust?

Petrologic and chemical evidence have not resolved the first problem. Hawkins (1977, 1979) and Saunders and others (1979) show that ocean-ridge basalts and marginal-basin basalts more or less overlap in their chemistry and petrography, and reliable distinguishing criteria have not yet come to light. Modern marginal basins have not yielded keratophyric or more silicic volcanic rocks; thus, the occurrence of keratophyric rocks within the ophiolite does not provide evidence for a marginal basin origin, as once supposed. We believe that the Coast Range ophiolite formed at a spreading ocean ridge that lay in *front* of an active island arc. The evidence, however, comes mostly from the sedimentary record, and our explanation is deferred to a later section.

Regarding the second question, there have indeed been later magmatic additions to the Middle Jurassic (pre-Oxfordian) oceanic crust now preserved in the Coast Ranges. Two kinds of renewed magmatic activity are documented, both in the early Tithonian some 20 m.y. after the oceanic crust had formed. One type formed the large oceanic volcano (seamount) at Stonyford (loc. 18), built of high FeTi, REE-rich pillow basalt.

The field relations, age, and chemistry of the Stonyford seamount are discussed in earlier parts of this chapter. The second type of post-ophiolite magmatic activity in the early Tithonian involves the intrusion of small amounts of highly siliceous magma. At Llanada (loc. 7) and Del Puerto Canyon (loc. 9), quartz keratophyre sills and dikes invade the volcanic member of the ophiolite and the overlying early Tithonian marine tuffaceous strata (Lotta Creek Tuff). Far to the south at Santa Cruz Island (loc. 1), a small plagioclase mass (Alamos pluton) dated at about 144 m.y. intrudes 164 m.y. old ophiolite (Fig. 14-7). In all three cases the volume of new magma is very small, and its composition is highly siliceous, Na-rich, and very K-poor; that is, it corresponds to oceanic plagiogranite (Coleman and Peterman, 1975). Blake and Jones (Chapter 12, this volume) interpret these intrusions at the Diablo Range localities as evidence of island-arc magmatism associated with Franciscan subduction. However, the volume of the intrusions is very small, and their age (early Tithonian) makes them much older than Franciscan subduction, which began in the early Cretaceous (Blake and Jones, 1974). They are about the same age as the Stonyford seamount, which also predates Franciscan subduction. Thus, the age, the composition, and the very small volume of the Diablo Range quartz keratophyres favor an oceanic origin.

We conclude that the main bulk of the Coast Range ophiolite is composed of oceanic lithosphere that formed at an oceanic spreading center (or centers) during the Middle Jurassic, and that off-axis magmatic activity has made local additions to the oceanic crust during the early Tithonian, approximately 20 m.y. later. The Stonyford seamount appears to have been a typical ocean-island volcano, perhaps related to mantle plume activity. The plagiogranite (quartz keratophyre) intrusions appear to be oceanic in character, but whether they are the product of ocean-ridge crest jumping (Menard, 1978) or some form of off-axis magmatic activity is unknown. Arc-type magmatic activity is not represented.

We turn now to the development of the main, Middle Jurassic stage of oceanic crust growth. An important problem here is to relate the volcanic rocks to processes going on in the underlying magma chambers (represented now by the plutonic rocks) and to link the magmatic processes with tectonics.

Magma that erupted to form the lavas can be related to magma that formed the cumulates by comparing their FeO*-MgO ratios. This ratio is an index of the fractionation of tholeiitic magma, and it has been widely applied to oceanic basalts (e.g., Bryan and others, 1976; Hekinian and others, 1976; Bryan and Moore, 1977) and to cumulates in layered igneous intrusions. The FeO*-Mg ratio of the basalt magma is estimated from the whole-rock composition, since glass is no longer preserved. This will not be greatly in error for the relatively aphyric Coast Range lavas, if the alteration is not too severe. The FeO*-MgO ratio of the liquids from which the cumulates crystallized may be estimated from the composition of the cumulus olivine and known value of the crystal-liquid partition coefficient (Roeder and Emslie, 1970; Church and Riccio, 1977; Stern, 1979). Figure 14-11 plots the FeO*-MgO of olivines from cumulate ultramafic rocks and gabbros at Point Sal (Hopson and Frano, 1977) and Del Puerto Canyon (Evarts, 1977) against the FeO*-MgO of the liquids from which they crystallized, using $K_D = 0.27$ determined for Mid-Atlantic Ridge basaltic glasses (Bryan, 1979; Rhodes and others, in press). These data show that the FeO*-MgO ratio of the Point Sal magma changed from 0.70 to 1.75 during the time its cumulates (ol, cpx + ol, pl + cpx + ol) were forming (Figs. 14–9, and 14–11). The FeO*-MgO ratio of the Del Puerto magma ranged from 0.76 to 1.95 during the time its ultramafic cumulates (ol, ol + cpx ± opx) were forming (Fig. 14–11). This ratio increased to still higher values in the Del Puerto magma while opx + cpx - pl were crystallizing and settling to form the gabbro cumulates (Evarts, 1977, Fig. 6), but specific values cannot be calculated when olivine is no longer a liquidus phase. Also plotted on Fig. 14–11 for comparison is the range of basalt (or basaltic glass) compositions for (1) the Coast Range ophiolite, (2) the Stonyford seamount, (3) the Mid-Atlantic Ridge (Bryan, 1979), (4) the Lau Basin in the western Pacific (Hawkins, 1977), and (5) Kilauea volcano in Hawaii. These are provided as examples of basaltic magma associated with a slow-spreading ocean ridge (3), a spreading interarc basin (4), and an oceanic-island volcano (5).

The significance of these data is as follows. First, the limited range of melt fractionation in the Point Sal magma chamber while olivine continuously crystallized and settled supports our earlier conclusion from the cryptic and phase variation that this was an open-system magma chamber that was unable to fractionate extensively because of repeated replenishment by new primitive magma. Such open-system conditions would exist at an oceanic spreading center where the magma chamber was open and was continuously replenished from the center, while spreading moved the two sides apart. Extreme iron enrichment (FeO*-MgO = 4.0), however, is shown by the noncumulus dioritic rocks at the top of the Point Sal plutonic sequence. This indicates that the magma chamber completed its crystallization and differentiation under closed-system conditions, probably when spreading had moved the sides of the magma chamber beyond the reach of replenishment from the center. Most of the Coast Range ophiolite *lavas* (Fig. 14-11) have FeO*-MgO ratios that correspond to the Point Sal magma during its main, open-system stage of crystallization. This suggests that these lavas were erupted along an active spreading axis, rather than off to the sides where late-stage residual magma had gone to closed-system crystallization and extreme fractionation. This agrees with observations at modern spreading ocean ridges, where the extrusion of new lava is limited to a relatively narrow zone close to the spreading axis (Bryan and Moore, 1977).

Fractionation in the Del Puerto magma chamber presents an important contrast to Point Sal. Here the FeO*-MgO ratio of the melt changed from 0.70 to 1.96 while only the basal ultramafic (ol, ol + cpx ± opx) cumulates were forming, and it continued to fractionate to still higher values while the gabbroic (pl + opx + cpx) cumulates developed. Most of the Coast Range ophiolite lavas correspond to the Del Puerto magma during its *earliest* stage of fractionation while the ultramafic cumulates were forming, but not to the later stages when the iron-rich gabbroic melts and even more fractionated dioritic melts evolved. This suggests that the Del Puerto magma changed from open- to closed-system fractionation quite early in its crystallization history. This may mean that active sea-floor spreading ceased or jumped to a new location, cutting off replenishment. The relative abundance of quartz-rich differentiates (including quartz keratophyre) at Del Puerto Canyon might be explained by this lack of replenishment, which probably is tectonically controlled.

Another clue to tectonic conditions during magmatic crystallization of the ophiolite comes from the amount of plutonic rock. It was pointed out earlier that the Coast Range ophiolite has very thick volcanic sequences, but relatively thin, perhaps even discontinuous, plutonic sequences. This equates with small magma chambers. The thickest plutonic sequence, about 1.5 km, is at Point Sal, but this is thin compared with up to 6 km of gabbro observed in the Samail ophiolite of Oman (Hopson and Pallister, 1979a). It has been shown from thermal calculations that the size of magma chambers beneath spreading ocean ridges is critically dependent upon spreading rate. Thus, fast-spreading ridges will have the largest magma chambers, while magma chambers cannot be continuously maintained if the half-spreading rate drops below about 0.5 cm/yr (Sleep, 1975; Kusznir and Bott, 1976). Thus, we infer from the rather small size and perhaps discontinuous character of the Coast Range ophiolite magma chambers that they were developed in relatively slow spreading oceanic crust. We have already deduced slow spreading or occasional cessation of spreading from the early change from open- to closed-system fractionation, illustrated by the Del Puerto Canyon plutonic sequence.

Coast Range Ophiolite Compared with Seismic Velocity Structure of Oceanic Crust

Despite evidence for an oceanic crustal origin, the stratigraphic thickness of the ophiolite above the peridotite contact seems too thin, relative to oceanic crust in seismic profiles for the deep oceans and interarc basins. The ophiolite volcanic zone corresponds quite well to oceanic layer 2, but the thin plutonic sequences (up to 1.5 km) do not begin to match the 3.5- to 7-km thicknesses for layer 3, determined from seismic refraction at sea (Raitt, 1963; Shor and others, 1970; Woollard, 1975). This problem is not peculiar to the Coast Range ophiolite, but holds for many of the world's well-known ophiolite complexes, which, from other evidence, are clearly oceanic crustal remnants (Moores and Jackson, 1974).

This "thickness problem" probably results from the assumption that the seismically defined oceanic crust-mantle boundary (M-discontinuity) coincides with the transition from gabbroic to ultramafic rocks. Alternatively, if the upper part of the peridotite layer were partly serpentinized, its seismic velocities would be lowered and M would lie at a deeper level, that is, at the boundary between serpentinized and fresh peridotite (Clague and Straley, 1977). Luyendyk and Nichols (1977) point out an additional important effect: the serpentinization of lower oceanic crustal and upper mantle rocks can produce a seismic velocity reversal. Two results of this velocity reversal are (1) the low-velocity layer goes undetected by conventional seismic refraction techniques, and (2) the increased delay time produced by the low-velocity layer puts the seismic M-discontinuity at a greater *apparent* depth than it really is. Two lines of evidence from the deep ocean basins suggest that this situation is actually realized: (1) recent seismic evidence shows that velocity reversals do occur at the base of the oceanic crust in some areas (Lewis and Snydsman, 1977; Meeder and others, 1977), and (2) dredging shows that serpentinized peridotite is locally abundant within the Mid-Atlantic Ridge and some other deep-ocean sites, demonstrating that serpentinization does take place in spreading ocean-ridge settings.

Thus, it now seems evident that layer 3 of some oceanic crustal sections can consist of thin gabbroic sequences underlain by partly serpentinized dunite and harzburgite, which pass downward (M-discontinuity) into fresh mantle harzburgite. This type of oceanic crustal profile would result from very small magma chambers, which solidify to thin plutonic sequences, and to sea-water penetration extending through this thin igneous crust into the underlying peridotite, producing a serpentinized upper layer.

The California Coast Range ophiolite evidently preserves a sample of this type of oceanic crust. The thick volcanic sequence, including basal dike and sill complex, adequately matches the thickness and seismic velocity of oceanic layer 2 (Hopson and others, 1975a; Nichols, 1977), and the thickness requirements of layer 3 are satisfied if a partly serpentinized peridotite layer (V_p = 5.5) approximately 1 km thick underlies the thin (1.5 km) plutonic sequence (Nichols, 1977; Nichols and others, in press). The assumption in the Nichols model that partial sepentinization occurred in situ beneath the ocean floor is strongly supported at ophiolite localities in the northern Coast Ranges, where stratigraphic evidence clearly shows that peridotitic oceanic basement has already extensively serpentinized (and melanged) before distal clastic sediments of the basal Great Valley Sequence were deposited on the deep ocean floor (locs. 20 to 23). Also, at other northern Coast Range localities (locs. 17, 18), uplifted oceanic basement shed serpentinite detritus into the lower part of the Great Valley Sequence while it was being deposited. Thus, Coast Range ophiolite peridotite was already partly serpentinized while it was still a part of the sea floor, even though additional serpentinization by low-temperature meteoric waters occurred much later, during or following its emplacement on land (Magaritz and Taylor, 1976). Finally, Sr-isotopic evidence indicates that sea water penetrated at least as deep as the ultramafic cumulate zone of the ophiolite at Point Sal (Lass and Davis, 1979; Davis and Lass, 1976), and it probably reached deeper. Deep penetration of heated sea-water solutions would explain the in situ serpentinization along the top of upper mantle peridotite, beneath thin oceanic crust.

In summary, the stratigraphy and thickness of the Coast Range ophiolite are compatible with the seismic velocity structure of modern oceanic crust where partly serpentinized ultramafic rock in the lower part of layer 3 produces a velocity reversal that overestimates the depth of M. Such crust is probably characteristic of slow-spreading ocean ridges that have only small magma chambers, which solidify to thin plutonic sequences.

JURASSIC SEDIMENTARY ROCKS ASSOCIATED WITH THE OPHIOLITE

The sedimentary rocks that are intercalated with the ophiolite lavas, and that have been built up upon them, have already been described in detail for the individual localities. Here we will briefly summarize the evidence for conclusions that bear on the tectonic history of the ophiolite.

Middle Jurassic Interpillow Sediments

Small pockets of sediment between lava pillows and injecting the base of lava flows are locally abundant in the Coast Range ophiolite. They are usually red radiolarian cherts or else light-colored, recrystallized limestones with rare faint remnants of coccoliths. These rocks were derived from biogenic pelagic oozes, which normally accumulate slowly. Since they are usually uncontaminated with terrigenous or volcaniclastic detritus, we infer a depositional environment in the open ocean, well removed from continental margins or active volcanic arcs.

MnFe-rich umbers, conspicuous as interpillow sediments in the Cyprus and Oman ophiolites, are not found in the Coast Range ophiolite lavas. This might suggest that the Coast Range ophiolite spreading system lacked an active ridge-crest hot spring system, like those found today along moderately fast spreading ridges such as the Galapagos and East Pacific rises. It would compare more closely in terms of sediment, to the slow-spreading Mid-Atlantic Ridge, where there is little hot spring activity and only biogenic calcareous ooze is accumulating between eruptions.

Arc-derived volcaniclastic materials are also lacking as interpillow sediment. This argues against formation of the ophiolite in a spreading interarc basin, where the lavas could hardly escape being blanketed, between eruptions, by airborne tephra and current-transported volcaniclastic detritus.

In conclusion, sedimentation during the Middle Jurassic (Fig. 14-8) in the region where the ophiolite formed consisted mainly of the exceedingly slow accumulation of siliceous and calcareous biogenic oozes. An open-ocean setting is inferred.

Late Jurassic Cherts and Volcaniclastic Sediments

Oxfordian to lower Tithonian pelagic and hemipelagic deposits blanket the upper surface of the ophiolite lavas (Fig. 14-8). These range from nearly pure radiolarian cherts through tuffaceous cherts to water-laid tuffs and coarser volcaniclastic sediments. There is a complete gradation between these end members, when the ophiolite remnants are considered collectively. This gradation is evident in the field and also under the microscope, where radiolaria, volcanic microlites, and pumice and shard outlines are all readily identified. The volcanic crystal assemblage consists chiefly of lathlike plagioclase, augite, hypersthene, hornblende, and magnetitie, typical of andesitic and dacitic tephra. The gradation from chert to volcanic ash (tuff) is also evident from whole-rock chemistry. Figure 14-12 plots SiO_2 versus Al_2O_3 for suites of these sediments from four ophiolite localities, revealing their intergradation. Other sediment components are not apparent; clastic sand grains, for example, are rarely seen. These deposits were evidently laid down in a deep marine environment in which radiolarian ooze was accumulating, but was mixed to varying degrees with airborne volcanic dust and ash that eventually settled to the bottom, or with sandy volcaniclastic detritus transported by turbidity currents. An active volcanic arc was the source of this calc-alkaline detritus.

Figure 14-13 shows how the thickness of these deposits is related to their compo-

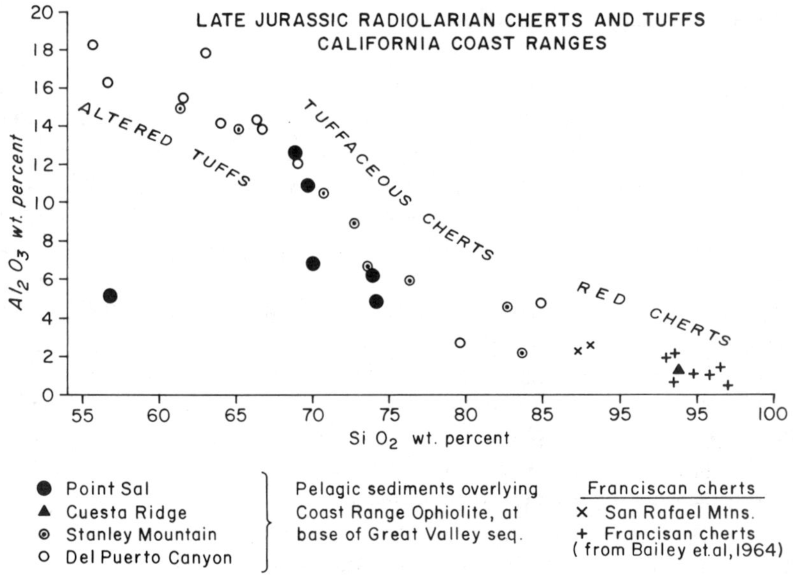

Fig. 14-12. Plot of alumina versus silica for Late Jurassic radiolarian cherts, tuffaceous cherts, and tuffs from ophiolite localities 2 to and 9 of Fig. 14-1.

sition. The radiolarian cherts with a low volcaniclastic component are generally less than 100 m thick, but the sections thicken markedly with an increase in volcaniclastic components. Moreover, the volcaniclastic sediments themselves coarsen in the direction of thickening. Thus, viewed collectively, the Oxfordian to early Tithonian sediment blanket that covers the Coast Range ophiolite forms a tapering wedge, from a thick proximal facies of andesitic to dacitic volcaniclastics, including bouldery submarine debris-flow deposits, to a thin distal facies of radiolarian chert with only traces of tuffaceous component. One step farther takes one to the tuff-free radiolarian cherts that cover Franciscan pillow lavas (Figs. 14-12 and 14-13).

It will not escape the reader's attention, viewing Fig. 14-8, that a time gap exists between the age of the ophiolite and the age of the lowest sedimentary deposit above it. This may partly be due to a poor correlation between the radiometric and fossil time scales for the Jurassic. However, there is also an abrupt change in sediment and faunal types at the ophiolite-basal sediment contact, suggesting a real time gap. At Point Sal, for example (Fig. 14-8, loc. 2), there is limestone (interpillow) just below the contact and chert above, no volcaniclastic component in the limestone but abundant tuff in the chert, and no radiolaria in the limestone but abundant well-preserved radiolaria in the chert. This suggests that a major submarine disconformity exists here, and that seafloor spreading had carried the oceanic crust from one environment to another during an extended period of nondeposition. Van Andel and others (1975) record submarine unconformities in the central equatorial Pacific that span up to about 30 m.y., so a hiatus of 10 to 20 m.y. between the ophiolite and overlying sediments is not exceptional.

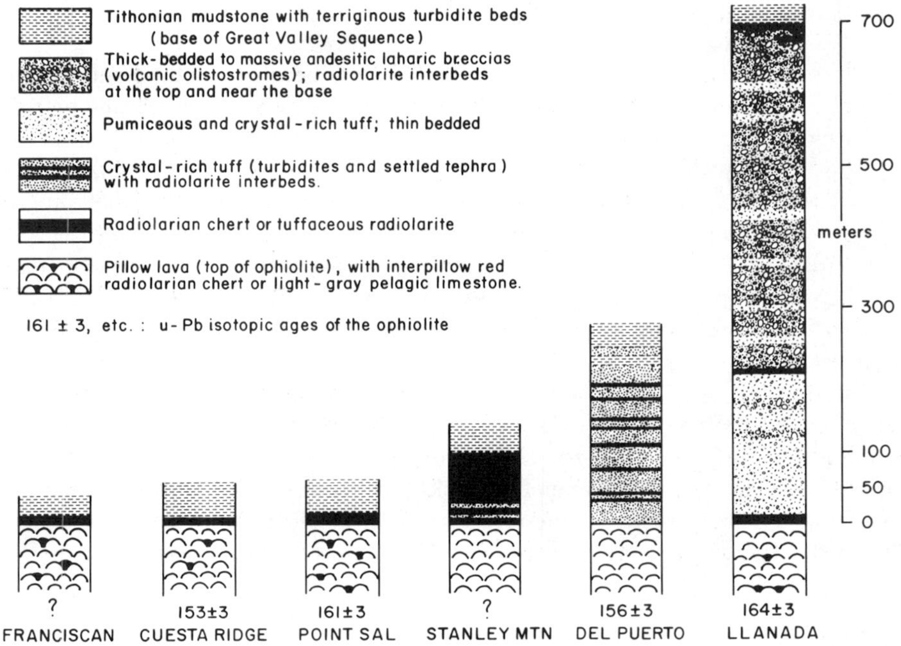

Fig. 14-13. Columnar sections comparing thickness and facies changes for Oxfordian to lower Kimmeridgian cherty and volcaniclastic strata that rest on the ophiolite at localities shown in Fig. 14-1. Shown for comparison is a columnar section of Franciscan pillow lava capped by red radiolarian chert.

Late Jurassic Ophiolite Breccias

Breccias with gravel- to boulder-sized clasts consisting entirely of ophiolite detritus occur between the top of the ophiolite and the basal mudstones of the Great Valley Sequence at localities in the Sonoma-Lake County area and along the western side of the Sacramento Valley (locs. 13 to 15, 22, 23). The breccias are polymictic, with angular unsorted fragments held together by iron oxide or by small amounts of iron-rich mudstone. Some of the breccias consist chiefly of fragments of the ophiolite volcanics, while others include abundant diabase and/or plutonic rock fragments, including pyroxenite. It was concluded earlier that the breccias were talus deposits that accumulated at the foot of submarine fault scarps that uplifted the igneous oceanic crust (see descriptions of localities 13, 14, 22, 23). Other workers interpret these breccias as ophiolitic olistostromes (Phipps and others, 1979). In either case, the ophiolitic breccias record the disruption and uplift of the igneous oceanic crust during Kimmeridgian or early Tithonian time, before deposition of middle Tithonian basal mudstones of the Great Valley Sequence.

Latest Jurassic Strata of the Great Valley Sequence

Middle to upper Tithonian mudstones, with thin interbeds of siltstone and sandstone and local conglomeratic lenses, cover all of the other deposits. The structure, petrology, and transport directions of these strata are well known from the studies of Irwin (1960), Ojakangas (1968), Dickinson (1971, 1974), Dickinson and Rich (1972), and many others. These strata represent the distal edges of large submarine fans that prograded southward (in the northern Coast Range) and westward across the deep ocean floor. Their source terranes were the igneous and metamorphic rocks of the Sierra Nevada and Klamath Mountains. A close modern analogy would be the Astoria and Nitinat fans that cover the Gorda-Juan de Fuca plate off the Oregon-Washington coast (Kulm and Fowler, 1974).

Interbedded with the lower Great Valley Sequence in northern California, especially toward its base, are beds of dark sandstone and microbreccia. These have been called basaltic sandstone (Brown, 1964), but they really are polymictic lithic wackes composed of a wide array of ophiolitic rocks plus radiolarian chert, slaty argillite, and so on. They evidently represent the recurring uplift of blocks of the oceanic basement, which shed detritus locally into the advancing terrigenous submarine fan.

Conclusion

The record of Jurassic sedimentation within and above the ophiolite presents a clear progression from open-ocean pelagic sedimentation, through a period of nondeposition and/or sea-floor erosion, to resumption of pelagic sedimentation (radiolarian ooze), but within reach of volcanic arc-derived tephra and volcaniclastic flowage deposits, and, finally, to cessation of volcanic sedimentation as an advancing wedge of terrigenous subsea fan deposits spread over the deep-sea floor. The tectonic consequences of this progression are explored in the next section.

TECTONIC HISTORY OF THE COAST RANGE OPHIOLITE

Here the tectonic history of the ophiolite is traced from its formation in the Middle Jurassic to its encounter with a continental margin in the latest Jurassic and Early Cretaceous.

Middle Jurassic Plate Accretion

The progression of sedimentary environments from Middle to Late Jurassic time, described in the previous section, is compatible with formation of the ophiolite at a spreading ocean ridge that was well removed from land, its subsequent transport by sea-floor spreading across an oceanic region where sedimentation was nil, its approach to an active volcanic arc, followed by encounter with a continental margin. However, the sedimentary progression (stratigraphic sequence) is definitely incompatible with an origin of the ophiolite in a spreading back-arc basin. For example, an interarc rift would initially be narrow and then gradually widen. Volcaniclastic sedimentation from the sides would accompany the formation of new oceanic crust, and then would bury it afterward. Recent deep drilling across the Mariana Trough (DSDP Leg 60), for example, found that the sediments are predominantly volcaniclastic material derived from Mariana arc volcanism since Miocene time.

Not only the nature of the sediments but also the time of their deposition is critical. Thus, Evarts (1977) postulated that the Del Puerto ophiolite remnant formed in an interarc basin setting because of its thick cover of volcaniclastic sediments. However, Fig. 14-8 shows that these volcaniclastics did not begin to accumulate until millions of years after the ophiolite had formed. The problem of explaining how a developing rift, adjacent to an active arc, is kept bare of sediment for millions of years seems insurmountable. The open-ocean ridge interpretation of the ophiolite is therefore adopted. The open-ocean ridge that gave rise to the Coast Range ophiolite was probably slow spreading, perhaps like the modern Mid-Atlantic Ridge, but situated nearer to land. The evidence for slow spreading, developed earlier, is summarized as follows:

1. The thin plutonic sequences indicate small magma chambers, which are a consequence of slow spreading (Kusznir and Bott, 1976).
2. The premature change from open- to closed-system fractionation, shown especially by the Del Puerto (loc. 9) plutonic sequence, means termination of replenishment. This may be linked to a tectonic control such as a spreading slow-down or the abandonment of a magma chamber.
3. The sheeted sill complexes developed between the plutonic and volcanic units appear to result from intrusion of magma into a slow-rifting (or nonrifting) roof.
4. The absence of MnFe-rich sediments (umbers) between the pillows may mean a slow or inactive hot-spring system, which suggests slow spreading.

The question of how far the ridge lay from land may be roughly estimated from the time interval between the formation of new oceanic crust and the beginning of its burial by the clastic sediments, if the spreading rate is known or can be estimated (Lanphere, 1971). More specifically, this time interval might measure (1) the length of time it took for sea-floor spreading to move a segment of newly formed crust from an accreting to a consuming plate boundary, or (2) the time it took for an actively growing continental rise to advance its front seaward over stationary (or very slowly spreading) sea floor to this crustal segment. In the present case, possibility (1) probably applies from the time the crust was formed until about the late Kimmeridgian, whereas possibility (2) holds for the early Tithonian. The time interval, for the Point Sal section, ranges between about 20 and 25 m.y., using the Van Hinte 1976a time scale (Fig. 14-8).

Using the sea-floor spreading model and assuming an average half-spreading rate of 3 cm/yr and a time interval of ~20 m.y., and further assuming that the ridge axis did not greatly change its position during that spreading period, the ridge axis would have lain approximately 600 km seaward of the continental rise. Alternatively, a slow half-spreading rate of 1 cm/yr would place the ridge axis only about 200 km out. The actual distance was probably somewhere between these extreme estimates.

Middle and Late Jurassic Sea-Floor Spreading and Deep-Sea Sedimentation

Following formation at a slow-spreading ocean ridge, the ophiolite was transported relatively passively toward the continent by sea-floor spreading. Initially, the ophiolite was evidently kept swept clean of sediments, as discussed earlier. Beginning in the Oxfordian, however, cherts, tuffaceous cherts, and tuffs were deposited on its surface. The thin, relatively pure chert sections (e.g., locs. 2, 4) must have been deposited in deep-ocean sites far from any active volcanic arc. Thicker, more tuff-rich sections were within the reach of airfall and turbidity current-derived volcanogenic sediments, but still beyond the influence of terrigenous clastic sedimentation. Thus, an oceanic environment still obtained for much of the Middle and Late Jurassic.

Late Jurassic (Kimmeridgian) Sea-Floor Disruption

The early stages of sea-floor spreading of the "Coast Range segment" of oceanic crust formed about 165 to 152 m.y. ago were evidently serene, from the sedimentary record discussed previously. However, beginning in the Kimmeridgian (Fig. 14-8), the widespread development of ophiolitic breccias on the upper surface of the ophiolite in the northern Coast Range terrane evidently marks the development of large fault scarps on the sea floor, which built up thick talus accumulations. Some scarps were sufficiently large to expose the deeper plutonic crustal layers, judging from the talus blocks of cumulate gabbro and pyroxenite. At the South Fork of Elder Creek (loc. 23), basement pyroxenite was brought all the way to the surface, for it is covered unconformably by ophiolitic breccia. The ophiolitic breccias at some localities contain such a mixture of different plutonic and volcanic rock types that it seems unlikely they came from a single scarp face. Transportation and mixing of fragments seems to be indicated.

Large-scale faulting at the surface was accompanied by strong tectonic disruption and mixing at depth in the ophiolite belt west of the northern part of Sacramento Valley. This is impressively documented by the serpentinite-matrix melanging, recorded especially at Paskenta (see loc. 22 description). It appears that oceanic crust and upper mantle were strongly broken and dismembered, allowing sea water to penetrate into the peridotite to cause extensive serpentinization. Further strong deformation mobilized the serpentinite, causing it to rise diapirically and to intrude the disrupted upper crust. The large-scale vertical faulting of the sea floor, resulting in the ophiolitic scree deposits, is evidently closely linked to the serpentinization and melanging at depth. The serpentinite melanging is not seen south of Tehama County, however, and neither the melanging nor the ophiolitic breccias are recognized south of Mount St. Helena (loc. 13).

Late Jurassic (Early Tithonian) Off-Ridge Intrusion and Volcanism

Early Tithonian time saw renewed igneous activity at some of the Coast Range ophiolite sites. At Santa Cruz Island (loc. 1), ophiolite (Willows Complex) about 164 m.y. old was intruded by the Alamos plagiogranite some 20 m.y. later (see description of locality 1 and the section on isotopic ages). Minor siliceous intrusions of similar age (based on radiolaria in closely associated sedimentary rocks) also occur in the Diablo Range. One such intrusion, a 30-m-thick sill of quartz keratophyre, intrudes the thick section of arc-derived pyroclastic rocks at Llanada (loc. 7). In a similar fashion, quartz keratophyre sills intrude the cherts and arc-derived volcaniclastic rocks of the Lotta Creek Tuff at Del Puerto Canyon (loc. 9). These Tithonian intrusive rocks are all similar chemically to the low-K_2O siliceous differentiates of the Middle Jurassic ophiolite. We conclude (p. 431, 494) that they are the product of off-ridge oceanic igneous processes, not of island-arc magmas (Jones and others, 1976).

A different kind of off-ridge magmatic activity is represented at Stonyford (loc. 18), where a 2- to 3-km-thick pile of TiO_2-rich basalt of early Tithonian age overlies a small remnant of harzburgite tectonite. The thick Stonyford volcanic pile is evidently the remains of a large Tithonian seamount that was built upon older oceanic crust (now mostly faulted out; see loc. 18 description).

Latest Jurassic (Tithonian) Sedimentation and Tectonism

Middle to Late Tithonian time was marked by the encroachment of subsea fans of terrigenous detritus advancing westward from the Sierran terrane and southward from the Klamath terrane out over the disrupted sea floor. This encroachment of continent-derived muds and distal turbidities began somewhat earlier at northern Coast Range localities, starting in the earliest Tithonian at Paskenta (Jones, 1975). During this time, sheets of ophiolitic sandy detritus spread out from uplifted masses of oceanic basement, which evidently included diapirs of serpentinized peridotite. Much of this material came from the west, where a topographic barrier was rising, according to Maxwell (1974, (1977). These ophiolitic sands indicate that the sea-floor disruption which began in Kimmeridgian time continued into the Tithonian.

Tectonic Summary and Proposed Model

At this stage it is perhaps worthwhile to briefly summarize the tectonic history of the ophiolite and integrate this history into a tectonic model. Formation of the ophiolite at an open-ocean ridge has been supported earlier and is a rather simple (although still controversial) concept. Tectonic disruption of the ophiolite, its relationship to active volcanic arc(s), its sources of terrigenous sediments, and its relation to the Franciscan Complex are considered in the framework of a simple model, as follows.

The strong tectonic disruption and melanging of the oceanic crustal sequence might be related to fracture-zone tectonics, as postulated for the Sierran serpentinite melanges by Saleeby (1979) or to subduction tectonics to the west as suggested by Maxwell (1977). We favor a pattern of tectonic deformation similar to the one affecting the Nazca plate today as it approaches the Peru-Chile trench. Figure 14-14 shows the disruption of this plate as reconstructed by Hussong and others (1975a,b). In applying this analog to the Coast Range ophiolite, we envision the following sequences of stages:

1. In Callovian and Oxfordian time sea-floor spreading was carrying the undisrupted oceanic plate eastward toward a subduction zone (or zones) in the Sierran Foothills arc-subduction complex (e.g., Schweickert and Cowan, 1975). Volcanism from that complex supplied the tuffaceous detritus that was deposited along with radiolarian ooze on the oceanic plate.

2. Collision within the arc complex (Kimmeridgian, Nevada orogeny) choked off subduction, bringing about the cessation of volcanism in and the uplift of the Sierran arc terrane, which then began to shed terrigenous detritus (Great Valley Sequence) that spread rapidly westward over the sea floor.

3. The choking off of the old subduction zone also forced the locus of subduction to jump to the west. But the subduction zone did not jump to a new site (the site of Franciscan subduction far to the west) in a single step, as shown, for example, by Schweickert and Cowan (1975, Fig. 3). Instead, we believe that the subduction zone jumped westward in a series of steps. Thus, as suggested in Fig. 14-14, the progressively westward stepping subduction zone, while failing to become (dare we say) "entrenched" at any of the intermediate positions, would nevertheless leave in its wake a legacy of severe tectonic disruption of the intervening oceanic crust. This disruption was likely responsible for the development of great scarps on the ocean floor from which ophiolitic talus (now ophiolite breccia) was shed. It may also have provided access for seawater to the deeper levels of the ophiolite, thus promoting the development of serpentinites. These later rose diapirically to be exposed, eroded, and recycled, along with other ophiolite detritus, into the sedimentary sequence atop other portions of the ophiolite.

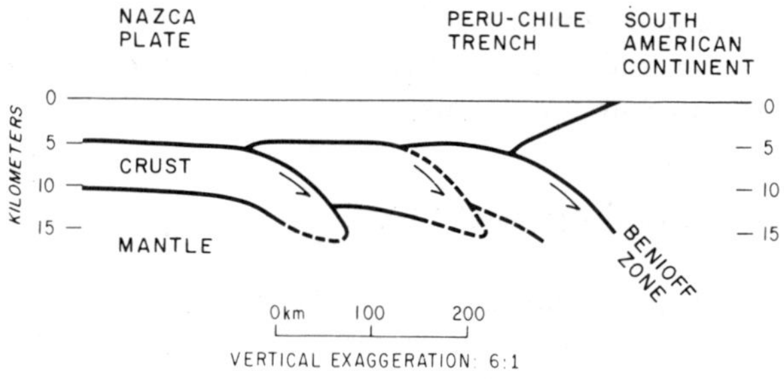

Fig. 14-14. Diagrammatic representation of crustal compression and underthrusting of a portion of the Nazca plate in front of the Peru-Chile trench (after Hussong and others, 1975a).

4. By latest Tithonian or earliest Cretaceous, the Franciscan subduction zone became established to the west of the Coast Range ophiolite, thus trapping the ophiolite against the continent. Continued long-term subduction then progessively built up the highly deformed accretionary wedge of the Franciscan complex.

The earliest (Tithonian-Valanginian) clastic sedimentary rocks within the Franciscan complex tend to be thicker bedded and coarser grained than contemporaneous flyschlike strata of the adjacent Great Valley Sequence that rest on the ophiolite. This raises the question of a source for these sediments, which obviously cannot be a simple distal facies of the already fine-grained Great Valley Sequence. Blake and Jones (1974) have postulated an island arc that lay between the Great Valley and Franciscan realms. Sediment shed westward from this inferred arc is thought to have supplied the Franciscan, while the distal Great Valley strata, to the east of the arc, were fed from the more distant Sierran-Klamath provinces to the east and north (Blake and Jones, 1974).

This postulated western arc poses major difficulties, however: an arc large enough to have shed vast volumes of sediment westward to the Franciscan realm would also have shed sediment eastward into the adjacent Great Valley depositional realm. Paleocurrent data for the Tithonian to Valanginian Great Valley strata, however, do not show transport from a western direction (Ojakangas, 1968). Furthermore, cherts from the Franciscan tend to be strikingly free from tuffaceous admixtures, whereas cherts from the Coast Range ophiolite-Great Valley sections contain variable and sometimes very large volcaniclastic contributions (see Figs. 14-12 and 14-13). This argues strongly against Franciscan proximity to an *active* arc. Finally, and most telling, no trace of the postulated western arc remains, even though it must have been huge to have supplied the vast volumes of sediment now found in the older part of the Franciscan complex.

These difficulties, taken together, make the presence of a Late Jurassic-Early Cretaceous volcanic arc to the west of the Great Valley realm highly unlikely. We suggest an alternate source for the thick-bedded sandy Franciscan strata: the proximal strata of Tithonian to Valanginian age of the Great Valley Sequence itself. Compression along the edge of the eastward-spreading oceanic (Farallon?) plate following termination of the subduction zone along the Sierran Foothills zone, but prior to its full reestablishment in the Franciscan subduction zone, may have resulted in extensive faulting and uplift of the intervening plate, as discussed earlier. Such uplift in the realm of proximal Great Valley Sequence deposition might have resulted in the delivery, via submarine canyons and in the form of submarine slides and grain flows, of large volumes of these unconsolidated sandy sediments through the continental shelf and slope (more distal Great Valley facies) and into a deepening Franciscan trench.

It may be asked, where does one find lower Great Valley Sequence strata that are thinned or missing, where they slid away to feed the Franciscan? Presumably, the remnants of this cannibalized proximal facies now lie buried beneath Cenozoic deposits of the Great Valley. The Diablo Range, however, is a region where much of the Upper Jurassic and Lower Cretaceous part of the Great Valley Sequence is missing.

The final aspect of the Coast Range ophiolite's tectonic evolution that we will consider briefly is the tectonic juxtaposition of the ophiolite, with its overlying strata, against the Franciscan Complex. The contact between the ophiolite-Great Valley Sequence and the Franciscan Complex has been regarded as a thrust fault of great magnitude (Bailey and others, 1964, 1970). In fact, the contact is widely held to be an exposed fossil Benioff zone, having resulted from underthrusting of the Franciscan beneath the Great Valley Sequence during subduction of the oceanic plate on the west (e.g., Ernst, 1970; Page, 1972). Strong dissenting views that challenge the existence of the Coast Range thrust have also been expressed (e.g., Raney, 1973; Maxwell, 1974).

From our own observations of the contacts between the base of the ophiolite and the Franciscan complex, we conclude that differential movement has been limited in most cases (e.g., see descriptions for locs. 19 and 20). We do not deny the need for a major structural discontinuity between the ophiolite-Great Valley Sequence on one hand and the Franciscan Complex on the other hand. Rather we believe that the *present* contact, in most cases, represents smaller scale Cenozoic faulting rather than the Coast Range thrust.

ACKNOWLEDGMENTS

Hopson's work on this project during 1974-1977 was partly funded by grants from the Academic Senate Research Committee, University of California, Santa Barbara. Mattinson was supported in part by National Science Foundation grant EAR78-12958. Pessagno's work was supported by grants from the National Science Foundation (GA-35094, DES72-01528, DES72-01528-AO1, EAR76-22029, EAR78-12923) and by funding from the Atlantic-Richfield Company, the Exxon Production Research Company, and the Mobil Oil Corporation. Steven Atlas assisted Pessagno in the field in 1975. The Oregon Department of Geology and Mineral Industries granted permission to reproduce figures 14-3, 14-9 and 14-10. Robert McLaughlin guided us through the ophiolite remnants at Geyser Peak and Harbin Springs. Marshall Maddock showed us the Lone Tree Creek ophiolite remnant. Ben Page first introduced us to the Cuesta Ridge ophiolite remnant. We heartily acknowledge the help of these and other colleagues who have enriched our knowledge of Coast Range geology. Foremost among these are Edgar H. Bailey, David L. Jones, M. Clark Blake, W. Porter Irwin, and Stephen Bezore, who first demonstrated the existence of Jurassic oceanic lithosphere in the California Coast Ranges.

Clarence R. Allen
Seismological Laboratory
California Institute of Technology
Pasadena, California 91125

15

THE MODERN SAN ANDREAS FAULT

ABSTRACT

Although the San Andreas fault was once viewed by most geologists as a grossly aberrant tectonic feature, it is now recognized as representing normal plate-tectonic processes and is, in fact, one of the simpler of the transform faults that cut the continents. The fault zone is continuous for at least 1100 km, but individual surface traces are en echelon segments of much more limited extent, and they represent fracture processes that currently are not fully understood. Studies of both the fault's physiographic expression and its recent earthquake history emphasize the remarkable temporal repeatability of seismic events along it; during Holocene time, earthquakes and their associated effects have repeated one another at given localities with almost uncanny reproducibility. Two great earthquakes occurred along the fault in 1857 and 1906, and although two remaining segments have not broken in large events during the historic record, the most likely next great earthquake is a repeat of the 1857 event. Seismographic stations in California now number perhaps 500, and the recent dramatic increases have shown that epicenters tend to markedly "pull in" toward the active fault trace in some segments but not others. In general, the parts of the fault showing concentrated alignments of small shocks are the same parts that display continuous or episodic surface creep. Currently locked segments of the fault, presumably those of highest seismic hazard, are characterized by very low, scattered seismicity. Short-term epicenter maps are thus not necessarily a good representation of long-term seismic hazard, which is much better portrayed by geologic studies of the fault's Holocene history, using radiometric dating of offset and disturbed strata. The 1857 and 1906 earthquakes represent the types of events we must be prepared for in the future, although their engineering effects were not as great as often imagined. The hazards of both surface fault displacements and heavy ground shaking are not yet fully understood, but appear to be problems capable of solution in the foreseeable future.

INTRODUCTION

Only a few years ago, California's San Andreas fault was looked upon by most geologists as some sort of aberrant feature of the earth's crust. Not only were the predominantly horizontal fault displacements that had been observed during major earthquakes such as that of 1906 thought to be rare by worldwide standards, but seemingly even more anomalous were the claims of total lateral displacements along the fault of several hundred kilometers (Hill and Dibblee, 1953; Crowell, 1962). It is now generally recognized, of course, that not only is the San Andreas fault completely compatible with world wide tectonism as embodied in the concepts of plate tectonics, but that features of this type are in fact demanded by relative plate motions in many parts of the world. And some of the same alleged displacements that were once criticized as being ridiculously large, such as the 25 miles suggested by Noble (1926), are now being viewed as perhaps too small (Woodburne, 1975). Indeed, it now appears that the San Andreas fault is among the *simplest* of the major transform faults that cut the continents. For example, Allen (1965) drew close comparisons between the San Andreas fault and the Alpine fault of New Zealand, the Atacama fault of Chile, and the Philippine fault. But it now turns out that each of these features has complications not shared by the San Andreas: the Atacama fault, although probably once a major transform fault, is now characterized by predominantly vertical displacements (Arabasz, 1971); the Alpine fault has very complex changes of tectonic regime with depth (Arabasz and Robinson, 1976); and the Philippine fault is evidently the result of rather complex oblique subduction (Fitch, 1972). Similarly, the North Anatolian fault of Turkey is situated anomalously within a highly complicated structural belt (McKenzie, 1972), and the spectacular strike-slip faults of western China are only indirectly related to plate-boundary processes (Molnar and Tapponnier, 1975). In contrast, the San Andreas fault appears to be a relatively straightforward transform that happens to cut across the corner of a continent, although dissenting views have been expressed (e.g., Hill, 1974). Its history, on the other hand, is anything but simple (Atwater, 1970), and many aspects of the geologic record remain to be worked out. The emphasis in this chapter will instead be on the modern San Andreas fault—its geological and seismological setting, its Holocene history, and its current seismic-hazard potential.

From its type locality at San Andreas Lake near San Francisco (Lawson, 1895), the fault can be followed as a distinct geologic feature north 200 km to Point Arena, and then for at least another 215 km as a predominantly submarine feature to the vicinity of Shelter Cove,[1] Humboldt County. Southeast from San Andreas Lake, one can easily trace the fault as a continuous feature for some 500 km to the Transverse Ranges, north of Los Angeles, where the fault frays out into a number of branches, all of which trend southeastward toward the Gulf of California and must be regarded as members of the overall fault system. The parent name has usually been associated with the easternmost member of the system, which trends through San Gorgonio Pass, Riverside County, and terminates as an identifiable single fault trace east of the Salton Sea. Although contemporary seismicity gives little reason for assigning the parent name to one branch or another, much the largest displacement during geologic history appears to have taken place on the easternmost break (Crowell, 1962, 1975a) rather than on the seemingly more continuous San Jacinto fault (Sharp, 1967). Thus assignment of the San Andreas nomenclature to the easternmost branch seems logical, and the fault may be assumed to terminate as a single, well-defined feature near Bertram, Imperial County. Still farther southeast in the Imperial Valley, individual faults within the overall system take on an en echelon pattern, separated by ridgelike segments, which is typical of the floor of the Gulf of California (Hill and others, 1975; Johnson, 1979).

The San Andreas thus extends as a well-defined single fault zone from Shelter Cove to Bertram (Fig. 15-1), a distance of some 1100 km. As such, it is certainly one of the longest identified continental strike-slip faults in the world, exceeded perhaps only by the Altyn Tagh fault of western China (Molnar and Tapponier, 1975).

[1]Most of the place names mentioned in the text are shown on Fig. 15-1.

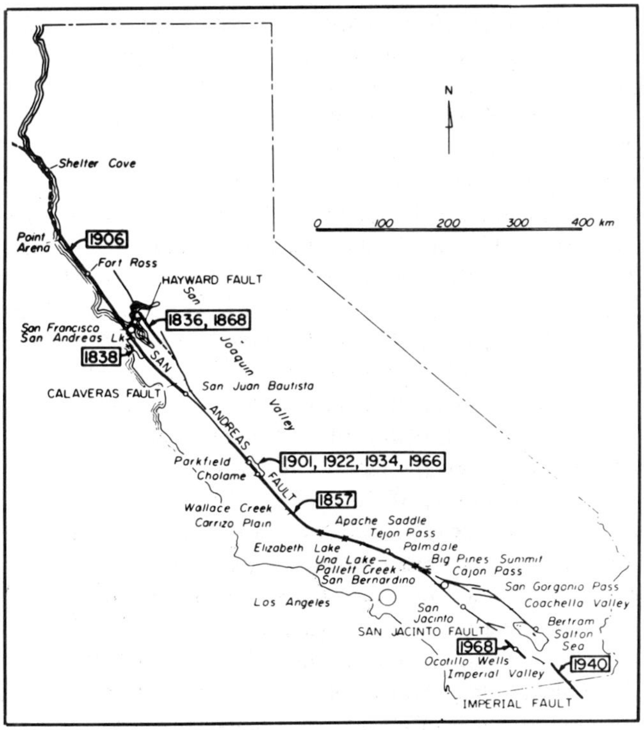

Fig. 15-1. Map showing extent of faulting during significant earthquakes on the San Andreas fault (heavy segments) and showing locations of most place names mentioned in text.

CONTINUITY OF THE FAULT

Just how continuous is the San Andreas fault between Shelter Cove and Bertram? Certainly it appears as a single line on a small-scale map of California, but no one continuous surficial break can be traced over this entire distance, regardless of what may happen at depth. It is clear from the detailed description of the 1906 faulting (Lawson and others, 1908), as well as from aerial photographic studies of Holocene fault features (e.g., Ross, 1969), that the ground rupture during an individual great earthquake cannot consist of a single uninterrupted surficial fracture extending for hundreds of kilometers. The longest surficial individual strands appear to be about 18 km in length (Wallace, 1973). The perpendicular distance between overlapping individual en echelon traces is often only a few tens or hundreds of meters, so that it is easy to visualize a continuous fracture at depth over much of the length of the fault, although Wallace (1973) and Bakun and others (1980) have pointed out that some individual en echelon segments retain their identity down at least through the seismogenic zone.

Even the entire width of the fault *zone*, comprising a number of en echelon or anastomosing traces, becomes very complicated in a few places. In San Gorgonio Pass, for example, unresolved problems remain as to which, if any, members of the zone have continuity through this structural "knot," where the San Andreas fault as traced from the southeast first intersects the older east-west structures of the Transverse Ranges (Allen, 1957; Dibblee, 1964). There can be little doubt that large earthquakes have taken place in this area within Holocene time, but probably a myriad of individual surface breaks occurred here, particularly along the many thrust faults, instead of throughgoing fractures. Similarly, it seems likely that the next great earthquake along the fault zone in the San Gorgonio Pass area will locally be associated with very complex faulting quite unlike that along the simpler and more typical parts of the fault to the northwest and southeast.

In the entire 1100-km length of the fault from Shelter Cove to Bertram, the most conspicuous en echelon interruption of the Holocene fault trace appears to be that near Cholame, San Luis Obispo County, where right-stepping parallel traces are about 2 km apart and are reflected in the "pull-apart" nature of the Cholame Valley itself (Dickinson, 1966b). Hypocentral locations of aftershocks of the 1966 Parkfield-Cholame earthquake (Eaton and others, 1970) indicate that the fault is continuous at depth, but apparently takes a gentle bend that leads to the more discontinuous en echelon breaks at the surface. (These very accurate hypocentral locations also provide, incidentally, one of the best lines of evidence supporting the near verticality of the fault surface, at least to depths of about 15 km.) Significantly, the Cholame area is also the point where the locked and slipping segments of the San Andreas fault meet, as well as the only point along the entire fault to have experienced a number of almost identical moderate earthquakes within the historic record. It is also the probable locus of foreshocks of the great 1857 earthquake (Sieh, 1978a). Clearly, the detailed surficial geometry of the fault trace is reflecting important mechanical processes at depth (Bakun and others, 1980).

In some major earthquakes, individual en echelon segments have limited the extent of fault rupture, and thus the ultimate magnitude. During the great Erzincan (Turkey) earthquake of 1939, for example, rupture was limited to a single 250-km-long en echelon segment of the much longer North Anatolian fault (Ketin and Roesli, 1953; Allen, 1969). And in the 1940 Imperial Valley (California) earthquake, rupture was similarly limited to a single en echelon segment within the longer fault zone. This cannot be assumed as a rule, however: In the 1968 Borrego Mountain (California) earthquake, rupture occurred along two en echelon segments of the San Jacinto fault zone (Clark, 1972), and Sieh (1978b) has argued that the 1857 earthquake on the San Andreas fault, although probably commencing near Cholame, involved rupture for several tens of kilometers to the north of the Cholame Valley as well as for 300 km to the south.

PHYSIOGRAPHIC EXPRESSION

On a relief map of California at almost any scale, the San Andreas fault appears as a distinct scar cutting across the state, for three principal reasons: (1) to some extent, one is looking at the primary relief of actual fault scarps along the fault trace; (2) probably more important, one is viewing relief caused by differential erosion of contrasting rocks that have been brought into juxtaposition by lateral displacements along the fault, producing fault-line scarps; and (3) certainly most important, one is observing the very significant effect of erosion of the crushed and altered materials within the fault zone, which gives rise to the marked linear valleys along the fault trace. The question often asked as to the width of the crushed and altered zone is difficult to answer not only because of great variability along the length of the fault, but also because the degree of modification is somewhat gradational as one moves away from the most active strands of the fault. One of the best estimates, nevertheless, comes from the excavation of the Elizabeth Lake Tunnel (Los Angeles County) through the fault zone in 1908 to 1911; Mulholland (1918) reported that that the rock was relatively coherent on both sides of the fault zone, bounding a highly sheared septum about three-quarters of a mile wide, which corresponds roughly to the width of the surficial alluviated valley at that point. Because of the nature of this rock within the fault zone, it is hardly surprising that good exposures of the fault are rare; only where the fault zone is undergoing rapid erosion, such as near Apache Saddle (Kern County) or Big Pines summit (Los Angeles County), are excellent natural exposures of rocks within the fault zone to be seen.

Although the fault zone itself is usually marked by a distinct linear valley, owing to differential erosion, it is nevertheless impressive that throughout most of central and southern California one must travel uphill to reach the fault zone. That is, the fault zone tends to be in mountainous or hilly country that stands high with respect to the surrounding terrain. This is perhaps not surprising in view of the compressional horizontal stresses that must be associated with strike-slip faulting. The current southern California uplift (Castle and others, 1976) may represent only the latest phase of regional uplift centered on the fault system and apparently related to strain accumulation thereon. Only in the Salton Sea region is the fault devoid of major relief, and this is logically related to the transition to the ocean-floor tectonic style that characterizes the fault system still farther southeast in the Gulf of California.

Even in a more detailed sense, localized uplift along the fault is obvious in areas such as Carrizo Plain (San Luis Obispo County), where the fault appears as a weltlike ridge bisected by minor stream valleys eroding along the most recent fault trace (Fig. 15-2). One must conclude that upward "squeezing out" of materials within the fault zone at depth has caused the localized welt in the very young gravels at the surface, similar to the mechanism proposed by Wallace (1949) for the center-trough ridges along the fault farther southeast.

Where one side of the fault is consistently higher than the other, such as along the north flank of the San Gabriel Mountains or along the south flank of the San Bernardino Mountains, the fault is characterized by thrust faults dipping into the adjacent ranges. These thrusts must steepen very rapidly with depth, however, in order to preserve linearity of the overall fault zone, and such steepening has in fact been demonstrated in the very excellent exposures along the San Jacinto fault where it cuts the Peninsular Ranges (Sharp, 1967). The low-angle faulting is apparently a consequence both of the

Fig. 15-2. San Andreas fault in Carrizo Plain area, showing weltlike nature of surface expression. Photo by C. R. Allen.

local vertical components of displacement (Allen, 1965) and of mass movements that tend to deform the fault surface (Wellman, 1955).

Detailed physiographic features of recent fault movement within the fault zone have been discussed by numerous authors (e.g., Lawson and others, 1908; Wallace, 1949, 1975; Sharp, 1954; Sieh, 1978b). With the recognition of major active strike-slip faults in other parts of the world, it has become clear that faults of this type are characterized by a truly unique set of geomorphic features that are remarkably similar among strike-slip faults even in widely differing climatic environments (Allen, 1965). Particularly impressive are the features associated with the overlapping ends of individual en echelon segments–typical "pull-apart" depressions where the sense of stepping is the same as that of the overall displacement (e.g., right-stepping segments on a right-lateral fault), and typical compressive upwarps where relations are the opposite (e.g., see Hall, Chapter 17 and Crowell, Chapter 18, this volume). Such features occur on all scales, from the major depressed basins such as the Salton Sea area (Lomnitz and others, 1970; Elders and others, 1972) or the San Jacinto Valley (Sharp, 1975), to the small sag ponds between en echelon segments that are only a few meters apart, such as at Una Lake near Palmdale (Ross, 1969). Uplifts associated with en echelon offsets are perhaps less obvious than depressions but are well expressed in some areas, such as the highly deformed Ocotillo badlands, caught between the overlapping ends of two en echelon segments of the San Jacinto fault zone in San Diego County (Sharp and Clark, 1972) or the hill on which the Fremont Civic Center (Alameda County) is located (Cluff and others, 1972). Indeed, with these geometric relationships in mind, it is easy to determine the sense of overall lateral displacement on many active strike-slip faults simply by reference to aerial photographs, even in the absence of directly offset features such as displaced stream channels. In this way, for example, the Garlock fault can be clearly identified as left-lateral on the basis of fault relationships at Koehn Lake (Kern County) or the Elsinore fault as right-lateral on the basis of the fault geometry at Lake Elsinore (Riverside County). Even at the scale of satellite images, such relationships are clearly visible and have been used to determine current horizontal fault motions in western China.

The mechanism leading to en echelon segments along the San Andreas fault remains an unsolved problem, and such features are common along other regional strike-slip faults as well, such as in Turkey (Arpat and Saroglu, 1975) and New Zealand (Clayton, 1966). Particularly troublesome is the fact that both right- and left-stepping segments may occur along the same fault (Wallace, 1973). One explanation is that they are related to places where the original linear fault trace has been slightly bent by subsequent deformation (Freund, 1971), and this certainly must be a factor in some locations. This would not explain, however, the seeming dominance, at least for long segments, of the "thrust shear" type (Morganstern and Tchalenko, 1967; Wallace, 1973), which displays the same sense of en echelon offset as that seen in ridge-transform-ridge patterns and seems to represent a more fundamental mechanical process.

REPEATABILITY OF SURFACE FAULTING

One cannot view aerial photographs of most segments of the San Andreas fault without being impressed by the fact that the most recent displacements along the fault have all occurred along almost exactly the same trace or within a very few meters thereof. In the Carrizo Plain area, for example (Fig. 15-3), sharp lateral offsets of stream channels vary from about 8 m, representing the great 1857 earthquake (Sieh, 1978b), to more than 300 m (Wallace, 1968; Hall and Sieh, 1977), and they all occur along essentially the same line. This can only mean that earthquake after earthquake has been associated with a fault line that very faithfully repeated that of earlier events. At Wallace Creek in the northern Carrizo Plain, Sieh (personal communication, 1979) considers that all major strike slip during the past 3400 years, and probably within the past 10,000 years, has taken place within a rupture zone less than 1 m wide, representing perhaps thirty individual large earthquakes during Holocene time. One must conclude, therefore, that the proba-

Fig. 15-3. Vertical aerial photograph of San Andreas fault in Carrizo Plain. Stream channels are offset from as little as 8 m (right), representing 1857 earthquake, to several hundred meters (left, indicating recurrent movements along same trace. Length of fault shown is 880 m. U.S. Geol. Survey photo.

bility of the next earthquake at this locality being associated with fault displacement along exactly the same trace is very high, as was emphasized by Wallace (1968). This is an important conclusion in formulating zoning and land-use policy. Certainly not all segments of the fault are as simple as that at Carrizo Plain, and in some areas two or more parallel traces appear to be equally active, but the ability to predict exactly where within the fault zone future ruptures will take place is high along most parts of the fault.

In the light of this apparently faithful repeatability of surface faulting, two questions arise: (1) how does one rationalize this observation with the fact that the fault zone at depth, or at least where exposed in basement rocks, is up to several kilometers wide, and (2) how is a new fault ever "born" if the preexisting trace faithfully repeats itself? Despite the tendency of the fault trace in many areas to remain fixed over periods of as long as 10,000 years, the overall history of the fault as reflected in the basement rocks is considerably greater than this, by at least 8 million years (Crowell, personal communication). Thus, despite the apparent repeatability of fault movements within Holocene time, this epoch represents in reality only a miniscule portion of the fault's long history, and abundant time exists during which the active fault trace could have migrated from place to place within the zone, either by suddenly changing its course or, more likely, by gradually extending splays until they become integrated into the system in ways to accommodate major motions. It is certainly possible that a major new branch could be "born" abruptly during a single earthquake, but it seems more likely that new faults grow by extending themselves intermittently from earthquake to earthquake. To the author's knowledge, there is no case in the worldwide historical record where an earthquake was associated with primary faulting along a fault that did not, for the most part, correspond to a preexisting fracture; and in virtually all the best-studied cases, these fractures had earlier histories of Holocene displacements (Allen, 1975).

Even more impressive than the tendency to repeat movements along the same trace is the tendency to repeat the same intricate details of fault displacement from earthquake to earthquake. During the 1968 Borrego Mountain earthquake on the Coyote Creek fault, for example (Fig. 15-1), wherever the fault trace lay at the base of a preexisting Holocene scarp, the displacement during the 1968 event was in the same sense as that demonstrated by the earlier scarp (Fig. 15-4), despite the scissoring movement along the fault (Clark, 1972). For a few kilometers north of Ocotillo Wells (San Diego County), for example, the displacement of both the Holocene and 1968 scarps was up on the west, whereas the reverse was true along the west edge of the Ocotillo badlands to the south. And still farther south, where the 1968 movement was almost entirely strike slip, the same sense of Holocene displacement was suggested by the absence of a pre-1968 scarp, although a groundwater barrier locally marked the Holocene trace. This almost uncanny repetition of earlier events, as though the earthquakes had a "memory" lasting hundreds of years, must represent a remarkable mechanical uniformity to the strain-buildup and strain-release process.

That repeatability has characterized the earthquake history elsewhere along the San Andreas system is suggested by descriptions of the 1906 faulting, particularly near Fort Ross (Sonoma County) (Lawson and others, 1908), as well as by field observations of J. P. Buwalda (unpublished field notes) following the 1940 Imperial Valley earthquake.

Fig. 15-4. Scarp of 1968 Borrego Mountain earthquake, at base of larger Holocene scarp. Movement in 1968 faithfully reproduced details of earlier Holocene displacements, including scissoring along fault. Photo by C. R. Allen.

The only significant vertical displacements observed in 1940 were at the north end of the Imperial fault trace, which is the only part of the trace marked by a Holocene scarp. Very likely the same phenomenon has characterized most worldwide earthquakes associated with surface faulting, but seldom has the faulting at the time of the earthquake been compared in detail to the preexisting Holocene scarps.

EARTHQUAKES ON THE SAN ANDREAS FAULT

Within the historic record, two great earthquakes have occurred on the San Andreas fault, those of 1857 in southern California and 1906 in northern California (Fig. 15-1). These two events were similar in many ways: both were associated with a few meters of right-lateral slip; both were associated with fault rupture lengths of several hundred kilometers; and both were felt over somewhat comparable areas (Agnew and Sieh, 1978). There were also differences: the 1857 event was preceded by a conspicuous foreshock series (Sieh, 1978a); fault displacements were seemingly more variable in 1857; and both the average and maximum displacements in 1857 were about one and a half times those of 1906, which, despite a somewhat shorter 1857 rupture length, suggest a slightly higher moment and surface-wave magnitude for the 1857 event (Sieh, 1978b). Nevertheless, these two earthquakes stand as the "type examples" of great earthquakes on the San Andreas fault, and they presumably represent the kind of events for which we must be prepared in the future.

One other historic San Andreas earthquake that may have approached the size of the 1857 and 1906 events was that of 1838 in the San Francisco Bay area. Surficial rupture during this earthquake extended along a trace of the San Andreas fault through much of the San Francisco Peninsula, where the shaking was obviously intense, but the faulting probably did not extend nearly so far to the north as in 1906 (Louderback, 1947). This earthquake is nevertheless significant, because it represents the only segment of the main San Andreas fault where large earthquakes are known to have repeated within the historic record and thus might give some clues to recurrence intervals; clearly this same segment broke again in 1906 (Lawson and others, 1908). The time interval from 1906 to the present is, of course, somewhat longer than that from 1838 to 1906, so one might infer that this segment of the fault is again capable of rupture. On the other hand, the many earthquakes that occurred in the Bay area between 1830 and 1906 may represent a clustering, as suggested by Wallace (1970), and the consistently smaller fault displacements measured in 1906 on the San Francisco Peninsula as compared to those farther north may, in fact, reflect the fact that some of the strain had already been relieved there in 1838. The interval between major clusters would thus remain in the hundreds of years (Wallace, 1970), as suggested by the geodetic data as well as the recent studies of Holocene fault displacements in southern California (Sieh, 1978c).

One other segment of the San Andreas fault has been characterized by a remarkable series of repeated earthquakes within the historic record, although none of these has exceeded magnitude 6.5. This is the 25-km-long segment between Parkfield (Monterey

County) and Cholame (San Luis Obispo County), which has experienced almost identical earthquakes of magnitude 5.5 to 6.5 in 1901, 1922, 1934, and 1966 (Fig. 15-1). Each of these four events was probably associated with surface displacements of a few centimeters along essentially the same segment (Byerly and Wilson, 1935; Townley and Allen, 1939; Brown and others, 1967), and at least the 1934 and 1966 events, as well as their foreshocks, wrote seismograms at several California stations that are astonishingly similar (Bakun and McEvilly, 1978). The surprising repeatability of surface faulting has already been emphasized in the preceding section, but in this case we see that the details of fracturing represented in seismograms are also faithfully repeated, again emphasizing the continuity of mechanical processes. Also mentioned earlier was the unique tectonic environment of the Parkfield-Cholame area, with the marked en echelon offset of the fault trace, as well as the meeting point of the slipping and locked segments of the fault. At least between 1934 and the present, and probably earlier, the part of the fault that ruptured during these earthquakes has also been characterized by episodic creep (Brown and others, 1967; Smith and Wyss, 1968; Goulty and Gilman, 1978), but the rate of creep is rapidly decreasing southward, and the creep zone seems to terminate near the southernmost point of the zone of fault rupture, at least as observed in 1966. In view of the foreshocks that occurred before the 1857 earthquake in the same area, which were also apparently of about magnitude 6, Sieh (1978a) proposes that after a few Parkfield-Cholame-type earthquakes over perhaps 100 years, the strain becomes so great at the end of the locked segment of the fault that the next such event in essence breaks the camel's back and becomes a foreshock that starts the cascading of ruptures along the fault that results in a great earthquake. Certainly the next magnitude 6 Parkfield-Cholame earthquake will be viewed with interest by seismologists!

In terms of the potential for future earthquakes, those segments of the San Andreas fault that *have not* broken during the historic record are fully of as much interest as those that *have*. Only two such major segments remain today (Fig. 15-5): (1) the 80-km-long segment between the southeast end of the 1906 rupture at San Juan Bautista (San Benito County) and the northwest end of the 1857 rupture, probably near the San Benito-Monterey County line (Sieh, 1978b), and (2) the 180-km-long southeast end of the fault beyond the segment that broke in 1857, roughly from Cajon Pass (San Bernardino County) to Bertram (Imperial County). These two segments will be discussed separately.

Despite the fact that the segment of the fault between the ends of the 1906 and 1857 ruptures has not broken during a large historic earthquake and might therefore be visualized as particularly highly stressed, the author has argued that large earthquakes in this segment are unlikely (Allen, 1968). Evidence gained since 1968 appears further to support this view, which is based on the following lines of evidence: (1) This is the very segment of the fault where creep was first recognized (Steinbrugge and Zacher, 1960) and appears to be taking place at a high and relatively constant rate. The geodetically measured creep rate of about 33 mm/yr (Savage and Burford, 1973) is probably sufficient to accommodate the local plate motion without intermittent large earthquakes, although the possibility of some elastic strain accumulation cannot be completely ruled out (Thatcher, 1979). (2) Creep along this segment of the fault has been taking place at a relatively constant rate for at least 70 years (Brown and Wallace, 1968), so it seems unlikely to be a temporal precursor to a large earthquake, as has been suggested as a possible model by Nason (1971). (3) Rock types in this segment of the fault, and particularly the serpentinite that is abundant within the fault zone, are different from those elsewhere along the fault and suggest that creep may be a permanent rather than a temporary phenomenon here.

Irwin and Barnes (1975) further emphasize the unique geology of the Cholame-San Juan Bautista segment, which is the only part of the fault that cuts off the massive nappe of the Great Valley Sequence, which has been thrust westward from the San Joaquin Valley region with a serpentinite-rich ophiolite at its base; they also emphasize the possible role of metamorphic fluids in the underlying Franciscan Complex migrating into the fault zone because of being capped by the less permeable serpentinites at the base of the nappe. It is interesting that north of the San Juan Bautista area, it is the Calaveras and Hayward faults that cut off the nappe (Irwin and Barnes, 1975), and it is on these faults rather than the San Andreas that creep continues through the Bay area (Nason, 1973). Moderately large earthquakes associated with surface faulting occurred on the Hayward fault in 1836 and 1868 (Fig. 15-1) (Lawson and others, 1908) so earthquakes of this size are certainly not incompatible with continuing creep; but these earthquakes are nevertheless a far cry from the truly great earthquakes of 1906 and 1857, which appear to be limited to those segments of the fault with distinct geologic characteristics. In particular, the concept that basement rock types can affect the mode of surficial fault strain release appears to be more and more valid.

The potential for a great earthquake at the southeastern end of the San Andreas fault represents a more perplexing problem. No creep has been observed along much of this segment, and unbroken concrete drainage ditches near San Bernardino suggest that none has taken place here for at least 50 years. Farther southeast, however, right-lateral creep, albeit minor, has been measured in the Coachella Valley, Riverside County (Keller and others, 1978), and parts of this segment showed displacements of about 1 cm triggered by the 1968 Borrego Mountain earthquake, possibly reflecting strain accumulation in the surficial sediments resulting from creep in the underlying basement rocks (Allen and others, 1972). Furthermore, both high micro-earthquake activity and moderate historic activity have characterized this entire region of branching and anastomosing faults, which stands in contrast to the relatively simple, seismically quiet segment of the fault northwest of Cajon Pass that broke during the great earthquake of 1857. Thus, Allen (1968) included this southeasternmost segment in his "southern California active area" (Fig. 15-5) with the implication that only moderate earthquakes were to be expected—no greater than magnitude 7.5. Wallace (1970) similarly concluded that the largest expectable earthquake on this segment was about magnitude 7.

Several arguments might now be invoked in opposition to this view: (1) A conspicuous 100-km-long seismic gap has developed since at least 1950 centered on the San Andreas fault in the Coachella Valley north of the Salton Sea (Fig. 15-6). Although a 100-km fault length is probably not sufficient in itself to generate a great earthquake, it is not hard to visualize a break continuing into the structurally very complex, and thus seismically more active, San Gorgonio Pass area to the northwest. Furthermore, the predominance of thrusting in the Pass area might call for a smaller fault length to pro-

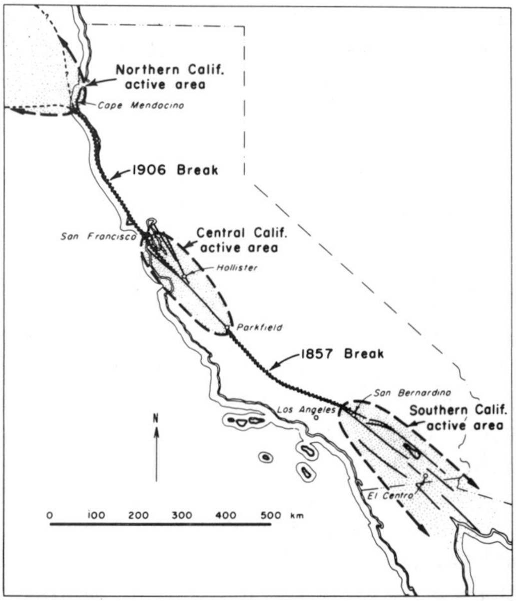

Fig. 15-5. Areas of contrasting seismic behavior along the San Andreas fault system. Reproduced from Allen (1968).

duce a large earthquake than would a simple, linear strike-slip fault; the magnitude 7.7 Kern County earthquake of 1952 was generated by a thrust-fault rupture of only 60 km length (Benioff, 1955). (2) The San Gorgonio Pass area, as previously mentioned, represents a major "structural knot" along the San Andreas fault, which could be considered analogous to the two major bends farther north that were proposed by Allen (1968) as helping to "lock" the fault during the time intervals between great earthquakes. (3) Sieh (1978b) has suggested that because the 1857 displacements along the southern third of the break were only about half those observed farther north, this could imply the earthquakes should occur twice as often in the southern sector, and he presents evidence that this may in fact be the case. Under these circumstances, Sieh further suggests (1978c) that perhaps the fault segment corresponding to the southern end of the 1857 trace breaks during great earthquakes alternately to the north, as it did in 1857, and to the south–into the "southern California active area" of Allen (1968). Thus the next great earthquake on the San Andreas fault might be caused by rupture from somewhere near Palmdale (Los Angeles County) southeast to Bertram, a distance of more than 300 km.

The author finds it difficult to choose between this scenario and that of his 1968 paper, but in view of the great complexity of faulting south of the Transverse Ranges, he is inclined to favor the concept of numerous somewhat smaller earthquakes in this region, although events as large as magnitude 7.5 can still be very damaging locally. Furthermore, strain accumulation appears to be distributed rather uniformly across the southernmost part of the state, rather than being concentrated close to any one branch of the fault system (Savage and others, 1978). Thus, in the opinion of the author, the next truly great earthquake on the San Andreas fault will most likely be a repeat of the 1857

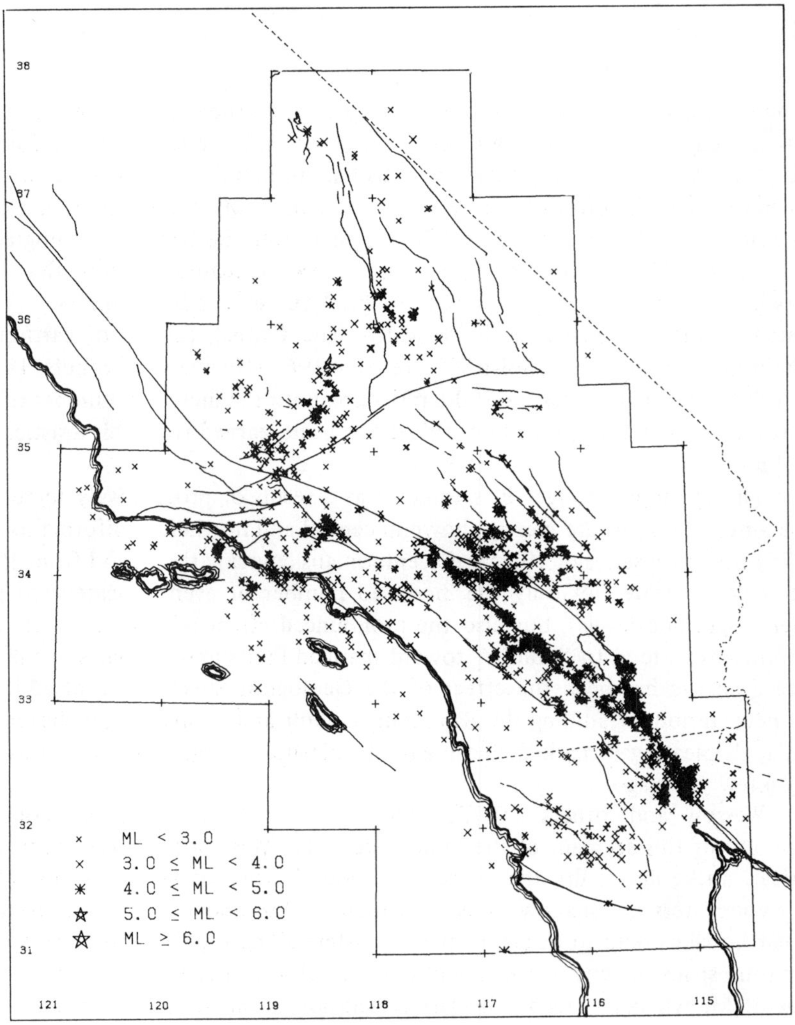

Fig. 15-6. Epicenters of earthquakes in southern California region for 6-month period from 1 January to 1 July 1978. Outlined area is region of coverage of Caltech-USGS Southern California Seismographic Network. Gulf of California (Mexico) at lower right; Nevada border (dashed) at upper right. Single lines are major faults. About 3000 individual epicenters are shown.

event, since this segment of the fault is now more highly strained than the only other segment, that of the 1906 break, that appears capable of generating truly great events.

CURRENT SEISMICITY

Current seismicity and its relationship to geologic structure have been summarized for northern California by Bolt and others (1968) and for southern California by Allen and others (1965). The primary emphasis herein, however, is on the changing concepts that have resulted from the greatly increased density of seismographic stations along the San Andreas fault in recent years. Whereas there were perhaps 35 seismographic stations operating in California in 1969, this number has increased to more than 500 only 10 years later, primarily owing to U.S. Geological Survey funding. Similarly, the number of located events has increased dramatically: The Caltech catalog of earthquakes in the southern California region shows 509 events in 1968 (Hileman and others, 1973), whereas the 1978 count is in excess of 7500, reflecting both the increased number of stations and improved data-analysis procedures. What has been learned from this drastically improved capability?

The apparent changes in seismicity as a result of progressively increasing number of recording stations are clearly shown in central and northern California. Bolt and Miller (1975) present a series of 10-year epicenter maps covering the period from 1910 to 1970, and they illustrate not only the increased number of events located (to progressively lower magnitude levels), but also the pronounced effect of more accurate locations, so that epicenters tend to "pull in" toward mapped faults in some parts of the region. Even more dramatic has been the effect of the Geological Survey's recent addition of some 150 new stations straddling the San Andreas fault and its major branches in central California, delineating a marked lineament of epicenters (Fig. 15-7) that was unsuspected only a few years ago.

Wesson and others (1977) point out the following conclusions concerning seismicity of this segment of the fault system: (1) Most earthquakes occur on mappable, vertical, strike-slip faults that have a previous history of Holocene displacements; (2) the hypocenters are shallow, most less than 10 km and virtually all less than 15 km; (3) earthquakes tend to occur in spatial clusters along the fault that are relatively stable over time spans of tens of years; and (4) the densely aligned epicenters occur mainly in areas of known fault creep. The last conclusion is particularly significant; virtually no small earthquakes are aligned along the fault north or south of the known creeping segment between Cholame and San Juan Bautista, including the great extent of the fault to the north that broke in 1906. Furthermore, the other principal alignment of epicenters in Fig. 15-7 is along the Calaveras-Hayward fault system, which is the other principal known area of fault creep in central California. In view of these findings, particular interest now centers on southern California, where the networks have been expanded more recently and where one wonders if the same conclusions hold.

An epicenter map of the southern California region made prior to 1970 appeared

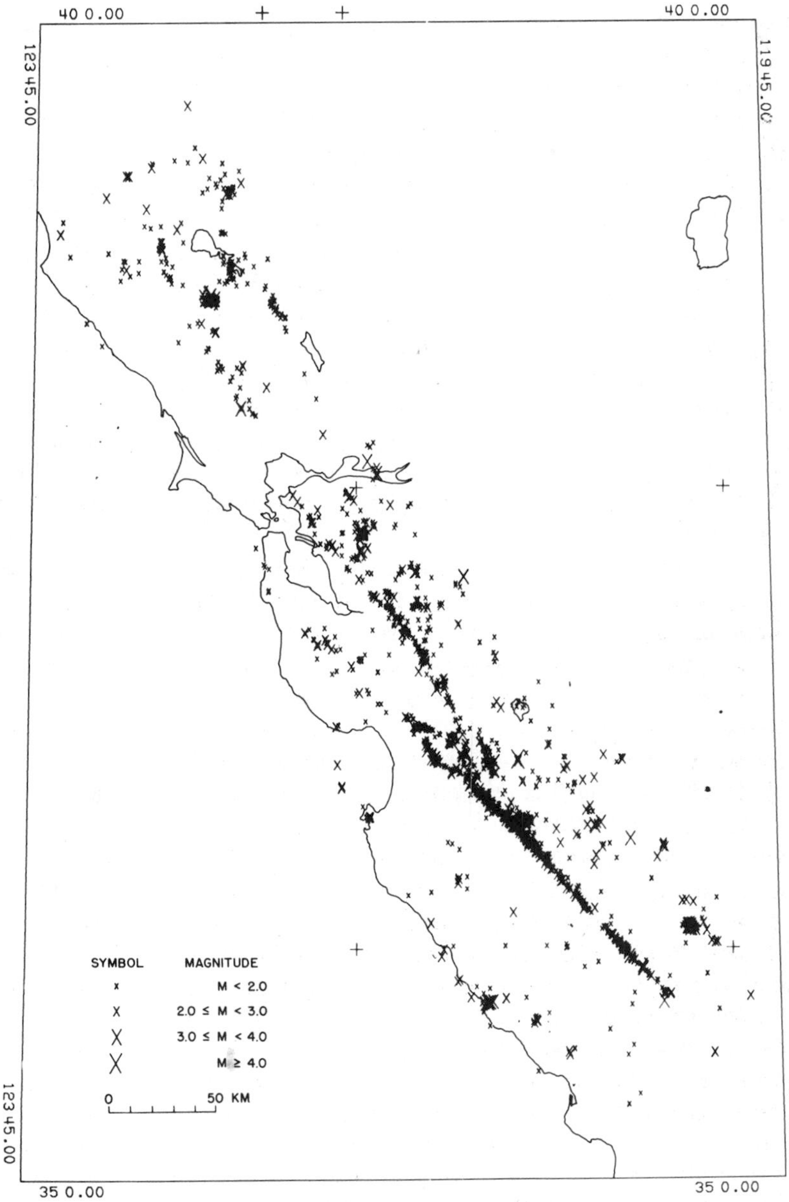

Fig. 15-7. Epicenters of earthquakes in central California for the year 1976, based on data from USGS Central California Network (McHugh and Lester, 1978). From Eaton (in press).

roughly as though someone had fired a shotgun at it, with no obvious alignments and only a few concentrated clusters marking the localities of individual moderate earthquakes and their aftershocks during the sample period (e.g., Hileman and others, 1973, Fig. 55). Major historic earthquakes in southern California have occurred along known active faults, but there are too few such events within the short historic record to delineate trends, except perhaps along the San Jacinto fault, and smaller earthquakes appeared to have occurred with a much more random distribution (Allen and others, 1965). With the drastically improved network capabilities, have epicenters tended to "pull in" toward the major active faults, as was observed in parts of central California? The answer is both yes and no, as can be seen from a recent 6-month map of some 3000 individual well-located epicenters (Fig. 15-6). In the Imperial Valley and adjacent parts of Mexico, there has indeed been a tendency for a linear concentration of activity, although even here, very detailed studies by Johnson (1979) show that the hypocenters do not by any means delineate simple fault surfaces. Similarly, the southeastern San Jacinto fault zone is marked by a rough alignment of epicenters that was not obvious on earlier maps. But the rest of the map *still* looks as though a shotgun had been fired at it, which has been a surprising development; despite the great abundance of well-delineated active faults in the southern part of the state, most small earthquakes are not obviously related to them.

In the evaluation of seismic hazard, a critical question is the degree to which a short-term seismicity map such as Fig. 15-6 is also representative of long-term seismicity, and therefore of long-term seismic hazard. That is, do the 3000 earthquakes shown on Fig. 15-6, which occurred over a 6-month period and include events as small as magnitude 1.0, show the same pattern of seismicity as would 3000 shocks of magnitude 4.5 and greater distributed over a 200-year period, representing about the same level of average seismicity? Allen and others (1965) have argued that the answer to this question is emphatically *no,* and that seismically inactive areas during one period of a few tens of years might well be active during others. One has only to note the current absence of seismicity on the segment of the San Andreas fault that broke in 1857 or the current quiescence along the Garlock fault, a major fault conjugate to the San Andreas that has not broken within the historic record but which has ruptured at least nine to seventeen times in the past 14,700 years (Burke and Clark, 1978). Although some areas such as the Imperial Valley are probably areas of continuing moderate activity, and this may also be true of the Transverse Ranges, many parts of the seismicity map will undoubtedly look grossly different when sampled during periods several tens of years apart. One must therefore be exceedingly cautious in evaluating long-term seismic hazard from a short-term seismicity map, regardless of how many individual earthquakes may be represented.

It is instructive to compare the four previously mentioned conclusions of Wesson and others (1977) from central California with tentative conclusions from the southern part of the state: (1) Although most major faults in southern California are vertical strike-slip features, and most large earthquakes have probably had strike-slip mechanisms, it is not clear that most small earthquakes necessarily share that mechanism. Certainly many small earthquakes close to the San Andreas fault are *not* of strike-slip origin (e.g., McNally and others, 1978). Furthermore, most events within the Transverse Ranges are of thrust origin (Lee and others, 1979; Pechmann, in press), which is not unexpected in view of the great bend in the San Andreas fault as it passes through this region. (2) Most earthquakes in southern California are in the same depth range as that for central California, although a few in San Gorgonio Pass reach 20 km. (3) We see no suggestion of the persistence of spatial clusters over periods of tens of years on individual faults, although, except on the Imperial fault, we have not looked as intensely as did Wesson and others (1977) in central California. (4) As is the case in central California, lineaments of epicenters in southern California appear to be limited to those fault segments where creep or episodic creep is taking place. Such creep has been well documented for the Imperial fault (Goulty and others, 1978) and is suggested for parts of the San Jacinto fault (Keller and others, 1978). But for those parts of the San Andreas fault where no creep is occurring, both in northern and southern California, there is no obvious alignment of small earthquakes; these are, of course, the very segments that broke in 1857 and 1906. At the micro-earthquake level, there may have been some increase in activity in recent years along the Palmdale-Tejon Pass segment of the fault in southern California, but the Carrizo Plain region remains virtually devoid of even the smallest shocks (Brune and Allen, 1967; Carlson and others, 1979).

Further differences in mechanical processes along different segments of the fault are illustrated by the aftershock patterns of the 1966 Parkfield-Cholame earthquake and the 1968 Borrego Mountain earthquake, both on major strike-slip elements of the fault system. Whereas the 1966 aftershocks showed remarkable alignment on virtually a single vertical fault surface (Eaton and others, 1970), investigation of the 1968 aftershocks using the same equipment and the same techniques revealed a spread of aftershocks over a wide oval-shaped area (Hamilton, 1972; Allen and Nordquist, 1972), although the focal mechanisms for the main shocks were almost identical.

Thus it is dangerous to draw sweeping conclusions about current seismicity that apply to the entire San Andreas fault system. Two patterns do seem to persist: (1) Those segments of the fault and its branches that have numerous small earthquakes aligned along the trace are the same segments that display continuous or episodic creep. In fact, the detailed seismicity might be used as a tool in the search for field evidence of surficial fault creep. (2) The fault segments that ruptured in 1857 and 1906 are generally characterized by very low current activity even in the micro-earthquake range. Since these appear to be the segments of greatest current hazard, this is nothing more than a restatement of the principle of the seismic gap (Fedotov, 1965).

HOLOCENE HISTORY

Probably the most exciting recent contributions to our knowledge of the San Andreas fault have come from studies of its history of Holocene displacement. Particularly important have been the results from trenches excavated across the fault, where strata that can be radiometrically dated are seen to be offset, and where older strata are seen to be offset more than younger strata, thus indicating continuing fault movements during the time interval in which the strata were deposited. In one of the earliest of such studies, Clark and others (1972) showed that a branch of the Coyote Creek fault, whose movement caused the 1968 Borrego Mountain earthquake (Fig. 15-1), displaced lake beds of 3080-year age by an amount about seventeen times that observed in 1968 at the same place. These data, together with measurements of offsets of intermediate amounts on strata of intermediate ages, indicated that 1968-type earthquakes ($M = 6.4$) would have had to occur on the average of about once every 200 years to give the observed relationships. Thus, for the first time, a recurrence interval was established for the fault, assuming that the 1968 event was typical of earlier earthquakes. This assumption seems reasonable, inasmuch as the style of surficial fault motion in 1968 was remarkably similar to that of earlier events, as was discussed previously, and because the total length of the fault suggests that earthquakes still larger than $M = 6.4$, and thus with longer recurrence intervals, are unlikely (Bonilla and Buchanan, 1970; Slemmons, 1977).

Considerably more diagnostic than the trenches across the 1968 break are a series of recent trenches across the main San Andreas fault in late Holocene marsh deposits at Pallett Creek, Los Angeles County (Fig. 15-1). Sieh (1978c) not only demonstrates here that progressively older strata are displaced by progressively greater amounts (Fig. 15-8), but he has also identified specific effects of sand liquefaction that he associates with the heavy shaking of individual prehistoric earthquakes. Some twenty-five radiocarbon dates were obtained on interlayered peat beds, and by piecing together the very complicated relationships, Sieh has identified eight large prehistoric earthquakes within the past 1500 years in addition to the known 1857 event (Fig. 15-9). Comparison of the effects of these events with those known to have occurred in 1857 leads to the conclusion that at least six of the eight prehistoric events were comparable in size to that of 1857, estimated by Sieh (1978b) as of magnitude 8.25+. The average recurrence interval between large earthquakes on this segment of the San Andreas fault during the past 1500 years has thus been about 160 years, although some have occurred as close together as 57 years and some as far apart as 275 years. Sieh (1978c) further points out that the intervals appear to have a crudely bimodal distribution, with pairs of large earthquakes separated by longer time spans. Not only is the average recurrence interval itself of considerable importance in engineering and land-use planning, but also the variations in it.

The ability to recognize individual prehistoric earthquakes in the manner elucidated by Sieh (1978c) represents an extremely powerful tool that should eventually permit a thorough understanding of the long-term seismic cycle along the San Andreas fault. The unique geologic conditions found at Pallett Creek, however, are not easily duplicated; it is an area of rapid and continuous sedimentation straddling the fault trace, with interlayered sands and peats in a marshlike environment, but one in which the regimen has recently changed so that the groundwater table has lowered to the point where trench excavation is now possible. Several attempts to reveal similar sequences at other locations along the fault have as yet not been very successful, but sooner or later they will be. With the dating of individual prehistoric earthquakes at a number of points along the fault, several important questions can be answered: What is the longest extent of faulting during a

Fig. 15-8. Trench wall at Pallett Creek, showing fault offset of older units (68, 61, etc.), creating a fault scarp that was then buried by unbroken younger units (71, 73, etc.) From K. E. Sieh (1978c). Units 68 and 71 thus bracket the time of this fault displacement, and radiocarbon dates on these units allow placement of the earthquake at 1470 ± 40 A.D. (event V on Figure 9). No subsequent displacements have occurred on this particular branch of the fault, although a younger branch (behind shoring) moved as recently as 1857. Photo by K. E. Sieh.

single earthquake, and thus the maximum magnitude? Is it possible that both Los Angeles and San Francisco might be heavily shaken during a single San Andreas event? Is there a repeated cycle of earthquakes along the fault, for example, alternating between north and south? To what extent do large earthquakes repeat their detailed characteristics? Will trenches in the presently creeping zone reveal a different style of offset (e.g., continuous rather than intermittent increases of displacement with age) that will prove once and for all whether large earthquakes can occur in these reaches? Is it true, as hypothesized by Sieh (1978c), that some parts of the "locked" segment are characterized by fewer but larger earthquakes than other parts? Are slip rates deduced from displacements during prehistoric earthquakes compatible with slip rates derived from geodetic observations or

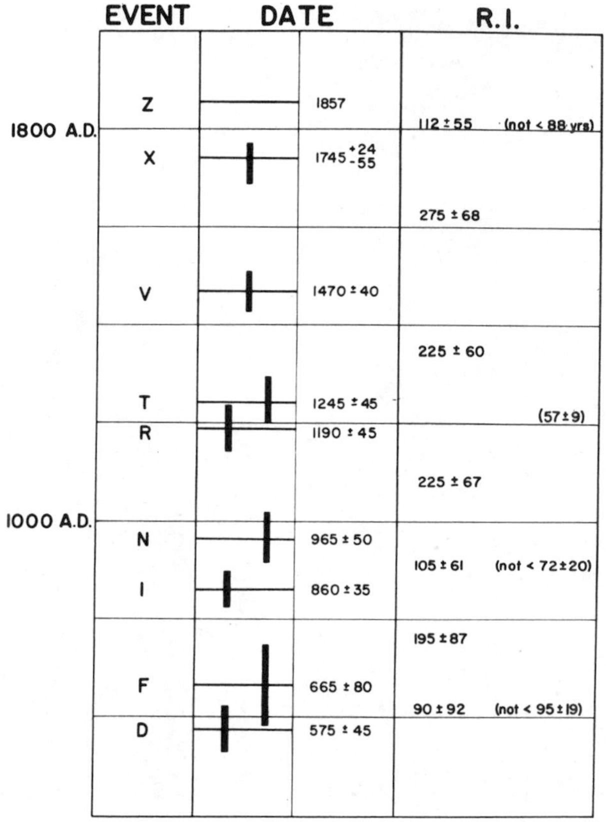

Fig. 15-9. Dates of events and recurrence intervals (RI) at Pallett Creek. Vertical bars in the second column indicate a statistical uncertainty of 1 standard deviation. From Sieh (1978c).

from plate-tectonics calculations? The importance of finding other localities along the fault similar to Sieh's Pallett Creek locality is very great indeed.

CAN CALIFORNIANS LIVE SAFELY WITH THE SAN ANDREAS FAULT?

Within the short historic record, two great earthquakes have taken place along the San Andreas fault in California, and the time has been sufficiently long since the last such event that most scientists agree that another one could occur tomorrow as no great surprise. Even without the evidence from strain-accumulation data, the recurrence interval of 160 years indicated by the Pallett Creek data (Sieh, 1978c) is so short that prudent engineers, land-use planners, and government officials must assume that a large event will take place within the lives of structures now being planned. Thus Californians must be prepared to "live" with major earthquakes on the San Andreas fault. Is this realistically possible?

Earthquake hazards arise from a number of sources, only two of which are of major concern along the San Andreas fault: (1) rupture through structures that are built squarely across the fault trace; and (2) widespread heavy shaking that is an indirect result of the fault displacement. The danger of fault rupture has been somewhat exaggerated in the public eye, and it generally represents only a very small proportion of total earthquake losses. Nevertheless, fault displacement through critical facilities such as dams or nuclear facilities could be disastrous, and avoiding active fault traces, particularly along features as obvious as is the San Andreas, seems relatively straightforward. It is disconcerting that many structures have been built directly athwart the most active trace of the San Andreas fault even in recent years. As was emphasized in an earlier section, fault movements tend to repeat themselves faithfully along previous Holocene traces, so identification of the most dangerous areas is generally within the state of the art of modern geologic practice. California's Alquist-Priolo Special Studies Zone Act of 1972 does, in fact, require special geologic studies aimed at the problem of surface fault displacement for new development of four or more housing units (Hart, 1977). The siting of very critical structures near faults has sometimes involved greater debate because of the concern over whether branch faults or older traces could be rejuvenated. Although it is generally agreed that most structures cannot practicably be designed to withstand large fault displacements through their foundations, some very special structures such as utility crossings have been designed to undergo as much as 10 m of strike-slip displacement without loss of function. Even some embankment dams in California have been designed on the assumption of several meters of fault displacement through the foundation (Sherard and others, 1974), including one squarely across the most recent trace of the San Andreas fault near Palmdale. In any case, the fault-displacement problem, while sometimes a matter of debate, appears to be perfectly solvable by existing geological and engineering techniques, usually by judicious placement of structures to avoid the most dangerous potential lines of rupture.

The quantitative specification of ground shaking to be expected during a great earthquake on the San Andreas fault represents a more difficult problem, because no strong-motion accelerograph records exist from sites close to great earthquakes anywhere in the world, and conclusions must be reached from theoretical calculations and from the indirect evidence of observed damage to structures, soil effects, and the like. The fact that a great earthquake on the San Andreas fault need not necessarily be as destructive as is sometimes imagined is indicated by several lines of evidence: (1) The number of major buildings in San Francisco that survived the 1906 earthquake without significant structural damage is much greater than generally appreciated. (2) Isoseismal maps of the 1906 earthquake (Lawson and others, 1908) give an undue impression of the severity of shaking close to the fault because of the arbitrary assignment of intensity X solely on the basis of fault rupture (Nason, 1978). (3) Recent studies by Jennings and Kanamori (1979) indicate that the local magnitude (M_L) of the 1906 earthquake was about 6.75

to 7, which is a better measure of the intensity of shaking in the frequency range of most buildings than is the accepted M_S of 8.25 (Gutenberg and Richter, 1954). Furthermore, this M_L can be compared to the even larger M_L of 7.2 for the 1952 Kern County earthquake (Kanamori and Jennings, 1978), which was not severely damaging to many well-built structures only 5 km from the fault trace in Arvin (Steinbrugge and Moran, 1955). (4) The sparse historical records of the great 1857 earthquake on the southern segment of the San Andreas fault suggest that, although the intensity may have reached IX along the fault trace, the severity of shaking in areas as far away as Los Angeles was not extremely high, and that "were the 1857 earthquake to be repeated today there would not be extensive damage to low-rise construction in the metropolitan Los Angeles area" (Agnew and Sieh, 1978). (5) Theoretical calculations by Kanamori (1979) of the ground shaking due to a repeat of the 1857 event, assuming a complex multiple rupture along the fault, suggest that the shaking in the Los Angeles area, at least in the 1- 5-s period range, would be roughly comparable to that empirically specified in the standard design earthquakes that are currently used in the design of well-built high-rise structures in the area.

There is probably greater confidence in assigning a realistic ground motion to points at some distance from the fault during a great earthquake than there is in specifying what might happen at given localities very close to the fault. This is because of the complexities of the fault-rupture process, which involves starting and stopping phases as the rupture progresses, which in turn may give rise to wide and unpredictable variations in high-frequency shaking (1 to 10 Hz) at different points along the fault. At some distance away, such effects tend to average out, particularly at longer periods. These near-field complexities are still the subject of intense research efforts and probably represent our greatest challenge in attempting to design adequately for earthquake shaking in seismic areas such as California. There is no reason to believe, however, that within the near future we cannot at least put realistic limits on what kinds of shaking are credible and incredible, so that structures can be designed with this in mind. Even very heavy shaking seems to be within the realm of behavior that can be dealt with in modern engineering practice, so that, in the long run, structures virtually anywhere in California should be capable of being built with adequate margins of seismic safety.

ACKNOWLEDGMENTS

Most of our knowledge of current seismicity in southern California comes from the Southern California Seismographic Network, which is mainly supported by the U.S. Geological Survey (Contract No. 14-08-0001-16719), with contributing support from the State of California (Division of Mines and Geology Agreement No. 5-9015) and the Caltech Earthquake Research Affiliates. The author appreciates the critical comments of John C. Crowell, Kerry E. Sieh, and Robert E. Wallace. Contribution Number 3286, Division of Geological and Planetary Sciences, California Institute of Technology, Pasadena, California 91125.

D. G. Howell and J. G. Vedder
U.S. Geological Survey
Menlo Park, California

16

STRUCTURAL IMPLICATIONS OF STRATIGRAPHIC DISCONTINUITIES ACROSS THE SOUTHERN CALIFORNIA BORDERLAND

ABSTRACT

Throughout mainland California, geologic events recorded in Jurassic and younger rocks indicate a nearly continuously active continental margin. By analogy, rocks on the California Continental Borderland suggest episodes of subduction and transform faulting similar to those on the mainland. Two types of basement rocks are juxtaposed: Upper Jurassic (?) through Lower Tertiary (?) melange and blueschist of a subduction or accretionary complex that are overthrust by Upper Jurassic ophiolite, arc-volcanogenic and forearc sedimentary rocks. The distribution of basement within the overriding plate, when compared with correlative rocks elsewhere in California, implies east-west foreshortening. The spatial relations of both the upper and lower plate rocks suggest northwest-directed dislocation of basement blocks in the borderland.

Cretaceous and Lower Tertiary strata are discontinuously exposed along the mainland coast and underlie much of the central part of the borderland. These thick clastic wedges are composed largely of turbidites that accumulated in forearc basins. Local middle Cretaceous and late Paleocene hiatuses in sedimentation and concurrent regional lapses in magmatism may represent times of transform faulting that interrupted subduction.

The modern topography of the borderland began to take shape in Oligocene and Miocene time, and ridges and basins grew in a wrench-tectonic setting within the evolving pliant margin between the Pacific and North American plates. Tholeiitic and calc-alkaline volcanism accompanied the early stages of development of typical borderland features. Miocene volcanogenic rocks commonly are present on ridges and knolls throughout the borderland, but it is improbable that these rocks are coextensive with basin floors.

INTRODUCTION

Location, Geomorphic Setting, and Scope

The California Continental Borderland is a predominantly subsea geomorphic province that extends from Point Conception to Cedros Island (Fig. 16-1). The northern part, which encompasses the Santa Barbara Channel and the northern Channel Islands, overlaps the western Transverse Ranges. Southward, the borderland narrows, and the numerous basins and ridges coalesce to form a single ridge-basin pair expressed by Cedros Island and Bahia Sebastian Vizcaino. In the north, the western edge is marked by the Patton Escarpment, in the south, by Cedros Deep (Moore, 1969). Ancestral parts of the borderland include the Ventura and Los Angeles basins of southern California and the Vizcaino basin of central Baja California.

With the exception of the west-trending Santa Barbara Channel, all the basins are elongate northwest, and many are ellipsoid to rhomboid in outline. Relief locally is as much as 2400 m, and slope gradients are as much as 15° along the flanks of some ridges, such as the escarpment northeast of San Clemente Island. The large ridges in the central

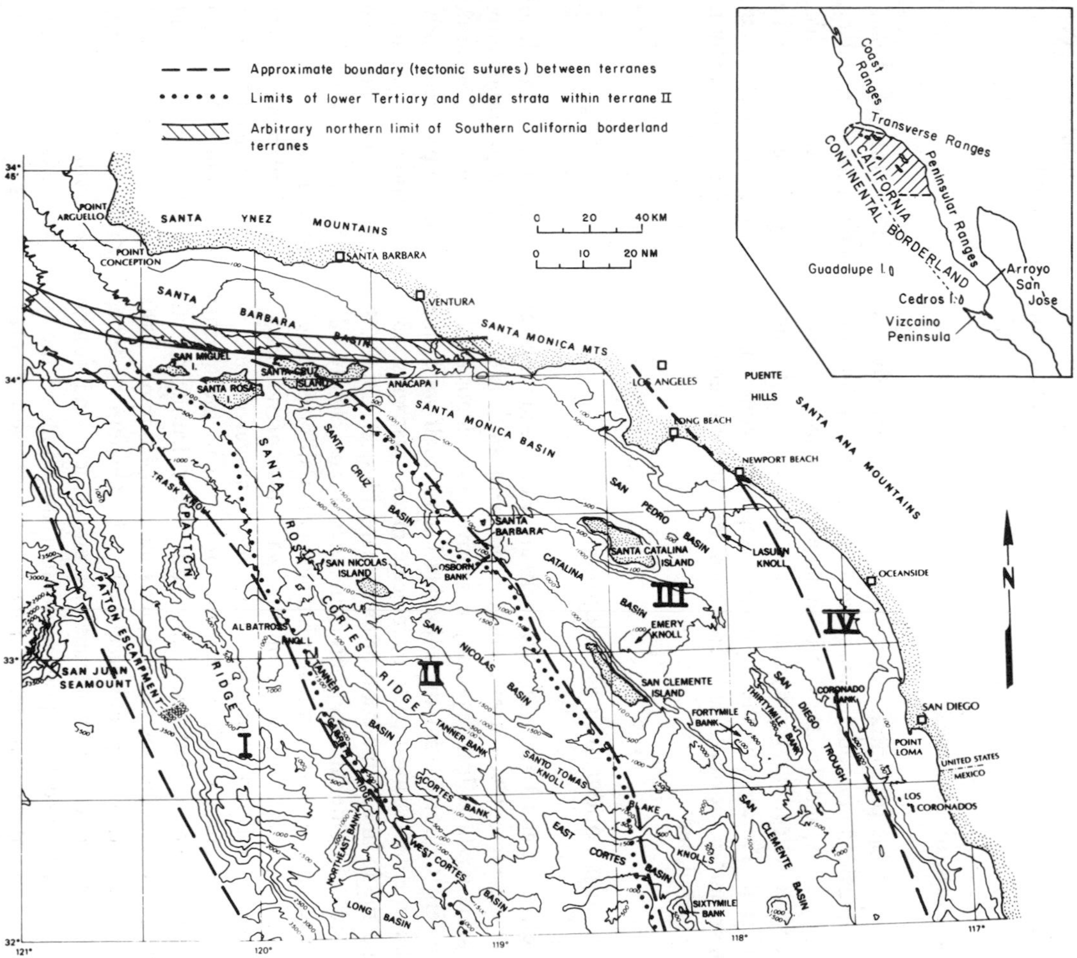

Fig. 16-1. Index map of the northern part of the California Continental Borderland showing generalized boundaries between four terranes (I to IV) with contrasting stratigraphic sequences. Area of this report is shown by hachures on the index map. Depths are in meters.

part of the borderland, however, typically have gently sloping flanks. Although most of the ridges are linear, a few are gently curved, and others tend to bifurcate southeastward.

This paper focuses on the geology of the northern part of the borderland (north of 32°N latitude). For the north limit, the Santa Barbara Channel is arbitrarily selected (Fig. 16-1), even though the borderland, by definition, extends to the mainland shore north of the channel. Although the data are sparse, the petrologic nature and known distribution of the main rock types in the borderland are used to infer kinematic models that may help to explain the evolution of this part of the California margin.

Borderland Terranes

On the basis of contrasting stratigraphic successions the southern California borderland is subdivided into four terranes, two of which extend into the coastal belt of the mainland (Figs. 16-1 and 16-2). The boundaries are indistinct, but by integrating topographic, acoustic-reflection, gravity, and rock sample data the limits of each terrane can be inferred. Presumably the boundaries represent zones of regional structural dislocation; their nature, however, is not well known, and our delineation of them undoubtedly is oversimplified.

Age and Topographic Development

Cretaceous and Lower Tertiary rocks are widespread in the central part of the borderland and are present at places along the northern and eastern shelves. These predominantly marine strata reflect regionally extensive basin deposition. In Oligocene time, much of western California and the adjoining Pacific shelf were emergent, and subsequent early Miocene depositional patterns indicate marine transgression. The inception of volcanism and the local accumulation of basin-margin breccia about 24 to 20 m.y. ago heralded the advent of borderland-style structure; these events closely followed the inferred shift from convergent to transform tectonism along the California margin about 30 m.y. ago.

Topographic relief probably was most pronounced on the borderland in late Miocene and early Pliocene time, when bathyal marine conditions prevailed in most of the basins and when bordering highlands to the north and east were contributing large volumes of detritus to the basin floors. During this time, episodes of uplift and subsidence, commonly at rates of as much as 1.0 m per 1000 years, effectively increased the preexisting ridge-and-basin relief. In Quaternary time, eustatic sea-level changes affected topography by sequential emergences and submergences, while diastrophic events continued to impose substantial alterations on the landscape, particularly in the region of intersection between west- and northwest-oriented structures. Extraordinarily rapid late Pleistocene and Holocene uplift rates of nearly 8.0 m per 1000 years are recorded in marine terrace deposits near Ventura (K. R. Lajoie, oral communication, 1978).

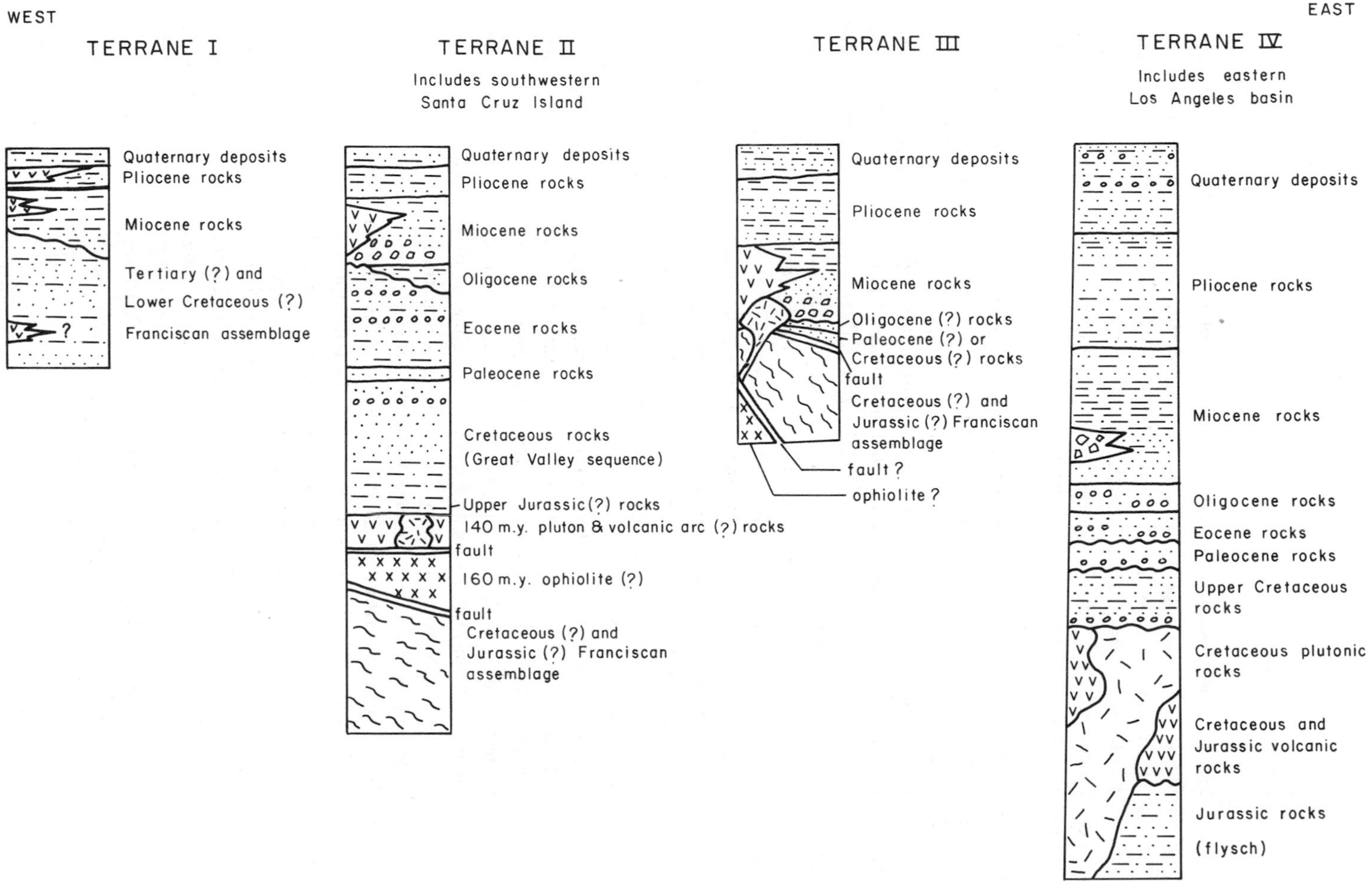

Fig. 16-2. Generalized stratigraphic columns for each of the four terranes of the southern California borderland. Thicknesses of rock units are relative and actual scale is not intended.

BASEMENT ROCKS

Regional Relations

The oldest known crystalline rocks from the borderland are on Santa Cruz Island and are of Middle and Late Jurassic age (Mattinson and Hill, 1976; Platt, 1976). Throughout California, Jurassic rocks lie within one of several roughly parallel terranes composed of either volcanic arc-batholithic, oceanic-crustal, forearc-basin, or accretionary epiclastic rocks (Jones and Irwin, 1975). A brief summary of these rocks is germane, for an understanding of the Mesozoic rocks of the borderland can be drawn by analogy with these better-known terranes.

In the western Sierran Foothills belt of central California, Upper Jurassic and older rocks have been tectonically accreted onto North America (Schweickert and Cowan, 1975; Jones and others, 1976; Saleeby and others, 1978: Schweickert, 1978). These rocks are dominantly of oceanic and arc affinities, although locally, quartzofeldspathic sandstones are present, implying a continental source in part (Behrman, 1978; Behrman and Parkison, 1978). In the western Sierran Foothills belt and in the western Klamath Mountains region of northern California and southern Oregon, the Jurassic terranes form imbricate nappes with westward vergence (Dott, 1971; Jones and Imlay, 1973; Jones and others, 1976; Irwin and others, 1978). Most of the pre-Tithonian Jurassic strata show a penetrative fabric, presumably reflecting the Late Jurassic Nevadan orogeny.

In the western Transverse Ranges, the Upper Jurassic Santa Monica Slate, which consists chiefly of metamorphosed epiclastic and volcaniclastic sandstone (Hoots, 1931), is correlative with and lithologically much like the Mariposa Formation of the western Sierran Foothills belt (Jones and others, 1976). Unlike the Santa Monica Slate, the Bedford Canyon Formation of the northern Peninsular Ranges contains Middle Jurassic fossils and consists largely of quartzofeldspathic sandstone and argillite; it generally is overturned and internally deformed (Criscione and others, 1978).

The petrology and distribution of Mesozoic rocks in Sonora and Baja California, Mexico, indicate a complex history of growth and collapse of volcanic arcs and marginal basins, but the rocks in that region do not have a penetrative fabric indicative of the Nevadan orogeny (Gastil and others, 1978; Rangin, 1978a). The section exposed on Cedros Island is noteworthy, for all the Mesozoic rocks inferred to underlie the borderland are displayed on the island in a relatively coherent sequence (Jones and others, 1976; Rangin, 1978a). The three primary lithologic units include (1) blueschist, serpentinite, graywacke melange, and broken formation at the lowest structural level, (2) dismembered ophiolite of gabbro, sheeted diabase sills, pillow basalt, and tuff overlain depositionally by an epiclastic sequence of Late Jurassic and Early Cretaceous age, and (3) Upper Cretaceous epiclastic rocks depositionally overlying an andesitic arc complex that is intruded by a 142 to 148 m.y. old tonalite pluton (the young epiclastic unit overlies the older epiclastic unit with local disconformity). In their respective order, these rocks correspond to (1) the Franciscan assemblage, (2) the Coast Range ophiolite and lower part of the Great Valley Sequence, and (3) rocks of the western Sierran Foothills belt west of the Melones fault and overlying Upper Cretaceous epiclastic rocks of the upper part of the Great Valley Sequence.

Figure 16-3 summarizes the chronology of lithologic terranes for the borderland and adjacent areas. From a plate-tectonic perspective it is apparent that throughout Mesozoic time the west margin of North America, now partly occupied by the borderland, was dominantly a convergent boundary that may have been subjected to brief episodes of transform faulting.

Southern California Borderland

Franciscan assemblage Blueschist-facies metagraywacke and metavolcanic rocks are exposed on Santa Catalina Island and in the Palos Verdes Hills (Woodford, 1924). On the island, these rocks are commonly associated with amphibolite- and greenschist-facies rocks (Platt, 1976). Blueschist and other schistose rocks also form subsea bedrock exposures south of San Clemente Island on Sixtymile Bank and the central ridge of Blake Knolls and southeast of Santa Catalina Island on Thirtymile Bank and possibly at Emery Knoll between these islands. The westernmost known body of glaucophane schist is on a narrow, unnamed subsea ridge 10 km southwest of the southern tip of Santa Rosa Island (Vedder and others, 1974). Other rocks dredged from this ridge include quartz-crossite schist, actinolite schist, and saussuritized gabbro (?). On northeastern Santa Cruz Island, the Union Gherini No. 1 well bottomed in melangelike rocks that include pillow basalt containing lawsonite (Howell and others, 1976). Beneath the Gulf of Santa Catalina, a buried schist ridge is inferred from acoustic-reflection data (Junger, 1974), as well as from sedimentologic studies of Miocene schist-breccia deposits (Stuart, 1976). Thus, it is apparent that blueschist and related high-pressure metamorphic basement rocks are widespread within terrane III of the borderland and locally present along the north part of terrane II.

In terrane I, a variety of zeolite-cemented volcanic and low-grade metamorphic rocks occurs at places on Patton Ridge, Albatross Knoll, and on the deeper parts of the Patton Escarpment. These rocks include unmetamorphosed altered basalt, laumontitic graywacke, argillite, phyllite, and greenchert. Serpentinite and ultramafic blocks have been sampled locally on the ridge and escarpment (Emery and Shepard, 1945; Winterer and others, 1969; Vedder and others, 1974, 1977; J. C. Taylor (unpublished data). Uchupi and Emery (1963) report glaucophane schist from the southern part of Patton Ridge, but their samples may represent erratics. It seems likely that the Patton Ridge-Albatross Knoll "basement" is part of the Franciscan assemblage, possibly analogous to the Coastal belt of northern California but dissimilar to the higher-grade metamorphic rocks in adjoining areas.

Coast Range ophiolite A dismembered ophiolite approximately 160 m.y. old occurs sporadically throughout central and northern California (Bailey and others, 1970). Just north of the borderland at Point Sal and at places in the San Rafael and southern Santa Lucia Mountains, the Franciscan assemblage typically is structurally overlain by the Coast Range ophiolite, which in turn is depositionally overlain by Upper Jurassic and

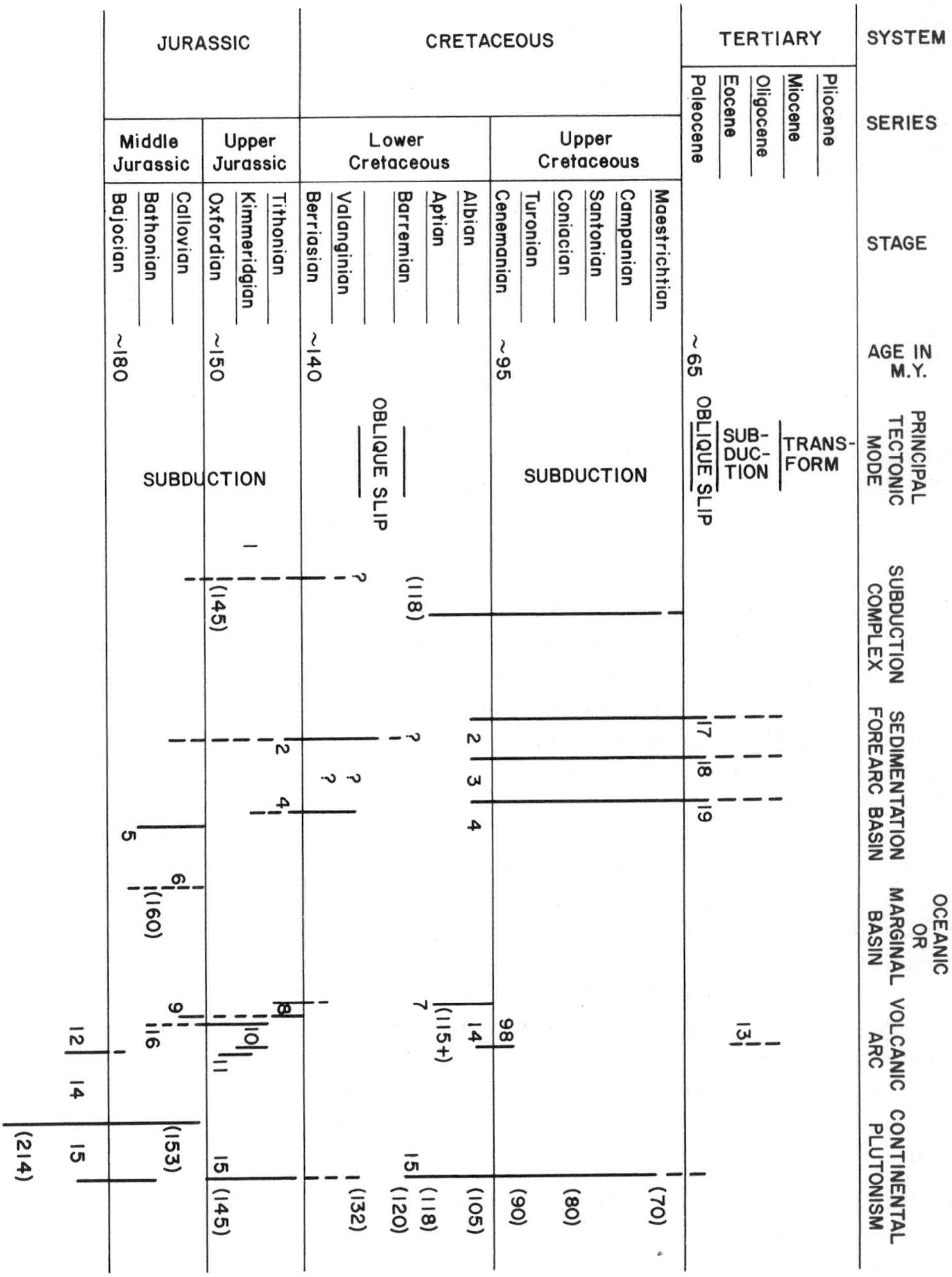
SYSTEM
SERIES
STAGE
AGE IN M.Y.
PRINCIPAL TECTONIC MODE
SUBDUCTION COMPLEX
SEDIMENTATION FOREARC BASIN
OCEANIC OR MARGINAL BASIN
VOLCANIC ARC
CONTINENTAL PLUTONISM
JURASSIC
CRETACEOUS
TERTIARY
Middle Jurassic
Upper Jurassic
Lower Cretaceous
Upper Cretaceous
Paleocene
Eocene
Oligocene
Miocene
Pliocene
Bajocian
Bathonian
Callovian
Oxfordian
Kimmeridgian
Tithonian
Berriasian
Valanginian
Barremian
Aptian
Albian
Cenemanian
Turonian
Coniacian
Santonian
Campanian
Maestrichtian
~180
~150
~140
~95
~65
SUBDUCTION
OBLIQUE SLIP
SUBDUCTION
OBLIQUE SLIP
SUB-DUC-TION
TRANS-FORM
(145)
(118)
(160)
(115+)
(153)
(214)
(145)
(132)
(120)
(118)
(105)
(90)
(80)
(70)

Fig. 16-3. Summary of inferred ages and geologic settings of rock sequences that are believed to be sensitive indicators of subduction processes. Correlation of radiometrically dated rocks with biostratigraphic zones is imprecise for the Jurassic. Cretaceous and Paleocene hiatuses in subduction may represent episodes of transform faulting, although confirmatory data are lacking. Numbers in parentheses on the chart are published radiogenic dates.

1. Franciscan assemblage and underlying tuff (San Francisco to Vizcaino Peninsula; Bailey and others, 1964; Lanphere, 1971).
2. Great Valley Sequence (Vizcaino Peninsula and vicinity; Jones and others, 1976).
3. Great Valley Sequence (Terrane II; Weaver, 1969; Vedder and others, 1974; Paul and others, 1976).
4. Great Valley Sequence (Coast Ranges, central California; Howell and others, 1977).
5. Bedford Canyon Formation (Peninsular Ranges, Imlay, 1964; Criscione and others, 1978).
6. Coast Range ophiolite (California Coast Ranges to Vizcaino Peninsula; Jones and others, 1976).
7. Alisitos Formation (Baja California; Allison, 1955).
8. Santiago Peak Volcanics (Fife and others, 1967).
9. Unnamed tuff (Vizcaino Peninsula and vicinity; Jones and others, 1976).
10. Santa Cruz Island Schist of Weaver (1969) (Mattinson and Hill, 1976).
11. Santa Monica Slate (Imlay, 1963).
12. Unnamed volcaniclastic strata (Arroyo San Jose, Baja California; Jones and others, 1976; Gastil and others, 1975).
13. Basin and Range Province volcanic rocks (Cross and Pilger, 1977).
14. Eastern Sierra Nevada rocks (Fiske and Tobisch, 1978).
15. Sierran and Peninsular Ranges plutonic rocks (Evernden and Kistler, 1970; Krummenacher and others, 1975).
16. Sierran Foothills Jurassic metamorphic rocks (Duffield and Sharp, 1975; Behrman and Parkison, 1978).
17. Paleogene strata (Vizcaino Peninsula; Minch and others, 1976).
18. Paleogene strata (southern California borderland; Howell, 1975a); Vedder and others, 1974).
19. Paleogene strata (Coast Ranges, central California; Clarke and others, 1975).

Other pertinent references for the Cretaceous time scale are Lanphere and Jones (in press), Obradovich and Cobban (1975), and Van Hinte (1976a, b).

Lower Cretaceous epiclastic rocks (lower part of the Great Valley Sequence). Nearly identical structural and stratigraphic relations are evident on Cedros Island and Vizcaino Peninsula at the south end of the borderland.

On Santa Cruz Island within terrane II, the Willows Diorite of Weaver (1969) consists dominantly of hornblende diorite with minor amounts of gabbro and ultramafic rocks (Hill, 1976). Zircons from this body have yielded nearly concordant ^{205}Pb–^{238}U ages of 162 ± 3 m.y. (Mattinson and Hill, 1976); both age and composition of the dioritic rocks suggest a correlation with the Coast Range ophiolite.

Elsewhere in the borderland, scattered occurrences of saussuritized gabbro (ophiolite) include a body on the south side of Santa Catalina Island in terrane III (Platt and Stuart, 1974), subsea exposures on northern San Clemente Ridge 12 to 20 km southeast of Osborn Bank in terrane III, and on the predominantly blueschist ridge in terrane II 20 km south of Santa Rosa Island. Samples from the crest of Trask Knoll in terrane II include serpentinite, fractured pyroxenite, and actinolite schist (?), which possibly represent the Coast Range ophiolite.

On the southwest side of Santa Cruz Island, a sequential change of clast types in Miocene breccia units indicates downward erosion of adjoining basement from saussuritized gabbro and diorite (ophiolite) into blueschist and greenschist (McLean and others, 1976). Clasts of these basement types abound in Miocene beds throughout terrane III of the borderland, but apparently are limited to the northwestern and southeastern parts of terrane II (Vedder and Howell, 1976). Similar breccia beds are widely distributed in terrane IV along the mainland coast (Woodford, 1925; Stuart, 1976). In a belt extending about 30 km northwest from Oceanside, clasts of metagabbro constitute as much as 50% of the assemblage in the basal part of the breccia sequence, but at higher levels blueschist and other Franciscan-derived clasts predominate. Farther northwest near Laguna Beach, this systematic change is not evident, although some breccia zones within the section contain as much as 40% metagabbro clasts.

A few occurrences of possible ophiolite basement on the borderland and the distribution of cobbles and boulders of possible ophiolitic debris mixed with, or stratigraphically near, blueschist clasts suggest a formerly broad areal extent and contiguity of the Franciscan assemblage and the Coast Range ophiolite. From the ordered sequence of clast types in the Miocene breccia on Santa Cruz Island and locally along the mainland coast and from observable thrust relations in the Coast Ranges, the most facile reconstruction places the ophiolite mass structurally over the Franciscan assemblage.

Jurassic arc The Santa Cruz Island Schist of Weaver (1969) contains low-grade greenschist-facies minerals (epidote, chlorite, quartz, albite, ± actinolite, ± muscovite) that indicate principally spilitic to quartz-keratophyric volcanic protoliths together with small amounts of epiclastic rock types (Hill, 1976). This schist mass has a preferred west-trending foliation. Intruding the Santa Cruz Island Schist is the 141 ± 3 m.y. old Alamos Tonalite of Weaver (1969), an irregularly shaped and unsheared body of leucocratic tonalite (Mattinson and Hill, 1976). The schist and the intrusive tonalite are much like the Jurassic volcanic-arc rocks and the tonalite body on Cedros Island (Jones and others, 1976). On Santa Cruz Island, however, the penetrative fabric of the volcanogenic schist is similar to that of arc-related rocks along the Sierran Foothills belt west of the Melones fault (Jones and Irwin, 1975). By analogy, we postulate that the Santa Cruz Island Schist was affected by the Nevadan orogeny and subsequently was intruded by the Alamos Tonalite. The North Valley Anchorage fault of Hill (1976) separates the schist and tonalite from the unsheared Willows Diorite (ophiolite) on Santa Cruz Island. Presumably, this fault is of great magnitude, similar to upper-plate nappe structures in southwestern Oregon, where thrust sheets consisting of ophiolite overlain by Upper Jurassic and Lower Cretaceous strata, or allochthonous metamorphosed Upper Jurassic volcanic-arc rocks, structurally overlie Franciscan-equivalent rocks (Jones and Imlay, 1973).

In the basal beds of the Miocene breccia northwest of Oceanside, greenschist clasts (epidote, chlorite, albite, quartz ± actinolite) accompany metagabbro (ophiolite) clasts as one of the main constituents. This suite of clasts closely resembles basement rocks on Santa Cruz Island and may indicate that the paired metamorphosed arc-unsheared ophiolite extended over a wide region, in particular, west and possibly south of the present site of the Bedford Canyon Formation. The age and mode of emplacement of the Santa Cruz Island Schist and possible correlative rocks are enigmatic, for along the mainland southeast of the Santa Monica Mountains, known Jurassic rocks do not show evidence of the Nevadan orogeny. To us, these relations suggest east-west foreshortening of the terranes.

SEDIMENTARY AND VOLCANIC ROCKS

Upper Jurassic (?) and Cretaceous Strata

The only subaerially exposed, fossiliferous Mesozoic rocks are at the north edge of terrane II, where Upper Cretaceous strata as much as 3000 m thick form the western seacliffs of San Miguel Island. Benthic foraminifers in this section are assigned to the upper Turonian through lower Maestrichtian stages. Within the same terrane, one exploratory well on the southern part of Santa Rosa Island, two on the west part of Santa Cruz, and several in the Dall-Tanner-Cortes Bank area have penetrated Cretaceous rocks. Fossil assemblages as old as Cenomanian have been identified from the stratigraphic test well drilled on southeastern Cortes Bank, where basement is inferred to be 300 m or more below the lowest beds drilled (Paul and others, 1976). On the northern Garrett Ridge, claystone containing a coccolithophorid assemblage is assigned a late Albian to early Turonian age. Unfossiliferous siltstone and sandstone of probable Cretaceous age are present at Nidever Bank, and Coniacian and Santonian siltstone and conglomerate are exposed on a shallow area near the northwest end of Cortes Bank. Along the base of the slope 15 km southeast of San Nicolas Island, dredged samples of argillite and sandstone resemble rock types of the lower part of the Great Valley Sequence.

Beneath the northwest end of Santa Cruz Island, north of the Santa Cruz fault, the Richfield Santa Cruz Island No. 1 well penetrated Upper Cretaceous strata below Miocene volcanic rocks; near the east end of the island, in the Union Gherini No. 1 well, the same volcanic rocks overlie Miocene and possibly older Tertiary clastic rocks and the Franciscan assemblage (McLean and others, 1976). South of the Santa Cruz fault, Upper Cretaceous strata in the subsurface lie below a thick Paleogene clastic sequence. These diverse subsurface sections imply different geologic histories and suggest probable allochthonous relations for the parts of the island block.

Acoustic-reflection data from the borderland indicate that Cretaceous strata are widely distributed beneath terrane II. Along the west edge of this terrane these strata apparently are truncated by a broad zone of high-angle faults; on the east, the same sequence of beds locally pinches out eastward by overlap onto basement rocks, which may include ophiolite. The maximum age and thickness of the Cretaceous and possible Upper Jurassic strata (Great Valley Sequence equivalents) are not known, although well cores and gravity anomalies suggest a minimum thickness of 3000 m (L. A. Beyer, oral communication, 1978). For comparison, Upper Jurassic and Cretaceous strata in the San Rafael Mountains north of the borderland range in thickness from approximately 4000 to 6500 m. Equivalent strata in the Vizcaino Peninsula are more than 1000 to 2000 m thick.

Along the east edge of the borderland in terrane IV (east of Newport-Inglewood fault zone), Upper Cretaceous strata are discontinuously exposed between the northwestern Santa Ana Mountains and Point Loma, as well as in northwestern Baja California. These strata are upper Turonian and younger and were deposited in nonmarine to westward-prograding subsea-fan environments. At the southeast end of Santa Catalina Island in terrane III, altered sedimentary rocks enveloped by Miocene hypabyssal rocks have lithologic affinities to Upper Cretaceous and lower Paleogene strata on the mainland (Vedder and others, 1979). Other than the Franciscan assemblage, these pervasively intruded beds on the island together with overlying Oligocene (?) red beds are the only known pre-Miocene rocks in terrane III. The stratigraphic and structural relations between the Franciscan and these isolated epiclastic rocks are obliterated by the Miocene intrusive rocks.

A regional basin analysis for the Cretaceous of the borderland has not been attempted because known Cretaceous strata are subaerially exposed only on San Miguel Island. The upper Turonian to lower Maestrichtian rocks on this island are dominantly coarse-grained turbidites, including a granitic- and volcanic-cobble conglomerate. By analogy with equivalent strata on the mainland, terrane II seems to represent part of a forearc basin that lay west of a Jurassic and Cretaceous arc complex. The near absence of Cretaceous rocks in terrane III may be an artifact of postdepositional erosion (Woodford and Gander, 1977). There are, however, very few sandstone clasts in the Miocene schist-breccia beds that lie within and along the margins of terrane III (Vedder and Howell, 1976), suggesting that Cretaceous sediments, if deposited in terrane III, were very thin or confined to basins of limited extent.

In northern and central California, proximal facies of Upper Cretaceous strata rest on the Sierra Nevada crystalline basement, and distal forearc basin deposits overlie older strata with an ophiolite basement, which in turn is structurally above the Franciscan assemblage. Because the contact is buried, the nature of the relations between the ophiolite and the Sierran crystalline rocks is not known. Some analogous relations are present in southern California, where postdepositional disruption is suggested. Within terrane IV, proximal Upper Cretaceous strata overlap crystalline basement, whereas in terrane II a nearly complete Great Valley Sequence is inferred to overlie ophiolite. The wide expanse of intervening Franciscan basement in terrane III implies tectonic dismemberment, probably involving both uplift of terrane III and strike slip along its east edge.

Tertiary Strata

Paleocene Nonmarine and shallow-marine Paleocene strata flank the northern Peninsular Ranges and parts of the west margin of Baja California, although they have not been reported from western San Diego County. Within the borderland, the known distribution of Paleocene rocks is restricted to terrane II, unless the small patch of feldspathic sedimentary rocks enveloped in hypabyssal rocks on Santa Catalina Island is Paleocene rather than Cretaceous.

Paleocene rocks are exposed on Santa Cruz and San Miguel Islands. On Santa Cruz Island, a 90-m-thick section consisting of predominantly medium to fine grained sandstone and siltstone is inferred to be shallow marine (Doerner, 1969). On San Miguel Island, Paleocene sandstone and siltstone turbidites form a section that may be as thick as 460 m. From well-log and acoustic-reflection data, Paleocene rocks are known to occur in the subsurface section of Santa Rosa Island and Cortes Bank and probably are present beneath a large part of Santa Rosa-Cortes Ridge. It is noteworthy that a thin zone of algal limestone of possible Paleocene age was penetrated in the stratigraphic test well on Cortes Bank.

These data are too few to reconstruct depositional patterns. In the borderland, however, known Paleocene sedimentary features, the unconformity at the top of the section on Santa Cruz Island, and the algal limestone beneath Cortes Bank suggest shoaling of the postulated Cretaceous forearc basin in Paleocene time, perhaps resembling depositional conditions in the San Rafael Mountains.

Eocene Eocene strata are discontinuously distributed along most of the east edge of the borderland in terrane IV; like the Cretaceous and Paleocene rocks, farther offshore they are virtually restricted to terrane II. Eocene strata probably are truncated by faults along the western edge of terrane II, and both faults and depositional wedge-outs inferred from acoustic-reflection data mark the limits of Eocene rocks along the eastern margin of this terrane (Vedder and others, 1974).

Silicified metavolcanic clasts typify the Eocene Poway Group of the San Diego district and other Eocene rock units on the borderland. These clasts represent a voluminous accumulation of exotic detritus possibly derived as far away as Sonora, Mexico (Merriam, 1972; Minch and others, 1976; Abbott and Smith, 1978). In the Poway Group, these distinctive clasts and associated arkosic sandstone occur in west-trending fluvial-channel, alluvial fan, coastal plain-fan delta, paralic, shelf, and submarine channel environments of deposition (Howell and Link, 1976).

Conglomerate is a small yet significant constituent in the Eocene rocks of the borderland, and the cobbles closely resemble those in the Poway Group (Howell, 1975a; Minch and others, 1976). Most of the enclosing flyschlike strata typically are medium- to coarse-grained sandstone and mudstone with an overall sandstone-mudstone and shale ratio of about 1:1. Paleocurrent directions on the islands (Fig. 16–4) indicate a west-facing depositional system that fans out from the Santa Cruz Island area toward the south, west, and northwest (Cole, 1975; Merschat, 1971; Howell, 1975a; Erickson, 1975). The petrographic uniformity, the flyschlike character, and the systematic flow patterns all suggest that the Eocene strata of the borderland are part of a single large subsea-fan system. Thick conglomerate zones in the northern Channel Islands and the lenticular sandstone and conglomerate body on Santa Cruz Island indicate that the apex of this fan probably lay near the northeastern embayment of Santa Cruz Basin (Howell, 1975a). This anomalous situation led Howell and others (1974) to propose an episode of middle Miocene right slip that moved this fan system northwest away from correlative shallower marine and nonmarine facies in the San Diego district.

From paleomagnetic data, Kamerling and Luyendyk (1977) suggest that Santa

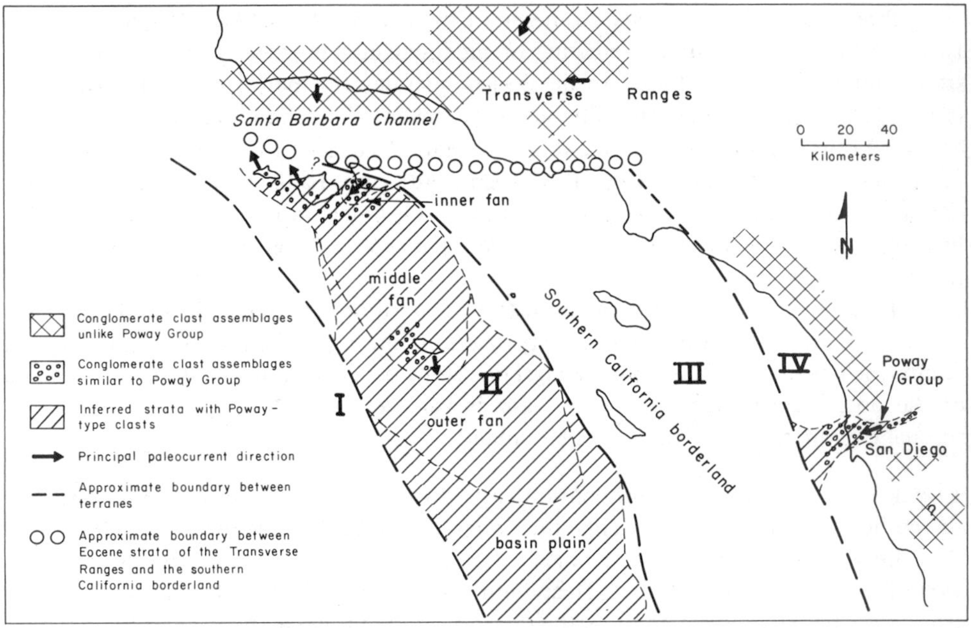

Fig. 16-4. Generalized distribution of Eocene strata in terranes I to IV and adjoining areas of southern California. Clast assemblages typical of the Poway Group conglomerates are restricted to the San Diego district and the southern California borderland. Subsea-fan lithofacies within terrane II suggest that possibly as much as 200 km of postdepositional right slip separated the northeast part of terrane II from an area west of San Diego (Howell, 1975a). The petrologic relation between Eocene strata of the borderland and the Transverse Ranges is unknown.

Cruz Island has been rotated more than 75° clockwise since middle Miocene time. A palinspastic correction for this amount of rotation reorients the Eocene channel at Santa Cruz Island to a southward flow direction, which is not imcompatible with the reconstructed geometry of the entire fan complex. However, such rotation must be local in extent, for a palinspastic correcion for all of terrane II would result in an unlikely juxtaposition of east-flowing distal fan facies at San Nicolas Island against nonmarine and west-flowing proximal fan facies of the mainland (Fig. 16-4).

Oligocene Throughout onshore southern California, strata of Oligocene age are predominantly nonmarine, although in the western San Ynez Mountains and possibly in the Santa Monica Mountains red beds intertongue westward with shallow-marine lithofacies.

On San Miguel and Santa Cruz islands, Miocene rocks lie unconformably on upper Eocene strata. Mudstone at the top of the Eocene section on Santa Cruz Island locally is variegated, suggesting that nonmarine beds may have been deposited nearby but subsequently were eroded (Doerner, 1969). On Santa Rosa Island, a nonmarine section at least 150 m thick unconformably overlies upper Eocene strata and is gradationally overlain by shallow-marine Miocene strata (Weaver and Doerner, 1969). Sediment transport was toward the north, and conglomerate beds composed of siliceous volcanic pebbles indicate reworking from emergent Eocene and possibly older strata lying to the south. Oligocene mudstone beds at Dall, Tanner, and Cortes banks show that marine deposition spread over much of the southwest part of terrane II. Correlative beds are thought to separate Eocene and Miocene strata on the platform northeast of San Nicolas Island.

Farther east, on Santa Catalina Island, a sequence of variegated beds consisting of mudstone, sandstone, and conglomerate lies stratigraphically between Miocene breccia and probably Paleogene or Upper Cretaceous strata (Vedder and others, 1979). Pebbles and cobbles comprising gneissic, quartzitic, plutonic, and siliceous volcanic rock types imply an eastern source terrane. On Patton Ridge in terrane I, bedded strata locally lie below a section of known late Oligocene and early Miocene siltstone (J. K. Crouch, written communication, 1977). The age of these underlying stratified rocks is not known, but they may be Oligocene or older.

From the available data, it seems likely that marine Oligocene strata principally are confined to former basins within the area of terrane II. Thus, some segments of the preexisting broad Cretaceous and Early Tertiary basin of the central borderland persisted through the regional marine regression of Oligocene time. In the vicinity of the northern island platform, however, parts of this ancestral basin apparently were emergent during the Oligocene.

Miocene The present configuration of borderland features probably began to evolve in early Miocene time. Initial stages of development are poorly known because of the fragmentary geologic record, and it is not yet possible to trace all phases of growth of the offshore basins. By the end of the Miocene, well-defined ridge and basin topography characterized most of the borderland (Vedder and others, 1979).

Lower Miocene strata in the offshore region seem to be restricted to terranes I and II and possibly the northernmost part of terrane IV; they are exposed on San Miguel, Santa Rosa, and Santa Cruz islands. These basal Miocene beds, and most equivalent strata elsewhere in southern California, mark the beginning of a regional marine transgression.

On Santa Rosa Island, recycled pebbles in lower Miocene conglomeratic sandstone beds reflect uplift and erosion of Paleocene (?) and Eocene strata to the south; on Santa Cruz Island, clasts in lower and middle Miocene breccia beds indicate denudation of both Paleogene beds and the local basement rocks, followed by erosion of a nearby blueschist terrane (Yeats, 1968a, b; Howell and others, 1974). Similar schist breccias were widely dispersed during the middle Miocene in the central and eastern parts of the borderland along the mainland coast from Point Mugu to Oceanside and locally near Tijuana (Woodford, 1925; Stuart, 1976; Vedder and Howell, 1976). The principal source area for the mainland schist breccias is inferred to have been a Miocene ridge system, now covered by younger rocks, within terrane III (Junger, 1974). Secondary sources were the schist ridge just south of Santa Rosa Island and several schist knobs in the vicinity of Blake Knolls and Sixtymile Bank. The widespread breccia beds in and surrounding terrane III indicate an episode of rapid differential uplift and erosion of basement blocks in that area.

At most places where lower Miocene transgressive marine beds occur, they are superseded by sandstone and mudstone beds that contain sublittoral and bathyal fossils, respectively. From Santa Cruz Island westward through Santa Rosa Island, middle Miocene strata are largely volcaniclastic and grade westward and southward into siliceous mudstone (Howell and McLean, 1976). Along the Santa Cruz-Catalina Ridge and at San Clemente Island in terrane III, similar volcaniclastic beds are interlayered and overlain by middle and late Miocene claystone and siltstone beds (Vedder and Howell, 1976; Vedder and others, 1979). This regional deepening and transgression apparently was interrupted by local development of high-standing topographic features in terrane II in the central part of the Santa Rosa-Cortes Ridge during middle Miocene time (Arnal, 1976, Fig. 8). Subsidence predominated in the borderland basins at the end of the epoch (Vedder and others, 1979).

Pliocene By Pliocene time, the sea floor probably closely resembled that of the modern borderland (Blake and others, 1978). Parts of Patton Ridge in terrane I may have been above sea level, as known Pliocene rocks are limited to the north and south ends, and Miocene and older rocks predominate elsewhere on the ridge. Pliocene shallow-marine and coastal dune deposits cap parts of Santa Rosa and Santa Cruz islands (Weaver and Meyer, 1969; Vedder and others, 1979). In contrast to the extraordinarily thick sections in the Ventura and Los Angeles basins and the nearshore basins, relatively thin accumulations of Pliocene sediment are present in the basins of terrane II. This seaward decrease in thickness suggests that submarine ridges and possibly islands to the east and north impeded the dispersal of terrigenous detritus, and that pelagic and hemipelagic debris and locally transported material from the barrier ridges were the primary sources of sediment.

The nearshore basins received large amounts of sediment from the north and east, yet the sea floor in the subsiding axial parts remained as deep as 1500 m as recently as late Pliocene time. Unconformities within the sections on the flanks of these basins demonstrate that adjoining areas were being structurally deformed (Fischer, 1976; Greene, 1976; Junger and Wagner, 1977). Late Pliocene marine strata now in water depths of 1250 m near the northwest end of San Clemente Ridge contain foraminiferal assemblages that imply water depths nearly twice that deep at the time of deposition, and early Pliocene beds now exposed on Santa Catalina Island probably were deposited at depths greater than 2000 m (R. E. Arnal, written communications, 1977, 1978). These depth changes indicate tectonic uplifts along the margins of terranes II and III that may have been as much as 2000 m since the early part of the epoch and 1000 m since the late part; other areas, such as Santa Rosa and Santa Cruz islands, seem to have remained relatively stable. Maximum rates of uplift are estimated to have been between 0.5 and 0.7 m per 1000 years.

Quaternary Strata

Pleistocene Marine and nonmarine sediments accumulated rapidly in the coastal basins during Pleistocene time, while in basins farther offshore, deposition of marine beds generally decreased with increasing distance from the mainland. Unconformities are recognizable on acoustic-reflection profiles in the Santa Monica and San Pedro Basins (Greene, 1976; Fischer, 1976; Junger and Wagner, 1977; Nardin and Henyey, 1978). In basins beyond the islands, interruptions in sedimentation are less distinct, suggesting that tectonism diminished seaward and that the contribution of pelagic material tended to obscure erosional gaps. Downslope transport of terrigenous sediment was obstructed by ridges and basin sills.

Eustatic sea-level changes are in part reponsible for the cutting of flights of terraces that are locally well displayed along the island and mainland coasts. Correlation of terraces throughout the borderland, based on matching altitudes, is not reliable because of the effects of disharmonic tectonic deformation (Vedder and others, 1979). Preliminary amino acid age determinations for terrace deposits at San Nicolas Island, Palos Verdes Hills, and San Joaquin Hills imply different rates of uplift for each of these structural blocks as well as rate changes during the last 500,000 years (Lajoie and Wehmiller, 1978). The highest estimated rate among these sites is slightly less than 1.0 m per 1000 years at Palos Verdes Hills. Intense deformation along the north half of the Santa Barbara Channel is manifested by uplifted and tilted marine terraces and north-dipping thrust faults.

Cenozoic Volcanic Rocks

Volcanism on the borderland spanned most of late Cenozoic time, although Quaternary activity seems to have been limited to the area south of 32°N latitude. This episode culminated in middle Miocene time and diminished sharply during the late Miocene. The rocks are chiefly flows with minor amounts of pyroclastics; compositionally, they range from rhyolitic to basaltic, but andesitic types predominate. Their total extent and volume are not known, but it is likely that they constitute large parts of the borderland south of 31°N latitude (Krause, 1965).

Little is known about the distribution and age of volcanic rocks in terrane I other than that they are concentrated in the southern part. Pliocene and older (?) submarine extrusives form most of Northeast Bank (Hawkins and others, 1971) and early (?) and middle (?) Miocene basaltic and andesitic rocks are present on knolls west and north of this bank and at the southern end of Patton Ridge. Hawkins (1970) reports tholeiitic basalt from the Patton Escarpment. Pyroclastic rocks, possibly of early Pliocene age, and basaltic rocks of unknown age have been dredged from the north end of the ridge.

Late Oligocene and/or early Miocene volcanism occurred in the western part of terrane II, where andesitic to dacitic volcanic rocks about 24 to 18 m.y. old form parts of San Miguel Island, ridges just west of this island (Palmer, 1965), and Cortes and Tanner banks (Paul and others, 1976; D. Bukry, written communication, 1976). Volcanic rocks are sparse in the northern part of the Santa Rosa-Cortes Ridge, although small amounts of pyroclastic material commonly are interbedded in strata of early (?) and middle Miocene age. Dikes such as those exposed on San Nicolas Island and Begg Rock probably are sporadically distributed along the same part of this ridge, but their exact age is unknown. Acoustic-reflection data indicate that neither Santa Cruz nor San

Nicolas basins have significant amounts of interlayered volcanic rocks in the underlying sedimentary sections. Magnetic data suggest that Tertiary volcanic rocks with high contrasts in magnetic susceptibility are not coextensive with the entire basin floors (L. A. Beyer, oral communication, 1978).

In terrane III, early igneous activity included a 19 m.y. old quartz diorite stock on Santa Catalina Island followed by extrusive events at Santa Cruz, Anacapa, Santa Barbara, Santa Catalina, and San Clemente islands, where ages range from 16 to 12 m.y. old. Large subsea tracts of volcanic rocks are spread along the Santa Cruz-Catalina Ridge, the Santa Barbara Island-San Clemente Ridge area, and the Thirtymile-Fortymile Bank area. Along the flanks of Santa Monica and San Pedro basins, ridgetop exposures of volcanic rocks extend downdip for an unknown distance beneath the basins (Junger and Wagner, 1977). Terrane IV probably contains middle Miocene basaltic extrusive rocks in its southernmost part and diabasic intrusive rocks in its northernmost part.

Preliminary chemical data indicate that the preponderance of volcanic rock on Santa Catalina islands is tholeiitic to calc-alkaline (Crowe and others, 1976). On these two islands, the rocks were largely subaerially emplaced, but on Santa Rosa, Santa Barbara, and San Clemente islands, submarine volcanism predominated. Figure 16-5 illustrates the variety and complex relations of igneous rock bodies on Santa Catalina Island; these include an intrusive stock, hypabyssal dike swarms, a near-surface dome, subaerial and submarine volcanic flows and breccias, as well as air-fall and reworked tephra deposits.

Thick accumulations of volcaniclastic debris were deposited along the flanks of the middle Miocene volcanic ridges. These rocks commonly are interbedded with lava flows, and on Santa Cruz and Santa Rosa islands they grade distally into subsea-fan facies (Howell and McLean, 1976). Away from these volcanic ridges, siliceous hemipelagic material blanketed much of the sea floor. Thin layers of hemipelagic debris also accumulated on the flanks of volcanoes during periods of quiescence; for example, mid-bathyal foraminiferal shale is intercalated with basalt flows on Santa Barbara Island, neritic foraminiferal shale is interbedded with andesitic and dacitic rocks on Santa Catalina Island, and neritic to bathyal foraminiferal and diatomaceous shale occur with dacite on San Clemente Island.

It is noteworthy that the greatest encroachment of seas and some of the deepest basins of late Cenozoic time in southern California developed near the end of Miocene time (Natland, 1957; Ingle, 1973). Perhaps this was in part a response to diminished igneous activity and concomitant subsidence that accompanied thermal cooling within the crust.

SUMMARY

The types and distribution of basement rocks in the southern California borderland are poorly known. However, a coherent pattern emerges (Figs. 16-6 and 16-7) if the limited data are used in direct comparison to relations established for similar rock types in northern California and Baja California.

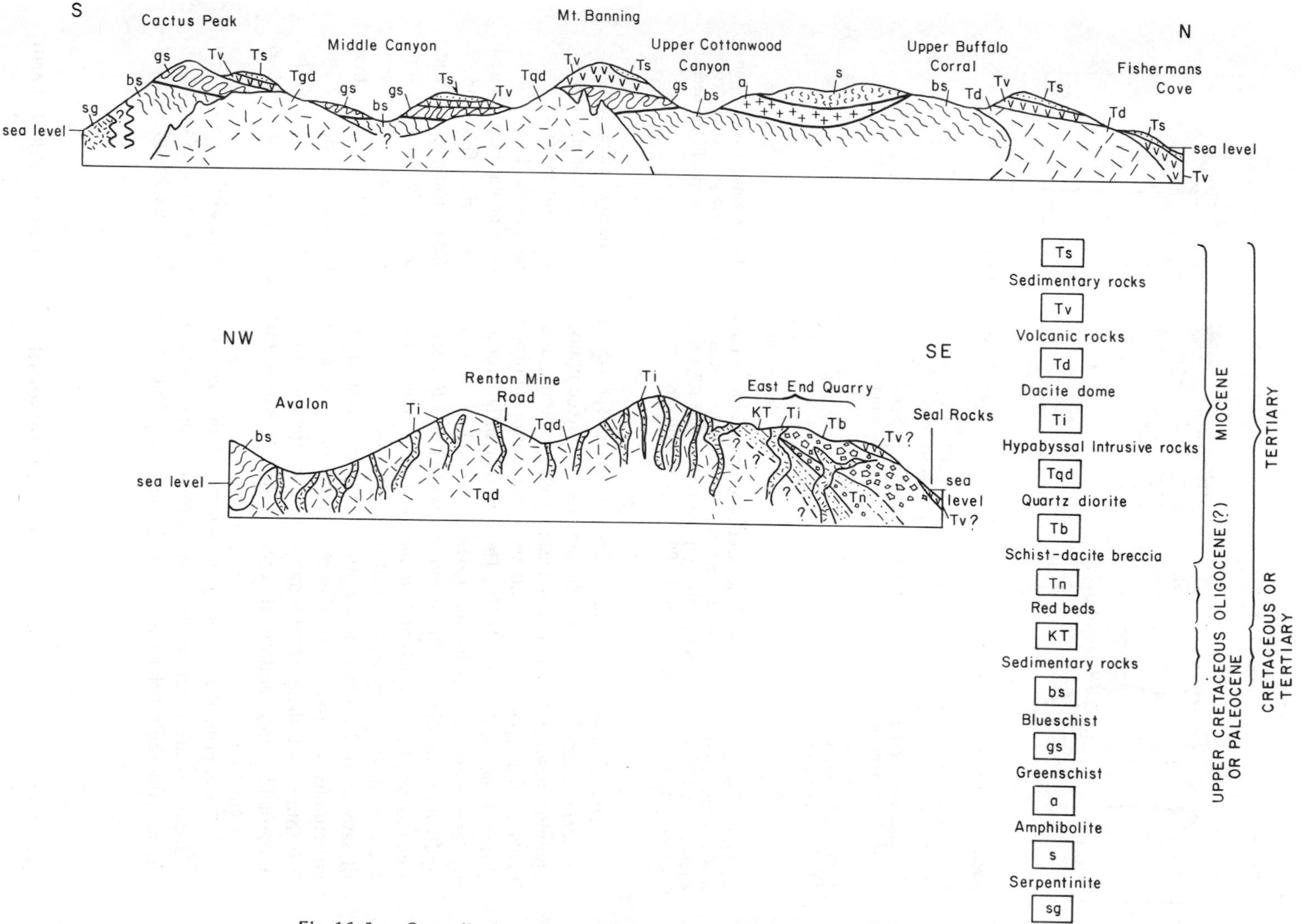

Fig. 16-5. Generalized sections through Santa Catalina Island illustrating the variety of relations between metamorphic basement and Neogene sedimentary and igneous rocks (plutonic rocks, hypabyssal dikes, domes, flows, and tephra).

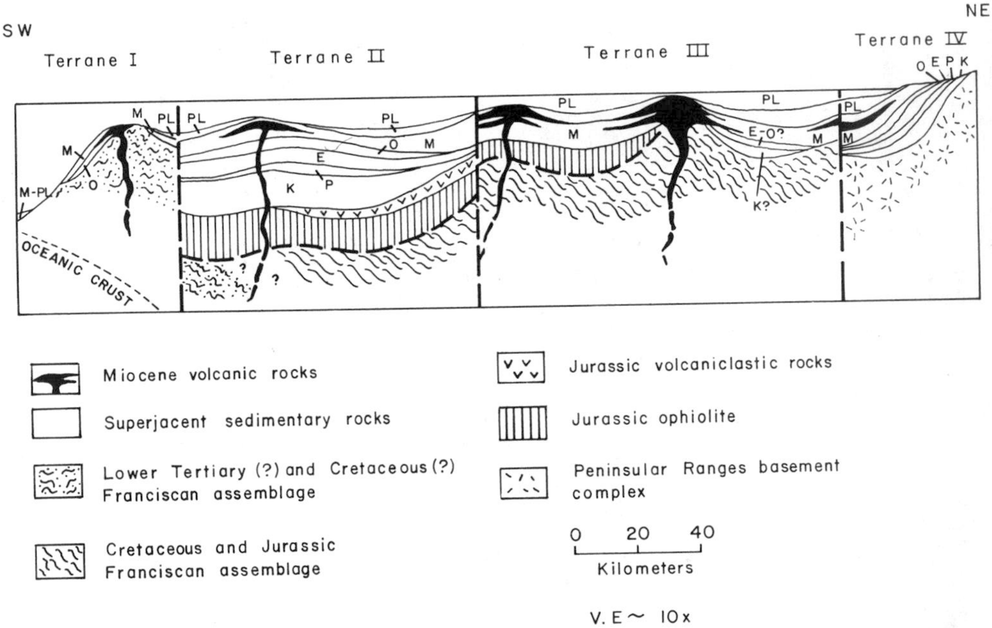

Fig. 16-6. Schematic east-west cross section of the southern California borderland. The relations shown in this figure are diagrammatic and largely speculative. Letter symbols indicate sedimentary rock sequences as follows: K, Upper Jurassic (?) and Cretaceous; K?, Cretaceous or Paleocene; P, Paleocene; E, Eocene; O, Oligocene; M, Miocene; Pl, Pliocene and younger.

The arrangement of late Mesozoic rocks (post-Nevadan orogeny) suggests that an accretionary prism of subducted material structurally underlies an upper plate of ophiolite basement and forearc-basin epiclastic strata, and that the fault separating the lower metamorphosed plate and the upper unmetamorphosed plate is the Coast Range thrust (Bailey and others, 1970). The original configuration of this thrust has been altered by later tectonism, including Neogene rebound of the subducted material. This kind of uplift is exemplified by the Diablo and Santa Lucia Ranges of the Coast Ranges (see Page, this volume, Chapter 13) and similar antiformal uplifts on the Vizcaino Peninsula. From these relations, it seems likely that the Willows Diorite and the scattered occurrences of saussuritized gabbro on the borderland are remnants of an upper-plate ophiolite that structurally overrode high-pressure metamorphic rocks such as the Catalina Schist. The distribution of these accretionary rocks in the eastern and westernmost parts of the borderland imply antiformal uplift, some of which may have occurred as early as Paleocene time.

To explain the present conjoinment of the basement rocks, additional structural dislocations are required. The first involves Upper Jurassic volcanic arc rocks (western Sierran Foothills belt) overriding the ophiolite (Fig. 16-7 Santa Cruz Island area). If

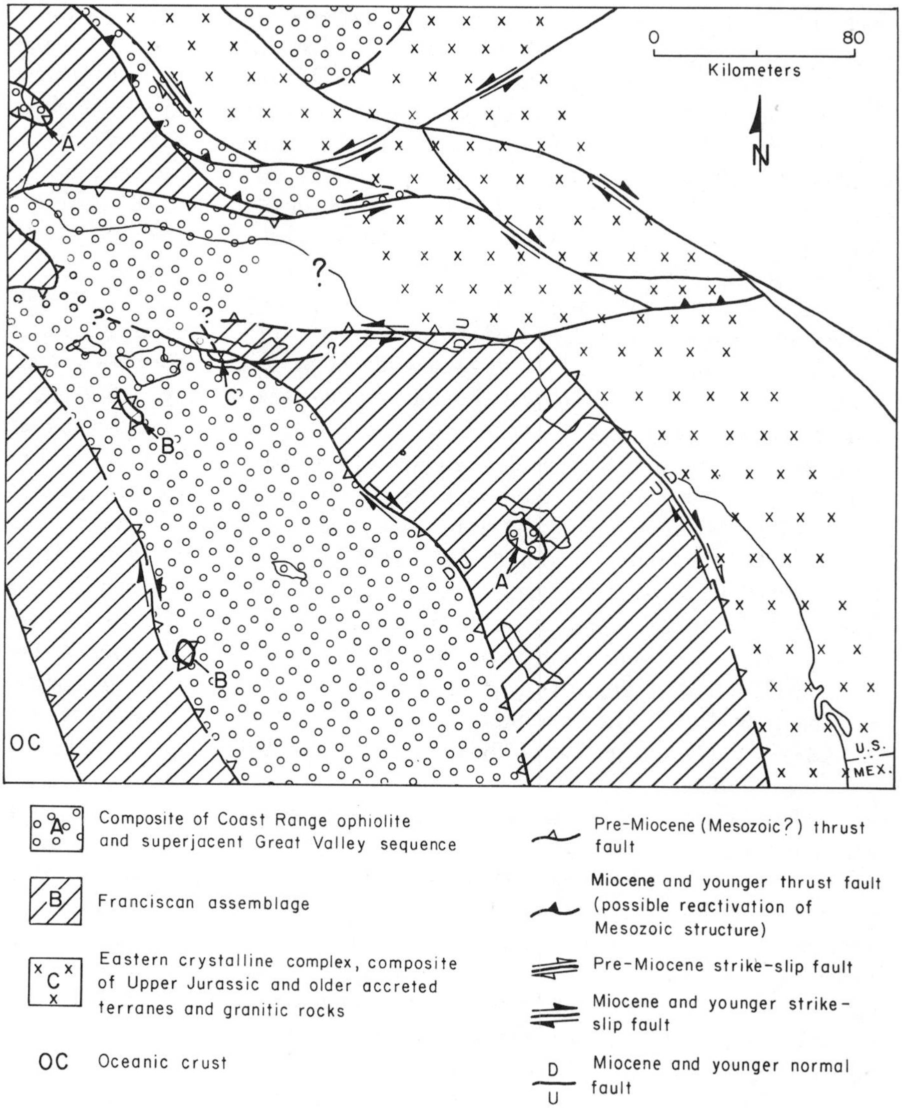

Fig. 16-7. Schematic map illustrating inferred distribution and possible structural arrangement of basement rocks in southwestern California. Multiple symbols on boundary faults indicate different episodes and types of movement.

clastic rocks in the borderland are no older than earliest Cretaceous, then the ophiolite could have been juxtaposed with a sheared arc after emplacement of the Alamos Tonalite in latest Jurassic time; the foreshortening of crystalline terranes may have been a late phase of the Nevadan orogeny. In the western Sacramento Valley and on Cedros Island, however, the part of the Great Valley Sequence that directly overlies the ophiolite is upper Kimmeridgian, coeval with or possibly older than the Nevadan orogeny (Jones and others, 1976). If correlative Upper Jurassic and younger Lower Cretaceous strata occur in the borderland, structural emplacement of the Santa Cruz Island Schist may have been as late as mid-Cretaceous, indicating that the schist has overridden the older epiclastic rocks in a fashion similar to that of inferred nappes in the western Klamath Mountains, where thrusting in large part is unrelated to the Nevadan orogeny (see Irwin, Chapter 2 and Burchfiel and Davis, Chapter 3, this volume).

A second structural dislocation requires juxtaposition of Franciscan rocks of the lower plate against crystalline batholithic rocks of the Peninsular Ranges along the ancestral Newport-Inglewood fault zone and the zone of deformation that extends southeastward from it at the west boundary of terrane IV (Figs. 16-1 and 16-7). A third is the change from a Paleonene high for terrane III to a basin-ridge system. The schist ridges of terrane III generally are ellipsoidal in plan and fault bounded on at least one side, and the overlying basin deposits are Miocene and younger. This pattern of structural modification reflects Neogene wrench tectonics. West of the region of schist ridges, in the central part of borderland terrane II, shallow areas are underlain by en echelon northwest-trending upwarps such as Cortes and Tanner banks and the San Nicolas Island platform. Junger (1976) attributed this structural configuration to wrench tectonics in a region of relatively thick anisotropic upper crust. The westernmost part of the borderland, terrane I, consists of ridges of diverse shapes and sizes and a basement that may be equivalent to the Coastal belt Franciscan of northern California. For the origin of basins and ridges in the Patton Ridge area, Crouch (1977) proposed strike-slip faulting as the primary cause.

West of terrane IV, each of the three seemingly distinct basement terranes has responded somewhat differently to a Neogene right-shear stress field (Figs. 16-7 and 16-8). The basement rocks are largely Mesozoic in age, and their spatial relations require recurrent episodes and different kinds of structural dislocation to account for their present position. Across the entire region comprising terranes I to III, accretionary subducted rocks presumably occur either at the surface or below an ophiolite. West of San Diego, this belt of rocks is nearly 300 km wide, far wider than any other known Late Jurassic to Early Tertiary accretionary terrane in North America; but at the south end of the borderland, near Cedros Island, the width of accretionary rock narrows to less than 100 km. This pattern suggests to us a possible northward reshuffling and outward stacking of the terranes with strike slip along their boundaries in a manner somewhat like that proposed by Crouch (1978).

Geotectonic Model

To explain the evolution of the borderland, we envision an orderly succession of tectonic events that is amenable to both the modern concepts of continental-margin development

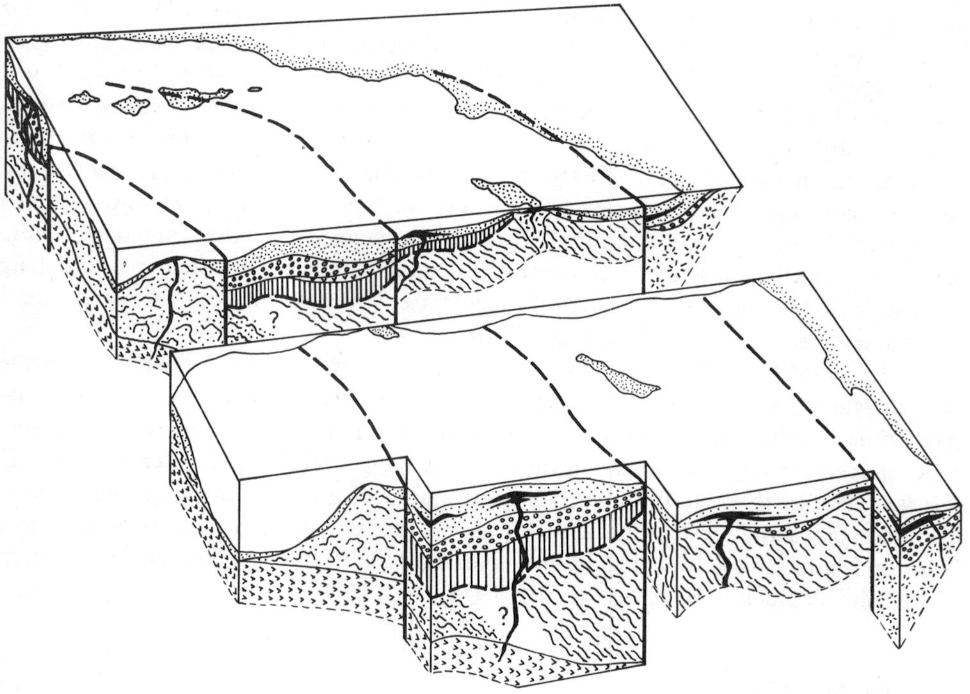

Fig. 16-8. Block diagram illustrating in simplified fashion the inferred response to right-shear stress and isostatic rebound among the different terranes of the borderland. For example, south of the northern Channel Islands: terrane IV is rigid, high standing, and tilted seaward; terrane III is highly ductile and high standing; terrane II is moderately ductile and low standing; and terrane I is warped and high standing. The front edge of the block shows apparent translocation of terranes and not actual slip; the middle slice is a vertical section similar to Fig. 16-6.

and the known distribution of rocks. Following the Nevadan orogeny, east-directed subduction stepped westward, trapping oceanic crust between the arc and the trench. The resulting basin was floored by the 160 m.y. old Coast Range ophiolite. Subduction was the dominant tectonic mode during the late Mesozoic and early Cenozoic, possibly interrupted by several episodes of crustal slivering due to oblique subduction (middle Cretaceous and Paleocene). By about 30 m.y. ago, compression gave way to right shear with the southward propagation of the Rivera triple junction and the juxtaposing of the North American and Pacific plates. Following the cessation of compression, the old subducted accretionary rock rebounded isostatically, and concurrently the margin of California and Baja California was subjected to major right-shear stress. These two independent dynamic forces resulted in the strain pattern that realigned the four basement terranes of the southern California borderland.

It is uncertain why the older accretionary terranes widened by outward stacking in the borderland. Perhaps the widening is related to Neogene extension of the Basin and Range province and westward movement of the Sierra Nevada block. Such large-scale crustal distension into west-central California presumably would have kinked the North American plate just north of the borderland to form a large bight that obstructed northwestward drift of Pacific-margin fault slivers in Neogene time. The tectonic relations in the northern part of this bight (Santa Barbara Channel and Transverse Ranges) are complex and involve west-trending left slip and west-directed thrusting sympathetic to Basin and Range province extension, right slip in response to North American-Pacific plate shear, north-south crustal shortening accompanying north-directed collision along the southern boundary of the western Transverse Ranges bight, and local (?) crustal rotation resulting from drag along major faults.

The combination of north-directed slivering and west-directed bulging along the west margin of North America may be recurrent. A similar occurrence in middle Cretaceous time fits the relations of the westward-migrated Klamath Mountains and the wide zone of accretionary rocks in the northern Coast Ranges (Blake and Jones, Chapter 12, this volume). In the Late Cretaceous, a northward slicing of a west-protruding bulge of the Mesozoic batholith by right-slip faults may explain the distribution of basement rocks in central and southern California and the Late Cretaceous depositional patterns of the Salinian block (Howell and Vedder, 1978).

ACKNOWLEDGMENTS

This paper summarizes the work of numerous individuals of the U.S. Geological Survey, universities, and the petroleum industry, who have shared their concepts of the geology of the southern California borderland. Early drafts of this paper have benefited from suggestions for improvement by S. B. Bachman, L. A. Beyer, W. G. Ernst, D. L. Jones, Hugh McLean, and other contributors to this volume.

Clarence A. Hall, Jr.
Department of Earth and Space Sciences
University of California, Los Angeles

17

EVOLUTION OF THE WESTERN TRANSVERSE RANGES MICROPLATE: LATE CENOZOIC FAULTING AND BASINAL DEVELOPMENT

ABSTRACT

Following the emplacement and deposition of Miocene volcanic intrusive and extrusive and pyroclastic rocks over a wide area of western California, the western Transverse Ranges drifted northwestward between major bounding faults. The volcanic and pyroclastic rocks may have been deposited along flanks of a submarine high in a linear zone from near Santa Barbara to Cambria. The Transverse Ranges microplate is that area south of the southern Coast Ranges, north of the Peninsular Ranges, and west of the San Gabriel fault. The major structural features bounding the microplate are the Lompoc-Solvang-Santa Ynez fault zone on the north and the Malibu Coast-Santa Monica-Raymond Hill fault zone on the south. The San Gabriel (and perhaps the Rinconada) fault may have acted as part of an eastern bounding fault zone during the northwestward migration of the Transverse Ranges microplate. The timing of northwestward migration and rotation of the Transverse Ranges microplate is largely dictated by the formation of the Lompoc-Santa Maria pull-apart basin. The latter is floored by Jurassic and possibly Cretaceous rocks and in general lacks marine Oligocene to lower Miocene sedimentary and middle Miocene volcanic and pyroclastic rocks present along the shoulders of the basin. Counterclockwise rotation of the western Transverse Ranges occurred during formation of the pull-apart structure; however, clockwise rotation, based on paleomagnetic data, of individual blocks probably occurred within the microplate between right lateral faults, during extension, or perhaps prior to the development of the pull-apart structure. The microplate has moved more than 70 km westward and northwestward since the late medial Miocene time.

The San Gregorio-San Simeon-Hosgri fault zone transects the western part of the Lompoc-Santa Maria basin, and formed following, or concurrent with, the formation of the basin. This strike-slip zone joins the San Andreas fault zone north of San Francisco and south of the present northern limit of Salinian granitic basement. Because the San Gregorio-San Simeon-Hosgri fault intersects the San Andreas fault, the total apparent offset of the Sierran-type granitic rocks along the San Andreas fault system is the sum of offsets along the San Andreas and the San Gregorio-Hosgri fault zone. The 80 to 115 km offset along the San Gregorio-Hosgri fault zone accounts for some of the displacement during Tertiary time between the Pacific and North American plate and could bring the calculated strike slip between these two plates during the last 4 to 10 m.y. more in line with the amounts measured along the fault's northern and southern trace.

INTRODUCTION

Perplexing geologic problems in southern California include (1) the possibility of a two-stage development of the San Andreas fault and (2) the east-west trend of the western Transverse Ranges lying between the southern Coast Ranges and the northern Peninsular Ranges, both of which have northwest trends. An understanding of the development of the western Transverse Ranges is related to an understanding of the San Gabriel-San Andreas fault system and the time or times of movement along a series of major faults

within the regions (e.g., Ozena, Rinconada, Sur-Nacimiento, Pine Mountain, Santa Maria River, Santa Ynez, Lompoc-Solvang, and Santa Monica-Malibu Coast-Raymond Hill faults). A review of the tectonic evolution of the Transverse Ranges and related faults is given by Jahns (1973), other papers in the proceedings volume edited by Kovach and Nur

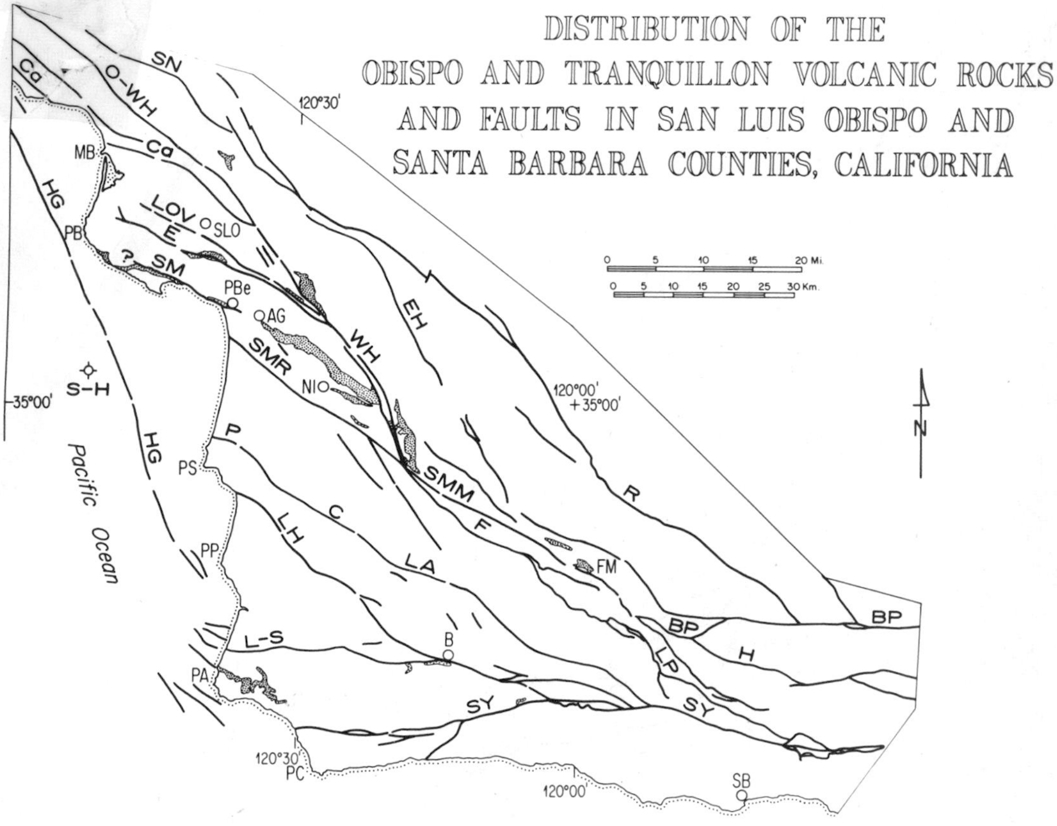

Fig. 17-1. Map showing the generalized distribution of middle Miocene volcanic and pyroclastic rocks (pattern) in central California, western San Luis Obispo County, and northern Santa Barbara County. Some of the faults in this region are also shown for purposes of comparison with Fig. 17-5. Only the larger outcrops of middle Miocene volcanic and pyroclastic rocks are shown because of the scale of the map. Middle Miocene pyroclastic and volcanic rocks (Obispo Formation and Tranquillon Volcanics) are absent or rare between the Santa Maria River-Foxen Canyon-Little Pine Fault (SMR-F-LP). Faults: **BP**, Big Pine; **C**, Casmalia; **Ca**, Cambria; **Cu**, Camuesa; **E**, Edna; **EH**, East Huasna; **F**, Foxen Canyon; **HG**, Hosgri; **LA**, Los Alamos; **LP**, Little Pine; **L-S**, Lompoc-Solvang; **LOV**, Los Osos Valley; **O**, Oceanic; **P**, Pezzoni; **PM**, Pine Mountain; **R**, Rinconada; **SM**, San Miguelito; **SSM**, Santa Maria Mesa; **SMR**, Santa Maria River; **SN**, Sur-Nacimiento; **SY**, Santa Ynez; **WH**, West Huasna. Other symbols: **B**, Buellton; **FM**, Figueroa Mountain; **MB**, Morro Bay; **NI**, Nipomo; **PA**, Point Arguello; **PB**, Point Buchon; **PBe**, Pismo Beach; **PC**, Point Conception; **PP**, Purisima Point; **PS**, Point Sal; **SB**, Santa Barbara; **SLO**, San Luis Obispo. **S-H** is Standard-Humble "Oceano" No. 1 exploratory well.

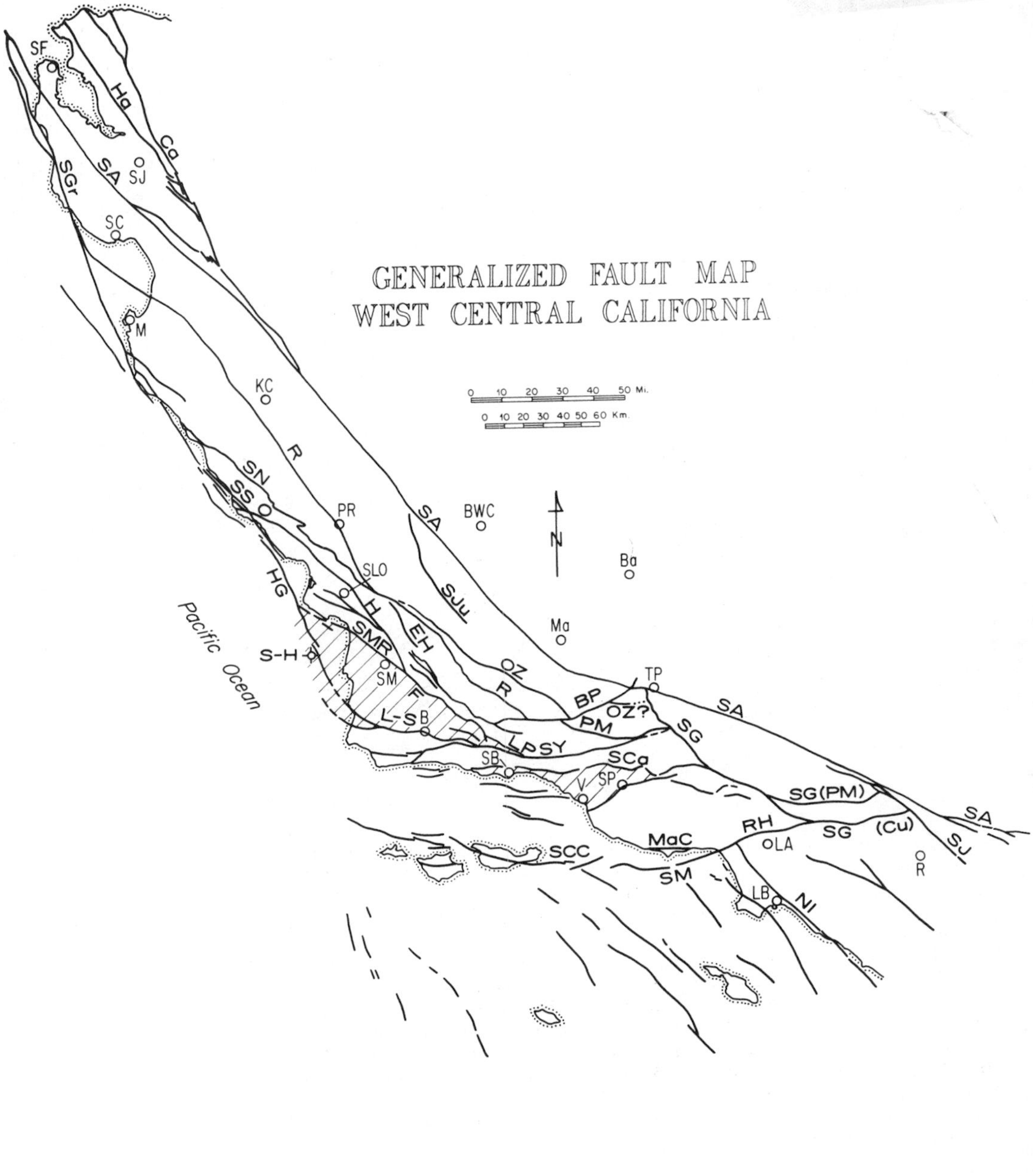
GENERALIZED FAULT MAP
WEST CENTRAL CALIFORNIA
0 10 20 30 40 50 Mi.
0 10 20 30 40 50 60 Km.
N
SF
Ha
Ca
SJ
SA
SGr
SC
M
KC
R
SN
SS
PR
SA
BWC
Ba
SLO
HG
H
SJu
Pacific Ocean
SMR
EH
Ma
S-H
SM
OZ
TP
R
F
BP
OZ?
SA
L-S
B
PM
SG
LP
SY
SB
SCa
V
SP
SG(PM)
SA
RH
SG
(Cu)
SJ
MaC
LA
R
SCC
SM
LB
NI

Fig. 17-2. Generalized fault map of western California. Modified from *Fault Map of California* (Jennings, 1975) and maps compiled by Buchanan-Banks and others (1978). Faults: **BP**, Big Pine; **Ca**, Calaveras; **Cu**, Cucamonga; **EH**, East Huasna; **F**, Foxen Canyon; **H**, West Huasna; **Ha**, Hayward; **HG**, Hosgri; **LP**, Little Pine; **MaC**, Malibu Coast; **NI**, Newport-Inglewood; **O**, Oceanic; **Oz**, Ozena; **PM**, Pine Mountain; **R**, Rinconada; **RH**, Raymond Hill; **SA**, San Andreas; **SCa**, San Cayetano; **SCr**, Santa Cruz Island; **SG**, San Gabriel; **SGr**, San Gregorio; **SJ**, San Jacinto; **SJu**, San Juan; **SM**, Santa Monica; **SMR**, Santa Maria River; **SN**, Sur-Nacimiento; **SS**, San Simeon; **SY**, Santa Ynez. Other symbols: B, Buellton; BWC, Blackwells Corners; KC, King City; LA, Los Angeles; LB, Long Beach; M, Monterey; Ma, Maricopa; MB, Morro Bay; Pa, Palmdale; PP, Purisima Point; PR, Paso Robles; PS, Point Sal; R, Riverside; SB, Santa Barbara; SC, Santa Cruz; SF, San Francisco; SLO, San Luis Obispo; SM, Santa Maria; SP, Santa Paula; TP, Tejon Pass; V, Ventura. Ruled areas are Lompoc-Santa Maria and Santa Clara Trough pull-apart basins. **S-H**, Standard-Humble "Oceano" No. 1 exploratory well.

Selected references for faults, from north to south, are San Gregorio fault, Graham (1977) and Graham and Dickinson (1978a, b); Rinconada fault, Dibblee (1976); Sur-Nacimiento and southern part of Rinconada faults, Page (1970a); San Simeon-Hosgri fault, Hall (1975); Big Pine fault, Dibblee (1976); Santa Maria River, Foxen Canyon, Little Pine, Lompoc-Solvang, Hall (1977, 1978a); the western extension of the Santa Maria River fault is probably that fault mentioned by Dept. of Water Resources (1970, p. 14); Santa Ynez (early right-lateral movement), Schmitka (1973); Santa Cruz Island fault and 32° counterclockwise rotation of southern part of Santa Cruz Island, Howell (1976a); San Cayetano-Oak Ridge faults, Crowell (1976); San Gabriel fault, Crowell (1975a); Cucamonga fault, Crowell (1973); Santa Monica fault, Sage (1973) and Campbell and Yerkes (1976); northern Peninsular Ranges faults, Crowell (1974b); San Jacinto fault, Sharp (1967); northern part of San Andreas fault, Clarke and Nilsen (1973) and Suppe (1970); southern part of San Andreas fault, Crowell (1962, 1973, 1975c); proto-San Andreas and Newport-Inglewood faults, Suppe (1970). A review of many of these faults in provided by Jahns (1973), and seismicity is discussed by Gawthrop (1975, 1978).

(1973), and by Campbell and Yerkes (1976). A clue to a relatively definitive time or times of movement along the system of faults in the southern Coast Ranges and in, or bounding, the western Transverse Ranges is provided by the Lompoc-Santa Maria pull-apart basin (Hall, 1977, 1978a) and its associated rocks. The history of this basin as suggested by the distribution of middle Miocene volcanic rocks and other Tertiary rocks provides temporal restrictions concerning the emplacement of the western Transverse Ranges microplate in its present position. This in turn places constraints on the timing of motion along several major faults within western southern California, and seems to agree with limits suggested by Campbell and Yerkes (1971, 1976).

Any model that attempts a fault and basinal developmental history in the southern Coast Ranges and the Transverse Ranges must include explanations for the distribution of middle Miocene volcanic and Tertiary and pyroclastic rocks, as well as sequences of Mesozoic rocks in the Point Sal and San Simeon areas, and the abrupt termination of a thick section of Eocene and Oligocene marine strata at the northern boundary of the western Transverse Ranges. Figure 17-1 shows the distribution of middle Miocene pyroclastic and volcanic rocks; however, not all volcanic rocks of this age are shown because of the limiting scale of the map. Figure 17-2 depicts the principal faults that are considered in this analysis.

To set the stage for the post-middle Miocene structural history of the southern Coast Ranges and Transverse Ranges, an abbreviated outline of several salient events that occurred in the region is first presented in Table 17-1. This is followed by a review of the history of the Santa Maria basin and, finally, by a summary of major structural events. The actual geologic history, largely for the middle and late Miocene, is more complex than presented here, but the outline of Table 17-1 should serve to illustrate the tectonic mosaic of interdependent structural features that characterize an evolving mobile continental margin.

TABLE 17-1. Outline of Some Major Post-Mesozoic Events, West Central California

Paleocene (?) or Oligocene (∿38 m.y.; Atwater, 1970)	Early-stage movement along either the northern part of San Andreas fault or a parallel proto-San Andreas and the proto-San Andreas fault in southern California
Oligocene-early Miocene	Local subsidence following deposition of nonmarine rocks of Sespe and Lospe formations; continued subsidence during late Oligocene through late early Miocene
Middle Miocene (15–17 m.y.; Turner, 1970)	Volcanism during late early and middle Miocene, and continued subsidence
Late middle and upper Miocene (8–14 m.y.)	Northwestward movement of Transverse Ranges (probably more than 70 km) and development of pull-apart structures (e.g., Santa Maria basin), largely following middle Miocene volcanic events; continued subsidence regionally, more rapid locally; e.g., Lompoc-Santa Maria basin
Miocene-Pliocene (4–8 m.y.)	Right-slip along the West Huasna-Los Osos fault and the western San Gabriel fault during late Miocene or Pliocene; Right slip along San Gregorio-San Simeon-Hosgri fault zone during some portion of the interval between latest Miocene and latest Pliocene
Late Pliocene-Pleistocene-Holocene (<4 m.y.; Crowell, Chapter 18, this volume)	Left slip along Big Pine and Santa Ynez faults; crustal shortening in southernmost western Coast Ranges and Transverse Ranges; initiation of movement along San Andreas fault south of Tejon Pass, producing about 250 km of right slip along San Andreas fault (Crowell, 1962; Crowell, Chapter 18, this volume)

DEVELOPMENT OF THE LOMPOC-SANTA MARIA BASIN

The origin and development of the Lompoc-Santa Maria basin has been discussed by Hall (1977, 1978a). Hall (1977) first proposed that the Lompoc-Santa Maria basin is a pull-apart structure bounded on the north and northeast by the Santa Maria River-Foxen Canyon-Little Pine fault zone and on the south and southwest by the Lompoc-Solvang and part of the Santa Ynez fault zone (Fig. 17-3). This latter fault zone was later named the Santa Ynez River fault by Sylvester and Darrow (1979) when they presented evidence for its presence in northern Santa Barbara County, California. The name Lompoc-Solvang fault is preferred because it is more descriptive of the geographic location of the fault and avoids confusion with the Santa Ynez fault. Of historic interest is the fact that some of the faults mentioned are shown, although unnamed, on a 1922 map compiled by Willis and Wood (1922). The Santa Maria River and Lompoc-Solvang zones, making up a fault system in northernmost western Santa Barbara County, constitute the shoulders of a basin that developed after deposition and emplacement of the Obispo and Tranquillon volcanic and pyroclastic rocks. These rocks have been radiometrically dated by Turner (1970) as 15 to 17 m.y. old, using K-Ar whole-rock methods. Miocene volcanic and pyroclastic rocks are largely absent from the pull-apart structure but are present at, and extend away from, the shoulders of the basin. Thus, the basin may have developed between 12 and 16 m.y. ago and continued during the Miocene and early Pliocene. The apparent absence or possible rare occurrence of the marine Oligocene and lower Miocene Vaqueros and Rincon formations in the basin and their presence along the shoulders of the basin and the fact that the basin is largely floored only by Late Jurassic ophiolite, chert and shale, Franciscan rocks, or Cretaceous sandstone are factors that support the hypothesis that the Lompoc-Santa Maria basin is a pull-apart structure.

The following is a brief history of the basin following the deposition of the non-marine Oligocene Sespe (Lospe) Formation. After the Sespe (Lospe) was deposited there was subsidence of the region, with the subsequent deposition of the shallow-water deposits of the Oligocene Vaqueros Sandstone, the Oligocene-Miocene bathyal or abyssal deposits of the Rincon Shale, and lower and middle Miocene Obispo-Tranquillon formations (also in part deep water). The Obispo and Tranquillon formations contain tuffaceous siltstone, tuff, and diabase or basalt; the tuffs are locally associated with marine fossils (Hall and others, 1966). The Lompoc-Santa Maria pull-apart structure began to develop along an existing fault system (Santa Maria River-Foxen Canyon-Little Pine-Lompoc-Solvang) between 12 and 14 m.y. ago, or somewhat later. Oblique rifting evidently occurred along this fault system in a manner described by Crowell (1974b) (see also Fig. 17-4) and analogous to the development of rhombochasms described by Carey (1958). The term "pull-apart" was first used by Burchfiel and Stewart (1966) to describe the formation of Death Valley California.

The development of the Lompoc-Santa Maria pull-apart structure can be visualized by a simple exercise using your hands. Place the palms of your hands together, palms flat. The imaginary plane between your hands represents the ancestral Santa Maria River-Lompoc fault complex or system, mentioned previously (Fig. 17-3). With your hands and fingers pointing away from you, imagine you are looking toward the southeast. As the flattened palms are rubbed together your hands move easily by one another, representing simple strike-slip faulting (Fig. 17-4). However, if you slightly cup your right hand and repeat the palm-rubbing motion, you will find that the tips of the fingers of your left hand will tend to catch in the palm of your right hand, and with continued pushing of your left hand away from you, the fingers of your left hand tend to bend, and your left hand pulls away from your right hand. At this point the area between your hands represents the pull-apart basin. Your right hand represents the region west of the early Santa Maria River-Lompoc system as you imagine you are looking to the southeast; your left

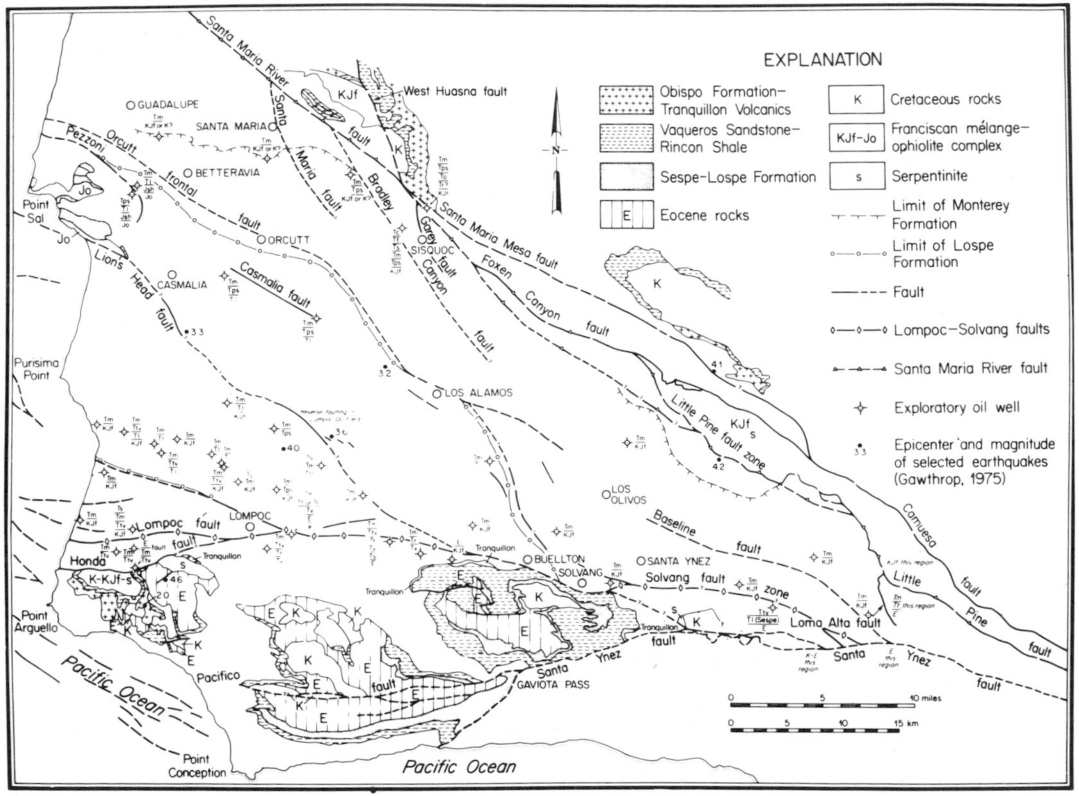

Fig. 17-3. Generalized pre-late Miocene (Monterey Formation) paleogeologic map of the Santa Maria-Lompoc region. Sites of selected exploratory wells are shown with stratigraphic section encountered in well. Stratigraphic symbols: Ts, Sisquoc Formation, upper Miocene and lower Pliocene; Tm, Monterey Formation, middle and upper Miocene; Tps, Point Sal Formation, middle Miocene; Tt, "Temblor Formation," middle Miocene; Ttv, Tranquillon Volcanics, early middle Miocene; Tr?, Rincon Shale?, early Miocene; Tv?, Vaqueros Sandstone?, Oligocene or early Miocene; Tl, Lospe Formation, Oligocene; Jsh, Jurassic Shale; Jo, Jurassic ophiolite; KJf, Franciscan melange.

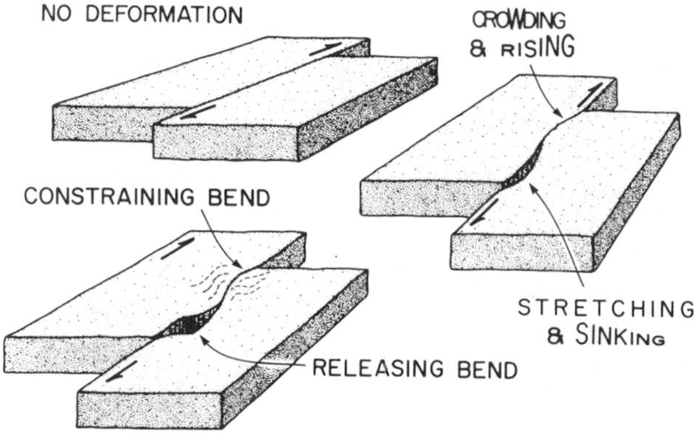

Fig. 17-4. Deformation along a strike-slip fault, from Crowell (1974b). Relatively little deformation of the two juxtaposed blocks takes place along a straight fault. If right slip occurs along a fault with a gentle bend, oblique shortening is produced at the restraining or constraining bend and extension at the releasing bend. In the case of a marked bend in the fault, there is pull-apart at the releasing bend and folds and thrust faults at the restraining bend. In the lower left-hand diagram, north would be toward the lower left, south in the upper right as a model for the Lompoc-Santa Maria pull-apart structure.

hand represents the rocks that today lie to the northeast of the Santa Maria River-Foxen Canyon-Little Pine fault; and the area between your hands represents the Lompoc-Santa Maria basin, a basin that for the most part consists of a floor of Franciscan melange or Jurassic ophiolite and shale overlain by later Miocene and Pliocene rocks (Fig. 17-3). Irregularities in the fault zone may be responsible for upthrust blocks and pull-apart basins as shown in Fig. 17-4. The basin must have formed for the most part immediately following the deposition and emplacement of the Obispo-Tranquillon pyroclastic, sedimentary, and volcanic complex during the middle Miocene. These rocks are almost completely absent from the basin, but occur on the flanks of the basin (your hands in the preceding analogy).

The basin-flooring Jurassic and Franciscan rocks failed through mass flowage because they were already highly sheared and were easily folded, faulted, and stretched. Because of the incompetent nature of the Franciscan rocks, very little or no igneous activity occurred in the basin during the pull-apart process. Regional subsidence continued through the late Oligocene to the late Miocene (Vaqueros, Rincon, Obispo-Tranquillon, Point Sal, Monterey formations), but was more pronounced in the central part of the basin during the late middle and late Miocene (Monterey Formation). The depth of the bottom of the Monterey in the basin is now between 3000 and 4500 m. The land east of the basin was probably rising and accompanying the development of the pull-apart structure. With the uplift to the east of the basin, there would have been an increase in slope and runoff so that, with the depression and development of the basin, portions of the section overlying the Jurassic or Franciscan rocks, that is, much of the late Oligocene and Miocene rocks, Sespe (Lospe) and all or most all of the Vaqueros, Rincon, and Obispo-Tranquillon rocks were sluiced out of the pull-apart structure down submarine canyons and carried to the west. Returning to the model using your hands, this part of the history of the basin can be visualized if the operations outlined are repeated; however, this time hold your thumbs over the site of the pull-apart structure. Your thumbs represent the preceding sequence of strata and should be thought of as continuous at first with the strata on the flanks of the basin, represented by your two index fingers. As you move your left hand away from you and past your cupped right hand, your left hand tends to pull away from the right thumb (the Santa Maria River-Foxen Canyon-Little Pine fault would be between your left hand and left side of your left thumb, and the Lompoc-Solvang fault would be between the right side of your right thumb and your right index finger). Not only would there be dip slip aloang the flanks of the deepening basin, but there would have been contemporaneous strike-slip faulting (~70 km). As the rocks, represented by your thumb, sink into the basin, fragmentation within the basin, increased runoff and increased slope development, and faulting would tend to destroy most of the Middle Tertiary section of rocks. Submarine erosion of the Oligocene and Miocene rocks within the basin would have uncovered the underlying Jurassic and Franciscan rocks, and the remnants of the Middle Tertiary section of rocks were in turn covered by the regionally transgressing Monterey sea as the area continued to subside. Significant right slip along the Santa Maria River-Lompoc fault zone accompanied dip slip as the late Miocene and Pliocene seas continued to flood the deepening basin.

Accompanying development of the basin, fragmentation of the trough, right-slip rifting, and rotational movement occurred that changed the trend of the western part of the fault system, that is, the Lompoc-Solvang fault or western shoulder of the basin. The change in trend would have been from the northwest to the west. This is one of the most significant aspects of the development of the Lompoc-Santa Maria basin because of its relation to the history of the western Transverse Ranges. The development of the pull-apart structure, counterclockwise rotation, and accompanying right slip along the fault system seem to account for the distribution of thick section (maximum 7000 m) Cretaceous, Eocene, Oligocene, and lower Miocene rocks in the western Transverse Ranges and their absence from the Lompoc-Santa Maria basin. It seems unlikely that the Santa Maria basin could have been a wedge-shaped "high" that persisted from the Cretaceous to the middle Miocene, as proposed by Dibblee (1978), and then rapidly became a deep-water basin during the mid-Miocene. The distribution of volcanic and pyroclastic rocks of the Tranquillon and Obispo formations and the deep-water-deposited sediments of the Rincon (lower Miocene) Shale on the flanks of the basin, the depth of the base of the Monterey, and the proximity of dissimilar rock sections along the Lompoc-Solvang fault (Fig. 17-3) are some points that cannot be reconciled with a long-standing Cretaceous to Miocene "high"—a wedge-shaped high over which a maximum thickness of 7000 m of rock is missing but present on the flanks.

Returning for the final time to your hands as a means of visualizing in a very generalized way how the basin and the western Transverse Ranges developed, place your hands together, palms flat, thumbs over the gap between two hands, and the fingers of right hand extended beyond those of the left. You are looking to the southeast. The rock relationships relative to your hands are as described previously with the addition that the ball of your right thumb and the joint between index finger and palm of the left hand represent the lateral extent (north or western limit) of Cretaceous, Eocene, and marine Oligocene rocks. Your left thumb represents Sespe (Lospe), Vaqueros, Rincon, and Obispo lying on older rocks. Cup your right hand, move your left hand, which is flat with tips of fingers in cup of right hand, away from you (southward in map view), while at the same time moving your right hand toward you (northwest and westward). Your left fingers will tend to bend, your right fingers will tend to straighten, and your right hand will tend to rotate and move to the right in a counterclockwise motion. As this operation is completed note the relative position of the ball of your right thumb and the index finger-palm joint of your left. Your right hand and right wrist represent the western Transverse Ranges (see Fig. 17-3) that have moved westward or northwestward.

Thus, the age of the Obispo-Tranquillon volcanic rocks and the late middle Miocene rocks overlying a Franciscan basin floor strongly suggest a late middle and late Miocene development of the Lompoc-Santa Maria basin, and this event in turn suggests the time of emplacement of the Transverse Ranges microplate, that is, perhaps between 12 and 16 m.y.b.p. Left slip along the Santa Ynez fault did not occur until the Quaternary and may have played only a relatively minor role in the structural evolution along its trace in the westernmost Transverse Ranges. The development of the Santa Maria pull-apart structure occurred before movement along the San Simeon-Hosgri fault. The sequence of strata in the basin near Point Sal, that is, the Jurassic ophiolite chert and shale and the Lospe, Monterey, and possibly the Pliocene rocks, was moved northward after the deposition of the Miocene Monterey Formation or Pliocene rocks; this event suggests that the basin must have been developing until the Pliocene or at least until the Miocene Monterey and Sisquoc formations had been deposited.

SAN GREGORIO-SAN SIMEON-HOSGRI FAULT

As mentioned previously, movement on the San Gregorio-San Simeon-Hosgri fault (Fig. 17-2) occurred after or during the development of the Lompoc-Santa Maria basin because the packages of rocks described by Hall (1975) from San Simeon and Point Sal areas represent post-pull-apart basinal development; that is, there is an absence of Vaqueros, Rincon, and Obispo-Tranquillon rocks at the two localities. The Lompoc-Solvang fault may curve to the north offshore and may parallel or be represented by the Hosgri fault zone to the north. Sequences of Jurassic ophiolite, Jurassic chert, Jurassic shale, Lospe, Monterey, Point Sal, Sisquoc, and Careaga formations occur in the vicinity of Point Sal and San Simeon, localities separated by approximately 100 km. Some of the lateral slip along the fault zone may be accommodated in folds and reverse slip splays at the southern end of the fault zone (Hamilton and Willingham, 1977, p. 429); however, there has been approximately 80 km of right slip along the fault zone after the development of the Santa

Maria basin (Hall, 1975, 1977), and perhaps even more (115 km) to the north, based on the well-documented work of Graham (1977, p. 424), Graham and Dickinson (1978a), and Clark and Brabb (1978) on the northern (San Gregorio) part of the fault zone. Other estimates of right slip along the San Gregorio-Hosgri fault zone are 10 to 25 km (Willingham, 1978), 15 to 20 km (Hamilton and Willingham, 1978), 30 to 35 km (Seiders, 1978, 1979), and 145 km (Payne and others, 1978a, b). However, if preslip restorations are made using these figures, dissimilar assemblages of rocks are in juxtaposition.

Rocks from the Standard-Humble "Oceano" No. 1 exploratory well (Howell and others, 1978) (Fig. 17-2) would have originally been on the south side of the Lompoc-Santa Maria basin if they have been moved 80 km or more to the northwest along the west side of the Hosgri fault during late Miocene or late Pliocene time. This assumes that the Hosgri fault veers inland near Purisima Point (Fig. 17-2). The "Oceano" No. 1 well encountered Sisquoc, Monterey, Obispo-Tranquillon, and Franciscan or ophiolite or Lospe rocks. Similar sections are known from near Lompoc, 50 to 60 km southwest of the "Oceano" well and situated on the north side of Hosgri fault (Fig. 17-3). The offshore "Oceano" section of rock is not correlative with the adjacent onshore section in the Santa Maria region, nor is it correlated with rocks at Point Sal. It cannot be correlated because of the presence of several hundred meters of pyroclastic rocks between the Monterey and Lospe or Jurassic ophiolite or Franciscan rocks in the offshore well. These pyroclastic and the upper volcanic rocks are probably the Obispo-Tranquillon volcanics, and they are absent in the on-land part of the Santa Maria basin and are not present at Point Sal. Although Miocene volcanic rocks are generally absent from the Lompoc-Santa Maria basin, there are local exceptions (Dibblee, 1950, and Fig. 14-3). However, no cross sections through the Santa Maria area shown by Woodring and Bramlette (1950) or cross sections drawn by Dibblee (1950) through the Lompoc-Santa Maria basin show Obispo or Tranquillon volcanic rocks above the Lospe Formation north of the site of the Lompoc-Solvang fault.

Information from the "Oceano" well neither negates nor supports large strike-slip along the San Gregorio-San Simeon-Hosgri fault; however, it provides an important reference point in terms of the late history of the Lompoc-Santa Maria basin. It is suggested here that the rocks in the vicinity of the "Oceano" No. 1 well were at one time in juxtaposition with a counterpart section on the northeast side of the San Gregorio-San Simeon-Hosgri fault system, and that the section of interest is in the Santa Maria basin north of the projection of the Hosgri fault on land between Purisima Point and Point Arguello. The rocks were moved along the San Simeon-Hosgri fault during the late Miocene to late Pliocene for a distance of 50 km or more (Figs. 17-1 to 17-3). During the Pleistocene there was crustal shortening in the basin, analogous to that described by Crowell (1976) for the Santa Clara trough and in the western Transverse Ranges (Yerkes and Lee, 1979). At this time all but the western part of the onshore part of the Hosgri fault, that is, that part of the fault southeast of Purisima Point and north of the Lompoc-Solvang fault, was probably obscured by crustal shortening; the offshore portion of the Hosgri fault, with its northern trend, was not markedly affected by the shortening. Left slip along the Santa Ynez, Baseline, Santa Rita, and Honda (the latter two near Lompoc) (Fig. 17-7), with associated en echelon folds, as postulated by Dibblee (1950) and Sylvester and Darrow (1979), occurred during the Quaternary and would have tended to obscure the southeastern onshore and perhaps nearshore part of the Hosgri fault; but right slip measured in tens of kilometers occurred along the Lompoc-Solvang-Santa Ynez fault at the time of northwestward movement of the Transverse Ranges microplate. Right slip along the southern part of the Hosgri fault may presently be taken up in folding, along the Lions Head fault, along en echelon or reverse faults in the central and northern part of the Santa Maria basin, and offshore from Point Arguello.

The interpretation of the offshore geophysical data relative to the southern part of the Hosgri fault south of Purisima Point is not completely resolved. Payne and others (1978a) suggest that the Hosgri probably traverses across a complex fold belt and comes onto land between Purisima Point and Point Arguello (Fig. 17-2), although this is not clear from the work of Payne and others (1978b, 1979).

Regarding the San Gregorio-San Simeon-Hosgri fault zone, there has been approximately 80 km of right slip along the San Simeon-Hosgri part of the zone since the late Miocene or late Pliocene. The three named faults (San Gregorio, San Simeon, and Hosgri) make up a zone and are not a single fault, at least at the surface; an analogy can be drawn with strands of the Garlock and San Andreas faults, each of which includes strands that are separated from one another by tens to hundreds of meters. A direct join at the surface between the Hosgri and San Simeon fault has not been proposed (Hall, 1975) and is probably precluded by aeromagnetic data (Hamilton, 1977). However, two assemblages of rocks, one located east of the Hosgri fault near Point Sal and the other west of the San Simeon fault near San Simeon, are distinctive and strikingly similar. These assemblages are separated by approximately 100 km and have apparently been offset along the San Simeon-Hosgri fault. The most pronounced similarities at these widely separated localities occur between gabbro, wehrlite, dunite, serpentinite, and sheeted dike and sill complex (Page, Chapter 13, and Hopson and others, Chapter 14, this volume) of the Jurassic ophiolite; grayish-green tuffaceous Jurassic chert, Jurassic shale; the Lospe Formation, which contains abundant clasts of ophiolite and several thin beds of volcanic ash or volcanic rocks and breccia; the Monterey Formation, particularly some horizons or members that include thick beds of black chert; and, most importantly, the similarity of sequences of rocks (Hall, 1975). Another consideration is the presence of hematized serpentinite in the Franciscan rocks of the San Simeon region that is seemingly absent elsewhere in that area. However, clasts of this distinctive rock type are present as clasts in the Lospe Formation near Point Sal. Pronounced dissimilarities occur between assemblages of rocks east and west of the San Simeon fault in the Cambria-San Simeon area (Hall, 1975). For example, a maximum thickness of 680 m of Vaqueros, Rincon, and Obispo formations is present east of the San Simeon fault in the Cambria-San Simeon region, but these rocks are absent to the west. The Lospe Formation consists of an ophiolite-clast conglomerate and volcanic rocks and interbedded pyroclastic rocks west of the San Simeon fault, but the formation is largely claystone and some conglomerate to the east. The conglomerate contains dacite clasts and Cambria felsite clasts of local derivation east of the fault in the Cambria-San Simeon region, but not west of the fault. The Monterey Formation consists of claystone, diatomaceous siltstone, and thickly bedded black chert west of the San Simeon, but thickly bedded black chert is absent or rare east of San Simeon fault in the Cambria-San Simeon region.

UNSLIPPING STRIKE-SLIP FAULTS: A GENERALIZED PALEOGEOLOGY OF THE WESTERN TRANSVERSE RANGES AND SOUTHERN COAST RANGES

The distribution of Mesozoic and Cenozoic rocks within the southern Coast Ranges and western Transverse Ranges seems to require significant lateral displacements along known faults in order to account for existing mismatches of assemblages of rocks on opposite

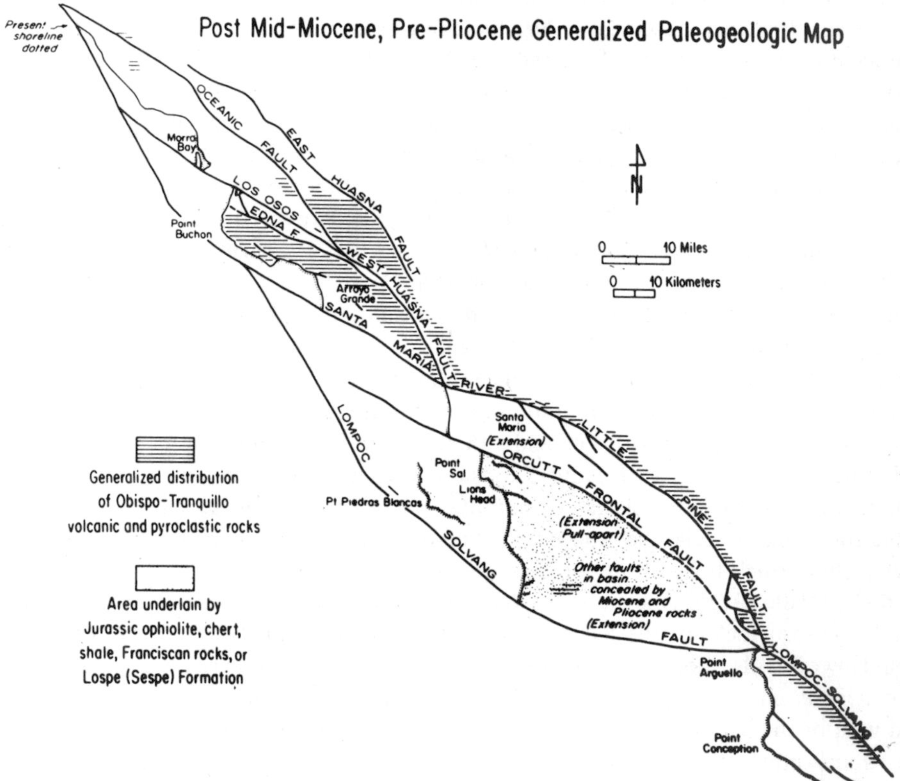

Fig. 17-5. Generalized post- mid-Miocene-pre-Pliocene paleogeologic map. The Santa Maria pull-apart structure is not completely restored to its preextension and pull-apart configuration. Such a preextension configuration aligns mid-Miocene Obispo and Tranquillon volcanic and pyroclastic rocks. Note that Miocene to Holocene rocks in the Santa Maria basin cover areas that are probably faulted in the subsurface. Pull-apart must have been accompanied by significant slivering and fragmentation with extension occurring between sets of buried faults. Several faults in the region of the Orcutt Frontal fault are included on this generalized map with the Orcutt Frontal fault.

sides of these faults. Some of the pre-mid-Miocene restorations that are apparently required are (Figs. 17-5 to 17-7) (1) unslipping the Piedras Blancas-San Simeon block southward along the San Simeon-Hosgri fault to near the latitude of Point Sal; (2) the Point Buchon block is moved southeastward along the Los Osos-West Huasna fault zone 3810 to 9140 m; (3) probable Franciscan or Cretaceous sandstone in the Santa Maria block (between the Orcutt frontal and Santa Maria River faults, Fig. 17-3) has moved an unknown distance from the southeast along the Santa Maria River-Little Pine fault; (4) Jurassic ophiolite, chert, and shale, along with the Lospe or Sespe formation, is moved an unknown distance from the southeast along the Orcutt frontal fault and either the Los Alamos-Baseline faults or an inferred fault along the southeastern edge of the subsurface limit of the Lospe Formation between Los Alamos and Solvang (Fig. 17-3); and (5) the western or northern part of the western Transverse Ranges requires 70 km or more of eastward and southeastward restoration along the Lompoc-Solvang fault. The last restoration will juxtapose Jurassic, Cretaceous, Eocene, Oligocene, and Miocene rocks with similar rocks in the vicinity of the Loma Alta fault (Figs. 17-1 to 17-3 and 17-5). The

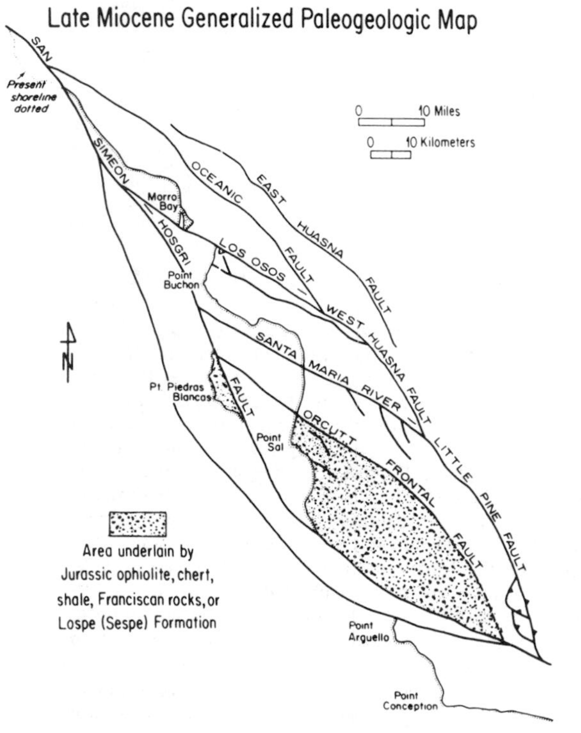

Fig. 17-6. Late Miocene generalized paleogeologic map. Northward slip of the Piedras Blancas block is constrained by the known distributional limits of the Jurassic ophiolite, chert, shale, and Lospe Formation, which are overlain by the Monterey and Sisquoc formations in both the Piedras Blancas and Point Sal blocks.

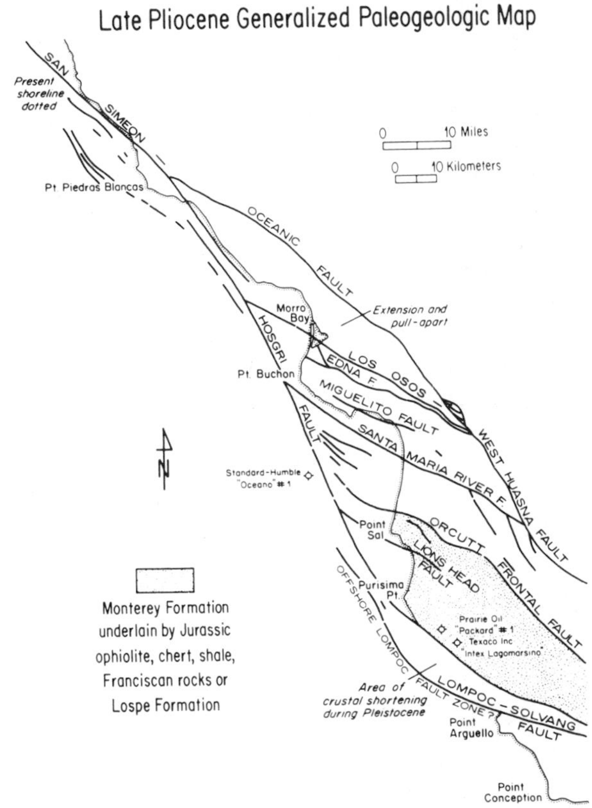

Fig. 17-7. Late Pliocene generalized paleogeologic map. Extension and rotation of the region between the Oceanic and Los Osos fault may have occurred at an earlier date; however, Pliocene rocks are cut by the Los Osos fault. Note that late Pliocene or early Pleistocene movement on the West Huasna fault took place along the zone of weakness along the Santa Maria River-Little Pine fault southeast of the junction of the West Huasna and Santa Maria River faults. The section of rock encountered in the Standard-Humble "Oceano" No. 1 well consists of Sisquoc Formation, Monterey Formation, Obispo or Tranquillon pyroclastic rocks, and either Franciscan rocks and ophiolite or the Lospe Formation. Prarie Oil "Packard" No. 1 encountered Monterey Formation, Tranquillon volcanics, and the Lospe Formation. Texaco, Inc., "Intex-Lagomarsino" encountered Monterey Formation, Tranquillon volcanics and Franciscan rocks. From Fig. 17-3 it can be seen that the presence of the Tranquillon volcanic rocks in this part of the basin or elsewhere in the basin is rare. Because of Pleistocene crustal shortening, a late Pliocene palinspastic reconstruction along the Lompoc-Solvang fault is uncertain. However, note that the area of crustal shortening is the area from which the rocks in the "Oceano" well would have moved prior to shortening.

restoration shown in Fig. 17-5 will roughly align Jurassic ophiolite, chert, and shale of the Point Sal block with Jurassic ophiolite, chert, and shale that crops out south and southeast of Figueroa Mountain, northwest of the Camuesa fault. Also aligned are the Tranquillon Volcanics near the western end of the Transverse Ranges (Point Arguello, Buellton) with the southeastern extension of the Obispo Formation (from near Santa Maria southeast to Figueroa Mountain to near the Camuesa fault. That is, pre-mid-Miocene unslipping and pre-pull-apart restoration would bring early mid-Miocene volcanic and pyroclastic rocks in the western Transverse Ranges into alignment with such rocks along an ancestral Hosgri-Santa Maria River-Foxen Canyon-Little Pine-Lompoc-Solvang fault (Fig. 17-5).

Extension (Figs. 17-5 and 17-7) probably occurred between the fault blocks mentioned previously and could have been more pronounced in the regions of known Franciscan melange, such as between Camuesa and Little Pine faults and along the southwestern part of the Santa Maria basin. These restorations suggest that the ophiolite clasts in the Lospe (Sespe) Formation of Oligocene age have been derived from the ophiolite in the Point Sal block, or in the case of the Sespe Formation near Solvang to have come from the Point Sal block or a southeastern extension of the ophiolite near Figueroa Mountain.

Strike slip alone cannot account for the distribution of rocks mentioned previously, but must have accompanied the development of the Santa Maria pull-apart structure, extension, and fault slivering of rocks in that structure. The eastern extension of the Solvang fault could be the eastern part of the Santa Ynez fault. Clearly, if the western Transverse Ranges have moved 70 km or more they must have moved along a fault from a region to the east, because similar rocks are known from that region. This aspect of the model is difficult to reconcile with what is known of the eastern part of the Santa Ynez fault. The Santa Ynez fault is terminated by the Pine Mountain-Aqua Blanca thrust and apparently does not reach the San Gabriel fault (Gordon, 1979). As mentioned previously, it is the San Gabriel fault that is envisaged as the eastern boundary of the western Transverse Ranges, and it is along this fault that the microplate would have moved north and northwestward (Figs. 17-8 and 17-9). Perhaps the middle Miocene trace of the eastern Santa Ynez fault is obscured by post-middle-Miocene thrusting along the Pine Mountain thrust or is covered by upper Miocene rocks near the San Gabriel fault. Proposed left slip of 6 km or more (Dibblee, 1978) along the Santa Ynez fault faces the same problem and must call upon significant shortening near the eastern termination of the Santa Ynez fault. Clearly, late Cenozoic left slip (Dibblee, 1978) along the Santa Ynez fault and crustal shortening of many kilometers in the western Transverse Ranges (Yerkes and Lee, 1979) have complicated and obscured the early history of the pull-apart structure and the Lompoc-Solvang fault.

Others have suggested alternative models to account for the distribution of Mesozoic and Cenozoic rocks in the region of the western Transverse Ranges and the southern Coast Ranges. For example, Dibblee (1978) suggests that the wedge-shaped Santa Maria basin between the western Transverse Ranges or Santa Ynez Mountains and the San Rafael Mountains (northeast of the Santa Maria River-Little Pine fault) was a persistent topographic high from the Jurassic or Cretaceous (Franciscan) until the deposition of the Miocene Monterey Formation. Such a proposal requires (1) that the topographic high

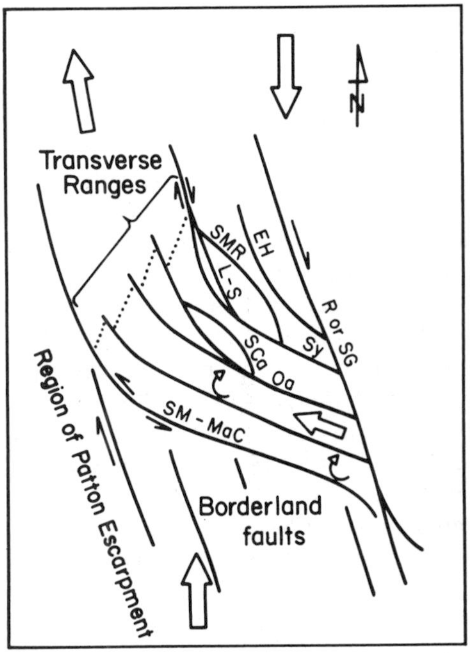

Fig. 17-8. Development of the Transverse Ranges during the late Miocene. R, Rinconada; SG, San Gabriel; EH, East Huasna, SMR, Santa Maria River; L-S, Lompoc-Solvang; SCa, San Cayetano; Oa, Oak Ridge; Sy, Simi; SM-MaC, Santa Monica-Malibu Coast faults. Movement of a series of blocks making up the Transverse Ranges began in the late middle Miocene and before movement along the present-day San Gabriel fault. The Rinconada fault (R) (or Sur-Nacimiento fault) was perhaps offset by the San Gabriel fault in the late Miocene. The eastern branch of the southern part of the San Gabriel fault is considered here to be the faulted-off part of the Pine Mountain fault east of the San Gabriel fault. Counterclockwise rotation of the Transverse Ranges province occurred as it moved westward between the proto-Santa Ynez and Santa Monica-Malibu Coastal-Raymond Hill faults. Pull-apart basins (between SMR and L-S, and SCa and Oa faults) developed in the Santa Maria, Ventura, and Los Angeles areas in response to north-south compression at northwest-southeast or east-west bends in northerly trending faults. Restoration to early middle Miocene paleogeology requires palinspastically slipping the Transverse Ranges province eastward approximately 70 to 90 km or more, moving the Peninsular Ranges westward by about the same amount and then moving the Transverse Ranges clockwise.

would have been wedge-shaped from Jurassic or Cretaceous to the present, and (2) would have been surrounded by deep water during a part of the Tertiary Period (Saucesian-Relizian, Rincon Shale); (3) there would have been mid-Miocene pyroclastic or volcanic rocks deposited along the margins of the wedge-shaped "high," but not on the wedge-shaped topographic high; and (4) the Oligo-Miocene sea gradually became deeper around the margins of the basin following the deposition of the nonmarine Lospe (Sespe) and during deposition of the shallow- to deep-water marine-deposited Vaqueros, Rincon, Tranquillon-Obispo, Point Sal, and Monterey formations, but the wedge-shaped "high" of Franciscan rocks and some Lospe Formation would have sunk rapidly during the deposition of the Monterey Formation.

Dibblee (1950, 1978) also suggests that the thick (~7000 m) Cretaceous, Eocene, Oligocene-lower Miocene stratigraphic sequence in the western Transverse Ranges thins rapidly northward near the Santa Ynez River (Dibblee, 1950), where the Monterey Formation lies on Franciscan basement. Such a suggestion seems to require an unreasonable rate of thinning. For example, there is present in the log of Marathon Oil Company's "Salsipuedes" No. 1 well (T. 6 N, R. 34W., sec. 2), near Lompoc, a section of rock that includes the Monterey Shale, Tranquillon Volcanics, Rincon Shale, Vaqueros Sandstone,

Fig. 17-9. Generailized stages of development of the Transverse Ranges and Lompoc-Santa Maria pull-apart structure. SA, San Andreas fault or proto-San Andreas fault; SG, San Gabriel fault; SY, Santa Ynez fault; MaC-SM, Malibu Coast-Santa Monica fault zone; L-S, Lompoc-Solvang fault zone. In (a) structures and textures are oriented nearly north-south and are depicted by the vertically ruled lines. A bend has developed in the San Andreas, proto-San Andreas, or in the Rinconada or early San Gabriel fault. In (b), as a reflection of the bending of the San Andreas or proto-San Andreas, the Santa Ynez fault (and related faults) along with the Malibu Coast-Santa Monica fault could have developed between what is now the southern Coast Ranges and the southern California borderland and northern Peninsular Ranges. The southern part of the western Transverse Ranges and Pacific Plate was apparently pushed northward, perhaps 5° to 10° in latitude (Kamerling and Luyendyk, 1977, 1979; Kamerling and others, 1978). In (c) is depicted the counterclockwise rotation of the Transverse Ranges microplate, clockwise rotation of individual blocks within the microplate to account for east-west trends in former north-south trending structures and textures, and pull-apart (crosshatched) in some areas.

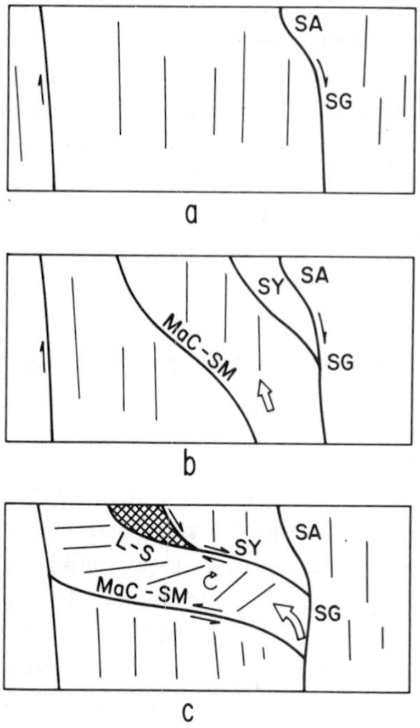

and Eocene and Cretaceous shale. However, only 2 miles north of this well the Texaco, Inc. "Irma Wilson" well (T. 7 N., R. 34 W., sec. 26) encountered Monterey Shale, Point Sal Formation, and Franciscan rocks. Near Solvang, the Exxon "Florence Selby" well (T. 6 N., R. 31 W., sec. 14) encountered Monterey Shale overlying "Knoxville" or Franciscan rocks, while approximately 1 mile south of this well Dibblee (1950) has mapped surface exposures of the Espada Shale, and Sespe, Vaqueros, Rincon, Tranquillon, and Monterey formations. Similar abrupt changes in stratigraphic section are present near the northern shoulder of the Santa Maria basin. For example, Cretaceous rocks, Rincon Shale, Obispo tuff, and Monterey Shale crop out north of the Santa Maria River, but the Monterey Shale lies on Franciscan or Cretaceous rocks immediately south of these outcrops (Hall, 1978b). All the abrupt changes in stratigraphy seem to require faults between wells and outcrop areas.

Other models to account for the present location of the western Transverse Ranges in part propose a northwestward translation of the western Transverse Ranges of 10° of latitude since early Miocene (Kamerling and Luyendyk, 1977, 1979; Crouch, 1979). Such proposals are not consistent with assemblages of rocks in widely separated areas. The Cretaceous-Eocene-Oligocene-Miocene section in the northwestern Transverse Ranges is unlike that near Los Angeles or Baja California, areas from which the Transverse Ranges are proposed to have been rotated and translated.

It is possible that the southern part of the Transverse Ranges, that is, south of Ventura, contains rocks that are closely related to those of southern California or Baja California, as suggested by Crouch (1979), and the large translations and rotations (Kammerling and Luyendyk, 1979) seemingly required by the paleomagnetic data may require translation of 10° of latitude. However, an alternative to large translations may be explained by right lateral shearing and accompanying right rotation (Greenhaus and Cox, 1979). Significant mid-Miocene left slip along the western Santa Ynez or Lompoc-Solvang faults is unsatisfactory for the same reason—the distinctive rock assemblages of the westernmost Transverse Ranges are not known from the offshore north of Point Arguello, the area from which these rocks would have had to come if there had been mid-Miocene left slip.

SUMMARY OF MAJOR STRUCTURAL EVENTS IN SOUTHERN CALIFORNIA

The tectonic history of central California may be placed in the context of the constant-motion plate-tectonic model proposed by Atwater (1970) and Atwater and Molnar (1973). The calculated amount of right slip along the modern San Andreas fault is 450 km in the last 10 m.y. (Atwater and Molnar, 1973) and exceeds the measured amount of 320 km in the last 4 m.y. (J. C. Crowell, personal communication, 1978) by perhaps as much as 100 to 150 km. Some of the relative motion, at least 80 km (Hall, 1977, 1978a) and perhaps 115 km (Graham and Dickinson, 1978a, b), in the last 4 or 5 to 13 m.y. (Hall, 1975) or 5 m.y. (Hall, 1978a) may have been taken up along the San Simeon-Hosgri fault or nearby faults (Payne and others, 1978a, b), thus further supporting the calculated amounts of right slip that have been made by Atwater and Molnar (1973).

From research and summaries of work provided in papers by authors of this volume and by Jahns (1973), Campbell and Yerkes (1976), and others, it is clear that it is not possible to treat or interpret the structural history of the Transverse Ranges without considering the neighboring provinces, because they appear to have interacted during deformation. Thus, a complete summary should include not only the southern Coast Ranges and Transverse Ranges, but all surrounding provinces or regions—the goal of this Rubey volume. Earlier chapters in the structural history of southern California are discussed by Page (1970a, Chapter 13, this volume), Suppe (1970), Dickinson and others (1972), Clarke and Nilsen (1973), and Howell (1975c; see also Chapter 16, this volume) for the late Mesozoic; and other reports deal with the Sur-Nacimiento fault and its possible southern extension (Newport-Inglewood fault, Jahns, 1973) and the proto-San Andreas (Nilsen and Clarke, 1975). The early history of the Rinconada fault is discussed by Dibblee (1976). Page (1970a) extends the Sur-Nacimiento fault south to the Big Pine fault, whereas Dibblee (1976) truncates the Sur-Nacimiento fault by the Rinconada fault near San Luis Obispo and continues the Rinconada to the Big Pine fault. However, it seems possible that there could have been reactivation, or continued movement, along the Sur-Nacimiento fault of Page (1970a) along the part lying south of San Luis Obispo. The segment of the Rinconada north of San Luis Obispo could have developed at the time of reactivation of the Sur-Nacimiento fault, perhaps analogous to the Calaveras fault where it branches from the San Andreas fault. It may have been during the Cretaceous-Tertiary transition or later that the Sierran-type granitic rocks were carried northward a few hundred kilometers (Crowell, 1973) in a first stage of movement along the San Andreas fault or a parallel fault (Campbell and Yerkes, 1976). The following review begins with the middle Miocene after all these important structural events have occurred.

1. Westward movement of the western Transverse Ranges province or microplate west of the San Gabriel fault occurred during the late middle Miocene or late Miocene along the Malibu Coast-Santa Monica-Raymond Hill fault zone and along an early stage Santa Ynez-Lompoc-Solvang-Santa Maria River-Little Pine fault system (Figs. 17-2 and 17-5). This event was concurrent with pull-apart basinal development in the Santa Maria region (middle and late Miocene) (Hall, 1977, 1978a) and possibly the Santa Clara trough (late Miocene and Pliocene) (Crowell, 1976) (Fig. 17-2, diagonally ruled). The western Transverse Ranges moved at least 70 to 90 km westward, following or perhaps in part during the emplacement or deposition of the Obispo-Tranquillon pyroclastic and volcanic rocks (15 to 17 m.y. old). The present 70-km-wide western Transverse Ranges province or microplate moved from a region north and possibly somewhat west of Los Angeles, where structures and textural features within the province had had a north or northwest trend, to the present westward trend. Seventy-five degrees of clockwise rotation of the Santa Monica block in the southern part of the western Transverse Ranges in southern California, deduced by Kamerling and Luyendyk (1977) from paleomagnetic data, could be consistent with a model of northwestward drift of the western Transverse Ranges microplate if the clockwise rotation is considered for individual blocks within the Transverse Ranges microplate (Figs. 17-8 and 17-9). The clockwise rotation could have occurred between sets of right lateral faults within the microplate in a fashion similar to that discussed by Beck (1976) or owing to extension, as described by Greenhaus and Cox (1979). The block containing the middle Miocene Conejo volcanic rocks, which Kamerling and Luyendyk (1977, 1979) and Kamerling and others (1978) report has rotated 70° clockwise, lies between faults in the Santa Monica Mountains and faults such as the Oak Ridge-San Cayentano fault system or faults in the vicinity of Ventura. However, more work must be done to resolve the differing suggestions for rotation of the Santa Monica Mountains and Santa Cruz Island within the Transverse Ranges. For example, Kamerling and others (1978) propose post-middle-Miocene clockwise rotation of 90° for Santa Cruz Island. Howell (1976a, p. 449), on the other hand, postulates 32° of counterclockwise rotation for the southern part of Santa Cruz Island. Jones and Irwin (1975) and Jones and others (1976) suggest major clockwise rotation of the Transverse Ranges (Santa Monica Slate). Kamerling and others (1978) propose 75° of clockwise rotation and 10° of northward transport of the Channel Islands and western Transverse Ranges. How these rotations fit into the major northwestward postulated translation and counterclockwise drift of the western Transverse Ranges microplate remains problematical. Perhaps a structural boundary lies between the northern and southern parts of the western Transverse Ranges (perhaps near Ventura); however, the distribution of Miocene pyroclastic rocks and the sedimentary sequences that are present in the Santa Maria basin and along the shoulders

of the basin, as well as the distribution of Mesozoic and Cenozoic rocks in the northern part of the western Transverse Ranges, are difficult to explain if regional clockwise rotation of the Transverse Ranges microplate or significant Miocene left slip along the Lompoc-Solvang is considered. These same relationships are also difficult to reconcile with the subduction model for the Transverse Ranges proposed by Hamilton (1978a) and Crouch (1978, 1979).

Counterclockwise rotation of the western Transverse Ranges could have taken place at the time of the formation of the Lompoc-Santa Maria Basin and as the microplate moved northwestward (Fig. 17-9). The western Transverse Ranges are approximately 70 km wide, and left-lateral displacement along the Santa Monica-Raymond Hill fault zone is at least 60 km (Sage, 1973) and perhaps more (Campbell and Yerkes, 1976). Thus, the width of the Transverse Ranges province and the width of the area from which it moved are palinspastically compatible (see Howell and Vedder, Chapter 16, Fig. 16-7, this volume).

2. Movement along the San Gabriel fault was taking place at the same time, or shortly after, as the westward drift of the western Transverse Ranges microplate. The proto-San Andreas was no longer the most important north- or northwestward-trending fault after the Transverse Ranges began their westward shift, but was supplanted by the San Gabriel fault. The San Gabriel-early San Andreas fault system was extant by about 12 m.y.b.p. (Crowell, 1973) or 4 to 8 m.y. (J. C. Crowell, personal communication, 1978) and was the main strand of the San Andreas fault zone until about 3 or 4 m.y. ago (Crowell, 1973).

3. Movement along the San Gregorio-San Simeon-Hosgri fault zone, and possibly joining with the Santa Ynez fault, commenced following the formation of the Santa Maria pull-apart structure, perhaps 8 to 13 m.y.b.p. or later (i.e., 4 to 5 m.y. ago) if the Pliocene rocks in the San Simeon area have been moved from the Santa Maria basin (Figs. 17-6 and 17-7). Consideration of right slip on the San Gregorio-San Simeon-Hosgri fault zone has prompted Graham (1977; Graham and Dickinson, 1978a, b) to suggest that the need for right slip on the proto-San Andreas fault is reduced by one-third to perhaps two-thirds. They propose that 535 to 625 km of total offset along the San Andreas fault based on the offset of Sierran-type granitic rocks is an integration of offsets of the connecting San Gregorio and the San Andreas faults (Fig. 17-2). They cite post-Oligocene right slip on the San Andreas as being 310 km; they then deduct 115 km of San Gregorio-San Simeon-Hosgri right slip from total offset of granitic basement and suggest 200 to 110 km (maximum and minimum) of right slip along the proto-San Andreas. If the interpretation by Graham and Dickinson (1978a, b) is correct, then most of the movement between the Pacific and North American plates has been accomplished by the San Andreas fault during at least the past 4 to 6 m.y., while between 4 to 6 m.y. and the early Miocene, plate motion was distributed between the San Andreas and San Gregorio-San Simeon-Hosgri fault zone. This leads to the further implication that the San Gregorio-Hosgri fault absorbed a large proportion of the relative motion between the Pacific and North American plates before the Pliocene. However, rocks dated by Hall (1975) as Pliocene in the San Simeon region were apparently moved northward along the San Gregorio-Hosgri fault after basinal formation. They are lithologically most similar to Pliocene rocks in the Santa Maria region, and dissimilar lithologically to the Miocene or Pliocene Pismo Formation

that crops out a few kilometers from them on the east side of the San Simeon fault. If the San Simeon Pliocene rocks have moved from the Santa Maria basin, then the San Gregorio-San Simeon-Hosgri fault zone could have absorbed a relatively large proportion of the relative motion between the Pacific and North American plates prior to the Pleistocene instead of prior to the Pliocene. Any interpretation that suggests pre-Pliocene movement along the San Gregorio-Hosgri requires that the San Gregorio-Hosgri fault be older than Pliocene or there has been multistage movement (Figs. 17-5 to 17-7).

4. Lateral movement along the San Gabriel fault lessened or nearly ceased during latest Pliocene or Pleistocene time, and the main strand of the southern part of the San Andreas fault zone became the present San Andreas fault south of Tejon Pass.

5. Movement along the Big Pine fault (Fig. 17-2) probably took place during the Quaternary (Dibblee, 1976, p. 44). The fault offsets the Ozena, Rinconada (Sur-Nacimiento), and Pine Mountain faults (Fig. 17-2) by 14 km to possibly more than 20 km. Perhaps left-lateral movement along this fault is taken up along the East Huasna fault or by folding within the Santa Maria basin. If the effects of movement along the San Gabriel fault and the Big Pine fault are removed, and a palinspastic map is constructed for the San Andreas, Ozena, Rinconada, and Santa Maria River faults, it can be shown that there is an arcuate pattern to all these faults, with maximum curvature near the latitude of the Big Pine fault (Figs. 17-2 and 17-5). This arcuate pattern possibly developed during and after the emplacement of the western Transverse Ranges microplate during the middle Miocene.

6. Basinal shortening probably occurred in the Santa Maria basin and Santa Clara trough at the time of faulting along the Big Pine fault and concurrent with strike slip along the extant strand of the San Andreas fault south of Tejon Pass, possibly 4 m.y.b.p. (Crowell, 1975a). This event obscures the early history of the San Gregorio-San Simeon-Hosgri fault where it evidently comes ashore near Purisima Point (Fig. 17-7).

7. Total right slip along the San Andreas fault has been about 530 km (Crowell, 1973) or 535 to 625 km (Graham and Dickinson, 1978a), but only about 260 km along the part south of Tejon Pass (TP) (Crowell, 1973) (Fig. 17-2). The two-stage history along the fault prompted Suppe (1970a) to suggest that the proto-San Andreas fault included a northern segment and perhaps the Newport-Inglewood fault. Crowell (1973) outlines the difficulties with this model.

However, a two-stage history for the San Andreas may be required (Suppe, 1970). Careful reconstruction of the pre-middle Miocene or middle Miocene paleogeology of the Transverse Ranges may provide a means for further testing the hypothesis; however, as suggested by Graham and Dickinson (1978a, b), early stage displacement on a proto-San Andreas need not be as large as postulated by early workers if the San Gregorio-San Simeon-Hosgri fault is considered.

To conclude, the Lompoc-Santa Maria basin is a pull-apart structure that developed during late middle Miocene time. Its formation provides a critical constraint for the principal time of northwestward movement of the Transverse Ranges; furthermore, development of the basin can account for the present distribution of Mesozoic and lower Cenozoic rocks in the western Transverse Ranges and their absence in the Lompoc-Santa

Maria basin. Late Tertiary faulting accounts for approximately 80 km of right slip along the Hosgri fault and perhaps as much as 115 km of right slip along the northern part of the San Gregorio-San Simeon-Hosgri fault zone.

ACKNOWLEDGMENTS

This paper represents a part of the ongoing research in the southern Coast Ranges and western Transverse Ranges. The manuscript and concepts have benefited from critical discussions with W. G. Ernst, J. C. Crowell, and W. R. Dickinson. The research was supported by the U.S. Geological Survey, the University of California Committee on Research, and by the National Science Foundation grant number EAR77-14676. To these institutions and scientists, the author expresses his thanks for assistance and suggestions.

John C. Crowell
Department of Geological Sciences
University of California
Santa Barbara, California 93106

18

AN OUTLINE OF THE TECTONIC HISTORY OF SOUTHEASTERN CALIFORNIA

ABSTRACT

In southeastern California a basement terrane of gneisses and granitic rocks of Precambrian and Mesozoic age is thrust upon a thick sequence of Mesozoic (?) graywacke and other clastic sediments containing basic volcanic rocks. This sequence, metamorphosed in association with thrusting, now constitutes the Orocopia Schist and lies beneath the folded and fragmented Orocopia-Chocolate Mountains thrust. Later regional folding of the thrust plate and strike slip measured in many tens of kilometers, along with deep erosion, accounts for the present outcrop pattern. The Orocopia Schist and the thrust zone, with its belt of mylonites and retrograded rocks in the overriding plate, are now exposed in several folded and faulted fensters so that the present dips of the thrust have no relation to its original dip nor to its original direction of movement. Another thrust of probable latest Cretaceous or Early Tertiary age, the Mule Mountains thrust, is exposed about 45 km to the northeast, dips southwest, and is presumably somewhat younger based on the tentative interpretation of regional relations. In fact, the somewhat older Orocopia-Chocolate Mountains thrust may have been transected and locally reactivated at the same time as movement developed on the Mule Mountains thrust.

Two possible tectonic explanations of these relations are: (1) the protolith of the schist was deposited in a back-arc or oblique-rift basin in mid-Mesozoic time, and then metamorphosed as a continental terrain (including a waning arc system) to the southwest was displaced northeastward across it; or (2) the protolith may have been deposited in a trench and then metamorphosed within a northeastward shallow-dipping subduction zone or accretionary prism. Later reactivation with overriding toward the northeast may account both for northeastern vergence of minor folds in the Orocopia Schist and for northeastern vergence of major folds in thick-cleaved sedimentary rocks of Cretaceous age beneath the Mule Mountains thrust. These latter strata are of very different sedimentary and metamorphic facies from the Orocopia Schist.

Following a long transitional period in the mid-Tertiary when basin-range faulting and volcanism occurred, southeastern California was disrupted by faults of the San Andreas transform system. Although right-slip fragmentation of the continental margin began to the southwest near the coast as early as 29 m.y. ago, it affected the belt along the present San Andreas-San Gabriel-San Jacinto faults only about 8 m.y. ago, or in late Miocene time. Displacements since then along these faults have amounted to about 330 km. Much of the right-slip displacement is associated with the opening of the Gulf of California, an event that began about 4 m.y. ago. Since then, about 250 km of right slip on the system has taken place, or at an average rate of about 6.25 cm/yr. As the divergent plate boundary occupied by the Gulf of California enlarged, slices of terrain are pictured as moving northwestward along a series of transform faults from a system of rhombic pull-aparts within the Gulf. These transform faults constitute branches of the San Andreas, San Jacinto, and Elsinore faults in particular. The rugged topography of southeastern California is the result of such tectonic movements, where the transform system meets the divergent plate boundary. Northwest of the Gulf and Salton Trough, movements on subparallel right-slip faults have squeezed upward some long blocks between faults to make mountain ranges, and other blocks have sagged downward to form deep valleys and basins.

INTRODUCTION

The tectonic history of the region including the eastern Transverse Ranges, the Salton Trough, and the terrane eastward to the Colorado River consists mainly of older convergent plate-boundary systems disrupted and overprinted by the transform San Andreas system where it meets the divergent Gulf of California rift. Emphasis here is placed on these post-Jurassic tectonic events, although the record of previous episodes is preserved here and there. The tectonic kaleidoscope really unfolds within the Cretaceous Period, with the convergence against the North American plate of a series of plates coming in from Pacific regions. Southeastern California has in fact been at or near the boundary between major lithospheric plates since well back in the Mesozoic Era, and at times even before that.

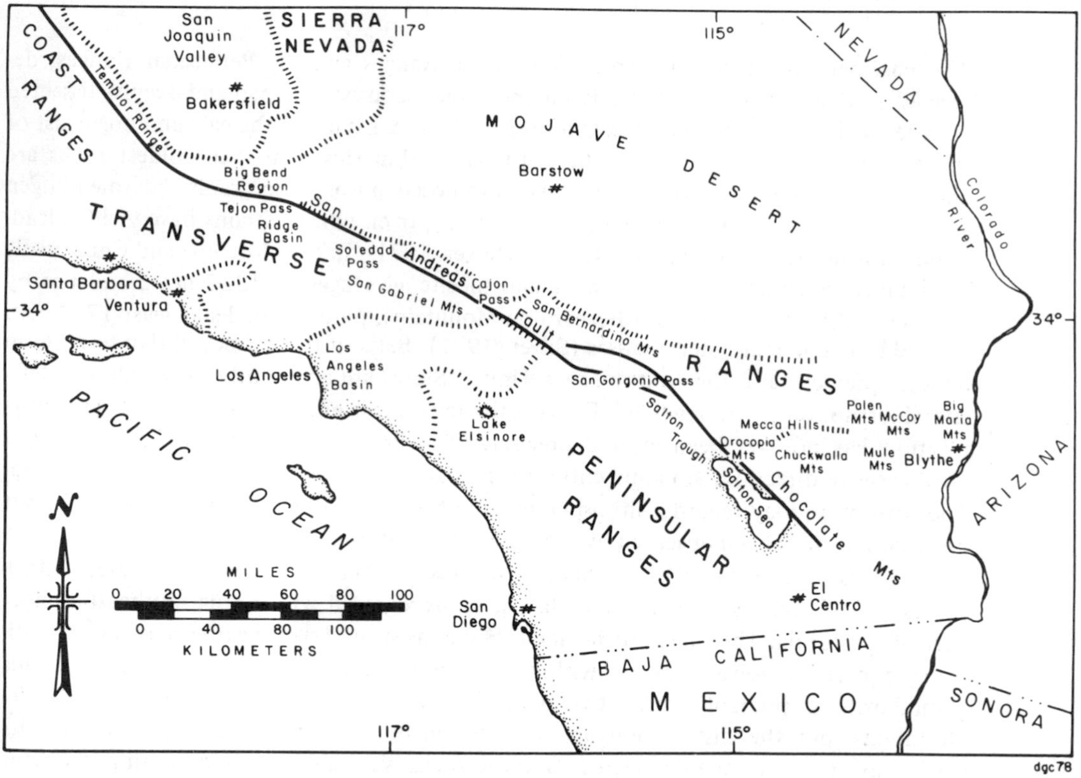

Fig. 18-1. Geographic location map, southern California.

Today the boundary between the Pacific and North American lithospheric plates lies within southern California and consists of a broad and splintered belt extending eastward from the continental edge to the region of the Salton Trough. The offshore borderland is made up of fault-block wedges that stand high between depressed elongate basins. Similar tectonic and topographic styles extend inland and are represented by the Los Angeles Basin, the northern Peninsular Ranges, and valleys and ranges elongated in a northwesterly direction (Fig. 18-1). These tectonic and topographic trends are truncated sharply on the north by the Transverse Ranges, where east-west trends lie athwart those extending to the southeast and northwest. The Transverse Ranges in turn meld gradually into the Coast Ranges farther to the northwest. Southeastward, the structure and resulting topography result primarily from the tectonic style where the San Andreas fault system joins with the Gulf of California Rift. Inasmuch as these tectonic features originated in late Miocene times, the record of older or pre-Miocene tectonics is held within the terranes crosscut by this younger system and reveals major thrusting and eroded volcanic belts.

PRE-SAN ANDREAS ROCKS AND STRUCTURES

The basement rocks of the central Transverse Ranges and the Peninsular Ranges, described in Chapters 10 to 12 of this volume, largely consist of broad and deeply dissected terranes of Precambrian and Mesozoic rocks. The rocks lying to the east and northeast of the Salton Trough are described here briefly. Within this sector, the oldest rocks are gneisses and granulites intruded by metamorphosed porphyritic granites, termed augen geneisses in the field (Crowell and Walker, 1962), from which zircons have yielded lead-uranium isotopic ages of 1670 ± 15 m.y. (Silver, 1971). In the Orocopia and Chuckwalla Mountains, synenites associated with anorthosite give ages of 1220 m.y., and nearby gneisses, 1425 m.y. In the San Bernardino Mountains, plutonic rocks at least 1750 ± 15 m.y. old intrude still older gneisses (Silver, 1971). Belts of these Precambrian terranes in Sonora, Mexico, and the southwestern United States are apparently offset along a left-lateral disruption extending S 50° E across the region (Silver and Anderson, 1974), but its position has not yet been located precisely. The Precambrian belts, which trend in a northeasterly direction, are apparently offset from 700 to 800 km in a left-lateral sense. The disruption may record a lithospheric plate boundary active during the Jurassic, but one that has since been obscured by later granitic intrusions.

Intrusive igneous rocks, probably emplaced within arc systems associated with a succession of convergent plate boundaries, are recognized within this southeasternmost part of California. The oldest so far identified consists of Triassic garnet-bearing granodiorite, exposed at several places within the Chocolate Mountains (Dillon, 1976). This granodiorite is probably correlative with the Lowe Granodiorite of the San Gabriel Mountains, but the distribution and orientation of the arc system responsible for its origin have not yet been delineated. Granitic rocks were also emplaced during two later periods, about 160 to 170 m.y. ago and between 75 and 90 m.y. ago. The later plutons are widely distributed within arc systems possessing a northwest-southwest trend, parallel to the Cretaceous continental margin.

Sedimentary rocks of Precambrian and Paleozoic age have not been recognized eastward from the Salton Trough to the Colorado River. They occur in the San Bernardino Mountains and adjacent ranges on the north, however, where the metasedimentary sequence lies unconformably upon dated Precambrian rocks (Stewart and Poole, 1975). On the northeast, highly deformed Paleozoic strata have been identified in the Big Maria Mountains (Hamilton, 1964a, b) and in western Arizona (Miller, 1970).

Orocopia-Chocolate Mountains Thrust

A major thrust, now folded and fragmented by later faulting, is exposed within the Orocopia and Chocolate Mountains and extends into southwesternmost Arizona. Mylonite and other coherent cataclastic rocks occur along the movement zone, although at places soft gouge indicates that the thrust was locally reactivated much later. The footwall sequence consists of the Orocopia Schist, a unit that before metamorphism was composed of graywacke, sandstone, mudstone, chert, and basic volcanic rocks of probably Mesozoic age (Dillon, 1976; Haxel; 1977, Haxel and Dillon, 1978). Quartzofeldspathic schist now predominates, displaying lithologic layering derived from sedimentary bedding. Here and there mesoscopic isoclinal folds suggest that bedding has been widely transposed into foliation. Metamorphic mineral assemblages belong to the albite-epidote-amphibolite facies and grade increases slightly upward to the Orocopia-Chocolate Mountains Thrust. Metamorphism of the schist therefore apparently occurred in close relationship, both spatially and temporally, with thrust movement, and with burial and heating as the thrust plate was emplaced. This overlying plate consists of Precambrian and younger plutonic and metamorphic rocks.

The crosscutting relations and ages of metamorphism of the Orocopia Schist indicate that thrusting took place in latest Cretaceous or early Paleocene time (Haxel and Dillon, 1978). The protolith is probably not older than Mesozoic inasmuch as Mesozoic plutons do not crosscut the schist. The protolith is inferred to have had an ensimatic origin; however, with only oceanic floor beneath it, there may have been deep sources for sialic intrusions.

The Orocopia-Chocolate Mountains thrust is a segment of the Vincent Thrust of the Transverse Ranges, where it is particularly well exposed (Ehlig, 1968). In all areas the thrust is characterized by (1) greenschist in the footwall that is petrographically nearly identical and that appears to have had the same protolith and history, including metamorphism, (2) cataclastic rocks and minor structures along the thrust surface that are strikingly similar, and (3) similar rocks within the upper plate, including a thick zone of retrograded granulites and granites. If the folding of the thrust pieces is "undone," and the fragments now distributed along the San Andreas and other faults are reassembled, a major thrust fault is indicated. Such reconstruction reveals that the thrust, with its associated schist in the footwall, lies in a belt with a WNW-ESE trend. But because the thrust has been folded and dismembered by later tectonic events, the present dip of thrust exposures has no relation to its original dip. In arriving at a tectonic model to account for the thick protolith of the schist, as well as for the metamorphism, and deformation to account for the thrust, we must look to both minor structures and regional relationships.

Some minor structures in the thrust zone indicate vergence and tectonic transport toward the NNE. Haxel and Dillon (1978) therefore propose that the protolith of the schist was deposited in a Mesozoic back-arc basin that was later closed by northeastward thrusting at the end of Cretaceous time (Fig. 18-2). As the arc within older sialic crust was carried northeastward upon thick sediments within the back-arc basin, the schist formed as the overridden terrane was depressed, transposed, and metamorphosed. Similarly, the basin behind or east of the arc may have been formed as an oblique rift zone tectonically like the Gulf of California today. Such a model might be associated with a proto-San Andreas fault, along a transform system between the Kula and Farallon plates (Haxel and Dillon, 1978). Both of these explanations involve the closing of a long ensimatic basin filled with sediment, lying behind or NNE of an arc system, and the origin of the thrust and metamorphism during the closing. In both explanations the direction of vergence is NNE, and the thrust dipped SSW during displacement.

Minor folds indicating tectonic transport toward the NNE affect the principal foliation in the schist and record a late stage in the deformation after the schistosity was largely formed. Little evidence yet indicates that these minor folds followed directly upon movements with the same orientations as those reponsible for the penetrative

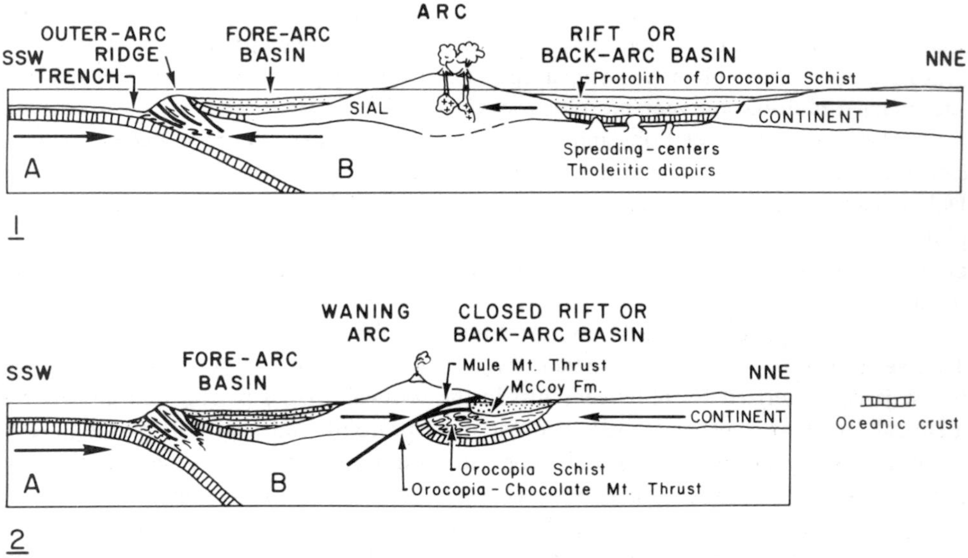

Fig. 18-2. Diagrammatic cross section through southeasterrmost California. 1: During Late Cretaceous time; 2: During earliest Tertiary time. A: Kula, Vancouver, or Farallon lithospheric plate (Menard, 1978). B: North American lithospheric plate. Sketches modified from Dillon (1976, Fig. 44) and Haxel and Dillon (1978). Refer to text for discussion.

schistosity. Data in hand permit that the principal movements were in other directions, including those associated with a NNE-dipping thrust system. Such a dip, with tectonic transport from the NNE toward the SSW, would fit deformation associated with a gently NNE dipping subduction zone. Under this explanation (Fig. 18-3), the protolith of the Orocopia Schist would constitute trench deposits, laid down upon oceanic crust, that were then carried down the subduction zone. The sediments now metamorphosed largely to the greenschist facies would constitute a facies, similar to the Franciscan Complex, that was subducted beneath the sialic North American plate. The schists make up a highly deformed wedge cut through by several thrusts where the uppermost thrusts bring slices of sialic basement above schist of ensimatic origin. As compression and shortening continued, hot basement beneath and around the arc is pictured as moving southwestward upon the distributive thrust, producing the schists with their characteristic inverse grade of metamorphism. Favorable to this hypothesis involving thrusting within a subduction

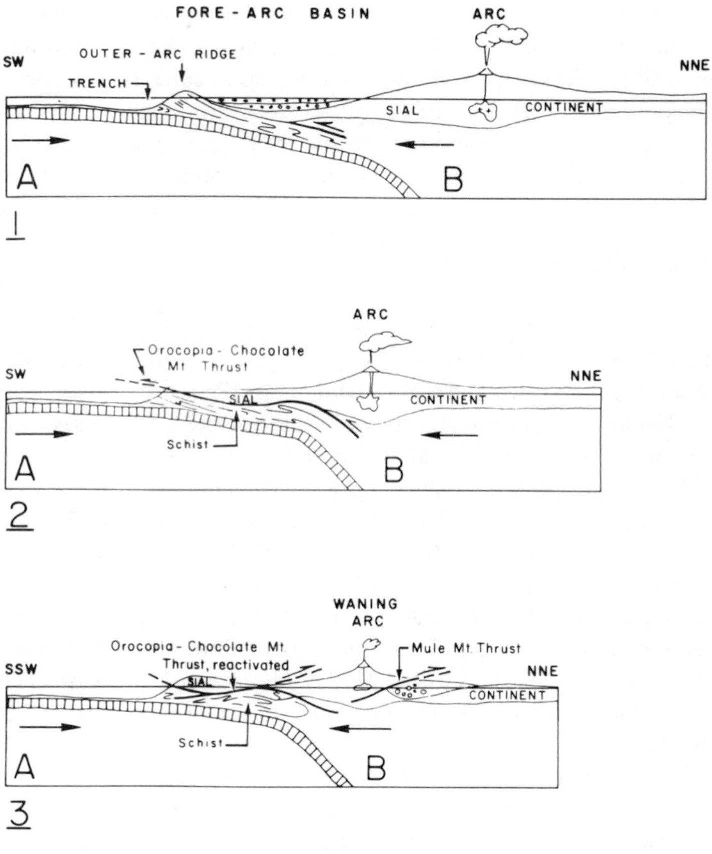

Fig. 18-3. Alternative diagrammatic cross section through southeastern California. 1: During Late Jurassic-Early Cretaceous time when lithospheric plates from the Pacific (A) converged with North American plate (B). Refer to text for discussion.

complex are observations that some mineral relationships in the Pelona Schist of the Transverse Ranges are transitional between blueschist and greenschist facies (Graham, 1975; Graham and England, 1976), inasmuch as high-pressure, low-temperature blueschist minerals are commonly associated with subduction zones (Ernst, 1970, 1971c).

Such a subduction zone would dip to the NNE, whereas a major thrust dipping to the SSW should crop out along strike somewhere in the Mojave-Sonoran Desert region. No northeastern leading edge or lip of a huge thrust sheet has yet been recognized, although the Mule Mountains thrust described next, east of the Salton Trough, has some of the needed characteristics. Nor has a locality been found where the northeastern edge of the Pelona-Orocopia Schist lies upon older rocks. It is therefore easy to picture that the thrust with the schist below it is plunging downward and out of sight on the northeast. Later folding, largely during the Miocene, might then have arched up the thin plate near its leading edge on the southwest to make the schist fensters. The late minor structures with NNE vergence, according to this speculation, would record only later events, perhaps those associated with NNE-directed late shortening across the region. Perhaps they were formed in sympathy with the development of the Mule Mountains thrust, described next briefly. The gently NNE dipping subduction zone, later deformed and reactivated, would presumably lie SSW of its calc-alkaline arc. Such an explanation would apparently require that the arc rocks now to the SSW were brought to this position by later faulting, mainly strike slip or are younger and not related.

Mule Mountains Thrust

A major thrust zone, now dipping gently southwestward with folds beneath it verging northeastward and with strong cleavage, is exposed in the northern Mule Mountains and in the southern tips of the Palen and McCoy Mountains (Pelka, 1973b). The zone consists of as much as 300 m of phyllonite and phyllite interspersed with lenses of plutonic and volcanic rocks originally formed as andesitic and quartz latite shallow intrusions and flows. In the Mule Mountains the hanging wall is made up of dark gneisses and porphyritic diorite, similar to gneisses invaded by dated Precambrian syenites 25 km to the west (L .T. Silver, 1974). These upper-plate rocks also include plutons recording a Cretaceous thermal event (Armstrong and Suppe, 1973). Beneath the thrust lies about 8000 m of cleaved sedimentary, volcaniclastic, and volcanic strata, now weakly metamorphosed so that sericite and epidote with subordinate quartz and albite have resulted from incipient recrystallization; they are assigned to the McCoy Mountains Formation (Pelka, 1973b). Facies changes within both the volcanic and epiclastic strata show that a volcanic source, perhaps an arc system, lay to the southwest. The formation is therefore provisionally inferred to be a back-arc accumulation. Fossil wood within this unit is no older than Late Cretaceous in age, so the thrusting and imposition of cleavage and large overturned folds verging northeastward is considered as formed during latest Cretaceous or Tertiary times. Although no volcanic intrusions have yet been recognized that crosscut the thrust zone. faults of presumed basin-range origin crosscut cleavage, so the thrusting is tentatively dated as earlier than 20 m.y. or early Miocene (Crowe and others, 1979). The upper plate of the Mule Mountains thrust apparently moved northeastward at about the same time (or somewhat later) as the Orocopia-Chocolate Mountains thrust and perhaps in a similar direction.

Because the Mule Mountains and Orocopia-Chocolate Mountains thrusts have many similar characteristics, are of about the same age, and lie within 40 km of each other, are they parts of the same thrust system? Data in hand suggest not. In the southern Chocolate Mountains, Haxel (1977) shows the Winterhaven Formation as unconformably overlying the Chocolate Mountains thrust, and not cut by it. This stratigraphic unit consists of about 350 m of low-grade metamorphic rocks that formerly were siltstone, sandstone, some conglomerate, and with some andesite and dacite at the base. In general aspect the sequence is similar to the McCoy Formation, but the age of the Winterhaven has not yet been determined. If the two formations are correlative, the Chocolate Mountains thrust is older than the Mule Mountains thrust by at least the time interval represented by the deposition of the Winterhaven and McCoy formations. The two thrusts may be nearly of the same age, however, and both perhaps resulted from the closing of a rift or back-arc basin lying to the NNE of an arc system. Perhaps the Chocolate Mountains thrust first overrode the Orocopia Schist, and then, in later stages of the closing, the Mule Mountains thrust overrode the McCoy-Winterhaven Formation. This is the arrangement sketched in Fig. 18-2. On the other hand, the two thrust systems are shown as unrelated in origin in Fig. 18-3, but with sympathetic distributive movements occurring on the Orocopia-Chocolate Mountains thrust during later deformations associated with the Mule Mountains thrust. At the moment, the second hypothesis is the one I prefer.

In summary, knowledge of the tectonic setting and regional relations of major thrust faults in southeastern California and adjoining areas is unsatisfactory. Understanding of movement directions through the study of minor structures and petrofabrics and of timing through correlation studies is needed especially. East-directed thrust sheets of late Mesozoic and Early Tertiary age are well known in southernmost Nevada and the Death Valley region, and similar thrusts have been traced even farther south (Burchfiel and Davis, 1972, 1975). The relationships of these faults to strong deformation, including major thrusting in the Big Maria Mountains (Hamilton, 1964a, b; Terry, 1975) and in west-central Arizona (Miller, 1970; Miller and McKee, 1971) is unclear. Many of these faults have east-directed displacements, similar to those inferred for the Mule Mountains thrust and perhaps the Vincent system. But west-directed thrusts of about the same age are recognized in northern and central California within the Klamath Mountains, Coast Ranges, and Sierra Nevada. Where is this system in southern California? Is it represented by the original movements on the Vincent system?

Other Paleogene Relations

Sedimentary and volcanic rocks younger than the Orocopia-Chocolate Mountains thrust and older than the inception of San Andreas displacements overlie the basement in southeasternmost California. These include the Eocene Maniobra Formation consisting of about 1460 m of marine shale, sandstone, and conglomerate exposed only in the northern

Orocopia Mountains (Crowell and Susuki, 1959; Crowell, 1962, 1975b; Howell, 1975a, b). The formation was laid down nearshore and was probably part of the sequence at the eastern margin of an Early Tertiary forearc basin, and has later been disconnected completely by displacements on the San Andreas fault system from correlative units in the central and northern Transverse Ranges. In the latter region the prism within the forearc basin includes thick Eocene and Paleogene strata and, even farther northwest, Cretaceous beds (Dibblee, 1967; Sage 1975).

In southeasternmost California some of the overlying volcanic sequences probably record as well the presence of an early Cenozoic subduction system along the Pacific continental margin. Here volcanism started again somewhat before 33 m.y. ago after a long Early Tertiary lapse (Dickinson and Snyder, 1979; see also, Cross and Pilger, 1978). These calc-alkaline rocks include basaltic to rhyodacitic lava flows and breccias dated by potassium argon methods as between 26 and 35 m.y. (Crowe and others, 1979). Rhyodacitic and rhyolitic flows, domes, volcaniclastic units, and ignimbrite sheets range in age from 22 to 28 m.y. and mafic lava flows from 13 to 29 m.y. Plutonic rocks with ages between 21 and 26 m.y. are locally exposed and may lie at depth beneath many of the volcanic centers. The northern boundary of the Cenozoic volcanic and plutonic province is marked by the east-west trending Salton Creek fault between the Orocopia and Chocolate Mountains, which was active during mid-Cenozoic time. This fault belongs with the group dealt with here as one of the transverse structures.

TRANSITIONAL AND TRANSVERSE STRUCTURES

Three regions with nearly east-west structures defining narrow basins are present near the San Andreas fault system and are truncated and offset by it (Crowell, 1960, 1962, 1975a; Bohannon, 1975, 1976). The northwesternmost of these Oligocene and early Miocene fault-bounded basins lies along the Big Pine fault in the north-central Transverse Ranges. Here sedimentary breccias and conglomerates of the Plush Ranch and Simmler formations grade rapidly away from fault scarps (Carman, 1964; Kahle, 1966). In the Soledad Pass region, three small basins containing similar facies and bounded also by fault scarps (Muehlberger, 1958) lie between the San Gabriel and San Andreas faults. These basins, filled with sediments and volcanics of the Vasquez Formation, are overlapped by mid-Miocene beds of the Tick Canyon and Mint Canyon formations (Bohannon, 1975, 1976). Similar facies and structures occur in the Orocopia Mountains far to the southeast (Crowell, 1975b). This series of basins with their bounding faults, which appear to be part of the same chain with ENE trend, are now displaced by the combined San Andreas and San Gabriel faults and form part of the evidence for a total displacement on the transcurrent system in southern California of the order of 300 km (Crowell, 1960, 1962, 1975a). Their significance in the context here is that, although their plate-tectonic origin is as yet obscure, they predate the San Andreas system. Perhaps they mark the place where the tip of the Pacific lithospheric plate first met the North American Plate and they may initially have had quite a different orientation. Did the triple-junction so formed somehow disrupt the leading edge of the North American Plate at a high angle? Did this disruption somehow form these transverse structures and basins? Were the Oligocene-Early Miocene basins oriented with more northerly trends and subsequently rotated (Luyendyk and others, 1980)?

SAN ANDREAS FAULT SYSTEM

The San Andreas fault system passes obliquely across the Transverse Ranges from the northern Coast Ranges to the Salton Trough. Within the Coast Ranges on the northwest, the modern and principal strand of the system is remarkably straight and is characterized by recent movement and seismicity (Scholz and others, 1969; Nason, 1973). Here it is apparently moving as the main transform boundary between the North American and Pacific plates. In the north-central part of the Transverse Ranges, however, it changes trend to form the Big trend near the intersection with several major faults, including the Big Pine, Garlock, and San Gabriel, and then takes a straight southeasterly course to Cajon Pass (Dibblee, 1967). From this vicinity the main modern strand, judging from its active seismicity as well as straight course, is the San Jacinto fault (Thatcher and others, 1975; Sharp, 1967, 1975). This fault, as well as strands of the system east of San Gorgonio Pass (Allen, 1957; Dibblee, 1970) and the Elsinore fault to the west, extend into the Salton Trough and the head of the Gulf of California on the southeast. A purpose of this chapter is to describe briefly the history of this fault system within the Transverse Ranges and on southeastward, but with a particular emphasis on its extension into the Gulf of California.

The San Andreas system originated as a result of the meeting of the Pacific plate with the North American plate during late Oligocene time, or about 29 m.y. ago. Previously, the Farallon plate formed an obliquely convergent boundary with the North American plate (Menard, 1978), and the geology of the continental margin was characterized by a trench, outer-arc ridge, forearc basin, and an arc system progressing eastward. The arrangement of these belts was orderly, and each had a general northwesterly trend until fragmentation in the early Miocene, following an interval after the first meeting of the Pacific and North American plates. The Sespe Formation along the coast was the last widespread stratigraphic unit deposited in the arc-trench system and laid down in late Oligocene and early Miocene time. It was deposited as a series of coalescing alluvial and piedmont fans that reached from a highland belt on the northeast toward the sea on the southwest. Toward the east, however, the Sespe apparently grades into the Simmler, Plush Ranch, Vasquez, and Diligencia formations, and is in part correlative. A transition in topography and tectonic style eastward is therefore implied during early Miocene times and following close upon the deposition of lower units of the Sespe Formation. By the end of the early Miocene, however, the regularity of these geographic belts was disrupted, and a broad region along the plate juncture evolved into a splintered transform boundary. Geologic features formed before this event were rearranged as a consequence. In fact, much of the anticipated success in working out the history of these later movements depends on recognizing displaced and rotated fragments of the previous orderliness associated with the geology of the convergent plate boundary. Reconstructions are in part based on reassembling pre-Miocene plate-convergent models.

In southeastern California, rock units and structures older than late Miocene, about 8 m.y. ago, are all displaced on the San Andreas fault the maximum amount, about 320 km, regardless of their antiquity. These include basement rocks as well as several Tertiary units with similar petrology and all arranged in the same sequence (Crowell, 1960, 1962, 1975a). Latest Miocene and younger units, however, originated during displacements and show successively lesser offsets as they become younger.

The San Andreas system is now active tectonically, and the rugged topography today is related to the tectonics of the broad transform boundary. This topography, which in turn controls the facies and distribution of sediments, is the result of the splintering of the plate margins. Blocks between faults within this transform regime are at places obliquely squeezed upward to form mountains and highlands and obliquely depressed downward at other places to form intervening basins (Crowell, 1974b; Yeats, 1976, 1978). This style of tectonics and sedimentation has prevailed in southern California since the birth of the San Andreas system when the Pacific plate first met the North American plate about 29 m.y. ago (Atwater, 1970; Atwater and Molnar, 1973; Blake and others, 1978). Between 29 m.y. ago and about 8 m.y. ago, however, it apparently operated to the west, and especially in the offshore borderland.

The youngest formations showing the maximum displacement along the present San Andreas-San Gabriel faults are the Mint Canyon and Caliente formations of late Miocene age (Ehlig and others, 1975b). The formations were formed as alluvial fans sloping basinward down a broad valley toward lakes and playas on the southwest from a distinctive volcanic province identified in the northern Chocolate Mountains. The medial and distal facies of these fan and playa deposits, containing characteristic volcanic detritus, lie downflow from this source as shown by paleocurrent data and facies changes. At present, however, the Caliente Formation is found southwest of the San Gabriel and San Andreas faults in the north-central Transverse Ranges, and the Mint Canyon Formation southwest of the San Andreas fault within the Soledad Pass region. Only when displacements on the San Andreas and San Gabriel are restored are paleocurrents, facies changes, and distinctive lithic clasts lined up sensibly downflow from their source area. Two conclusions follow: (1) An interval followed the origin of the transverse chain of basins and structures when the regional slope toward the southwest again prevailed; and (2) the date of inception of the San Andreas system in this region began after their deposition, that is, no earlier than latest Miocene. According to recent revisions of the Neogene time scale, this would date the inception of the San Andreas at about 8 m.y. ago (Berggren and van Couvering, 1974). The average rate of displacement since inception is therefore about 3.75 cm/yr on the combined San Gabriel-San Andreas system.

Several sedimentary formations display distinct facies changes into coarse sedimentary breccias and conglomerates against fault scarps of the San Andreas system and demonstrate that the faults were active during the time of deposition. The Violin Breccia in Ridge Basin documents convincingly that the San Gabriel fault began activity during late Mohnian times, about 8 m.y. ago (Crowell, 1975c). In the southern Temblor Range in the southeastern Coast Ranges, the Santa Margarita Formation, also latest Miocene in age, includes coarse breccias laid down along the base of a San Andreas scarp (Huffman, 1972; Fletcher, 1967; Vedder, 1975). Coarse formations in the Mecca Hills, such as the Mecca Conglomerate, show similar relations (Sylvester and Smith, 1976).

Other nonmarine formations now exposed in patches along the San Andreas in southeastern California, do not so clearly show deposition in basins whose shapes were formed by the fault system. These include the Anaverde and Punchbowl formations along the northern flank of the San Gabriel Mountains (Woodburne, 1975), slices of the Mill Creek Formation (lower Pliocene or upper Miocene) along the southern margin of the San Bernardino Mountains (Gibson, 1971), and the Coachella Fanglomerate in the eastern San Gorgonio Pass region (Peterson, 1975). These formations along with several others have been crosscut by the fault system and portions displaced.

On the southeast the San Andreas fault system, constituting a transform lithospheric plate boundary, merges with the Gulf of California, a divergent plate boundary. Although the tectonic arrangement of the two types of plate boundaries is complex in detail, the style, much simplified, can be diagrammed. Figure 18-4 shows the generalized orientation of the San Andreas transform and the rifted borders of the Salton Trough and Gulf of California. The shape of basins within the Gulf and of the coastline on either side suggests a rhombic shape for pull-apart basins. Linear spreading centers are drawn as following this shape, whereas in reality they probably are quite irregular and not located symmetrically with respect to the borders. In the diagram, an intact continent is illus-

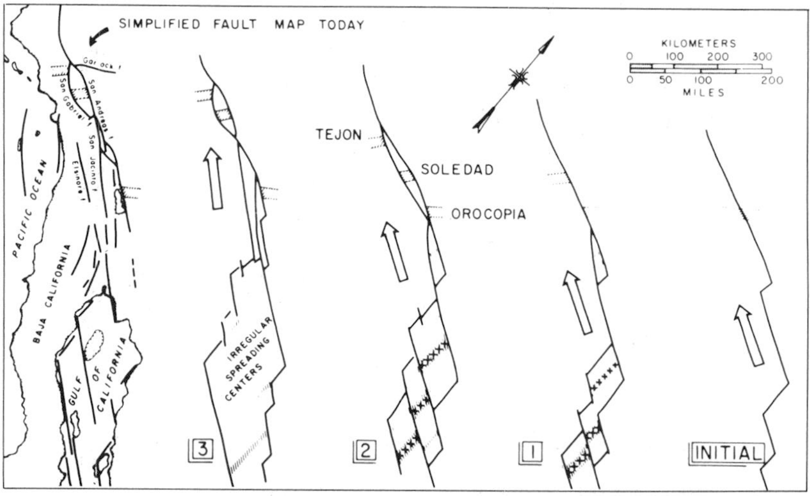

Fig. 18-4. Diagram to illustrate style of displacement of Baja and western California from mainland North America, beginning in late Miocene time. Initially, the Tejon-Soledad-Orocopia terrane was intact with an east-west trend, and then separated, somewhat as shown. Rhombic spreading centers are envisioned in the Gulf of California and account for the opening of the rift. Displacement northwestward occurs on a system of transform faults, both continental and oceanic. The diagram is very much oversimplified. Refer to text for discussion.

trated on the east and north from which the peninsula of Baja California, along with north California west of the San Andreas fault system, moved away from the continent. The rate of moving is equal to double the spreading rate, and the spreading ridge itself is moving in the same direction at the spreading rate.

Two types of transforms are depicted (Crowell, 1974a, Fig. 3). Oceanic transforms confined to the new floor of the Gulf do not cut into the continental rocks on either side. Such transform faults serve as boundaries between two areas of spreading geometry. Continental transforms, in contrast, such as the San Andreas itself, transect ancient rocks that are part of the sialic crust. Linear features with their piercing points within the prerifting and pretransform rocks reveal the total displacement (Crowell, 1962). Both oceanic and continental transform faults are varieties of growth faults.

The geology of this region where the San Andreas transform system meets the Gulf of California rift is complicated by several active faults, all moving more or less at the same time geologically (Fig. 18-5). The principal active faults now are the San Jacinto,

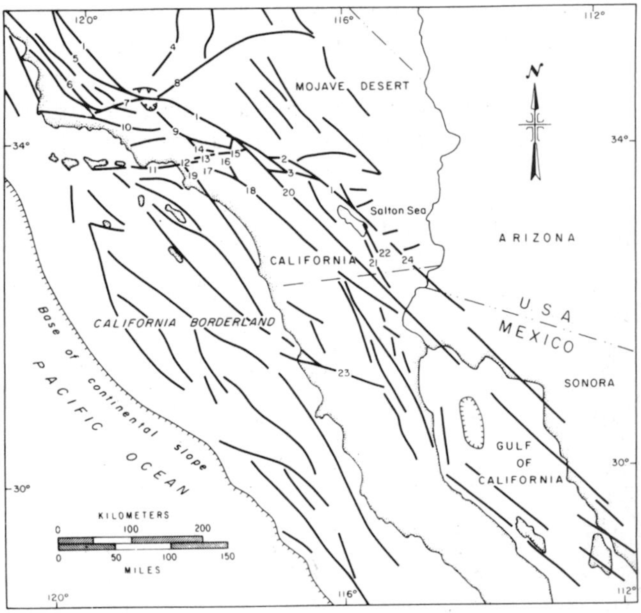

Fig. 18-5. Simplified fault map of southern California and northwestern Mexico. Fault names: 1, San Andreas; 2, northern branch of San Andreas or Mission Creek; 3, southern branch of San Andreas or Banning; 4, White Wolf; 5, Red Hills-San Juan-Chimineas; 6, Nacimiento; 7, Big Pine; 8, Garlock; 9, San Gabriel; 10, Santa Ynez; 11, Malibu Coastal; 12, Santa Monica; 13, Raymond Hill; 14, Sierra Madre; 15, Cucamonga; 16, Chino; 17, Whittier; 18, Elsinore; 19, Newport-Inglewood; 20, San Jacinto; 21, Imperial; 22, Brawley; 23, Agua Blanca; 24, Sand Hills-Algodones. Base redrawn from King, 1969. Refer to text for discussion.

Elsinore, Imperial, and Brawley. The San Andreas itself to the southeast of San Gorgonio Pass is pictured as less involved at present in the transform movement between the major lithospheric plates, but is nonetheless active at places. The course of the San Andreas fault can be traced from the Salton Sea southeastward into Sonora, but along this trend it does not cut recent alluvium (Olmsted and others, 1973; Merriam, 1965, 1972; Gastil and Krummenacher, 1974, 1977). Southeast of the Salton Sea, it has apparently been abandoned, when a pull-apart basin formed to sidestep the major plate motions to the San Jacinto by way of the Brawley and Imperial faults. A spreading center beneath the southern end of the Salton Sea that fits this interpretation is suggested by seismic activity, geothermal areas, modern volcanic centers, and gravity and magnetic anomalies (Muffler and White, 1969; Elders and others, 1972; Hill and others, 1975; Robinson and others, 1976; Johnson and Hadley, 1976; Hill, 1977; Sharp, 1976, 1977). Similar tectonic features, outlining an active rhombic pull-apart, probably occur farther southeast within the Gulf of California (Bischoff and Henyey, 1974).

Southeastward from the Salton Trough the dominant tectonic feature is the divergent plate boundary, the Gulf of California Rift (Carey, 1958; Hamilton, 1961; Elders and others, 1972). When the East Pacific Rise occupied the region about 4.5 m.y. ago, continental rocks constituting western Mexico were separated along an irregular saw-toothed fracture, and the terrane to the west began to move northwestward away from the divergent boundary (Larson and others, 1968; Moore and Buffington, 1968; Karig and Jensky, 1972). The movement was and is oblique, however, and parallel to the trend of many transform faults that are marked in the bathymetry of the gulf floor. The pattern of transform faults and rhombic pull-apart basins is especially clear in the southern part of the gulf where sediments fail to blanket completely the topographic patterns formed by lateral rifting. In the northern part of the gulf the sedimentary blanket is thick so that the tectonic topography is obscure. Here the Colorado River and its immediate ancestors have brought enormous volumes of sediment into the opening gash within the continental rocks and have filled the widening rift to near sea level as the pull-apart grew.

The southeastern portion of the San Andreas fault system therefore consists of a series of near-parallel major right-slip faults that originated in the late Miocene and together perform a transform function at the splintered boundary between the North American and Pacific lithospheric plates. In California the set of transform faults on the southeast is now more or less moving simultaneously, with perhaps the San Jacinto fault being the most active at present. It was probably born in late Pliocene time as the San Andreas zone in the San Gorgonio Pass region was bent into a constraining bend. Here the northern branch of the San Andreas fault (Mission Canyon fault) and the southern branch (Banning fault) were slowly rotated into positions that in part inhibited their movement or so that they took on thrust or oblique-slip displacements. This constraining, rotation, and reorientation of principal slip directions is envisioned as gradual and continuous. The constraining during the twisting may have resulted also in raising the blocks to the north and south of San Gorgonio Pass and in depressing the narrow block between them. It is the result of deformation through time of a soft and yieldable crust.

The Elsinore fault zone, lying about 35 km to the southwest of the San Jacinto fault, appears to be of about the same age as the San Jacinto. It is now active as shown by fault topography. In addition, the size and shape of the Lake Elsinore depression suggests that it originated as a pull-apart with about 6 km of right slip where the fault-zone trend locally forms a releasing bend (Crowell, 1974b, Fig. 3). The total slip on the Elsinore zone is probably about 30 km as indicatd by the offset of facies-change lines in Paleocene strata (Lamar, 1959; Sage, 1973; Campbell and Yerkes, 1976). The tectonic relations of the Elsinore fault zone with the older (in part) Whittier fault to the northwest, the Chino fault to the north, and the origin of the Los Angeles Basin are still unclear.

The San Jacinto fault, with about 24 km of right slip (Sharp, 1967), and the Elsinore, with about 30 km, apparently are the result of a coastal or southwestward stepping of the main transform boundary beginning in the late Pliocene Epoch. In southernmost California it appears that the San Andreas system consists of subparallel transforms lengthening from the divergent widening in the Gulf of California. There is no evidence here that faults of the San Andreas system are now moving eastward, as may be the situation in the Mojave region to the north (Garfunkel, 1974), although over a longer time span they have moved inland from the coast.

The straight segment of the San Andreas proper northwest of San Gorgonio Pass, to the vicinity of the Tejon Pass region, is probably also a recently born strand of the system. Although the time of origin is not completely known, it is probably younger than about 4 m.y. ago when the San Gabriel fault was abandoned as the principal through-going fault of the system (Crowell, 1975a). In the southern Big Bend region the San Gabriel fault is overlapped by beds of late Pliocene age near the top of the Ridge Basin sequence (Crowell, 1975c). Apparently, as the Big Bend of the San Andreas began to grow, a series of subparallel faults to the northeast originated sequentially. These include the Clearwater and branches of the Liebre system, faults that lie between the San Andreas and San Gabriel at the northwest. The main fault, however, developed as the present San Andreas in this region. In approximate terms, because the critical data are still obscure, the main strand of the active San Andreas fault here is probably about 4 m.y. old with a total right slip of about 250 km. This gives an average rate of displacement of 6.25 cm/yr.

Understanding of the complex movement history in the eastern San Gabriel Mountains and in the Cajon Pass-San Gorgonio Pass region must await more data, however. Tectonic relations here are especially clouded because in this vicinity the enigmatic structures of roughly east-west trend enmesh with those of the San Andreas system. The geometry and history of the Cucamonga, Sierra Madre, San Gabriel, Raymond Hill, Santa Monica, and Malibu Coastal faults are still quite obscure, and of especial concern is the manner in which they terminate on the east. The tectonic discontinuity along this east-west trend is certainly one of the most profound in southern California and still evades explanation. In fact, the Transverse Ranges appear to lack deep roots to support them isostatically as suggested by the interpretation of earthquake waves and gravity studies (Hadley and Kanamori, 1977; Grannell, 1971), so perhaps they are held up by plate-tectonic compression across the region.

In summary, the San Andreas fault system from the north-central Transverse Ranges to the Gulf of California consists of a series of subparallel transform splays. The oldest recognized fault of the system, born about 8 m.y. ago, is the San Gabriel fault. It was abandoned about 4 m.y. ago and has since been somewhat warped, bent, and rotated. In the middle of its course, it branches into two splays: a northern and a southern branch. The southern melds with the Sierra Madre fault, which has subsequently assumed thrust slip (Proctor and others, 1970). The northern branch is offset by younger faults in the eastern San Gabriel Mountains. In the complex region of Cajon Pass the identity of the San Gabriel fault system is lost locally, but this ancient strand somehow made its way to the northeastern flank of the Salton Trough and on southeastward, because terranes are offset along this stretch. From this region on into Mexico, if it ever extended that far, the course of this mid-Pliocene fault has not yet been identified. More likely it ended at the corner of a rhombic pull-apart basin at the head of the developing Gulf of California.

This episode of strike-slip movement on the San Gabriel fault apparently preceded the opening of the Gulf of California. Moreover, the right slip during this 4 m.y. interval from about 4 to 8 m.y. ago amounted to only about 50 km. The San Gabriel fault may therefore have developed as an onshore part of a complex transform and pull-apart system that was operating most vigorously in the California Borderland and which formed the Los Angeles, Ventura, and Maricopa basins. Only later, with the opening of the Gulf when the East Pacific Rise was sited at its mouth, did Gulf transforms truly work northwestward to join with faults of somewhat different origin. In short, the San Andreas system in southern California, as a whole, has a multiple origin and represents the melding of at least two tectonic schemes.

First, fault elements originated as the continental edge was splintered in the late Miocene in response to tectonic events involved in basin formation and in the elevation of ranges in the California borderland and northeastward on the mainland within the transverse Ranges; these displacements were apparently associated with large rotations of tectonic units in a clockwise sense (Kamerling and Luyendyk, 1977). Second, transform faults associated with opening of the Gulf of California reached northwestward and joined with those continental transforms already formed. During the opening, some new faults originated, such as the San Jacinto and the southeastern reaches of the Elsinore. These faults fragmented and displaced parts of the terrane now underlying the Peninsular Ranges. The San Jacinto extended northwestward into the San Gabriel Mountains and is now in the process of joining the San Andreas fault on the northeastern flank of that range. The great right-slip faults of southern California truly have a multiple and complex history.

ACKNOWLEDGMENTS

I am particularly indebted to a stimulating group of former graduate students who have recently contributed significantly to the understanding of the tectonics of southeastern California: R. G. Bohannon, B. M. Crowe, J. T. Dillon, G. B. Haxel, G. J. Pelka, and O. G. Sage, Jr. Many field trips with these and other colleagues, and especially with P. L.

Ehlig, have been rewarding. My recent research on California tectonic problems has been in part supported by U.S. National Science Foundation grant DES 71-00498, U.S. Geological Survey grants 395 and 473, and the University of California, Santa Barbara.

W. G. Ernst
Department of Earth and Space Sciences
University of California
Los Angeles, California 90024

19

SUMMARY OF THE GEOTECTONIC DEVELOPMENT OF CALIFORNIA

ABSTRACT

Although geologic relationships within each province attest to relatively unique plate-tectonic histories, in general the collage of lithotectonic belts within the state bears testimony to five principal developmental stages: (1) a middle-late Precambrian calc-alkaline volcanic + plutonic convergent continental margin or island arc; (2) a latest Precambrian-early Paleozoic rifted western North American margin; (3) a late Paleozoic-early Mesozoic island arc + marginal basin adjacent to the continent; (4) a late Mesozoic-early Cenozoic convergent margin; and (5) a late Cenozoic complex continental transform system along the continental margin. The subduction zones apparently involved important periods typified by low-angle oblique convergence and/or local tectonic erosion through subduction.

The assembly of lithotectonic belts was mainly accomplished by (1) more or less continuous calc-alkaline volcanism + plutonism sited above subduction zones, with concomitant sedimentation on the adjacent shelf, continental slope, rise, and trench environments; (2) relatively shallow water deposition along passive strike-slip and pull-apart margins; and (3) episodic collision of oceanic and/or sialic fragments along conservative, oblique, and consumptive plate boundaries. Thus the Klamath, Sierran, Peninsular Range, and Franciscan complexes, and probably the southern California borderland as well, show clear evidence of landward tectonic imbrication and seaward accretion (i.e., younging) with time. Tertiary-Quaternary basins in the Coast Ranges also exhibit roughly a westward progradation.

Periods of continental truncation along convergent, transform, or rifted margins have offset, scattered, and completely removed sections of the western North American crust. Specific examples include (1) the latest Precambrian rifting of western North America prior to deposition of the miogeoclinal Eocambrian strata in easternmost California; (2) the Permocarboniferous truncation and removal of the eugeoclinal + miogeoclinal + platform section in southwesternmost Nevada and southern California; (3) the problematic early Mesozoic left-lateral north-south trending megashear in the Mojave Desert + southern Sierra Nevada region; (4) the possible oblique subduction and obliteration of the western portion of the Salinian batholithic terrane and outboard arc-trench gap Great Valley units in the southern Coast Ranges; and (5) the late (possibly also early?) Cenozoic transection of the Mesozoic continental margin, and 300 to 550-km northwestward transportation and fragmentation of the Salinia-Nacimiento block in westernmost California.

The present-day geology of California, an intricate lithotectonic collage, is seen to reflect the complex interplay between semicontinuous igneous, sedimentary, and tectonic constructional processes and the episodic attendance of destructive plate motions that have truncated, pulled apart, dispersed, and carried away segments of the continental margin.

INTRODUCTION

What should be quite clear from the preceding chapters is that numerous constructive and destructive processes have been at work in shaping the geology of the state. Moreover, the nature of the plate-tectonic interactions themselves is still disputed. Because of this controversy, the reader may feel that plate-tectonic theory is to some extent invalidated. On the contrary, it seems evident to me that the application of such scenarios to the lithotectonic history and present-day geologic configuration of California for the first time provides a reasonably coherent rationalization of the origin of this complicated mountain belt. Debatable interpretations provide a measure of the inadequacies of the geologic-geophysical information and the gross oversimplification of existing plate-tectonic models. Extensions and refinements to both should be forthcoming as a result of continued work, such as that reported in this volume.

CONTINENTAL MARGINS AND CALIFORNIAN GEOLOGY

The principal varieties of lithospheric plate boundaries involving continental margins have been described by many authors. The four chief models, very much simplified, are illustrated diagramatically by Dickinson (Chapter 1, Fig. 1-1, this volume). Of course, the geology of each margin is a function of its unique history and represents a complex interplay among diverse constructional and destructional processes. The Atlantic type is a rifted margin produced by divergent plate motion. In contrast, both Andean and Japanese types are convergent plate junctions, whereas the "Californian" type is characterized by a broad zone of strike slip, and is therefore a conservative plate boundary. The convergent models may involve pure "head on" subduction, but oblique underflow is apparently more common. In either case, convergent motion may result in accretion of reworked sialic material adjacent to and landward of the trench; alternatively, tectonic erosion of segments of the continental margin (not shown in Dickinson's illustration) may occur.

The plate-tectonic processes that operated in California must be inferred from the lithologic products preserved in the continental margin. Some of these interpretations, described in the previous eighteen chapters, will now be summarized briefly, based on intervals of time rather than geologic provinces.

PRECAMBRIAN HISTORY OF CALIFORNIA

As pointed out by Dickinson in Chapter 1 of this volume, the record of plate-tectonic interactions within the state is both most widespread and apparent for the youngest rock systems. As one tries to look backward in time, the preserved lithologic sections are more fragmentary and the deduced history consequently is more obscure. Nonetheless, we know that continental crust, hydrosphere, and atmosphere had evolved to an essentially modern aspect by about 2000 m.y.b.p. at the latest (Windley, 1977), with the result that crustal sections from that time onward are in large part correlatable with currently active earth-building processes. A notable exception involves the apparent lack of ophiolites *sensu stricto* prior to Phanerozoic time.

Occurrences of rocks of Precambrian age are very limited in California. Although their presence has been hypothesized in several batholithic terranes, they are known only with certainty from the San Gabriel and San Bernardino Mountains, the Mojave-Sonoran Desert, and the Basin and Range provinces.

Portions of the San Gabriel Complex have been fragmented and reassembled at different times. This terrane probably is still being modified due to ongoing translational motion along the San Andreas system. The oldest rocks belong to the Mendenhall or pre-Mendenhall quartzofeldspathic gneiss-amphibolite complex of 1715 ± 30 m.y. age, intruded by 1670 ±15 m.y. old granitic gneisses. The San Gabriel anorthosite-gabbro-syenite complex was emplaced at about 1220 ± 10 m.y.b.p. The thermal aureole produced by this differentiated mafic pluton caused the amphibolitization of relict granulites in Mendenhall gneisses on the south and southeast margins of the anorthositic complex, but whether an early stage of metamorphism attending intrusion produced the granulites (Ehlig, Chapter 10, this volume) or whether they formed prior to emplacement of the anorthositic complex, at about 1440 m.y. ago (Silver and others, 1963), is unclear. In any case, lithologies preserved in the San Gabriel Mountains bear testimony to deep crustal levels of continental metamorphism and addition of both mafic and felsic calc-alkaline igneous material.

Isolated masses of Precambrian basement, in part covered by and faulted against largely Mesozoic granitic and volcanosedimentary units, protrude through a veritable sea of alluvium in the Mojave-Sonoran Desert. The limited lateral continuity hampers regional correlations, but the situation is somewhat alleviated by the fact that, where outcrops occur, they are virtually 100% exposed. Two major belts of ancient crystalline rocks have been recognized by Silver and Anderson (1974), based on geochronologic data for cogenetic zircon suites in granites and their country rocks: the older terrane on the northwest is characterized by a metamorphic basement 1820 m.y. old invaded by 100 m.y. younger granitic plutons; the younger, more southeasterly terrane consists of 1720 m.y. old country rocks intruded by 1650 m.y. old granites. Juxtaposition of these ancient crystalline belts in early Mesozoic time defines an hypothesized left-lateral megashear transecting the Mojave Desert in a roughly north-south direction. Like the San Gabriels, this Precambrian terrane reflects deep-level sialic additions to the continental crust.

Similar metamorphosed rocks crop out to the south in the San Bernardino and Chocolate + Orocopia Mountains. They are also found as far north as Death Valley, where they are unconformably overlain by a feebly recrystallized sedimentary series of latest Precambrian age. These strata, and the overlying Eocambrian + lower Paleozoic section, represent well-sorted, chemically mature sediments produced along an Atlantic-type continental margin (Stewart, 1972; Nelson, Chapter 8). The initiation of rifting resulted locally in formation of the Amargosa aulacogen, site of the chiefly terrigenous Pahrump Group. Thus the latest Precambrian and early Paleozoic time interval was dominated by shallow-water, passive margin-type deposition in this portion of California.

Rocks of Paleozoic age are present in large tracts of eastern California and crop out as far west as the Klamath Mountains in the north, possibly in Salinia in the west-central portion of the state, and in the Mojave Desert and probably the San Gabriel Mountains to the south. Judging from isopach and sedimentary facies trends, the ancient Pacific margin of North America ran roughly north-south, from southernmost Idaho and central Nevada into southeastern California prior to the Antler orogeny (Burchfiel and Davis, 1972; Stewart and Suczek, 1977; Dickinson, Chapter 1, Fig. 1-3). The onlapping sediments aggregate about 1 km thickness on the southeast near the shelf edge, but thicken northwestward to nearly 10,000 m of section (Dickinson, Chapter 1). To the west lay one or more early Paleozoic island arcs in the Klamath Mountains and northwestern Sierra Nevada.

Eastward-directed overthrusting along the Roberts Mountains thrust in latest Devonian-earliest Mississippian time brought volcanogenic argillites of a subduction complex over an autochthonous miogeoclinal sequence (Stewart and Poole, 1974). The Antler orogeny signaled the closure of an intervening marginal basin and the suturing of these island arcs onto the western perimeter of North America. The Shoo Fly Complex in the northern Sierra Nevada bears testimony to this event. Dickinson (1977) and Schweickert and Snyder (Chapter 7) have presented the case for westward consumption of the marginal basin, whereas Burchfiel and Davis (1975) explained the suturing event through a process of eastward plate descent involving uncoupling and overthrusting (obduction) of the island-arc superstructure. At the latitude of the Sierra Nevada, renewed subduction offshore allowed the construction of a late Paleozoic volcanic arc lying west of a newly formed marginal basin. The latter was presumably the site of accumulation of the Calaveras Complex. Unlike the present-day Japanese archipelago (perhaps the type section for an island arc), which faces the subducting oceanic crust-capped lithospheric plate, these Paleozoic island arcs of central California apparently faced North America, reflecting westward underflow of the marginal basin. This volcanosedimentary unit constituted the leading edge of the continental crust-capped slab. During the Sonoma orogeny at the end of Permian to earliest Triassic time, this lithologic packet, too, moved eastward along the Golconda thrust toward the North American craton.

Farther north, in the Klamath province, a general westward younging and tectonic imbrication of mid + upper Paleozoic and Mesozoic units along eastward-dipping thrusts (see Irwin, 1960, 1977, Chapter 2, this volume, Fig. 2-6) suggests that, subsequent to the Antler orogeny, the paleopacific plate descended to the east beneath the accreting continental margin. Here the east-dipping Trinity thrust has juxtaposed the eastern Paleozoic + Triassic eugeoclinal section—a back-arc basin assemblage (?) that appears to correlate in part with the northern Sierran Shoo Fly—and underlying Trinity Ophiolite against the structurally lower, more westerly central metamorphic belt. The age of metamorphism of the latter (~380 to 400 m.y.) suggests a Devonian event; however, the time of final assembly in the eastern Klamath Mountains, judging by late Paleozoic-earliest Mesozoic depositional ages, indicates that late stages of the thrusting probably should be correlated with a phase of the Sonoma orogeny. The plate-tectonic interpretation of the central metamorphic belt is uncertain: whereas Hamilton (1969) regarded it as a microcontinental fragment rafted into the subduction zone, others have suggested that it is the remnant of an island arc. In any case, it is a relatively thin flap of Paleozoic metamorphic rocks resting tectonically on the Calaveras equivalent "western Paleozoic and Triassic" belt, which contains Jurassic strata (Irwin, Chapter 2, this volume; Irwin and others, 1977, 1978).

Upper (?) Paleozoic strata of southern and southeastern California, including alleged equivalents in Salinia, are characterized by platform carbonates and orthoquartzites, and lack abundant volcanogenic detritus; hence this widespread but relatively thin sequence is thought to have been deposited on the continental shelf and slope (Burchfiel and Davis, 1972, 1975, Chapter 9, this volume). This miogeoclinal sequence passes to the northwest into a eugeoclinal facies, of which fragments are preserved in the Sierra Nevada batholith north of the latitude of the White Mountains and throughout the Western Foothills belt as the Shoo Fly, Kings-Kaweah Ophiolite and perhaps older portions of the Calaveras units.

Paleozoic miogeoclinal-eugeoclinal facies trends are abruptly terminated on the southwest by north-northwest striking Mesozoic lithotectonic belts that roughly parallel the present continental margin. Because of this profound truncation of the Paleozoic system, Hamilton (1969) and Burchfiel and Davis (1972) have postulated a major rifting event at the end of Paleozoic time. The latter authors (Chapter 3, this volume) correlate the Klamath "western Paleozoic + Triassic" belt with the Sierran Calaveras Formation, and suggest that the rifted margin lay on the continental side of these units. On the other hand, Dickinson (Chapter 1, this volume) has related this pronounced depositional offset to an original configuration of the late Precambrian rifted margin, now somewhat obscured due to Cenozoic strike slip on the San Andreas system. Tectonic erosion during subduction is yet another possibility (e.g., see Scholl and Vallier, 1979).

To summarize, the Paleozoic Era was characterized in much of eastern and southern California by deposition along a rifted continental margin that shoaled southeastward. Offshore island arcs were accreted to the continent by the (westward?) consumption of intervening back-arc basins and eastward-directed overthrusting along the Roberts Mountains (latest Devonian-earliest Mississippian) and Golconda (latest Permian-earliest Triassic) thrusts in Nevada. In the Klamath Mountains of Oregon and northern California, however, the age relations and juxtapositions of the various lithic belts suggest that the eastern Paleozoic and Triassic province plus underlying Trinity Ophiolite were emplaced to the west over the subducted and recrystallizing central metamorphic belt at the end of Devonian time. Effects of the Sonoma orogeny are confined to the ultimate juxtaposition of these two terranes. Renewed thrusting along the east-dipping Siskiyou thrust along the western margin of the central metamorphic belt in Triassic and later time reflects normal Pacific-type underflow and westward continental growth.

Igneous, metamorphic, and sedimentary rocks of Mesozoic age are both widespread and abundant throughout all portions of the state. In aggregate, they bear testimony to a very active stage of continental accretion that evidently attended Pacific-type lithospheric plate convergence. Oblique subduction seems to have been the dominant process in central California. Three principal tectonic environments are preserved in the lithologic record: the subduction zone, the arc-trench gap, and the volcanic-plutonic arc. There is evidence supporting the possible existence of at least one marginal basin. In addition, oblique rifting + continental fragmentation, large-scale strike-slip faulting, and probably subcrustal erosion as well have caused the truncation of preexisting structural trends.

The Klamath Mountains mark the site of a seaward prograding continental margin. The late Paleozoic process of tectonic imbrication along east-dipping thrust faults continued throughout Mesozoic time, with the western Paleozoic and Triassic belt (North Fork, Hayford, and Rattlesnake Creek units), the western Jurassic (Galice + Rogue formations), and the Franciscan terrane occupying successively more oceanward positions along the western margin of North America. Ophiolitic slabs and slices, such as the Josephine ultramafic complex (possibly correlative with the Smartville block of the northern Sierra Nevada), occupy the suture zones along which these lithotectonic belts have been juxtaposed. The Franciscan Complex, which is confined to the Coast Ranges, lies tectonically beneath the South Fork Mountain thrust and represents voluminous late Mesozoic episodic offscrapings from the downgoing slab. Exotic (allochthonous) sedimentary + igneous masses appear to have been brought into the continental margin by oblique convergence (e.g., see Alvarez and others, 1979). In contrast, the two older, more easterly Klamath belts each contain telescoped lithic remnants of probable subduction zone, arc-trench gap, and volcanic island-arc regimes (Irwin, 1972; Burchfiel and Davis, Chapter 3, this volume). The entire Klamath province has been punctuated by discrete, pre-Cretaceous calc-alkaline plutons; isolated tectonic blocks of blueschist 214 to 223 m.y. old occurring in the Hayfork + North Fork, and of unknown age in the eastern Paleozoic + Triassic belts, indicate subduction zone recrystallation. Thus, additions to the Klamath continental crust here involved both surficial accretion and deep-seated emplacement.

Farther to the southeast, the Sierra Nevada and White-Inyo Ranges are dominated by coalescing granitic plutons (see Bateman and Eaton, 1966; Bateman, 1974, Chapter 4, this volume). Although the earliest bodies are as old as about 210 m.y., the major batholithic units are Cretaceous in age. Plutons on the east (e.g., White-Inyo granitic rocks) tend to be more silicic and potassic (Bateman and Dodge, 1970), suggesting possibly the admixture of components of deeper mantle (?) levels of derivation, compared to those in the central and western Sierra Nevada. Five major intrusive episodes in the time span Late Triassic-Late Cretaceous have been postulated by Kistler and others (1971); these epochs, each of which involved intrusive events over a 10 to 15 m.y. interval, apparently are separated in time by about 30 m.y. In addition, the initial ^{87}Sr-^{86}Sr ratios of the granitic rocks appear to reflect the relative proximity to the Mesozoic continental margin (Kistler and Peterman, 1973); evidently, the magmas interacted isotopically with the preexisting crust during assimilation or anatexis, or the melts were derived from isotopically distinct portions of the upper mantle. All the batholithic rocks possess textures indicating emplacement as crystal mushes; hence it is difficult to see how the magmas could represent uncontaminated, high-temperature liquids direct from the mantle.

Remnants of the volcanic carapace into which the plutons were intruded, such as the rhyodacitic Ritter sequence (Bateman, Chapter 4), are confined to central and eastern Sierra pendants. Eastward-dipping reverse faults in the Western Foothills belt have juxtaposed oceanic crust against the continental margin (Saleeby, 1978a, b; 1979; Chapter 6, this volume; Saleeby and others, 1978). This tectonic regime contains subduction zone olistostromes, ophiolites, and calc-alkaline volcanic arc rocks (andesites + dacites), as well as eugeoclinal sedimentary units. Dominantly of early and mid-Mesozoic age, this Logtown Ridge-Mariposa-Smartville complex is equivalent to the Rogue-Galice-Josephine assembly in the Klamaths, and like the latter lies outboard of the more easterly, predominantly Paleozoic section. In general, tops face eastward in both younger and older western Sierran terranes (Bateman, Chapter 4; Schweickert, Chapter 5, this volume). Although some older ophiolitic units such as the Kings-Kaweah complex seem to have been transported thousands of kilometers northward during early Mesozoic sea-floor spreading, the northwest-trending tectonostratigraphic and batholithic belts in the Sierra Nevada appear to reflect stages involving large components of lithospheric plate convergence. The Mariposa belt lies outboard of the Sierran Foothills suture, which constitutes the contact between the Mesozoic accretionary terrane and the more easterly, largely Paleozoic basement. The Klamaths and the Sierra Nevada are correlatable in terms of their petrotectonic histories (e.g., Davis, 1969), but the latter terrane appears to represent exposure of somewhat deeper structural levels than the former. Westward movement of the Klamath complex relative to the Sierra Nevada belt appears to have begun by the end of Early Cretaceous time (Jones and Irwin, 1971); aggregate left-lateral strike slip on a series of Great Valley tear faults was approximately 100 km.

The Great Valley-Franciscan couplet lies on the Pacific Ocean side of the Klamath-Sierran volcanic-plutonic belt. Although its plate-tectonic setting has been recently reinterpreted as a strike-slip collage (Jones and others, 1978a; Blake and Jones, Chapter 12, this volume), this pair of lithologic belts has been widely recognized as a classically developed arc-trench gap section and trench melange, respectively (Dickinson, 1970; Page, 1978, Chapter 13, this volume). Both assemblages consist chiefly of poorly sorted, first-cycle clastic sediments derived in large part from the landward adjoining calc-alkaline volcanic-plutonic arc (Dickinson and Rich, 1972; Jacobson, 1978). At least the Cretaceous portion of the Great Valley Sequence represents a miogeoclinal wedge laid down on the westernmost Sierra + Klamath + Salinia + Peninsular Range basement and immediately seaward oceanic crust, all generally regarded a portions of the North American lithospheric plate. In contrast, the Franciscan coherent turbidites and sedimentary melanges, representing mid-fan and olistostromal units, respectively, were deposited exclusively on an allochthonous oceanic substrate, the capping crust of one or more paleopacific plates. Unlike the rather orderly Great Valley Sequence, much of the Franciscan group has been thoroughly tectonized. The two terranes are juxtaposed along the east-dipping Coast Range thrust and related faults (Bailey and others, 1970). Movement along this system may have occurred at various times in the Cretaceous and Early Tertiary subsequent to initiation of the subduction zone about 150 m.y. ago. Prior to the westward stepping of this junction, the convergent lithospheric plate boundary lay roughly 100 km to the east in the vicinity of the Sierran Western Foothills belt. The time of step-out must have followed generation of the Great Valley ophiolite at 155 ± 5 m.y. (Lanphere, 1971; Hopson and others, 1975a, b; Hopson and others, Chapter 14, this volume), but before deposition of the overlying Tithonian cherts 140 m.y. ago (Pessagno, 1973). Alternatively, Schweickert and Cowan (1975) have suggested that the ophiolite represents oceanic crust flooring a marginal basin lying immediately to the west of an east-facing western Sierran island arc. Although large components of northward movement of the oceanic lithosphere have been postulated (Blake and Jones, Chapter 12, this volume), the contrast in deformation and in inferred pressures of metamorphic recrystallization across the Coast Range thrust system requires periods of rapid convergence and profound underflow of the blueschistic Franciscan terrane prior to decoupling from the downgoing slab (Ernst, 1965, 1971a, 1977 Suppe, 1972).

Most of the Mesozoic volcanic and granitic rocks of the Mojave Desert (including the San Bernardino Mountains), the granitic sliver that constitutes the Salinian block, and the Peninsular Ranges batholith represent southward prolongations of the Klamath-Sierra Nevada volcanic-plutonic arc. Cenozoic translations along the San Andreas transform system have displaced segments of this terrane 300 to as much as 550 km northwest from their original positions (see Page, Chapter 13), but it is clear that these units mark the mid to late Mesozoic Andean-type continental margin of North America. As in the Sierra Nevada, more mafic portions of the batholith, possessing low initial ^{87}Sr-^{86}Sr ratios and exhibiting unfractionated rare-earth-element patterns, in general occur on the west side of the Peninsular Ranges, with more silicic, more potassic, higher ^{87}Sr-^{86}Sr types situated progressively farther east. In addition, the age of post-Jurassic granitic emplacement and crystallization tends to decrease toward the craton. Plutons on the south and west are characterized by very high K-Rb ratios, analogous to isotopically primitive island arcs; this ratio decreases in the northern and eastern Peninsular Range granites (L. T. Silver, personal communication, 1978). Gastil and others, 1978, Chapter 11, this volume) have proposed that the western, more mafic portion of this terrane was an island arc constructed on oceanic basement, whereas the more easterly volcanic-plutonic belt was sited near the North American continental margin (Andean-type arc); eastward consumption along two parallel subduction zones is presumed to have obliterated the hypothesized intervening marginal basin, with collision and final suturing occurring in earliest to mid-Cretaceous time.

The California borderland, including the islands of Santa Catalina, Cedros and the San Benitos, as well as Palos Verdes Peninsula and San Sebastian Vizcaino Peninsula, contains the southern equivalents of the Great Valley-Franciscan couplet. Although poorly exposed and complexly faulted offshore, their presence supports the late Mesozoic existence of a convergent boundary marking the western termination of the North American lithospheric plate. The great width of this terrane suggests tectonic duplication accompanying strike-slip faulting (Howell and Vedder, Chapter 16, this volume).

The Transverse Ranges in general, and the San Gabriel Mountains in particular, contain some of the most puzzling Mesozoic rocks exposed in the state. The highly alkalic 220 ± 10 m.y. old Lowe Grandiorite has invaded the Precambrian and Paleozoic (?) sections. Similar to this San Gabriel occurrence, plutons of Early Triassic age have been emplaced in the Mojave Desert, San Bernardino Mountains, and White-Inyo Range (C. F. Miller, 1977b, 1978), possibly signaling the onset of Mesozoic lithospheric plate convergence and inferred deep-seated partial fusion of subducted, eclogitized oceanic crust. According to Burchfiel and Davis (Chapter 9), this Mesozoic volcanic-plutonic terrane passed eastward to a (shallow) back-arc sedimentary basin in the eastern Mojave. The San Gabriels are also the site of several Cretaceous granitic intrusions, which have induced a metamorphic overprinting on the adjacent country rocks.

About 55 to 60 m.y. ago, this calc-alkaline plutonic arc complex apparently moved en masse northeastward over the ensimatic Pelona Schist terrane along the gently west dipping Vincent thrust (Ehlig, Chapter 10; Graham and England, 1976). Its areal and structural disposition and both relict volcanoclastic and new, relativley high-pressure metamorphic mineral assemblages suggest that the Pelona (Orocopia) protolith was deposited in a late Mesozoic back-arc basin (perhaps a southwesterly, more oceanic extension of the shallow depositional basin of the eastern Mojave continental regime), then deformed and recrystallized during southwestward subduction accompanying a latest Cretaceous closure and destruction of the depositional trough (Haxel and Dillon, 1978). Fragments of this hypothesized marginal basin sequence are dispersed along the San Andreas fault, extending from the eastern Transverse Ranges southeast to the Salton Sea and into southwesternmost Arizona. A contrasting tectonic model has been advanced by Yeats 1968a, b) and by Burchfiel and Davis (Chapter 9, this volume): they have interpreted the Pelona Schist as a Franciscan equivalent that initially lay on the west, but was subducted to relatively shallow depths beneath the calc-alkaline plutonic arc, then displaced upward to reappear far to the east in windows framed by the overlying granitic terrane.

Whatever the origin and sense of movement on the Vincent-Orocopia thrust system, the rocks of both upper and lower plates seem to have reached their metamorphic culminations nearly in the present locations relative to one another; hence recrystallization evidently outlasted thrusting. In contrast, the Coast Range thrust (with which this system is often compared) juxtaposed rocks of different metamorphic ages and P-T conditions of formation; that is, thrusting outlasted metamorphism.

CENOZOIC HISTORY OF CALIFORNIA

Cenozoic volcanic rocks, predominantly of calc-alkaline affinities, were erupted in much of the eastern portion of the state, but crop out sporadically in the California Coast Ranges and western Transverse Ranges as well. The Modoc Plateau consists of extrusive units representing the southern terminus of the currently active Cascade Mountain Range and its Tertiary volcanic arc + basaltic plateau precursors. Continental tholeiites, high-Al basalts, andesites, and dacites are widespread in this northeastern portion of California. They probably represent the superjacent carapace of a continental margin calc-alkaline plutonic series and back-arc lavas generated over the eastward-descending Farallon lithospheric plate.

Father to the southeast, in the block faulted Basin and Range and Mojave Desert regions, alkalic basalts have joined the volcanic assemblage, indicating the derivation of magmas from different upper mantle (and crustal?) levels. This change in character is evidently a consequence of the fact that Early Tertiary convergence gave way in Miocene time to the ongoing stage of complex northwest-trending right-lateral transform motion involving east-west crustal extension as well as local compression (Wise, 1963). Thinning of the crust and associated volcanism along a northern extrapolation of the East Pacific Rise (Menard, 1964) may reflect distributory back-arc spreading in the Basin and Range (Stewart, 1978). The top of the low-velocity zone in the mantle (asthenosphere) is at the base of this attenuated crust and, along with high heat flow measurements, indicates the present-day existence of a buried heat source beneath the Basin and Range. Overriding of the East Pacific Rise can be inferred from an examination of Fig. 1 of the Preface.

The Coast Ranges and western portion of the Transverse Ranges bear evidence of widespread basaltic and calc-alkaline volcanism and hypabyssal intrusion from late Oligocene time to the present. These rocks are of secondary importance, however, compared to the voluminous clastic sedimentary units. Igneous activity appears to be related to a thermal event associated with the progressively greater overriding of the East Pacific Rise heat source commencing approximately 29 m.y. ago (Atwater, 1970; Hawkins, 1970). The shallow level of derivation of the melts is attested to by the absence of alkali basalts and hyperalkalic felsic plugs.

Cenozoic marine depositional basins covered most of the western half of the state, but are currently much reduced in areal extent to the present-day narrow continental shelves and the moderately broad southern California borderland. In contrast, continental deposition has proceeded at high local rates in the Basin and Range, and in parts of the Great Valley, the Coast Ranges, the Mojave Desert, and the Transverse Ranges.

Both onshore and offshore areas are characterized by horst and graben structures; the California Coast Ranges and borderland are also typified by moderately intense folding, reflecting local compression. Moreover, associated with the San Andreas strike-slip system in western California, rotations and pull-apart structures are increasingly being recognized (Crowell, 1974b; Hall, 1978a, Chapter 17, this volume). Positive topographic areas along this shear system reflect "locking bends" where excess mass has accumulated; the San Gabriel Mountains represent a prime example. Because the transform boundary transects the continental margin at a low angle, northwestward transportation of the Salinia-Nacimiento block relative to the North American plate has duplicated the volcanic-plutonic arc, arc-trench gap, and subduction zone in the central Coast Ranges and borderland (Crouch, 1979).

Although the present-day kinematic picture is complex, movement along the San Andreas shear system in general appears to have stepped eastward with time. Whereas earlier stages of slip were accommodated chiefly along the San Simeon-Hosgri, southern California borderland complex, and San Gabriel faults (e.g., see Hall, 1975, Chapter 17, this volume; Crowell, 1975a, c, Chapter 18, this volume; Howell and Vedder, 1978, Chapter 16, this volume) during slivering of the Salinian Block, movement has gradually intensified on later breaks such as the San Jacinto and Hayward strands (Allen, 1957, Chapter 15, this volume; Garfunkel, 1974). This eastward migration of the active strike-slip zone probably is a result of the progressive overriding of the East Pacific Rise by the North American continental crust-capped lithospheric plate.

Measurements of the remnant magnetism of Miocene volcanic rocks from the western Transverse Ranges indicate eruptions at about 10° lower magnetic latitude, and subsequently an approximate 75° clockwise rotation of the units during transportation to their present location (Kamerling and Luyendyk, 1977, 1979). Such rotation suggests that individual segments of the western Transverse Ranges may have behaved as coherent blocks caught in the master right-lateral San Andreas shear system.

A final point of interest involves fluctuations in sea level. Studying the bathymetry of mid-oceanic rift systems, Parsons and Sclater (1977) recognized that water depths are roughly proportional to the age of the oceanic lithosphere; hence as the mean age of the ocean basins increases, sea water tends to drain from the continents. Dickinson points out (Chapter 1) that worldwide marine regressions should be characteristic of times in which young oceanic crust-capped lithosphere is preferentially subducted, as would be the case when North America encroached upon the East Pacific Rise. Thus the local obliteration of this ridge system along the California coast in mid-Cenozoic time could have been responsible for the ubiquitous hiatus in shallow-marine deposition on the continents and the flood of terrestrial sedimentation during Oligocene time. It also seems likely that regional elevation of the lithosphere would accompany an overriding of the thermal anomaly by the leading edge of the continent.

To summarize, easternmost California during Cenozoic time was the site of continental margin subduction zone calc-alkaline igneous activity (locally extinguished by the cessation of convergent lithospheric plate motion), block faulting + extension, uplift (especially in the batholithic terranes), and local continental alluviation. Western California was subjected to widespread but retreating marine embayments, restricted igneous activity of exclusively shallow derivation, local compression, and oblique rifting —all associated with an early stage of oblique subduction, gradually succeeded by the North American–Pacific transform system. This major strain system, although simply expressed within the ocean basin, has taken on a very complex aspect in the medium of sialic crust.

CONCLUDING STATEMENT

The geologic growth and evolution of California may be more fully appreciated through an interpretation of the various lithotectonic regimes in the light of plate-tectonic theory. Judging by the various chapters of this book, and as briefly summarized, California has experienced a long and eventful plate-tectonic development. Clearly, it presents a well-studied and reasonably well documented case history of an evolving continental margin. How typical a case history it represents can only be ascertained through comparison with other accreting plate margins. It is certain, however, that each terrane possesses unique aspects that are a composite function of the local geometry and previous history, during interacting and changing plate motions.

A rapid perusal of this history provides the reader with a comfortable feeling of appreciation for the processes that led to the present geological configuration of the state. However, a probing study of this book raises more questions than it answers. We are far from a detailed understanding of the orogenic assembly of this segment of the North American continental margin. Perhaps this is due in part to a faulty geological data base or, more likely, to misinterpretation of the significance of key portions of the admittedly fragmentary data. These latter have been employed to support markedly contrasting scenarios for the geologic evolution of a specific terrane. Simple plate-tectonic models in their present forms seem incapable of explaining many of the complexities, and even some of the fundamental problems such as the direction of lithospheric underflow, the exotic versus autochthonous site of origin, and the nature of the boundaries between pairs of terranes. It is also certain that the concepts of the new global tectonics are most readily applied to young, relatively rigid oceanic crust-capped lithosphere. Thick, ancient continental crust is the composite product of numerous earlier tectonic interactions, and consequently tends to behave in a less competent, more complex fashion that is difficult to analyze correctly, plate-tectonic theory notwithstanding. Obviously, study of the geotectonic development of California is far from finished.

REFERENCES

Aalto, K. R., 1976, Sedimentology of a melange: Franciscan of Trinidad, California: *Jour. Sed. Petrology,* v. 46, p. 913–929.

Abbott, E. W., 1971, Structural Geology of the Southern Silurian Hills, San Bernardino County, California: Unpublished M.S. thesis, Rice University, Houston, Tex., 48pp.

Abbott, P. L., and Smith, T. E., 1978, Eocene conglomerate trace element comparison of clasts in southwestern California and northwestern Mexico: *Jour. Geology,* v. 86, p. 753–762.

Acosta, M. G., 1966, Geology of the Bahía Soledad Embayment, Baja California, Mexico: Unpublished M.S. thesis, San Diego State University, San Diego, Calif., 93 pp.

——, 1970, Soledad–Punta China area, in *Pacific Slope Geology of Northern Baja California and Adjacent Alta California:* Pacific Section, Am. Assoc. Petroleum Geologists, Soc. Econ. Paleontologists and Mineralogists, and Soc. Explor. Geophys., p. 30–36.

Agnew, D. C., and Sieh, K. E., 1978, A documentary study of the felt effects of the great California earthquake of 1857: *Seismol. Soc. America Bull.,* v. 68, p. 1717–1730.

Albers, J. P., 1967, Belt of sigmoidal bending and right-lateral faulting in the western Great Basin: *Geol. Soc. America Bull.,* v. 78, p. 143–156.

——, and Robertson, J. F., 1961, Geology and ore deposits of East Shasta copper-zinc district, Shasta County, California: *U.S. Geol. Survey Prof. Paper 338,* 107 pp.

Alencaster, G., and others, 1961, Paleontología del Triásico Superior de Sonora (I–IV): Publicación Instituto de Geologia, U.N.A.M., Paleontología Mexicana, no. 11.

Allen, C. R., 1957, San Andreas fault zone in San Gorgonio Pass, southern California: *Geol. Soc. America Bull.,* v. 68, p. 315–350.

——, 1962, Circum-Pacific faulting in the Philippines-Taiwan region: *Jour. Geophys. Res.,* v. 67, 4795–4812.

——, 1965, Transcurrent faults in continental areas: *Royal Soc. [London] Philos. Trans.,* v. 258A, p. 82–89.

——, 1968, The tectonic environments of seismically active and inactive areas along the San Andreas fault system: *Stanford Univ. Pub. Geol. Sci.,* v. 11, p. 70–82.

——, 1969, Active faulting in northern Turkey: *Calif. Inst. Tech., Div. Geol. Sci., Contr. No. 1577,* 32 pp.

——, 1975, Geological criteria for evaluating seismicity: *Geol. Soc. America Bull.,* v. 86, p. 1041–1057.

——, and Nordquist, J. M., 1972, Foreshock, main shock, and larger aftershocks of the Borrego Mountain earthquake: *U.S. Geol. Survey Prof. Paper 787,* p. 16–23.

——, and others, 1960, Agua Blanca fault–a major transverse structure of northern Baja California, Mexico: *Geol. Soc. America Bull.,* v. 71, p. 457–482.

——, and others, 1965, Relationship between seismicity and geologic structure in the southern California region: *Seismol. Soc. America Bull.,* v. 55, p. 753–797.

——, and others, 1972, Displacements on the Imperial, Superstition Hills, and San Andreas faults triggered by the Borrego Mountain earthquake: *U.S. Geol. Survey Prof. Paper 787,* p. 87–104.

Allison, E. C., 1955, Middle Cretaceous gastropoda from Punta China, Baja California, Mexico: *Jour. Paleontology,* v. 29, p. 400–432.

——, 1964, Geology of areas bordering Gulf of California, *in* van Andel, T. H., and Shor, G. G., Jr., eds., Marine Geology of the Gulf of California–symposium: *Am. Assoc. Petroleum Geologists Mem. 3,* p. 3–29.

——, 1974, The type Alisitos Formation (Cretaceous, Aptian-Albian) of Baja California and its bivalve fauna in *Geology of Peninsular California:* Pacific Section, Am. Assoc. Petroleum Geologists, Soc. Econ. Petrologists and Mineralogists, and Soc. Explor. Geophys. Guidebook, p. 29–59.

Alt, J. N., 1979, Analysis of repeated geodetic leveling data for the Livermore Valley area, Alameda County, California [abstract]: *Geol. Soc. America Abstracts with Programs,* v. 11, p. 65.

Altamirano, R. F. J., 1972, Tectonica de la porción meridional de Baja California Sur: *Soc. Geol. Mexicana, Memoria II Convencion Nacional,* p. 113–114.

Alvarez, W., and others, 1979, Franciscan limestone deposited at 17° south paleolatitude [abstract] : *Geol. America Abstracts with Programs,* v. 11, p. 66.

Anderson, D. L., 1971, The San Andreas fault: *Sci. American,* v. 255, no. 5, p. 53–66.

Anderson, T. B., and others, 1974, Geology of a Late Devonian fossil locality in the Sierra Buttes Formation, Dugan Pond, Sierra City quadrangle, California [abstract] : *Geol. Soc. America Abstracts with Programs,* v. 6, p. 139.

Anderson, T. H., and Silver, L. T., 1969, Mesozoic magmatic events of the northern Sonora coastal region, Mexico [abstract] : *Geol. Soc. America Abstracts with Programs,* v. 1, p. 3–4.

——, and others, 1972, Observaciones geochronologicas sobre los complejos cristalinos de Sonora y Oaxaca, Mexico: *Soc. Geol. Mexicana, Mem. II Convencion Nacional,* p. 115–122.

Ando, C. J., 1977, Disrupted ophiolitic sequence in the south central Klamath Mountains, California [abstract] : *Geol. Soc. America Abstracts with Programs,* v. 9, p. 380.

——, and others, 1976, Structural and stratigraphic equivalence of the Stuart Fork, North Fork, and Hayfork terranes, central Klamath Mountains, California [abstract] : *Geol. Soc. America Abstracts with Programs,* v. 8, p. 349–350.

——, and others, 1977, Geologic summary and road log of portions of the central Klamath Mountains, California, in *Geology of the Klamath Mountains, Northern California:* Geol. Soc. America, Cordilleran Section Meeting, Fieldtrip Guidebook, p. 134–156.

Arabasz, W. J., 1971, Geological and geophysical studies of the Atacama fault zone in northern Chile: Unpublished Ph.D. dissertation, California Institute of Technology, Pasadena, 175 pp.

——, and Robinson, R., 1976, Microseismicity and geologic structure in the northern South Island, New Zealand: *New Zealand Jour. Geology and Geophysics,* v. 19, p. 569–601.

Armstrong, R. L., 1968, Sevier orogenic belt in Nevada and Utah: *Geol. Soc. America Bull.,* v. 79, p. 429–458.

——, 1974, Magmatism, orogenic timing and orogenic diachronism in the Cordillera from Mexico to Canada: *Nature,* v. 247, p. 348–351.

——, 1975, Precambrian (1500 m.y. old) rocks of central Idaho—the Salmon River Arch and its role in Cordilleran sedimentation and tectonics: *Am. Jour. Sci.,* v. 275A, p. 437–467.

——, 1978, The pre-Cenozoic Phanerozoic time scale—a computer file of critical dates and consequences of new and in-progress decay constant revisions: *Am. Assoc. Petroleum Geologists,* Stud. Geol., No. 6, p. 73–91.

——, and Dick, H. J., 1974. A model for the development of thin overthrust sheets of crystalline rock: *Geology*, v. 2, p. 35–40.

——, and Suppe, John, 1973, Potassium-argon geochronometry of Mesozoic igneous rocks in Nevada, Utah, and southern California: *Geol. Soc. America Bull.*, v. 84, p. 1375–1392.

Arnal, R. E., 1976, Miocene paleobathymetric changes of the Santa Rosa-Cortes Ridge area, California Continental Borderland, *in* Howell, D. G., ed., *Aspects of the Geologic History of the California Continental Borderland:* Pacific Section, Am. Assoc. Petroleum Geologists Misc. Publ. 24, p. 60–79.

Arpat, E., and Saroglu, F., 1975, Some recent tectonic events in Turkey [in Turkish]: *Turkiye Jeol. Kurumu Bull.*, v. 18, p. 91–101.

Atwater, Tanya, 1970, Implications of plate tectonics for the Cenozoic tectonic evolution of western North America: *Geol. Soc. America Bull.*, v. 81, p. 3513–3536.

——, and Molnar, P., 1973, Relative motion of the Pacific and North American plates deduced from sea-floor spreading in the Atlantic, Indian, and South Pacific oceans: in Kovach, R. L., and Nur, A., eds., Proceedings, Conf. on Tectonic Problems of the San Andreas Fault System: *Stanford Univ. Pubs. Geol. Sci.*, v. 13, p. 136–148.

Ave'Lallemant, H. G., 1976, Structure of the Canyon Mountain (Oregon) ophiolite and its implication for sea-floor spreading: *Geol. Soc. America Spec. Paper 173,* 49 pp.

Axelrod, D. I., 1957, Late Tertiary floras and the Sierra Nevada uplift: *Geol. Soc. America Bull.*, v. 68, p. 19–45.

——, 1962, Post-Pliocene uplift of the Sierra Nevada, California: *Geol. Soc. America Bull.*, v. 73, p. 183–198.

Bachman, S. B., 1978, A Cretaceous and early Tertiary subduction complex, Mendocino Coast, northern California, *in* Howell, D. G., and McDougall, K. A., eds., *Mesozoic Paleogeography of the Western United States:* Pacific Section, Soc. Econ. Paleontologists and Mineralogists, Pacific Coast Paleogeography Symposium 2, p. 419–430.

Bailey, E. H., and Blake, M. C., 1969, Tectonic development of western California during the late Mesozoic: *Geotektonika,* no. 3, p. 17–30; no. 4, p. 24–34.

——, and Blake, M. C., 1974, Major chemical characteristics of Mesozoic Coast Range ophiolite in California: *U.S. Geol. Survey Jour. Res.*, v. 2, p. 637–656.

——, and Irwin, W. P., 1959, K-feldspar content of Jurassic and Cretaceous graywackes of the northern Coast Ranges and Sacramento Valley, California: *Am. Assoc. Petroleum Geologists Bull.*, v. 43, p. 2797–2809.

——, and Jones, D. L., 1973a, Preliminary lithologic map, Colyear Springs quadrangle, California: *U.S. Geol. Survey Misc. Field Studies Map MF-516,* scale 1:48,000.

——, and Jones, D. L., 1973b, Metamorphic facies indicated by vein minerals in basal beds of the Great Valley sequence, northern California: *U.S. Geol. Survey Jour. Res.*, v. 1, no. 4, p. 383–385.

——, and others, 1964, Franciscan and related rocks and their significance in the geology of western California: *Calif. Div. Mines and Geology Bull. 183,* 177 pp.

——, and others, 1970, On-land Mesozoic oceanic crust in California Coast Ranges: *U.S. Geol. Survey Prof. Paper 700-C,* p. C70–C81.

Baird, A. K., 1962, Superposed deformation in the central Sierra Nevada foothills east of the Mother Lode: *Univ. Calif. Pubs. Geol. Sci.*, v. 42, p. 1–70.

——, and others, 1974, Transverse Range Province: A unique structural-petrochemical belt across the San Andreas fault system: *Geol. Soc. America Bull.*, v. 85, p. 163–174.

Bakun, W. H., and McEvilly, T. V., 1978, Comparison of Parkfield earthquakes (1934 through 1936) [abstract] : *Trans. Am. Geophys. Union,* v. 59, p. 1126.

——, and others, 1980. Implication of seismicity for failure of a section of the San Andreas fault: *Seismol. Soc. America Bull.*

Ballard, R. D., and Moore, J. G., 1977, *Photographic Atlas of the Mid-Atlantic Ridge Rift Valley:* New York, Springer-Verlag, 114 pp.

——, and others, 1978, Amar 78 preliminary results IV: Amar rift valley—Evidence for cycles in the evolution of rift valleys [abstract] : *Trans. Am. Geophys. Union,* v. 59, no. 12, p. 1198–1199.

——, and others, in press, The Galapagos rift at 86°W: 3. Sheet flows, collapse pits, and lava lakes of the rift valley: *Jour. Geophys. Res.*

Bamba, T., 1974, Ophiolite from Ergani Mining District, southeastern Turkey: *Mining Geology,* v. 24, p. 297–305.

Bandy, O. L., and Ingle, J. C., 1970, Neogene planktonic events and radiometric scale, Calif.: *Geol. Soc. America Spec. Paper 124,* 172 pp.

Barker, D. S., 1977, Northern Tran-Pecos magmatic province: Introduction and comparison with the Kenya Rift: *Geol. Soc. America Bull.,* v. 88, p. 1421–1427.

Barnes, I., and O'Neil, J. R., 1969, The relationship between fluids in some fresh alpine-type ultramafics and possible modern serpentinization, western United States: *Geol. Soc. America Bull.,* v. 80, p. 1947–1960.

Barthelmy, D. A., 1975, Geology of the El Arco—Calmalli area, Baja California, Mexico: Unpublished M.S. thesis, San Diego State University, San Diego, Calif., 130 pp.

Bateman, P. C., 1962, Geology, *in* Schumacher, Genny, ed., *Deepest Valley, Guide to Owens Valley and Its Mountain Lakes, Roadsides, and Trails:* San Francisco, Sierra Club, p. 100–122.

——, 1965a, Geology and tungsten mineralization of the Bishop District, California: *U.S. Geol. Survey Prof. Paper 470,* 208 pp.

——, 1965b, Geologic map of the Blackcap Mountain quadrangle, Fresno County, California: *U.S. Geol. Survey Geol. Quad. Map GQ-428,* scale 1:62,500.

——, 1974, Model for the origin of Sierra granites: *Calif. Geology,* v. 27, no. 1, p. 3–5.

——, 1979, Cross section of the Sierra Nevada from Madera to the White Mountains: *Geol. Soc. America Map and Chart Series MC-28E.*

——, and Chappell, B. W., 1979, Crystallization, fractionation and solidification of the Tuolumne intrusive series, Yosemite National Park, California: *Geol. Soc. America Bull.,* v. 90, no. 5, p. 465–482.

——, and Clark, L. D., 1974, Stratigraphic and structural setting of Sierra Nevada Batholith, California: *Pacific Geology,* v. 8, p. 79–89.

——, and Dodge, F. C. W., 1970, Variations of major chemical constituents across the central Sierra Nevada batholith: *Geol. Soc. America Bull.,* v. 81, p. 409–420.

——, and Eaton, J. P., 1967, Sierra Nevada batholith: *Science,* v. 158, p. 1407–1417.

——, and Moore, J. G., 1965, Geologic map of the Mount Goddard quadrangle, Fresno and Inyo Counties, California: *U.S. Geol. Survey Geol. Quad. Map GQ-429,* scale 1:62,500.

——, and Nokleberg, W. J., 1978, Solidification of the Mount Givens Granodiorite, Sierra Nevada, California: *Jour. Geology,* v. 86, p. 59–75.

——, and Wahrhaftig, Clyde, 1966, Geology of the Sierra Nevada, *in* Bailey, E. A., ed., Geology of Northern California: *Calif. Div. Mines and Geology Bull. 190,* p. 107–172.

——, and Wones, D. R., 1972, Geologic map of the Huntington Lake quadrangle, central Sierra Nevada, California: *U.S. Geol. Survey Geol. Quad. Map GQ-987,* scale 1:62,500.

——, and others, 1963, The Sierra Nevada batholith—A synthesis of recent work across the central part: *U.S. Geol. Survey Prof. Paper 414D,* p. D1–D46.

——, and others, 1971, Geologic map of the Kaiser Peak quadrangle, central Sierra Nevada, California: *U.S. Geol. Survey Geol. Quad. Map GQ-894,* scale 1:62,500.

Bauder, J. M., and Liou, J. G., 1979, Tectonic outlier of Great Valley sequence in Franciscan terrain, Diablo Range, California: *Geol. Soc. America Bull.,* v. 90, p. 561–568.

Beauvais, L., and Stump, T. E., 1976, Corals, mollusks and paleogeography of Late Jurassic strata of the Poso Serna region, Sonora, Mexico: *Paleogeography, Paleoclimatology, Paleoecology,* v. 19, p. 275–301.

Beck, M. E., Jr., 1976, Discordant paleomagnetic pole positions as evidence of regional shear in the western Cordillera of North America: *Am. Jour. Sci.,* v. 276, p. 694–712.

——, and Plumley, P. W., 1979, Late Cenozoic subduction and continental-margin truncation along the northern Middle America Trench: Discussion: *Geol. Soc. America Bull.,* v. 90, p. 792–793.

Behrman, P. G., 1978, Pre-Callovian rocks west of the Melones Fault Zone, central Sierra Nevada foothills, *in* Howell, D. G., and McDougall, K. A., eds., *Mesozoic Paleogeography of the Western United States:* Pacific Section, Soc. Econ. Paleontologists and Mineralogists, Pacific Coast Paleogeography Symposium 2, p. 337–348.

——, and Parkison, G. A., 1978, Paleogeographic significance of the Callovian to Kimmeridgian strata central Sierra Nevada foothills, California, *in* Howell, D. G., and McDougall, K. A., eds., *Mesozoic Paleogeography of the Western United States:* Pacific Section, Soc. Econ. Paleontologists and Mineralogists, Pacific Coast Paleogeography Symposium 2, p. 349–360.

Benioff, H., 1955, Mechanism and strain characteristics of the White Wolf fault as indicated by the aftershock sequence: *Calif. Div. Mines Bull. 171,* p. 199–202.

Bennett, J. D., 1978, Tectonics and metamorphism of northern Sumatra Transect: Unpublished abstract, 2 pp.

Berg, H. C., and others, 1978, Map showing pre-Cenozoic tectonostratigraphic terranes of southeastern Alaska and adjacent areas: *U.S. Geol. Survey Open File Report 78-1085.*

Berger, W. H., and Winterer, E. L., 1974, Plate stratigraphy and the fluctuating carbonate line *in* Hsu, K. J., and Jenkyns, H. C., eds., *Pelagic Sediments on Land and Under the Sea:* Internat. Assoc. Sedimentologists Special Publ. No. 1, p. 11–48.

Berggren, W. A., and van Couvering, J. A., 1974, The Late Neogene, biostratigraphy, geochronology, and paleoclimatology of the last 15 million years in marine and continental sequences: *Paleogeography, Paleoclimatology, and Paleoecology,* v. 16, p. 1–216.

Berkland, J. O., 1973, Rice Valley outlier—new sequence of Cretaceous-Paleocene strata in northern Coast Ranges, California: *Geol. Soc. America Bull.,* v. 84, p. 2389–2406.

——, and others, 1972, What is Franciscan? *Am. Assoc. Petroleum Geologists Bull.,* v. 56, p. 2295–2302.

Best, Myron, 1963, Petrology of the Guadalupe igneous complex, southwestern Sierra Nevada foothills, California: *Jour. Petrology,* v. 4, no. 2, p. 223–259.

Best, M. G., 1963, Petrology and structural analysis of metamorphic rocks in the southwestern Sierra Nevada Foothills, California: Univ. Calif. Pubs., *Geol. Sci.,* v. 42, p. 111–157.

Beutner, E. C., 1977, Evidence and implications of a late Cretaceous Paleogene island arc and marginal basin along the California coast [abstract]: *Geol. Soc. America Abstracts with Programs,* v. 9, no. 4, p. 389.

Bezore, S. P., 1969, The Mount Saint Helena ultramafic-mafic complex of the northern Coast Ranges [abstract]: *Geol. Soc. America Abstracts with Programs,* v. 1, no. 3, p. 5.

——, 1971, Ophiolitic and associated rocks near Mount St. Helena, p. 23–28 *in* Lipps, J. H., and Moores, E. M., eds., *Geologic Guide to the Northern Coast Ranges, Point Reyes Region, California:* Geol. Soc. Sacramento, Annual Field Trip Guidebook for 1971, 135 pp.

Bischoff, J. L., and Henyey, T. L., 1974, Tectonic elements of the central part of the Gulf of California: *Geol. Soc. America Bull.,* v. 85, p. 1893–1904.

Bishop, C. C., 1963, Geologic map of California, Needles sheet: *Calif. Div. Mines and Geology,* scale 1:250,000.

——, 1970, Upper Cretaceous stratigraphy of the west side of the northern San Joaquin Valley, Stanislaus and San Joaquin Counties, California: *Calif. Div. Mines and Geology Spec. Report 104,* 29 pp.

Bishop, D. G., 1977, South Fork Mountain Schist at Black Butte and Cottonwood Creek, northern California: *Geology,* v. 55, p. 595–599.

Blake, M. C., Jr., and Jones, D. L., 1974, Origin of Franciscan melanges in northern California, *in* Dott, R. H., Jr., and Shaver, R. H., eds., *Modern and Ancient Geosynclinal Sedimentation:* Soc. Econ. Paleontologists and Mineralogists Spec. Pub. 19, p. 345–357.

——, and Jones, D. L., 1977, Plate tectonic history of the Yolla Bolly Junction, northern California: *Geol. Soc. America, Guide for 73rd Annual Cordilleran Section Meeting, Sacramento, Calif.*

Blake, M. C., and Jones, D. L., 1978, Allochthonous terranes in northern California?—A reinterpretation, *in* Howell, D. G., and McDougall, K. A., eds., *Mesozoic Paleogeography of the Western United States:* Pacific Section, Soc. Econ. Paleontologists and Mineralogists, Pacific Coast Paleogeography Symposium 2, p. 397–400.

——, and Wright, R. H., 1976, Petrology of Franciscan and graywacke in the north San Francisco Bay region [abstract]: *Geol. Soc. America Abstracts with Programs,* v. 8, no. 3, p. 356–357.

——, 1967, Upside-down metamorphic zonation, blueschist facies, along a regional thrust in California and Oregon: *U.S. Geol. Survey Prof. Paper 575–C,* p. 1–9.

——, and others, 1971, Preliminary geologic map of western Sonoma County and northernmost Marin County, California: *U.S. Geol. Survey Open-File Map,* scale 1:62,500.

——, and others, 1974, Preliminary geologic map of Marin, San Francisco, and parts of adjacent counties, California: *U.S. Geol. Survey Misc. Field Studies Map MF-573,* scale 1:62,500.

——, and others, 1975, Trace element characteristics of Mesozoic Coast Range ophiolite in California [abstract] : *Trans. Am. Geophys. Union,* v. 56, no. 12, p. 1079–1080.

——, and others, 1978, Neogene basin formation in relation to plate-tectonic evolution of San Andreas fault system, California: *Am. Assoc. Petroleum Geologists Bull.,* v. 62, no. 3, p. 344–372.

Bloxam, T. W., 1960, Jadeite-rocks and glaucophane-schists from Angel Island, San Francisco Bay, California: *Am. Jour. Sci.,* v. 258, p. 555–573.

Bodenlos, A. J., 1950, Geology of the Red Mountain magnesite district, Santa Clara and Stanislaus Counties, California: *Calif. Jour. Mines and Geology,* v. 46, p. 223–278.

Bohannon, R. G., 1975, Mid-Tertiary conglomerates and their bearing on Transverse Range tectonics, southern California, *in* Crowell, J. C., ed., *The San Andreas Fault in Southern California:* Calif. Div. Mines and Geology Special Report 118, p. 75–82.

——, 1976, Mid-Tertiary nonmarine rocks along the San Andreas fault in southern California: Unpublished Ph.D. dissertation, University of California, Santa Barbara, 309 pp.

Bolt, B., and Miller, R. D., 1975, *Catalogue of Earthquakes in Northern California and adjoining areas, 1 January 1910–31 December 1972:* University of California, Berkeley, 567 pp.

——, and others, 1968, Seismological evidence on the tectonics of central and northern California and the Mendocino escarpment: *Seismol. Soc. America Bull.,* v. 58, p. 1725–1767.

Bond, G. C., and others, 1977, *Paleozoic-Mesozoic Rocks of the Northern Sierra, Field Guide:* Geol. Soc. America, 73rd Annual Meeting, Cordilleran Section, 35 pp.

Bonilla, M. G., and Buchanan, J. M., 1970, Interim report on worldwide historic surface faulting: *U.S. Geol. Survey Open File Report,* 32 pp.

Bonneau, M., 1971, Una nueva area Cretacica fosilifera en el Estado de Sinaloa: *Bol. Soc. Geol. Mexicana,* v. 32, no. 2, p. 159–167.

Bostick, N. H., 1974, Phytoclasts as indicators of thermal metamorphism, Franciscan assemblage and Great Valley sequence (Upper Mesozoic), California: *Geol. Soc. America Spec. Paper 153,* p. 1–17.

Boucot, A. J., and Potter, A. W., 1977, Middle Devonian orogeny and biogeographical relations in areas along the North American Pacific Rim: *Univ. Calif. Riverside Mus. Contrib.* 4, p. 210–219.

Boudier, F., and Coleman, R. G., in press, Cross section through the peridotite in the Samail ophiolite, southeastern Oman Mountains: *Jour. Geophys. Res.,* v. 86

Bowen, O. E., 1954, Geology and mineral deposits of Barstow quadrangle, San Bernardino Couny, California: *Calif. Div. Mines Bull. 165,* 208 pp.

Bradley, W. C., and Griggs, G. B., 1976, Form, genesis, and deformation of central California wave-cut platforms: *Geol. Soc. America Bull.,* v. 87, p. 433–449.

Brook, C. A., 1977, Stratigraphy and structure of the Saddlebag Lake roof pendant, Sierra Nevada, California: *Geol. Soc. America Bull.*, v. 88, p. 321–331.

——, and others, 1974, Nature of the angular unconformity between the Paleozoic metasedimentary rocks and the Mesozoic metavolcanic rocks in the eastern Sierra Nevada, California: *Geol. Soc. America Bull.*, v. 85, p. 571–576.

——, and others, 1979, Fossiliferous upper Paleozoic rocks and their structural setting in the Ritter Range and Saddlebag Lake roof pendants, central Sierra Nevada, California [abstract] : *Geol. Soc. America Abstracts with Programs*, v. 11, p. 71.

Brookfield, M. E., 1977, The emplacement of giant ophiolite nappes. I. Mesozoic-Cenozoic examples: *Tectonophysics*, v. 37, p. 247–303.

Brown, J. A., 1968a, Probable thrust contact between Franciscan Formation and Great Valley Sequence northeast of Santa Maria, California [abstract] : *Geol. Soc. America Spec. Paper 115, Abstracts for 1967*, p. 313–314.

——, 1968b, Thrust contact between Franciscan Group and Great Valley Sequence northeast of Santa Maria, California: Unpublished Ph.D. dissertation, University of Southern California, Los Angeles, 236 pp.

Brown, R. D., Jr., 1964, Geologic map of the Stonyford, quadrangle, Gleen, Colusa, and Lake Counties, California. *U.S. Geol. Survey Mineral Inv. Field Studies Map MF-279*, scale 1:48,000;

——, and Wallace, R. E., 1968, Current and historic fault movement along the San Andreas fault between Paicines and Camp Dix, California: *Stanford Univ. Pub. Geol. Sci.*, v. 11, p. 22–41.

——, and others, 1967. The Parkfield-Cholame, California, earthquakes of July–August 1966–surface geology effects, water-resources aspects, and preliminary seismic data: *U.S. Geol. Survey Prof. Paper 579*, 66 pp.

Brune, J. N., and Allen, C. R., 1967, A micro-earthquake survey of the San Andreas fault system in southern California: *Seismol. Soc. America Bull.*, v. 57, p. 277–296.

Bryan, W. B., 1972, Morphology of quench crystals in submarine basalts: *Jour. Geophys. Res.*, v. 77, p. 5812–5819.

——, 1979, Regional variation and petrogenesis of basalt glasses from the FAMOUS area, mid-Atlantic Ridge: *Jour. Petrology*, v. 20, p. 293–325.

——, and Moore, J. G., 1977, Compositional variations of young basalts in the mid-Atlantic Ridge rift valley near lat. 36°49′N: *Geol. Soc. America Bull.*, v. 88, p. 566–570.

——, and others, 1976, Inferred geologic settings and differentiation in basalts from the Deep-Sea Drilling Project: *Jour. Geophys. Res.*, v. 81, p. 4285–4304.

Buchanan-Banks, J. M., and others, 1978, Preliminary map showing recency of faulting in coastal south-central California: *U.S. Geol. Survey Misc. Field Studies Map 910.*

Buckley, C. P., 1972, Structural geology of the northern Silver Peak Mountains, Esmeralda County, Nevada [abstract] : *Geol. Soc. America Abstracts with Programs*, v. 4, p. 132.

——, 1974, Interpretation of Mesozoic displacement along the Furnace Creek Fault [abstract] : *Geol. Soc. America Abstracts with Programs*, v. 6, p. 149–150.

——, and others, 1975, The Jurassic Flysch of the Santa Ana Mountains: an example of obduction? [abstract] : *Geol. Soc. America Abstracts with Programs*, v. 7, p. 300.

Buer, K. Y., 1977, Stratigraphy, structure and petrology of a portion of the Smartsville complex, northern Sierra Nevada, California [abstract] : *Geol. Soc. America Abstracts with Programs,* v. 9, no. 4, p. 394.

Burch, S. H., 1968, Tectonic emplacement of the Burro Mountain ultramafic body, Santa Lucia Range, California: *Geol. Soc. America Bull.,* v. 78, p. 527–544.

Burchfiel, B. C., and Davis, G. A., 1968, Two-sided nature of the Cordilleran orogen and its tectonic implications: *Int. Geol. Congress, XXIII Sess. Rept., Proc. of Sect. 3,* p. 175–184.

——, and Davis, G. A., 1971, Clark Mountain thrust complex in the Cordillera of southeastern California, Geologic Summary and Field Trip Guide: *Univ. Calif. Riverside Mus. Contr.,* v. 1, p. 1–28.

——, and Davis, G. A., 1972, Structural framework and evolution of the southern part of the Cordilleran orogen, western United States: *Am. Jour. Sci.,* v. 272, p. 97–118.

——, and Davis, G. A., 1975, Nature and controls of Cordilleran orogenesis, western United States: extensions of an earlier synthesis: *Am. Jour. Sci.,* v. 275–A, p. 363–396.

——, and Davis, G. A., 1977, Geology of the Sagamore Canyon-Slaughterhouse Spring Area, New York Mountains, California: *Geol. Soc. America Bull.,* v. 88, p. 1623–1640.

——, and Stewart, J. H., 1966, "Pull-apart" origin of the central segment of Death Valley, California: *Geol. Soc. America Bull.,* v. 77, p. 439–442.

——, and others, 1970, An early Mesozoic deformation belt in south-central Nevada-southeastern California: *Geol. Soc. America Bull.,* v. 81, p. 211–215.

——, and others, 1974, Geology of the Spring Mountains, Nevada: *Geol. Soc. America Bull.,* v. 85, p. 1013–1022.

Burford, R. O., 1965, Strain analysis across the San Andreas fault and Coast Ranges of California: *Proc. 2nd Internat. Symposium on Recent Crustal Movements,* Aulanko, Finland, p. 100–110.

Burke, D. B., and Clark, M. M., 1978, Late Quaternary activity along the Garlock fault at Koehn Lake, Fremont Valley, California [abstract] : *Trans. Am. Geophys. Union,* v. 59, p. 1126.

Burrett, C. F., 1974, Plate tectonics and the fusion of Asia: *Earth Planet. Sci. Lett.,* v. 21, p. 181–189.

Bushee, J., and others, 1963, Lead-alpha dates for some basement rocks of southwestern California: *Geol. Soc. America Bull.,* v. 74, p. 803–806.

Byerly, P., and Wilson, J. T., 1935, The central California earthquakes of May 16, 1933, and June 7, 1934: *Seismol. Soc. America Bull.,* v. 25, p. 223–246.

Cady, J. W., 1975, Magnetic and gravity anomalies in the Great Valley and western Sierra Nevada Metamorphic belt, California: *Geol. Soc. America Spec. Paper 168,* 56 pp.

Callender, J. C., 1975, Geology of the York Mountain area, southern Coast Ranges, California: Unpublished Ph.D. dissertation, Harvard University, Cambridge, Mass., 293 pp.

Cameron, S. C., 1978, Geology of the Potasi Mountain Area, southeastern Nevada: Unpublished M.S. thesis, Rice University, Houston, Tex., 85 pp.

Cameron, W. E., and others, 1979, Petrographic dissimilarities between ophiolitic and ocean-floor basalts [abstract]: *Cyprus Geol. Survey Dept., Abstracts Internat. Ophiolite Symposium, Nicosia,* p. 108–109.

Campbell, R. H., and Yerkes, R. F., 1971, Cenozoic evolution of the Santa Monica Mountains-Los Angeles basin area: II. Relation to plate tectonics of the northeast Pacific Ocean [abstract]: *Geol. Soc. America Abstracts with Programs,* v. 3, no. 2, p. 92.

——, and Yerkes, R. F., 1976, Cenozoic evolution of the Los Angeles basin area—relation to plate tectonics: *Pacific Section Am. Assoc. Petrol. Geologists and Petrologists Misc. Pub. 24,* p. 541–558.

Carder, D. S., 1973, Trans-California seismic profile, Death Valley to Monterey Bay; *Seismol. Soc. America Bull.,* v. 63, p. 480–493.

Carey, S. W., 1958, A tectonic approach to continental drift, p. 177–355, *in* Carey, S. W., ed., *Continental Drift: A Symposium:* Hobart, Australia, University of Tasmania, Geology Dept., 375 pp.

Carillo, M., 1971, La geologia de la hoja San Jose de Gracia, Sinaloa: Tesis professional, Univ. Nac. Autónoma de México.

Carlson, R., and others, 1979, A survey of microearthquake activity along the San Andreas fault from Carrizo Plains to Lake Hughes: *Seismol. Soc. America Bull.,* v. 69, p. 177–186.

Carman, M. F., Jr., 1964, Geology of the Lockwood Valley area, Kern and Ventura counties, California: *Calif. Div. Mines Spec. Report 81,* 62 pp.

Carmichael, I. S. E., and others, 1974, *Igneous Petrology:* New York, McGraw-Hill Book Co., 739 pp.

Carr, M. D., 1977, Stratigraphy, timing, and nature of emplacement of the Contact thrust plate in the Goodsprings District, southern Nevada [abstract]: *Geol. Soc. America Abstracts with Programs,* v. 9, no. 4, p. 397–398.

——, 1978, Geology of the Goodsprings Area, southeastern Nevada: Unpublished Ph.D. dissertation, Rice University, Houston, Tex., 150 pp.

Carter, Bruce, and Silver, L. T., 1971, Post-emplacement structural history of the San Gabriel anorthosite complex [abstract]: *Geol. Soc. America Abstracts wtih Programs,* v. 3, p. 92–93.

——, and Silver, L. T., 1972, Structure and petrology of the San Gabriel anorthosite-syenite body, California: *24th Internat. Geol. Congress, Section 3,* p. 303–311.

Casey, T. A. L., and Dickinson, W. R., 1976, Sedimentary serpentinite of the Miocene Big Blue Formation near Cantua Creek, California, *in* Fritsche, A. E., and others, eds., *The Neogene Symposium:* Pacific Section, Soc. Econ. Paleontologists and Mineralogists, p. 65–74.

Cashman, P. H., 1974, Cross-section of a portion of the western Paleozoic and Triassic subprovince, Salmon River, Klamath Mountains, California, in *Geologic Guide to the Southern Klamath Mountains:* Geol. Soc. Sacramento, Ann. Field Trip Guidebook, p. 62–68.

Cashman, S. M., 1977a, Correlation of the Duzel Formation with the Central Metamorphic Belt, northeastern Klamath Mountains, California [abstract]: *Geol. Soc. America Abstracts with Programs,* v. 9, p. 398.

——, 1977b, Structure and petrology of part of the Duzel Formation and related rocks in the Klamath Mountains southwest of Yreka, California: Unpublished Ph.D. dissertation, University of Washington, Seattle, 94 pp.

Castle, R. O., and others, 1976, Aseismic uplift in southern California: *Science,* v. 192, p. 251–253.

Cebull, S. E., 1972, Sense of displacement along Foothills fault system: new evidence from the Melones fault zone, western Sierra Nevada, California: *Geol. Soc. America Bull.,* v. 83, p. 1185–1190.

Charlton, Doug, 1978, An upward-coarsening volcaniclastic submarine fan constructed on the Rattlesnake Creek ophiolite, Ironside Mountain quadrangle, Trinity County, Klamath Mountain, California [abstract] : *Geol. Soc. America Abstracts with Programs,* v. 10, no. 3, p. 99.

Chayes, F., 1969, The chemical composition of Cenozoic andesite, *in* McBirney, A. R., ed., *Proceedings of the Andesite Conference:* Oregon Dept. Geology and Mineral Industries Bull., v. 65, p. 1–20.

Chen, J. H., 1977, Uranium-lead isotopic ages from the southern Sierra Nevada batholith and adjacent areas, California: Unpublished Ph.D. dissertation, University of California, Santa Barbara, 138 pp.

——, and Tilton, G. T., 1978, Lead and strontium isotopic studies of the southern Sierra Nevada batholith, California [abstract] : *Geol. Soc. America Abstracts with Programs,* v. 10, p. 99–100.

Chesterman, C. W., 1975, Geologic map of the Matterhorn Peak quadrangle, Mono and Tuolumne Counties: *Calif. Div. Mines and Geology Geo. Map Sheet 22,* scale 1:48,000.

Chhibber, H. L., 1934, *The Geology of Burma:* London, Macmillan and Co., 538 pp.

Chipping, D. H., 1972, Early Tertiary paleogeography of central California: *Am. Assoc. Petroleum Geologists Bull.,* v. 56, p. 480–493.

Christensen, M. N., 1963, Structure of metamorphic rocks at Mineral King, California: *Univ. California Pubs. Geol. Sci.,* v. 42, no. 4, p. 159–198.

——, 1965, Late Cenozoic deformation in the central Coast Ranges of California: *Geol. Soc. America Bull.,* v. 76, 1105–1124.

Christiansen, R. L., and Lipman, P. W., 1972, Cenozoic volcanism and plate-tectonic evolution of the western United States. II. Late Cenozoic: *Roy. Soc. London Phil. Trans.,* v. 271, p. 249–284.

Church, W. R., and Riccio, L., 1977, Fractionation trends in the Bay of Islands ophiolite of Newfoundland: polycyclic cumulate sequences in ophiolites and their classification: *Canadian Jour. Earth Sci.,* v. 14, p. 1156–1165.

Churkin, Michael, Jr., 1974a, Paleozoic marginal ocean basin volcnic arc systems in the Cordilleran foldbelt, *in* Dott, R. H., Jr., and Shaver, R. H., eds., *Modern and Ancient Geosynclinal Sedimentation:* Soc. Econ. Paleontologists and Mineralogists Spec. Pub. No. 19, p. 174–192.

——, 1974b, Deep-sea drilling for landlubber geologists—the southwest Pacific, an accordion plate tectonics analog for the Cordilleran geosyncline: *Geology,* v. 2, p. 339–342.

——, and Eberlein, G. D., 1977, Ancient borderland terranes of the North American Cordillera: correlations and microplate tectonics: *Geol. Soc. America Bull.,* v. 88, p. 769–786.

——, and Kay, M., 1967, Graptolite-bearing Ordovician siliceous and volcanic rocks, northern Independence Range, Nevada: *Geol. Soc. America Bull.,* v. 78, p. 651–668.

Clague, D. A., and Straley, P. F., 1977, Petrologic nature of the oceanic Moho: *Geology,* v. 5, p. 133–136.

Clark, J. C., and Brabb, E. E., 1978, Stratigraphic contrasts across the San Gregorio fault, Santa Cruz Mountains, west central California: *Calif. Div. Mines and Geology Spec. Report 137,* p. 3–12.

Clark, K. F., and others, 1977, Posición estratigráfia en tiempo y espacio de mineralización en la provincia de la Sierra Madre Occidental, en Durango, Mexico: *Asociación Ingeneria Minera, Metalogia, y Geologia de Mexico, 12th Convención Nacional, Memoria,* p. 197–244.

Clark, L. D., 1960, Foothills fault system, western Sierra Nevada, California: *Geol. Soc. America Bull.,* v. 71, p. 483–496.

——, 1964, Stratigraphy and structure of part of the western Sierra Nevada metamorphic belt, California: *U.S. Geol. Survey Prof. Paper 410,* 70 pp.

——, 1970, Geology of the San Andreas 15-minute quadrangle, California: *Calif. Div. Mines and Geology Bull. 195,* 23 pp.

——, 1976, Stratigraphy of the north half of the western Sierra Nevada metamorphic belt, California: *U.S. Geol. Survey Prof. Paper 923,* 26 pp.

——, and others, 1962, Angular unconformity between Mesozoic and Paleozoic rocks in the northern Sierra Nevada, California: *U.S. Geol. Survey Prof. Paper 450B,* p. B15–B19.

Clark, M. M., 1972, Surface rupture along the Coyote Creek fault: *U.S. Geol. Survey Prof. Paper 787,* p. 55–86.

——, and others, 1972, Holocene activity of the Coyote Creek fault as recorded in the sediments of Lake Cahuilla: *U.S. Geol. Survey Prof. Paper 787,* p. 112–130.

Clarke, S. H., Jr., and Nilsen, T. H., 1973, Displacement of Eocene strata and implications for the history of offset along the San Andreas fault, central and northern California, *in* Kovach, R. L., and Nur, Amos, eds., *Proceedings of the Conference on Tectonic Problems of the San Andreas Fault System:* Stanford Univ. Publ. Geol. Sci., v. 13, p. 358–367.

——, and others, 1975, Paleogene geography of California, *in* Weaver, D. W., and others, eds., *Conference on Future Energy Horizons of the Pacific Coast, Paleogene Symposium and Selected Technical Papers:* Pacific Section, Am. Assoc. Petroleum Geologists, Annual Meeting 1975, p. 124–154.

Clayton, L., 1966, Tectonic depressions along the Hope fault, a transcurrent fault in North Canterbury, New Zealand: *New Zealand Jour. Geology Geophys.,* v. 9, p. 95–104.

Cloos, Ernst, 1932, "Feather joints" as indicators of the direction of movement on faults, thrusts, joints and magmatic contacts: *Proc. National Academy of Sciences,* v. 18, p. 387–395.

Cluff, L. S., and others, 1972, Site evaluation in seismically active regions—an interdisciplinary team approach: *Proc. Internat. Conf. on Microzonation [Seattle],* p. 957–987.

Colburn, I. P., 1970, The trench concept as a model for central California Cretaceous basin of deposition [abstract] : *Geol. Soc. America Abstracts with Programs,* v. 2, no. 2, p. 82–83.

Cole, M. R., 1975, Eocene sedimentation and paleocurrents, San Nicolas Island, California: *Geol. Soc. America, Cordilleran Section, Guidebook,* Los Angeles, March 1975, 32 pp.

Coleman, R. G., 1967, Low-temperature reaction zones and alpine ultramafic rocks of California, Oregon, and Washington: *U.S. Geol. Survey Bull. 1247,* 49 pp.

——, 1971a, Plate tectonic emplacement of upper mantle periodotites along continental edges: *Jour. Geophys. Res.,* v. 76, no. 5, p. 1212–1222.

——, 1971b, Petrological and geophysical nature of serpentinites: *Geol. Soc. America Bull.,* v. 82, p. 897–918.

——, 1972, The Colebrooke schist of southwestern Oregon and its relation to the tectonic evolution of the region: *U.S. Geol. Survey Bull., 1339,* 61 pp.

——, 1977, *Ophiolites:* New York, Springer-Verlag, 229 pp.

——, and Keith, T. E., 1971, A chemical study of serpentinization–Burro Mountain, California: *Jour. Petrology,* v. 12, p. 311–328.

——, and Lanphere, M. A., 1971, Distribution and age of high-grade blueschists, associated eclogites, and amphibolites from Oregon and California: *Geol. Soc. America Bull.,* v. 82, p. 2397–2412.

——, and Peterman, Z. E., 1975, Oceanic plagiogranites: *Jour. Geophys. Revs.,* v. 80, p. 1099–1108.

——, and others, 1976, The amphibolite of Briggs Creek: a tectonic slice of metamorphosed oceanic crust in southwestern Oregon [abstract] : *Geol. Soc. America Abstracts with Programs,* v. 8, no. 3, p. 363.

Compton, R. R., 1955, Trondhjemite batholith near Bidwell Bar, California: *Geol. Soc. America Bull.,* v. 66, p. 9–44.

——, 1960, Charnockitic rocks of Santa Lucia Range, California: *Am. Jour. Sci.,* v. 258, p. 609–636.

——, 1966a, Granitic and metamorphic rocks of the Salinian block, California Coast Ranges, in *Geology of Northern California:* Calif. Div. Mines and Geology Bull. 190, p. 277–287.

——, 1966b, Analysis of Plio-Pleistocene deformation and stresses in northern Santa Lucia Range, California: *Geol. Soc. America Bull.,* v. 77, p. 1361–1380.

Condie, K. C., and Snansieng, Sathian, 1971, Petrology and geochemistry of the Duzel (Ordovician) and Gazelle (Silurian) Formations, northern California: *Jour. Sed. Petrology,* v. 41, no. 3, p. 741–751.

Coney, P. J., 1972, Cordilleran tectonics and North America plate motion: *Am. Jour. Sci.,* v. 272, p. 603–628.

——, 1978, Mesozoic-Cenozoic Cordilleran plate tectonics, *in* Smith, R. B., and Eaton, G. P., eds., *Cenozoic Tectonics and the Regional Geophysics of the Western Cordillera:* Geol. Soc. America Mem. 152, p. 33–50.

Conrad, R. L., and Davis, T. E., 1977, Rb/Sr Geochronology of cataclastic rocks of the Vincent thrust, San Gabriel Mountains, southern California [abstract] : *Geol. Soc. America Abstracts with Programs,* v. 9, no. 4, p. 403–404.

Cooper, A. K., and others, 1976, Plate tectonic model for the evolution of the eastern Bering Sea Basin: *Geol. Soc. America Bull.,* v. 87, p. 1119–1126.

Cooper, G. A., and others, 1965, Fauna Permica de el Antimonio, oeste de Sonora, Mexico: *Instituto de Geologia Univ. Nac. Autónoma Mexico, Bol. 58, Part 3,* 122 pp.

Cotton, W. R., 1972, Preliminary geologic map of the Franciscan rocks in the central part of the Diablo Range, Santa Clara and Alameda Counties, California: *U.S. Geol. Survey Misc. Field Studies Map MF-343.*

Cowan, D. S., 1974, Deformation and metamorphism of the Franciscan subduction zone complex northwest of Pacheco Pass, California: *Geol. Soc. America Bull.*, v. 85, p. 1623–1634.

——. 1978, Origin of blueschist-bearing chaotic rocks in the Franciscan Complex, San Simeon, California: *Geol. Soc. America Bull.*, v. 89, p. 1415–1423.

——, and Page, B. M., 1975, Recycled Franciscan material in Franciscan melange west of Paso Robles, California: *Geol. Soc. America Bull.*, v. 86, p. 1089–1095.

——, and Silling, R. M., 1978, A dynamic, scaled model of accretion at trenches and its implications for the tectonic evolution of subduction complexes: *Jour. Geophys. Res.*, v. 83, p. 5389–5396.

Cox, D. P., 1967, Reconnaissance geology of the Helena quadrangle, Trinity County, California, in *Short Contributions to California Geology:* Calif. Div. Mines and Geology Spec. Report 92, p. 43–55.

——, and Pratt, W. P., 1973, Submarine chert-argillite slide-breccia of Paleozoic age in the southern Klamath Mountains, California: *Geol. Soc. America Bull.*, v. 84, p. 1423–1438.

Crawford, K. E., 1975, The Geology of the Franciscan tectonic assemblage near Mount Hamilton, California: Unpublished Ph.D. dissertation, University of California, Los Angeles, 137 pp.

——, 1976, Reconnaissance geologic map of the Eylar Mountain quadrangle, Santa Clara and Alameda Counties, California: *U.S. Geol. Survey Misc. Field Studies Map MF-764.*

Creely, R. S., 1965, Geology of the Oroville quadrangle, California: *Calif. Div. Mines and Geology Bull. 184,* p. 1–86.

Crickmay, C. H., 1933, The structural connection between the Coast Range of British Columbia and the Cascade Range of Washington: *Geol. Mag.*, v. 67, p. 482–491.

Crippen, R. A., Jr., 1951, Nephrite jade and associated rocks of the Cape San Martin region, Monterey County, California: *Calif. Div. Mines Spec. Report 10-A,* 14 pp.

Criscione, J. J., and others, 1978, The age and sedimentation/diagenesis for the Bedford Canyon Formation and the Santa Monica Formation in southern California: a Rb/Sr evaluation, *in* Howell, D. G., and McDougall, K. A., eds., *Mesozoic Paleogeography of the Western United States:* Pacific Section, Soc. Econ. Paleontologists and Mineralogists, Pacific Coast Paleogeography Symposium 2, p. 385–396.

Cross, T. A., 1973, Implications of igneous activity for the Early Cenozoic tectonic evolution of the western United States [abstract] : *Geol. Soc. America Abstracts with Programs,* v. 5, no. 7, p. 587.

——, and Pilger, R. H., 1977, Influence of "absolute" plate motion on Cenozoic igneous activity in the Cordillera [abstract] : *Geol. Soc. America Abstracts with Programs,* v. 9, no. 4, p. 406–407.

——, and Pilger, R. H., Jr., 1978, Constraints on absolute motion and plate interaction inferred from Cenozoic igneous activity in the western United States: *Am. Jour. Sci.*, v. 278, p. 865–902.

Crouch, J. K., 1977, Structure of the outer California continental borderland and a possible analogue in the region between the San Andreas and San Gabriel faults [abstract] : *Geol. Soc. America Abstracts with Programs,* v. 9, no. 4, p. 407.

——, 1978, Neogene tectonic evolution of the California Continental Borderland and western Transverse Ranges: *U.S. Geol. Survey Open-File Report 78-606,* 24 pp.

——, 1979, Neogene tectonic evolution of the California Continental Borderland and western Transverse Ranges: *Geol. Soc. America Bull.*, v. 90, p. 338–345.

Crowder, D. F., and Ross, D. C., 1970, Permian(?) to Jurassic(?) metavolcanic and related rocks that mark a major structural break in the northern White Mountains, California-Nevada: *U.S. Geol. Survey Prof. Paper 800-B*, p. B195–B203.

——, and Sheridan, M. F., 1972, Geologic map of the White Mountain Peak quadrangle, Mono County, California: *U.S. Geol. Survey Geol. Quad. Map GQ-1012*, scale 1:62,500.

——, and others, 1973, Granitic rocks of the White Mountains area, California-Nevada: Age and regional significance: *Geol. Soc. America Bull.*, v. 84, no. 1, p. 285–296.

Crowe, B. M., and others, 1976, Petrography and major element chemistry of the Santa Cruz Island volcanics, *in* Howell, D. G., *Aspects of the Geologic History of the California Continental Borderland:* Pacific Section, Am. Assoc. Petroleum Geologists Misc. Pub. 24, p. 196–215.

——, and others, 1979, Regional stratigraphy, K-Ar ages, and tectonic implications of Cenozoic volcanic rocks, southeastern California: *Am. Jour. Sci.*, v. 279, p. 186–216.

Crowell, J. C., 1952, Probable large lateral displacement on San Gabriel fault, southern California: *Am. Assoc. Petroleum Geologists Bull.*, v. 36, p. 2026–2035.

——, 1954, Strike-slip displacement of the San Gabriel fault, southern California, *in* Jahns, R. H., ed., *Geology of Southern California:* Calif. Div. Mines Bull. 170, Ch. 4, Contrib. 6, p. 49–52.

——, 1960, The San Andreas fault in southern California: *Report 21st Internat. Geol. Congress, Copenhagen,* part 18, p. 45–52.

——, 1962, Displacement along the San Andreas fault, California: *Geol. Soc. America Spec. Paper 71*, 61 pp.

——, 1973, Problems concerning the San Andreas fault system in southern California: *Stanford Univ. Pub. Geol. Sci.*, v. 13, p. 125–135.

——, 1974a, Sedimentation along the San Andreas fault, California, *in* Dott, R. H., Jr., and Shaver, R. H., eds., *Modern and Ancient Geosynclinal Sedimentation*: Soc. Econ. Paleontologists and Mineralogists Spec. Pub. 19, p. 292–303.

——, 1974b. Origin of late Cenozoic basins in southern California, *in* Dickinson, W. R., ed., *Tectonics and Sedimentation:* Soc. Econ. Paleontologists and Mineralogists Spec. Pub. 22, 204 pp.

——, 1975a, The San Andreas fault in southern California, p. 7–27 *in* Crowell, J. C., ed., *San Andreas Fault in Southern California:* Calif. Div. Mines and Geology Spec. Report 118, 272 pp.

——, 1975b, Geologic sketch of the Orocopia Mountains, southeastern California, p. 99–110 *in* Corwell, J. C., ed., *San Andreas Fault in Southern California:* Calif. Div. Mines and Geology Spec. Report 118, 272 pp.

——, 1975c, The San Gabriel Fault and Ridge Basin, p. 223–233 *in* Crowell, J. C., ed., *The San Andreas Fault in Southern California:* Calif. Div. Mines and Geology Spec. Report 118, 272 pp.

——, 1976, Implications of crustal stretching and shortening of coastal Ventura basin, California: *Pacific Section, Am. Assoc. Petroleum Geologists Misc. Publ. 24,* p. 365–382.

——, and Frakes, L. A., 1970, Phanerozoic glaciation and the causes of ice ages: *Am. Jour. Sci.*, v. 268, p. 193–224.

——, and Susuki, Takeo, 1959, Eocene stratigraphy and paleontology, Orocopia Mountains, southeastern California: *Geol. Soc. America Bull.*, v. 70, p. 581–592.

——, and Walker, J. W. R., 1962, Anorthosite and related rocks along the San Andreas fault, southern California: *Univ. California Pub. Geol. Sci.*, v. 40, no. 4, p. 219–288.

Curray, J. R., and Moore, D. G., 1974, Sedimentary and tectonic processes in the Bengal deep sea fan and geosyncline, *in* Burk, C. A., and Drake, C. L., eds., *Continental Margins:* New York, Springer-Verlag, p. 617–627.

——, and others, 1979, Tectonics of the Andaman Sea and Burma, *in* Watkins, J. S., and others, eds., *Geological and Geophysical Investigations of Continental Margins:* Am. Assoc. Petroleum Geologists Mem. 29, p. 189–198.

Curtis, G. H., and others, 1958, Age determination of some granitic rocks in California by the potassium-argon method: *Calif. Div. Mines and Geology Spec. Report 54,* 16 pp.

Cyamex Scientific Team, 1978, First submersible study of the East Pacific Rise: RITA (Riversa-Tamayo) Project 21°N [abstract]: *Trans. Am. Geophys. Union,* v. 59, no. 12, p. 1198.

d'Allura, J. A., and others, 1977, Paleozoic rocks of the northern Sierra Nevada: their structural and paleogeographic implications, *in* Stewart, J. H., and others, eds., *Paleozoic Paleogeography of the Western United States:* Pacific Section, Soc. Econ. Paleontologists and Mineralogists, Pacific Coast Paleogeography Symposium 1, p. 395–408.

Dalziel, I. W. D., and others, 1974, Fossil marginal basin in the southern Andes: *Nature,* v. 250, p. 291–294.

Danner, W. R., 1976, The Tethyan realm and the Paleozoic Tethyan province of western North America [abstract]: *Geol. Soc. America Abstracts with Programs,* v. 8, no. 6, p. 827.

Daviess, S. N., 1971, Barbados: a major submarine gravity slide: *Geol. Soc. America Bull.,* v. 82, p. 2593–2602.

Davis, G. A., 1966, Metamorphic and granitic history of the Klamath Mountains, *in* Bailey, E. H., ed., *Geology Northern California:* Calif. Div. Mines and Geology Bull. 190, p. 39–50.

Davis, G. A, Anderson, J. L, Frost, E. G, and Shackelford, T. J., 1979, Regional Miocene detachment faulting and early Tertiary (?) mylonitization, Whipple-Buckskin Rawhide Mountains, southeastern California and western Arizona, *in* Abbott, P. L (ed.), *Geological Excursions in the Southern California Area*: San Diego State Univ., p. 75–108.

——, 1968, Westward thrusting in the south-central Klamath Mountains, California: *Geol. Soc. America Bull.,* v. 79, p. 911–933.

——, 1969, Tectonic correlations, Klamath Mountains and western Sierra Nevada, California: *Geol. Soc. America Bull.,* v. 80, p. 1095–1108.

——, 1973, Relations between the Keystone and Red Springs thrust faults, eastern Spring Mountains, Nevada: *Geol. Soc. America Bull.,* v. 84, p. 3709–3716.

——, 1980, Problems of intraplate extensional tectonics, western United States, p. 84–95, *in Continental Tectonics:* National Research Council, Washington, D.C. 197 p.

——, and Burchfiel, B. C., 1973, Garlock fault: an intracontinental transform structure, southern California: *Geol. Soc. America Bull.*, v. 84, p. 1407–1422.

——, and Lipman, P. W., 1962, Revised structural sequence of pre-Cretaceous metamorphic rocks in the southern Klamath Mountains, California: *Geol. Soc. America Bull.*, v. 73, no. 12, p. 1547–1552.

——, and others, 1965, Structure, metamorphism, and plutonism in the south-central Klamath Mountains, California; *Geol. Soc. America Bull.*, v. 76, no. 8, p. 933–966.

——, and others, 1978, Mesozoic construction of the Cordilleran "collage," central British Columbia to central California, *in* Howell, D. G., and McDougall, K. A., eds., *Mesozoic Paleogeography of the Western United States:* Pacific Section, Soc. Econ. Paleontologists and Mineralogists, Pacific Coast Paleogeography Symposium 2, p. 1–32.

Davis, T. E., and Lass, G. L., 1976, Strontium isotopic composition of plutonic and volcanic rocks from the Jurassic Point Sal ophiolite [abstract] : *Trans. Am. Geophys. Union,* v. 57, p. 160.

Day, D., 1977, The petrology of a mafic dike complex, Yuba County, California: Unpublished M.S. thesis, University of California, Davis, 113 pp.

Department of Water Resources, State of California, The Resources Agency, 1970, Seawater intrusion: Pismo Guadalupe area; *Dept. Water Resources Bull. 63–3,* 76 pp.

Dewey, J. F., 1975, Finite plate implications: some implications for the evolution of rock masses at plate margins: *Am. Jour. Sci.*, v. 275-A, p. 260–284.

——, and Burke, Kevin, 1974, Hot spots and continental breakup: implications for collisional orogeny: *Geology,* v. 2, p. 57–60.

Dibblee, T. W., Jr., 1950, Geology of southwestern Santa Barbara County, California: *Calif. Div. Mines Bull. 150,* 95 pp.

——, 1964, Geologic map of the San Gorgonio Mountain quadrangle, San Bernardino and Riverside Counties, California. *U.S. Geol. Survey Misc. Geol. Inv. Map I–431.*

——, 1966, Geology of the Palo Alto quadrangle, Santa Clara and San Mateo counties, California: *Calif. Div. Mines and Geology Map Sheet 8.*

——, 1967, Areal geology of the western Mojave Desert, California: *U.S. Geol. Survey Prof. Paper 522,* 153 pp.

——, 1970, Regional geologic map of San Andreas and related faults in eastern San Gabriel Mountains, San Bernardino Mountains, western San Jacinto Mountains and vicinity: *U.S. Geol. Survey Open File Map,* scale 1:125,000.

——, 1973, Stratigraphy of the southern Coast Ranges near the San Andreas fault from Cholame to Maricopa, California: *U.S. Geol. Survey Prof. Paper 764,* 45 pp.

——, 1976, The Rinconada and related faults in the southern Coast Ranges, California, and their tectonic significance: *U.S. Geol. Survey Prof. Paper 981,* 55 pp.

——, 1977, Sedimentology and diastrophism during Oligocene time relative to the San Andreas fault system, *in* Nilsen, T. H., ed., *Late Mesozoic and Cenozoic Sedimentation and Tectonics in California:* San Joaquin Geol. Soc. Short Course, Bakersfield, Calif., p. 99–108.

——, 1978, Analysis of geologic-seismic hazards to Point Conception LNG Terminal Site: *County of Santa Barbara, California, March 1978, Report,* 70 pp.

Dick, H. J. B., 1973, K-Ar dating of intrusive rocks in the Josephine peridotite and Rogue Formation west of Cave Junction, southwestern Oregon [abstract] : *Geol. Soc. America Abstracts with Programs,* v. 5, no. 1, p. 33–34.

——, 1977, Partial melting in the Josephine peridotite, I, The effect on mineral composition and its consequence for geobarometry and geothermometry: *Am. Jour. Sci.*, v. 277, p. 801–832.

Dickinson, W. R., 1965, Tertiary stratigraphy of the Church Creek area, Montery County, California: *Calif. Div. Mines and Geology Spec. Report 86,* p. 25–44.

——, 1966a, Table Mountain serpentinite extrusion in California Coast Ranges; *Geol. Soc. America Bull.*, v. 77, p. 451–472.

——, 1966b, Structural relationships of San Andreas fault system, Cholame Valley and Castle Mountain Range, California: *Geol. Soc. America Bull.,* v. 77, p. 707–726.

——, 1970, Relations of andesite, granites, and derivative sandstones to arc-trench tectonics: *Rev. Geophys. Space Phys.*, v. 8, p. 813–860.

——, 1971, Clastic sedimentary sequences deposited in shelf, slope and trough settings between magmatic arcs and associated trenches: *Pacific Geology,* v. 3, p. 15–30.

——, 1972, Evidence for plate-tectonic regimes in the rock record: *Am. Jour. Sci.,* v. 272, p. 551–576.

——, 1974, Sedimentation within and beside ancient and modern magmatic arcs, *in* Dott, R. H., Jr., and Shaver, R. H., eds., *Modern and Ancient Geosynclinal Sedimentation:* Soc. Econ. Paleontologists and Mineralogists Spec. Pub. 19, p. 230–239.

——, 1975, Potash-depth (K-h) relations in continental margin and intraoceanic magmatic arcs: *Geology,* v. 3, p. 53–56.

——, 1976, Sedimentary basins developed during evolution of Mesozoic-Cenozoic arc-trench system in western North America: *Canadian Jour. Earth Sci.*, v. 13, p. 1268–1287.

——, 1977, Paleozoic plate tectonics and the evolution of the Cordilleran continental margin, *in* Stewart, J. H., and others, eds., *Paleozoic Paleogeography of the Western United Sates:* Pacific Section, Soc. Econ. Paleontologists and Mineralogists, Pacific Coast Paleogeography Symposium 1, p. 137–156.

——, and Rich, E. I., 1972, Petrologic intervals and petrofacies in the Great Valley Sequence, Sacramento Valley, California: *Geol. Soc. America Bull.,* v. 83, p. 3007–3024.

——, and Seely, D. R., 1979, Structure and stratigraphy of forearc regions: *Am. Assoc. Petroleum Geologists Bull.*, v. 63, no. 1, p. 2–31.

——, and Snyder, W. S., 1978, Plate tectonics of the Laramide Orogeny: *Geol. Soc. America Mem. 151,* 370 pp.

——, and Snyder, W. S., 1979, Geometry of triple junctions related to San Andreas transform: *Jour. Geophys. Res.*, v. 84, p. 561–572.

——, and Suczek, C. A., in press, Plate tectonics and sandstone compositions: *Jour. Sed. Petrology.*

——, and Thayer, T. P., 1978, Paleogeographic and paleotectonic implications of Mesozoic stratigraphy and structure in the John Day inlier of central Oregon, *in* Howell, D. G., and McDougall, K. A., eds., *Mesozoic Paleogeography of the Western United States:* Pacific Section, Soc. Econ. Paleontologists and Mineralogists, Pacific Coast Paleogeography Symposium 2, p. 147–161.

——, and others, 1969, Burial metamorphism of the late Mesozoic Great Valley Sequence, Cache Creek, California: *Geol. Soc. America Bull.*, v. 80, p. 519–526.

——, and others, 1972, Test of new global tectonics–discussion: *Am. Assoc. Petroleum Geologists Bull.*, v. 56, p. 375–384.

Diller, J. S., 1892, Geology of the Taylorsville region, California: *Geol. Soc. America Bull.*, v. 3, p. 370–394.

——, 1895, Description of the Lassen Peak sheet (California): *U.S. Geol. Survey Geol. Atlas, Folio 15,* 4 pp.

——, 1905, The Bragdon formation: *Am. Jour. Sci.,* v. 4, no. 19, p. 379–387.

——, 1906, Description of the Redding quadrangle (California): *U.S. Geol. Survey Geologic Atlas, Redding Folio, no. 138,* 14 pp.

——, 1908, Geology of the Taylorsville region, California: *U.S. Geol. Survey Bull. 353,* 128 pp.

——, and Kay, G. F., 1909, Mineral resources of the Grants Pass quadrangle and bordering districts, Oregon: *U.S. Geol. Survey Bull. 380,* p. 48–70.

Dillon, J. T., 1976, Geology of the Chocolate and Cargo Muchacho Mountains, southeasternmost California: Unpublished Ph.D. dissertation, University of California, Santa Barbara, 380 pp.

——, and Haxel, G., 1975, The Chocolate Mountain-Orocopia-Vincent thrust system as a tectonic element of late Mesozoic California [abstract] : *Geol. Soc. America Abstracts with Programs,* v. 7, p. 311–312.

Dodge, F. C. W., 1972a, Trace-element contents of some plutonic rocks of the Sierra Nevada batholith: *U.S. Geol. Survey Bull. 1314-F,* p. F1–13.

——, 1972b, Variation of ferrous-ferric ratios in the central Sierra Nevada batholith, U.S.A.: *34th Internat. Geol. Cong. Proc., Sec. 10, Montreal,* 1972, p. 12–19.

Doe, B. R., and Delavaux, M. H., 1973, Variations in lead-isotopic compositions in Mesozoic granitic rocks of California: a preliminary investigation: *Geol. Soc. America Bull.,* v. 84, p. 3513–3526.

Doerner, D. P., 1969, Lower Tertiary biostratigraphy of southwestern Santa Cruz Island (California), *in* Weaver, D. W., and others, eds., *Geology of the Northern Channel Islands:* Pacific Sections, Am. Assoc. Petroleum Geologists and Soc. Econ. Paleontologists and Mineralogists, Special Pub., p. 17–29.

Dott, R. H., Jr., 1971, Geology of the southwestern Oregon coast west of the 124th meridian: *Oregon Dept. Geology and Mineral Industries Bull. 69,* 63 pp.

Douglass, R. C., 1967, Permian Tethyan fusulinids from California: *U.S. Geol. Survey Prof. Paper 583-A,* p. 7–43.

Drewes, Harald, 1959, Turtleback faults of Death Valley, California: a reinterpretation: *Geol. Soc. America Bull.,* v. 70, p. 1497–1508.

——, 1971, Mesozoic stratigraphy of the Santa Rita Mountains, southeast of Tucson, Arizona: *U.S. Geol. Survey Prof. Paper 658-C,* 81 pp.

——, 1976, Plutonic rocks of the Santa Rita Mountains, southeast of Tucson, Arizona: *U.S. Geol. Survey Prof. Paper 915,* 75 pp.

——, 1978, The Cordilleran orogenic belt between Nevada and Chihuahua: *Geol. Soc. America Bull.,* v. 89, no. 5, p. 641–657.

Duffield, W. A., and Sharp, R. V., 1975, Geology of the Sierra foothills melange and adjacent areas, Amador County, California: *U.S.Geol. Survey Prof. Paper 827,* 30 pp.

Dunne, G. C., 1977, Geology and structural evolution of Old Dad Mountain, Mojave Desert, California: *Geol. Soc. America Bull.,* v. 88, p. 737–748.

——, and Gulliver, R. M., 1976, Superposed synbatholithic deformations in eastern wallrocks, Sierra Nevada batholith, California [abstract] : *Geol. Soc. America Abstracts with Programs,* v. 8, no. 6, p. 846.

——, and others, 1978, Mesozoic evolution of rocks of the White, Inyo, Argus and Slate ranges, eastern California, *in* Howell, D. G., and McDougall, K. A., eds., *Mesozoic Paleogeography of the Western United States:* Pacific Section, Soc. Econ. Paleontologists and Mineralogists, Pacific Coast Paleogeography Symposium 2, p. 189–208.

——, and others, 1975, The Bean Canyon Formation of the Tehachapi Mountains, California: An early Mesozoic arc-trench gap deposit? [abstract] : *Geol. Soc. America Abstracts with Programs,* v. 7, no. 3, p. 314.

Durrell, Cordell, 1940, Metamorphism in the southern Sierra Nevada northeast of Visalia, California: *Univ. California Publs. Bull. Dept. Geol. Sci.,* v. 25, no. 1, p. 1–118.

——, and d'Allura, Jad, 1977, Upper Paleozoic section in eastern Plumas and Sierra Counties, northern Sierra Nevada, California: *Geol. Soc. America Bull.,* v. 88, p. 844–852.

——, and Proctor, P. D., 1948, Iron-ore deposits near Lake Hawley and Spencer Lakes, Sierra County, California: *Calif. Div. Mines Bull. 129,* part L, p. 165–192.

Easton, W. H., and Imlay, R. W., 1955, Upper Jurassic fossil localities in Franciscan and Knoxville Formations in southern California: *Am. Assoc. Petroleum Geologists Bull.,* v. 39, p. 2336–2340.

Eaton, J. P., 1966, Crustal structure in southern and central California from seismic evidence: in Geology of Northern California, ed: E. H. Bailey, p. 419–426, *California Division Mines and Geology, Bull.* 190, 508 pp.

Eaton, J. P., in press, Temporal variation in the pattern of seismicity in central California: *Proc. UNESCO Internat. Symposium on Earthquake Prediction* [1979].

——, and others, 1970, Aftershocks of the 1966 Parkfield-Cholame earthquake: A detailed study: *Seismol. Soc. America Bull.,* v. 60, p. 1151–1197.

Eberly, L. D., and Stanley, T. B., Jr., 1978, Cenozoic stratigraphy and geologic history of southwestern Arizona: *Geol. Soc. America Bull.,* v. 89, p. 901–920.

Echeverria, L. M., 1977, Oceanic basalt magmas in clasted wedges [abstract] : *Geol. Soc. America Abstracts with Programs,* v. 9, p. 963.

Ehlert, K. W., and Ehlig, P. L., 1977, The "polka-dot" granite and the rate of displacement on the San Andreas fault in southern California [abstract] : *Geol. Soc. America Abstracts with Programs,* v. 9, p. 415–416.

Ehlig, P. L., 1958, Geology of the Mount Baldy region of the San Gabriel Mountains, California: Unpublished Ph.D. dissertation, University of California, Los Angeles, 153 pp.

——, 1968, Causes of distribution of Pelona, Rand, and Orocopia Schist along the San Andreas and Garlock faults, *in* Dickinson, W. R., and Grantz, Arthur, eds., *Proceedings of the Conference on Geologic Problems of San Andreas Fault System:* Stanford Univ. Pub. in Geol. Sci., v. 11, p. 294–305.

——, 1975, Basement rocks of the San Gabriel Mountains, south of the San Andreas fault, southern California: *Calif. Div. Mines and Geology Spec. Report 118,* p. 177–186.

——, and Ehlert, K. W., 1972, Offset of Miocene Mint Canyon Formation from volcanic source along San Andreas Fault, southern California [abstract] : *Geol. Soc. America Abstracts with Programs,* v. 4, no. 3, p. 154.

——, and Joseph, S. E., 1977, Polka dot granite and correlation of La Panza quartz monzonite with Cretaceous batholithic rocks north of Salton Trough, *in* Howell, D. G., and others, eds., *Cretaceous Geology of the California Coast Ranges West of the San Andreas Fault:* Pacific Section, Soc. Econ. Paleontologists and Mineralogists, Pacific Coast Paleogeography Field Guide 2, p. 91–96.

——, and others, 1975a, Tectonic implications of the cooling ages of the Pelona schist [abstract] : *Geol. Soc. America Abstracts with Programs,* v. 7, no. 3, p. 314–315.

——, and others, 1975b, Offset of Upper Miocene Caliente and Mint Canyon Formations along the San Gabriel and San Andreas faults, *in* Crowell, J. C., ed., *San Andreas Fault in Southern California:* Calif. Div. Mines and Geology Spec. Report 118, p. 83–92.

Ehrenberg, S. M., 1975, Feather River ultramafic body, northern Sierra Nevada, California: *Geol. Soc. America Bull.,* v. 86, p. 1235–1243.

Elders, W. A., and others, 1972, Crustal spreading in southern California: *Science,* v. 178, p. 15–24.

Elliott, M. A., and Bostwick, D. A., 1973, Occurrence of *Yabeina* in the Klamath Mountains, Siskiyou County, California [abstract] : *Geol. Soc. America Abstracts with Programs,* v. 5, no. 1, p. 38.

Emery, K. O., and Shepard, F. P., 1945, Lithology of the sea floor off southern California: *Geol. Soc. America Bull.,* v. 56, p. 431–478.

Engel, A. E. J., 1963, Geologic evolution of North America: *Science,* v. 140, p. 143–152.

Enos, P., 1963, Jurassic age of Franciscan Formation south of Panoche Pass, California: *Am. Assoc. Petroleum Geologists Bull.,* v. 47, no. 1, p. 158–163.

——, 1965, Geology of the Western Vallecitos syncline, San Benito County, California: *Calif. Div. Mines and Geology Map Sheet 5,* scale 1:31,680.

Eric, J. H., and others, 1955, Geology and mineral deposits of the Angels Camp and Sonora quadrangles, Calaveras and Tuolumne Counties, California: *Calif. Div. Mines and Geology Spec. Report 41,* 55 pp.

Erickson, J. W., 1975, Sedimentology of the South Point Formation (Eocene), Santa Rosa Island, California, *in* Weaver, D. W., and others, eds., *Paleogene Symposium:* Pacific Section, Soc. Econ. Paleontologists and Mineralogists, p. 169–190.

Ernst, W. G., 1965, Mineral parageneses of Franciscan metamorphic rocks, Panoche Pass, California: *Geol. Soc. America Bull.,* v. 76, p. 879–914.

——, 1970, Tectonic contact between the Franciscan melange and the Great Valley Sequence, crustal expression of a Late Mesozoic Benioff Zone: *Jour. Geophys. Res.,* v. 75, p. 886–902.

——, 1971a, Do mineral parageneses reflect unusually high-pressure conditions of Franciscan metamorphism?: *Am. Jour. Sci.,* v. 270, p. 81–108.

——, 1971b, Petrologic reconnaissance of Franciscan metagraywackes from the Diablo Range, central California Coast Ranges: *Jour. Petrology,* v. 12, no. 2, p. 413–437.

——, 1971c, Metamorphic zonations on presumably subducted lithospheric plates from Japan, California, and the Alps: *Contrib. Mineralogy Petrology,* v. 34, p. 43–59.

——, 1973, Blueschist metamorphism and P-T regimes in active subduction zones; *Tectonophysics,* v. 17, p. 255–272.

——, 1975, Systematics of large-scale tectonics and age progressions in Alpine and circum-Pacific blueschist belts: *Tectonophysics,* v. 26, p. 229–246.

——, 1977, Mineral parageneses and plate tectonic settings of relatively high-pressure metamorphic belts: *Fortschrift für Mineralogie,* v. 54, p. 192–222.

——, and Hall, C. A., 1974, Geology and petrology of the Cambria Felsite, a new Oligocene formation, west-central California Coast Ranges: *Geol. Soc. America Bull.*, v. 85, p. 523–532.

——, and others, 1970, Comparative study of low-grade metamorphism in the California Coast Ranges and the outer metamorphic belt of Japan: *Geol. Soc. America Mem. 124*, 276 pp.

Evarts, R. C., 1977, The geology and petrology of the Del Puerto ophiolite, Diablo Range, central California Coast Ranges, *in* Coleman, R. G., and Irwin, W. P., eds., *North American Ophiolites:* Oregon Department of Geology and Mineral Industries Bull. 95, p. 121–140.

——, 1978, The Del Puerto ophiolite: structural and petrologic evolution: Unpublished Ph.D. dissertation, Stanford University, Stanford, Calif.

Evenson, W. A., 1973, Geology of the southern Kilbeck Hills and an adjacent portion of the Old Woman Mountains, eastern Mojave Desert, San Bernardino County, California: Unpublished M.S. thesis, University of Southern California, Los Angeles, 51 pp.

Evernden, J. F., and Kistler, R. W., 1970, Chronology of emplacement of Mesozoic batholithic complexes in California and western Nevada: *U.S. Geol. Survey Prof. Paper 623*, 42 pp.

Evitt, W. R., and Pierce, S. T., 1975, Early Tertiary age from the coastal belt of the Franciscan complex, northern California: *Geology*, v. 3, p. 433–436.

Ewing, A. H., 1976, Depositional and post-depositional features in cumulate rocks, Vourinos ophiolite complex, northern Greece [abstract]: *Trans. Am. Geophys. Union*, v. 57, no. 12, p. 1027.

Fairbanks, H. W., 1896, The geology of Point Sal: *Univ. Calif. Dept. Geology Bull.*, v. 2, no. 1, p. 1–92.

——, 1904, San Luis, California: *U.S. Geol. Survey Geol. Atlas*, Folio 101.

Farley, T., and Ehlig, P. L., 1977, Displacement on the Punchbowl fault based on occurrence of "polka-dot" granite clasts [abstract]: *Geol. Soc. America Abstracts with Programs*, v. 9, p. 419.

Fedotov, S. A., 1965, Regularities in the distribution of strong earthquakes in Kamchatka, the Kurile Islands, and northeastern Japan [in Russian]: *Akad. Nauk SSSR Inst. Fiziki Zemli Trudy*, v. 36, p. 36.

Ferguson, H. G., and others, 1951a, Geology of the Winnemucca quadrangle, Nevada: *U.S. Geol. Survey Geol. Quad. Map GQ-11.*

——, and others, 1951b, Geology of the Mount Moses quadrangle, Nevada: *U.S. Geol. Survey Geol. Quad. Map GQ-12.*

——, and others, 1952, Geology of the Golconda quadrangle, Nevada: *U.S. Geol. Survey Geol. Quad. Map GQ-15.*

Fife, D. L., 1968, Geology of the Bahia Santa Rosalia quadrangle, Baja California, Mexico: Unpublished M.S. thesis, San Diego State University, San Diego, Calif., 100 pp.

——, and others, 1967, Late Jurassic age of the Santiago Peak Volcanics, California: *Geol. Soc. America Bull.*, v. 78, p. 299–303.

Finch, J. W., and Abbott, P. J., 1977, Petrology of a Triassic marine section Vizcaino Peninsula, Baja California Sur, Mexico: *Sedimentary Geology*, v. 19, p. 253–273.

Fischer, P. J., 1976, Late Neogene-Quaternary tectonics and depositional environments of the Santa Barbara Basin, California, *in* Fritsche, A. E., and others, eds., *The Neogene Symposium:* Pacific Section, Soc. Econ. Paleontologists and Mineralogists, p. 33–52.

Fiske, R. S., and Tobisch, O. T., 1978, Paleogeographic significance of volcanic rocks of the Ritter Range pendant, central Sierra Nevada, California, *in* Howell, D. G., and McDougall, K. A., eds., *Mesozoic Paleogeography of the Western United States:* Pacific Section, Soc. Econ. Paleontologists and Mineralogists, Pacific Coast Paleogeography Symposium 2, p. 209–222.

——, and others, 1963, Geology of Mount Rainier National Park, Washington: *U.S. Geol. Survey Prof. Paper 444,* 93 pp.

——, and others, 1977, Minarets Caldera: A Cretaceous volcanic center in the Ritter Range pendant, central Sierra Nevada, California [abstract]: *Geol. Soc. America Abstracts with Programs,* v. 9, no. 7, p. 975.

Fitch, T. J., 1970, Earthquake mechanisms in the Himalayan, Burmese, and Andaman regions and continental tectonics in Central Asia: *Jour. Geophys. Res.,* v. 75, p. 2699–2709.

——, 1972, Plate convergence, transcurrent faults, and internal deformation adjacent to southeast Asia and the western Pacific: *Jour. Geophys. Res.,* v. 77, p. 4432–4460.

Fleck, R. J., 1967, Structural significance of the contact between Franciscan and Cenozoic rocks—southern San Francisco Peninsula, California: Unpublished M.S. thesis, Stanford University, Stanford, Calif.

——, 1970, Tectonic style, magnitude, and age of deformation in the Sevier orogenic belt in southern Nevada and eastern California: *Geol. Soc. America Bull.,* v. 81, p. 1705–1720.

Fletcher, G. L., 1967, Post Late Miocene displacement along the San Andreas fault zone, central California, in *Guidebook, Gabilan Range and Adjacent San Andreas Fault:* Pacific Section, Am. Assoc. Petroleum Geologists, p. 74–80.

Freund, R., 1971, The Hope fault—a strike slip fault in New Zealand: *New Zealand Geol. Survey Bull.,* n.s. 86, 49 pp.

Fritz, D. M., 1974, Potassium-argon ages, chemistry, and structure across the Coast Range fault zone, west of Paskenta, northern California [abstract]: *Geol. Soc. America Abstracts with Programs,* v. 6, no. 7, p. 745.

——, 1975, Ophiolite belt west of Paskenta, northern California Coast Range: Unpublished M.S. thesis, University of Texas at Austin, 63 pp.

Fyfe, W. S., and Zardini, R., 1967, Metaconglomerates in the Franciscan Formation near Pacheco Pass, California: *Am. Jour. Sci.,* v. 265, p. 819–830.

Gabrielse, Hubert, 1972, Younger Precambrian of the Canadian Cordillera: *Am. Jour. Sci.,* v. 272, p. 521–536.

Galehouse, J. S., 1967, Provenance and paleocurrents of the Paso Robles Formation, California: *Geol. Soc. America Bull.,* v. 78, p. 951–978.

Garcia, M. O., 1976, Rogue River island arc complex, Western Jurassic belt, Klamath Mountains, Oregon [abstract]: *Geol. Soc. America Abstracts with Programs,* v. 8, no. 3, p. 375.

——, 1978, Criteria for the identification of ancient volcanic arcs: *Earth Sci. Reviews,* v. 14, p. 147–165.

Garfunkel, Z., 1973, History of the San Andreas fault as a plate boundary: *Geol. Soc. America Bull.,* v. 84, p. 2409–2430.

——, 1974, Model for the late Cenozoic tectonic history of the Mojave Desert, California, and for its relation to adjacent regions: *Geol. Soc. America Bull.,* v. 85, p. 1931–1944.

Gastil, R. G., 1977, Subduction, accretion, and batholithic emplacement, in *Plutonism in*

Relation to Volcanism and Metamorphism: Internat. Geol. Correlations Program Circum-Pacific Plutonism Project, 7th Meeting, Toyawa, Japan.

——, and Allison, E. C., 1966, An Upper Cretaceous fault-line coast [abstract] : *Am. Assoc. Petroleum Geologists Bull.,* v. 50, p. 647–648.

——, and Krummenacher, Daniel, 1974, Reconnaissance geologic map of coastal Sonora: *Geol. Soc. America Map and Chart Series, MC–16.*

——, and Krummenacher, Daniel, 1977, Reconnaissance geology of coastal Sonora between Puerto Lobos and Bahia Kino: *Geol. Soc. America Bull.,* v. 88, p. 189–198.

——, and Phillips, R. P., 1974, Neogene tectonic evolution of the Salinian Block, west-central California: Comment: *Geology,* v. 2, p. 391.

——, and Rowley, G. M., 1978, Subduction, accretion and batholith emplacement [abstract] : *Geol. Soc. America Abstracts with Programs,* v. 10, p. 106–107.

——, and others, 1972, The reconstruction of Mesozoic California: *Proc. 24th Internat. Geol. Congress, Section 3,* p. 217–229.

——, and others, 1973, Permian fusulinids from near San Felipe, Baja California: *Am. Assoc. Petroleum Geologists Bull.,* v. 57, p. 746–747.

——, and others, 1974, The west Mexico batholith belt: *Pacific Geology,* v. 8, p. 73–78.

——, and others, 1975, Reconnaissance geology of the State of Baja California: *Geol. Soc. America Mem. 140,* 170 pp.

——, and others, 1976, La zona batolitica del sur de California y el occidente de Mexico: *Bol. Soc. Geol. Mexicana* XXXVII, p. 84–90.

——, and others, 1978, Mesozoic history of peninsular California and related areas east of the Gulf of California and related areas east of the Gulf of California, *in* Howell, D. G., and McDougall, K. A., eds., *Mesozoic Paleogeography of the Western United States:* Pacific Section, Soc. Econ. Paleontologists and Mineralogists, Pacific Coast Paleogeography Symposium 2, p. 107–116.

Gastil, R. G., and others, 1979a, Reconnaissance geologic map of the central part of the state of Nayarit, Mexico: *Geological Society of America Map and Chart Series,* no. MC–24, scale 1:200,000, summary: *Geologic Society of America Bulletin,* v. 80, p. 15–18.

Gastil, R. G., and others, 1979b, The record of Cenozoic volcanism around the Gulf of California: *Geological Society of America Bulletin Part 1,* v. 90, p. 819–857.

Gawthrop, W., 1975, Seismicity of the central California coastal region: *U.S. Geol. Survey Open-File Report 75–134,* 87 pp.

——, 1978, Seismicity and tectonics of the central California coastal region: *Calif. Div. Mines and Geology Spec. Report 137,* p. 33–56.

Gealy, W. K., 1951, Geology of the Healdsburg quadrangle, California: *Calif. Div. Mines Bull. 161,* 50 pp.

George, R. P., Jr., 1978, Structural petrology of the Olympus ultramafic complex in the Troodos ophiolite, Cyprus: *Geol. Soc. America Bull.,* v. 89, p. 845–865.

Ghent, E. D., 1965, Glaucophane-schist facies metamorphism in the Black Butte area, northern Coast Ranges, California: *Am. Jour. Sci.,* v. 263, no. 5, p. 385–400.

Gibson, R. C., 1971, Non-marine turbidites and the San Andreas Fault, San Bernardino Mountains, California, *in* Elders, W. A., ed., *Geological Excursions in Southern California:* University of California, Riverside, Campus Museum Pubs., no. 1, 181 pp.

Gilbert, W. G., 1973, Franciscan rocks near Sur fault zone, northern Santa Lucia Range, California: *Geol. Soc. America Bull.,* v. 84, p. 3317–3328.

——, 1974, Franciscan rocks near Sur fault zone, northern Santa Lucia Range, California: Reply (to discussion by L. A. Raymond): *Geol. Soc. America Bull.,* v. 85, p. 1826.

——, and Dickinson, W. R., 1970, Stratigraphic variations in sandstone petrology, Great Valley Sequence, central California Coast: *Geol. Soc. America Bull.,* v. 81, p. 949–954.

Gilluly, J., 1956, General geology of central Cochise County, Arizona: *U.S. Geol. Survey Prof. Paper 281,* 169 pp.

——, 1967, Geologic map of the Winnemucca quadrangle, Pershing and Humboldt Counties, Nevada: *U.S. Geol. Survey Geol. Quad. Map GQ-656.*

——, and Gates, O., 1965, Tectonic and igneous geology of the northern Shoshone Range, Nevada: *U.S. Geol. Survey Prof. Paper 465,* 53 pp.

Girty, G. H., and Schweickert, R. A., 1979, Preliminary results of a detailed study of the "lower" Shoo Fly, Bowman Lake, northern Sierra Nevada, California [abstract]: *Geol. Soc. America Abstracts with Programs,* v. 11, p. 79.

Gobbett, D. J., 1972, Geological map of the Malay Peninsula: Kuala Lumpur, *Geol. Soc. Malaysia,* scale 1:1,000,000.

——, 1973, Permian Fusulinacea, *in* Hallam, A., ed., *Atlas of Paleobiogeography:* Amsterdam, Elsevier, p. 151–158.

Goff, F. E., and McLaughlin, R. J., 1976, Geology of the Cobb Mountain-Ford Flat geothermal area, Lake County, California: *U.S. Geol. Survey Open File Map 76-221.*

Gordon, S. A., 1979, Relations between the Santa Ynez fault zone and the Pine Mountain thrust fault system, Piru Mountain, California [abstract]: *Geol. Soc. America with Programs,* v. 11, p. 80.

Goulty, N. R., and Gilman, R., 1978, Repeated creep events on the San Andreas fault near Parkfield, California, recorded by a strainmeter array: *Jour. Geophys. Res.,* v. 83, p. 5415–5419.

——, and others, 1978, Large creep events on the Imperial fault, California: *Seismol. Soc. America Bull.,* v. 68, p. 517–521.

Graham, C. M., 1975, Inverted metamorphic zonation and mineralogy of the Pelona Schist, Sierra Pelona, Transverse Ranges [abstract]: *Geol. Soc. America Abstracts with Programs,* v. 7, p. 321–322.

——, and England, P. C., 1976, Thermal regimes and regional metamorphism in the vicinity of overthrust fault—An example of shear heating and inverted metamorphic zonation from southern California: *Earth Planet. Sci. Lett.,* v. 31, p. 142–152.

Graham, S. A., 1977, Apparent offsets of on-land geologic features across the San Gregorio-Hosgri fault trend [abstract]: *Geol. Soc. America Abstracts with Programs,* v. 9, no. 4, p. 424.

——, 1978, Role of the Salinian Block in the evolution of the San Andreas fault system: *Am. Assoc. Petroleum Geologists Bull.,* v. 62, p. 2214–2231.

——, and Dickinson, W. R., 1978a, Evidence for 115 kilometers of right slip on the San Gregorio-Hosgri fault trend: *Science,* v. 199, p. 179–181.

——, and Dickinson, W. R., 1978b, Apparent offsets of on-land geologic features across the San Gregorio-Hosgri fault trend: *Calif. Div. Mines and Geology Spec. Report 137,* p. 13–24.

——, and others, 1975, Himalayan-Bengal model for flysch dispersal in the Appalachian-Ouachita system: *Geol. Soc. America Bull.,* v. 86, p. 273–286.

Grannell, R. B., 1971, A regional gravity survey of the San Gabriel Mountains, California [abstract] : *Geol. Soc. America Abstracts with Programs,* v. 3, p. 127.

Green, D. H., and Ringwood, A. E., 1967, The stability fields of aluminous pyroxene peridotite and garnet peridotite and their relevance in upper mantle structure: *Earth Planet. Sci. Lett.,* v. 3, p. 151–160.

Greene, H. G., 1976, Late Cenozoic geology of the Ventura basin, California, *in* Howell, D. G., ed., *Aspects of the Geologic History of the California Continental Borderland:* Am. Assoc. Petroleum Geologists Misc. Pub. 24, p. 499–529.

——, 1977, Geology of the Monterey Bay region: *U.S. Geol. Survey Open-File Report 77–718,* 347 pp.

Greenhaus, M. R., and Cox, A., 1979, Paleomagnetism of the Morro Rock-Islay Hill complex as evidence for crustal block rotations in central coastal California: *Jour. Geophys. Res.,* v. 84, p. 2393–2400.

Griffin, W. L., 1967, Provenance, deposition, and deformation of the San Benito Gravels, *in* Marks, J. G., chairman, *Gabilan Range and Adjacent San Andreas Fault:* Pacific Sections, Am. Assoc. Petroleum Geologists and Soc. Econ. Paleontologists and Mineralogists, Field Trip Guidebook, p. 61–73.

Grose, L. T., 1959, Structure and petrology of the northeast part of the Soda Mountains, San Bernardino County, California: *Geol. Soc. America Bull.,* v. 70, p. 1509–1548.

Grow, J. A., 1973, Implications of deep sea drilling, Sites 186 and 187 on island arc structure, *in* Creager, J. S., and others, eds., *Initial Reports of the Deep Sea Drilling Project,* v. 19: Washington, D.C., U.S. Government Printing Office, p. 799–801.

——, and Bowin, C. O., 1975, Evidence for high-density crust and mantle beneath the Chile Trench due to the descending lithosphere: *Jour. Geophys. Res.,* v. 80, p. 1449–1458.

Gucwa, P. R., 1975, Middle to Late Cretaceous sedimentary melange, Franciscan Complex, northern California: *Geology,* v. 3, p. 105–108.

Gulliver, R. M., 1976, Regional deformations distinguished by cross-cutting relationships in the Talc City Hills, eastern California [abstract] : *Geol. Soc. America Abstracts with Programs,* v. 8, no. 6, p. 896.

Gutenberg, B., and Richter, C. F., 1954, *Seismicity of the Earth and Associated Phenomena:* Princeton, N.J., Princeton University, 310 pp.

Hadley, David, and Kanamori, Hiroo, 1977, Seismic structure of the Transverse Ranges, California: *Geol. Soc. America Bull.,* v. 88, p. 1469–1478.

Hall, C. A., 1973a, Geology of the Arroyo Grande quadrangle, California: *Calif. Div. Mines and Geology Map Sheet 24.*

——, 1973b, Geologic map of the Morro Bay South and Port San Luis quadrangles, San Luis Obispo County, California: *U.S. Geol. Survey Misc. Field Studies Map MF–511.*

——, 1974, Geologic map of the Cambria region, San Luis Obispo County, California: *U.S. Geol. Survey Misc. Field Studies Map MF–599.*

——, 1975, San Simeon-Hosgri fault system, coastal California: economic and environmental implications: *Science,* v. 190, p. 1291–1294.

——, 1976, Geologic map of the San Simeon-Piedras Blancas region, San Luis Obispo County, California: *U.S. Geol. Survey Misc. Field Studies Map MF–784.*

——, 1977, Origin and development of the Lompoc-Santa Maria pull-apart basin and its relation to the San Simeon-Hosgri fault, California [abstract] : *Geol. Soc. America Abstracts with Programs,* v. 9, no. 4, p. 428.

——, 1978a, Origin and development of the Lompoc-Santa Maria pull-apart basin and its relation to the San Simeon-Hosgri strike-slip fault, western California: *Calif. Div. Mines and Geology Spec. Report 137,* p. 25–31.

——, 1978b, Geologic map of Twitchell Dam, parts of Santa Maria and Tepusquet quadrangles, Santa Barbara County, California: *U.S. Geol. Survey Misc. Field Studies Map MF 933,* 2 sheets, scale 1:24,000.

——, and Corbató, C. E., 1967, Stratigraphy and structure of Mesozoic and Cenozoic rocks, Nipomo quadrangle, southern Coast Ranges, California; *Geol. Soc. America Bull.,* v. 78, p. 559–582.

——, and Prior, S. W., 1975, Geologic map of the Cayucas-San Luis Obispo region, San Luis Obispo County, California: *U.S. Geol. Survey Misc. Field Studies Map MF–686.*

——, and others, 1966, Potassium-argon age of the Obispo Formation with *Pecten lompocensis* Arnold, southern Coast Ranges, California: *Geol. Soc. America Bull.,* v. 77, p. 443–445.

Hall, J. M., and Robinson, P. T., 1979, Deep crustal drilling in the North Atlantic Ocean: *Science,* v. 204, no. 4393, p. 573–586.

Hall, N. T., and Sieh, K. E., 1977, Late Holocene rate of slip on the San Andreas fault in the northern Carrizo Plain, San Luis Obispo County, California [abstract] : *Geol. Soc. America Abstracts with Programs,* v. 9, p. 428.

Hamilton, D. L., 1977, Neogene and Holocene slip along northwest-trending faults of the central California region: Appendix 41A–1 to FSAR for Diablo Canyon Nuclear Power Plant, AEC Docket Nos. 50–275 and 50–323.

——, and Willingham, C. R., 1977, Hosgri fault zone: structure, amount of displacement and relationship to structures of the western Transverse Ranges [abstract] : *Geol. Soc. America Abstracts with Programs,* v. 9, no. 4, p. 429.

——, and Willingham, C. R., 1978, Evidence for a maximum of 20 km of Neogene right slip along the San Gregorio fault zone of central California [abstract] : *Trans. Am. Geophys. Union,* v. 59, no. 12, p. 1210.

Hamilton, R. M., 1972, Aftershocks of the Borrego Mountain earthquake from April 12 to June 12, 1968: *U.S. Geol. Survey Prof. Paper 787,* p. 31–54.

Hamilton, Warren, 1961, The origin of the Gulf of California: *Geol. Soc. America Bull.,* v. 72, p. 1307–1318.

——, 1963, Metamorphism in the Riggins region, western Idaho: *U.S. Geol. Survey Prof. Paper 436,* 95 pp.

——, 1964a, Geologic map of the Big Maria Mountains NE Quadrangle, Riverside County, California and Yuma County, Arizona: *U.S. Geol. Survey Geol. Quad. Map GQ–350,* scale 1:24,000.

——, 1964b, Nappes in southeastern California [abstract] : *Geol. Soc. America Spec. Paper 75,* p. 274.

——, 1969, Mesozoic California and the underflow of the Pacific mantle: *Geol. Soc. America Bull.,* v. 80, p. 2409–2430.

——, 1971, Tectonic framework of southeastern California [abstract] : *Geol. Soc. America Abstracts with Programs,* v. 3, no. 2, p. 130–131.

——, 1973, Tectonics of the Indonesian region: *Geol. Soc. Malaysia Bull.,* v. 6, p. 3–10.

——, 1977, Subduction in the Indonesian region, *in* Talwani, M., and Pitman, W. C., III, eds., *Island Arcs, Deep Sea Trenches, and Backarc Basins:* Am. Geophys. Union, Maurice Ewing Ser. 1, p. 15–31.

——, 1978a, Mesozoic tectonics of the western United States, *in* Howell, D. G., and McDougall, K. A., eds., *Mesozoic Paleogeography of the Western United States:* Pacific Section, Soc. Econ. Paleontologists and Mineralogists, Pacific Coast Paleogeography Symposium 2, p. 33–70.

——, 1978b, Tectonic map of the Indonesian region: *U.S. Geol. Survey Map 1-875-D.*

——, and Myers, W. B., 1966, Cenozoic tectonics of the western United States: *Rev. Geophys.,* v. 4, p. 509–549.

——, and Myers, W. B., 1967, The nature of batholiths: *U.S. Geol. Survey Prof. Paper 554-C.*

——, and Myers, W. B., 1968, Cenozoic tectonic relationships between the western United States and the Pacific Basin, *in* Dickinson, W. R., and Grantz, Arthur, eds., *Proceedings of Conference on Geologic Problems of San Andreas Fault System:* Stanford Univ. Pubs. Geol. Sci., v. 11, p. 342–357.

Hansen, Edward, 1967, Methods of deducing slip-line orientations from the geometry of folds: *Carnegie Inst. Washington Year Book 65,* p. 387–405.

Harding, T. P., 1974, Petroleum traps associated with wrench faults: *Am. Assoc. Petroleum Geologists Bull.,* v. 58, 1290–1304.

——, 1976, Tectonic significance and hydrocarbon trapping consequences of sequential folding synchronous with San Andreas faulting, San Joaquin Valley, California: *Am. Assoc. Petroleum Geologists Bull.,* v. 60, p. 356–378.

Hardy, L. R., 1972, Geology of an allochthonous Jurassic sequence in the Sierra de Santa Rosa, northwest Sonora, Mexico: Unpublished M.S. thesis, San Diego State University, San Diego, Calif., 92 pp.

Harland, W. B., 1971, Tectonic transpression in Caledonian Spitsbergen: *Geol. Mag.,* v. 108, p. 27–42.

Harper, G. D., 1978, Preliminary report on the western Jurassic belt, Klamath Mountains, vicinity of the Smith River, northwestern California [abstract] *Geol. Soc. America Abstracts with Programs,* v. 10, no. 3, p. 108.

——, 1979, "Anomalous" ophiolite underlying Late Jurassic metasedimentary rocks of the Galice Formation, western Jurassic belt, northwestern California [abstract]: *Geol. Soc. America Abstracts with Programs,* v. 11, no. 3, p. 82.

Hart, E. W., 1959, Geology of limestone and dolomite deposits in the south half of the Standard 7½-minute quadrangle, Tuolumne County, California: *Calif. Div. Mines and Geology Spec. Report 58,* 25 pp.

——, 1977, Fault hazard zones in California: *Calif. Div. Mines and Geology Spec. Pub. 72,* 24 pp.

Hawkins, J. W., Jr., 1970, Petrology and possible tectonic significance of late Cenozoic volcanic rocks, southern California and Baja California: *Geol. Soc. America Bull.,* v. 81, no.11, p. 3323–3338.

——, 1977, Petrologic and geochemical characteristics of marginal basin basalta, *in* Talwani, M., and Pitman, W. C., III, eds., *Island Arcs, Deep Sea Trenches, and Back-arc Basins:* Am . Geophys. Union, Maurice Ewing Ser. 1, p. 355–365.

——, 1979, Geology of marginal basins and their significance to the origin of ophiolites [abstract]: *Cyprus Geol. Survey Dept., Abstracts Internat. Ophiolite Symposium, Nicosia,* p. 119–120.

——, and others, 1971, Volcanic petrology and geologic history of Northeast Bank, southern California borderland: *Geol. Soc. America Bull.,* v. 82, no. 1, p. 219–228.

Haxel, Gordon, 1977, The Orocopia Schist and the Chocolate Mountain thrust in the Picacho-Peter Kane Mountain area, southeasternmost California: Unpublished Ph.D. dissertation, University of California, Santa Barbara, 277 pp.

——, and Dillon, John, 1978, The Pelona-Orocopia Schist and Vincent-Chocolate Mountain thrust system, southern California, *in* Howell, D. G., and McDougall, K. A., eds., *Mesozoic Paleogeography of the Western United States:* Pacific Section, Soc. Econ. Paleontologists and Mineralogists, Pacific Coast Paleogeography Symposium 2, p. 453–469.

Hayes, P. T., 1970, Cretaceous paleogeography of southeastern Arizona and adjacent areas: *U.S. Geol. Survey Prof. Paper 658-B,* 42 pp.

Hazzard, J. C., and Mason, J. F., 1936, Middle Cambrian formations of the Providence and Marble Mountains, California: *Geol. Soc. America Bull.,* v. 47, p. 229–240.

Healy, J. H., 1963, Crustal structure along the coast of California from seismic refraction measurements: *Jour. Geophys. Res.,* v. 68, p. 5777–5787.

——, and Peake, L. G., 1975, Seismic velocity structure along a section of the San Andreas fault near Bear Valley, California: *Seismol. Soc. America Bull.,* v. 65, p. 1177–1197.

Hekinian, R., and others, 1976, Volcanic rocks and processes of the mid-Atlantic Ridge rift valley near 36°49′N: *Contrib. Mineralogy Petrology,* v. 58, p. 83–110.

Helmstaedt, Herwart, and Doig, Ronald, 1975, Eclogite nodules from kimberlite pipes of the Colorado Plateau—samples of subducted Franciscan-type oceanic lithosphere: *Physics Chemistry Earth,* v. 9, p. 95–111.

Helwig, James, 1974, Eugeosynclinal basement and a collage concept of orogenic belts: *Soc. Econ.Paleontologists and Mineralogists Spec. Pub. 19,* p. 359–376.

Henry, C. D., 1975, Geology and geochronology of the granitic batholithic complex, Sinaloa, Mexico [abstract]: *Geol. Soc. America Abstracts with Programs,* v. 7, p. 172.

Hewett, D. G., 1956, Geology and mineral resources of the Ivanpah quadrangle, California and Nevada: *U.S. Geol. Survey Prof. Paper 275,* 172 pp.

Hickey, R., and Frey, M. A., 1979, Petrogenesis of high-Mg andesites: geochemical evidence [abstract] : *Trans. Am. Geophys. Union,* v. 60, no. 18, p. 413.

Hietanen, Anna, 1973, Geology of the Pulga and Bucks Lake quadrangles, Butte and Plumas Counties, California: *U.S. Geol. Survey Prof. Paper 731,* 66 pp.

——, 1976, Metamorphism and plutonism around the Middle and South Forks of the Feather River, California: *U.S. Geol. Survey Prof. Paper 920,* 30 pp.

——, 1977, Paleozoic-Mesozoic boundary in the Berry Creek quadrangle, northwestern Sierra Nevada, California: *U.S. Geol. Survey Prof. Paper 1027,* 22 pp.

Higgs, D. V., 1954, Anorthosite and related rocks of the western San Gabriel Mountains, southern California. *Univ. Calif. Pubs. Geol. Sci.,* v. 30, p. 171–222.

Hilde, T. W. C., and others, 1977, Evolution of the western Pacific and its margin: *Tectonophysics,* v. 28, p. 148–165.

Hileman, J. A., and others, 1973, *Seismicity of the Southern California Region, 1 January 1932 to 31 December 1972:* Pasadena, Calif. Inst. Technology, 497 pp.

Hill, D. J., 1976, Geology of the Jurassic basement rocks, Santa Cruz Island, California and correlation with other Mesozoic basement terranes in California, *in* Howell, D. G., ed., *Aspects of the Geologic History of the Southern California Continental*

Borderland: Pacific Section, Am. Assoc. Petroleum Geologists Misc. Pub. 24, p. 16–46.

Hill, D. P., 1977, A model for earthquake swarms: *Jour. Geophys. Res.,* v. 84, p. 1347–1353.

——, and others, 1975, Earthquakes, active faults, and geothermal areas in the Imperial Valley, California: *Science,* v. 188, p. 1306–1308.

Hill, M. L., 1954, Tectonics of faulting in southern California, *in* Jahns, R. H., ed., *Geology of Southern California:* Calif. Div. Mines Bull. 170, Chapter 4, Part 1, p. 5–13.

——, 1974, Is the San Andreas a transform fault?: *Geology,* v. 2, p. 535–536.

——, and Dibblee, T. W., 1953, San Andreas, Garlock, and Big Pine faults—a study of the character, history, and significance of their displacements: *Geol. Soc. America Bull.,* v. 64, p. 443–458.

——, and Troxel, B. W., 1966, Tectonics of Death Valley region, California: *Geol. Soc. America Bull.,* v. 77, p. 435–438.

Hillhouse, John, 1977, Paleomagnetism of the Triassic Nikolai Greenstone, McCarthy quadrangle, Alaska: *Canadian Jour. Earth Sci.,* v. 14, no. 11, p. 2578–2592.

Himmelberg, G. R., and Coleman, R. G., 1968, Chemistry of primary minerals and rocks from the Red Mountain-Del Puerto ultramafic mass, California, *in Geological Survey Research 1968:* U.S. Geol. Survey Prof. Paper 600–C, p. C18–C26.

Holcombe, C. J., 1977, How rigid are the lithospheric plates? Fault and shear rotations in southeast Asia: *Jour. Geol. Soc. London,* v. 134, p. 325–342.

Hollenbaugh, K. M., 1970, Geology of a portion of the north flank of the San Bernardino Mountains, California [abstract] : *Geol. Soc. America Abstracts with Programs,* v. 2, no. 2, p. 103.

Hoots, H. W., 1931, Geology of the eastern part of the Santa Monica Mountains, Los Angeles County, California: *U.S. Geol. Survey Prof. Paper 164,* p. 3–134.

Hopson, C. A., 1975, Features of igneous cumulates in ophiolites formed at spreading ocean ridges [abstract] : *Trans. Am. Geophys. Union,* v. 56, no. 12, p. 1079.

——, 1976, Sheeted sill complexes in ophiolites: their occurrence and tectonic significance [abstract] : *Trans. Am. Geophys. Union,* v. 57, no. 12, p. 1026.

——, and Frano, C. J., 1977, Igneous history of the Point Sal ophiolite, southern California, *in* Coleman, R. G., and Irwin, W. P., eds., *North American Ophiolites:* Oregon Dept. Geology and Mineral Industries Bull. 95, p. 161–183.

——, and Pallister, J. S., 1978, Gabbro sections in Samail ophiolite, southeastern Oman Mountains [abstract] : *Geol. Soc. America Abstracts with Programs,* v. 10, no. 7, p. 424.

——, and Pallister, J. S., 1979, Samail ophiolite magma chamber: I, Evidence from gabbro phase variation, internal structure and layering: *Cyprus Geol. Survey Dept., Abstracts Internat. Ophiolite Symposium, Nicosia,* p. 37.

——, and Pallister, J. S., 1980, Crystal fractionation and magma mixing during growth of the Samail ophiolite, eastern Oman [abstract] : *Trans. Am. Geophys. Union,* v. 61, p. 67.

——, and others, 1973, Late Jurassic ophiolite at Point Sal, Santa Barbara County, California [abstract] : *Geol. Soc. America Abstracts with Programs,* v. 5, no. 1, p. 58.

——, and others, 1975a, Preliminary report and geologic guide to the Jurassic ophiolite near Point Sal, southern California coast: *Cordilleran Section, Geol. Soc. America, 71st Annual Meeting, Field Trip No. 5,* p. 36.

——, and others, 1975b, Record of Late Jurassic sea-floor spreading, California Coast Ranges [abstract] : *Geol. Soc. America Abstracts with Programs,* v. 7, p. 326.

Hotz, P. E., 1971, Plutonic rocks of the Klamath Mountains, California and Oregon: *U.S. Geol. Survey Prof. Paper 684–B,* 20 pp.

——, 1973, Blueschist metamorphism in the Yreka-Fort Jones area, Klamath Mountains, California: *U.S. Geol. Survey Jour. Res.,* v. 1, no. 1, p. 53–61.

——, 1974, Preliminary geologic map of the Yreka quadrangle, California: *U.S. Geol. Survey Misc. Field Studies Map MF-568.*

——, 1977, Geology of the Yreka quadrangle, Siskiyou County, California: *U.S. Geol. Survey Bull. 1436,* 72 pp.

——, and Willden, R., 1964, Geology and mineral deposits of the Osgood Mountains quadrangle, Humboldt County, Nevada: *U.S. Geol. Survey Prof. Paper 431,* 128 pp.

——, and others, 1977, Triassic blueschist from northern California and north-central Oregon: *Geology,* v. 5, p. 659–663.

Howard, A. D., 1973, Modified contour-generalization procedure as applied to the Santa Lucia Range, California: *Geol. Soc. America Bull.,* v. 84, p. 3415–3428.

Howell, D. G., 1975a, Middle Eocene paleogeography of southern California, *in* Weaver, D. W., Hornaday, G. R., and Tipton, Ann, eds., *Conference on Future Energy Horizons of the Pacific Coast, Paleogene Symposium and Selected Technical Papers:* Pacific Section, Am. Assoc. Petroleum Geologists, p. 272–293.

——, 1975b, Early and middle Eocene shoreline offset by the San Andreas fault, southern California, *in* Crowell, J. C., ed., *San Andreas Fault in Southern California:* Calif. Div. Mines and Geology Spec. Report 118, p. 69–74.

——, 1975c, Hypothesis suggesting 700 km of right slip in California along northwest-oriented faults: *Geology,* v. 3, p. 81–83.

——, 1976a, Late-Miocene counterclockwise rotation of the south half of Santa Cruz Island: *Pacific Section, Am. Assoc. Petroleum Geologists Misc. Pub. 24,* p. 449–454.

——, 1976b, Hypothesis suggesting 700 km of right slip in California along northwest oriented faults: Reply: *Geology,* v. 4, no. 10, p. 632–633.

——, and Link, M. H., 1976, Conglomerate facies of Eocene fluvial to bathyal fan deposits, San Diego to the Channel Islands, California [abstract] : *Geol. Soc. America Abstracts with Programs,* v. 8, no. 6, p. 930.

——, and McLean, Hugh, 1976, Middle Miocene paleogeography, Santa Cruz and Santa Rosa Island, *in* Howell, D. G., ed., *Aspects of the Geologic History of the California Continental Borderland:* Pacific Section, Am. Assoc. Petroleum Geologists Misc. Pub. 24, p. 266–293.

——, and Vedder, J. G., 1978, Late Cretaceous paleogeography of the Salinian block, California, *in* Howell, D. G., and McDougall, K. A., eds., *Mesozoic Paleogeography of the Western United States:* Pacific Section, Soc. Econ. Paleontologists and Mineralogists, Pacific Coast Paleogeography Symposium 2, p. 107–116.

——, and others, 1976, Cenozoic tectonism on Santa Cruz Island, *in* Howell, D. G., ed., *Aspects of the Geologic History of the California Continental Borderland:* Pacific Section, Am. Assoc. Petroleum Geologists Misc. Pub. 24, p. 392–417.

——, and others, 1974, Possible strike-slip faulting in the southern California borderland: *Geology,* v. 2, no. 2, p. 93–98.

——, and others, 1977, Review of Cretaceous geology, Salinian and Nacimiento blocks, Coast Ranges of central California, *in* Howell, D. G., and others, eds., *Cretaceous Geology of the California Coast Ranges, West of the San Andreas Fault:* Pacific Section, Soc. Econ. Paleontologists and Mineralogists, Pacific Coast Paleogeography Field Guide 2, p. 1–46.

——, and others, 1978, General geology, petroleum appraisal, and nature of environmental hazards eastern Pacific Shelf Latitude 28° to 38° North: *U.S. Geol. Survey Circular 786,* 29 pp.

Hsü, K. J., 1955, Granulites and mylonites of the region about Cucamonga and San Antonio Canyons, San Gabriel Mountains, California: *Univ. Calif. Pub. Geol. Sci.,* v. 30, p. 223–324.

——, 1968, Principles of melanges and their bearing on the Franciscan-Knoxville paradox: *Geol. Soc. America Bull.,* v. 79, p. 1063–1074.

——, 1969, Preliminary report and geologic guide to Franciscan melanges of the Morro Bay-San Simeon area, California: *Calif. Div. Mines and Geology Spec. Pub. 35,* 46 pp.

——, 1974, Melanges and their distinction from olistostromes, *in* Dott, R. H., Jr., and Shaver, R. H., eds., *Modern and Ancient Geosynclinal Sedimentation:* Soc. Econ. Paleontologists and Mineralogists Spec. Pub. 19, p. 321–333.

Huber, N. K., 1968, Geologic map of the Shuteye Peak quadrangle, Sierra Nevada, California: *U.S. Geol. Survey Geol. Quad. Map GQ-728,* scale 1:62,500.

——, and Rinehart, C. D., 1965, Geologic map of the Devils Postpile quadrangle, Sierra Nevada, California: *U.S. Geol. Survey Geol. Quad. Map GQ-437,* scale 1:62,500.

Hudson, F. S., 1922, Geology of the Cuyamaca region of California, with special reference to the origin of the nickeliferous pyrrhotite: *Univ. Calif. Dept. Geol. Sci. Bull.,* v. 13, p. 175–252.

Huffman, O. F., 1970, Miocene and post-Miocene offset on the San Andreas fault in central California [abstract]: *Geol. Soc. America Abstracts with Programs,* v. 2, p. 104–105.

——, 1972, Lateral displacement of Upper Miocene rocks and the Neogene history of offset along the San Andreas fault in central California: *Geol. Soc. America Bull.,* v. 83, p. 2913–2946.

——, and others, 1975a. Compressional faulting of the oceanic crust prior to subduction in the Peru-Chile Trench: *Geology,* v. 3, p. 601–604.

——, and others, 1975b, Crustal structure of the Peru-Chile trench, 8°12′S latitude, *in* Sutton, G. H., and others, eds., *The Geophysics of the Pacific Ocean Basin and Its Margin* (Woollard Volume): Am. Geophys. Union Geophys. Monogr. 19.

——, and others, 1978, Leg 60 ends in Guam: *Geotimes,* v. 23, p. 19–22.

Imlay, R. W., 1959, Succession and speciation of the Pelecypod Aucella: *U.S. Geol. Survey Prof. Paper 314-G,* p. 155–169.

——, 1961, Late Jurassic ammonites from the western Sierra Nevada, California: *U.S. Geol. Survey Prof. Paper 374-D,* p. D1–D30.

——, 1963, Jurassic fossils from southern California: *Jour. Paleontology,* v. 37, no. 1, p. 97–107.

——, 1964, Middle and Upper Jurassic fossils from southern California: *Jour. Paleontology*, v. 38, no. 5, p. 505–509.

Ingersoll, R. V., 1978, Paleogeography and paleotectonics of the late Mesozoic forearc basin of northern and central California, *in* Howell, D. G., and McDougall, K. A., eds., *Mesozoic Paleogeography of the Western United States:* Pacific Section, Soc. Econ. Paleontologists and Mineralogists, Pacific Coast Paleogeography Symposium 2, p. 471–482.

Ingle, J. C., Jr., 1973, Biostratigraphy and paleoecology of early Miocene through early Pleistocene benthonic and planktonic Foraminifers, San Joaquin Hills-Newport Bay-Dana Point area, Orange County, California, in *Miocene Sedimentary Environments and Biofacies, Southeastern Los Angeles Basin:* Guidebook, Soc. Econ. Paleontologists and Mineralogists, p. 18–38.

——, and others, 1976, Evidence and implications of worldwide late Paleogene climatic and eustatic events [abstract]: *Geol. Soc. America Abstracts with Programs*, v. 8, p. 934–935.

Irvine, T. N., 1965, Chromian spinel as a petrogenetic indicator—Part 1, Theory: *Canadian Jour. Earth Sci.*, v. 2, p. 648–672.

——, 1970, Crystallization sequences in the Muskox intrusion and other layered intrusions. I, olivine-pyroxene-plagioclase relations: *Geol. Soc. South Africa Spec. Pub. 1*, p. 441–476.

Irwin W. P., 1960, Geologic reconnaissance of the northern Coast Ranges and Klamath Mountains, California, with a summary of the mineral resources: *Calif. Div. Mines Bull. 179*, 80 pp.

——, 1963, Preliminary geologic map of the Weaverville quadrangle, California: *U.S. Geol. Survey Mineral Inv. Field Studies Map MF-275*, scale 1:62,500.

——, 1964, Late Mesozoic orogenies in the ultramafic belts of northwestern California and southwestern Oregon, in *Geological Survey Research, 1964:* U.S. Geol. Survey Prof. Paper 501-C, p. C1–C9.

——, 1966, Geology of the Klamath Mountains province: *Calif. Div. Mines and Geology Bull. 190*, p. 19–38.

——, 1972, Terranes of the western Paleozoic and Triassic belt in the southern Klamath Mountains, California, in *Geological Survey Research, 1972:* U.S. Geol. Survey Prof. Paper 800-C, p. C103–C111.

——, 1973, Sequential minimum ages of oceanic crust in accreted tectonic plates of northern California and southern Oregon [abstract]: *Geol. Soc. America Abstracts with Programs*, v. 5, no. 1, p. 62–63.

——, 1977, Review of Paleozoic rocks of the Klamath Mountains, *in* Stewart J. H., and others, eds., *Paleozoic Paleogeography of the Western United States:* Pacific Section, Soc. Econ. Paleontologists and Mineralogists, Pacific Coast Paleogeography Symposium 1, p. 441–454.

——, and Barnes I., 1975, Effect of geologic structure and metamorphic fluids on seismic behavior of the San Andreas fault system in central and northern California: *Geology*, v. 3, p. 713–716.

——, and Galanis, S. P., Jr., 1976, Map showing limestone and selected fossil localities in the Klamath Mountains, California and Oregon: *U.S. Geol. Survey Misc. Field Studies Map MF-749*, scale 1:500,000.

——, and others, 1978, Radiolarian from pre-Nevadan rocks of the Klamath Mountains,

California and Oregon, *in* Howell, D. G., and McDougall, K. A., eds. *Mesozoic Paleogeography of the Western United States:* Pacific Section, Soc. Econ. Paleontologists and Mineralogists, Pacific Coast Paleogeography Symposium 2, p. 303–310.

——, and others, 1977, Significance of Mesozoic radiolarians from the pre-Nevadan rocks of the southern Klamath Mountains, California: *Geology,* v. 5, p. 557–562.

——, and Lipman, P. W., 1962, A regional ultramafic sheet in eastern Klamath Mountains, California, in *Geological Survey Reserach, 1962:* U.S. Geol. Survey Prof. Paper 450–C, p. C18–C21.

Jackson, E. D., 1971, The origin of ultramafic rocks by cumulus processes: *Fortschr. Mineralogie,* v. 48, p. 128–174.

——, and Thayer, T. P., 1972, Some criteria for distinguishing between stratiform, concentric, and alpine peridotite-gabbro complexes: *Internat. Geol. Congr., 24th, Montreal,* sect. 2, p. 289–296.

Jacobson, M. I., 1978, Petrologic variations in Franciscan sandstone from the Diablo Range, California, *in* Howell, D. G., and McDougall, K. A., eds., *Mesozoic Paleogeography of the Western United States:* Pacific Section, Soc. Econ. Paleontologists and Mineralogists, Pacific Coast Paleogeography Symposium 2, p. 401–417.

Jahns, R. H., 1973, Tectonic evolution of the Transverse Ranges Province as related to the San Andreas Fault System: *Stanford Univ. Pub. Geol. Sci.,* v. 13, p. 149–170.

Janda, R. J., 1965, Quaternary alluvium near Friant, California: *Internat. Quaternary Assoc. Guidebook for Field Conf. 1, Northern Great Basin and California,* p. 128–133.

Jeffrey, D. H., 1978, Volcanics and intrusives of northern Sumatra [unpublished abstract] : Workshop on the Sumatra Transect, 1 p.

Jennings, C. W., 1958, Geologic map of California, San Luis Obispo sheet: *Calif. Div. Mines and Geology,* scale 1:250,000.

——, 1959, Geologic map of California, Santa Maria sheet: *Calif. Div. Mines and Geology,* scale 1:250,000.

——, 1975, Fault map of California, with locations of volcanoes, thermal springs and thermal wells: *Calif. Div. Mines and Geology, Calif. Geol. Data Map Series, Map No. 1.*

——, 1977, Geologic map of California: *Calif. Div. Mines and Geology,* scale 1:750,000.

——, and Burnett, J. L., 1961, San Francisco sheet, Geologic map of California, Olaf P. Jenkins edition: *Calif. Div. Mines and Geology.*

——, and Strand, R. G., 1958, Santa Cruz Sheet, Geologic Map of California, Olaf P. Jenkins edition: *Calif. Div. Mines and Geology.*

——, and Strand, R. G., 1960, Ukiah sheet, Geologic Map of California: *Calif. Div. Mines and Geology,* scale 1:250,000.

——, and others, 1963, Trona sheet, Geologic Map of California: *Calif. Div. Mines and Geology,* scale 1:250,000.

Jennings, P. C., and Kanamori, H., 1979, Determination of local magnitude, M_L, from seismoscope records: *Seismol. Soc. America Bull.,* v. 69, p. 1267–1288.

Jensky, W. A., 1975, Reconnaissance geology and geochronology of the Bahia de Banderas area, Nayarit and Jalisco, Mexico: Unpublished M.S. thesis, University of California, Santa Barbara, 80 pp.

Johnson, A. M., and Page, B. M., 1976. A theory of concentric, kink, and sinusoidal folding and of monoclinal flexuring of compressible, elastic multilayers–VII, Development of folds within the Huasna syncline, San Luis Obispo County, California: *Tectonophysics,* v. 33, p. 97–143.

Johnson, C. E., 1979, Seismotectonics of the Imperial Valley of southern California: Unpublished Ph.D. dissertation, California Institute of Technology, Los Angeles, part 2, 332 pp.

——, and Hadley, D. M., 1976, Tectonic implications of the Brawley earthquake swarm, Imperial Valley, January, 1975: *Seismol. Soc. America Bull.,* v. 66, p. 1133–1144.

Johnson, J. D., and Normark, W. R., 1974a, Neogene tectonic evolution of the Salinian Block, west-central California: *Geology,* v. 2, p. 11–14.

——, and Normark, W. R., 1974b, Neogene tectonic evolution of the Salinian Block, west-central California: Reply [to discussion]: *Geology,* v. 2, p. 391, 394.

Johnston, B. K., 1957, Geology of a part of the Manly Peak Quadrangle, southern Panamint Range, California: *Univ. Calif. Pub. Geol. Sci.,* v. 30, p. 353–423.

Jones, D. L., 1973, Structural significance of upper Mesozoic biostratigraphic units in northern California and southwestern Oregon [abstract]: *Geol. Soc. America Abstracts with Programs,* v. 5, no. 7, p. 684–685.

——, 1975, Discovery of *Buchia rugosa* of Kimmeridgian age from the base of the Great Valley Sequence [abstract]: *Geol. Soc. America Abstracts with Programs,* v. 7, no. 3, p. 330.

——, and Bailey, E. H., 1973, Preliminary biostratigraphic map, Colyear Springs quadrangle, California: *U.S. Geol. Survey Misc. Field Studies Map MF-517,* scale 1: 8,000.

——, and Imlay, R. E., 1973, Structure of Upper Jurassic and Lower Cretaceous rocks in the Riddle area, southwestern Oregon [abstract]: *Geol. Soc. America Abstracts with Programs,* v. 5, no. 1, p. 64.

——, and Irwin, W. P., 1971, Structural implications of an offset Early Cretaceous shoreline in northern California: *Geol. Soc. America Bull.,* v. 82, p. 815–822.

——, and Irwin, W. P., 1975, Rotated Jurassic rocks in the Transverse Ranges, California [abstract]: *Geol. Soc. America Abstracts with Programs,* v. 7, p. 330–331.

——, and Moore, J. G., 1973, Lower Jurassic ammonite from the south-central Sierra Nevada, California: *U.S. Geol. Survey Jour. Research,* v. 1, no. 4, p. 453–458.

——, and others, 1969, Structural stratigraphic significance of the *Buchia* zones in the Colyear Springs-Paskenta area, California: *U.S. Geol. Survey Prof. Paper 647-A,* 24 pp.

——, and others, 1972, Southeastern Alaska–a displaced continental fragment? *U.S. Geol. Survey Prof. Paper 800-B,* p. B211–B217.

——, and others, 1976, The four Jurassic belts of northern California and their significance to the geology of the southern California borderland, *in* Howell, D. G., ed., *Aspects of the Geologic History of the California Continental Borderland:* Pacific Section, Am. Assoc. Petroleum Geologists Misc. Pub. 24, p. 343–362.

——, and others, 1977, Wrangellia–A displaced terrane in northwestern North America: *Canadian Jour. Earth Sci.,* v. 14, no. 11, p. 2565–2577.

——, and others, 1978a, Distribution of late Mesozoic melanges along the Pacific Coast of North America: *Tectonophysics,* v. 47, p. 207–222.

——, and others, 1978b, Microplate tectonics of Alaska–Significance for the Mesozoic history of the Pacific Coast of North America, *in* Howell, D. G., and McDougall, K. A., eds., *Mesozoic Paleogeography of Western United States:* Pacific Section, Soc. Econ. Paleontologists and Mineralogists, Pacific Coast Paleogeography Symposium 2, p. 71-74.

——, and others, 1978c, Revised ages of chert in the Roberts Mountains allochton, northern Nevada [abstract]: *Geol. Soc. America Abstracts with Programs,* v. 10, p. 111.

Jones, M., Van der Voo, R., Churkin, M., Jr., and Eberlein, G. D., 1977, Paleozoic paleomagnetic results from the Alexander terrane of southeastern Alaska [abstract]: *Trans. Am. Geophys. Union,* v. 58, p. 743-744.

Joseph, S. E., and Davis, T. E., 1977, $^{87}Sr/^{86}Sr$ correlation of rapakivi-textured porphyry to measure offset on the San Andreas fault [abstract]: *Geol. Soc. America Abstracts with Programs,* v. 9, no. 4, p. 443.

——, and others, 1978, Rb/Sr geochronology and geochemistry of the lower granodiorite, central San Gabriel Mountains, California [abstract]: *Geol. Soc. America Abstracts with Programs,* v. 10, no. 3, p. 111.

Junger, Arne, 1974, Source of the San Onofre Breccia [abstract]: *Geol. Soc. America Abstracts with Programs,* v. 6, no. 3, p. 199-200.

——, 1976, Tectonics of the southern California borderland, *in* Howell, D. G., ed., *Aspects of the Geologic History of the California Continental Borderland:* Pacific Section, Am. Ass. Petroleum Geologists Misc. Pub. 24, p. 486-498.

——, and Wagner, H. C., 1977, Geology of the Santa Monica and San Pedro basins, California Continental Borderland: *U.S. Geol. Survey Misc. Field Studies Map MF-820.*

Kahle, J. E., 1966, Megabreccias and sedimentary structures of the Plush Ranch Formation, northern Ventura County, California: Unpublished M.A. thesis, University of California, Los Angeles, 125 pp.

Kaizuka, S., 1975, A tectonic model for the morphology of arc-trench systems, especially for the echelon ridges and mid-arc faults: *Japanese Jour. Geol. Geog.,* v. 45, p. 9-28.

——, and others, 1973, Quaternary tectonic and recent seismic crustal movements in the Arauco Peninsula and its environs, central Chile: *Geogr. Rep. Tokyo Metropol. Univ.,* no. 8, p. 1-49.

Kamerling, M. J., and Luyendyk, B. P., 1977, Tectonic rotation of the Santa Monica Mountains in southern California [abstract]: *Trans. Am. Geophys. Union,* v. 58, p. 1126.

——, and Luyendyk, B. P., 1979, Tectonic rotations of the Santa Monica Mountains region, western Transverse Ranges, California, suggested by paleomagnetic vectors: *Geol. Soc. America Bull.,* v. 90, p. 331-337.

——, Luyendyk, B. P., and Marshall, M., 1978, Paleomagnetism and tectonic rotation of parts of the Transverse Ranges, California [abstract]: *Trans. Am. Geophys. Union,* v. 59, no. 12, p. 1058.

Kanamori, H., 1979, A semi-empirical approach to prediction of long-period ground motions, from great earthquakes: *Seismol. Soc. America Bull.,* v. 69, p. 1645-1670.

——, and Jennings, P. C., 1978, The determination of local magnitude, M_L, from strong-motion accelerograms: *Seismol. Soc. America Bull.,* v. 68, p. 471-485.

Karig, D. E., 1970, Ridges and basins of the Tonga-Kermadec island arc system: *Jour. Geophys. Res.,* v. 75, p. 239-254.

——, 1971, Origin and development of marginal basins in the western Pacific: *Jour. Geophys. Res.*, v. 76, p. 2542–2561.

——, 1972, Remnant arcs: *Geol. Soc. America Bull.*, v. 83, p. 1057–1068.

——, 1974, Tectonic erosion at trenches: *Earth Planet. Sci. Lett.*, v. 21, p. 209–212.

——, 1980, Material transport within accretionary prisms and the "knocker" problem: *Jour. Geology*, v. 88, p. 27–40.

——, and Jensky, Wallace, 1972, The proto-Gulf of California: *Earth Planet. Sci. Lett.*, v. 17, p. 169–174.

——, and Moore, G. F., 1975a, Tectonic complexities in the Bonin arc system: *Tectonophysics,* v. 27, p. 97–118.

——, and Moore, G. F., 1975b, Tectonically controlled sedimentation in marginal basins: *Earth Planet. Sci. Lett.*, v. 26, p. 233–238.

——, and Sharman, G. F., 1975, Subduction and accretion in trenches: *Geol. Soc. America Bull.*, v. 86, p. 377–389.

——, and others, 1978, Late Cenozoic subduction and continental margin truncation along the northern middle America trench: *Geol. Soc. America Bull.*, v. 89, p. 265–276.

——, and others, 1979, Structure and Cenozoic evolution of the Sunda arc in the central Sumatra region, *in* Watkins, J. S., and others, eds., *Geological and Geophysical Investigations of Continental Margins:* Am. Assoc. Petroleum Geologists Mem. 29, p. 223–238.

Kay, M., 1951, North American geosynclines: *Geol. Soc. America Mem. 48,* 143 pp.

Keith, S. B., 1978, Paleosubduction geometries inferred from Cretaceous and Tertiary magmatic patterns in southwestern North America: *Geology,* v. 6, p. 516–521.

Keller, R. P., and others, 1978, Monitoring slip along major faults in southern California: *Seismol. Soc. America Bull.*, v. 68, p. 1187–1190.

Kemp, W. R., 1975, Petrochemical associations within the foothill copper-zinc belt, Sierra Nevada, California [abstract]: *Geol. Soc. America Abstracts with Programs,* v. 7, p. 332.

——, 1976, The foothill copper-zinc belt, Sierra Nevada, California, a volcanogenic massive sulfide province [abstract]: *Geol. Soc. America Abstracts with Programs,* v. 8, p. 387–388.

——, and Payne, A. L., 1975, Petrochemical associations within the foothill copper-zinc belt, Sierra Nevada, California [abstract]: *Geol. Soc. America Abstracts with Programs,* v. 7, p. 332.

Kempner, W. C., 1977, The magnetic properties of the Point Sal ophiolite: A comparison with oceanic crust: Unpublished M.A. thesis, University of California, Santa Barbara, 134 pp.

Ketin, I., and Roesli, F., 1953, Makroseismische Untersuchungen uber das nordwestanatolischen Beben vom 18 Marz 1953: *Eclogae Geol. Helvetiae,* v. 46, p. 187–208.

Kilmer, F. H., 1977, Reconnaissance geology of Cedros Island, Baja California, Mexico: *Southern Calif. Acad. Sci. Bull.*, v. 91–98.

King, P. B., 1969, Tectonic map of North America: *U.S. Geol. Survey,* scale 1:5,000,000.

King, R. E., 1939, Geological reconnaissance in northern Sierra Madre Occidental of Mexico: *Geol. Soc. America Bull.*, v. 50, p. 1625–1727.

——, and others, 1944, Geology and paleontology of the Permian area northwest of

Delicias, southwestern Coahuila, Mexico: *Geol. Soc. America Spec. Paper 52,* 34 pp.

Kinkel, A. R., Jr., Hall, W. E., and Albers, J. P., 1956, Geology and base-metal deposits of West Shasta copper-zinc district, Shasta County, California: *U.S. Geol. Survey Prof. Paper 285,* 156 pp.

Kistler, R. W., 1966a, Geologic map of the Mono Craters quadrangle, Mono and Tuolumne Counties, California: *U.S. Geol. Survey Geol. Quad. Map GQ–462,* scale 1:62,500.

——, 1966b, Structure and metamorphism in the Mono Craters quadrangle, Sierra Nevada, California: *U.S. Geol. Survey Bull. 1221-E,* p. E1–E53.

——, 1973, Geologic map of the Hetch Hetchy Reservoir quadrangle, Yosemite National Park, California: *U.S. Geol. Survey Geol. Quad. Map GQ–1112,* scale 1:62,500.

——, 1978, Mesozoic paleogeography of California: A viewpoint from isotope geology, *in* Howell, D. G., and McDougall, K. A., eds., *Mesozoic Paleogeography of the Western United States:* Pacific Section, Soc. Econ. Paleontologists and Mineralogists, Pacific Coast Paleogeography Symposium 2, p. 75–84.

——, and Bateman, P. C., 1966, Stratigraphy and structure of the Dinkey Creek roof pendant in the central Sierra Nevada, California: *U.S. Geol. Survey Prof. Paper 524–B,* 14 pp.

——, and Peterman, Z. E., 1973, Variations in Sr, Rb, K, Na, and initial $^{87}Sr/^{86}Sr$ in Mesozoic granitic rocks and intruded wall rocks in central California: *Geol. Soc. America Bull.,* v. 84, no. 11, p. 3489–3512.

Kistler, R. W. and Peterman, Z. E., 1978, Reconstruction of crustal blocks of California on the basis of initial strontium isotopic compositions of Mesozoic granitic rocks: *U.S. Geol. Survey Prof. Paper 1071,* 17 p.

——, and others, 1965, Isotopic ages of minerals from granitic rocks of the central Sierra Nevada and Inyo Mountains, California: *Geol. Soc. America Bull.,* v. 76, no. 2, p. 155–164.

——, and others, 1971, Sierra Nevada plutonic cycle: Part I, Origin of composite granitic batholiths: *Geol. Soc. America Bull.,* v. 82, p. 853–868.

——, and others, 1973, Strontium isotopes and the San Andreas fault, *in* Kovach, R. L., and Nur, A., eds., *Proceedings of the Conference on Tectonic Problems of the San Andreas Fault System:* Stanford Univ. Pubs. Geol. Sci., v. 13, p. 339–347.

Kittleman, L. R., 1973, Mineralogy, correlation, and grain-size distribution of Mazama tephra and other postglacial pyroclastic layers, Pacific Northwest: *Geol. Soc. America Bull.,* v. 84, p. 2957–2980.

Klein, C. W., 1977, Thrust plates of the north-central Klamath Mountains near Happy Camp, California: *Calif. Div. Mines and Geology Spec. Report 129,* p. 23–26.

Kleist, J. R., 1974, Deformation by soft-sediment extension in the Coast Belt, Franciscan Complex: *Geology,* v. 2, p. 501–504.

Krogh, T. E., 1973, A low-contamination method for hydrothermal decomposition of zircon and extraction of U and Pb for isotopic age determination: *Geochim. et Cosmochim. Acta,* v. 37, p. 485–494.

Knopf, Adolph, and Thelen, P., 1905, Sketch of the geology of Mineral King, California: *Univ. Calif. Pubs. Dept. Geology Bull.,* v. 4, no. 12, p. 227–262.

Koch, J. G., 1966, Late Mesozoic stratigraphy and tectonic history, Port Orford-Gold Beach area, southwestern Oregon coast: *Am. Assoc. Petroleum Geologists Bull.,* v. 50, p. 25–71.

Kovach, R. L., and Nur, A., 1973, *Proceedings of the Conference on Tectonic Problems of the San Andreas Fault System:* Stanford Univ. Pubs. Geol. Sci., v. 13, 494 pp.

Krause, D. C., 1965, Tectonics, bathymetry and geomagnetism of the southern continental borderland west of Baja California, Mexico: *Geol. Soc. America Bull.,* v. 76, no. 6, p. 617-650.

Krauskopf, K. B., 1971, Geologic map of the Mt. Barcroft quadrangle, California-Nevada: *U.S. Geol. Survey Geol. Quad. Map GQ-960,* scale 1:62,500.

Krummenacher, D., and others, 1975, K-Ar apparent ages, Peninsular Ranges batholith, southern California and Baja California: *Geol. Soc. America Bull.,* v. 86, no. 6, p. 760-768.

Kulm, L. D., and Fowler, G. A., 1974, Cenozoic sedimentary framework of the Gorda-Juan de Fuca plate and adjacent continental margin—a review, *in* Dott, R. H., Jr., and Shaver, R. H., eds., *Modern and Ancient Geosynclinal Sedimentation:* Soc. Econ. Paleontologists and Mineralogists Special Pub. No. 19, p. 212-229.

——, and others, 1973, Site 173, *in* Kulm and others, *Initital Reports of the Deep Sea Drilling Project,* v. 18: Washington, D.C., U.S. Governement Printing Office, p. 15-30.

Kuno, H., 1966, Lateral variation of basalt magma type across continental margins and island arcs: *Bull. Volc.,* v. 29, p. 195-222.

Kupfer, D. H., 1960, Thrust-faulting and chaos structure, Silurian Hills, San Bernardino County, California: *Geol. Soc. America Bull.,* v. 71, p. 181-214.

Kusznir, N. J., and Bott, M. H. P., 1976, A thermal study of the formation of oceanic crust: *Geophys. Jour. Roy. Astronom. Soc.,* v. 47, p. 83-95.

Labotka, T. C., and Albee, A. L., 1977, Late Precambrian depositional environment of the Pahrump Group, Panamint Mountains, California: *Calif. Div. Mines and Geology Spec. Report 129,* p. 93-100.

Lachenbruch, A. H., 1968, Preliminary geothermal model of the Sierra Nevada: *Jour. Geophys. Res.,* v. 73, p. 6977-6989.

——, and others, 1976, Geothermal setting and simple heat conduction models for Long Valley caldera: *Jour. Geophys. Res.,* v, 81, no. 5, p. 769-784.

Lajoie, K. R., and Wehmiller, J. F., 1978, Quaternary uplift rates, southern California borderland [abstract], in *A Multidisciplinary Symposium on the California Islands:* Santa Barbara Museum of Natural History, Santa Barbara, Calif.

——, and others, 1972, Marine terrace deformation: San Mateo and Santa Cruz Counties, in *Progress Report on USGS Quarternary Studies in San Francisco Bay Area, An Informal Collection of Preliminary Papers:* Friends of the Pleistocene Guidebook, p. 100-113.

Lamar, D. L., 1959, Geology of the Corona area, Orange, Riverside, and San Bernardino Counties, California: Unpublished M.A. thesis, University of California, Los Angeles, 95 pp.

Lambert, R. St. J., 1971, The pre-Pleistocene time-scale—a review, *in* Harland, W. P., and Francis, E. H., eds., *The Phanerozoic Time-Scale—A Supplement:* Geol. Soc. London Special Pub. No. 5, p. 9-34.

Langseth, M., and others, 1978, Transects begun: *Geotimes,* v. 23, p. 22-36.

Lanphere, M. A., 1964, Geochronologic studies in the eastern Mojave Desert, California: *Jour. Geology,* v. 72, no. 4, p. 381-399.

——, 1971, Age of the Mesozoic oceanic crust in the California Coast Ranges: *Geol. Soc. America Bull.,* v. 82, p. 3209-3212.

——, and Jones, D. L., in press, Cretaceous time scale from North America: *Am. Assoc. Petroleum Geologists Spec. Pub.*

——, and others, 1963, Redistribution of strontium and rubidium isotopes during metamorphism, World Beater Complex, Panamint Range, California, *in* Craig, Harmon, and others, *Isotopic and Cosmic Chemistry:* Amsterdam, North-Holland, p. 269–320.

——, and others, 1968, Isotopic age of the Nevadan orogeny and older plutonic and metamorphic events in the Klamath Mountains, California: *Geol. Soc. America Bull.,* v. 79, p. 1027–1052.

——, and others, 1975, Early Cretaceous metamorphic age of the South Fork Mountain Schist in the northern Coast Ranges of California [abstract]: *Geol. Soc. America Abstracts with Programs,* v. 7, no. 3, p. 340.

——, and others, 1978, Early Cretaceous metamorphic age of the South Fork Mountain Schist in the northern Coast Ranges of California: *Am. Jour. Sci.,* v. 278, p. 798–815.

Larsen, E. S., Jr., 1948, Batholith and associated rocks of Corona, Elsinore, and San Luis Rey quadrangles, southern California: *Geol. Soc. America Mem. 29,* 182 pp.

Larson, R. L., and Chase, C. G., 1972, Late Mesozoic evolution of the western Pacific Ocean: *Geol. Soc. America Bull.,* v. 83, p. 3627–3644.

——, and Pitman, W. C., 1972, World-wide correlation of Mesozoic magnetic anomalies, and its implications: *Geol. Soc. America Bull.,* v. 83, p. 3645–3662.

——, and others, 1968, Gulf of California: A result of ocean-floor spreading and transform faulting: *Science,* v. 161, p. 781–784.

Lass, G. L., and Davis, T. E., 1979, Effects of hydrothermal contamination on strontium isotope systematics in rocks and minerals of the Point Sal ophiolite, Santa Barbara County, California, U.S.A. [abstract]: *Cyprus Geol. Survey Dept., Internat. Ophiolite Symposium Abstracts,* p. 43–44.

Lawson, A. C., 1895, Sketch of the geology of the San Francisco Peninsula: *U.S. Geol. Survey 15th Annual Report,* p. 405–476.

——, and others, 1914, Description of the San Francisco district: Tamalpais, San Francisco, Concord, San Mateo, and Hayward quadrangles: *U.S. Geol. Survey Geol. Atlas,* Folio 193, 24 pp.

——, and others, 1908, The California earthquake of April 18, 1906: Report of the State Earthquake Investigation Commission: *Carnegie Inst. Washington Pub. 87,* 2 v.

Lawton, J. E., 1956, Geology of the north half of the Morgan Valley quadrangle and the south half of the Wilbur Springs quadrangle, California: Unpublished Ph.D. dissertation, Stanford University, Stanford, Calif.

Lee, W. H. K., and others, 1979, Recent earthquake activity and focal mechanisms in the western Transverse Ranges, California: *U.S. Geol. Survey Circ. 799-A,* 37 pp.

Leith, C. J., 1949, Geology of the Quien Sabe quadrangle, California: *Calif. Div. Mines Bull. 147,* 60 pp.

Lewis, B. T. R., and Snydsman, W. E., 1977, Evidence for a low-velocity layer at the base of the oceanic crust: *Nature,* v. 266, no. 5600, p. 340–344.

Lewis, J. F., 1972, Petrology of ejected plutonic blocks of Soufriere volcano, St. Vincent West Indies: *Jour. Petrology,* v. 14, p. 81–112.

Lindsley-Griffin, Nancy, 1973, Lower Paleozoic ophiolite of the Scott Mountains, eastern Klamath Mountains, California [abstract]: *Geol. Soc. America Abstracts with Programs,* v. 5, no. 1, p. 71–72.

——, 1977, Paleogeographic implications of ophiolites: The Ordovician Trinity complex,

Klamath Mountains, California, *in* Stewart, J. H., and others, eds., *Paleozoic Paleogeography of the Western United States:* Pacific Section, Soc. Econ. Paleontologists and Mineralogists, Pacific Coast Paleogeography Symposium 1, p. 409–420.

——, and others, 1974, Geology of the Lovers Leap area, eastern Klamath Mountains, California, *in Geologic Guide to the Southern Klamath Mountains:* Geol. Soc. Sacramento, Annual Field Trip Guidebook, p. 82–100.

Link, M. H., and Howell, D. G., 1976, Conglomerate facies of Eocene fluvial to shelf submarine deposits, San Diego County, California [abstract]: *Geol. Soc. America Abstracts with Programs,* v. 8, no. 6, p. 930.

Liou, J. G., and others, 1977, The east Taiwan ophiolite, its occurrence, petrology, metamorphism and tectonic setting: *Mining Research Service Organization, Republic of China, Special Report No. 1,* 212 pp.

Lipman, P. W., 1964, Structure and origin of an ultramafic pluton in the Klamath Mountains, California: *Am. Jour. Sci.,* v. 262, p. 199–222.

——, and others, 1971, Evolving subduction zones in the western United States as interpreted from igneous rocks: *Science,* v. 174, p. 821–823.

Locke, A., and others, 1940, Sierra Nevada tectonic pattern: *Geol. Soc. American Bull.,* v. 51, p. 513–539.

Lockwood, J. P., and Lydon, P. A., 1975, Geologic map of the Mount Abbot quadrangle central Sierra Nevada, California: *U.S. Geol. Survey Geol. Quad. Map GQ–1155,* scale 1:62,500.

Lomnitz, C., and others, 1970, Seismicity and tectonics of the northern Gulf of California region, Mexico: preliminary results: *Geofisica Internac.,* v. 10, p. 37–48.

Loney, R. A., and Himmelberg, G. R., 1977, Geology of the Josephine Peridotite, Vulcan Peak area, southwestern Oregon: *U.S. Geol. Survey Jour. Res.,* v. 5, no. 6, p. 761–781.

Loney, R. A., and others, 1971, Structure and petrology of the alpine-type peridotite at Burro Mountains, California, U.S.A.: *Jour. Petrology,* v. 12, p. 245–309.

Longwell, C. R., 1926, Structure studies in southern Nevada and western Arizona: *Geol. Soc. America Bull.,* v. 37, p. 551–584.

——, 1973, Structural studies in southern Nevada and western Arizona: A correction: *Geol. Soc. America Bull.,* v. 84, p. 3717–3720.

——, and others, 1965, Geology and mineral deposits of Clark County, Nevada: *Nevada Burea of Mines Bull. 62,* 218 pp.

Loomis, A. A., 1966, Contact metamorphic reactions and processes in the Mt. Tallac roof remnant, Sierra Nevada, California: *Jour. Petrology,* v. 7, p. 221–245.

Louderback, G. D., 1947, Central California earthquakes of the 1830's: *Seismol. Soc. America Bull.,* v. 37, p. 33–74.

Lowe, G. D., 1974, Terrestrial paleoenvironmental implications of the middle Eocene Ardath flora (Ardath shale, La Jolla group), San Diego, California: Unpublished M.S. thesis, San Diego State University, San Diego, Calif., 90 pp.

Lowell, J. D., 1960, Ordovician miogeosynclinal margin in central Nevada: *21st Internat. Geol. Cong. Report,* part 7, p. 7–17.

——, 1972, Spitsbergen Tertiary orogenic belt and the Spitsbergen fracture zone: *Geol. Soc. America Bull.,* v. 83, p. 3091–3102.

Luyendyk, B. P., 1970, Dips of downgoing lithospheric plates beneath island arcs: *Geol. Soc. America Bull.,* v. 81, p. 3411–3416.

——, and Nichols, Jr., 1977, Petrologic nature of the oceanic Moho: Comment: *Geology,* v. 5, 1980, p. 578–579.

——, and others, 1980, Geometric model for Neogene crustal rotations in southern California, *Geol. Soc. Amer. Bull.,* v. 91, p. 211–217.

Macdonald, G. A., 1941, Geology of the western Sierra Nevada between the Kings River and the San Joaquin River, California: *Univ. Calif. Pubs. Geol. Sci. Bull.,* v. 26, p. 215–286.

Mack, S., and others, 1979, Origin and emplacement of the Academy pluton, Fresno County, California: *Geol. Soc. America Bull.,* v. 90, no. 4, p. 321–323.

MacLeod, N. S., and Pratt, R. M., 1973, Petrology of volcanic rocks recovered on Leg 18, *in* Kulm, L. D., and others, eds., *Initial Reports of the Deep Sea Drilling Project,* v. 18: Washington, D.C., U.S. Government Printing Office, p. 935–945.

MacMillan, J. R., 1972, Late Paleozoic and Mesozoic tectonic events in west-central Nevada: Unpublished Ph.D. dissertation, Northwestern University, Evanston, Ill., 146 pp.

Maddock, M. E., 1964, Geology of the Mt. Boardman quadrangle, Santa Clara and Stanislaus Counties, California: *Calif. Div. Mines and Geology, Map Sheet 3.*

Magaritz, M., and Taylor, H. P., 1976, Oxygen, hydrogen, and carbon isotopic studies of the Franciscan formation, Coast Ranges, California: *Geochim. Cosmochim. Acta,* v. 40, p. 215–234.

Malpas, J., and Strong, D. F., 1975, A comparison of chrome-spinels in ophiolites and mantle diapirs of Newfoundland: *Geochim. Cosmochim. Acta,* v. 39, p. 1045–1060.

Malpas, P., and Talkington, R. W., 1979, The significance of the critical zone in the petrogenesis of the Bay of Islands ophiolite, western Newfoundland: *Cyprus Geol. Survey Dept., Abstracts Internat. Ophiolite Symposium, Nicosia,* p. 98.

Malpica, C. R., 1972, Rocas marinas del Paleozoico tardio en el area de San Jose de Gracia, Sinaloa: *Soc. Geol. Mexicana, Mem. II Convención Nacional,* p. 174–175.

Mansfield, C. F., III, 1972, Petrography and sedimentology of the late Mesozoic Great Valley Sequence, near Coalinga, California: Unpublished Ph.D. dissertation, Stanford University, Stanford, Calif.

Marsh, O. T., 1960, Geology of the Orchard Peak area, California: *Calif. Div. Mines and Geology Spec. Report 62,* 42 pp.

Marshak, R. S., and Karig, D. E., 1977, Triple junctions as a cause for anomalously near-trench igneous activity between the trench and volcanic arc: *Geology,* v. 5, p. 233–236.

Marvin, R. F., and others, 1973, Radiometric ages of igneous rocks from Pima, Santa Cruz, and Cochise Counties, southeastern Arizona: *U.S. Geol. Survey Bull. 1379,* 27 pp.

Matthews, R. A., and Burnett, J. C., 1965, Fresno sheet, Geol. Map of California: *Calif. Div. Mines and Geology,* scale 1:250,000.

Matthews, Vincent, 1973, Pinnacles-Neenach correlations: A restriction for models of the origin of the Transverse Ranges and the big bend in the San Andreas fault: *Geol. Soc. America Bull.,* v. 84, p. 683–688.

——, 1976, Correlation of Pinnacles and Neenach Volcanic Formations and their bearing on San Andreas fault problem: *Am. Assoc. Petroleum Geologists Bull.,* v. 60, p. 2128–2141.

Matti, J. C., and McKee, E. H., 1977, Silurian and Lower Devonian paleogeography of the outer continental shelf of the Cordilleran miogeocline, central Nevada, *in* Stewart J. H., and others, eds., *Paleozoic Paleogeography of the Western United States:* Pacific Section, Soc. Econ. Paleontologists and Mineralogists, Pacific Coast Paleogeography Symposium 1, p. 181–216.

Mattinson, J. M., and Hill, D. J., 1976, Age of plutonic basement rocks, Santa Cruz Island, California, *in* Howell, D. G., ed., *Aspects of the Geologic History of the California Continental Borderland:* Pacific Section, Am. Assoc. Petroleum Geologists Misc. Pub. 24, p. 53–59.

——, and Hopson, C. A., 1972, Paleozoic ophiolite complexes in Washington and northern California: *Carnegie Inst. Washington Year Book 71,* p. 578–583.

——, and others, 1972, U-Pb studies of plutonic rocks of the Salinian block, California: *Carnegie Inst. Washington Year Book 71,* p. 571–576.

Maxwell, J. C., 1974, Anatomy of an orogen: *Geol. Soc. America Bull.,* v. 85, p. 1195–1204.

——, 1976, Possible serpentinite diapirism, Lake and Colusa Counties, California [abstract]: *Geol. Soc. America Abstracts with Programs,* v. 8, no. 3, p. 394.

——, 1977, Depositional environment of Jurassic-lower Cretaceous sediments, western margin of Sacramento Valley [abstract]: *Geol. Soc. America Abstracts with Programs,* v. 9, no. 4, p. 461–462.

May, J. C., and Hewitt, R. L., 1948, The basement complex in well samples from the Sacramento and San Joaquin Valleys, California: *Calif. Div. Mines Jour. Mines and Geology,* v. 44, no. 2, p. 129–158.

Mayo, E. B., 1941, Deformation in the interval Mount Lyell-Mount Whitney, California: *Geol. Soc. America Bull.,* v. 52, p. 1001–1084.

McCulloh, T. H., 1952, Geology of the southern half of the Lane Mountain quadrangle, California: Unpublished Ph.D. dissertation, University of California, Los Angeles, 182 pp.

McCulloch, T. H., 1954, Problems of the metamorphic and igneous rocks of the Mojave Desert: *Calif. Div. of Mines Bull. 170,* Sec. 2, Chapt. 7.

McEldowney, R. C., 1970a, Geology of the northern Sierra Pinta, Baja California, Mexico: Unpublished M.S. thesis, San Diego State Univeristy, San Diego, Calif., 78 pp.

——, 1970b, An occurrence of Paleozoic fossils in Baja California, Mexico [abstract]: *Geol. Soc. America Abstracts with Programs,* v. 2, p. 117.

McFall, C. C., 1968, Reconnaissance geology of the Concepcion Bay area, Baja California, Mexico: *Stanford Univ. Pub. Geol. Sci.,* v. 10, no. 5, 25 pp.

McHugh, C. A., and Lester, F. W., 1978, Catalog of earthquakes along the San Andreas fault system in central California, for the year 1976: *U.S. Geol. Survey Open File Report 78-1051,* 91 pp.

McJunkin, R. D., and others, 1979, An isotopic age for Smartville ophiolite and the obduction of metavolcanic rocks in the northwestern Sierran foothills, California [abstract]: *Geol. Soc. America Abstracts with Programs,* v. 11, no. 3, p. 91.

McKee, B., 1962, Widespread occurrence of jadeite, lawsonite, and glaucophane in central California: *Am. Jour. Sci.,* v. 260, p. 596–610.

McKee, E. H., 1971, Tertiary igneous chronology of the Great Basin of western United States—implications for tectonic models: *Geol. Soc. America Bull.,* v. 82, p. 3497–3502.

——, and Nash, D. B., 1967, Potassium-argon ages of granitic rocks in the Inyo batholith, east-central California: *Geol. Soc. America Bull.,* v. 78, no. 5, p. 669–680.

——, and Nelson, C. A., 1967, Geologic map of the Soldier Pass quadrangle, California and Nevada: *U.S. Geol. Survey Geol. Quad. Map GQ-654,* scale 1:62,500.

McKenzie, D., 1972, Active tectonics of the Mediterranean region: *Roy. Astron. Soc. Geophys. Jour.,* v. 30, p. 109–185.

McLaughlin, R. J., 1974, Preliminary geologic map of the Geysers steam field and vicinity, Sonoma County, California: *U.S. Geol. Survey Open-File Map 74-238.*

——, 1975, Preliminary field compilation of in-progress geologic mapping in the Geysers geothermal area, California: *U.S. Geol. Survey Open-File Map 75-198.*

——, 1976, Significance of age relationships of rocks above and below upper Jurassic ophiolite in the Geysers-Clear Lake area, California [abstract]: *Geol. Soc. America Abstracts with Programs,* v. 8, p. 394–395.

——, 1977, The Franciscan assemblage and Great Valley Sequence in the Geysers-Clear Lake region of northern California, in *Field Trip Guide to the Geysers-Clear Lake Area:* Cordilleran Section, Geol. Soc. America Annual Meeting, 1977, p. 3–24.

——, 1978, Preliminary geologic map and structural sections of the central Mayacamas Mountains and the Geysers steamfield, Sonoma, Lake, and Mendocino Counties, California: *U.S. Geol. Survey Open-File Report 78-389,* 1 map, scale 1:24,000, explanation and cross sections, 2 sheets.

——, and Pessagno, E. A., Jr., 1978, Significance of age relations of rocks above and below Upper Jurassic ophiolite in the Geysers-Clear Lake area, California: *U.S. Geol. Survey Jour. Res.,* v. 6, p. 715–726.

——, and Stanley, W. D., 1976, Pre-Tertiary geology and structural control of geothermal resources, the Geysers steam field, California: *United Nations Symposium on Development and Use of Geothermal Resources, 2d Proc., San Francisco,* v. 1, p. 475–485.

McLean, Hugh, and others, 1976, Miocene strata on Santa Cruz and Santa Rosa Islands—a reflection of tectonics in the southern California borderland, *in* Howell, D. G., ed., *Aspects of the Geologic History of the California Continental Borderland:* Pacific Section, Am. Assoc. Petroleum Geologists Misc. Pub. 24, p. 241–255.

McMath, V. E., 1966, Geology of the Taylorsville area, northern Sierra Nevada, California: *Calif. Div. Mines and Geology Bull.,* v. 190, p. 173–183.

McNally, K. C., and others, 1978, Earthquake swarm along the San Andreas fault near Palmdale, southern California, 1976 to 1977: *Science,* v. 201, p. 814–817.

Meeder, C. A., and others, 1977, The structure of the oceanic crust off southern Peru determined from an ocean bottom seismometer. *Earth Planet. Sci. Lett.,* v. 37, p. 13–28.

Menard, H. W., 1964, *Marine Geology of the Pacific:* New York, McGraw-Hill, 271 pp.

——, 1978, Fragmentation of the Farallon Plate by pivoting subduction: *Jour. Geology,* v. 86, p. 99–110.

Menzies, M., and others, 1977a, Rare earth and trace element geochemistry of a fragment of Jurassic seafloor, Point Sal, California: *Geochim. Cosmochim. Acta,* v. 41, p. 1419–1430.

——, and others, 1977b, Rare earth and trace element geochemistry of metabasalts from the Point Sal ophiolite, California: *Earth Planet. Sci. Lett.,* v. 37, p. 203–215.

Merriam, C. W., 1961, Silurian and Devonian rocks of the Klamath Mountains, California, in *Geological Survey Research, 1961:* U.S. Geol. Survey Prof. Paper 424-C, p. C188–C190.

——, 1963, Geology of the Cerro Gordo Mining District, Inyo County, California: *U.S. Geol. Survey Prof. Paper 408,* 83 pp.

——, and Anderson, C. A., 1942, Reconnaissance survey of the Roberts Mountains, Nevada: *Geol. Soc. America Bull.,* v. 53, p. 1675–1727.

Merriam, R. H., 1965, San Jacinto Fault in northwestern Sonora, Mexico: *Geol. Soc. America Bull.,* v. 76, p. 1051–1054.

——, 1972, Reconnaissance geologic map of Sonoyta Quadrangle, northwest Sonora, Mexico: *Geol. Soc. America Bull.,* v. 83, p. 3533–3536.

Merschat, W. R., 1971, Lower Tertiary paleocurrent trends, Santa Cruz Island, California: Unpublished M.A. thesis, Ohio University, Athens.

Miller, C. F., 1977a, Alkali-rich monzonites, California: Origin of near silica saturated alkaline rocks and their significance in a calc-alkaline batholithic belt: Unpublished Ph.D. dissertation, University of California, Los Angeles, 283 pp.

——, 1977b, Early alkalic plutonism in the calc-alkalic batholithic belt of California: *Geology,* v. 5, p. 685–688.

——, 1978, An early Mesozoic alkalic magmatic belt in western North America, *in* Howell, D. G., and McDougall, K. A., eds., *Mesozoic Paleogeography of the Western United States:* Pacific Section, Soc. Econ. Paleontologists and Mineralogists, Pacific Coast Paleogeography Symposium 2, p. 163–187.

Miller, E. L., 1977, Geology of the Victorville region, California: Unpublished Ph.D. dissertation, Rice University, Houston, Tex., 226 pp.

——, 1978, The Fairview Valley Formation: A Mesozoic intraorogenic deposit in the southwestern Mojave Desert, *in* Howell, D. G., and McDougall, K. A., eds., *Mesozoic Paleogeography of the Western United States:* Pacific Section, Soc. Econ. Paleontologists and Mineralogists, Pacific Coast Paleogeography Symposium 2, p. 277–282.

——, and Carr, M. D., 1978, Recognition of Aztec equivalent sandstones and associated Mesozoic metasedimentary deposits within the Mesozoic magmatic arc in the southwestern Mojave Desert, California, *in* Howell, D. E., and McDougall, K. A., eds., *Mesozoic Paleogeography of the Western United States:* Pacific Section, Soc. Econ. Paleontologists and Mineralogists, Pacific Coast Paleogeography Symposium 2, p. 283–290.

Miller, F. K., 1970, Geologic map of the Quartzite quadrangle, Yuma County, Arizona: *U.S. Geol. Survey Geol. Quad. Map GQ-841.*

——, and McKee, E. H., 1971, Thrust and strike-slip faulting in the Plomosa Mountains, southwestern Arizona: *Geol. Soc. America Bull.,* v. 82, p. 717–722.

——, and Morton, D. M., 1977, Comparison of granitic intrusions in the Pelona and Orocopia schists, southern California: *U.S. Geol. Survey Jour. Res.,* v. 5, p. 643–649.

Miller, W. J., 1931, Anorthosite in Los Angeles County, California: *Jour. Geology,* v. 39, p. 331–344.

——, 1934, Geology of the western San Gabriel Mountains of California: *Univ. Calif. Pub. Math. and Phys. Sci.,* v. 1, no. 1, p. 1–114.

——, 1944, Geology of the Palm Springs-Blythe strip, Riverside County, California: *Calif. Jour. Mines and Geology,* v. 40, p. 11–72.

Minch, J. A., 1969, A depositional contact between the pre-batholithic Jurassic and Cretaceous rocks in Baja California, Mexico [abstract]: *Geol. Soc. America Abstracts with Programs for 1969,* part 3, p. 42–43.

——, and others, 1976a, Geology of the Vizcaino Peninsula, *in* Howell, D. G., ed., *Aspects of the Geologic History of the California Continental Borderland:* Pacific Section, Am. Assoc. Petroleum Geologists Misc. Pub. 24, p. 136–195.

——, and others, 1976b, Clast populations in Sespe and Poway conglomerates and their possible bearing on the tectonics of the southern California borderland, *in* Howell, D. G., ed., *Aspects of the Geologic History of the California Continental Borderland:* Pacific Section, Am. Assoc. Petroleum Geologists Misc. Pub. 24, p. 256–265.

Minster, J. B., and Jordan, T. H., 1978, Present-day plate motions: *Jour. Geophys. Res.,* v. 83, p. 5331–5354.

Miyashiro, A., 1961, Evolution of metamorphic belts: *Jour. Petrology,* v. 2, p. 277–311.

——, 1967, Orogeny, regional metamorphism, and magmatism in the Japanese Islands: *Medd. Dansk Geol. Foren.,* v. 17, p. 390–446.

——, 1973, Paired and unpaired metamorphic belts: *Tectonophysics,* v. 17, p. 241–254.

Moiseyev, A. N., 1966, Geology and geochemistry of the Wilbur Springs quicksilver deposits, Lake and Colusa Counties, California: Unpublished Ph.D. dissertation, Stanford University, Stanford, Calif.

——, 1968, The Wilbur Springs quicksilver district (California): Example of a study of hydrothermal processes by combining field geology and theoretical geochemistry: *Econ. Geology,* v. 63, p. 169–181.

——, 1970, Late serpentinite movements in the California Coast Ranges: New evidence and its implications: *Geol. Soc. America Bull.,* v. 81, p. 1721–1732.

Molnar, P., and Tapponnier, P., 1975, Cenozoic tectonics of Asia: Effects of a continental collision: *Science,* v. 189, p. 419–426.

Monger, J. W. H., 1977, Upper Paleozoic rocks of the western Canadian Cordillera and their bearing on Cordilleran evolution: *Canadian Jour. Earth Sci.,* v. 14, p. 1832–1859.

——, and others, 1972, Evolution of the Canadian Cordillera: A plate-tectonic model: *Am. Jour. Sci.,* v. 272, p. 577–602.

Moody, J. D., and Hill, M. J., 1957, Wrenchfault tectonics: *Geol. Soc. America Bull.,* v. 67, p. 1207–1246.

Moore, D. E., 1977, Sedimentary, deformational, and metamorphic history of Franciscan conglomerates of the Diablo Range, California: Unpublished Ph.D. dissertation, Stanford University, Stanford, Calif., 261 pp.

——, and Liou, J. G., 1977, Discovery of some detrital blueschist pebbles in two Franciscan metaconglomerates near Red Mountain, northeast Diablo Range, California [abstract]: *Geol. Soc. America Abstracts with Programs,* v. 9, p. 1099.

Moore, D. G., 1969, Reflection profiling studies of the California Continental Borderland: *Geol. Soc. America Spec. Paper 107,* 138 pp.

——, 1973, Plate-edge deformation and crustal growth, Gulf of California structural province: *Geol. Soc. America Bull.,* v. 84, p. 1883–1906.

——, 1978, Submarine slides, *in* Voight, B., *Rockslides and Avalanches: Natural Phenomena,* Pt. 1 (Dev. in Geotech. Engr. Ser., Vol. 14a): Elsevier, Chapter 16, Amsterdam.

——, and Buffington, E. C., 1968, Transform faulting and growth of the Gulf of California since the Late Pliocene: *Science,* v. 161, p. 1238–1241.

Moore, G. F., and Karig, D. E., 1976, Deformed trench and trench-slope deposits on Nias Island, Indonesia [abstract]: *Geol. Soc. America Abstracts with Programs,* v. 8, p. 1017.

Moore, G. W., 1976, Basin development in the California borderland of the Basin and Range province, in Howell, D. G., ed., *Aspects of the Geologic History of the California Continental Borderland:* Pacific Section, Am. Assoc. Petroleum Geologists Misc. Pub. 24, p. 383–392.

Moore, J. C., and Connelly, W., 1979, Tectonic history of the continental margin of southwestern Alaska: Late Triassic to earliest Tertiary, *in* Watkins, J. S., and others, eds., *Geol. and Geophys. Invest. of Continental Margins:* Am. Assoc. Petroleum Geologists Mem. 29, p. TN 860/851/M.

——, and Karig, D. E., 1976, Sedimentology, structural geology, and tectonics of the Shikoku subduction zone, southwestern Japan: *Geol. Soc. America Bull.,* v. 87, p. 1259–1268.

——, and others, 1979, Middle America Trench: *Geotimes,* v. 24, no. 9, p. 20–22.

Moore, J. G., 1959, The quartz diorite boundary line in the western United States: *Jour. Geology,* v. 67, p. 197–210.

——, 1962, K/Na ratio of Cenozoic igneous rocks of the western United States: *Geochim. Cosmochim. Acta,* v. 26, p. 101–130.

——, 1963, Geology of the Mount Pinchot quadrangle, southern Sierra Nevada, California: *U.S. Geol. Survey Bull. 1130,* 152 pp.

——, 1965, Petrology of deep-sea basalt near Hawaii: *Am. Jour. Sci.,* v. 263, p. 40–52.

——, 1970, Water content of basalt erupted on the ocean floor: *Contrib. Mineral. Petrol.,* v. 28, p. 272–279.

——, and Dodge, F. C., 1962, Mesozoic age of metamorphic rocks in the Kings River area, southern Sierra Nevada, California, in *Geological Survey Research 1962:* U.S. Geol. Survey Prof. Paper 450–B, p. B19–B21.

——, and du Bray, E., 1978, Mapped offset of the right-lateral Kern Canyon fault, southern Sierra Nevada, California: *Geology,* v. 6, p. 205–208.

——, and Hopson, C. A., 1961, The Independence dike swarm in eastern California: *Am. Jour. Sci.,* v. 259, p. 241–259.

——, and Marks, L. Y., 1972, Mineral resources of the High Sierra primitive area: *U.S. Geol. Survey Bull. 1371–A,* p. 1–39.

——, and Schilling, J. G., 1973, Vesicles, water, and sulfur in Reykjanes Ridge basalts: *Contrib. Mineral. Petrol.,* v. 41, p. 105–118.

Moore, J. G., and others, 1979, Geologic guide to the Kings Canyon highway, central Sierra Nevada, California: *Geol. Soc. America,* Cordilleran Section, 75th Ann. Mtg., San Jose, Calif., 33 p.

Moore, J. N., 1976, Depositional environments of the Lower Cambrian Poleta Formation and its stratigraphic equivalents, California and Nevada: *Brigham Young Univ. Geology Series,* v. 23, pt. 2, p. 23–38.

Moore, J. N., and Foster, C. T., Jr., 1980, Lower Paleozoic metasedimentary rocks in the east-central Sierra Nevada, California: Correlations with Great Basin formations: *Geol. Soc. America Bull.,* v. 91, p. 37–43.

Moore, S. C., in preparation, Rooted thrust faults of the eastern Sierra Nevada batholithic margin.

Moores, E. M., 1970, Ultramafics and orogeny, with models of the U.S. Cordillera and the Tethys: *Nature,* v. 228, p. 837–842.

——, 1972, Model for Jurassic island arc-continental margin collision in California [abstract]: *Geol. Soc. America Abstracts with Programs,* v. 4, no. 3, p. 202.

——, and Jackson, E. D., 1974, Ophiolites and ocean crust: *Nature,* v. 250, p. 136–139.

——, and others, 1979, The Nevadan orogeny, northern Sierra Nevada: An abrupt arc-arc collision [abstract]: *Geol. Soc. America Abstracts with Programs,* v. 11, no. 3, p. 118.

Moran, A. I., 1976, Allochthonous carbonate debris in Mesozoic flysch deposits in Santa Ana Mountains, California. *Am. Assoc. Petroleum Geologists Bull.,* v. 60, no. 11, p. 2038–2043.

Morgan, B. A., 1973, Tuolumne River ophiolite, western Sierra Nevada, California [abstract]: *Geol. Soc. America Abstracts with Programs,* v. 5, p. 83.

——, 1976, Geology of Chinese Camp and Moccasin quadrangles, Tuolumne County, California: *U.S. Geol. Survey Misc. Field Studies Map MF-840,* scale 1:24,000.

——, and Rankin, D. W., 1972, Major structural break between Paleozoic and Mesozoic rocks in the eastern Sierra Nevada, California: *Geol. Soc. America Bull.,* v. 83, p. 3739–3744.

——, and Stern, T. W., 1977, Chronology of tectonic and plutonic events in the western Sierra Nevada between Sonora and Mariposa, California [abstract]: *Geol. Soc. America Abstracts with Programs,* v. 9, no. 4, p. 471–472.

Morganstern, N. R., and Tchalenko, J. S., 1967, Microstructural observations on shear zones for slips in natural clays: *Geotech. Conf., Oslo, 1967, Proc.,* v. 1, p. 147–152.

Morrell, R. P., 1978, Geology and mineral paragenesis of Franciscan metagraywacke near Paradise Flat, northwest of Pacheco Pass, California: Unpublished M.S. thesis, Stanford University, Stanford, Calif., 73 pp.

Moscoso, B. A., 1967, A thick section of "flysch" in the Santa Ana Mountains, southern California: Unpublished M.S. thesis, San Diego State University, San Diego, Calif., 106 pp.

Muehlberger, W. R., 1958, Geology of northern Soledad Basin, Los Angeles County, California: *Am. Assoc. Petroleum Geologists Bull.,* v. 42, p. 1812–1844.

Muffler, L. J. R., and White, D. E., 1969, Active metamorphism of Upper Cenozoic sediments in the Salton Sea Geothermal Field and the Salton Trough, southeastern California: *Geol. Soc. America Bull.,* v. 80, p. 157–182.

Mulholland, W., 1918, Earthquakes in their relation to the Los Angeles aqueduct: *Seismol. Soc. America Bull.,* v. 8, p. 13–19.

Mullan, H. S., 1975, Unpublished manuscript submitted to the Circum-Pacific Plutonism Project, 3rd Meeting, San Diego State University, San Diego, Calif., 16 pp.

Muller, S. W., Ferguson, H. W., and Roberts, R. J., 1951, Geology of the Mount Tobin quadrangle, Nevada: *U.S. Geol. Survey Geol. Quad. Map GQ-7.*

Mullineaux, D. R., and others, 1975, Widespread late glacial and postglacial tephra deposits from Mount St. Helens volcano, Washington: *U.S. Geol. Survey Jour. Res.,* v. 3, p. 329–335.

Murphy, M. A., and others, 1969, Geology of the Ono quadrangle, Shasta and Tehama Counties, California: *Calif. Div. Mines and Geology Bull. 192,* 28 pp.

Nardin, T. R., and Henyey, T. L., 1978, Plio-Pleistocene diastrophism in the area of the Santa Monica and San Pedro shelves, California Continental Borderland: *Am. Assoc. Petroleum Geologists Bull.,* v. 62, no. 2, p. 247–272.

Nason, R. D., 1971, Measurement and theory of fault creep slippage in central California: *Roy. Soc. New Zealand Bull. 9,* p. 181–187.

——, 1973, Fault creep and earthquakes on the San Andreas Fault, *in* Kovach, R. L., and

Nur, Amos, eds., *Proceedings, Conference on Tectonic Problems of the San Andreas Fault System:* Stanford Univ. Pubs. Geol. Sci., v. 13, p. 275–285.

——, 1978, Seismic intensities in the 1906 earthquake fault zone [abstract]: *Earthquake Notes,* v. 49, p. 1.

Natland, M. L., 1957, Paleoecology of west coast Tertiary sediments, *in* Ladd, H. S., ed., *Treatise on Marine Ecology and Paleoecology:* Geol. Soc. America Mem. 67, p. 543–571.

Nelson, C. A., 1966a, Geologic map of the Waucoba Mountain quadrangle, Inyo County, California: *U.S. Geol. Survey Geol. Quad. Map GQ–528,* scale 1:62,500.

——, 1966b, Geologic map of the Blanco Mountain quadrangle, Inyo and Mono Counties, California: *U.S. Geol. Survey Geol. Quad. Map GQ–529,* scale 1:62,500.

——, 1976, Late Precambrian-Early Cambrian stratigraphic and faunal succession of eastern California and the Precambrian-Cambrian boundary, *in* Moore, J. N., and Fritsche, A. E., eds., *Depositional Environments of Lower Paleozoic Rocks in the White-Inyo Mountains, Inyo County, California:* Pacific Section, Soc. Econ. Paleontologists and Mineralogists, Pacific Coast Paleogeography Field Guide 1, p. 31–42.

——, and Sylvester, A. G., 1971, Wall rock decarbonation and forcible emplacement of the Birch Creek pluton, southern White Mountains, California: *Geol. Soc. America Bull.,* v. 82, p. 2891–2904.

——, and others, 1978, Geologic map of Papoose Flat pluton, Inyo Mountains, California: *Geol. Soc. America Map Series MC–20.*

Nichols, G., 1977, The seismic structure of the Point Sal ophiolite and its relationship to oceanic crustal structure: Unpublished M.A. thesis, University of California, Santa Barbara.

Nichols, J. and others, in press, Seismic velocity structure of the ophiolite at Point Sal, southern California, determined from laboratory measurements: *Geophys. Jour. Roy. Astron. Soc.* v. 60.

Nicolas, A., and Jackson, E. D., 1972, Répartition en deux provinces des péridotites des chaines alpines longeant la Méditerranée: Implications tectoniques: *Bull. Suisse Min. Petrog.,* v. 52, p. 479–495.

——, and others, 1980, Interpretation of peridotite structures from ophiolitic and oceanic environments: *Am. Jour. Sci.,* E. D. Jackson volume v. 280–A, p. 192–210.

Nilsen, T. H., 1977, Paleogeography of Mississippian turbidites in south-central Idaho, *in* Stewart, J. H., and others, eds., *Paleozoic Paleogeography of the Western United States:* Pacific Section, Soc. Econ. Paleontologists and Mineralogists, Pacific Coast Paleogeography Symposium 1, p. 275–300.

——, 1978, Late Cretaceous geology of California and the problem of the proto-San Andreas fault, *in* Howell, D. G., and McDougall, K. A., eds., *Mesozoic Paleogeography of the Western United States:* Pacific Section, Soc. Econ. Paleontologists and Mineralogists, Pacific Coast Paleogeography Symposium 2, p. 559–573.

——, and Clarke, S. H., Jr., 1975, Sedimentation and tectonics in the early Tertiary continental borderland of central California: *U.S. Geol. Survey Prof. Paper 925,* 64 pp.

——, and Dibblee, T. W., 1979, Geology of the central Diablo Range between Hollister and New Idria, California: *Cordilleran Section, Geol. Soc. America Meeting, San Jose, Calif., April 9–11, 1979, Field Trip Guidebook.*

Noble, D. C., 1962, Mesozoic geology of the southern Pine Nut Range, Douglas County, Nevada: Unpublished Ph.D. dissertation, Stanford University, Stanford, Calif., 251 pp.

Noble, L. F., 1926, The San Andreas rift and some other active faults in the desert region of southeastern California: *Carnegie Inst. Washington Year Book 25,* p. 415–428.

——, 1941, Structural features of the Virgin Spring area, Death Valley, California: *Geol. Soc. America Bull.,* v. 52, p. 941–1000.

Nockolds, S. R., 1954, Average chemical composition of some igneous rocks: *Geol. Soc. America Bull.,* v. 65, p. 1007–1032.

Nokleberg, W. J., 1970, Geology of the Strawberry Tungsten Mine roof pendant: Unpublished Ph.D. dissertation, University of California, Santa Barbara, 156 pp.

——, 1975, Structural analysis of a collision between an oceanic plate and a continental plate along the lower Kings River in the Sierra Nevada [abstract]: *Geol. Soc. America Abstracts with Programs,* v. 7, p. 357–358.

——, and Kistler, R. W., 1977, Mesozoic deformations in the central Sierra Nevada, California [abstract]: *Geol. Soc. America Abstracts with Programs,* v. 9, no. 4, p. 475.

——, and Kistler, R. W., 1980, Paleozoic and Mesozoic deformations, central Sierra Nevada, California, *U.S. Geol. Surv. Prof. Paper 1145,* 24 p.

Novitsky, J. M., and Burchfiel, B. C., 1973, Pre-Aztec (Upper Triassic(?)-Lower Jurassic) thrusting, Cowhole Mountains, southeastern California [abstract]: *Geol. Soc. America Abstracts with Programs,* v. 5, no. 7, p. 755.

Novitsky-Evans, J. M., 1978, Geology of the Cowhole Mountains, California: Unpublished Ph.D. dissertation, Rice University, Houston, Tex., 102 pp.

Nutt, C. J., 1977, The Escondido mafic-ultramafic complex, a concentrically zoned body in the Santa Lucia Range, California: Unpublished M.S. thesis, Stanford University, Stanford, Calif., 90 pp.

Oakeshott, G. B., 1937, Geology and mineral deposits of the western San Gabriel Mountains, Los Angeles County: *Calif. Div. Mines, Calif. Jour. Mines and Geology,* v. 33, no. 3, p. 215–249.

——, 1954, Geology of the western San Gabriel Mountains, Los Angeles County, *in* Jahns, R. H., ed. *Geology of Southern California:* Calif. Div. Mines Bull. 170, map sheet 9.

——, 1958, Geology and mineral deposits of San Fernando quadrangle, Los Angeles County, California: *Calif. Div. Mines Bull. 172,* 147 pp.

Obradovich, J. D., and Cobban, W. A., 1975, A time-scale of the Late Cretaceous of the western interior of North America, *in* Caldwell, W. G. E., ed., *Cretaceous System in the Western Interior of North America:* Geol. Assoc. of Canada Spec. Paper 13, p. 31–54.

O'Connor, J. T., 1965, A classification for quartz-rich igneous rocks based on feldspar ratios, in *Geological Survey Research, 1965:* U.S. Geol. Survey Prof. Paper 525–B, p. B79–B84.

O'Hara, M. J., 1977, Geochemical evolution during fractional crystallization of a periodically refilled magma chamber: *Nature,* v. 266, no. 5602, p. 503–507.

Ojakangas, R. W., 1968, Cretaceous sedimentation, Sacramento Valley, California: *Geol. Soc. America Bull.,* v. 79, p. 973–1008.

Oliver H. W., 1977, Gravity and magnetic investigations of the Sierra Nevada batholith, California: *Geol. Soc. America Bull.,* v. 88, no. 3, p. 445–461.

——, and Robbins, S. L., 1973, Complete Bouguer anomaly map of the Mariposa $1^\circ \times 2^\circ$ and part of the Goldfield $1^\circ \times 2^\circ$ quadrangles: *U.S. Geol. Survey Open-File Map,* scale 1:250,000.

——, and Robbins, S. L., 1975, Preliminary complete Bouguer gravity map of Fresno 1° X 2° quadrangle, California: *U.S. Geol. Survey Open File Report 74-187.*

Olmsted, F. H., and others, 1973, Geohydrology of the Yuma area: *U.S. Geol. Survey Prof. Paper 486-H,* 227 pp.

Olson, J. C., and others, 1954, Rare earth mineral deposits of the Mountain Pass District, San Bernardino County, California: *U.S. Geol. Survey Prof. Paper 261.*

Oxburgh, E. R., and Turcotte, D. L., 1971, Origin of paired metamorphic belts and crustal dilation in island arc regions: *Jour. Geophys. Res.,* v. 76, p. 1315–1327.

Packer, D. R., and Stone, D. B., 1974, Paleomagnetism of Jurassic rocks from southern Alaska and the tectonic implications: *Canadian Jour. Earth Sci.,* v. 11, no. 7, p. 976–997.

Page, B. M., 1966, Geology of the Coast Ranges of California, *in* Bailey, E. H., ed., *Geology of Northern California:* Calif. Div. Mines and Geology Bull. 190, p. 255–276.

——, 1969, Relation between ocean floor spreading and the structure of the Santa Lucia Range, California [abstract]: *Geol. Soc. America Abstracts with Programs for 1969,* part 3, p. 51–52.

——, 1970a, Sur-Nacimiento fault zone of California—continental Margin tectonics: *Geol. Soc. America Bull.,* v. 81, p. 667–690.

——, 1970b, Time of completion of underthrusting of Franciscan beneath Great Valley rocks west of Salinian block, California: *Geol. Soc. America Bull.,* v. 81, p. 2825–2834.

——, 1972, Oceanic crust and mantle fragment in subduction complex near San Luis Obispo, California: *Geol. Soc. America Bull.,* v. 83, p. 957–972.

——, 1978, Franciscan melanges compared with olistostromes of Taiwan and Italy: *Tectonophysics,* v. 47, p. 223–246.

——, and Tabor, L. L., 1967, Chaotic structure and decollement in Cenozoic rocks near Stanford University California: *Geol. Soc. America Bull.,* v. 78, p. 1–12.

——, and others, 1979, Geologic cross section of the continental margin off San Luis Obispo, the southern Coast Ranges, and the San Joaquin Valley, California: *Geol. Soc. America Map and Chart Series MC-28G.*

Page, P. G. N., 1978, Serpentinites of northern Sumatra [unpublished abstract]: *Workshop on the Sumatra Transect,* 1 p.

Pallister, J. S., and Hopson, C. A., 1979, Samail ophiolite magma chamber: II, Evidence from the cryptic variation and mineral chemistry: *Cyprus Geol. Survey Dept., Abstracts Internat. Ophiolite Symposium, Nicosia,* p. 38.

Palmer, H. D., 1965, Geology of Richardson Rock, northern Channel Islands, Santa Barbara County, California: *Geol. Soc. America Bull.,* v. 76, no. 10, p. 1197–1202.

Parkison, G. A., 1976, Tectonics and sedimentation along a Late Jurassic(?) active continental margin, western Sierra Nevada foothills, California: Unpublished M.S. thesis, University of California, Berkeley, 160 pp.

Parsons, Barry, and Sclater, J. H., 1977, An analysis of the variation of ocean floor bathymetry and heat flow with age: *Jour. Geophys. Res.,* v. 82, p. 803–827.

Paul, R. C., and others, 1976, Geological and operational summary, southern California deep stratigraphic test ocs-CAL 75-70 no. 1, Cortes Bank area offshore California: *U.S. Geol. Survey Open-File Report 76-232,* 65 pp.

Payne, C. M., and others, 1978a, Evidence for post-Wisconsin displacement on the Hosgri fault near Point Sal, California [abstract]: *Trans. Am. Geophys. Union,* v. 59, no. 12, p. 1210.

——, and others, 1978b, Investigation of the Hosgri fault offshore southern California, Point Sal to Point Conception: Prepared for U.S. Geol. Survey by Fugro, Inc., Los Angeles, Calif., 17 pp.

——, and others, 1979, Quaternary faulting in the offshore region between Point Sal and Point Conception, California [abstract]: *Geol. Soc. America Abstracts with Programs,* v. 11, p. 121.

Pechmann, J. C., in press, Tectonic implications of small earthquakes in the central Transverse Ranges, California: *U.S. Geol. Survey Prof. Paper.*

Peck, D. L., 1964, Preliminary geologic map of the Merced Peak quadrangle, California: *U.S. Geol. Survey Mineral Inv. Field Studies Map MF-281,* scale 1:48,000.

——, and others, 1964, Geology of the central and northern parts of the western Cascade Range in Oregon: *U.S. Geol. Survey Prof. Paper 449,* 56 pp.

Pelka, G. J., 1973a, Geology of the McCoy and Palen Mountains, southeastern California [abstract]: *Geol. Soc. America Abstracts with Programs,* v. 5, no. 1, p. 89–90.

——, 1973b, Geology of the McCoy and Palen Mountains, southeastern California: Unpublished Ph.D. dissertation, University of California, Santa Barbara, 162 pp.

Pelton, P. J., 1966, Mississippian rocks of southwestern Great Basin, Nevada and California: Unpublished Ph.D. dissertation, Rice University, Houston, Tex., 99 pp.

Perfit, R., and Heezen, B. C., 1978, The geology and evolution of the Cayman Trench: *Geol. Soc. America Bull.,* v. 89, p. 1155–1174.

Peselnick, Louis, and others, 1977, Anisotropic elastic velocities of some upper mantle xenoliths underlying the Sierra Nevada batholith: *Jour. Geophys. Res.,* v. 82, no. 14, p. 2005–2010.

Pessagno, E. A., Jr., 1973, Age and significance of radiolarian cherts in the California Coast Ranges: *Geology,* v. 1, p. 153–156.

——, 1976, Radiolarian zonation and stratigraphy of the upper Cretaceous portion of the Great Valley Sequence, California Coast Ranges: *Micropaleontology,* Spec. Pub. 2, p. 1–95.

——, 1977a, Upper Jurassic radiolaria and radiolarian biostratigraphy of the California Coast Ranges: *Micropaleontology,* v. 23, p. 56–113.

——, 1977b, Lower Cretaceous radiolarian biostratigraphy of the Great Valley Sequence and Franciscan complex, California Coast Ranges: *Cushman Found. for Foraminiferal Research, Spec. Pub. 15,* 87 pp.

——, and Newport, R. L., 1972, A technique for extracting radiolaria from radiolarian cherts: *Micropaleontology,* v. 18, no. 2, p. 231–234.

——, and Poisson, A., in press, Lower Jurassic radiolaria from the Gümüslü allochthon of southwestern Turkey (Taurides Occidentales): *Bull. of Mineral Res. and Explor. Institute of Turkey:* 55 ms pp.

Peterman, Z. E., and others, 1967, $^{87}Sr/^{86}Sr$ ratios in some eugeosynclinal sedimentary rocks and their bearing on the origin of granitic magma in orogenic belts: *Earth Planet. Sci. Lett.,* v. 2, p. 433–439.

Peterson, M. S., 1975, Geology of the Coachella fanglomerate, *in* Crowell, J. C., ed., *San Andreas Fault in Southern California:* Calif. Div. Mines and Geology Spec. Report 118, p. 119–126.

Phipps, S. P., and others, 1979, Ophiolitic olistostromes in the basal Great Valley Sequence, Napa County, California Coast Ranges [abstract]: *Geol. Soc. America Abstracts with Programs,* v. 11, no. 3, p. 122.

Pike, J. E. N., 1974, Intrusions and intrusive complexes in the San Luis Obispo ophiolite: A chemical and petrologic study: Unpublished Ph.D. dissertation, Stanford University, Stanford, Calif., 212 pp.

——, 1976, Bulk chemistry of the San Luis Obispo ophiolite: Genetic implications [abstract] : *Trans. Am. Geophys. Union,* v. 57, no. 12, p. 1026.

Pilger, R. H., Jr., 1978, A closed Gulf of Mexico, pre-Atlantic Ocean plate reconstruction and the early rift history of the Gulf and North Atlantic: *Gulf Coast Assoc. of Geol. Societies, Trans.,* v. 28, p. 385–393.

——, and Henyey, T. L., 1977, Plate tectonic reconstruction and the Neogene tectonic and volcanic evolution of the California Borderland and Coast Ranges [abstract] : *Geol. Soc. America Abstracts with Programs,* v. 9, p. 482.

——, and Henyey, T. L., 1979, Mutual constraints on neogene volcanism in coastal California and the continental borderland, Pacific-North American plate interaction, and the development of the San Andreas fault system: *Tectonophysics,* v. 57, p. 189–210.

Piper, D. J. W., and others, 1973, Late Quaternary sedimentation in the active eastern Aleutian Trench: *Geology,* v. 1, p. 19–22.

Pischke, G. M., and others, 1979, Miocene paleomagnetic study of the tectonic history of Baja California, Mexico [abstract] : *Geol. Soc. America Abstracts with Programs,* v. 11, no. 3, p. 122.

Pitcher, W. S., 1978, The anatomy of a batholith: *Jour. Geol. Soc. London,* v. 135, p. 158–182.

Pitman, W. C., III, and others, 1974, The age of the ocean basins: *Geol. Soc. America Map MC-6,* scale 1:39,000,000.

Platt, J. P., 1975, Metamorphic and deformational processes in the Franciscan Complex, California: Some insights from the Catalina schist terrane: *Geol. Soc. America Bull.,* v. 86, p. 1337–1347.

——, 1976, The significance of the Catalina Schist in the history of southern California borderland, *in* Howell, D. G., ed., *Aspects of the Geological History of the California Continental Borderland:* Pacific Section, Amer. Assoc. Petroleum Geologists Misc. Pub. 24, p. 47–52.

——, and Stuart, C. J., 1974, Newport-Inglewood fault zone, Los Angeles Basin, California: Discussion: *Am. Assoc. Petroleum Geologists Bull.,* v. 58, no. 5, p. 877–898.

——, and others, 1976, Franciscan blueschist-facies metaconglomerate, Diablo Range, California: *Geol. Soc. America Bull.,* v. 87, p. 581–591.

Poole, F. G., 1974, Flysch deposits of Antler foreland basin, western United States, *in* Dickinson, W. R., ed., *Tectonics and Sedimentation:* Soc. Econ. Paleontologists and Mineralogists Spec. Pub. 22, p. 58–82.

——, and Sandberg, C. A., 1977, Mississippian paleogeography and tectonics of the western United States, *in* Stewart, J. H., and others, eds., *Paleozoic Paleogeography of the Western United Sates:* Pacific Section, Soc. Econ. Paleontologists and Mineralogists, Pacific Coast Paleogeography Symposium 1, p. 67–86.

——, and others, 1977, Silurian and Devonian paleogeography of the western United States, *in* Stewart, J. H., and others, eds., *Paleozoic Paleogeography of the Western*

United States: Pacific Section, Soc. Econ. Paleontologists and Mineralogists, Pacific Coast Paleogeography Symposium 1, p. 39–65.

Potter, A. W., and others, 1977, Stratigraphy and inferred tectonic framework of lower Paleozoic rocks in the eastern Klamath Mountains, northern California, *in* Stewart, J. H., and others, eds., *Paleozoic Paleogeography of the Western United States:* Pacific Section, Soc. Econ. Paleontologists and Mineralogists, Pacific Coast Paleogeography Symposium 1, p. 421–440.

Presnall, D. C., and Bateman, P. C., 1973, Fusion relationships in the system $NaAlSi_3O_8$–$CaAl_2Si_2O_8$–$KAlSi_3O_8$–SiO_2–H_2O and generation of granitic magmas in the Sierra Nevada batholith: *Geol. Soc. America Bull.,* v. 84, no. 10, p. 3181–3202.

Prestvik, T., 1979, A Caledonian ophiolite complex of Leka, north central Norway: *Cyprus Geol. Survey Dept., Abstracts Internat. Ophiolite Symposium, Nicosia,* p. 99–100.

Proctor, R. J., and others, 1970, Crossing the Sierra Madre fault zone in the Glendora Tunnel, San Gabriel Mountains, California: *Engineering Geology,* v. 4, p. 5–8, 57–63.

Raitt, R. W., 1963, The crustal rocks, *in* Maxwell, A. E., ed., *The Sea, Vol. 3:* New York, John Wiley and Sons, p. 85–102.

Ramberg, Hans, 1967, *Gravity, Deformation and the Earth's Crust:* London and New York, Academic Press, 214 pp.

——, 1970, Model studies in relation to intrusion of plutonic bodies, *in* Newall, G., and Rast, N., eds., *Mechanisms of Igneous Intrusion:* Geological Jour. Special Issue No. 2, Seel House Press, Liverpool, p. 261–286.

——, 1972, Theoretical models of density stratification and diapirism in the earth: *Jour. Geophys. Res.,* v. 77, p. 877–889.

Ramp, Lenin, 1975, Geology and mineral resources of the Upper Chetco drainage area, Oregon, including the Kamliopsis Wilderness and Big Craggies Botanical areas: *Oregon Dept. Geology and Mineral and Industries Bull. 88,* 49 pp.

Raney, J. A., 1973, Great Valley-Coast Range contact–an alternate hypothesis: [abstract] : *Geol. Soc. America Abstracts with Programs,* v. 5, no. 1, p. 93.

——, 1974, Geology of the Elk Creek-Stonyford area, northern California: Unpublished Ph.D. dissertation, University of Texas at Austin.

Rangin, Claude, 1978a, Speculative model of Mesozoic geodynamics, central Baja California to northeastern Sonora (Mexico), *in* Howell, D. G., and McDougall, K. A., eds., *Mesozoic Paleogeography of the Western United States:* Pacific Section, Soc. Econ. Paleontologists and Mineralogists, Pacific Coast Paleogeography Symposium 2, p. 85–106.

——, 1978b, Consideraciones sobre la evolucion geologica de la parte septentrional del Estado de Sonora, *in* Roldan, Q. J., and Salas, G. A., eds., *Libreto Guia, Primer Simposio Sobre la Geologia y Potencial Minero en el E Trado de Sonora:* Inst. Geologia, Universidad, Nacional Autónoma de Mexico, Hermosilo, Sonora, Mexico, Mayo, 1978, p. 35–56.

——, and Roldan, J., 1978, Preliminary report on middle Jurassic volcanics in northern Sonora, Mexico [abstract] : *Geol. Soc. America Abstracts with Programs,* v. 10, no. 3, p. 143.

Raymond, L. A., 1970, Cretaceous sedimentation and regional thrusting, northeastern Diablo Range, California: *Geol. Soc. America Bull.,* v. 81, p. 2123–2128.

——, 1973a, Franciscan geology of the Mount Oso area, California: Unpublished Ph.D. dissertation, University of California, Davis, 180 pp.

——, 1973b, Tesla-Ortigalita fault, Coast Range thrust fault, and Franciscan metamorphism, northeastern Diablo Range, California: *Geol. Soc. America Bull.*, v. 84, p. 3547–3562.

——, 1974, Possible modern analogs for rocks of the Franciscan Complex, Mount Oso area, California: *Geology*, v. 2, p. 143–146.

Reed, R. D., 1933, *Geology of California:* Tulsa, Okla., Am. Assoc. Petroleum Geologists, 355 pp.

——, and Hollister, J. S., 1936, *Structural Evolution of Southern California:* Tulsa, Okla., Am. Assoc. Petroleum Geologists, 157 pp.

Rhodes, J. M., and others, in press, Magma mixing at mid-ocean ridges: Evidence from basalts drilled near 22°N on the Mid-Atlantic Ridge: *Tectonophysics.*

Rich, E. I., 1971, Geologic map of the Wilbur Springs quadrangle, Colusa and Lake Counties, California: *U.S. Geol. Survey Misc. Geol. Inv. Map I-538,* scale 1:48,000.

Rich, Mark, 1971, Fusulinids from central Mojave Desert and adjoining area, California: *Jour. Paleontology,* v. 45, p. 1022–1207.

——, 1977, Pennsylvanian paleogeographic patterns in the western United States, *in* Howell, D. G., and McDougall, K. A., eds., *Mesozoic Paleogeography of the Western United States:* Pacific Section, Soc. Econ. Paleontologists and Mineralogists, Pacific Coast Paleogeography Symposium 2, p. 277–282.

Rindosh, M. C., 1977, Geology of the Tylerhorse Canyon pendant, southern Tehachapi Mountains, Kern County, California: Unpublished M.S. thesis, University of Southern California, Los Angeles, 80 pp.

Rinehart, C. D., and Ross, D. C., 1957, Geologic map of the Casa Diablo Mountain quadrangle, California: *U.S. Geol. Survey Geol. Quad. Map GQ-99,* scale 1:62,500.

——, and Ross, D. C., 1964, Geology and mineral deposits of the Mount Morrison quadrangle, Sierra Nevada, California, *with a section on* A gravity study of Long Valley, by L. C. Pakiser: *U.S. Geol. Survey Prof. Paper 385,* 106 pp.

——, and others, 1959, Paleozoic fossils in a thick stratigraphic section in the eastern Sierra Nevada, California: *Geol. Soc. America Bull.*, v. 70, p. 941–946.

Ringwood, A. E., 1975, *Composition and Petrology of the Earth's Mantle:* New York, McGraw-Hill, 618 pp.

Roberts, R. J., 1964, Stratigraphy and structure of the Antler Peak quadrangle, Humboldt and Lander Counties, Nevada: *U.S. Geol. Survey Prof. Paper 459A,* 93 pp.

——, 1968, Tectonic framework of the Great Basin: *Jour. Univ. Missouri, Rolla,* ser. 1, no. 1, p. 101–119.

——, and others, 1958, Paleozoic rocks of north-central Nevada: *Am Assoc. Petroleum Geologists Bull.*, v. 42, p. 2813–2857.

Robinson, J. W., 1974, Structure and stratigraphy of the northern Vizcaino Peninsula, Territory of Baja California, Mexico: Unpublished M.S. thesis, San Diego State University, San Diego, Calif., 114 pp.

Robinson, P.T., and others, 1976, Quaternary volcanism in the Salton Sea geothermal field, Imperial Valley, California: *Geol. Soc. America Bull.*, v. 87, p. 347–360.

Rodriquez-Torres, Rafael, 1972, Itinerario Geologicao Torreon, Coahuila to Durango, Durango, *in* Cordoba, D. A., and others, eds., *Memoria II Convencion Nacional:* Sociedad Geologia Mexicana, p. 17–77.

Roeder, P. L., and Emslie, R. F., 1970, Olivine-liquid equilibrium: *Contrib. Mineral. Petrol.,* v. 29, p. 275–289.

Rogers, J. J. W., and others, 1974, Paleozoic and Lower Mesozoic volcanism and continental growth in the western United States: *Geol. Soc. America Bull.,* v. 85, p. 1913–1924.

Rogers, T. H., 1966, San Jose sheet, Geologic Map of California: *Calif. Div. Mines and Geology,* scale 1:250,000.

Rogers, T. H., 1967, San Bernardo sheet, Geological Map of California: *Calif. Div. Mines and Geology,* scale 1:250,000.

Ross, D. C., 1958, Igneous and metamorphic rocks of parts of Sequoia and Kings Canyon National Parks, California: *Calif. Div. Mines and Geology Spec. Report 53,* 24 pp.

——, 1962, Correlation of granitic plutons across faulted Owens Valley, California: *U.S. Geol. Survey Prof. Paper 450-D,* p. 86–88.

——, 1967a, Generalized geologic map of the Inyo Mountains region, California: *U.S. Geol. Survey Misc. Geol. Inv. Map I-506,* scale 1:125,000.

——, 1967b, Geologic map of the Waucoba Wash quadrangle, Inyo County, California: *U.S. Geol. Survey Geol. Quad. Map GQ-612,* scale 1:62,500.

——, 1969, Recently active breaks along the San Andreas fault between Tejon Pass and Cajon Pass, southern California: *U.S. Geol. Survey Misc. Geol. Inv. Map I-553,* scale 1:24,000.

——, 1970, Quartz gabbro and anorthositic gabbro: Markers of offset along the San Andreas fault in the California Coast Ranges: *Geol. Soc. America Bull.,* v. 81, p. 3647–3662.

——, 1972a, Geologic map of the pre-Cenozoic basement rocks, Gabilan Range, Monterey and San Benito Counties, California: *U.S. Geol. Survey Misc. Field Studies Map MF-357,* scale 1:125,000.

——, 1972b, Petrographic and chemical reconnaissance study of some granitic and gneissic rocks near the San Andreas fault from Bodega Head to Cajon Pass, California: *U.S. Geol. Survey Prof. Paper 698,* 92 pp.

——, 1974, Map showing basement geology and location of wells drilled to basement, Salinian block, central and southern Coast Ranges, California: *U.S. Geol. Survey Misc. Field Studies Map MF-588,* scale 1:500,000.

——, 1976a, Reconnaissance geologic map of the pre-Cenozoic basement rocks, northern Santa Lucia Range, Monterey County, California: *U.S. Geol. Survey Misc. Field Studies Map MF-750,* scale 1:125,000.

——, 1976b, Metagraywacke in the Salinian block, central Coast Ranges, California—and a possible correlative across the San Andreas fault: *U.S. Geol. Survey Jour. Res.,* v. 4, p. 683–696.

——, 1977a, Pre-intrusive metasedimentary rocks of the Salinian block, California—a paleotectonic dilemma, *in* Stewart, J. H., and others, eds., *Paleozoic Paleogeography of the Western United States:* Pacific Section, Soc. Econ. Paleontologists and Mineralogists, Pacific Coast Paleogeography Symposium 1, p. 371–380.

——, 1977b, Maps showing sample localities and ternary plots and graphs showing modal and chemical data for granitic rocks of the Santa Lucia Range, Salinian block, California Coast Ranges: *U.S. Geol. Survey Misc. Field Studies Map MF-799.*

——, 1978, The Salinian block—a Mesozoic granitic orphan in the California Coast Ranges *in* Howell, D. G., and McDougall, K. A., eds., *Mesozoic Paleogeography of the West-*

ern United States: Pacific Section, Soc. Econ. Paleontologists and Mineralogists, Pacific Coast Paleogeography Symposium 2, p. 509–522.

——, and McCulloch, D. S., 1979, Cross section of the Southern Coast Ranges and San Joaquin Valley from offshore of Point Sur to Madera, California: *Geol. Soc. America Map and Chart Series MC-28H.*

——, and others, 1973, Cretaceous mafic conglomerate near Gualala offset 350 miles by San Andreas fault from oceanic crustal source near Eagle Rest Peak, California: *U.S. Geol. Survey Jour. Res.,* v. 1, p. 45–52.

Ruetz, J. W., 1976, Paleocene sedimentation in the northern Santa Lucia Range: Unpublished M.S. thesis, Stanford University, Stanford, Calif., 104 pp.

Russell, L. R., and Cebull, S. E., 1977, Structural-metamorphic chronology in a roof pendant near Oakhurst, California: Implications for the tectonics of the western Sierra Nevada: *Geol. Soc. America Bull.,* v. 88, p. 1530–1534.

Russell, S. J., and Nokelberg, W. J., 1977, Superimposition and timing of deformations in the Mount Morrison roof pendant and in the central Sierra Nevada, California: *Geol. Soc. America Bull.,* v. 88, p. 335–345.

Saad, A. H., 1969, Paleomagnetism of Franciscan ultramafic rocks from Red Mountain, California: *Jour. Geophys. Res.,* v. 74, p. 6567–6578.

Sage. O. G., Jr., 1973, Paleocene geography of southern California: Unpublished Ph.D. dissertation, University of California, Santa Barbara, 250 pp.

——, 1975, Sedimentological and tectonic implications of the Paleocene San Francisquito Formation, Los Angeles County, California, *in* Crowell, J. C., ed., *San Andreas Fault in Southern California:* Calif. Div. Mines and Geology Spec. Report 118, p. 162–169.

Salas, G. A., 1968, Areal geology and petrology of the igneous rocks of the Santa Ana region, northwest Sonora: *Soc. Geol. Mexicana Bol.,* v. 31, p. 11–63.

Saleeby, J. B., 1975, Structure, petrology and geochronology of the Kings-Kaweah mafic-ultramafic belt, southwestern Sierra Nevada foothills, California: Unpublished Ph.D. dissertation, University of California, Santa Barbara, 286 pp.

——, 1976, Onset of Cretaceous magmatism, southern Sierra Nevada batholith, California [abstract] : *Geol Soc. America Abstracts with Programs,* v. 9, no. 7, p. 1084.

——, 1977a, Fieldtrip guide to the Kings-Kaweah suture, southwestern Sierra Nevada foothills, California: *Geol. Soc. America, Cordilleran Section 73d Annual Meeting, Fieldtrip Guidebook,* 47 pp.

——, 1977b, Fracture zone tectonics, continental margin fragmentation, and emplacement of the Kings-Kaweah ophiolite belt, southwest Sierra Nevada, California, *in* Coleman, R. E., and Irwin, W. P., eds., *North American Ophiolites:* Oregon Dept. Geology and Mineral Industries Bull. 95, p. 141–160.

——, 1978a, Kings River Ophiolite, southwest Sierra Nevada foothills, California: *Geol. Soc. America Bull.,* v. 89, p. 617–636.

——, 1978b, Paleo-basement geology of the Sierra Nevada foothill metamorphic belt and its implications on the Nevadan orogeny [abstract] : *Geol. Soc. America Abstracts with Programs,* v. 10, no. 3, p. 145.

——, 1979, Kaweah serpentinite melange, southwest Sierra Nevada foothills, California: *Geol. Soc. America Bull.,* v. 90, p. 29–46.

Saleeby, J. B., and Behrman, P. G., in press, Igneous and metamorphic chronology of ophiolitic basement - Central and Southern Sierra Nevada Foothill Metamorphic Belt: *Contrib. Mineral and Petrology.*

——, and Chen, J., 1978, Preliminary report on initial lead and strontium isotopes from ophiolitic and batholithic rocks, southwest Sierra Nevada [abstract] : *Abstracts of Internat. Conf. on Geochronology, Cosmochronology and Isotope Geology.*

——, and Moores, E. M., 1979, Zircon ages on northern Sierra Nevada ophiolite remnants and some possible regional correlations [abstract] : *Geol. Soc. America Abstracts with Programs,* v. 11, no. 3, p. 125.

——, and Sharp, W. D., 1977, Jurassic igneous rocks emplaced along the Kings-Kaweah suture–southwestern Sierra Nevada foothills, California [abstract] : *Geol. Soc. America Abstracts with Programs,* v. 9, p. 494–495.

——, and Sharp, W. D., 1980, Geochronological development of the Sierra Nevada foothills between the Kings and Tule Rivers, California: *Geol. Soc. America Bull.,* v. 91, p. 317–332.

——, and Williams, Howell, 1978, Possible origin for California Great Valley gravity-magnetic anomalies [abstract] : *Trans. Am. Geophys. Union,* v. 59, p. 1189.

——, and others, 1978, Early Mesozoic paleotectonic-paleogeographic reconstruction of the southern Sierra Nevada region, *in* Howell, D. G., and McDougall, K. A., eds., *Mesozoic Paleogeography of Western United States:* Pacific Section, Soc. Econ. Paleontologists and Mineralogists, Pacific Coast Paleogeography Symposium 2, p. 311–336.

——, and others, 1979, Regional ophiolite terranes of California–vestiges of two complex ocean floor assemblages [abstract] : *Geol. Soc. America Abstracts with Programs,* v. 11, no. 7, p. 509.

Sanborn, A. F., 1960, Geology and paleontology of the southwest quarter of the Big Bend quadrangle, Shasta County, California: *Calif. Div. Mines Spec. Report 63,* 26 pp.

Santillán, M., and Barrera, T., 1930, Los Posibilidades petroliferas en la costa occidental de la Baja California, entre los paralelos 30 y 32 de latitud norte: *Mexico Instituto Geologia Anales,* v. 5, p. 1–37.

Sarna-Wojcicki, A., 1977, Correlation and age of a widespread Pleistocene ash bed in northern California and western Nevada [abstract] : *Geol. Soc. America Abstracts with Programs,* v. 9, p. 1155.

Saunders, A. D., and others, 1979, Ophiolites as ocean crust or marginal basin crust: A geochemical approach: *Cyprus Geol. Survey Dept., Abstracts Internat. Ophiolite Symposium, Nicosia,* p. 124–125.

Savage, J. C., and Burford, R. O., 1973, Geodetic determination of the relative plate motion in California: *Jour. Geophys. Res.,* v. 78, p. 832–845.

——, and others, 1978, Strain in southern California: Measured uniaxial north-south regional contraction: *Science,* v. 202, p. 883–885.

Schilling, F. A., Jr., 1962, The Upper Cretaceous stratigraphy of the Pecheco Pass quadrangle, California: Unpublished Ph.D. dissertation, Stanford University, Stanford, Calif., 146 pp.

Schmitka, R. O., 1973, Evidence for major right-lateral separation of Eocene rocks along the Santa Ynez fault, Santa Barbara and Ventura Counties, California [abstract] : *Geol. Soc. America Abstracts with Programs,* v. 5, no. 1, p. 104.

Scholl, D. W., and Marlow, M. S., 1974, Sedimentary sequences in modern Pacific trenches and the deformed circum-Pacific eugeosyncline, *in* Dott, R. H., and Shaver, R. H., eds., *Modern and Ancient Geosynclinal Sedimentation:* Soc. Econ. Paleontologists and Mineralogists Spec. Pub. 19, p. 193–211.

——, and Vallier, T. L., 1979, Tectonic erosion at a convergent margin, a mechanism that could have contributed to the truncation of Cordilleran geosynclinal trends [abstract] : *Geol. Soc. America Abstracts with Programs,* v. 11, p. 126.

——, and others, 1977, Sediment subduction and offscraping at Pacific margins, *in* Talwani, Manik, and Pitman, W. C., III, eds., *Island Arcs, Deep Sea Trenches, and Back-Arc Basins:* Am. Geophys. Union, Maurice-Ewing Series, v. 1, p. 199–210.

Scholz, C. H., and others, 1969, Seismic and aseismic slip on the San Andreas fault: *Jour. Geophys. Res.,* v. 74, p. 2049–2069.

Schweickert, R. A., 1974, Probable Late Paleozoic thrust fault near Sierra City, California [abstract] : *Geol. Soc. America Abstracts with Programs,* v. 6, p. 251.

——, 1976a, Early Mesozoic rifting and fragmentation of the Cordilleran orogen in the western U.S.A.: *Nature,* v. 260, p. 586–591.

——, 1976b, Shallow-level plutonic complexes in the eastern Sierra Nevada, California and their tectonic implications: *Geol. Soc. America Spec. Paper 176,* 58 pp.

——, 1976c, Lawsonite blueschist within the Melones fault zone, northern Sierra Nevada, California [abstract] : *Geol. Soc. America Abstracts with Programs,* v. 8, p. 409.

——, 1977, Major pre-Jurassic thrust fault between the Shoo Fly and Calaveras complexes, Sierra Nevada, California [abstract] : *Geol. Soc. America Abstracts with Programs,* v. 9, p. 497.

——, 1978, Triassic and Jurassic paleogeography of the Sierra Nevada and adjacent regions, California and western Nevada, *in* Howell, D. G., and McDougall, K. A., eds., *Mesozoic Paleogeography of the Western United States:* Pacific Section, Soc. Econ. Paleontologists and Mineralogists, Pacific Coast Paleogeography Symposium 2, p. 361–384.

——, 1979, Structural sequence of the Calaveras Complex between the Stanislaus and Tuolumne Rivers [abstract] : *Geol. Soc. America Abstracts with Programs,* v. 11, p. 127.

——, in press, Early Mesozoic magmatic arc in eastern California and western Nevada: *Tectonophysics,*

——, and Cowan, D. S., 1975, Early Mesozoic tectonic evolution of the western Sierra Nevada, California: *Geol. Soc. America Bull.,* v. 86, p. 1329–1336.

——, and others, 1977, Paleotectonic and paleogeographic significance of the Calaveras Complex, western Sierra Nevada, California, *in* Stewart J. H., and others, eds., *Paleozoic Paleogeography of the Western United States:* Pacific Section, Soc. Econ. Paleontologists and Mineralogists, Pacific Coast Paleogeography Symposium 1, p. 381–394.

——, and others, 1980, Lawsonite blueschist in the northern Sierra Nevada, California: *Geology,* v. 8, p. 27–31.

Sclater, J. G., and Fisher, R. L., 1974, Evolution of the east central Indian Ocean, with emphasis on the tectonic setting of the Ninetyeast Ridge: *Geol. Soc. America Bull.,* v. 85, p. 683–702.

——, and others, 1971, Elevation of ridges and evolution of the central eastern Pacific: *Jour. Geophys. Res.,* v. 76, p. 7888–7915.

Sears, D. H., 1953, Origin of Amargosa chaos, Virgin Spring area, Death Valley, California: *Jour. Geology,* v. 61, p. 182–186.

Seely, D. R., and others, 1974, Trench slope model, *in* Burk, C. A., and Drake, C. L., eds., *The Geology of Continental Margins:* New York, Springer-Verlag, p. 249–260.

Seiders, V. M., 1978, Onshore stratigraphic comparisons across the San Simeon and

Hosgri faults, California [abstract]: *Geol. Soc. America Abstracts with Programs,* v. 10, no. 3, p. 146.

——, 1979, San Gregorio-Hosgri fault zone south of Monterey Bay, California: A reduced estimate of maximum displacement: *U.S. Geol. Survey Open-File Report 79-385,* 10 pp.

——, and others, 1979, Radiolarians and conodonts from pebbles in Franciscan assemblage and Great Valley Sequence rocks of the California Coast Ranges: *Geology,* v. 7, p. 37–40.

Seyfert, C. K., 1974, Geology of the Sawyers Bar quadrangle, in *Geologic Guide to the Southern Klamath Mountains:* Geol. Soc. Sacramento, Annual Field Trip Guidebook, p. 69–81.

Sharaskin, A. Y., and Dobretsov, N. L., 1979, Marianites: The clinoenstative bearing pillow lavas associated with ophiolites of Mariana Island arc: *Cyprus Geol. Survey Dept., Abstracts Internat. Ophiolite Symposium, Nicosia,* p. 71–72.

Sharp, R. P., 1954, Physiographic features of faulting in southern California: *Calif. Div. Mines and Geology Bull. 170,* Chap. 5, p. 21–28.

Sharp, R. V., 1967, San Jacinto fault zone in the Peninsular Ranges of southern California: *Geol. Soc. America Bull.,* v. 78, p. 705–730.

——, 1975, En echelon fault patterns of the San Jacinto fault zone: *Calif. Div. Mines and Geology Spec. Report 118,* p. 147–152.

——, 1976, Surface faulting in Imperial Valley during the earthquake swarm of January-February, 1975: *Seismol. Soc. America Bull.,* v. 66, p. 1145–1154.

——, 1977, Map showing Holocene surface expression of the Brawley Fault, Imperial County, California: *U.S. Geol. Survey Misc. Field Studies Map MF-838.*

——, and Clark, M. M., 1972, Geologic evidence of previous faulting near the 1968 rupture on the Coyote Creek fault: *U.S. Geol. Survey Prof. Paper 787,* p. 131–140.

——, and others, 1972, The Borrego Mountain earthquake of April 9, 1968: *U.S. Geol. Survey Prof. Paper 787,* 207 pp.

Sharp, W. D., and Saleeby, J. B., 1979, The Calaveras Formation and syntectonic mid-Jurassic plutons between the Stanislaus and Tuolumne Rivers, California [abstract]: *Geol. Soc. America Abstracts with Programs,* v. 11, no. 3, p. 127.

Shelton, J. S., 1955, Glendora volcanic rocks, Los Angeles basin, California: *Geol. Soc. America Bull.,* v. 66, p. 45–90.

Sherard, J. L., and others, 1974, Poentially active faults in dam foundations: *Geotechnique,* v. 24, p. 367–428.

Shor, G. G., Jr., and others, 1970, Structure of the Pacific basin, *in* Maxwell, A. E., ed., *The Sea,* Vol. 4: New York, John Wiley and Sons, p. 3–28.

Shulz, S. S., and others, 1976, Catalog of creepmeter measurements in central California from 1973 through 1975: *U.S. Geol. Survey Open-File Report 77-31.*

Sieh, K. E., 1978a, Central California foreshocks of the great 1857 earthquake: *Seismol. Soc. America Bull.,* v. 68, p. 1731–1749.

——, 1978b, Slip along the San Andreas fault associated with the great 1857 earthquake: *Seismol. Soc. America Bull.,* v. 68, p. 1421–1448.

——, 1978c, Prehistoric large earthquakes produced by slip on the San Andreas fault at Pallett Creek, California: *Jour. Geophys. Res.,* v. 83, p. 3907–3939.

Sigit, S., 1962, Geologic map of Indonesia: *U.S. Geol. Survey Misc. Geol. Inv. Map I-414,* scale 1:2,000,000.

Silberling, N. J., 1973, Geologic events during Permian-Triassic time along the Pacific margin of the United States, *in* Logan, A., and Hills, L. V., eds., *The Permian and Triassic Systems and Their Mutual Boundary:* Alberta Soc. Petroleum Geologists Mem. 2, p. 345–362.

——, 1975, Age relationships of the Golconda thrust fault, Sonoma Range, north-central Nevada: *Geol. Soc. America Spec. Paper 163,* 28 pp.

——, and Roberts, R. J., 1962, Pre-Tertiary stratigraphy and structure of northwestern Nevada: *Geol. Soc. America Spec. Paper 72,* 58 pp.

——, and others, 1961, Upper Jurassic fossils from Bedford Canyon Formation, southern, California: *Am. Assoc. Petroleum Geologists Bull.,* v. 45, p. 1746–1748.

Silver, E. A., 1974, Structural interpretation from free-air gravity on the California continental margin, 35° to 40°N [abstract] : *Geol. Soc. America Abstracts with Programs,* v. 6, p. 253.

——, 1975, Geophysical studies and tectonic development of the continental margin off the western United States [abstract] : *Geol. Soc. America Abstracts with Programs,* v. 7, p. 1273.

——, and Moore, J. C., 1978, The Molucca Sea collision zone, Indonesia: *Jour. Geophys. Res.,* v. 83, p. 1681–1691.

Silver, L. T., 1966, Preliminary history of the crystalline complex of the central Transverse Ranges, Los Angeles County, California [abstract] : *Geol. Soc. America Abstracts with Programs,* v. 8, p. 201.

——, 1971, Problems of crystalline rocks of the Transverse Ranges [abstract] : *Geol. Soc. America Abstracts with Programs,* v. 3, p. 193–194.

——, 1974, Precambrian and "Precambrian(?)" crystalline rocks of the southwestern basin and range province of California and Arizona [abstract] : *Geol. Soc. America Abstracts with Programs,* v. 6, p. 253–254.

——, and Anderson, T. H., 1974, Possible left-lateral early to middle Mesozoic disruption of the southwestern North American craton margin [abstract] : *Geol. Soc. America Abstracts with Programs,* v. 6, p. 955–956.

——, and others, 1963, Precambrian age determination in the western San Gabriel Mountains, California: *Jour. Geology,* v. 71, p. 196–214.

——, and others, 1975, Petrological, geochemical, and geochronological asymmetries of the Peninsular Ranges batholith [abstract] : *Geol. Soc. America Abstracts with Programs,* v. 7, p. 375–376.

——, and others, 1977, Chronostratigraphic elements of the Precambrian rocks of the southwestern and far western United States [abstract] : *Geol. Soc. America Abstracts with Programs,* v. 9, no. 7, p. 1176.

Simpson, R. W., and Cox, A., 1977, Paleomagnetic evidence for tectonic rotation of the Oregon coast range: *Geology,* v. 5, p. 585–589.

Sinton, J. M., 1977, Equilibration history of the basal alpine-type periodotite, Red Mountain, New Zealand: *Jour. Petrology,* v. 18, p. 216–246.

Sleep, N. E., 1975, Formation of oceanic crust: Some thermal constraints: *Jour. Geophys. Res.,* v. 80, p. 4037–4042.

Slemmons, D. B., 1977, Faults and earthquake magnitude, Report 6, Misc. Paper S–73–1, State-of-the-art for Assessing Earthquake Hazard in the United States: U.S. Army Engineer Waterways Experiment Station, 166 pp.

Smith, D. P., 1977, San Juan-St. Francis fault—hypothesized major middle Tertiary right-lateral fault in central and southern California: *Calif. Div. Mines and Geology Spec. Report 129,* p. 41–50.

Smith, G. I., and Ketner, K. B., 1970, Lateral displacement on the Garlock fault, southeastern California, suggested by offset section of similar metasedimentary rocks: *U.S. Geol. Survey Prof. Paper 700-D,* pp. D1–D9.

Smith, G. W., and Ingersoll, R. V., 1978, The Cambria slab, San Luis Obispo County, California: A Late Cretaceous trench slope basin deposit [abstract]: *Geol. Soc. America Abstracts with Programs,* v. 10, p. 147.

Smith, S. W., and Wyss, Max, 1968, Displacement on the San Andreas fault initiated by the 1966 Parkfield earthquake: *Seismol. Soc. America Bull.,* v. 58, p. 1955–1973.

Snoke, A. W., 1977, A thrust plate of ophiolitic rocks in the Preston Peak area, Klamath Mountains, California: *Geol. Soc. America Bull.,* v. 88, p. 1641–1659.

Snyder, W. S., 1977, Origin and exploration for ore deposits in upper Paleozoic chert-greenstone complexes of northern Nevada: Unpublished Ph.D. dissertation, Stanford University, Stanford, Calif., 159 pp.

——, and Dickinson, W. R., 1979, Geometry of triple junctions related to San Andreas transform: *Jour. Geophys. Res.,* v. 84, p. 561–572.

——, and others, 1976, Tectonic implications of space-time patterns of Cenozoic magmatism in the western United States: *Earth Planet, Sci. Lett.,* v. 32, p. 91–106.

Soliman, S. M., 1965, Geology of the east half of the Mount Hamilton quadrangle, California: *Calif. Div. Mines and Geology Bull. 185,* 32 pp.

Speed, R. C., 1971a, Premo-Triassic continental margin tectonics in western Nevada [abstract]: *Geol. Soc. America Abstracts with Programs,* v. 3, p. 199.

——, 1971b, Golconda thrust, western Nevada: Regional extent [abstract]: *Geol. Soc. America Abstracts with Programs,* v. 3, p. 199–200.

——, 1974, Record of continental margin tectonics in the Candelaria Formation, Nevada [abstract]: *Geol. Soc. America Abstracts with Programs,* v. 6, p. 259.

——, 1977, Island-arc and other paleogeographic terranes of Late Paleozoic age in the western great basin, *in* Stewart, J. H., and others, eds., *Paleozoic Paleogeography of the Western United States:* Pacific Section, Soc. Econ. Paleontologists and Mineralogists, Pacific Coast Paleogeography Symposium 1, p. 349–362.

——, 1978, Paleogeographic and plate tectonic evolution of the early Mesozoic marine province of the western Great Basin, *in* Howell, D. G., and McDougall, K. A., eds., *Mesozoic Paleogeography of the Western United States:* Pacific Section, Soc. Econ. Paleontologists and Mineralogists, Pacific Coast Paleogeography Symposium 2, p. 253–270.

——, 1979, Collided Paleozoic platelet in the western United States: *Jour. Geology,* v. 87, no. 3, p. 279–292.

——, and Kistler, R. W., 1977, Correlations between pre-Tertiary rocks of the Great Basin and the Sierra Nevada [abstract]: *Geol. Soc. America Abstracts with Programs,* v. 9, no. 4, p. 505–506.

Spooner, E. T. C., and Fyfe, W. S., 1973, Sub-sea-floor metamorphism, heat and mass transfer: *Contrib. Mineral. and Petrology, v. 42,* p. 287–304.

Standlee, L. A., 1977, Structure and stratigraphy of the Shoo Fly Formation, northern Sierra Nevada, California [abstract]: *Geol. Soc. America Abstracts with Programs,* v. 9, p. 507.

Stanley, K. O., 1971, Tectonic and sedimentary history of Lower Jurassic Sunrise and

Dunlap Formations, west-central Nevada: *Am. Assoc. Petroleum Geologists Bull.*, v. 55, no. 3, p. 454.

——, and others, 1971, New hypothesis of Early Jurassic paleogeography and sediment dispersal for western United States: *Am. Assoc. Petroleum Geologists Bull.*, v. 55, p. 10–19.

——, and others, 1977, Depositional setting of some eugeosynclinal Ordovician rocks and structurally interleaved Devonian rocks in the Cordilleran mobile belt, Nevada, *in* Stewart, J. H., and others, eds., *Paleozoic Paleogeography of the Western United States:* Pacific Section, Soc. Econ. Paleontologists and Mineralogists, Pacific Coast Paleogeography Symposium 1, p. 259–274.

Stein, S., and Okal, E. A., 1978, Seismicity and tectonics of the Ninetyeast ridge area: Evidence for internal deformation of the Indian plate: *Jour. Geophys. Res.*, v. 83, p. 2233–2245.

Steinbrugge, K. V., and Moran, D. F., 1955, Structural damage to buildings, *in* Oakeshott, G. B., ed., *Earthquakes in Kern County During 1952:* Calif. Div. Mines Bull. 171, p. 259–270.

——, and Zacher, E. G., 1960, Creep on the San Andreas fault–fault creep and property damage: *Seismol. Soc. America Bull.*, v. 50, p. 389–396.

Stern, C., 1979, Open and closed system igneous fractionation within two Chilean ophiolites and the tectonic implication: *Contrib. Mineral. Petrol.*, v. 68, p. 243–258.

Stern, T. W., and others, in press, Isotopic Pb-U ages of zircon from the granitoids of the central Sierra Nevada batholith: *Jour. Geology.*

Stevens, C. H., and Ridley, A. P., 1974, Middle Paleozoic off-shelf deposits in southeastern California: Evidence for proximity of the Antler orogenic belt: *Geol. Soc. America Bull.*, v. 85, p. 27–32.

Stewart, J. H., 1967, Possible large right-lateral displacement along fault and shear zones in the Death Valley-Las Vegas area, California and Nevada: *Geol. Soc. America Bull.*, v. 78, p. 131–142.

——, 1970, Upper Precambrian and Lower Cambrian strata in the southern Great Basin, California and Nevada: *U.S. Geol. Survey Prof. Paper 620,* 206 pp.

——, 1971, Basin and Range structure: A system of horsts and grabens produced by deep-seated extension: *Geol. Soc. America Bull.*, v. 82, p. 1019–1044.

——, 1972, Initial deposits in the Cordilleran geosyncline: Evidence of a late Precambrian (< 850 m.y.) continental separation. *Geol. Soc. America Bull.*, v. 83, p. 1345–1360.

——, 1976, Late Precambrian evolution of North America: Plate tectonic implication: *Geology*, v. 4, p. 11–15.

——, 1978, Basin-range structure in western North America–A review, *in* Smith, R. B., and Eaton, G. P., eds., *Cenozoic Tectonics and Regional Geophysics of the Western Cordillera:* Geol. Soc. America Mem. 152, p. 7–24.

——, and Carlson, J. E., 1976, Geologic map of north-central Nevada: *Nevada Bureau of Mines and Geology Map 50.*

——, and Poole, F. G., 1974, Lower Paleozoic and uppermost Precambrian Cordilleran miogeocline, Great Basin, western United States, *in* Dickinson, W. R., ed., *Tectonics and Sedimentation:* Soc. Econ. Paleontologists and Mineralogists Special Publ. 22, p. 28–57.

——, and Poole, F. G., 1975, Extension of the Cordilleran miogeosynclinal belt to the San Andreas fault, southern California: *Geol. Soc. America Bull.*, v. 86, p. 205–212.

——, and Suczek, C. A., 1977, Cambrian and latest Precambrian paleogeography and tectonics in the western United States, *in* Stewart, J. H., and others, eds., *Paleozoic Paleogeography of the Western United States:* Pacific Section, Soc. Econ. Paleontologists and Mineralogists, Pacific Coast Paleogeography Symposium 1, p. 1–18.

——, and others, 1966, Last Chance thrust—a major fault in the eastern part of Inyo County, California: *U.S. Geol. Survey Prof. Paper 550-D, p. D23–D34.*

——, and others, 1968, Summary of regional evidence for right-lateral displacement in the western Great Basin: *Geol. Soc. America Bull.,* v. 79, p. 1407–1414.

——, and others, 1974, Geologic map of the Piper Peak quadrangle, Nevada-California: *U.S. Geol. Survey Geol. Quad. Map GQ-1186,* scale 1:62,500.

——, and others, 1977, Deep-water upper Paleozoic rocks in north-central Nevada—a study of the type area of the Havallah Formation, *in* Stewart, J. H., and others, eds., *Paleozoic Paleogeography of the Western United States:* Pacific Section, Soc. Econ. Paleontologists and Mineralogists, Pacific Coast Paleogeography Symposium 1, p. 337–348.

Stewart, R., and Peseknick, L., 1978, Systematic behavior of compressional velocity in Franciscan rocks at high temperature and pressure: *Jour. Geophys. Res.,* v. 83, p. 831–839.

Stewart, S. W., 1968, Preliminary comparison of seismic travel times and inferred crustal structure adjacent to the San Andreas fault in the Diablo and Gavilan Ranges of central California, *in* Dickinson, W. R., and Grantz, Arthur, eds., *Proceedings of Conference on Geologic Problems of San Andreas Fault System:* Stanford Univ. Pub. Geol. Sci., v. 11, p. 218–230.

Stone, D. B., and Packer, D. R., 1977, Tectonic implications of Alaska Peninsula paleomagnetic data: *Tectonophysics,* v. 37, p. 183–201.

Strand, R. G., 1959, San Luis Obispo sheet, Geologic Map of California: *Calif. Div. Mines,* scale 1:250,000.

——, 1967, Mariposa sheet, Geologic Map of California: *Calif. Div. Mines and Geology,* scale 1:250,000.

Streitz, Robert, and Stinson, M. C., 1977, Death Valley sheet, Geologic Map of California: *Calif. Div. Mines and Geology,* scale 1:250,000.

Stuart, C., 1976, Source terrane of the San Onofre breccia—Preliminary notes, *in* Howell, D. G., ed., *Aspects of the Geologic History of the California Continental Borderland:* Pacific Section, Am. Assoc. Petroleum Geologists Misc. Pub. 24, p. 309–325.

Suczek, C. A., 1977a, Sedimentology and petrology of the Cambrian Harmony Formation of north-central Nevada [abstract]: *Geol. Soc. America Abstracts with Programs,* v. 9., p. 510.

——, 1977b, Tectonic relations of the Harmony Formation, northern Nevada: Unpublished Ph.D. dissertation, Stanford University, Stanford, Calif. 96 pp.

Suppe, John, 1970, Offset of Late Mesozoic basement terrains by the San Andreas fault system: *Geol. Soc. America Bull.,* v. 81, p. 3253–3258.

——, 1972, Interrelationships of high-pressure metamorphism, deformation, and sedimentation in Franciscan tectonics, U.S.A.: *Reports of 24th Internat. Geol. Cong., Montreal, Sec. 3,* p. 552–559.

——, 1973, Geology of the Leech Lake Mountain-Ball Mountain region, California: *Univ. Calif. Publ. Geol. Sci.,* v. 107, p. 1–81.

——, 1977, The Coast Range décollement and post-subduction compression in northern California [abstract]: *Geol. Soc. America Abstracts with Programs,* v. 5, no. 4, p. 510–511.

——, 1978, Structural interpretation of the southern part of the northern Coast Ranges and Sacramento Valley: Summary: *Geol. Soc. America Bull.,* v. 90, p. 327–330 (map summary, GSA Map and Chart Series, MC-28B).

——, and Armstrong, R. L., 1972, Potassium-argon dating of Franciscan metamorphic rocks: *Am. Jour. Sci.,* v. 272, p. 217–233.

——, and Foland, K. A., 1978, The Goat Mountain schists and Pacific Ridge complex: A redeformed but still intact late Mesozoic schuppen complex, *in* Howell, D. G., and McDougall, K. A., eds., *Mesozoic Paleogeography of the Western United States:* Pacific Section, Soc. Econ. Paleontologists and Mineralogists, Pacific Coast Paleogeography Symposium 2, p. 431–451.

Sutter, J. F., 1968, Geochronology of major thrusts southern Great Basin, California: Unpublished M.S. thesis, Rice University, Houston, Tex. 32 pp.

Swe, W., and Dickinson, W. R., 1970, Sedimentation and thrusting of late Mesozoic rocks in the Coast Ranges near Clear Lake, California: *Geol. Soc. American Bull.,* v. 81, p. 165–188.

Sylvester, A. G., and Babcock, J. W., 1975, Significance of multiphase folding in the White-Inyo Range, eastern California [abstract] : *Geol. Soc. America Abstracts with Programs,* v. 7, p. 1289.

——, and Darrow, A. C., 1979, Structure and neotectonics of the western Santa Ynez fault system in southern California, *in* Whitten, C. A., and others, eds., *Recent Crustal Movements,* 1977 (Proc. 6th Internat. Symposium on Recent Crustal Movements, Stanford Univ., Palo Alto, Calif., July 25–30, 1977); *Tectonophysics,* v. 52, no. 1–4, p. 389–406.

——, and Smith, R. R., 1976, Tectonic transpression and basement-controlled deformation in San Andreas fault zone, Salton Trough, California: *Am. Assoc. Petroleum Geologists Bull.,* v. 60, 2081–2102.

——, and others, 1978a, Monzonites of the White-Inyo Range, California and their relation to the calc-alkaline Sierra Nevada batholith: *Geol. Soc. America Bull.,* v. 89, no. 11, p. 1677–1687.

——, and others, 1978b, Papoose Flat pluton: A granitic blister in the Inyo Mountains, California: *Geol. Soc. America Bull.,* v. 89, 1205–1219.

Taliaferro, N. L., 1942, Geologic history and correlation of the Jurassic of southwestern Oregon and California: *Geol. Soc. America Bull.,* v. 53, p. 71–112.

Taliaferro, N. L., 1943a, Geologic history and structure of the central Coast Ranges, California: *Calif. Div. Mines and Geology Bull. 118,* p. 119–163.

——, 1943b, Franciscan-Knoxville problem: *Am Assoc. Petroleum Geologists Bull.,* v. 27, p. 102–219.

——, 1944, Cretaceous and Paleocene of Santa Lucia Range, California: *Am. Assoc. Petroleum Geologists Bull.,* v. 28, p. 449–521.

Tarney, J., and others, 1976, Marginal basin "Rocas Verdes" complex from southern Chile: A model for Archaean greenstone belt formation, *in* Windley, B. F., ed., *Early History of the Earth:* London, Wiley, p. 131–147.

Teissere, R. F., and Beck, M. E., 1973, Divergent Cretaceous paleomagnetic pole positions for the southern California batholith, U.S.A.: *Earth Planet. Sci. Lett.,* v. 18, p. 296–300.

Telleen, K. E., 1977, Paleocurrents in part of the Franciscan Complex, California: *Geology,* v. 5, p. 49–51.

Terry, A. H., 1975, Thrust faulting with associated mylonites in the Chemehuevi Mountains, southeastern California [abstract]: *Geol. Soc. America Abstracts with Programs,* v. 7, p. 380.

Thatcher, Wayne, 1979, Systematic inversion of geodetic data in central California: *Jour. Geophys. Res.,* v. 84, p. 2283–2295.

——, and others, 1975, Seismic slip distribution along the San Jacinto fault zone, southern California, and its implications: *Geol. Soc. America Bull.,* v. 86, p. 1140–1146.

Tinsley, J. C., 1975, Quaternary geology of northern Salinas Valley, Monterey County, California: Unpublished Ph.D. disseration, Stanford University, Stanford, Calif.

Tobisch, O. T., and Fiske, R. S., 1976, Significance of conjugate folds and crenulations in the central Sierra Nevada, California: *Geol. Soc. America Bull.,* v. 87, p. 1411–1420.

——, and others, 1977, Strain in metamorphosed volcaniclastic rocks and its bearing on the evolution of orogenic belts: *Geol. Soc. America Bull.,* v. 88, p. 23–40.

Townley, S. D., and Allen, M. W., 1939, Descriptive catalog of earthquakes of the Pacific coast of the United States, 1769 to 1928: *Seismol. Soc. America Bull.,* v. 29, p. 1–297.

Trask, P. D., 1926, Geology of the Point Sur quadrangle, California: *Univ. Calif. Dept. Geol. Sci. Bull.,* v. 16, no. 6, p. 119–186.

Troughton, G. H., 1974, Stratigraphy of the Vizcaino Peninsula near Asuncion Bay, Territorio de Baja California, Mexico: Unpublished M.S. thesis, San Diego State University, San Diego, Calif., 83 pp.

Troxel, B. W., and Gunderson J. N., 1970, Geology of the Shadow Mountains and northern part of the Shadow Mountains SE Quadrangles, western San Bernardino County, California: *Calif. Div. Mines and Geology Preliminary Report 12.*

Turner, D. L., 1969, K-Ar ages of California Coast Range volcanics–implications for San Andreas fault displacement [abstract]: *Geol. Soc. America Abstracts with Programs,* v. 1, p. 70.

——, 1970, Potassium-argon dating of Pacific Coast Miocene foraminiferal stages, *in* Bandy, O. L., ed., *Radiometric Dating and Paleontologic Zonation:* Geol. Soc. America Spec. Paper 24, p. 91–129.

——, and others, 1970, The Obispo Formation and associated volcanic rocks in central California Coast Ranges–K-Ar ages and biochronologic significance [abstract]: *Geol. Soc. America Abstracts with Programs,* v. 2, p. 155.

Turner, H. W., 1893, Some recent contributions to the geology of California: *American Geologists,* v. 11, p. 307–324.

Tyler, D. L., 1975, Stratigraphy and structure of the Late-Precambrian-Early Cambrian clastic metasedimentary rocks of the Baldwin Lake area, San Bernardino Mountains, California: Unpublished M.S. thesis, Rice University, Houston, Tex., 50 pp.

Uchupi, Elazar, and Emery, K. O., 1963, The continental slope between San Francisco, California and Cedros Island, Mexico: *Deep-Sea Research,* v. 10, no. 4, p. 397–448.

Underwood, M. B., 1977a, Chaotic Franciscan rocks at Pfeiffer Beach, Point Sur quadrangle, California [abstract]: *Geol. Soc. America Abstracts with Programs,* v. 9., p. 519.

——, 1977b, The Pfeiffer Beach slab deposits, Monterey County, California: Possible trench-slope basin, in *Cretaceous Geology of the California Coast Ranges West of the San Andreas Fault:* Pacific Section, Soc. Econ. Paleontologists and Mineralogists, Pacific Coast Paleogeography Field Guide 2, p. 57–70.

Uyeda, S., 1977, Some basic problems in the trench-arc-back arc system, *in* Talwani, M., and Pitman, W. C., III, eds., *Island Arcs, Deep Sea Trench and Back Arc Basins:* Am. Geophys. Union, Maurice Ewing Series 1, p. 1–14.

Vail, S. G., and Dasch, E. J., 1977, Jurassic ophiolitic rocks in the Klamath Mountains, southwest Oregon and implications for pre-Nevadan paleogeography [abstract]: *Geol. Soc. America Abstracts with Programs,* v. 9, no. 4, p. 519–520.

Van Andel, T. H., and others, 1975, Cenozoic history and paleogeography of the central equatorial Pacific Ocean: *Geol. Soc. America Memoir 143,* 134 pp.

Van der Voo, R., and French, R. B., 1974, Apparent polar wandering for the Atlantic-bordering continents: Late Carboniferous to Eocene: *Earth Sci. Reviews,* v. 10, p. 99–119.

——, and others, 1977, Permian-Triassic continental configuration and the origin of the Gulf of Mexico: *Geology,* v. 5, p. 177–180.

Van Hinte, J. E., 1976a, A Jurassic time scale: *Am. Assoc. Petroleum Geologists Bull.,* v. 60, no. 4, p. 489–497.

——, 1976b, A Cretaceous time scale: *Am.Assoc. Petroleum Geologists Bull.,* v. 60, no. 4, p. 498–516.

Vedder, J. G., 1975, juxtaposed Tertiary strata along the San Andreas fault in the Temblor and Caliente Ranges, California, *in* Crowell, J. C., ed., *San Andreas Fault in Southern California:* Calif. Div. Mines and Geology Spec. Report 118, p. 234–240.

——, and Brown, R. D., 1968, Structural and stratigraphic relations along the Nacimiento fault in the southern Santa Lucia Range and San Rafael Mountains, California, *in* Dickinson, W. R., and Grantz, A., eds., *Proceedings of Conference on Geologic Problems of San Andreas Fault System:* Stanford Univ. Pub. Geol. Sci., v. 11, p. 242–259.

——, and Howell, D. G., 1976, Review of the distribution and tectonic implications of Miocene debris from the Catalina schist, California Continental Borderland and adjacent coastal areas, *in* Howell, D. G., ed., *Aspects of the Geologic History of the California Continental Borderland:* Pacific Section, Am. Assoc. Petroleum Geologists Misc. Pub. 24, p. 326–342.

——, and others, 1967, Reconnaissance geologic map of the central San Rafael Mountains and vicinity, Santa Barbara County, California: *U. S. Geol. Survey Misc. Geol. Inv. Map I–487.*

——, and others, 1974, Preliminary report on the geology of the continental borderland of southern California: *U.S. Geol. Survey Misc. Field Studies Map MF–624.*

——, and others, 1977, Description of pre-Quaternary samples, R/V ELLEN B. SCRIPPS, Sept. 1976, Patton Ridge to Blake Knolls, California Continental Borderland: *U.S. Geol. Survey Open-File Report 77–474.*

——, and others, 1979, Miocene rocks and their relation to older strata and basement, Santa Catalina Island, California, *in* Armantrout, John, and others, eds., *Cenozoic Paleogeography of the Western United States:* Pacific Section, Soc. Econ. Paleontologists and Mineralogists, Pacific Coast Paleogeography Symposium 3, 239–256.

Von Huene, Roland, 1974, Modern trench sediments, *in* Burk, C. A., and Drake, C. L., *The Geology of Continental Margins:* New York, Springer-Verlag, p. 207–211.

——, and others, 1963, Indian Wells Valley, Owens Valley, Long Valley, and Mono Basin: *Guidebook for Seismological Study Tour, 1963 Meeting of Internat. Union of Geodesy and Geophysics, U.S. Naval Ordnance Test Station, China Lake, Calif.*, p. 57–105.

——, and others, 1978, Japan Trench transected: *Geotimes,* v. 23, p. 16–21.

——, and others, 1979, Cross section, Alaska Peninsula-Kodiak Island-Aleutian Trench: *Geol. Soc. America Map and Chart Series, MC-28A.*

Wachs, D., and Hein, J. R., 1975, Franciscan limestones and their environments of deposition: *Geology,* v. 3, p. 29–33.

Walcott, C.D., 1895, The Appalachian type of folding in the White Mountain range of Inyo County, California: *Am. Jour. Sci.,* v. 49, p. 169–174.

Walker, R. G., 1976, Submarine-fan conglomerates and sandstones, Jurassic of southwestern Oregon, not associated with continental margin [abstract]: *Am. Assoc. Petroleum Geologists Bull.,* v. 60, no. 4, p. 730.

——, and Mutti, E., 1973, Turbidite facies and facies associations, in *Turbidites and Deep-Water Sedimentation:* Pacific Section, Soc. Econ. Paleontologists and Mineralogists, Short Course, Anaheim, Calif., p. 119–157.

Wallace, A. R., 1977, Geology and ore deposits, Kennedy Mining District, Pershing County, Nevada: Unpublished M.S. thesis, University of Colorado, Boulder, 87 pp.

Wallace, R. E., 1949, Structure of a portion of the San Andreas rift in southern California: *Geol. Soc. America Bull.,* v. 60, p. 781–806.

——, 1968, Notes on stream channels offset by the San Andreas fault, southern Coast Ranges, California: *Stanford Univ. Pub. Geol. Sci.,* v. 11, p. 6–21.

——, 1970, Earthquake recurrence intervals on the San Andreas fault: *Geol. Soc. America Bull.,* v. 81, 2875–2890.

——, 1973, Surface fracture patterns along the San Andreas fault: *Stanford Univ. Pub. Geol. Sci.,* v. 13, p. 248–250.

——, 1975, The San Andreas fault in the Carrizo Plain-Temblor Range region, California: *Calif. Div. Mines and Geology Spec. Report 118,* p. 241–250.

Wanless, H. R., and Shepard, F. R., 1936, Sea level, and climatic changes related to late Paleozoic cycles: *Geol. Soc. America Bull.,* v. 47, p. 1177–1206.

Wasserburg, G. J. F., and others, 1959, Ages in the Precambrian terrane of Death Valley, California: *Jour. Geology,* v. 67, p. 702.

Weaver, D. W., 1969, The pre-Tertiary rocks, *in* Weaver, D. W., and others, eds., *Geology of the Northern Channel Islands (California):* Pacific Sections, Am. Assoc. Petroleum Geologists, and Soc. Econ. Paleontologists and Mineralogists, Special Pub., p. 11–13.

——, and Doerner, D. P., 1969, Lower Tertiary stratigraphy, San Miguel and Santa Rosa Island, *in* Weaver, D. W., and others, eds., *Geology of the Northern Channel Islands (California):* Pacific Sections, Am. Assoc. Petroleum Geologists, and Soc. Econ. Paleontologists and Mineralogists, Special Pub., p. 30–47.

——, and Meyer, G. L., 1969, Stratigraphy of northeastern Santa Cruz Island, *in* Weaver, D. W., and others, eds., *Geology of the Northern Channel Islands (California):* Pacific Sections, Am. Assoc. Petroleum Geologists, and Soc. Econ. Paleontologists and Mineralogists, Special Pub., p. 95–104.

Weber, G. E., and Lajoie, K. R., 1977, Late Pleistocene and Holocene tectonics of the San

Gregorio fault zone between Moss Beach and Point Año Neuvo, San Mateo County, California [abstract] : *Geol. Soc. America Abstracts with Programs,* v. 9, p. 524.

Weisenberg, C. W., and Ave'Lallemant, H., 1977, Permo-Triassic emplacement of the Feather River ultramafic body, northern Sierra Nevada Mountains, California [abstract] : *Geol. Soc. America Abstracts with Programs,* v. 9, p. 525.

Wellman, H. W., 1955, The geology between Bruce Bay and Haast River, South Westland: *New Zealand Geol. Survey Bull. 48* (n.s.), 2nd ed., 46 pp.

Wells, F. G., and Walker, G. W., 1953, Geologic map of the Galice quadrangle, Oregon: *U.S. Geol. Survey Geol. Quad. Map GQ-25,* scale 1:62,500.

——, and others, 1949, Preliminary description of the geology of the Kerby quadrangle, Oregon: *Oregon Dept. Geology and Mineral Industries Bull.,* v. 40, 23 pp.

——, and others, 1959, Upper Ordovician (?) and Upper Silurian formations of the northern Klamath Mountains, California: *Geol. Soc. America Bull.,* v. 70, no. 5, p. 645–649.

Welsh, J. L., 1978, Emplacement of ultramafic rocks and metamorphism in the Marble Mountains area of the Klamath Mountains [abstract] : *Geol. Soc. America Abstracts with Programs,* v. 10, p. 153.

Wenner, D. B., and Taylor, H. P., 1971, Temperature of serpentinization of ultramafic rocks based on $^{18}O/^{16}O$ fractionation between co-existing serpentine and magnetite: *Contrib. Mineral Petrol.,* v. 32, p. 165–185.

Wesson, R. L., and others, 1977, Search for seismic forerunners to earthquakes in central California: *Tectonophysics,* v. 42, p. 111–126.

Whetten, J. T., and others, 1978, Ages of Mesozoic terranes in the San Juan Islands, Washington, *in* Howell, D. G., and McDougall, K. A., eds., *Mesozoic Paleogeography of the Western United States:* Pacific Section. Soc. Econ. Paleontologists and Mineralogists, Pacific Coast Paleogeography Symposium 2, p. 117–132.

——, and others, 1980, Allochthonous Jurassic ophiolite in Northwest Washington: *Geol. Soc. America Bull.,* Part I, v. 91, p. 359–368.

White, A. J. R., and Chappell, B. W., 1976, Ultrametamorphism and granitoid genesis: *Tectonophysics,* v. 43, p. 7–22.

White, E. H., and Guiza, R., 1948, Antimony deposits at El Antimonio district, Sonora, Mexico: *U.S. Geol. Survey Bull. 962-B,* p. 81–116.

White, R. S., 1977, Recent fold development in the Gulf of Oman: *Earth Planet. Sci. Lett.,* v. 36, p. 85–91.

——, and Klitgord, K., 1976, Sediment deformation and plate tectonics in the Gulf of Oman: *Earth Planet. Sci. Lett.,* v. 32, p. 199–209.

Whitebread, D. H., 1978, Preliminary geologic map of the Dun Glen quadrangle, Pershing County, Nevada: *U.S. Geol. Survey Open-File Report 78–407.*

Wiebe, R. A., 1970, Pre-Cenozoic tectonic history of the Salinian block, western California: *Geol. Soc. America Bull.,* v. 81, p. 1837–1842.

Wilcox, R. E., and others, 1973, Basic wrench tectonics: *Am Assoc. Petroleum Geologists Bull.,* v. 47, p. 74–96.

Williams, Howell, 1942, The geology of Crater Lake National Park, Oregon, with a reconnaissance of the Cascade Range southwest to Mt. Shasta: *Carnegie Inst. Washington Pub. 540,* 162 pp.

——, and Curtis, G. H., 1977, The Sutter Buttes of California: *Univ. Calif. Pub. Geol. Sci.*, v. 116, p. 1–56.

——, and McBirney, A. R., 1968, *Geologic and Geophysical features of Calderas:* Eugene, Ore., Center for Volcanology, University of Oregon, 87 pp.

——, and McBirney, A. R., 1979, *Volcanology:* San Francisco, Freeman, Cooper, 397 pp.

——, and others, 1958, *Petrography:* San Francisco, W. H. Freeman, 406 pp.

Willingham, C. R., 1978, Geophysical evidence for 15 km of right lateral displacement on the San Gregorio fault zone, California: *Trans. Am. Geophys. Union,* v. 59, no. 12, p. 1210.

Willis, Bailey, and Wood, H. O., 1922, Fault map of the State of California: *Seismol. Soc. America,* scale 1:506,880.

Wilson, J. T., 1965, A new class of faults and their bearing on continental drift: *Nature,* v. 207, p. 343–347.

Windley, B. F., 1977, *The Evolving Continents*: New York, John Wiley and Sons, 385 pp.

Winterer, E. L., and others, 1969, Inverted metamorphic zonation of Franciscan rocks of the California Borderland–possible relation to Mesozoic continental underthrusting [abstract] : *Geol. Soc. America Abstracts with Programs,* v. 1, no. 7, p. 240–241.

Wise, D. U., 1963, An outrageous hypothesis for the tectonic pattern of the North American Cordillera: *Geol. Soc. America Bull.,* v. 74, p. 357–362.

Wollenberg, H. A., and Smith, A. R., 1968, Radiogeologic studies in the central part of the Sierra Nevada batholith, California: *Jour. Geophys. Res.,* v. 73, no. 4, p. 1481–1495.

——, and Smith, A. R., 1970, Radiogenic heat production in prebatholithic rocks of the central Sierra Nevada: *Jour. Geophys. Res.,* v. 75, no. 2, p. 431–438.

Woodburne, W. O., 1975, Cenozoic stratigraphy of the Transverse Ranges and adjacent areas, southern California: *Geol. Soc. America Spec. Paper 162,* 91 pp.

Woodford, A. O., 1924, The Catalina metamorphic facies of the Franciscan series: *Univ. Calif. Dept. of Geol. Studies Bull.,* v. 15, no. 3, p. 49–68.

——, 1925, The San Onofre breccia: Its nature and origin: *Univ. Calif. Dept. of Geol. Sci. Bull.,* v. 15, no. 7, p. 159–280.

——, and Gander, Craig, 1977, Los Angeles erosion surface of middle Cretaceous age: *Am. Assoc. Petroleum Geologists Bull.,* v. 61, no. 11, p. 1979–1990.

Woodring, W. P., and Bramlette, M. N., 1950, Geology and paleontology of the Santa Maria District, California: *U.S. Geol. Survey Prof. Paper 222,* 185 pp.

Woollard, G. P., 1975, The interrelationships of crustal and upper mantle parameter values in the Pacific: *Rev. Geophys. Space, Phys.,* v. 13, p. 87–137.

Worrall, D. M., 1978, The Franciscan-South Fork Mountain Schist contact near South Yolla Bolly Peak, northern California Coast Ranges [abstract] : *Geol. Soc. America Abstracts with Programs,* v. 10, no. 3, p. 155.

Wright, J. E., 1979, Tectonic correlation between the south-central Klamath Mountains and Sierra Nevada foothill belt [abstract] : *Geol. Soc. America Abstracts with Programs,* v. 11, no. 3, p. 136.

Wright, L. A., and others, 1974a, Turtleback surfaces of Death Valley viewed as phenomena of extensional tectonics: *Geology,* v. 2, p. 53–54.

——, and others, 1974b, Precambrian sedimentary environments of the Death Valley region. California and Nevada in *Guidebook: Death Valley Region, California and*

Nevada, Field Trip No. 1: Geol. Soc. America, 70th Annual Meeting, March 29–31, 1974, Las Vegas Nevada: Shoshone, Calif., Death Valley Pub. Co., p. 27–35.

——, and others, 1976, Precambrian sedimentary environments of the Death Valley region, eastern California, *in* Troxel, B. W., and Wright, L. A., eds., *Geologic Features, Death Valley, California:* Calif. Div. Mines and Geology Spec. Report 106, p. 7–15.

Wright, W. H., III, and Schweickert, R. A., 1977, Tectonics and stratigraphy of the Calaveras complex central Sierra Nevada Foothills: *Fieldtrip Guidebook, 73rd Annual Meeting, Cordilleran Section, Geol. Soc. America,* 16 pp.

Xenophontos, C., and Bond, G. C., 1978, Petrology, sedimentation, and paleogeography of the Smartville terrane (Jurassic)—bearing on the genesis of the Smartville ophiolite, *in* Howell, D. G., and McDougall, K. A., eds., *Mesozoic Paleogeography of the Western United States:* Pacific Section, Soc. Econ. Paleontologists and Mineralogists, Pacific Coast Paleogeography Symposium 2, p. 291–302.

Yamashita, P. A., and Burford, R. O., 1973, Catalog of preliminary results from an 18-station creepmeter network along the San Andreas fault system in central California for the time interval June 1969 to June 1973: *U.S. Geol. Survey Open File Report.*

Yancy, T. E., 1975, Permian, marine biotic provinces in North America: *Jour. Paleontology,* v. 49, p. 758–766.

Yeats, R. S., 1968a, Rifting and rafting in the southern California borderland, *in* Dickinson, W. R., and Grantz, Arthur, eds., *Proceedings of Conference on Geologic Problems of San Andreas Fault System:* Stanford Univ. Pub. Geol. Sci., v. 11, p. 307–322.

——, 1968b, Southern California structure, sea floor spreading, and history of the Pacific basin: *Geol. Soc. America Bull.,* v. 79, p. 1693–1702.

——, 1976, Neogene tectonics of the central Ventura basin, California: *Proc. Neogene Symposium, Pacific Section, Soc. Econ. Paleontologists and Mineralogists,* p. 19–32.

——, 1978, Neogene acceleration of subsisdence rates in southern California: *Geology,* v. 6, p. 456–460.

Yerkes, R. F., and Lee, W. H. K., 1979, Seismicity and late Quaternary deformation in the Western Transverse Ranges [abstract] : *Geol. Soc. America Abstracts with Programs,* v. 11, p. 136–137.

Young, J. C., 1974, Geologic road log along U.S. Highway 295, between Gray Falls Campground and Berry Summit, Trinity and Humboldt Counties, California, *in* McGeary, D.. R. R., ed., *Geologic Guide to the Southern Klamath Mountains:* Geol. Soc. Sacramento, p. 116–120.

INDEX